Clinical Physiology

Clinical Physiology

EDITED BY

E.J. Moran Campbell
Professor of Medicine,
McMaster University,
Hamilton, Ontario,
Canada

C.J. Dickinson
Professor of Medicine and
Chairman, Department of
Medicine, St. Bartholomew's
Hospital Medical College
London

J.D.H. Slater
Physician,
The Middlesex Hospital
Senior Lecturer
The Middlesex Hospital
Medical School, London

C.R.W. Edwards
Professor of Clinical Medicine,
Edinburgh University
Chairman, Department of
Medicine, Western General
Hospital, Edinburgh, Scotland

K. Sikora
Director, Ludwig Institute for
Cancer Research
Consultant Radiotherapist
and Oncologist
Addenbrooke's Hospital,
Cambridge and Hinchingbrooke
Hospital, Huntingdon

FIFTH EDITION

BLACKWELL SCIENTIFIC PUBLICATIONS

OXFORD LONDON EDINBURGH

BOSTON PALO ALTO MELBOURNE

© 1963, 1968, 1974, 1984 by
Blackwell Scientific Publications
Editorial offices:
Osney Mead, Oxford, OX2 OEL
8 John Street, London, WCIN 2ES
9 Forrest Road, Edinburgh, EHI 2QH
52 Beacon Street, Boston
    Massachusetts 02108, USA
706 Cowper Street, Palo Alto
    California 94301, USA
99 Barry Street, Carlton
    Victoria 3053, Australia

First published 1960
Reprinted 1961
Second edition 1963
Reprinted 1965
Third edition 1968
Reprinted 1970
Fourth edition 1974
Fifth edition 1984

Set by Santype Ltd, Salisbury and
printed and bound in Great Britain
by Butler & Tanner Ltd
Frome and London

DISTRIBUTORS

USA
    Blackwell Mosby Book Distributors
    11830 Westline Industrial Drive
    St Louis, Missouri 63141

Canada
    Blackwell Mosby Book Distributors
    120 Melford Drive, Scarborough
    Ontario MIB 2X4

Australia
    Blackwell Scientific Book Distributors
    31 Advantage Road, Highett
    Victoria 3190

British Library
Cataloguing in Publication Data

Clinical physiology.—5th ed.
    1. Physiology, Pathological
    I. Campbell, E.J.
    616.07   RB113

    ISBN 0-632-00912-8

Contents

Contents

Contributors

J. BIENENSTOCK MD, MRCP, FRCP(C)
Professor of Medicine and Pathology, Department of Pathology, McMaster University,
Hamilton, Ontario, Canada
Immunology

J.F. CADE MD, PhD, FRACP, FCCP
Director of Intensive Care, The Royal Melbourne Hospital, Grattan Street, Parkville, Victoria,
Australia
Respiration

E.J.M. CAMPBELL BSc, PhD, MD, FRCP, FRCP(C), FRSC
Professor of Medicine, McMaster University, Hamilton, Ontario, Canada
Hydrogen Ion (Acid:Base) Regulation; Respiration

R.D. COHEN MA, MD, FRCP
Professor of Medicine, The London Hospital Medical College, London
Body Fluids

J.A. DENBURG MD, FRCP(C)
Associate Professor of Medicine, McMaster University, Hamilton, Ontario, Canada
Immunology

C.J. DICKINSON BSc, MA, DM, FRCP
Professor and Chairman, Department of Medicine, St Bartholomew's Hospital Medical College,
London
Circulation

C.R.W. EDWARDS MA, MD, FRCP, FRCPEd
Professor and Chairman, Department of Medicine, Western General Hospital, Edinburgh
Pituitary

J.N.A. HAMER MD, PhD, FRCP
Consultant Cardiologist, Department of Clinical Pharmacology, St Bartholomew's Hospital,
London
Circulation

C. HARDISTY MD, MRCP
Consultant Physician, Northern General Hospital, Sheffield
Thyroid

J. HIRSH MD, FRCP(C), FRACP
Professor and Chairman, Department of Medicine, McMaster University, Hamilton, Ontario,
Canada
Haemostasis

vii

H.S. JACOBS BA, MD, FRCP
Professor of Reproductive Endocrinology, The Middlesex Hospital, Mortimer Street, London
Sex

J.G. KELTON MD, FRCP(C)
Associate Professor, Department of Medicine, McMaster University, Hamilton, Ontario,
Canada
Haemostasis

B.R.F. LECKY MA, MD, MRCP
Senior Registrar, The National Hospital for Nervous Diseases, Queen Square, and Guy's
Hospital, London
Muscle

A.S. MEE MD, MRCP
Consultant Physician and Gastroenterologist, Battle Hospital, Reading
The Gut

A.J. McCOMAS BSc, MB, BS, FRCP(C)
Professor of Medicine (Neurology), Faculty of Health Sciences, McMaster University,
Hamilton, Ontario, Canada
Movement

G.J.R. McHARDY MA, BSc, BM, FRCP, FRCPE
Consultant Clinical Respiratory Physiologist, Lothian Health Board and Senior Lecturer,
Department of Respiratory Medicine, University of Edinburgh
Hydrogen Ion (Acid:Base) Regulation

J.A. MORGAN-HUGHES MD, FRCP
Consultant Physician, The National Hospital for Nervous Diseases, Queen Square, London
Muscle

D.S. MUNRO MD, FRCP
Professor of Medicine, Clinical Sciences Centre, Northern General Hospital, Sheffield
Thyroid

J.L.H. O'RIORDAN BSc, MA, DM, FRCP
Professor of Medicine and Deputy Director, Medical Unit, The Middlesex Hospital Medical
School, London
Bone

S.E. PAPAPOLOUS MD
Department of Clinical Endocrinology, University Hospital, Rijnsburgerweg 10, Holland
Bone

K. SIKORA MA, PhD, MRCP, FRCR
Director, Ludwig Institute for Cancer Research, Cambridge; Consultant Radiotherapist and
Oncologist, Addenbrooke's Hospital, Cambridge
Blood

J.D.H. SLATER MA, MB, FRCP
Physician and Senior Lecturer, The Middlesex Hospital Medical School, London
Adrenals

G.D. SWEENEY MB, ChB, PhD
Professor, Department of Medicine, Faculty of Health Sciences, McMaster University,
Hamilton, Ontario, Canada
Liver

C.J. TOEWS MD, PhD, FRCP(C)
Associate Professor, Department of Medicine, McMaster University, Hamilton, Ontario,
Canada
Energy

R.M. WINTER MB, BSc, MRCP
Consultant Clinical Geneticist, Kennedy–Galton Centre, Harperbury Hospital, and Clinical
Research Centre, Division of Inherited Metabolic Disease, Northwick Park Hospital
Genetics

C.S. WILCOX MA, BM, BCh, PhD, MRCP(UK)
Assistant Professor of Medicine, Harvard Medical School, Brigham and Women's Hospital,
Boston, Massachusetts 02115, USA
The Kidney

Preface to the Fifth Edition

For the first time this book has five editors. The two new ones are Professor C.R.W. Edwards and Dr K. Sikora, who have also revised, respectively, the chapters on the pituitary gland and that on blood.

One of the problems in preparing another edition of an established textbook is to avoid the stagnation inevitable when the same author is given the task of revising his previous accounts. To prevent this we have brought in a large number of new contributors; but we should like at the same time to thank all those who have contributed to the four previous editions.

We welcome many new contributors to this edition. Dr J. Hamer has helped revise the account of heart and circulation. Dr C.S. Willcox has taken over and revised the renal chapter, and Dr G.J.R. McHardy the account of hydrogen ion and acid-base regulation. Immunology has been comprehensively rewritten by Dr J. Bienenstock and Dr Denburg, and Dr Kelton has helped to update the chapter on haemostasis and thrombosis. Dr Papapoulos has jointly revised the chapter on bone and calcium metabolism, and Dr Lecky the previous account of skeletal muscle. Dr A. McComas has revised the chapter on the nervous system in the control of movement. Dr Mee has brought up-to-date the account of gut, and Dr G.D. Sweeney has rewritten the chapter on the liver. Dr Hardisty has jointly revised the chapter on the thyroid and its disorders, and Dr Winter has taken over and revised the account of genetics.

We are very grateful to our previous contributors who have undertaken very extensive revisions of their previous contributions—Professor R.D. Cohen, Dr J.F. Cade, Dr J. Hirsh, Dr J.L.H. O'Riordan, Dr J.A. Morgan Hughes, Dr C.J. Toews and Professor H.S. Jacobs.

This edition has fewer pages, but a larger format, than the last, and it has been more difficult than ever to be selective enough in all these rapidly expanding fields to avoid the book reaching an unwieldy size. As in all previous editions we have tried to select and emphasize those aspects of normal structure and function which are relevant to clinical practice, and all the contributors are, as they have always been, clinicians as well as physiologists. The style remains as didactic as before, and we have continued to resist the temptation to fill up the book with references, preferring to provide a few key references for further study at the end of each chapter. The book makes no pretentions to provide the last word on any subject. Where subjects are controversial we have tried to indicate this, but some readers may nevertheless feel that we have been at times too dogmatic. However, an account in which every statement is hedged around with escape clauses makes for dull reading.

As before, this book covers no special syllabus. It was originally written to give medical students more background to their studies than textbooks of medicine or of 'pure' physiology commonly provide. We hope that it will continue to furnish such a background. In addition, the bridge it provides between basic sciences and clinical medicine makes it suitable for postgraduates training for the various specialties as well as for practising physicians wishing to review the scientific basis of their practice.

MORAN CAMPBELL   CHRIS EDWARDS
JOHN DICKINSON   KAROL SIKORA
WILLIAM SLATER

Preface to the First Edition

One of the most striking changes in medicine in recent years has been the increasing use of physiology and biochemistry, not only to provide greater diagnostic accuracy, but also to guide treatment. In return, clinicians are making extensive use of their unique opportunities to observe disordered function in disease, and are thereby advancing basic physiological knowledge. For these and other reasons the contributors to this book believe that a good knowledge of physiology is becoming increasingly important in the practice of medicine. Applied physiology and functional pathology have, as yet, little place in teaching, and although there are many good textbooks of clinical biochemistry, there are few dealing with clinical physiology. This book, written by practising clinicians, is an attempt to fill the gap. We have not tried to cover the entire subject but have chosen rather to discuss those aspects which can profitably be presented from a more clinical standpoint than that of the academic physiologist, having in mind the interests of the senior student and postgraduate. Reluctantly, and only after much consultation, we decided not to include a chapter on neurology. This branch of medicine is, of course, firmly based on physiology, and neuro-physiology is rapidly expanding in many directions. Unfortunately the time has not yet come when the newer knowledge can be encompassed in a short account designed for the general reader.

Each chapter is divided into four sections. The first and second sections deal with normal and disordered function. The third is an account of the physiological principles underlying tests and measurements used in modern practice. The fourth section, 'Practical Assessment', is essentially a summary of the three preceding sections, to show how the information can be used in diagnosis and assessment. We hope that this section will prove useful in clinical practice by showing how evidence can be built up starting with clinical information and then proceeding to generally available procedures and, if necessary, to special techniques. In some chapters the connection between physiology and practice is so clear that it has been possible to summarize 'Practical Assessment' in almost tabular form. Technical details of tests have not been included, because this book does not pretend to be a manual of 'function testing'. One of the happy results of increased physiological knowledge is that many symptoms, signs and tests which were formerly empirical can now be rationally explained, thereby increasing the reliance that can be placed on clinical evidence and often decreasing the need for laboratory investigations.

References have not been included in the text. A selection of monographs, reviews

xiii

and key papers is given at the end of each chapter. We share the belief that students should be encouraged to use the library and we realize also that some more experienced readers will be irritated not to have some statements supported by references in the text. It is unfortunately not practicable to document the text to a degree suitable for both the beginner and the expert. The beginner will find plenty of further reading in the references and the expert should have little difficulty in tracing the source of any point. The style of presentation of the references has been chosen to help both types of readers, the title and length of all works being stated.

Although each contributor has been responsible for the preliminary writing of the section dealing with his special interest, there has been extensive consultation between contributors and editors.

The Middlesex Hospital   MORAN CAMPBELL
February, 1960   JOHN DICKINSON

Acknowledgements

We are grateful to our contributors, new and old, for their co-operation in this fifth edition. We also thank Mrs D. Blake for preparing the index, and Oxford Illustrators for preparing most of the figures. We acknowledge with thanks the permission of authors and publishers to reproduce those illustrations whose sources are cited in the text. As editors we are again indebted to Mr Per Saugman and the staff of Blackwell Scientific Publications for their help and forbearance.

1

Body Fluids

Normal physiology

WATER

Water is the major constituent of living cells and it is therefore important to consider the special properties that suit it for this role. These properties confer on living organisms an element of temperature stability which is essential if the rates of biochemical reactions are to be controlled, and allow substances of importance in these reactions to be held in solution.

The special features of water are due to the shape of the water molecule, which allows links between the hydrogen atoms of one water molecule and the oxygen atoms of neighbouring ones (so-called 'hydrogen bonding'). One consequence of these links is that, compared with other liquids, an unusually large amount of heat energy has to be absorbed in order to raise the temperature by a given amount. This means that water itself acts as a buffer against temperature changes. Similarly, an unusually large amount of heat has to be used to convert a unit weight of water into vapour. Thus, a given amount of cooling of the animal by evaporation of surface water (e.g. sweat) involves the loss from the body of only a relatively small amount of fluid. The same properties of the water molecule which give rise to hydrogen bonding also confer on the molecule electrical polarity, i.e. the molecule has a positive and negative end. Such substances have a high dielectric constant, which means that they are capable of reducing the electrical attraction between parts of molecules of other substances, e.g. Na^+ and Cl^- in crystals of sodium chloride—water is thus a good solvent for salts. It also dissolves other 'polar' substances (e.g. compounds with hydroxyl or carboxyl groups) by forming hydrogen bonds with them, but does not dissolve 'non-polar' compounds or groups, such as hydrocarbon chains. This principle has particular importance in determining the structure and permeability of cell membranes.

Some biologically important properties of solutions

Most biochemical reactions take place in solution and the speed of such reactions is related to the concentration of the reactants in solution. The relation to concentration is not a straightforward one because particles of a substance in solution, particularly if they are ions, interact electrostatically with each other, usually so as to reduce the

I

effective concentration. The factor by which the actual concentration has to be multiplied to obtain the effective concentration is known as the activity coefficient, and the effective concentration is known as the activity of the solute. 'Activity' is an apt term because it is closely related to the amount of 'drive' ('chemical potential') that a solute has for taking part in chemical and physical processes. The most important determinant of the activity coefficient in a solution of ions in water is the ionic strength (which is calculated from the concentrations and the valencies of the individual ions), but other factors are involved.

When in clinical practice we measure the amount of a constituent in, for example, plasma, in some instances we obtain actual concentrations and in others activities. Thus, when plasma sodium is measured by flame photometry, the actual concentration is measured. On the other hand, when pH is measured, using the potential generated at a glass electrode, it is the actual 'driving force' or activity of the hydrogen ion that is responsible for the reading obtained.

An important consequence of the concept of activity is the mechanism of osmotic pressure. When, for instance, sodium chloride is dissolved in water, not only is the activity of sodium and chloride ions lowered but so is that of the water itself, and the greater the concentration of sodium chloride the greater the lowering of the activity of water. Thus, if two sodium chloride solutions of different concentrations are separated by a membrane which impedes the movement of the sodium and chloride ion but not that of water, water will flow from the solution of higher water activity into that of lower activity, i.e. from the less concentrated solution into the more concentrated— unless the movement is opposed by a mechanical force equal to the difference in 'osmotic pressure' between the two solutions. These phenomena are fundamental in determining the passage of water across biological membranes.

For most solutes in biological fluids, the solute itself takes up a negligible volume of the solution. This is not so, however, for proteins. In plasma, for instance, the plasma proteins, which are present at a concentration of approximately 70 g/litre, occupy about 5 per cent of the volume of the plasma; thus, a concentration of sodium of 140 mmol/l plasma represents a concentration of 147·3 mmol/l plasma water. This becomes of clinical importance in conditions when plasma protein concentrations are grossly altered, as for example in myeloma or hyperlipoproteinaemia. In these conditions the plasma sodium may appear markedly lowered, with the plasma water sodium remaining normal.

BODY COMPARTMENTS

As the higher animals evolved from unicellular organisms, control of the physical and chemical composition of the cells' external environment became essential. This was not only because of the narrow limits of conditions under which life would be sustained, but also a consequence of cell differentiation. Thus, as organs of different functions within the animal developed it became essential for them to be able to communicate and interact with each other. Thus, the extracellular compartment evolved, comprising both a circulation and an interstitial space—that portion of the extracellular compartment which lies outside the circulation and lies functionally and anatomically between the circulation and the cells. The approximate water content of the intra- and extracellular compartments in a man weighing 70 kg are shown in Table 1.1.

Table 1.1. Average sizes of body compartments in an adult man

	% of body weight	Litres
Total body water	60	42
Intracellular water	40	28
Total extracellular water	20	14
Interstitial water	16	11·2
Plasma water	4	2·8

Not included in this table is a small amount of water that has been transferred across epithelial membranes into specialized spaces, such as the cerebrospinal fluid space, the lumen of the gut and the space between the osteoblast membrane and the surface of the bony trabeculum. The compositions of all of these transepithelial fluids are modified in varying degrees from ordinary extracellular fluid.

The intracellular compartment is delineated by the cell membranes, but because of the diversity of intracellular organelles is heterogeneous in composition. The volume and composition of each of the body compartments is normally maintained within narrow limits both by passage of water and solutes between the compartments and by interaction with the outside environment of the organism, mainly via the gut, kidneys, lungs and skin.

Factors affecting the relative composition of intra- and extracellular compartments

When the concentrations of substances in cells and in the extracellular compartment are unchanging this might be due to impermeability of cell membranes to the solute concerned. Thus no interchange between the two compartments would be possible, other than by specialized exo- and endocytotic mechanisms involving vesicle formation, which are not considered here. However, when the membrane is not impermeable to a particular solute, constancy of extra and intracellular concentrations must be due to one of two thermodynamically distinct types of distribution—these are known as 'equilibrium' and 'steady state' distributions.

The term 'equilibrium state' as applied to the cells and their environment means that the flux of a substance (e.g. an ion) into the cell is equal to the flux of that substance out of the cell *and that no input of energy is needed to maintain this situation.* The term 'steady state' distribution means that concentrations of a substance on either side of a membrane remain constant, *but are not in thermodynamic equilibrium.* The difference is well illustrated by reference to sodium, which is characterized by a far higher concentration in extracellular than intracellular fluid, in spite of the fact that both the concentration gradient and the direction of the potential across the cell membrane favour sodium entry into the cell. Since by and large the concentrations of sodium remain constant the fluxes of sodium into and out of cells must be equal. Nevertheless this does *not* represent an equilibrium state because the situation is maintained by the well-known 'sodium pump' in cell membranes, which continually extrudes sodium which has leaked in. In order to do so the pump needs extraneous energy, supplied in the form of adenosine triphosphate (ATP) by cell metabolism. This distribution is therefore a 'steady-state' as defined above. Interference with the sodium pump by treatment with ouabain or curtailing the supply of ATP will cause the intracellular sodium to rise at the expense of extracellular sodium until a new and lower concentration gradient is reached which is a little closer to a true equilibrium. Another type of steady-state is where cell metabolism is producing a metabolite which is lost from the cell, but to which the cell membrane is not very highly permeable. Thus in the case of the muscle cell producing lactate ions during exercise the intracellular lactate may rise to a level which drives sufficient lactate through the cell membrane to match the metabolic production rate of lactate. The cell lactate concentration then remains steady. Removal of the source of lactate production immediately results in a fall of intracellular lactate. In the case of carbon dioxide produced by tissue respiration the permeability of the cell membrane for CO_2 is so high that, even though CO_2 is being continually produced, under most conditions the cell partial pressure of CO_2 is probably not detectably different from its equilibrium value, which is in the case of CO_2 the same as that in the extracellular space. Of course, the cell PCO_2 must be *slightly* higher than the extracellular value, since otherwise there would be no net escape of CO_2 from the cell.

Appreciation of the contrast between equilibrium states and steady states is crucial to the understanding of the distribution of different substances across cell membranes. These distributions depend on the mechanisms by which those substances cross the membrane.

Distribution of water

As has been discussed water moves across cell membrane under the influence of the osmotic gradient. There is substantial evidence under normal conditions in man and vertebrates that equilibrium is attained with the cells being isosmotic with the extracellular space. The two compartments are not always necessarily isosmotic. Under certain transient conditions there may be temporary inequalities, and in man these produce clinical disturbances. Furthermore, quite apart from such transients, situations are seen not infrequently in clinical practice in which movement of water between the intra- and extracellular compartments may not be predictable on the basis of the cells behaving like perfect osmometers, i.e. with no change in the solute content of the cell under any conditions. Under some circumstances there are substantial movements of solute into and out of the cell. In other circumstances, osmotically active particles appear to be generated within the cell itself. Alternatively, cell solute may become less osmotically active. Because of the need to state general principles, the subsequent discussion of clinical states in this chapter takes no account of this type of event except in the case of brain cells, but its occurrence should nevertheless be borne in mind.

Distribution by simple diffusion

A number of substances of major importance distribute themselves between the intra- and extracellular compartments by simple diffusion, moving passively along concentra- tion, or, more precisely, activity gradients. Urea diffuses rapidly across cell membranes; so do the respiratory gases, carbon dioxide and oxygen. The equilibrium condition for all these substances is that of equal activities (usually represented as concentrations or partial pressures) on either side of the membrane. For CO_2 and urea at least these conditions are usually present, but under certain circumstances urea may be removed faster from the extracellular space than it can be replaced by diffusion from the cells, and concentration gradients of clinically significant magnitude may occur.

Distribution of electrolytes

Since electrolytes are charged the membrane potential plays a major part in their distribution across cell membranes. If equilibrium is reached the membrane potential E (positive on the outside) is related to the external and internal concentrations (more properly, activities) $[X^+]_e$ and $[X^+]_i$ of an ion X^+ by the equation

$$E \text{ (volts)} = \frac{RT}{nF} \log_e \frac{[X^+]_e}{[X^+]_i} = 0.061 \log_{10} \frac{[X^+]_e}{[X^+]_i} \text{ (at } 37° \text{ C)} \tag{1}$$

where T is the absolute temperature, R the gas constant, F the charge (faraday) and n the valency of the ion. This relationship is a fundamental one and is known as the Nernst equation. In many tissues this relationship is obeyed for the chloride ion and in skeletal muscle it is nearly obeyed for the potassium ion. In other words if for either of these ions one measures separately the membrane potential and the distribution ratio they are found to be related by the Nernst equation. It follows that in skeletal muscle

$$\frac{[K^+]_i}{[K^+]_e} \simeq \frac{[Cl^-]_e}{[Cl^-]_i} \tag{2}$$

This type of distribution is known as the Gibbs-Donnan equilibrium. Note that since the chloride ion is negatively charged the ratio of intra/extracellular concentration is inverted with respect to that for potassium. Thus for a membrane potential of 90 mV in skeletal muscle, if extracellular chloride has its normal value of about 100 mmol/l, intracellular chloride is just over 3 mmol/l. On similar reasoning, with $[K^+]_i$ about 150 mmol/l, $[K^+]_e$ is 4·5 mmol/l. Other cells which have lower membrane potentials have higher intracellular chloride; thus the liver cell, which has a membrane potential of approximately 40 mV, contains about 20 mmol/l chloride, and in the erythrocyte, which has a membrane potential as low as 8 mV, intracellular chloride is approximately 70 mmol/l.

So far in this section we have described equilibrium situations in which the membrane potential just balances the concentration gradient of the ion; there is said to be no 'electrochemical' gradient. But for many ions this is not so; the most well known example being that of sodium which is in far higher concentration outside the cell (e.g. 142 mmol/l) and far lower inside the cell (approximately 10 mmol/l for skeletal muscle) than would be expected from the membrane potential if equilibrium existed. There is thus a large electrochemical gradient trying to drive sodium into the cell, and this is exactly counteracted by the sodium pump which thereby maintains a steady state. Another example is that of the hydrogen ion which is actively pumped out of most mammalian cells so that although the cells are slightly more acid than the extracellular fluid, they are considerably less acid than would be the case if electrochemical equilibrium had been reached. In human erythrocytes the distribution of chloride, sulphate and hydrogen ion approximately obeys the Donnan equilibrium. But the intracellular content of potassium in erythrocytes is much higher than would fit with the Donnan equilibrium. The probable explanation is that the extrusion of sodium from the cells is linked with entry of potassium. In the human erythrocyte this pumping in of potassium is high in relation to the ability of the potassium ion to diffuse out passively across the red cell membrane, so the Donnan equilibrium is never reached.

The energy-requiring pumps for Na^+ and K^+ are examples of *active transport*. The absorption of glucose and amino acids from the gut and renal tubules are also good examples of this process. The mechanisms by which active transport is achieved are incompletely understood but they usually rapidly deteriorate when the energy supply is depleted by anoxia or metabolic poisons. There are however specific transporting mechanisms in cell membranes which do not require energy and merely facilitate passive transfer of ions across the membrane. Thus in erythrocytes a 'general anion transporter' has been described which facilitates the movements of chloride, bicarbonate and some other anions across the membrane; this transporter has been shown to consist of a protein of molecular weight approximately 95,000 embedded in the erythrocyte membrane.

Distribution of weak acids and bases

Although many ions are distributed according to the principles outlined above, certain weak acids and bases, which exist partially in the ionized and partially in the unionized form, have an entirely different method of distribution. This arises from the fact that cell membranes are permeable to the unionized form but very much less so to the ionized form. This is not particularly surprising since cell membranes have a very high lipid content and it would be expected that the unionized, and thus uncharged and less polar, form would be more soluble in cell membranes. Distribution equilibrium is, therefore, reached when the concentration of the *unionized* form is equal in both compartments. Since the degree of ionization in each compartment depends on the pH in each compartment, the relative total concentration (i.e. ionized plus unionized fraction) of the substances inside and outside the cells depends on the relative pH of the two compartments. The average pH within most cells is about 7·0, i.e. rather more acid than the extracellular pH of about 7·4, and it is not difficult to see that in the case of a weak acid distributing itself as described above, the total concentration in the cells at equilibrium would be less than in the extracellular space; the opposite would be true for a weak base.

The actual distribution of a weak acid (represented by HA, dissociating into A^- and H^+) *at equilibrium* is given by the formula

$$\frac{[T]_i}{[T]_e} = \frac{[H^+]_e}{[H^+]_i} \cdot \frac{([H^+]_i + K_a)}{([H^+]_e + K_a)} \tag{3}$$

Here i and e refer to the intra- and extracellular spaces, [T] represents the *total* concentration of the weak acid (i.e. $[HA] + A^-$) and K_a is the acid dissociation constant. This relationship has some interesting properties. Firstly, if K_a is very much greater than either $[H^+]_e$ or $[H^+]_i$, T is virtually identical to A^- and the formula reduces to

$$\frac{[A^-]_i}{[A^-]_e} = \frac{[H^+]_e}{[H^+]_i} \tag{4}$$

which means that for every 1 unit of pH difference between the two compartments, $[A^-]_i/[A^-]_e$ changes tenfold. Equation (4) has a superficial resemblance to a Donnan equilibrium (see eqn. (2)) but this would only be so if the distribution of $[H^+]$ was at equilibrium, which, as pointed out above, is not true for most cells. The second consequence of equation (3) is that if K_a is very much less than either $[H^+]_e$ or $[H^+]_i$ the formula reduces to

$$\frac{[T]_i}{[T]_e} = 1$$

i.e. there is no concentration gradient despite the pH difference. The formula for a weak base ($BH^+ \rightleftharpoons B + H^+$) corresponding to equation (3) is

$$\frac{[T]_i}{[T]_e} = \frac{[H^+]_i + K_a}{[H^+]_e + K_a}$$

This method of distribution, which is known as *non-ionic* diffusion applies not only to the interface between cells and extracellular fluids, but also to that between the extracellular fluids and 'transepithelial' compartments, such as the urine, the lumen of the gut and the cerebro-spinal fluid. Examples of metabolites known largely to obey non-ionic diffusion principles are ammonia (weak base) and urobilinogen (weak acid). Thus the concentration of ammonia plus ammonium is much higher in acid than in alkaline urine (non-ionic diffusion taking place across the renal tubular epithelium) and the concentration of ammonia and ammonium in the CSF relative to plasma has also been shown to be pH dependent. The amount of ammonia entering the blood stream from the gut is also controlled by the pH gradient; acidification of the gut contents is employed in liver failure to diminish absorption of ammonia and thus help to prevent hepatic coma.

Many drugs, including salicylates and barbiturates (weak acids) and mecamylamine (weak base) are also distributed by non-ionic diffusion and advantage is taken of this in the treatment of salicylate and some types of barbiturate poisoning by alkalinization of the urine. The pK of acetylsalicylic acid is 3·5 and a tenfold increase in urinary excretion should therefore be achieved for each increase of one unit of urine pH over blood pH. However, during the passage of urine down the tubules insufficient time occurs for full equilibrium between the blood and the tubular lumen to be reached and the advantage is therefore considerably less than theoretical; it is nevertheless a highly useful benefit. Only phenobarbitone and barbitone have pKs in the range in which significant benefit can be achieved by alkalinization of the urine in barbiturate poisoning. Some weak acids traverse cell membranes passively in both ionized and undissociated forms. Under these circumstances the equilibrium distribution is a complex compromise between the dictates of non-ionic diffusion and Donnan equilibrium.

Active control of permeability

The permeability of cell membranes is often controlled by hormonal or other means. Thus antidiuretic hormone causes concentration of urine by increasing the permeability of the distal renal tubule to water. The effective permeability of adipose and some other tissue cells to glucose is enhanced by insulin. This effect is now thought to be due to insulin-stimulated movement of glucose transporters from the endoplasmic reticulum to the cell membrane. During the passage of the action potential along skeletal muscle fibres there are striking changes in the permeability of the cell membrane to sodium, potassium and calcium.

Conclusions

The principle of isosmolality and the various mechanisms described above by which solutes traverse cell membranes are two major determinants of the relative electrolyte composition of the extra- and intracellular spaces. Another is the principle of electroneutrality; this means that the total ionic charge on solutes on either side of the membrane must be zero. Table 1.2 compares extracellular electrolytes with those of intracellular water in skeletal muscle.

Most of the values have been derived from actual measurements but, since it is not at present practical to measure precisely all substances contributing to charge and

Table 1.2. Comparison of the concentrations of the main electrolytes of whole serum and skeletal muscle intracellular fluid

	Whole serum mEq/l	I.C. fluid mEq/kg water
Sodium	142	11
Potassium	4	161
Calcium	5	4×10^{-4}
Magnesium	2	26
Total cations	153	198
Chloride	101	3
Bicarbonate	27	10
Phosphate	2	100
Sulphate	1	20
Organic acids	6	—
Proteins	16	65
Total anions	153	198

osmotic pressure, together with their valencies, activities and degree of binding (if any) in an inactive form to proteins, some of the quantities given are to some extent 'reasonable deductions'. It should be noted that in this example, taken from Maxwell and Kleeman (1962), the values have been expressed as milliequivalents per litre, in order to make the total of cations and anions the same, i.e. this is a 'charge' balance sheet. A similar 'osmotic pressure' balance sheet could be constructed; in that case the amounts of the individual constituents could be expressed as mmol/l, since osmotic pressure is dependent mainly on the total number of particles in solution. Substantial variations in composition occur between the different organs but the general principle is the same, with sodium and chloride the main extracellular ions and potassium, protein and phosphate the main intracellular ions.

TOTAL BODY SODIUM AND POTASSIUM

The total amount of sodium in the body is about 4000 mmol in an average-sized adult. Of this, just over one-third is in bone, but only 25–60 per cent of the bone sodium is freely exchangeable with the remainder of body sodium. About 97 per cent of body sodium is extracellular, either in plasma, interstitial fluid, 'transepithelial' fluid or bone; the remainder is intracellular, at very low concentration. The potassium content of the average person is about 3400 mmol, of which 90 per cent is in a freely exchangeable form. Over 95 per cent of this is intracellular.

ELECTROLYTE COMPOSITION OF 'TRANSEPITHELIAL' FLUIDS

The composition of 'transepithelial' fluids—such as intestinal secretions, sweat and cerebrospinal fluid, is modified from that of the extracellular fluid during secretion.

Table 1.3. Average electrolyte concentrations in gastrointestinal fluids of normal adults (mmol/l)

	Na	K
Gastric juice (mixed parietal and non-parietal)	19–70	6–17
Pancreatic	139–143	6–9
Bile (hepatic)	131–167	3–12
Whole small-intestinal fluid (from recent ileostomy)	105–144	6–29

Table 1.3 shows clinically important features of ionic composition of some intestinal fluids. The low concentration of sodium in gastric juice is due to replacement by hydrogen ions. The concentration of potassium is substantially higher than in plasma in some types of gastrointestinal secretion. The composition of sweat depends very much on the rate and duration of sweating. Sweat is very much hypotonic to plasma; the main ions are sodium and chloride, the concentrations of which become progressively lower during adaptation to a hot climate.

EXCHANGE OF WATER AND SOLUTES BETWEEN THE BLOOD AND INTERSTITIAL SPACE

The interstitial space, which is placed anatomically between the cells and the circulation in the capillaries, consists of a network of collagen fibres supporting a gel containing about 1 per cent of hyaluronic acid and other complex polysaccharides. Measurements (p. 45) suggest that the pressure within the interstitial space is slightly subatmospheric. This negative pressure combines with the hydrostatic pressure within the capillary lumina at the arterial end to overcome the osmotic pressure of the plasma proteins (the colloid osmotic, or 'oncotic' pressure) and thus fluid passes out of the capillaries into the interstitial spaces to come into contact with cells. At the venous end of the capillary the intracapillary hydrostatic pressure is lower, the osmotic pressure of the plasma proteins a little higher and fluid thus moves in the opposite direction. The movement of fluid into and out of the interstitial space from the capillaries is rapid, about three-quarters of the plasma exchanging with the interstitial fluid every minute.

In addition to these bulk movements of fluid under osmotic and hydrostatic forces, diffusion of small molecules may also play a part in the exchange between capillaries and interstitial space. Whatever the means of movement the capillaries are highly permeable to all small molecules, but much less to proteins. Under these circumstances a Donnan equilibrium is set up across the capillary membrane. This results for instance in the chloride concentration in interstitial water being about 5 per cent greater than in plasma water, and the sodium concentration 5 per cent less.

Although the fluid moving through the capillary wall is relatively protein free, it is by no means entirely so and albumin does slowly pass into the interstitial space, eventually mostly to be returned to the circulation via the lymphatics. Capillaries in different organs vary in their permeability to albumin, the hepatic sinusoids being particularly permeable. Although the concentration of albumin is low in interstitial fluid compared with plasma, just over half the body pool of albumin is extravascular (i.e. in the interstitial space) since the volume of the interstitial compartment is three to four times greater than the plasma volume.

Since the interstitial fluid pressure is negative, some special means of transferring fluid into and along the lymphatics must exist. An effective lymphatic 'pump' is probably provided by a combination of contractile elements in the walls of the terminal lymphatics, general muscular contractions and lymphatic valves.

CONTROL OF THE BODY WATER CONTENT

Water balance

The body gains water through the diet and from the oxidation of foodstuffs to carbon dioxide and water. It loses water through the kidneys, skin, lungs and faeces. Under normal conditions the renal loss is varied so that the overall loss matches the gains within very close limits; the osmolality of body fluid compartments is thus kept within a few mosmol/kg of the mean normal value of 282 mosmol/kg. A typical balance sheet is as follows for a normal adult on an ordinary diet in a temperate climate:

Intake (ml)		*Output (ml)*	
Water of diet	2250	Urine	1500
Water of metabolism	500	Skin	500
		Lungs	700
		Faeces	50
	2750		2750

These overall figures conceal very much larger exchanges which take place within the body between the extracellular fluid and the lumina of the gastrointestinal tract and renal tubules respectively. The following volumes are secreted daily into the gut of an adult and are virtually entirely reabsorbed: saliva 1500 ml, gastric juice 2500 ml, bile 500 ml, pancreatic juice 700 ml, succus entericus 3000 ml, total 8200 ml. 170 litres of glomerular filtrate are formed daily and all but 1·5–2 l. reabsorbed. Skin loss consists of two elements—the first is water lost by diffusion through the layers of the skin and subsequent vaporization. The rate of this loss depends on the skin temperature and the humidity of the air in contact with the skin, and in temperate climates amounts to 300–500 ml/day. The second element is sweat, the secretion of the sweat glands; this is very small in cool environments (100 ml/day), but may rise to several litres in hot climates and in severe exercise. The latent heat of vaporization required to vaporize both forms of skin water loss constitutes a major pathway of heat loss from the body.

Adjustment of water balance

Under normal conditions the osmolality of the body fluids is regulated extremely tightly—to within about ± 2 per cent of the mean plasma osmolality of 282 mosm/kg. This is achieved by adjustment of water intake, by means of the thirst mechanism, and urinary output.

Thirst. The sensation of thirst is generated by the hypothalamic thirst centre, which is stimulated when plasma osmolality exceeds about 290 mosm/kg. (It should be noted that if the plasma osmolality is raised by the presence of a substance such as urea which rapidly equilibrates between the extra- and intracellular compartments the thirst centre is not stimulated.) Because of variations in availability, emotional factors and social circumstances, the water intake is not always such as to just satisfy thirst and for this reason renal adjustment of water retention is an essential complement to thirst. In addition to extracellular hypertonicity, thirst is also stimulated by loss of extracellular fluid volume, as in haemorrhage or gastrointestinal fluid loss. This reaction is homeostatically inappropriate since it tends to result in replacement of the losses by water alone, without the other constituents of extracellular fluid.

Urinary output. About 20 ml/min of the glomerular filtrate (approx. 120 ml/min) reaches the collecting ducts. The amount of urine which is finally excreted can be varied between 20 ml/min and somewhat less than 0·5 ml/min. The plasma concentration of the antidiuretic hormone (ADH), which in man is arginine vasopressin, is the major determinant of the final excreted volume. ADH is elaborated in the hypothalamus and stored in the posterior pituitary. Its mode of action on the kidney is discussed in Chapter 4; here we are primarily concerned with the control of its release and effect on renal water reabsorption. In normal man graded increases in plasma osmolality stimulate the osmoreceptors on the supraoptic, paraventricular and suprachiasmatic nuclei of the hypothalamus and excite graded increases in the rate of ADH secretion from the posterior pituitary. There is a *threshold* plasma osmolality, about 280 mosm/kg, below which ADH secretion is virtually absent; above this threshold plasma ADH rises linearly with plasma osmolality to a level of 5–10 ng/l. These phenomena can be described by the simple equation

$$\text{plasma ADH (ng/l)} = 0.38 \text{ (plasma osmolality} - 280)$$

Here 0·38 is of course the slope of the linear rise of ADH and is a measure of the *sensitivity* of the osmoreceptors to changes in plasma osmolality. There is quite a wide normal variation of sensitivity. Urinary osmolality rises linearly with plasma ADH; a plasma concentration of 5 ng/l is sufficient to produce maximal water reabsorption in the collecting tubules. The 'gain' of the system is such that a change of plasma osmolality of only 1 mosm/kg will cause a change in urine osmolality of nearly 100 mosm/kg. It follows from the above considerations that an increase in total body water of only 1 per cent would be expected to decrease plasma osmolality by 2·9 mosm/kg, reduce plasma ADH by approximately fifty per cent and double the rate of urine output. Thus the ADH mechanism represents an extremely sensitive servomechanism regulating the osmolality of body fluids by adjusting renal water conservation.

Under normal conditions the small changes in plasma osmolality which excite or depress ADH secretion are due mainly to changes in the concentration of plasma sodium and equivalent anions, these being the major contributors to extracellular osmolality. Under pathological conditions variations in plasma osmolality may be due to other solutes, for example urea and glucose. Urea, which equilibrates across the blood brain barrier and cell membranes very rapidly has little or no effect on ADH secretion. Infusion of hypertonic glucose solutions, for reasons which are not entirely clear, results in suppression of ADH secretion. Thus in the presence of uraemia or hyperglycaemia the quantitative relationships between plasma ADH and measured plasma·osmolality break down. In other words the 'osmoreceptors' do not respond to absolute osmotic pressure *per se* in the extracellular space. It has been widely suggested, though not proven, that change in the volume of the osmoreceptor cells provides the actual signal for changes in ADH secretion. This theory would account for the lack of effect of a readily permeable substance like urea. It would also allow a situation where the osmotic pressure of both osmoreceptors and extracellular space would be lower than normal (and equal), in which the osmoreceptors were of *normal* volume, and therefore signalling the secretion of normal amounts of ADH; it would appear as if the 'threshold' had been reset to a low level but the 'sensitivity' was unaltered. It will be seen later that there are well described clinical examples of this situation.

Finally, in order to understand what follows later concerning disorders of water balance, there is a matter of definition to be discussed. It should be noted that there is a distinction between 'osmolality' and 'tonicity'. 'Osmolality' is a purely physical concept and refers to a value derived from measurement of freezing point depression; thus this value includes a contribution from solutes such as urea which are highly permeable to cell membranes and therefore do not contribute to the osmotic pressure gradient between the two compartments except in special circumstances (see p. 19). 'Tonicity' excludes such solutes; thus plasma tonicity may be derived from plasma osmolality by subtracting the value of plasma urea (in mmol/l). Osmotic disorders are therefore frequently referred to as hypo- or hypertonic states (rather than hypo- or hyperosmolar).

There are a considerable number of stimuli other than the degree of tonicity which alter the secretion of ADH. One of the most important is a change in blood volume, a fall resulting in increased secretion of ADH; under hypovolaemic conditions the osmotic threshold for ADH secretion is reduced and the osmotic sensitivity increased. The effect of volume is mediated reflexly, at least partly through receptors in the wall of the left atrium, and, like the response of the thirst centre to volume changes, appears somewhat inappropriate to the body needs under these circumstances. Acute lowering of arterial blood pressure is a potent stimulus to ADH release, mediated through the carotid baroreceptors; this effect is seen both in haemorrhage and in simple vaso-vagal fainting attacks. Other mechanisms which alter ADH release are the ingestion of alcohol (inhibitory), the administration of nicotine (stimulatory) and a number of pathological disturbances to be discussed later.

In conditions of water deprivation the kidney is limited in the extent to which it can reabsorb filtered water by its incapacity to produce a urine of concentration greater than approximately 1400 mosmol/kg, representing a specific gravity of about 1·035. Thus, the degree of reduction of urine volume depends on the amount of solute which has to be excreted, which, in its turn, is largely dependent on dietary intake of electrolytes and metabolic production of urea. This minimal 'obligatory' urine volume thus varies between about 500 and 1000 ml/day.

CONTROL OF BODY SODIUM AND EXTRACELLULAR FLUID VOLUME

A normal man usually ingests between 50 and 300 mmol sodium per day. Of this, approximately 2–5 mmol appear in the stools and a similar quantity is lost from the skin in the absence of sweating. The remainder is excreted in the urine and although small daily fluctuations occur, sodium balance remains unaltered in the long term. By far the most important active regulator of body sodium content is the kidney; alteration in appetite for salt is too crude a mechanism to contribute in any precise way to sodium homeostasis.

Body sodium and extracellular fluid volume are intimately linked in their homeostasis and are, for this reason, considered together. The reason for this association can be understood by considering the distribution of a volume of isotonic sodium chloride solution (sodium and chloride concentration both 150 mmol/l) infused intravenously. Since the infusion is isotonic, there is no reason why any transfer of fluid between the extra and intracellular compartments should take place. Thus, a virtually pure expansion of the extracellular compartment occurs. This behaviour contrasts with that seen after the infusion of 5 per cent dextrose solution. The dextrose (which is present to make the infusate isotonic, so that local haemolysis does not occur) is rapidly metabolized, leaving, in effect, an addition of water to the extracellular compartment, lowering its tonicity. Water, therefore, moves into the intracellular compartment until equality of osmotic pressure is restored. The load is shared between the body fluid compartments (in contrast to the effect of infused isotonic sodium chloride) and consequently the expansion of extracellular fluid volume is much less. Further linkages between sodium and extracellular fluid volume occur in relation to conditions giving rise to gastrointestinal fluid loss. These fluids are approximately isotonic and their main cation is sodium. Thus, the loss of a litre of intestinal fluid in the vomit in intestinal obstruction amounts approximately to the loss of a litre of extracellular fluid volume. Lastly, in the kidney, reabsorption of sodium and water are linked in the proximal tubule and loop of Henlé and commonly (but not always) associated in the distal tubule.

In addition to these relationships, there is also an association between extracellular fluid volume and blood volume; if a small quantity of isotonic sodium chloride is infused it will initially reduce the colloid osmotic (oncotic) pressure of the plasma and thus will pass into the interstitial space until the mean capillary and interstitial oncotic and hydrostatic pressures again balance out. Similarly, loss of isotonic extracellular fluid in disease is shared between the plasma and interstitial compartments.

The maintenance of a normal blood volume is of crucial importance, since the blood volume is a major factor in determining cardiac filling pressure and thus cardiac output. Sensitive and powerful homeostatic mechanisms directed towards the excretion or retention of sodium and water come into effect when the blood volume, and under many (but not all) circumstances the extracellular fluid volume (ECFV), exceed or fall below their normal magnitudes. It has become traditional to talk in terms of 'volume receptors' to describe the mechanisms that apparently sense alterations in fluid volumes, but there is no real evidence that compartmental volumes are sensed at all. The virtue in having a normal blood volume is that, in the presence of normal tone in the venous side of the circulation, it produces a cardiac filling pressure which results in adequate cardiac output and arterial blood pressure. It is not therefore surprising that certain pressure receptors, when stimulated or suppressed by changes in pressure, cause changes in the renal retention of sodium and water which result in alterations in extracellular fluid volume, including blood volume, and thus influence cardiac filling and output and arterial pressure. It may therefore appear that volume is being regulated, but it is not really correct to refer to 'volume receptors' or to specific control of extracellular fluid compartment volumes.

The best understood of these mechanisms is the renin-angiotensin-aldosterone system. The initial stimulus to renin release is low pressure in the renal glomerular afferent arterioles (a) on a pressure receptor in the juxtaglomerular apparatus and (b) in altering the amount of sodium flowing through the distal tubule at the point where the macula densa abuts on the granular cells of the juxtaglomerular apparatus. The sympathetic nervous system, acting through the renal nerves and circulatory catecholamines via adrenergic receptors is also involved in the control of renin release. Once released, renin acts enzymatically upon a polypeptide substrate in the plasma to produce a decapeptide, angiotensin I, which is cleaved by the 'converting enzyme' to an octapeptide, angiotensin II. (Angiotensin is described in greater detail in Chapters 2 and 4.) Although angiotensin II was originally described as a vasopressor hormone, it is now clear that in subpressor doses it stimulates aldosterone secretion by the adrenals and, furthermore, has a direct sodium retaining effect on the kidney. Aldosterone stimulates reabsorption of sodium from the tubular fluid of the distal part of the nephron in exchange for potassium ions. It probably also increases sodium reabsorption in the proximal tubule; it is not yet clear whether an exchange mechanism is involved at this site. Stimulation of the renin-angiotensin-aldosterone system is responsible for the rapid reduction in urinary sodium excretion which occurs when the dietary intake of sodium is abruptly reduced in normal subjects to 10 mmol/day; under these circumstances the urinary sodium falls below the level of sodium intake within 2–4 days. The extra sodium reabsorbed in the proximal tubule is accompanied by a quantity of water which makes the reabsorbed fluid isotonic; the retention of water in association with sodium reabsorbed elsewhere than in the proximal tubule may be effected by secretion of ADH when the plasma osmolality is increased by the sodium reabsorption. Thus the blood volume, extracellular fluid volume and cardiac performance are maintained. These vital quantities are also defended during the administration of diuretics, the degree of sodium and water depletion induced being limited by compensatory hypersecretion of aldosterone, due to rises in plasma renin and angiotensin.

Other receptors important in the present context lie in the walls of the left atrium, and are responsive to changes in cardiac filling. Distension of the atrium stimulates these receptors, whose afferent pathway appears to be the vagus. Such stimulation results in inhibition of ADH release and consequent water diuresis. The role of these receptors in circulatory control is poorly understood, especially as water diuresis seems a less appropriate response than a combined sodium and water diuresis. However, recent evidence suggests that receptors in the cardiopulmonary region may also influence renin release and thus sodium balance via a reflex involving the sympathetic innervation of the kidney.

A number of other mechanisms of possible importance in the control of renal sodium excretion have recently come to notice. Increasing evidence has been found in the past two decades of a hormone which induces salt loss—'the natriuretic hormone' or 'third factor'. This hormone appears to be secreted from a source as yet uncertain in response to overloading of the circulation or extracellular space, e.g. after saline infusion; its site of action in the renal tubule is as yet unclear. A factor of this sort may explain the phenomenon of 'aldosterone escape' which is observed in normal subjects persistently injected with aldosterone. Sodium retention surprisingly only occurs for a few days. A mild diuresis ensues and the subject ends up with a total body sodium somewhat greater than normal but usually insufficient to produce oedema. The role of the natriuretic factor in physiological and pathological responses is not yet established.

The renal handling of sodium and water is further discussed in Chapter 4.

CONTROL OF BODY AND PLASMA POTASSIUM

The amount of potassium which is normally consumed daily is 50–200 mmol. 5–15 mmol appear in the stools, less than 5 mmol are lost through the skin and the remainder is excreted in the urine. Although all but 2–3 per cent of body potassium is intracellular there is a rapid flux of potassium to and fro across the cell membranes. Thus, if an intravenous injection of isotopically labelled potassium is given, virtually all the isotope is cleared from the blood entering an organ in a single passage through it and is replaced by non-isotopic potassium from the cells.

The kidney is the only mechanism through which any substantial active regulation of total body potassium content is effected. The renal excretion of potassium is discussed in detail in Chapter 4, but some salient points are mentioned here. Potassium loss tends to be increased when aldosterone secretion is elevated and when an unusually large proportion of sodium escapes reabsorption in the proximal tubule, especially if accompanied by a relatively poorly reabsorbable anion such as bicarbonate. Renal potassium handling is also markedly affected by changes in acid-base status (see p. 27).

The renal excretion of potassium is also controlled by the level of serum potassium. High levels of serum potassium stimulate adrenal secretion of aldosterone, which results in increased secretion of potassium; a low serum potassium has the opposite effect. When the dietary intake of potassium is severely restricted, the serum level of potassium falls somewhat and the urinary excretion of potassium becomes much reduced. In contrast to the rapidity with which urinary sodium falls during restriction of sodium intake, it takes many days for the urinary potassium to fall when potassium intake is restricted. The mechanisms for renal retention of potassium are also less powerful than those for sodium. Whereas during the accumulation of oedema or during severe restriction of sodium intake daily urinary sodium excretions of less than 1 mmol are frequently seen, the excretion of potassium during severe restriction of potassium intake seldom falls below 5–10 mmol/day. In addition, faecal potassium under these conditions, though reduced compared with normal, contribute a substantial fraction of the overall potassium loss. For these reasons, considerable deficits of body potassium may be fairly rapidly incurred during severe restriction of intake.

Besides control by variation of renal excretion, plasma potassium may also be regulated in man by circulating catecholamines. There is a β_2-adrenergic receptor mechanism which when stimulated mediates the movement of potassium from the extracellular fluid into cells. This mechanism is responsible for the lowering of plasma potassium often observed during severe stress, for example, after myocardial infarction.

Disordered function

PURE WATER DEFICIENCY AND OTHER
HYPERTONIC STATES

Hypertonicity of the body fluids may be caused either by water deficiency or by accumulation of solutes. Osmotic shrinkage of intracellular volume occurs in both cases, and as will be seen, the consequences depend not only on the severity of the disturbance, but on the speed with which it develops and the nature of the solute accounting for the hypertonicity.

The classical example of pure water deficiency is pituitary diabetes insipidus (p.

561), due to destruction, either partial or complete, of the ADH secreting mechanism by disease or trauma. In terms of the description of the osmoreceptor mechanism given earlier in this chapter there is a marked reduction of the sensitivity of the mechanism to increases in plasma osmolality. Pituitary diabetes insipidus is a relatively infrequent condition and it is important to be aware of the comparatively common occurrence of relatively pure water depletion in seriously ill patients who are unable to indicate their desire for water, either because of disturbance of consciousness, or because of curarization in connection with artificial positive pressure respiration. Some other circumstances in which water deficiency may occur are sheer unavailability of water and resistance of the renal tubules to ADH (nephrogenic diabetes insipidus). The latter phenomenon may be the result of an inborn error of metabolism, or much more commonly, due to potassium deficiency, hypercalcaemia, urinary obstruction or the administration of lithium salts in the treatment of depression.

It should be pointed out that, given unimpeded access to water, patients with disorders of water turnover such as diabetes insipidus, may ingest and excrete enormous quantities of water and remain in perfect health until some event upsets the delicate balance of input and output; such a patient, rendered unconscious and unable to drink by, for example, a road accident, is in a very hazardous situation.

The initial effect of water deficiency is thirst, due to the rise in tonicity of the extracellular fluid (ECF). Water is withdrawn osmotically from the intracellular fluid (ICF) into the ECF, leading to a new equilibrium where the two compartments are of equal but higher tonicity than normal. Since the deficit is shared between the two compartments, signs of ECF volume and blood volume depletion (see later) occur much later in the development of the syndrome than is the case in circumstances where sodium and water loss occur together. The percentage contraction of all body compartments is the same and it can therefore be calculated theoretically that only about one-twelfth of the overall water deficit comes from the intravascular volume. In fact somewhat less than this fraction comes from the circulation, since rising plasma oncotic pressure tends to maintain plasma volume at the expense of interstitial volume. When more than 6–10% of the body water has been lost, drowsiness, confusion and eventually coma ensue.

When a hypertonic state is due to accumulation of solute the responsible substance is usually sodium (+ anions) or glucose. In this type of hypernatraemic state the excess sodium usually arises from therapeutic administration either by accident or design. Hypertonic sodium bicarbonate is routinely administered in the treatment of cardiac arrest in order to counteract metabolic acidosis. Inadvertent administration of hypertonic sodium chloride, salt poisoning in infants and the drinking of sea water are other causes. The immediate effect is to withdraw water rapidly from the intracellular space thus expanding the extracellular fluid volume. It should be noted that the major physiological distinction between the hypernatraemic state due to water depletion on the one hand and sodium excess on the other is in the effect on the extracellular volume which is decreased in water depletion and increased in sodium excess. The increase in extracellular volume in the latter disorder includes the circulation and pulmonary oedema due to cardiac overload may occur.

But particularly important is osmotic brain shrinkage which occurs in acute hypernatraemia. This may be sufficiently severe to rupture the vascular attachments and lead to intracranial haemorrhage. Clinically there is rapid onset of drowsiness followed by convulsions, deepening coma and death. The mortality of this acute syndrome is said to be over 90 per cent when the plasma osmolality exceeds 350 mosmol/kg.

Hypertonicity due to glucose is usually due either to uncontrolled diabetes mellitus or intravenous feeding with dextrose solutions in circumstances when insulin secretion is insufficient to deal with the glucose load. Though acute severe hyperglycaemia has in general the same major physiological and clinical consequences as described above for hypernatraemia there are certain important differences. Plasma sodium may be raised, normal or lowered, depending on numerous factors. The mechanisms tending to lower plasma sodium are movement of water osmotically from the intra- to extracellular compartment, loss of sodium in vomit and in the urine, because of osmotic diuresis and as an obligatory counterion to hydroxybutyrate and acetoacetate in diabetic ketoacidosis, and stimulation of ADH secretion because of eventual contraction of blood volume. Furthermore stimulation of thirst may result in replacement of sodium and water losses by water alone. On the other hand, loss of water because of osmotic diuresis and the effect of hyperglycaemia in suppressing ADH secretion tend to raise plasma sodium.

The above descriptions of syndromes due to solute excess are applicable to circumstances where these syndromes have developed rapidly and the principal events are the consequence of acute brain dehydration. The brain is however unique in that within a few hours of the start of dehydration new solute appears intracellularly. This new solute has the effect of counteracting the shrinkage and may be viewed as a protective mechanism. Some of the new solute represents sodium, potassium and chloride that has moved into the cell—but at least half the new solute has appeared from within the cell itself—so called 'idiogenic osmoles'. The nature of these 'idiogenic osmoles' is not clear though under some circumstances aminoacids may contribute. The consequence of this mechanism is that in experimental chronic hypernatraemia brain volume has returned to normal within seven days; in experimental hyperglycaemia this process only takes a few hours. Though this mechanism has clear protective advantages it gives rise to considerable dangers in treatment of hypertonic states in that if extracellular hypertonicity is lowered too quickly, before the formation of new solute in brain cells has had time to reverse, brain volume may be increased above normal ('cerebral oedema'). The clinical consequences (see below) are almost as serious as in brain dehydration, and great care is thus required in the management of hypertonic states. For instance, too rapid or vigorous lowering of the plasma glucose with insulin in diabetic hyperglycaemic precoma may be associated with a marked rise in cerebrospinal fluid pressure and a deterioration of consciousness. Even in the case of a highly diffusible substance such as urea it is possible to create a similar 'dis-equilibrium' state.

If uraemic patients are subject to too rapid haemodialysis then plasma urea concentration falls much more quickly than does the concentration on the brain side of the blood brain barrier. Confusion, disorientation and even coma may be the outcome of the consequent osmotic shift of water into the brain compartment.

WATER EXCESS

Again, the disturbance of body water is shared between both extracellular and intracellular compartments. This means that there is cellular swelling and, since the cranial cavity is of fixed volume, the predominant clinical manifestations of 'water intoxication' are due to raised intracranial pressure—nausea, vomiting, headaches, drowsiness, fits, coma and non-specific and changing neurological signs (e.g. extensor plantar responses) are characteristic. However, such manifestations are only seen in *acute* water intoxication, since in more chronic states brain-cells appear to reduce their content of osmotically active solute, thus allowing restoration of normal cell volume. This is the counterpart of the opposite effect described above in hypertonic states. In both acute and chronic water excess plasma sodium and osmolality may be strikingly reduced, levels of plasma sodium being not uncommonly as low as 100–110 mmol/l.

The most simple cause of water excess is acute oliguric renal failure, where ingested water cannot be excreted. The fluid intake is, therefore, usually restricted in these circumstances to 500 ml per day plus the equivalent volume of any urine passed. The 500 ml represents respiratory, skin and faecal loss, less the water produced during the metabolism of food stuffs.

Apart from this simple cause, one of the earliest types of water intoxication to achieve recognition was that associated with anterior pituitary failure. The excretion of ingested water at a normal rate requires the presence of adequate quantities of circulating glucocorticoids, which in pituitary failure are deficient because of partial or complete absence of adrenocorticotrophic hormone. The action of glucocorticoids in this respect is incompletely understood, but two factors which are thought to be involved are the effect of glucocorticoids in maintaining a normal glomerular filtration rate and a possible role in maintaining impermeability of the distal and collecting tubules to water in the absence of ADH. There is also some evidence that circulating ADH levels may be inappropriately high for the state of hydration, a situation which is discussed in some detail below. The high ADH levels are suppressible by the administration of glucocorticoids. A similar tendency to water intoxication exists in Addison's disease (primary adrenal failure), but here the situation is complicated by sodium deficiency and a contracted extracellular fluid volume; again, there is also some evidence in Addison's disease of high plasma ADH levels, which are suppressible by administration of glucocorticoids. A number of patients with hypopituitarism due to lesions which, in addition to the anterior pituitary, also affect the posterior pituitary and base of the hypothalamus develop diabetes insipidus when replacement steroid therapy is commenced.

In recent years a further group of disorders characterized by excessive water retention has been described. In these disorders high levels of antidiuretic hormone circulate in spite of low plasma sodium and osmolality; the osmolality of the urine is inappropriately high for the serum osmolality. In some cases, the excess ADH is derived from certain neoplasms of non-endocrine tissues, particularly the lung; the tumours synthesize and secrete the hormone and are not subject to the same control by plasma osmolality as is ADH secretion from its normal source. Inappropriate secretion may also be associated with a diverse selection of disorders such as the metabolic response to surgery (see p. 31), head injury, chest infections, meningitis, acute porphyria and myxoedema.

Measurements of plasma ADH in these patients under different states of plasma osmolality reveal a number of patterns of behaviour. Firstly, there may be continually high fluctuating levels of plasma ADH unrelated to plasma osmolality; secondly, the 'threshold' may be lowered, but the sensitivity unaltered, i.e. the 'reset osmoreceptor' situation described on p. 13. Thirdly, the only abnormality may be that ADH secretion fails to 'switch off' completely at low plasma osmolality; there is therefore no threshold. The nature of the stimuli to inappropriate ADH secretion in many of the conditions mentioned above is unclear, but they are non-osmotic in character. In many of these conditions the renal responsiveness to ADH becomes somewhat diminished; this allows a temporary steady state to be achieved with low plasma osmolality and sodium concentration, inappropriately high plasma ADH, and lower urine osmolality than expected from the level of plasma ADH. Certain commonly used drugs cause inappropriate ADH secretion; the most important clinically is the antidiabetic agent chlorpropramide, and there have been recorded instances of water intoxication syndromes due to this drug. Chlorpropramide also augments the sensitivity of renal tubules to ADH.

The pathophysiology of inappropriate secretion of ADH can be reproduced in many respects by chronic administration of ADH to normal subjects to an ordinary fluid intake. Water is retained, the plasma sodium and osmolality falls and the urine is concentrated. After two to three days, a sodium diuresis occurs. Although the glomerular filtration rate usually increases, the natriuresis is unrelated to the filtered load of sodium. It was originally thought that it was due to suppression of aldosterone secretion consequent upon the extracellular fluid volume's share of the increase in total body water. However, the natriuresis also occurs in normal subjects taking, in addition to ADH, doses of mineralcorticoid sufficient to swamp any possible changes in endogenous aldosterone secretion. Some mechanism other than suppression of aldosterone secretion may, therefore, be involved; stimulation of release of the 'natriuretic factor' is clearly a possibility. Both the natriuresis and the water retention contribute to the production of hyponatraemia, but frequently are not sufficient to account for the degree of lowering of plasma sodium that is observed; this is a situation where intracellular osmotic 'inactivation' of solute (see p. 20) has been proposed. In the inappropriate secretion syndrome the severity varies from a mild symptomless hyponatraemia to fully developed water intoxication. Attempts to correct the hyponatraemia with sodium chloride merely result in an extremely brisk renal excretion of the administered sodium. The symptoms and abnormal biochemical findings can, however, be reversed by restriction of fluid intake.

Water overload is occasionally due to disturbances of the thirst mechanism. Personality disturbances sometimes result in compulsive water drinking, ten or more litres of water per day being consumed. These patients may be usually distinguished from those with true diabetes insipidus by the history, the observation of a slightly lower than normal plasma osmolality (rather than higher), and by the fact that water deprivation produces a more highly concentrated urine than the administration of ADH; the opposite is the case in diabetes insipidus. It should be noted that prolonged high water intake in normal individuals results in a diminution of maximal concentrating activity in the presence of ADH; compulsive water drinkers thus do not concentrate as well as normal either on water deprivation or on administration of ADH.

Patients with hypothalamic lesions sometimes exhibit excessive thirst. This may be hazardous if ACTH secretion is also impaired.

SODIUM DEFICIENCY

From what has been said earlier about the close link between sodium and extracellular fluid volume homeostasis, it is not surprising that the major effects of sodium depletion are those of a low extracellular fluid volume. The earliest objective manifestation is a postural hypotension with a fall of systolic blood pressure of more than 10 mmHg on standing. This is presumably due to failure of venoconstriction in the face of low blood volume to maintain cardiac filling pressure on assuming the upright position, and to inadequate arteriolar vasoconstriction. Syncope may occur in more marked cases. Thirst and oliguria are also fairly early manifestations. Later, more striking effects appear—tachycardia, followed by low recumbent blood pressure, cold, sweating extremities, inelastic skin, dry tongue, low intraocular pressure and a rising blood urea level due to poor renal perfusion. When the cardiac output is markedly reduced, metabolic acidosis occurs due to increased production of lactic acid by hypoxic tissues.

Because of the poor peripheral blood flow, venous blood is much more deoxygenated than normal; this feature, together with rise in packed cell volume because of loss of plasma volume, is responsible for the peripheral cyanosis which is a marked feature. A 70 kg man showing most of the above features has lost at least 4 of his 14 litres of extracellular fluid.

This syndrome occurs in its purest form when fluid is lost from the gastrointestinal tract, as, for instance, in vomiting from intestinal obstruction, severe diarrhoea and biliary and pancreatic fistula. The fluids lost are approximately isotonic to extracellular fluid and have nearly the same sodium concentration. In chronic pyloric stenosis due to ulcer or neoplasm the vomitus is also usually of high sodium content, since hydrogen ion secretion tends to be suppressed by gastritis in the later stages of this condition. However, in many circumstances under which sodium is lost the eventual outcome is that sodium is lost *in excess of water*. This commonly happens in gastro-intestinal fluid loss either because the patient drinks and retains water in response to thirst and ADH release due to hypovolaemia, or because inappropriate treatment with effectively hypotonic fluids (such as 5 per cent dextrose or 4 per cent dextrose with 0·18 per cent sodium chloride) is given instead of isotonic (0·9 per cent) sodium chloride. This results in only a small expansion of the extracellular space since, because of osmotic considerations, most of the ingested or administered hypotonic fluid passes into the intracellular compartment, with resulting cellular overhydration as in water intoxication.

Sodium loss in excess of water also arises in other situations. A certain proportion of patients with chronic renal failure are not capable of reducing urinary sodium excretion when their sodium intake falls below a certain level. Loss of ECFV results, with further nitrogen retention due to the consequent fall in renal vascular perfusion. These patients are resistant to the action of aldosterone on distal tubular sodium reabsorption; this is partly accountable for by a large osmotic diuresis in the few remaining functional nephrons, due to the high level of plasma urea. Over-enthusiastic use of diuretics in oedematous states may also result in sodium loss in excess of water. In adrenocortical failure the lack of aldosterone results in renal salt loss and that of glucocorticoids in water retention. In addition, in adrenocortical failure, a significant amount of sodium moves from the ECF into the cells. A further cause of sodium loss in excess of water is seen in tropical climates and in heavy labourers when sweat losses are replaced by water without salt. Muscular cramps, possibly due to cellular overhy-dration, are a common feature of this type of electrolyte disturbance.

Serious depletion of ECF also occurs in uncontrolled diabetes mellitus, in association with both ketoacidosis and the less common non-ketotic hyperglycaemic coma. In ketoacidosis, sodium and water are lost because of both vomiting and renal loss. The latter occurs for two reasons—firstly, as a result of the osmotic diuresis due to the associated hyperglycaemia and secondly, because of the excretion of hydroxybutyrate and acetoacetate in the urine. In diabetic ketosis, these compounds are generated as the acids, in excess of the liver's catabolic capacity; their peripheral metabolism is also reduced. In an attempt to excrete the excess hydrogen ions the urine is maximally acidified, i.e. the urinary pH reaches its lower limit of about 4·6. Unfortunately, at this pH these fairly strong acids are approximately half dissociated and thus only half of the amount to be eliminated can be excreted as the undissociated acid. The remainder must be excreted in company with cations to preserve electroneutrality; sodium and potassium are for this reason lost in substantial amounts. Because of the osmotic diuresis water is generally lost in excess of sodium in ketoacidosis and even more so in the non-ketotic variety of hyperglycaemic coma, in which quite marked hyper-natraemia often occurs.

Even small degrees of extracellular fluid depletion bring into action the various homeostatic mechanisms discussed earlier. Renin is released and consequent increased secretion of aldosterone results in nearly complete disappearance of sodium from the urine when the gastrointestinal tract or the skin is the source of loss; when the primary source of sodium and water loss is the kidney, the compensatory effects of aldosterone may have only a variable effect in cutting down the loss. In the patient with normal kidneys, the mechanisms inducing sodium retention are, as previously mentioned, very powerful, the mechanisms for ensuring control of ECF and blood volume often overriding other homeostatic systems—for instance that of acid-base status; thus if a patient is both sodium and water depleted and alkalotic—as in pyloric stenosis—the distal tubule will reabsorb sodium in exchange for hydrogen ions, with the result that an acid urine will be produced in spite of the alkalosis. However, in some patients with persistent loss of gastric juice due to vomiting the plasma bicarbonate level rises more rapidly than the capacity of the tubules to reabsorb bicarbonate. Abnormally large amounts of sodium bicarbonate reach the distal tubule where some but not all of the sodium is reabsorbed in exchange for potassium and hydrogen ions. The urine in these patients contains substantial amounts of sodium and potassium, but very little chloride, and is alkaline.

SODIUM EXCESS AND OEDEMA FORMATION

Excess body sodium is nearly always accompanied by an equivalent (isotonic) volume of water, so the main effects are those of expansion of the extracellular compartment. As the interstitial space is expanded, its compliance becomes progressively greater compared with that of the vascular compartment so that the fraction of the total excess ECF volume in the interstitial space steadily increases. When the initially negative interstitial pressure reaches a level slightly above atmospheric pressure, the capacity of the interstitial gel for taking up water is exceeded, and freely movable fluid is present. This is seen as 'pitting oedema' in subcutaneous tissues. Potential cavities bounded by the pleura, peritoneum and pericardium are also frequent sites of free fluid accumulation.

The factors responsible for the genesis of oedema during sodium retention can be deduced from Starling's principles governing the exchanges between capillaries and the interstitial space. Thus, high intracapillary hydrostatic pressure due to venous obstruction, hypoalbuminaemia, increased permeability to proteins, and failure of removal of interstitial proteins due to lymphatic disease are the main reasons at the capillary level for the development of oedema. However, sodium and water retention are essential ingredients of the process, since in their absence the development of oedema would cease as soon as sufficient fluid had moved out of the vascular compartment to lower the hydrostatic gradient or raise the plasma protein concentration to normal levels.

The mechanism of salt and water retention in some hypoalbuminaemic states such as the nephrotic syndrome is relatively well understood. The movement of fluid out of the capillaries because of low oncotic pressure tends to lower the circulating blood volume and the renin-angiotensin-aldosterone mechanism is stimulated, resulting in sodium and water retention. The oncotic pressure is kept low by the retained fluid, which, therefore, moves out into the interstitial space, thus increasing the oedema. Unlike normal individuals, such patients do not 'escape' from the action of aldosterone (see p. 16); it may be that since the vascular volume is not raised above normal in the nephrotic syndrome, no stimulus to release of the natriuretic hormone occurs. In contrast, patients suffering from an aldosterone-secreting tumour of the adrenal behave as do normal subjects repeatedly injected with aldosterone, the small increase in body sodium and water content which is present usually being insufficient to result in oedema.

Salt and water retention is also a prominent feature of cirrhosis and of right-sided heart failure; in these conditions the mechanism of sodium retention is not quite so well defined. In cirrhosis one view is that there is some ill-defined stimulus to renin secretion and thus to hyperaldosteronism, the resulting sodium and water retention expanding the vascular volume and 'overflowing' into the peritonial and interstitial compartments. The other view is that the effective vascular volume is low, due to hypoalbuminaemia and portal hypertension, and that renin and aldosterone over-production, with ensuing sodium and water retention are secondary effects of the hypovolaemia. Whatever the truth of the matter, once sodium and water retention has commenced the amount of it which appears in the peritoneal cavity as ascites is determined by a number of factors, including the height of the portal venous pressure, and the degree of hypoalbuminaemia. The obstruction to blood flow through the liver in cirrhosis may be either pre- or post-sinusoidal. If the former predominates most of the ascites will be formed from the transudation through the general visceral peritoneal surface, excluding the liver. If the major obstruction is post-sinusoidal, much of the ascites forms on the surface of the liver, and tends to be protein-rich because of the high permeability of the sinusoids to protein and the direct entry of newly synthesized albumin into the hepatic interstitial space. The hepatic lymphatics are grossly dilated in cirrhosis with portal hypertension, but even so cannot cope with the increased production of interstitial fluid when the obstruction is post-sinusoidal; under these circumstances the ascites largely represents hepatic lymph 'spill-over'.

In right heart failure the intravascular volume is increased (p. 56) and the possible mechanisms will be further discussed in Chapter 2. Venous thrombosis, commonly in the lower limbs, results in striking local oedema. In other types of oedema, such as that associated with inflammation and allergy (angioneurotic oedema) increased capillary permeability is mainly responsible.

Under certain circumstances, sodium and water retention, if they result in expansion of the extracellular fluid volume, appear to play a role in the determination of the arterial blood pressure, by inducing a secondary rise of peripheral vascular resistance (see p. 78).

POTASSIUM DEPLETION

Potassium depletion arises because of either excessive gastrointestinal or excessive renal loss. Inadequate intake frequently compounds the situation produced by excessive losses. Since potassium is the main intracellular cation, loss of non-fatty tissue ('lean body mass') is bound to result in loss of potassium from the body and under these circumstances the negative potassium balance bears a relatively constant quantitative relationship to the simultaneous negative nitrogen balance; $2 \cdot 7$–$3 \cdot 0$ mmol of potassium are lost with each gram of nitrogen. In this section the principal concern is with situations in which potassium is lost in excess of nitrogen, so that the intracellular concentration of potassium, relative to both nitrogen and water, is reduced.

The commonest cause of renal potassium loss seen in clinical practice is the administration of thiazide and certain other diuretics in the treatment of oedematous states. Many of these drugs cause more sodium than is usual to appear in the distal tubule where it may enter into exchange with potassium; furthermore, any reduction of circulating blood volume caused by the diuretic as well as increased renin release by the diuretic itself may stimulate aldosterone production and enhance the exchange. ('Exchange' does not necessarily imply a tightly coupled mechanism whereby sodium absorption and potassium secretion use the same mechanism. It is more probable that the often observed reciprocal relationships is dictated by the need to maintain electroneutrality). Any situation in which aldosterone secretion is excessive may have the same result. This is seen in its purest form in primary aldosteronism, but also is a prominent feature when aldosterone secretion is secondarily increased, e.g. in renal artery stenosis and accelerated hypertension. In addition, pronounced hypersecretion of cortisol, which has some mineralocorticoid effects, may cause renal loss of potassium; in Cushing's syndrome due to adrenal adenoma or pituitary-driven hyperplasia the cortisol excess is not usually sufficient to produce this effect, but in adrenal carcinoma and in adrenal hyperplasia due to ectopic corticotrophin-producing tumours potassium deficiency is a striking feature.

Another occasional cause of severe potassium depletion due to renal loss is Bartter's syndrome, a condition in which hypokalaemic alkalosis is associated with high plasma renin, angiotensin and raised aldosterone secretion, normal blood pressure, resistance of peripheral vasculature to the vasoconstrictor effects of infused angiotensin and hypertrophy of the juxtaglomerular apparatus. The cause is controversial; defects in tubular sodium potassium or chloride handling have been postulated. Fairly recently it has been discovered that patients with Bartter's syndrome excrete large quantities of prostaglandin E and metabolites of prostacyclin in the urine, and that the low plasma potassium can be partially corrected by administration of aspirin or indomethacin, inhibitors of prostaglandin synthesis. Though it is very far from clear that the disturbance of prostaglandin metabolism is the primary lesion rather than a secondary effect, this rare condition is important because it may point to a role for prostaglandins in potassium and sodium homeostasis.

The importance of the tubular sodium-potassium exchange is well illustrated in patients with primary aldosteronism; in this condition a high sodium diet will speed up the potassium loss and a low sodium diet reduce it; in contrast, in normal subjects aldosterone secretion would be suppressed by a high sodium diet, and thus, although more sodium might be presented to the exchange mechanism, the lack of circulating aldosterone would prevent much increase in potassium loss. The exchange mechanism is also responsible in a somewhat different way for the perpetuation of potassium loss once potassium depletion has been established. In potassium depleted states the plasma chloride is often low, either because the primary cause of the potassium deficit (e.g. vomiting or diuretics) has resulted in chloride loss, or because of excessive renal excretion of chloride, which is an effect of potassium depletion *per se*. Chloride is the main anion together with which sodium is reabsorbed in the renal tubules; a low filtered load of chloride due to a depressed plasma chloride concentration thus results in a fall of the fraction of sodium which is reabsorbed with chloride. The sodium which would have been reabsorbed with chloride is instead reabsorbed in exchange with potassium (and hydrogen) ions to preserve electroneutrality; thus potassium and hydrogen ions are lost into the urine and the potassium depletion perpetuated. In many types of potassium deficiency, it is not possible to repair the depletion completely by potassium administration unless the potassium salt employed is the chloride.

Renal potassium loss is also a feature of certain primary disturbances of acid-base status. In metabolic alkalosis, stimulation of distal tubular secretion of potassium ion results in excessive potassium loss; the evidence suggests that high pH in the distal tubular cells stimulates the entry of potassium into the cell and raises intracellular potassium, resulting in increased potassium movement into the tubular lumen down the electrochemical gradient. It has been pointed out previously how the obligatory excretion of hydroxybutyrate and acetoacetate anions in diabetic ketosis demands the simultaneous excretion of cations; urinary potassium loss is thus an important feature of this condition.

Chronic diarrhoea (due to ulcerative colitis or, for example, to laxative addiction), vomiting due to intestinal obstruction and loss through fistulae, are the usual sources of potassium loss through the gut. Diarrhoea fluid may contain potassium at a concentration as high as 40–70 mmol/l. In addition, in paralytic ileus, considerable amounts of potassium may be sequestered in the fluid contained in dilated loops of intestine; when the ileus resolves, much of the potassium is reabsorbed. In general, a patient whose plasma potassium level has fallen to 3 mmol/l or less because of gut loss would be expected to excrete less than 20–30 mmol/l potassium in the urine per 24 hours. If a greater amount is excreted, then primary renal loss must be suspected.

Effects of potassium depletion

Potassium depletion has profound effects on neuromuscular, cardiac, cerebral and renal function and on acid-base status. In most cases, it is not clear whether the effects are due to the lowered level of plasma potassium which is frequently but not always present, to the reduced intracellular potassium or to an altered ratio of extra- to intracellular potassium concentration (which determines cell membrane potential).

The most dangerous effects of potassium depletion are on cardiac rhythm. Many types of abnormality may be seen, including multiple ectopic beats, tachycardias and ventricular fibrillation. If the patient is receiving digitalis the effects of the drug are enhanced, including the toxic ones. The electrocardiogram in hypokalaemia may show depression of the S-T segment, prolonged Q-T intervals, U waves and tall P waves. Weakness, hypotonia and paralysis of skeletal muscle are often present. In the rare condition of familial periodic paralysis, hypokalaemia is due to movement of extra-cellular potassium into the cells; in this situation there is no whole body potassium depletion and it is, therefore, fairly certain that a low plasma potassium alone can result in paralysis, perhaps through hyperpolarization of muscle cell membranes. Pot-assium depletion also depresses the contractility of smooth muscle and may result in precipitation or prolongation of paralytic ileus—a situation commonly seen after abdominal operation. Tetany may be seen in potassium depletion; surprisingly, it may also be observed during the course of repletion. The rather variable effects on excitable tissues seen in potassium depletion might possibly be explained by variations in the relative concentrations of potassium on either side of the cell membrane.

The effects of potassium deficiency on acid-base status are complex. An extracellu-lar metabolic alkalosis usually occurs (revealed by high plasma bicarbonate concentration), unless the potassium depletion is itself part of a syndrome, such as renal tubular acidosis or chronic diarrhoea, which results in relative retention of hydrogen ions. In addition, in many, but not all circumstances, there appears to be an intracellular *acidosis*. The generation of the extracellular alkalosis involves two types of mechanism. The first is renal; potassium deficiency appears to have a direct effect in increasing bicarbonate reabsorption, and, as mentioned above, associated chloride deficiency, by increasing tubular sodium 'exchange' with hydrogen ions, has a similar effect. If ECF volume depletion is present (e.g. during diuretic therapy), this also appears to contribute directly to further bicarbonate reabsorption. These factors are respon-sible for the urine in potassium depletion usually being neutral, or mildly acid, rather than alkaline, which would be more appropriate to the blood acid-base status; the persistence of paradoxical aciduria has also been partly attributed to the presumed (but unproven) lowered pH within the distal tubular cells in potassium deficiency. The second mechanism responsible for extracellular alkalosis involves movements of ions between cells and ECF. Sodium and hydrogen ions appear to move into the cells, superficially to replace the lost potassium, but the actual mechanism may be that active transport systems which normally extrude sodium and hydrogen ions from cells are less efficient in potassium depletion. These events might be expected to result in intra-cellular acidosis, and this indeed can be demonstrated under certain experimental conditions. Figure 1.1 summarizes the factors discussed in this paragraph.

Full restoration of the alkalosis to normal requires the administration of potassium as the chloride salt, in the same way as does restoration of the K^+ deficit; adminis-

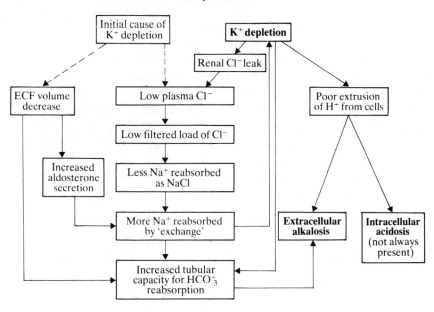

Figure 1.1. *Factors in the development of the acid-base disturbance in potassium depletion.*

tration of chloride as the *sodium* salt can partly, and, in mild cases, completely restore the extracellular bicarbonate to normal.

Potassium deficiency also affects renal function in a number of ways. For reasons which are unknown, glomerular filtration rate is often reduced. Vacuoles are often seen in the cells of the proximal tubules. The sensitivity of the distal and collecting tubules to ADH is reduced; thus polyuria and thirst are frequent symptoms. In spite of the usually slightly acid urine, the kidneys are unable to produce a urine of normally low pH when confronted with an acid load. This may be another manifestation of poor efficiency of hydrogen ion active transport under conditions of potassium deficiency.

Potassium depleted patients also show glucose intolerance, i.e. like diabetic subjects they cannot remove glucose loads from the circulation as quickly as can normal subjects. This may be due to their relatively poor secretion of insulin after a glucose load.

Potassium deficiencies of up to 10 per cent of the body potassium may be well tolerated. In severe cases, up to half of the body potassium may be lost.

POTASSIUM EXCESS

The capacity of cells for storing extra potassium is very small. Normally, large intakes of potassium are rapidly excreted in the urine, but if defective renal function is present the plasma potassium level rises rapidly. Cardiac arrest occurs usually when the level has risen to 7–9 mmol/l. This event is preceded by a characteristic series of changes in the electrocardiogram, consisting of peaking of T waves, abolition of P waves and widening of the QRS complex. This sequence of events occurs principally in oliguric renal failure, particularly when potassium intake is not limited or where tissue break-down is contributing K^+ to the circulation. The degree of hyperkalaemia in renal failure is enhanced by the acidosis which is frequently present (see p. 196). Muscle weakness is a less important feature of potassium overload. The deleterious effects of hyperkalaemia on cardiac muscle can be counteracted by increasing the extracellular concentration of calcium and sodium ions, phenomena which are frequently made use of in the emergency treatment of hyperkalaemia.

SOME SPECIAL SYNDROMES INVOLVING BODY FLUID COMPARTMENTS

Haemorrhage

A sudden reduction of the circulating blood volume sets off a large number of physio-logical mechanisms which seem primarily designed to maintain the blood supply to vital organs such as the brain and heart. The acute loss of blood initially lowers the central venous pressure; the potential effect of this in diminishing cardiac output may be initially counteracted by venoconstriction and reflex tachycardia. If this is ineffective and cardiac output does fall then arterial blood pressure is initially maintained by reflex sympathetic vasoconstriction in the skin and splanchnic areas, and further reflex tachycardia, the afferent signals arising from baroreceptors in the aortic arch and carotid sinus; if the haemorrhage is severe enough the blood pressure falls. Poor renal perfusion, either due to reflex renal vasoconstriction or to low blood pressure itself results in renin release, angiotension generation and increased aldosterone secretion. ADH is also released, perhaps via the left atrial receptor mechanism. These two hor-monal mechanisms cause sodium and water retention and a reduction of urinary output. The increased angiotension and vasopressin levels probably both sufficiently high to contribute significantly to maintenance of blood pressure by vasoconstriction. Cortisol secretion from the adrenal cortex is increased, as a result of pituitary ACTH secretion.

The clinical counterpart of these mechanisms is well known—cold, sweating, vaso-constricted extremities, collapsed superficial veins, tachycardia, hypotension, syncope and oliguria. This picture, loosely referred to as 'shock' is not specific for haemorrhage, but may arise whenever tissue perfusion is inadequate—as for instance in severe ECF volume depletion due to intestinal fluid loss or in 'cardiogenic shock' due to low cardiac output following cardiac infarction.

Haemorrhage is followed by a characteristic series of changes in the haemoglobin level. Immediately after the haemorrhage, there is no change in haemoglobin, but over the following 24–36 hours the level steadily falls, owing to the movement of interstitial fluid into the vascular compartment, because of capillary hypotension, and renal retention of sodium and water. The lost circulating albumin is rapidly resynthesized by the liver. All these events tend to restore the blood volume at the expense of the haemoglobin level, which is gradually regained over many days through increased erythropoiesis. The level of plasma urea may rise substantially after a haemorrhage, as a result of poor renal perfusion, protein breakdown as a part of the metabolic response to injury (see below) and, if the haemorrhage has been into the gut, absorption of the products of protein digestion.

Metabolic response to injury

Any injury (including surgical operation) of more than a trivial nature is followed by a characteristic sequence of events, which include sodium and water retention, hyponatraemia and potassium and nitrogen loss. Water retention, independently of sodium, is probably largely due to increased ADH secretion and this phase only lasts for about 24 hours. In contrast, sodium retention, partly related to increased circulating levels of aldosterone and cortisol, may last for several days and is of particular importance in patients whose cardiac function is suspect or inadequate; in these circumstances, the post-operative administration of more than minimal quantities of sodium may precipitate or worsen heart failure. In spite of sodium retention, post-operative hyponatraemia is often quite marked; possible reasons include inappropriate ADH secretion and movement of water out of cells because of potassium loss. The negative nitrogen balance is short-lived in uncomplicated cases; it is accompanied by a similar short-lived potassium loss, which is, however, in excess of that which would be expected from the nitrogen loss.

Principles of clinical tests and measurements

PLASMA SODIUM

A raised plasma sodium level is due either to water depletion or to excessive administration of sodium salts. The interpretation of a low plasma sodium is more difficult; there are three main possibilities:

1 Sodium depletion in excess of water (e.g. intestinal loss with inappropriate replacement by hypotonic fluids; adrenal cortical insufficiency).

2 Water overload (e.g. hypopituitarism; most types of inappropriate secretion of ADH).

3 'Reset' of osmoreceptors.

It is of great importance to distinguish between these possibilities. Frequently, hyponatraemia is regarded as an indication for administration of sodium; this is, of course, appropriate for possibility (1), but in (2) or (3) will make little difference to the hyponatraemia, and if any salt-retaining condition, such as heart failure, is present, only make matters worse. The first possibility may be distinguished from the others by the evidence of extracellular volume depletion that will be present—the earliest manifestation is postural hypotension, but later a raised or rising plasma urea and other signs associated with ECF volume depletion will be evident. In (2) the urine is usually persistently hyperosmolar to plasma and cannot be appropriately diluted; the plasma urea is usually lower than normal. In (3) the urine can be concentrated and diluted normally, but in response to a lower level of plasma osmolality than normal.

The plasma sodium as measured by flame photometry may appear falsely low because of raised plasma protein or lipid concentrations (see p. 2). Under these circumstances, water comprises an abnormally small fraction of a unit volume of plasma.

PLASMA POTASSIUM

The level of plasma potassium is frequently used as an indicator of whole body potassium deficiency but, in fact, only correlates rather poorly with body potassium stores and under some circumstances moves in an opposite direction to that anticipated. Thus in systemic alkalosis, during glucose entry into cells under the influence of insulin, in familial periodic paralysis and in response to high circulating catecholamines the cells take up potassium; the resulting hypokalaemia is unrelated to the total body potassium. Nevertheless, in the absence of such factors, a low level of plasma potassium is a useful indicator that a degree of potassium depletion is present.

As might be expected from the effect of alkalosis in shifting potassium into cells, systemic acidosis frequently results in movement in the opposite direction and hyperkalaemia is a common feature of both respiratory and metabolic acidosis. The mechanism by which acid-base changes affect potassium movements across cell membranes is uncertain. Hyperkalaemia is a common occurrence in both chronic and acute renal failure, where the effect of acidosis is augmented by failure to excrete both ingested potassium and potassium derived from tissue breakdown. Failure of potassium excretion, together with exit of potassium from the cells is the basis of the hyperkalaemia seen in primary adrenal insufficiency.

PLASMA CHLORIDE

Alterations in plasma chloride levels generally reflect those of plasma sodium, except where changes in acid-base status are also present, when plasma chloride and bicarbonate tend to change in a roughly reciprocal fashion, except when increased quantities of 'unmeasured anions' are present (see 'anion gap' below). For reasons outlined above plasma chloride is low in a large number of patients with potassium deficiency.

PLASMA BICARBONATE

The factors determining the plasma concentration of this anion are discussed in Chapter 5.

ANION GAP

It is often useful to calculate the 'anion gap'. This is done by adding together the concentrations of the principal plasma cations ($Na^+ + K^-$) and subtracting from the result the main plasma anions ($Cl^- + HCO_3^-$). The difference, which is normally 10–16 mmol/l, is due to unmeasured anions, mainly negatively charged proteins in health, but phosphate, sulphate, lactate, pyruvate, 3-hydroxybutyrate, acetoacetate and other known and unknown anions contribute. The anion gap may be increased in uraemia (due to the accumulation of non-volatile acids), in diabetic ketoacidosis (due mainly to 3-hydroxybutyrate, acetoacetate), in salicylate poisoning (due to salicylate, lactate and ketoacids) and in lactic acidosis. The main value of the anion gap is in the diagnosis and management of metabolic acidoses; as an example, the presence in a patient with metabolic acidosis of a large anion gap in the absence of the first three causes listed above is strong evidence for lactic acidosis. Patients who have metabolic acidosis due to causes other than the accumulation of organic acids (e.g. renal tubular bicarbonate leak) have a normal anion gap, in which the fall in plasma bicarbonate is compensated for by a rise in plasma chloride. The anion gap may also be lowered; this may be due to hypoproteinaemia, but may also alert the physician to the presence of a positively charged paraprotein in the plasma, as may occur in myelomatosis.

PLASMA AND URINE OSMOLALITY

Osmolality is a measure of the osmotic pressure of a solution and is related to the molality (or combined molality of a mixture of solutes) by the activity coefficient(s). Osmolality is measured in mosmoles/kg solvent, in contradistinction to osmolarity, which is measured in mosmoles per litre of whole solution. Osmolality is the relevant measurement when considering osmotic equilibrium between body compartments.

Sodium and its main accompanying anions, chloride and bicarbonate are under normal circumstances by far the major contributors to plasma osmolality. Thus, since all these ions are monovalent, the concentration of serum sodium correlates well with the plasma osmolality. When an osmometer is not available, plasma osmolality can be approximated by the following expression:

$$2(Na^+ + K^+) + \text{plasma glucose} + \text{plasma urea}$$

(all measured in mmol/l). It can be seen from this expression that in hyperglycaemia and uraemia the simple relationship of measured plasma osmolality to the plasma sodium is upset.

Urine osmolality is not consistently related to any one solute and is, therefore, a more indispensable measurement than plasma osmolality. A normal adult should be able to achieve a value of at least 900 mosmol/kg after 18 hours of water deprivation. Measurement of urine osmolality under such conditions is of value in the assessment of failure of urinary concentration, whether of renal or non-renal origin.

PLASMA UREA AND CREATININE CONCENTRATIONS

The following are the main factors affecting the level of plasma urea:
1 Renal function.
2 The intake of protein.
3 The rate of catabolism of endogenous protein.
4 The rate of reincorporation of urea nitrogen into amino acids.

Plasma urea is usually measured as a test of renal function, but obviously can only be regarded as such if factors 2–4 are of normal magnitude. Similarly, a change in plasma urea can only be regarded as reflecting change in renal function if the other factors remain unaltered. Thus, in renal failure, high levels of plasma urea can be greatly reduced by lowering the protein intake, especially if most of the intake consists of essential amino acids in the optimal proportion for efficient utilization, thereby reducing catabolism of muscle protein to supply essential amino acids. It has been shown that, in normal and uraemic subjects, urea, usually regarded as an inert end-product, may be reconverted into amino acids, especially under conditions of reduced protein intake. When endogenous protein catabolism is high, as in the metabolic response to injury or infection, increased endogenous urea production raises the level of plasma urea. Plasma urea level rises progressively during sodium and water depletion because of lowered glomerular filtration rate. Hyponatraemia is unlikely to be due to sodium depletion unless the plasma urea is raised or rising.

Because of the many factors affecting plasma urea, the measurement of plasma creatinine, which is closely related to glomerular filtration rate, is preferable as an indicator of renal function.

PACKED CELL VOLUME (PCV)

The PCV measures the fraction of a volume of blood which is occupied by red cells. Thus, changes in plasma volume which are not accompanied by proportional changes in red cell mass may be recognized from serial measurements of PCV. Such observations are, in practice, principally of value in recognizing dehydration due to sodium and water loss, where the volume depletion is shared between the vascular and interstitial compartments. One obvious difficulty is lack of knowledge of the value before the start of the illness. Thus, if the patient was initially anaemic or polycythaemic, entirely misleading conclusions could be drawn.

PLASMA PROTEINS

Because albumin is a relatively small molecule, it usually accounts for over 90 per cent of the colloid osmotic pressure, although it only contributes about 60 per cent of the total plasma protein concentration in grams per litre. Thus, plasma albumin concentration rather than total protein concentration is the principal osmotic factor determining the distribution of fluid between the vascular and interstitial spaces. Although it is often said that oedema appears when the plasma albumin concentration is below 3 g per 100 ml, this relationship is, in practice, extremely variable. This might be expected since hydrostatic forces in the capillary and interstitial space are also important factors in determining fluid partition and may themselves be varying.

A low plasma albumin concentration is usually a result of either poor synthesis, as in starvation or liver disease or increased loss, as in the nephrotic syndrome or protein-losing enteropathy. High levels of albumin usually indicate dehydration, but may be artefactual, due to prolonged congestion of the arm during venepuncture.

CENTRAL VENOUS PRESSURE (CVP)

This is a standard clinical measurement achieved by inserting a catheter through a forearm or neck vein into the right atrium or superior vena cava, and attaching it to a saline manometer, the reference point of which is selected as the level of the centre of the right atrium. Normal CVP ranges from 0–11 cm H_2O (supine). Its main use is in the assessment of the degree of filling of the venous system. Thus in haemorrhage or fluid loss, the CVP may be low and appropriate fluids may be administered until the CVP is restored to the normal range. In over-transfusion or in myocardial failure (often due to iatrogenic blood or fluid overloading) the CVP is raised. Caution is, however, necessary in making these simple deductions. The CVP may be low in spite of a normal blood volume because of low venous tone, or normal in the presence of low blood volume because of high venous tone. Furthermore, the CVP is not necessarily directly related to the filling pressure of the *left* side of the heart and it is, therefore, wise in the assessment of fluid overload to take into account other evidence of left ventricular performance as well.

MEASUREMENT OF FLUID COMPARTMENT VOLUMES

The volumes of the various body fluid compartments can only be measured by dilution. A known quantity of an appropriate substance, thought to be largely confined to the space whose volume is required, is injected intravenously (or taken by mouth if intestinal absorption is complete) and its concentration measured in the plasma (or preferably the plasma water) when even distribution within the space has occurred. Ideally, multiple samples of plasma should be assayed to determine the equilibrium concentration when all the mixing processes are complete but in practice somewhat arbitrary equilibration periods are generally taken, depending on the space being measured. The volume of distribution is given by the amount of indicator remaining in the body (i.e. the amount given minus the amount that has left the body during the period of equilibration) divided by the equilibrium concentration.

Plasma is the only fluid space that can be sampled directly so that only the volume of the plasma, extracellular fluid and total body water can be measured directly; the volumes of the interstitial cell fluid and the intracellular fluid have to be assessed by subtraction. An inadequate equilibration period tends to produce an underestimation of the volume, and leakage out of the space in question tends to produce an overestimation. When all losses during the period of equilibration cannot be measured, the volume of distribution may be obtained by extrapolation from the slope of the plasma concentrations after mixing is complete. This makes the additional assumption that losses are taking place at a steady rate and there are the inherent difficulties of determining the precise slope.

Total body water is generally measured from the volume of distribution either of tritiated water (^3H_2O), which is radioactive, or of 'heavy water' (D_2O). Urea, ethanol and antipyrine can also be used.

Approximations to the extracellular fluid volume can be made from the volume of distribution of thiocyanate, inulin, sulphate, chloride, bromide or radioactive sodium. Inulin, a relatively large molecule, is partly excluded from the interstices in the meshes of mucopolysaccharide molecules in ground substance, so that the values obtained underestimate the ECF volume; sodium, bromide and chloride inevitably penetrate across cell membranes to some degree so their volumes of distribution overestimate the ECF volume.

The blood volume can be measured from the volume of distribution of albumin (labelled with a dye such as Evans' blue or with radioactive iodine), or from the volume of distribution of suitably labelled erythrocytes (usually labelled with radioactive chromium). In both cases, the estimation of final concentration is often made on whole blood; the answers obtained with the two markers are not quite identical, mainly because of variations in PVC between the blood obtained from peripheral veins and that in certain organs. Labelled albumin disappears from the circulation at varying rates and misleading values are often obtained if the extrapolation technique is not used. Blood volume measurements are quite frequently used in conjunction with observations of pulse rate, arterial pressure and CVP to gauge the rate and volume of transfusion in the treatment of haemorrhage.

BODY SODIUM AND POTASSIUM CONTENT

The total *exchangeable* sodium and potassium in the body are measured by the same principles of dilution used for body space measurements except that the quantity of marker in the body at equilibrium is divided by the ratio of labelled to unlabelled electrolyte (i.e. the so-called specific activity), instead of the ratio of the labelled marker to water (i.e. the concentration); the same limitations apply. The proportion of *total* body sodium and potassium which the total exchangeable values represent is discussed on p. 9.

A very small but constant proportion of the whole body potassium consists of the naturally occurring radioisotope ^{40}K. This natural radioactivity can be measured in a very sensitive and suitably calibrated whole body counter, and is a direct estimate of the total body potassium. This method, though technically difficult, has considerable potential, since repeated observation can be made without injections of radioisotopes, and there are no equilibration problems.

EXTERNAL BALANCE STUDIES

In this technique, the intake and total output of the substances under study (e.g. Na or K) is measured and it can, therefore, be deduced whether net gain or loss of the substance is occurring. If carried out over long periods, small errors due, for example, to the impossibility of measuring losses through the skin, may accumulate and cause a major error. Nevertheless, the technique is of considerable clinical value in some circumstances—for instance in gauging sodium and potassium replacements during prolonged ileus after abdominal surgery.

Practical assessment

HAEMORRHAGE

Clinical observations

Loss of less than 10 per cent of blood volume: no definite signs. Increasing losses cause, in roughly the following sequence: pale, cold skin; postural hypotension; empty veins; tachycardia; small pulse pressure; hypotension even when lying flat; sweating; restlessness; hyperventilation; mental confusion.

Routine methods

Immediately after haemorrhage: reduction in CVP.
Within a few hours: falling PCV and Hb; rising plasma urea, particularly with blood loss into the bowel; rising plasma creatinine.

Special techniques

Volume of distribution of ^{131}I-albumin or Evans' blue (plasma volume); cardiac output.

WATER DEPLETION

Clinical observations

Thirst; dry mouth; oliguria. Later coma when it may be difficult to distinguish cause from effect. No physical signs until late. Then, loss of skin turgor, tachycardia; hypotension; fever (particularly in infants).

Routine methods

Concentrated urine; hypernatraemia; increased plasma osmolality.

HYPEROSMOLAR STATES DUE TO SOLUTE EXCESS

Clinical observations

Disorientation, fits, coma.

Routine methods

Elevated plasma glucose, plasma sodium, plasma osmolality.

SODIUM DEPLETION

Clinical observations

Lethargy; postural hypotension; thirst; inelastic subcutaneous tissues. Later, signs of circulatory failure, often with vomiting. Cramps with sodium depletion in excess of water. Physical signs not prominent until fluid loss > 6 per cent of body weight, but postural hypotension appears earlier.

Routine methods

Low CVP, weight loss, raised PCV and Hb; rising plasma urea and creatinine; plasma sodium concentration normal or low. Urine concentration of sodium and chloride very low except when renal sodium loss is the primary cause.

Special techniques

Total exchangeable sodium; balance studies; plasma volume; ECF volume (e.g. bromide or inulin space).

POTASSIUM DEPLETION

Clinical observations

Weakness and reduced tone of limb muscles; cardiac arhythmias; constipation; paralytic ileus. Sometimes carpo-pedal spasm (or positive Trousseau sign). Thirst; polyuria. But often no clinical evidence.

Routine methods

Plasma potassium concentration often, but not necessarily, low. Raised plasma bicarbonate and low plasma chloride concentration, unless depletion due to diarrhoea or renal tubular acidosis; arterial pH, Pco_2, ECG. Urine potassium to determine route of depletion.

Special techniques

Total exchangeable potassium; balance studies; tests for renal functional lesions of potassium depletion (p. 29). Whole body ^{40}K counting.

WATER OVERLOAD

Clinical observations

Restlessness; nausea; vomiting; headache; variable neurological signs; fits; coma.

Routine methods

Lowered plasma sodium. Comparison of urinary and plasma osmolality. Plasma urea normal or low.

Special techniques

Total body water.

OVERLOADING WITH BLOOD, PLASMA OR SALINE

Clinical observations

Raised jugular venous pressure; pulmonary and/or peripheral oedema; effusions.

Routine methods

Increased central venous pressure.

Special techniques

As for body fluid deficits.

References

BARTTER F.C. & SCHWARTZ W.B. (1967) The syndrome of inappropriate secretion of antidiuretic hormone. *Amer. J. Med.* **42,** 79.

EARLEY L.E. & DOUGHARTY T.M. (1969) Sodium metabolism. *New Eng. J. Med.* **281,** 72–86.

EDSALL J.T. & WYMAN J. (1958) *Biophysical Chemistry*, Chapter 2. New York, Academic Press.

GAUER O.H., HENRY J.P. & BEHN C. (1970) The regulation of extracellular fluid volume. *Ann. Rev. Physiol.* **21,** 547–94.

KASSIRER J.P., BERKMAN P.M., LAURENZ D.R. & SCHWARTZ W.B. (1965) The critical role of chloride in the correction of hypokalaemic alkalosis in man. *Amer. J. Med.* **209,** 655–8.

LUKE R.G. & LEVITIN H. (1967) Impaired renal conservation of chloride and the acid-base changes associated with potassium depletion in the rat. *Clin. Sci.* **32,** 511–26.

LEDINGHAM J.M. (1971) Blood pressure regulation in renal failure. *J. Roy. Coll. Phys. London.* **5,** 103–34.

MORGAN D.B. & THOMAS T.H. (1979) Water balance and hyponatraemia. *Clin. Sci.* **56,** 517–522.

MAXWELL M.H. & KLEEMAN C.R. (1962) *Clinical Disorders of Fluid and Electrolyte Metabolism.* New York, McGraw Hill.

ORLOFF M.J. (1970) Pathogenesis and surgical treatment of intractable ascites associated with alcoholic cirrhosis. *Ann. N.Y. Acad. Sci.* **170,** 213–36.

ROBERTSON G.L., SHELTON R.L. & ATHAR S. (1976) The osmoregulation of vasopressin. *Kidney International* **10,** 28–37.

ROBINSON J.R. (1960) Metabolism of intracellular water. *Physiol. Rev.* **40,** 112–49.

SELDIN D.W. & RECTOR F.C. (1972) The generation and maintenance of metabolic alkalosis. *Kidney International,* **1,** 306–21.

SNASHALL P.D., LUCAS J., GUZ A. & FLOYER M.A. (1971) Measurement of the interstitial 'fluid' pressure by means of a cotton wick in man and animals: an analysis of the origin of the pressure. *Clin. Sci.* **41,** 35–53.

SQUIRES R.D. & HUTH E.J. (1959) Experimental potassium depletion in normal human subjects. I. Relation of ionic intakes to the renal conservation of potassium. *J. Clin. Invest.* **38,** 1134–1148.

STOCKIGT J.R. (1977) Potassium homeostasis. *Austral. & New Zealand J. Med.* **7,** 66–77.

SUMMERSKILL W.H.J. (1969) Ascites: the kidney in liver disease. In SCHIFF L. (ed.). *Diseases of the Liver,* p. 355. Philadelphia, Lippincott.

WALLACE M., RICHARDS P., CHESSER E. & WRONG O.M. (1968) Persistent alkalosis and hypokalaemia caused by surreptitious vomiting. *Quart. J. Med.* **37,** 577–88.

2

Heart and Circulation

Normal function

THE HEART

All cardiac muscle has the intrinsic capacity for rhythmic excitation. The contractile muscle fibres have different rates of spontaneous rhythmic contraction; but their coupling together as a syncytium and through specialized conducting tissue, so that the activation process can spread from cell to cell, normally leads to coordinated contraction. This starts at the sino-atrial node, the tissue with the fastest natural rate and which is under autonomic control. The tendency to spontaneous activity of most fibres is normally suppressed by the faster rate of the sino-atrial node, but the atria also contain specialized fibres which can be provoked into spontaneous rhythmic contraction. Ventricular muscle fibres can also contract on their own at a slow rate, but are normally excited only through the specialized conducting tissue.

Activation of the atria from the sino-atrial node at the junction of the right atrium and superior vena cava spreads rapidly producing an almost simultaneous activation of the two atria. Activation of the right atrium slightly precedes the left, and the two corresponding components of the P wave can be distinguished on the ECG. The atrioventricular node at the base of the atrial septum acts as the transmitting station to feed the activation process to the ventricles. It is also responsible, by slow conduction between the cells, for the delay of about 0·20 s between atrial and ventricular activation, thus allowing atrial contraction to finish ventricular filling before the ventricles contract. Fast conducting Purkinje fibres carry the impulse down the bundle of His and its branches to reach all parts of the ventricle almost at once, thus ensuring the coordinated contraction necessary for efficient systole. The processes of excitation by depolarization, coupling by entry of calcium ions, contraction itself, and the energy pathways are essentially similar to those of skeletal muscle.

In a young adult whose cardiac nervous connections and response to circulating catecholamines have been blocked by drugs the heart contracts about 110 times/min. This intrinsic rate falls with age and is about 80 beats/min at the age of 70 years. The intrinsic rate is modified by the central nervous system, which can slow the heart by means of impulses transmitted in the vagi, and speed it up by impulses in the sympathetic cardioaccelerator nerves and by catecholamines released from the adrenal medulla. In a supine young adult at rest the heart is under predominantly vagal influence and contracts about 70 times/min, but can, for example, with exercise accelerate to 200 beats/min or more. This is an important mechanism for increasing cardiac output when venous return increases on exercise. By the age of 60 the maximum heart rate is about 170, presumably a reflection of the fall in intrinsic heart rate with age.

The healthy heart obeys Starling's law, which states that the force of cardiac contraction increases with increasing diastolic stretch of the cardiac muscle fibres. If neurohormonal influences do not change, the volume of blood expelled at each beat (the stroke volume) depends on cardiac filling; the atrial boost to filling just before ventricular systole becomes particularly important under stress. At a heart rate of 70 beats/min in a young adult at rest, each ventricle contains at the end of diastole about 120 ml of blood, 70 ml of which is expelled at each systole. The resting cardiac output is therefore approximately 5 l/min.

The force of cardiac contraction is continuously modified by the ionic environment, especially by the concentrations of potassium and calcium in the blood. Vagal inhibitory nerve impulses reduce heart rate and also slightly reduce the force of atrial contraction. Vagal stimulation also slows conduction at the atrio-ventricular node, but vagal influences do not extend to the ventricular muscle, whereas circulating catecholamines and sympathetic cardioaccelerator nerve activity increase heart rate, improve conduction and can greatly increase the force of atrial and ventricular contraction. However, the cardiac output can only be increased some 10% even by maximal activation of the sympatho-adrenal system unless venous return is augmented at the same time. Venous return in the absence of skeletal muscle contraction is essentially a passive process.

THE PULMONARY CIRCULATION

The normal mean pulmonary capillary pressure is probably only a few mm Hg above mean alveolar pressure, which in a normal subject breathing quietly is about − 10 mm Hg. Therefore little extra pressure is needed to drive blood into the lungs, and the pulmonary artery pressure is normally about 20/8 with a mean of about 12 mm Hg. The mean left atrial pressure is about 4 mm Hg, the left ventricular pressure about 120/5, and the aortic pressure 120/80, with a mean of about 93 mm Hg. All these pressures are referred to the level of the tricuspid valve (in practical terms approximately mid-chest in a horizontal subject), where the mean right atrial pressure is about atmospheric, i.e. 0 mm Hg. At the sternal angle, semi-recumbent, (often used as a base-line clinically), the measured pressures are some 5 mm Hg lower. Since pressures are higher on the left side, communication at any level produces a shunt from left to right unless the pulmonary vessels are abnormal. Pulmonary capillary pressure is normally about 7 mm Hg, but varies in different parts of the lung according to posture. This normally low pressure favours reabsorption of fluid by colloid osmotic forces (see below) and protects against pulmonary oedema (p. 63).

Because of the low cardiac filling pressure, changes in intra-thoracic pressure have a big influence on venous return and hence on cardiac filling and cardiac output. During deep inspiration the intrathoracic pressure may be -25 mm Hg, and during diastole most of this pressure is transmitted to the lax right ventricle, which sucks blood in. The diastolic volume of the right ventricle therefore increases and the right ventricle has a greater amount of blood to eject. Expiration on the other hand, squeezes blood out of the lungs so that the output from the left ventricle increases. Right ventricular contraction begins just before the left but finishes later, so that the pulmonary second sound (marking pulmonary valve closure) follows the aortic second sound (marking aortic valve closure). The first sound is associated with atrioventricular valve closure. The tricuspid and mitral components of the sound may be heard separately as a split first sound.

THE CARDIAC CYCLE

The rhythmic activity of the heart produces alternate contraction and relaxation (systole and diastole) and in the case of the ventricles leads to pulsatile filling of the arterial tree alternating with periods of ventricular filling. Atrial contraction is timed just before ventricular systole to add a final boost to ventricular filling (Fig. 2.1). Closure and opening of the atrio-ventricular valves (mitral and tricuspid) and the semilunar valves (aortic and pulmonary) in relation to the pressure changes leads to further subdivisions of ventricular systole. Isometric (more properly 'isovolumic', as some change in ventricular shape may occur) contraction occurs after the atrioventricular valves close. This produces the first heart sound which occurs before the semilunar valves open, allowing ventricular ejection. A further isometric relaxation phase occurs after the semilunar valves close, creating the second heart sound. Then follows atrio-ventricular valve opening, giving an opening snap in the more rigid valve of mitral stenosis.

Ventricular filling is rapid at first, and when filling is nearly complete a soft low-pitched third heart sound is heard in normal children and adolescents, but is lost in adult life, to return under conditions of enhanced blood flow or in the presence of ventricular muscle disease. In the succeeding portion of diastole there is little flow and atrial and ventricular pressures equilibrate (diastasis). Atrial systole adds a final boost to ventricular filling and if accentuated in ventricular muscle disease may produce a fourth (or atrial sound) of similar quality to the third heart sound (i.e. low-pitched).

REGULATION OF CARDIAC OUTPUT

Some 5 l of blood fill the heart and circulation. The system is relatively stiff and removal or addition of relatively small volumes of blood, such as 1 pint (600 ml), have a profound effect on venous filling. In the absence of compensatory factors such manoeuvres would change venous return, and hence cardiac output, over the range of about 2 to 8 l/min. In practice cardiac output is stabilized by a number of compensatory mechanisms, which are listed in order of rapidity of action, though this is not necessarily the order of importance for long-term adjustment.

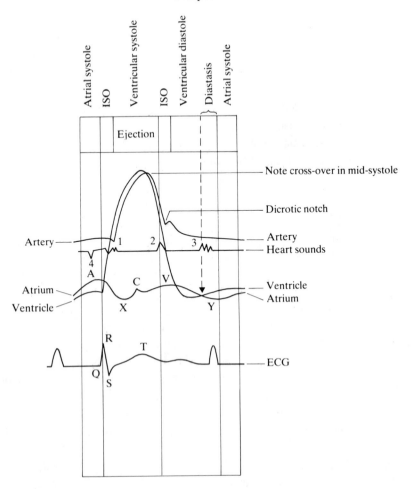

Figure 2.1. *The Cardiac Cycle*

A schematic diagram of ventricular, atrial and arterial pressure pulses to show the phases of the cardiac cycle in relation to the triggering events of the electrocardiogram (ECG) and the heart sounds. Aortic and left ventricular pressure have been shown to cross-over in mid-systole as the ventricle changes from active ejection to passive emptying. 'ISO' marks the isometric phases of systole and diastole, when both atrioventricular and semilunar valves are closed. The period of equilibration of ventricular and atrial diastolic pressures after the third heart sound is known as 'diastasis'.

The sympathetic system. Receptors in the great veins and atrial walls detect the degree of filling and stretch of these organs. Ventricular wall tension can also probably be detected by mechanoreceptors in the inner surface of the ventricular wall. In addition, a primary change in cardiac output inevitably changes systemic arterial pressure, which can be detected by other stretch receptors in the walls of the large arteries, especially the carotid sinuses. A fall in blood volume decreases central venous pressure, cardiac filling, and systemic arterial pressure, and nervous reflexes mediated through the hind brain and sympathetic system rapidly restore cardiac filling by mobilizing blood from venous reservoirs, especially in the abdomen, where the liver has the largest mobilizable blood pool. Traditionally increased venous return has been attributed to constriction of veins, but it now seems that constriction of arteries supplying the splanchnic region may be the major factor. Clamping the aorta of a dog above the coeliac axis actually increases cardiac output, because the compliant splanchnic bed is less distended by arterial pressure and therefore empties much of its pooled blood into the general circulation.

Capillary fluid exchanges. The circulation normally contains some 3 l of fluid (as plasma) and there are about 11 l in the interstitial space (p. 3). The partition of this extracellular fluid depends on a dynamic equilibrium at the capillaries—between hydrostatic pressure tending to force fluid out and osmotic pressure of non-diffusible colloids tending to retain fluid in the plasma (Starling's capillary law). Any loss of blood, which must initially reduce the mean capillary pressure (normally about 15 mm Hg) tends to disturb the equilibrium and sucks fluid from the interstitial space (normally at a negative pressure of about −7 mm Hg) into the circulation. Quantitatively this mechanism is probably not very important in stabilizing cardiac output, because the compliance of the interstitial space is normally very low. Removal of a small volume of fluid lowers interstitial pressure greatly. Since water in the interstitial fluid is contained in a mucopolysaccharide gel, its free movement is much restricted, except in the presence of oedema. After the loss of 600 ml of blood from the circulation, for example, approximate calculations suggest that at best only about 250 ml of fluid could be transfered from the interstitial space into the blood stream before a new equilibrium was established at a more negative interstitial pressure. Thus this mechanism could immediately correct only about one-third of any blood volume deficit. Further replenishment of intravascular fluid comes either from fluid ingestion or from intracellular water.

Renal regulation of cardiac filling. The kidney plays a central role in cardiac output regulation, because it is the final arbiter of sodium and water excretion rate, and hence of extracellular fluid volume and circulatory filling. Regardless of other nervous or hormonal mechanisms, any reduction of renal perfusion pressure which might initially result from a fall of cardiac output reduces the volume and sodium concentration of the urine, thus conserving extracellular fluid. Afferent arteriolar constriction mediated by the sympathetic system has a similar effect, and probably plays a most important part in conserving extracellular fluid. It is notable that renal blood flow falls sharply after even slight blood loss, though there may be scarcely any change in systemic arterial pressure. In addition to the intrinsic and nervous mechanisms, a fall in venous return and cardiac output after haemorrhage is a powerful stimulus to the release of vasopressin from the posterior pituitary gland and hence to the conservation of water.

This stimulus can overrule the normal control of vasopressin secretion rate in accordance with plasma osmolality.

It is obvious that these renal mechanisms can compensate for an increase in cardiac output by increasing sodium and water excretion but they cannot in themselves compensate for reduced cardiac output. They can only prevent it getting worse by making the kidney retain fluid. However, thirst increases after haemorrhage, and increased fluid and continued salt ingestion in association with diminished renal excretion can eventually replace any loss of water and sodium.

Another renal mechanism which helps to protect cardiac output is the renin-angiotensin system. Reduced cardiac filling provokes the release of renin from the kidneys. The stimulus for the release of renin after moderate haemorrhage is not the fall in systemic arterial pressure itself, which may be scarcely measurable in the presence of intact vasomotor nervous reflexes, but rather increased sympathetic nervous impulses to the kidneys. There is sympathetic innervation of the juxta-glomerular apparatus, which is the site for renin storage. Renin release after slight haemorrhage is prevented by denervating the kidneys. In addition, the renal nerves may act indirectly by bringing about afferent arteriolar vasoconstriction and a fall of glomerular perfusion pressure. Severe systemic hypotension reduces glomerular capillary pressure *pari passu*. It is generally believed that this pressure, or some function of it, controls the rate of renin secretion. Renin catalyses the production of angiotensin (p. 49), which is a powerful vasoconstrictor, acting not only directly on arterioles, but also indirectly through the nervous system, to produce a sympathetically-mediated vasoconstriction. Over longer periods of time, the stimulation of aldosterone secretion by angiotensin will contribute to renal sodium retention, and in yet another way help to correct plasma volume depletion.

THE PERIPHERAL CIRCULATION

The systemic vasculature comprises the aorta and central arteries; the peripheral muscular arterial tree feeding the arterioles (arteries less than 150 μ internal diameter); the capillary bed; and the veins.

Aorta and the central arteries

The left ventricle ejects about two-thirds of its stroke volume in the first third of systole. Ejection performance is mainly determined in this early part of systole when the left ventricle is forcing blood into the aorta. Later in systole ventricular emptying is largely passive, following the initial kick given to the ventricular contents by muscular contraction; and pressure in the aorta comes to be slightly greater than that in the left ventricle. In diastole, after momentary retrograde flow, the aortic wall recoils, releasing the potential energy created by its distension in systole, and accelerating the forward flow of blood. The process is continued in the larger muscular arteries such as the iliac and brachial, and these arteries also contribute a small amount to the total peripheral arterial resistance when flow is greatly increased, e.g. during muscular exercise. The aorta and major central arteries act as a pressure chamber, converting the intermittent emptying of the left ventricle to an almost continuous flow of blood into the tissues.

Arterioles

An arteriole has a muscular wall which is thick in comparison with its lumen. The arterioles contribute most of the systemic vascular resistance, but once opened at a critical pressure they offer little further resistance to flow—the waterfall effect. In many tissues the rate of flow appears to be almost independent of pressure at the distal (capillary) end, provided that this is less than tissue pressure. It is likely that the number of arterioles open at any one time, rather than the average calibre of all the arterioles normally governs the vascular resistance offered by any tissue.

Capillaries

The total cross-sectional area of the capillary bed is more than 600 times that of the aorta but not all capillaries are open at any one time. In skeletal muscle at rest only 2 or 3 capillaries may be open in each mm^3 of tissue, whereas this may go up to 200 to 300 during exercise. Capillaries have no power of independent contraction, and are either fully open or fully closed according to the state of the feeding arterioles. In many tissues, notably skin, there are contractile arterio-venous anastomoses which allow a variable amount of blood to by-pass the capillary bed altogether. Some blood normally passes along these by-pass channels during rest, playing no known physiological role except in the case of the skin, in which it regulates heat loss.

Veins

The small veins have a large total cross-sectional area and are the main component of the circulatory volume, holding much more blood then the large diameter major arteries and veins. In a supine subject at rest, return of blood from the capillary bed is entirely a passive process, and depends on the difference between the mean capillary pressure (probably about 15 mm Hg) and right atrial pressure (about 0 mm Hg). In a normal erect subject passive distention of veins in the lower parts of the body would accommodate so much blood that cardiac filling would almost cease. Venous return during standing is sustained by at least three mechanisms. One is the pumping action of the muscles which compresses segments of peripheral veins isolated by valves, and propels blood towards the heart. A second is the reflex sympathetic constriction of the splanchnic arterioles and venous reservoirs. A third is the intrinsic, as well as reflex, shutting down of arteriolar sphincters in dependent limbs, greatly reducing the rate of flow of blood into the limbs.

SYMPATHO-ADRENAL VASCULAR RECEPTORS

The main peripheral vascular resistance lies in the arterioles (see above) whose innervation is dominantly sympathetic, adrenergic and vasoconstrictor. Although a cholinergic vasodilator system exists in several animals, it is doubtful whether there is one in man. Most circulatory reflexes, except the faint reaction, appear mainly to involve changes in sympathetic adrenergic vasoconstrictor activity.

The sympathetic receptors responsible for arteriolar constriction are generally termed α-adrenergic and can be stimulated by noradrenaline, which is the normal transmitter at the postganglionic sympathetic nerve endings; the vasoconstrictor action on skeletal muscle brought about by locally infused noradrenaline can be blocked by substances such as phentolamine and phenoxybenzamine, which are described as α-adrenergic blocking agents. Probably because noradrenaline is released from nerve terminals applied extremely close to the cell membrane, no locally infused blocking agent can completely prevent the vasoconstrictor response to adrenergic nerve stimulation. There are presynaptic α-receptors which when activated by released noradrenaline promote further depolarization. Thus blocking drugs may act presynaptically as well as on the effector cell, and a distinction has been made between α_1 and α_2-blocking agents. By contrast, the local infusion of adrenaline into the arterial supply of a skeletal muscle results in vasodilatation, which is described as a β-adrenergic response, and can be prevented by a different type of drug such as propranolol, which is described as a 'β-blocking' agent. The adrenergic receptors responsible for sympathetically mediated increases of rate and force of cardiac contraction are also blocked by β-blocking drugs. Normally the predominant action of the α-adrenergic vasoconstrictor effect completely swamps the β-adrenergic vasodilator action on general sympathetic stimulation, although both processes are mediated by the natural transmitter, noradenaline; general β-adrenergic vasodilatation is evident only in the presence of an α-blocker.

Because both adrenaline and noradrenaline possess both α- and β-stimulating activity, which varies in different species and in different organs of the same species, it is more satisfactory to define the nature of a receptor as α- or β- in terms of the effects upon it of α- and β- blocking drugs rather than in terms of its response to nordrenaline and adrenaline. Similarly two types of β-receptor (β_1 and β_2- predominantly cardiac or peripheral) have been recognized by different responses to different β-blocking drugs.

Veins, like most arterioles, possess an adrenergic vasoconstrictor innervation and constrict to noradrenaline; but they too can dilate in response to circulating adrenaline.

The total effect of a massive discharge of sympathetic nervous impulses and catecholamine release is to bring about a redistribution of blood flow, which is reduced in the skin and splanchnic regions, and increased in skeletal and cardiac muscle and in the brain. In animals deprived of their adrenal medullas cardiovascular adjustments involving sympathetic hyperactivity are impaired. In such animals cardiac output does not rise to a normal extent on exercise, for example.

FACTORS MODIFYING SYSTEMIC ARTERIAL PRESSURE

Sympatho-adrenal activity

Alterations in sympathetic nervous activity play an important part in controlling systemic arterial pressure. The systemic arterial baroreceptors in the large arteries form the afferent limb of a feedback loop whose integrating centres are in the medulla oblongata. Sympathetic activity is also influenced by higher centres, especially in the hypothalamus, and by the general level of activity in the reticular activating system. Decreased sympathetic nervous activity together with vagal inhibition of the heart are mainly responsible for the fall of blood pressure during sleep. The tonic action of the sympathetic nervous system also depends on the chemical environment of the medulla, and in some species on the spinal cord as well. Normally the cerebral blood flow is held constant and the chemical environment of the central nervous system does not change; but severe medullary ischaemia causes a massive sympatho-adrenal discharge which may act as a last ditch defence of the blood supply to vital centres.

Angiotensin

Arteriolar resistance can be altered by several vasoactive materials conveyed by the blood stream. The precise role of many of them is not fully known. This even applies to angiotensin II, the most powerful vasoconstrictor material known. It is likely that normally circulating concentrations of angiotensin exert a small vasoconstrictor effect. The effect of angiotensin II in maintaining normal blood pressure is most evident in states of sodium depletion and in the erect posture. A major action of angiotensin II is to control sympathetic stimulation by a powerful direct effect on the medullary centres near the area postrema where there is a local deficiency in the generally impermeable blood-brain barrier. The concentration of angiotensin increases rapidly if the renal perfusion pressure falls. It seems likely that the initial rise of systemic arterial pressure after putting a constricting clip on the main renal artery of an animal is due largely to the release of renin, liberating angiotensin in the blood stream. The factors governing renin release are more fully discussed in Chapter 4.

Blood volume

The state of filling of the circulation has an important part to play in regulating cardiac output and hence potentially in controlling systemic arterial pressure. Subjects given sodium-retaining steroids for a long period initially expand their plasma volume and cardiac output. After a few weeks the cardiac output tends to fall towards normal and at the same time the systemic arterial pressure rises. Roughly opposite changes occur after the long-term administration of diuretic agents which remove sodium and water from the body. These and other observations suggest that the control of sodium and water excretion by the kidney may influence long-term regulation of systemic arterial pressure. Although the systemic arterial baroreceptors control systemic arterial pressure over short periods of time, it seems unlikely that they exert any important influence on long-term stability of blood pressure. There is so far no general agreement about the means by which blood pressure is stabilized over long periods of time although most investigators give the kidney a dominant role, because of its ability to regulate sodium balance.

REGULATION OF TISSUE BLOOD FLOW

The regulation of blood flow to tissues such as the heart, brain and lungs will be considered separately; but it seems that almost all organs except the skin possess some intrinsic ability to control their blood flow in accordance with metabolic needs—a property known as 'autoregulation', though strictly this word is used to describe stability of blood flow despite changes in perfusion pressure. In organs such as the kidney which have a constraining capsule, autoregulation could be to some extent a passive process; but in most other situations the calibre of arterioles in an organ is probably governed mainly by its metabolic needs and its perfusion pressure. Even powerful extrinsic vasoconstrictor influences such as those of the sympatho-adrenal system cannot indefinitely reduce the blood flow to most organs because autoregulatory mechanisms eventually overrule the extrinsic system, which provides only short term control. In some situations, changes in perfusion pressure alone can alter arteriolar calibre, apparently by a local myogenic response of arteriolar smooth muscle; but in most organs some metabolite used or produced by the tissues appears to exert a controlling role. In skeletal muscle oxygen supply is the most important factor, and anoxaemia powerfully dilates muscle vessels independently of any nervous connections to the tissue and of changes in circulating vasoactive substances. In the cerebral circulation carbon dioxide tension is probably the main determining factor. Thus different vascular beds may depend on different control systems which preserve tissue blood flow despite changes in systemic arterial pressure.

Flow through some organs may be more dependent on nervous and circulatory chemical stimuli than on local metabolic influences. In particular the kidneys and skin have in common a vascular control system which is largely independent of metabolic needs. Indeed, the kidney appears to adjust its metabolic rate to its blood flow rather than the other way about. In both cases the normal blood flow must be enormously in excess of the minimal metabolic needs, thus enabling the kidney to operate flexibly in the control of water and electrolyte excretion, and the skin to regulate body temperature.

THE CORONARY CIRCULATION

Blood flow through the coronary arteries is continuous throughout the cardiac cycle, two-thirds occurring during diastole, as the capillary bed is occluded by the contracting muscle during systole. However the inner layers of the myocardium are also compressed by the high cavity pressure and receive no blood flow during systole. This is probably why they are more subject to ischaemic damage. The coronary sinus drains 90 per cent of the venous blood from the left ventricular muscle. The remainder, with blood from the right ventricle, drains into the other cardiac veins. The proportion of oxygen extracted by the heart muscle is remarkably constant at the relatively large value of around 150 ml O_2/l of blood flow. The volume of oxygen consumed at rest by the heart is 40 ml/min or 15 per cent of the total body oxygen consumption, and the normal coronary flow is about 250 ml/min or 5 per cent of the resting cardiac output. The main determinant of coronary flow is the oxygen consumption of the myocardium, which is directly related to the development of tension, determined by the pressure generated within the ventricular cavity and the radius of the chamber. An increase in heart rate increases the oxygen requirement of the myocardium. Sympathetic nerve stimulation probably has a slight direct vasoconstrictor effect on the coronary vessels, and the increased coronary blood flow produced by increased sympathetic nervous stimulation is mainly due to the action on the myocardium and the metabolic effect of the resulting increase in cardiac work. Vagal stimulation has no direct effect on the coronary vasculature; angiotensin and vasopressin are the only known naturally occurring substances having a vasoconstrictor action on the coronary arteries. Adrenaline has a slight β-stimulating dilator effect.

THE CEREBRAL CIRCULATION

The cerebral blood flow of man is normally 55 ml/100 g brain/min, which for a brain of 1400 g gives a total cerebral blood flow of about 750 ml/min, i.e. 15 per cent of the resting cardiac output. Consciousness is lost at cerebral blood flows of less than about 30 ml/100 g/min, and irreversible damage occurs at values sustained at less than 10 ml/100 g/min. Most general anaesthetics reduce cerebral blood flow in proportion to their depression of cerebral metabolic rate.

Until recently the cerebral circulation was thought to be largely independent of autonomic control. At systemic arterial pressures greater than about 75 mm Hg, the brain has the ability to regulate ('autoregulate') its intrinsic vascular resistance in such a way that blood flow remains almost constant despite changes in systemic arterial pressure. It probably does this partly by virtue of its enclosure within a rigid box. Patients with large skull defects tend to faint when standing up. There is probably also some degree of *myogenic* autoregulation; i.e. changes in vascular calibre induced by changes in distension pressure of the resistance vessels. However, stability of cerebral blood flow is probably achieved mainly by control of vascular calibre by intracellular hydrogen ion activity in smooth muscle cells, of which arterial carbon dioxide tension (P_{CO_2}) is an important determinant. If cerebral blood flow begins to fall, carbon dioxide accumulates and the increased P_{CO_2} brings about compensatory cerebral vaso-dilation. In the presence of high arterial P_{CO_2} almost all autoregulative capacity disappears. Hypoxaemia also dilates the cerebral vessels, possibly by local production of lactic acid. Recent work suggests that many of these local reactions are mediated through the vascular autonomic innervation, but the detailed mechanisms remain obscure.

The apparent overall stability of cerebral blood flow masks enormous local changes which are related to the amount of activity in different brain regions. A visual signal, for example, increases blood flow in the occipital cortex within a few seconds. Presumably the stimulus is initially metabolic, but local nervous connections to cerebral vessels might be involved.

Not only has the brain itself an intrinsic ability to control its own blood flow; the carotid and arterial baroreceptors lie on the path of blood from heart to brain, and their reflex action tends to maintain a constant systemic blood pressure over short periods of time, thus protecting the brain from acute changes in perfusion pressure.

THE PULMONARY CIRCULATION

The pulmonary arterial tree is perfused at a lower pressure than the systemic so that the effects of gravity are proportionally greater. In the erect posture the arterial pressure in the lung increases below the level of the hilum and decreases from the hilum towards the apex. Pulmonary capillary blood flow depends on the relative values of the alveolar, arteriolar and pulmonary venous pressures. At the apex of the lung the arteriolar pressure is reduced and the pulmonary capillaries are collapsed by the surrounding alveolar pressure. At the base of the lung the arteriolar and venous pressures exceed the alveolar pressure and the capillaries are patent throughout their length. In the midzone arteriolar pressure just exceeds alveolar pressure. However, alveolar pressure exceeds the local venous pressure, so that the arteriolar end of the capillary is open and the venous end closed. Blood flow measurements in the human lung show a progressive reduction of blood flow per unit of lung tissue from the base of the lung to the apex. This difference is reduced on exercise and abolished by lying down. Pulmonary hypertension (raised pulmonary arterial or pulmonary venous pressures), or a large left to right shunt of blood through the lungs also alters the usual pattern and flow in the lower parts of the lungs is relatively reduced, with a change to more even flow at base and apex as apical flow increases.

The principal function of the pulmonary circulation is gas exchange, and it has an appropriately large capillary surface area which contains approximately 20 per cent of the 400 ml blood in the pulmonary circulation. Local reduction in alveolar oxygen tension with or without hypercapnia appears to cause local vaso- and broncho-constriction, and teleologically serves to shut down flow in unventilated regions. This process tends to redistribute both blood and gas-flow to more appropriate areas. Systemic anoxaemia produces generalized pulmonary vasoconstriction; this has no obvious role in the adult but in the foetus diverts blood into the systemic circulation and hence to the placenta via the ductus arteriosus. A general response of this type is partly responsible for the pulmonary arterial hypertension, right ventricular hypertrophy and (ultimately) congestive heart failure of diffuse pulmonary disease, such as chronic bronchitis and emphysema—a condition known as 'cor pulmonale'.

The pulmonary circulation nourishes the alveolar walls, and in the event of pulmonary arterial occlusion there is inactivation of the surfactant lining material which normally reduces alveolar surface tension to prevent collapse. Consequently, atelectasis (lung collapse) commonly follows pulmonary arterial occlusion. The pulmonary circulation also filters particles from the mixed venous blood and in addition acts as a reservoir of blood, particularly during changes of posture. The lungs are also partly responsible for the conversion of angiotensin I to angiotensin II, the production of the platelet aggregation inhibitor, prostacyclin, and possibly for other metabolic reactions.

THE SPLANCHNIC CIRCULATION

The portal circulation will be considered in Chapter 13; but the amount of blood flowing through the liver has an important influence upon the circulating concentrations of hormones as well as upon general metabolism. For example the much reduced liver blood flow in severe haemorrhage contributes to the high blood concentrations of aldosterone (which is removed almost entirely by a single passage of blood through the liver) and also of vasopressin. The importance of the large labile blood pool in the liver has previously been mentioned. Its size in man is not exactly known, but the liver of the rabbit contains 30 per cent of all the circulating red cells.

LYMPHATIC CIRCULATION

Albumin molecules and other protein and lipid substances leaked from the capillaries are recovered from the extravascular tissues and returned to the systemic circulation by the lymphatics. The thoracic duct lymph flow in 24 hours is approximately equal to the blood volume and over the same period more than half the total circulating plasma protein is returned to the central veins through the thoracic duct. Although all protein and lipid components of the blood may be identified in lymph the detailed composition of the fluid varies with the region drained. Hepatic lymph is protein-rich and protein synthesis in the liver plays an important role in the maintenance of normal concentrations of most of the plasma proteins. Lymphatic capillaries are either open-ended or very permeable, allowing the free entry of macromolecules. Larger lymphatics have walls impermeable to substances of molecular weights greater than 6000, thus preventing loss of macromolecules. The pumping mechanisms of skeletal muscle contraction arterial pulsation, and respiration determine lymph movement and the vessels (like the veins) contain valves which prevent retrograde flow. Although peripheral lympho-venous communications may be demonstrated they have little significance in health. Foreign particles removed from the extravascular tissue are filtered off by the lymphatic nodes.

GENERAL CARDIOVASCULAR RESPONSES

Exercise

At the onset of exercise, the increased metabolic needs of the muscles are largely met by anaerobic mechanisms, but there is a rapid increase in blood flow to the active muscles and after 30 to 60 sec aerobic mechanisms predominate. At rest the mixed venous blood is about 70 per cent saturated with oxygen, but during severe exercise the saturation may fall as low as 25 per cent and the total oxygen consumption may increase 20-fold. In health the cardiac output may increase during severe exercise to 30 l/min, with a heart rate of about 190 beats/min (depending on age—see p. 41) and a stroke volume of over 150 ml/beat. The systemic arterial blood pressure only increases moderately and it is therefore obvious that peripheral resistance falls. However, the decreased resistance only occurs in active muscle. The peripheral resistance in inactive limbs and in the splanchnic vasculature rises and blood flow to these regions falls. The dominant requirement for heat loss in the skin causes cutaneous vasodilatation.

As the heart rate goes up, diastole gets shorter, but the heart also completes its contraction a little more rapidly, so that systole is a little shortened. The predominant effect of more frequent systoles as rate increases is to curtail the time available for ventricular filling between each beat. Filling is normally nearly complete in the first third of diastole and as increased venous pressure increases the flow rate, filling is well-maintained in the normal heart and the stroke volume (cardiac output per beat) is kept up at the maximal level during exercise. Thus the progressive rise in cardiac output is achieved by a corresponding increase in heart rate on more strenuous exercise. The opposite situation is present in some heart diseases such as mitral stenosis, where ventricular filling is impaired. The tachycardia of exercise fails to achieve an increase in cardiac output because stroke volume falls with the reduced duration of diastole producing reduced ventricular filling in spite of a gross rise in pulmonary venous pressure which may lead to pulmonary oedema. Any tachycardia is badly tolerated in this situation.

In health, the peak flow to an exercising limb occurs during the period of activity, with a period of hyperaemia followed by a rapid return to pre-exercising flow values in the recovery period. The magnitude of the post-exercise hyperaemia in a limb is directly proportional to the severity and duration of the preceding exercise up to a certain point, after which there is no further increase in flow. Physical training reduces the post-exercise hyperaemia probably by more effective capillary perfusion of the muscles during exercise; suggesting that accumulation of metabolites is responsible.

Postural changes

Sudden changes in posture affect the cardiac output by changing the filling pressure of the heart. Circulatory reflexes normally react quickly to restore the situation. On tipping a subject from the horizontal to the vertical position there is an immediate pooling of about 800 ml of blood in the veins of the lower limbs. The cardiac output falls and there is increased peripheral vasoconstrictor activity in arterioles and veins, preventing a reduction in blood pressure and protecting the capillary beds of the lower limb from the increased hydrostatic pressure which would otherwise result in massive transudation of fluid at capillary level. Active standing or tensing the leg muscles on tipping reduces the volume of blood pooled in the lower limbs and there is less reduction in cardiac output and a smaller increase in heart rate. After a very short time sympathetically-mediated venoconstriction diminishes, and a fresh equilibrium is attained in the new posture.

Temperature regulation

The skin circulation is under the control of temperature-regulating centres in the hypothalamus. In a hot environment the skin blood vessels dilate, the blood flow through them increases and heat is lost. The circulation through the skin can be so rapid as to be acting virtually as an arterio-venous fistula, so that the blood in the superficial veins may become almost fully 'arterialized' if the arm is placed in hot water. The cardiac output, pulse pressure and heart rate all increase to meet the demand for increased flow in response to heat, and the extra load on the heart is comparable to that in mild exercise. Hyperaemic conditions of the skin (e.g. extensive exfoliative dermatitis) may have the same effect, and sometimes cause a 'high output' stress on the heart.

In a cold environment there is peripheral vasoconstriction and the skin flow decreases. Provided that increased muscular activity, in the form of shivering does not occur, the cardiac output falls. The environmental temperature must be taken into account in attempting to assess the cardiac output clinically, for in a warm room the output may well be twice the lowest resting value.

Hypoxaemia and hypercapnia

The effect of moderate hypoxia on the circulation as a whole is to cause a rise in blood pressure and a rise in cardiac output; if severe it eventually causes circulatory collapse. Moderate carbon dioxide excess increases cardiac output usually with little change in blood pressure. The modes of action of anoxia and carbon dioxide excess are complex.

Both cause dilatation of most blood vessels by a direct local action, but also cause an increase in nervous vasoconstrictor tone. Hypoxia achieves this by stimulating the vasomotor centre reflexly through the sino-aortic chemoreceptors: carbon dioxide stimulates the vasomotor centre both reflexly and directly. The effects of hypoxia and carbon dioxide excess on the local blood flow in any organ and on the peripheral resistance as a whole are variable and depend on the balance between local and central actions.

Disordered function

HEART FAILURE

Heart failure might be defined in physiological terms as a failure of the heart to pump blood normally, and to maintain an adequate cardiac output. It would be difficult, though not impossible, to establish standard conditions of measurement of, say, a Starling curve relating filling pressure to cardiac output, and then to establish a normal range of variation in the population. One might then define heart failure as a condition in which this measure of cardiac performance fell short of some arbitrary criterion of normality, such as the lower 1 or 5 percentile for the population. This definition has the advantage of bringing heart failure into line with, for example, renal failure or respiratory failure, which can be specified in terms of deviations of blood urea concentration, or blood oxygen and carbon dioxide pressures, from an agreed normal range.

This is not what the physician means by heart failure. The term is not used to describe the results of sudden cessation of the circulation from failure of the heart beat, cardiac arrest from ventricular fibrillation or asystole, or sudden obstruction, as from a massive pulmonary embolism. The acute reduction of the cardiac output from massive left ventricular damage as in myocardial infarction is usually described as 'cardiogenic shock'. The term 'heart failure' is used as a shorthand expression to describe the compensated state of inadequate cardiac output which is associated with fluid retention producing vascular congestion and oedema, as fluid spills into the interstitial tissue in either the systemic or pulmonary vascular beds. Excess iatrogenic fluid administration or severe renal excretory failure can cause fluid overload which produces similar features of congestion in the presence of a normal heart. The description as 'congestive' heart failure enhances this view and is appropriate so long as it does not disguise the fact that the moderate congestion of the lungs may be clinically or even radiologically invisible.

This definition recognizes that many different conditions which cause failure of the heart to expel its contents adequately result in a common and easily recognizable clinical syndrome. The definition embraces cases in which all the clinical features of heart failure coexist with a cardiac output which may be normal or even greater than normal at rest, because of an excessive demand which may be metabolic or mechanical. Some cases of severe anaemia, thyrotoxicosis, cor pulmonale, beri-beri and arteriovenous fistula fall into this category. The definition implies that the symptoms and signs of heart failure are brought about by the same mechanism, whether the primary failure lies with the heart as a pump, or whether the needs of the tissues for blood are so much increased, as in severe anaemia, for example, that even a normal heart cannot keep up with them. Fundamentally similar mechanisms are operating in high and low output states.

When oedema occurs predominantly in the lungs and when left atrial pressure is raised, the heart failure is often described as 'left-sided'; when it occurs in the systemic circulation it is described as 'right-sided'. The distinction is in one sense trivial, because the fundamental disturbance is in the regulation of extracellular fluid volume, which is increased in all kinds of heart failure. The site at which oedema appears, usually either at the lung bases or at the ankles of an erect subject, depends on the relative degrees of impairment of performance of the two ventricles, and on the relative mechanical hindrances to the flow of blood in different parts of the central circulation. Nevertheless the distinction is useful as there are certain special problems in pulmonary oedema, which will be discussed more fully in a later section.

Heart failure used to be considered in terms of inadequate 'forward flow' and too much 'back pressure'. This concept of 'backward failure' seemed obvious to the early pathologists seeing congestion and dilatation in the part of the heart or venous system behind an obstructed valve or diseased ventricle, and may occur as a transient phenomenon acutely from shift of fluid from arterial and capillary systems or tissue fluid into the veins. However, the concept of backward failure is naive in explaining the extensive fluid retention which is better considered as a pathological retention of salt and water by the kidneys, secondary to inadequate cardiac performance. The 'forward failure' hypothesis thus suggests fluid retention as a response to inadequate renal perfusion. In an attempt to rescue the now relegated 'backward failure' hypothesis much work has gone into a search for volume receptors on the venous side of the circulation which could reflexly initiate fluid retention. Such receptors are well-known in the stretch receptors of the great veins and atria. Yet in heart failure these organs are demonstrably overstretched, and in acute experiments overstretching of the volume receptors is a stimulus for increased excretion rather than retention of fluid. Such a mechanism has been invoked to explain the diuresis characteristically associated with prolonged attacks of paroxysmal tachycardia. Attempts have been made to involve receptors in the 'backward failure' theory of heart failure by supposing that they respond only to a pulsatile stimulus, and that in heart failure the chronic overstretching of the great veins and atria reduces the pulsatile stimulation of the receptors. The 'forward failure' view seems a more likely explanation on the basis that the maintenance of an adequate systemic arterial pressure is the prime function of the heart, and suggests that any failure of the heart to sustain an adequate arterial pressure brings into action the many mechanisms under central nervous control which can increase cardiac output. By this hypothesis heart failure represents an attempt by the body to sustain arterial pressure by means which, though effective, eventually prove disadvantageous. An attractive feature of this hypothesis is that it explains why hypoxaemia, hypercapnia, anaemia, thyrotoxicosis, and arteriovenous fistula (all of which increase the demands of the tissues for the available blood, and lower peripheral systemic resistance) should provoke the same clinical syndrome as primary failure of the heart as a pump.

The concept of 'backward failure' remains useful in determining the distribution of the retained fluid as 'left-sided' or 'right-sided' failure. In fact the commoner heart diseases, such as hypertension, coronary artery disease, mitral or aortic valve disease involving the left heart lead to left-sided congestion. Isolated right-sided failure is rare but may occur in response to pulmonary hypertension, as in chronic lung disease. A common sequence of events is that left-sided failure leads on to right-sided failure, sometimes by the secondary development of pulmonary hypertension, but more often from the occurrence of atrial fibrillation which seems to interfere with tricuspid valve closure—the lack of atrial systole, and more important, atrial relaxation allowing tricuspid incompetence and putting an added load on the right ventricle. In addition, distension of the left ventricle may push the septum to the right, and interfere with right ventricular filling. Redistribution of retained fluid towards the right side may reduce left-sided congestion with a relative improvement in symptoms of pulmonary congestion (e.g. the improvement in breathlessness which is sometimes seen in long-standing mitral stenosis). A similar situation is found when both right and left sides of the heart are similarly affected. In constrictive pericarditis, for instance, both ventricles may be equally restricted by the tight pericardial sac, yet fluid retention and oedema are practically confined to the right side often with gross systemic oedema, congested liver and ascites (free fluid in the peritoneal cavity). The greater compliance and easier filling of the open systemic venous system than the tight closed pulmonary venous compartment may be responsible.

Any degree of acute failure of the heart as a pump will lead to proportionate reduction in mean capillary pressure, tending to disturb the balance of hydrostatic and osmotic forces across capillaries, and leading to fluid entering the circulation from the interstitial space and producing expansion of plasma volume. The 'forward failure' hypothesis explains the continued accumulation of fluid in the plasma and interstitial space. Oedema appears when the intercellular gel can hold no more fluid, so that spaces filled with freely mobile fluid appear between cells. It is clear that the excess fluid filling the interstitial spaces in heart failure is due to active retention of sodium and water by the kidney. Up to a point this mechanism is a valuable homeostatic process which allows the mean pressure filling the circulation to be increased, and for venous return to be enhanced. The increased filling of the heart stretches the heart muscle, and leads, according to Starling's law, to an increase in output. This will obviously tend to compensate in some degree for any primary failure of the heart to pump blood adequately, and is expressed graphically as elevation of the ventricular function curve relating cardiac output to filling pressure (Fig. 2.2).

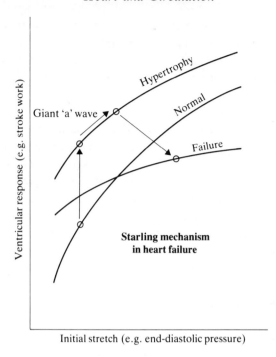

Figure 2.2 *The effects of heart failure demonstrated by ventricular function (Starling) curves.*
The hypothetical Starling curves shows the ventricular response as stroke work in relation to initial stretch of muscle fibre, related to end-diastolic pressure. The normal curve shows a progressive increase of response with increasing filling pressure. Hypertrophy of the ventricle produces a parallel higher curve, so that a greater response is obtained for a given filling pressure (vertical arrow). A similar functional response is produced by sympathetic stimulation and at least transiently by stimulating drugs, such as digitalis. The body uses forceful atrial contraction, producing a giant 'a' wave to boost ventricular filling and obtain additional stretch of the muscle fibres without any general rise in filling pressure (oblique upwards arrow). Impaired myocardial performance in failure produces a lower flatter curve, so that a higher filling pressure is needed to give the initial stretch required even for a lesser ventricular response (oblique downwards arrow), with the net effect of a persisting inadequate ventricular response in spite of a greatly raised filling pressure.

Heart failure as a disturbance of homeostasis

Salt and fluid retention in response to an inadequate output is a normal homeostatic mechanism leading to an improved cardiac output which corrects the situation. In heart failure this mechanism becomes disadvantageous, and heart failure can be regarded as a disturbance of a physiological feed-back control process. The key to the situation is that the fluid retention fails to correct the inadequate cardiac output, so there is a persisting error signal continuing the fluid retention. Several mechanisms contribute to the persistence of the error signal:

The formation of oedema. Loss of retained fluid from the circulation to increase the volume of interstitial fluid prevents the expected increase in ventricular filling. When interstitial fluid pressure exceeds atmospheric pressure, the freely mobile fluid is evident as oedema and free fluid also begins to accumulate in the serous cavities (e.g. pleura and peritoneum).

Ventricular dilatation. Dilatation of the affected ventricle has long been known as a cardinal feature of heart failure and is generally assumed to be associated with augmented performance by the Starling mechanism. However, the dilated chamber suffers two major disadvantages. Firstly, by the law of Laplace, much greater force is needed in the ventricular wall in proportion to the dilatation to generate the cavity pressure needed to open the aortic valve and achieve ejection. Secondly, the ventricular wall is not a simple elastic structure but consists of individual fibres which slide over one another as the chamber dilates, so that the degree of stretch transmitted to each fibre is less than expected. Although these factors are difficult to quantify it seems clear that the response of the dilated ventricle is less than expected, contributing to the flat ventricular function (Starling) curve in failure.

A-V valve incompetence. As a result of ventricular dilatation the A-V valve ring become stretched and the chordae tendineae are pulled down leading to mitral or tricuspid valve incompetence so that some of the ventricular output is lost as regurgitant flow.

Ventricular muscle disease. The compensatory process of ventricular hypertrophy behaves in some ways like hepatic cirrhosis with the development of patches of fibrosis, presumably due to local hypertrophy outrunning the available capillary bed. The loss of muscle function is associated with increasing stiffness impairing relaxation, and, in coronary artery disease, areas of fresh fibrosis contribute to the situation. Functional changes such as left bundle branch block producing incoordinate contraction can add to the problem. Although no specific metabolic disturbance has been identified in the myocardium in heart failure, the net effect of all these changes is a reduction in myocardial performance, evident as a low flat Starling curve, so that increasing fluid retention and rising filling pressure produce no corresponding rise in cardiac output and the error signal persists.

Compensatory mechanisms

As a response to the difficulties in fulfilling the requirements for a raised cardiac output in heart failure, reserve mechanisms used to increase cardiac output on exercise are brought into play and other responses play a part in maintaining cardiac output in the face of the adverse effects outlined above.

Tachycardia. An increase in heart rate is the main basis of the increase in cardiac output on exercise and is usual in heart failure as part of the general effect of sympathetic stimulation. A reflex or direct effect of the raised venous pressure may play a part. In pericardial tamponade, due, for instance, to an intrapericardial haemorrhage, ventricular filling and hence stroke volume are sharply limited and tachycardia remains the only compensatory mechanism. In other circumstances an undue tachycardia may accentuate failure by an adverse effect on ventricular filling. This is most often seen with uncontrolled atrial fibrillation in mitral stenosis, where a long diastolic filling time is needed because of the resistance to left ventricular filling, and the tachycardia with little change in the duration of systole greatly limits diastole, so that pulmonary congestion develops very quickly. Other hearts with impaired ventricular filling may behave similarly.

Sympathetic stimulation. An early part of the response to inadequate cardiac output is an increase in sympathetic tone, presumably as a reflex attempt to maintain arterial pressure by vasoconstriction in the face of a fallng cardiac output. This sympathetic effect boosts myocardial contractility in addition to increasing rate and plays a part in the renal changes responsible for 'forward failure'. In established failure sympathetic stimulation becomes extreme with very high plasma noradrenaline levels and has the undesirable effect of reducing the sensitivity of myocardial receptors, contributing to the disturbance of myocardial performance described above, and probably also inhibits myocardial noradrenaline synthesis.

Ventricular hypertrophy. The increased load on the ventricular muscle which eventually leads to heart failure, produces at an early stage a response of hypertrophy in the affected myocardium. The stimulus of an increased load requiring greater force of contraction, whether from an increase in cavity pressure, as in aortic stenosis or hypertension, or from an increase in stroke volume as in aortic incompetence or persistent ductus arteriosus with great run-off of left ventricular output to the pulmonary circulation, leads to myocardial hypertrophy until the load on the individual muscle fibres is returned to normal. Hypertrophy has the effect of giving the ventricle a new, higher Starling curve, so that a greater response is obtained for a given filling pressure (Fig. 2.2). An inescapable effect of ventricular hypertrophy is increased stiffness of the ventricular muscle in diastole, so that impaired filling may contribute to the problems and help to localise fluid retention to the affected side of the circulation. This is particularly seen with the concentric hypertrophy of pressure-load e.g. aortic stenosis, but is also a feature of some specific heart muscle diseases e.g. amyloid. In addition as described above diffuse fibrosis generally accompanies prolonged hypertrophy and adds to the problems of diastolic stiffness.

Accentuated atrial contraction. A forceful atrial boost obtains greater Starling response by a critical addition to ventricular filling with only a brief rise in venous pressure for a fraction of each cardiac cycle. This mechanism plays a major part in the early stages of the response to increased load, but may be very important in the presence of a stiff ventricle. Its loss, from atrial fibrillation, in aortic stenosis can be a disaster. Forceful atrial contraction produces characteristic physical signs, useful in the diagnosis of severe heart disease, not yet in failure. In the right heart the sharp 'a' wave in the jugular venous pulse, timing with the first heart sound is useful evidence of right ventricular stress, often lost in more severe disease in the large systolic wave of tricuspid valve incompetence. In the left heart an 'a' wave is not visible, but the boost to left ventricular filling produces a local low-pitched presystolic extra heart sound (the fourth or atrial sound). Often with the patient turned to the left the corresponding outward thrust is palpable as a presystolic outwards thrust before the main apex beat. As the fourth sound is so low-pitched, the palpable thrust is a useful clue; in difficult cases graphical demonstration in the apex cardiogram is useful.

The compensated state. In some patients a new equilibrium may be reached on the basis of the compensatory processes with a continued low cardiac output and a raised filling pressure on the affected side of the heart. If pulmonary congestion is dominant (left-sided failure) pulmonary oedema may be quickly fatal, but dominant right-sided failure may progress to gross oedema, hepatic congestion and ascites as in tricuspid valve disease and constrictive pericarditis, as these effects are not a direct threat to life.

Possible mechanisms of renal sodium and water retention in heart failure

The 'forward failure' hypothesis of heart failure (p. 57) requires that salt and fluid retention should occur as a response to an inadequate cardiac output. The scene is set by the activation of the sympathetic mechanism for redistribution of blood flow evident normally on exercise. This diverts flow to skeletal muscle at the expense of the renal and visceral circulation. Skin flow is reduced until the effect is overcome by the over-riding demand for heat loss; cerebral blood flow is relatively maintained. This mechanism is activated in response to the threat to blood arterial pressure as cardiac output falls. The continued evidence of sympathetic stimulation as failure progresses indicates that this process continues and there is evidence of some reduction in renal blood flow. Various factors may lead to salt and fluid retention on this basis:

Reduction in glomerular filtration. A slight fall in renal perfusion might be expected to lead to salt retention, and is not unusual in generalized kidney disease; but in heart failure the changes do not seem great enough to account for the rapid progression and severity of the fluid accumulation.

Activation of the renin mechanism. Renin release is mediated by β-adrenergic receptors, and is probably initiated as part of the general sympathetic stimulation of heart failure, but a fall in glomerular perfusion pressure may stimulate renin release directly from the juxta-glomerular apparatus. Renin releases angiotensin I in the plasma and forms the powerful pressor substance angiotensin II which plays a part in maintaining the peripheral resistance in the face of a falling cardiac output in heart failure. Angiotensin II stimulates the adrenal cortex to produce the sodium retaining hormone aldosterone which acts on a Na/K exchange mechanism in the distal tubule. Again the quantitative hormonal disturbances do not seem adequate to explain the fluid retention in failure, although aldosterone is of undoubted importance in producing potassium loss.

'Third factor'. It has been suggested that lack of a natriuretic 'third factor' may cause fluid retention in heart failure. At one time changes in plasma protein osmotic pressure alongside the renal tubules was suggested, but this theory is now discredited. A current explanation is that there is an intra-renal redistribution of blood flow, perhaps as part of the sympathetic response with an increased flow to deeper (juxta-medullary) nephrons of greater sodium retaining capacity; such a mechanism would be a logical extension of a homeostatic response to fluid loss.

Vasopressin release. Although not generally a contributing factor in heart failure, vasopressin is often considered as part of the normal responses of the body to changes in fluid balance. Vasopressin is normally released in response to a rise in osmotic pressure and causes water retention in the kidneys to correct the situation; loss of production in diabetes insipidus produces great water loss, and an increased water retention. The fluid retention of heart failure is generally iso-osmotic, but in severe cases plasma osmotic pressure falls, suggesting interference with free water-handling in the kidney. In very severe dried-out states on treatment, reduction in the effective plasma volume may provoke release of vasopressin through a second emergency mechanism and the resulting water retention may cause a fall in plasma osmolality 'inappropriate' to the vasopressin levels as these are normally suppressed by a fall in osmolality.

Pulmonary oedema ('left-sided failure')

Pathological retention of sodium and water in the body can eventually lead to pulmonary oedema for purely hydrostatic reasons, e.g. when pulmonary venous pressure rises in left ventricular failure or mitral stenosis, when pulmonary capillary pressure exceeds plasma colloid osmotic pressure, or when the drainage capacity of the pulmonary lymphatics is exceeded. At an early stage breathlessness is produced by increased stiffness of the lungs from oedema of the interstitial tissue. However, pulmonary oedema, especially its acute form, which typically occurs during sleep, also has an important sympathetic reflex component. Characteristically in pulmonary oedema the blood pressure and heart rate are increased above normal, and there is sweating and pupillary dilatation. Such changes occurring during sleep may be trigger factors in susceptible subjects with serious left heart disease. It is even possible in the experimental animal to produce acute pulmonary oedema without any pre-existing retention of sodium and water (e.g. by the injection of protein solutions into the cerebral ventricles). Acute pulmonary oedema may occasionally be precipitated in man by subarachnoid haemorrhage, suggesting that central sympathetic stimulation may be responsible. However, myocardial disturbance from scattered necrotic foci is found in states of intensive sympathetic stimulation and may be the underlying cardiac cause in these situations. In most cases there is some existing underlying heart disorder with expansion of the extracellular fluid, and the sympathetic reflex component is superimposed upon this and triggers off an attack, probably partly by peripheral venous constriction shifting blood into the thorax. The nature of the stimulus to the sympatho-adrenal system is not always apparent, but the autonomic accompaniments of rapid eyeball movement (REM) sleep may be a factor in some apparently spontaneous attacks.

Pulmonary oedema has also been observed to occur in some previously fit men when suddenly exposed to altitudes in excess of 9000 feet. It appears to be related to exertion and inadequate acclimatization and it has been said to occur with normal values for pulmonary capillary and venous pressure. There is no generally agreed explanation, but increased capillary permeability after hypoxic arteriolar constriction may play some part.

Constrictive pericarditis

Constrictive pericarditis produces many of the features of heart failure. In this condition the heart is encased in a rigid casing, which limits its diastolic filling. Thus although the heart itself may be normal, increased filling pressure increases cardiac output very little. The heart fails because it cannot fill and not because it cannot empty. A similar situation may occur in some forms of cardiomyopathy in which there is myocardial fibrosis, and in constrictive endocarditis such as that due to African endomyocardial fibrosis. By interfering with cardiac filling these produce an identical clinical picture, which is characterized by the signs of right-sided congestive cardiac failure usually unaccompanied by pulmonary congestion or orthopnoea. There are generally many physical signs of the restricted diastolic filling; e.g., an abrupt halt to the fall in jugular venous pressure in diastole the 'y' descent, a palpable thrust in the precordium, and a sharp third heart sound.

CENTRAL CIRCULATORY OBSTRUCTION

One of the important disturbances of heart disease is that due to obstruction of the circulation at some point at which the whole, or most of the cardiac output must pass. The symptoms depend on whether the obstruction is left or right-sided, but some effects tend to be similar whatever the site of obstruction. The cardiac output is reduced, and cannot increase normally on exercise. There is usually a small pulse, low blood pressure, cold extremities and peripheral cyanosis in any severe obstruction. A cyanotic facial blush is characteristic of sustained low cardiac output, because of vaso-dilatation in the blush area. Evidence of a high cardiac output rules out severe obstruction.

Left-sided central circulatory obstruction

When central circulatory obstruction is on the left side of the heart, it may cause congestion and oedema of the lungs, which alter pulmonary mechanics and gas exchange and characteristically cause breathlessness. These features are similar to left-sided heart failure and have the same basis of disordered physiology without myocardial disturbance, at least in the earlier stages.

Aortic stenosis. Resistance to left ventricular outflow by valve stenosis is countered by concentric left ventricular hypertrophy; the distensibility of the ventricle is reduced as a result and increased force of atrial contraction becomes of importance in ventricular filling. Resting cardiac output is maintained at normal levels until left ventricular failure supervenes, at which time mean left atrial and pulmonary venous and arterial pressures rise, and there may be pulmonary oedema. In severe aortic stenosis stroke volume fails to increase with exercise and the associated muscular vasodilation with reduction in blood flow elsewhere may produce syncope or compromise the coronary circulation, leading to angina. The cardinal feature of aortic stenosis is the small, slow-rising pulse, often called a 'plateau pulse', due to slow left ventricular ejection. The turbulent flow produces a loud ejection-type systolic murmur, which may produce a palpable thrill. The aortic component of the second sound is soft and may if audible be delayed beyond the pulmonary component (reversed splitting). There is left ventricular hypertrophy often with little dilatation, so the heart size may be normal on X-ray, though a dilated ascending aortic from the effects of the jet through the narrow valve, valve calcification or the features of pulmonary interstitial oedema may be present.

Hypertrophic obstructive cardiomyopathy (HOCM). In some forms of subvalvular stenosis the outflow of blood from the heart is not restricted throughout the whole period of systole, and a delayed obstruction results from muscular hypertrophy of the ventricular outflow tract. Such a lesion may be part of a general obstructive cardiomyopathy affecting the right and left ventricles, either singly or together. In this situation a high intraventricular pressure is developed during isometric systole, and when the pulmonary or aortic valves open there is an initial rapid ejection of blood which becomes interrupted by the closure of the anterior cusp of the mitral valve against the hypertrophied septal segment of the outflow tract, cutting off the sharp initial rise in the pulse. In severe stenosis stroke volume fails to increase with exercise and the associated reduction in blood flow may produce syncope or angina.

Mitral stenosis. Progressive obstruction to left ventricular inflow by the fibrosing valve in rheumatic mitral stenosis, produces gradually increasing breathlessness, due to left heart failure, at first evident only with the raised output of exercise, but later evident at rest and even progressing to pulmonary oedema. Enhanced contraction of the hypertrophied left atrium becomes of major importance in maintaining ventricular filling, and the onset of atrial fibrillation often marks the onset of rapid clinical deterioration. As mitral obstruction becomes more severe the left atrial pressure rises, and enough passive pulmonary arterial hypertension develops to maintain the pulmonary arteriovenous pressure gradient, until the left atrial pressure reaches about 25 mm Hg. There is then, sometimes, a disproportionate rise in pulmonary artery pressure, which may reach systemic levels. This active pulmonary hypertension is due to vasoconstriction followed by intimal thickening and muscular hypertrophy of the pulmonary arterioles, probably as a response to interstitial pulmonary oedema. By reducing cardiac output it tends to protect the pulmonary capillary bed from excessive venous pressure. However, this response imposes a strain on the right ventricle, and eventually right heart failure occurs. Pulmonary oedema occurs characteristically in the early stage of mitral stenosis and in patients not showing the pulmonary hypertensive reaction. Interstitial oedema makes the lungs stiffer and is probably the most important cause of shortness of breath in pulmonary congestion, which probably acts by stimulating vagal 'J' (juxta-pulmonary-capillary) receptors in the lung interstitium.

The major features of mitral stenosis are evident on auscultation. The first heart sound (mainly mitral valve closure) is accentuated by a combination of slight delay in mitral valve closure from the raised left atrial pressure so that the valve is exposed to a faster-rising left ventricular pressure and a slight increase in stiffness of the valve cusps. Just after the second sound, mitral valve opening is audible as a sharp snap. These features (loud first sound and opening snap) are lost in late severe disease when the valve tissue has become fibrotic and heavily calcified. The turbulent flow produces a low-pitched, rumbling diastolic murmur, delayed in onset until after the opening snap and accentuated up to the first heart sound if there is normal atrial contraction (the presystolic murmur).

Coarctation of the aorta. A congenital narrowing of the aorta at the point of insertion of the ductus arteriosus gives obstruction to flow to the lower part of the body, usually with high pressures proximally. There is accentuated pulsation of the arteries in the upper part of the body with a reduced and very delayed pulse below the obstruction. Lower body flow is maintained by collateral arteries mostly around the scapula, although a large internal mammary artery is mainly responsible for a systolic murmur which is delayed in relation to the heart sounds to fall in late systole. Tortuous intercostal arteries may notch the ribs on chest X-ray. Although the high proximal aortic pressure in severe coarctation may overload the heart and cause cardiac failure, coarctation is often seen as symptomless hypertension in young adults, until cerebral haemorrhage or aortic rupture occur. The normal cerebral blood flow in coarctation cannot be taken to represent abnormal cerebral vasoconstriction, because it can be fully explained by cerebral autoregulation. The elevation of proximal aortic pressure could be the simple mechanical result of a mismatch between cardiac contractile force and proximal large vessel resistance. However, the proximal hypertension could have a renal origin being functionally equivalent to bilateral renal artery stenosis.

Right-sided central circulatory obstruction

Pulmonary arteriolar obstruction, by reducing blood flow generally or to parts of the lungs, causes impaired gas exchange and hyperventilation presenting with central cyanosis and breathlessness. The enormous capacity of the right ventricle for hypertrophy often allows the maintenance of a normal resting cardiac output despite a many-fold increase in resistance in the pulmonary bed, until right-heart failure occurs.

The physical signs of severe pulmonary hypertension are very characteristic with a palpable pulmonary artery in the second left intercostal space and a palpable right ventricle in the third and fourth spaces close to the sternum. There is often a prominent 'a' wave in the jugular venous pulse. The pulmonary second sound is accentuated and the first sound is followed by a loud click due to sudden distension of the lax dilated pulmonary artery (ejection click). Frequent causes of pulmonary hypertension are mitral stenosis (see above) and septal defects (see below) which may reverse the direction of flow and produce central cyanosis from a right-to-left shunt; this situation is often called the Eisenmenger syndrome.

Pulmonary stenosis has similar effects but rarely produces as great a rise in right ventricular pressure as severe pulmonary hypertension. Tricuspid valve stenosis produces a low cardiac output with great elevation of the jugular venous pressure; it is usually part of a picture of severe multiple rheumatic valve disease.

Massive pulmonary embolism. In the normal subject small pulmonary emboli are rapidly absorbed, but in a congested lung (such as in mitral stenosis) they give rise to pulmonary infarcts, wedges of consolidated lung with inflammation on the pleural surface, presenting as chest pain or haemoptysis, and are a common complication in patients kept in bed for heart failure.

More serious is massive pulmonary embolus, usually from leg vein thrombosis in the hypercoagulable inactive state after a major operation or after cardiac infarction, which may cause sudden death in these situations. Patients that survive the initial incident present as acute shock with sudden fall in cardiac output due to massive circulatory obstruction—usually a big clot at the bifurcation of the main pulmonary artery. Pulmonary hypertension is not a feature as there is no time for the right ventricle to hypertrophy.

The cold clammy extremities with peripheral cyanosis are evidence of the low cardiac output. As there is no pulmonary hypertension (flow is very low) the pulmonary second sound may be normal. Yet the right ventricle is dilated, showing in the electrocardiogram as right bundle branch block, often with an unusual electrical axis, and the jugular venous pressure is raised, reflecting acute right ventricular failure.

The natural processes removing the clot in the pulmonary artery are so powerful that complete resolution may follow in survivors and persistent obstruction does not occur. The clot is absorbed to the arterial wall and may appear at necropsy as an atheromatous plaque. Persisting pulmonary arterial obstruction may follow repeated small emboli in hypercoagulable states, as in patients on the contraceptive pill, and can produce severe pulmonary hypertension (see below).

VALVAR INCOMPETENCE

Valvar incompetence is generally accommodated by dilatation of the chambers on either side of the leaking valve. With incompetence of the aortic and pulmonary valves the degree of backflow depends not only on the valve itself but also on the mean diastolic pressure in the great vessel concerned. Because pulmonary diastolic pressure is generally low, pulmonary incompetence is usually unimportant, unless it complicates severe pulmonary hypertension as described above in mitral stenosis or pulmonary arteriolar obstruction of other cause. Aortic incompetence is more serious, and may in itself be enough to cause heart failure from the large volume load on the left ventricle, which must have a big enough stroke volume to maintain forward flow while allowing for the regurgitant flow lost backward through the incompetent aortic valve; this produces a very large and collapsing arterial pulse. Mitral incompetence similarly gives a volume load to the left ventricle as the low resistance to regurgitant flow into the low-pressure left atrium allows regurgitation to begin before the aortic valve opens; the murmur therefore begins with first heart sound (pan-systolic).

The sudden development of valvar incompetence such as may follow perforation of an aortic or mitral valve cusp as a result of infective endocarditis, or disturbance of the papillary muscle/chordae tendineae system in the same condition or after myocardial infarction, tends to be most dramatic in its clinical severity. The main haemodynamic problem lies in the normal size of the unprepared left ventricle in the case of acute regurgitation. As a result a relatively slight degree of regurgitation produces a sudden and severe rise in ventricular diastolic pressure which is transmitted back through the left atrium to the pulmonary venous bed with the production of pulmonary oedema. The rise in ventricular diastolic pressure may be so rapid in severe acute aortic incompetence that there is almost immediate equilibration of aortic and ventricular pressures during diastole so that the murmur of regurgitation is absent, and the mitral valve may close prematurely early in diastole with loss of the normal first heart sound. Turbulent flow in the brief period the mitral valve is open may create a mitral diastolic murmur. In the case of acute mitral regurgitation the volume of blood regurgitated into the previously normal-sized left atrium produces a similar rise in pressure and the abrupt onset of pulmonary oedema with a relatively small heart on chest X-ray.

Tricuspid incompetence is rare in the absence of right-sided heart failure and rarely occurs without associated mitral incompetence. Tricuspid incompetence seems often to be a functional defect resulting from over stretching of the right ventricle in several conditions which have in common an increased right ventricular pressure. When tricuspid incompetence is gross the systemic veins and the liver show exaggerated systolic venous pulsation.

CARDIOVASCULAR SHUNTS

If there is an abnormal communication between two parts of the circulation, blood flows through the defect at a rate depending on the pressure difference across the shunt and the size of the defect. If the shunt is from the high pressure left side of the heart to the low pressure right side, the situation is physiologically similar to that found in valvar incompetence, in that part of the cardiac output passes through part of the heart twice and the work of the appropriate ventricle must increase if the systemic flow is to be maintained. Shunts may occur between any of the chambers of the heart or great vessels that are anatomically related to one another, but are commonly due to congenital defects in the atrial (ASD) or ventricular septum (VSD) or persistence of the ductus arteriosus (PDA). Many small ventricular septal defects demonstrated in the first few months of life are now known to close spontaneously.

The effect of a left-to-right shunt on the heart varies with the level of the shunt, as may be seen from Table 2.1.

Table 2.1. Left-to-right shunts

Level	Ventricle with increased flow	Specific signs
Atrial (ASD)	Right	Wide fixed split of second sound
Ventricle (VSD)	Left (and right)	Pan-systolic murmur at left sternal edge
Aorta (PDA)	Left	Continuous murmur below left clavicle.

The great increase in pulmonary blood flow produces dilatation of the main pulmonary artery and its branches are dilated, and the upper zone pulmonary veins are also dilated. The chambers of the heart affected by the increased flow are dilated (Table 2.1) giving a characteristic appearance for each anomaly. In VSD and PDA turbulent flow through the defect produces a specific murmur (Table 2.1), pan-systolic in the case of a ventricular defect as flow is a jet from the high pressure left ventricle throughout systole, and continuous in the case of a persistent ductus as flow can go on throughout the cardiac cycle, tending to peak around the time of the second heart sound. In an atrial defect, the relatively large defect and low atrial pressures lead to great left-to-right flow without a murmur. The great filling of the right ventricle delays pulmonary valve closure, so the second sound is widely split and the usual respiratory variation is lost in the common venous return to the two ventricles so the splitting does not change with respiration, producing the "fixed split second sound" which is a useful diagnostic feature. With all these shunts the work of the ventricle carrying the shunt flow is increased and dilatation and hypertrophy follow.

A right-to-left shunt may occur if there is an increase in the pressure on the right side of the heart, owing to some obstruction in the right heart itself such as pulmonary stenosis or, from pulmonary arteriolar obstruction. Such hypertrophy may be regarded teleologically as an attempt to protect the lungs from too large a blood flow, but in contrast to the similar situation in mitral stenosis does not recover when the underlying lesion is corrected surgically. The anatomical situation differs at each level; in persistent ductus arteriosus the blood enters the aorta from the pulmonary artery only if there is increased pulmonary vascular resistance, and only the lower half of the body (and sometimes the left arm as well) may be cyanosed, since the admixture of venous blood takes place distal to the origins of the innominate and left carotid arteries. At the other end of the scale is atrial septal defect across which the flow right to left may be produced by any disease of the right heart. The commonest example of a right-to-left shunt in adult life is Fallot's tetralogy in which the aorta arises from both the left and right ventricles over a ventricular septal defect, and there is pulmonary stenosis. The resistance to right ventricular outflow causes mixed venous blood to be ejected into the aorta. The characteristic effect of a right-to-left shunt is arterial unsaturation; cyanosis at rest is not obvious until the proportion of shunted blood in the peripheral arteries is over 30 per cent.

Pulmonary vascular hypertension

In the presence of large left-to-right shunts the pulmonary artery pressure is high because of the huge pulmonary blood flow, although the pulmonary vascular resistance remains low. In some children there appears to be persistence of the fetal type of muscular arterioles in the lungs. In adult life patients with large left-to-right shunts develop atheromatous changes in the intima of the pulmonary arteries, and eventually there is progressive pulmonary hypertension and right heart failure (Eisenmenger syndrome).

DISORDERS OF CARDIAC RHYTHM

When an initiating wave of depolarization reaches an area of depressed excitability there may be an abnormal local delay in conduction (block) which allows the excitation wave ot only to go forward by an alternative route, but also to return and reactivate previously blocked regions, and set up a second impulse shortly after the initiating depolarization. This 're-entry' phenomenon may account for *ventricular ectopic beats*, which are characterized by an abnormal sequence of ventricular activation as they arise eccentrically in the ventricles. Ectopic beats may be looked on as the building blocks from which more complex dysrhythmias are constructed. Repeated re-entry in the whole network of ventricular muscle fibres with focal areas of impaired conduction from disease or sympathetic stimulation may lead to the total breakdown of organized ventricular activation *Ventricular fibrillation*, with incoordinate electrical activity, allows no effective contraction; the heart looks like 'a bag of worms' at operation. It is a form of cardiac arrest and is fatal unless corrected by an electrical shock. In *atrial fibrillation* the conducting system is stimulated at a faster rate than the atrio-ventricular (AV) node can respond and the ventricular rate is usually inconveniently fast and irregular; appropriate inhibition of AV node conduction with vagotonic drugs such as digitalis glycosides will control the situation. *Heart block* with interference in atrio-ventricular conduction may result from any lesion of the atrioventricular conducting system and can be divided into block at the atrioventricular node, open to autonomic effects, or in the His-Purkinje system of the atrioventricular bundle or its branches, by detecting the His bundle spike on the intra-cardiac electrogram. Re-entry may also explain the *Wenckebach phenomenon* of progressively prolonged P–R interval so characteristic of inhibition of conduction at the atrio-ventricular node, which can be demonstrated even in normal subjects if the right atrium in stimulated at fast enough rates. Either the atrio-ventricular node or part of the bundle branch system is partially depolarized by retrograde excitation so that it fails to respond to the next excitation wave arriving from the usual direction.

At either AV nodal or His-Purkinje system levels, three degrees of block of increasing severity are recognized; 1. a long P–R interval (conduction is impaired but continues consistently with each P wave), 2. dropped beats (some P waves fail to conduct and are not followed by a QRS complex), 3. complete block (no P waves are conducted to the ventricle, which is controlled by an 'escape' rhythm at a slow rate from ventricular cells with pacemaker potential). Sudden cessation of the escape rhythm, usually temporary, is the usual cause of *Adams-Stokes attacks* (see below).

SYNCOPE

The simple faint

In a simple faint there is a sudden reflex fall in peripheral resistance, in filling pressure and in cardiac output. This causes faintness or loss of consciousness owing to insufficient cerebral blood flow. The sudden fall in peripheral resistance is caused mainly by relaxation of the vascular bed of skeletal muscle, which produces a big increase in muscle blood flow. The afferent limb of the reflex is not known, but many different sensory inputs may be causative in sensitive subjects. There is usually reflex bradycardia at first but a compensatory tachycardia quickly follows during recovery. Fainting attacks almost always occur in people who are standing still, and when they fall to the ground recovery starts, because venous return and cardiac output tend to return to normal. The patient usually has premonitory symptoms of nausea, sweating and dizziness. Recovery is never immediate, often taking several minutes to be complete.

Low output syncope

Syncope due to failure to maintain an adequate cardiac output occurs on effort when there is severe obstruction to the circulation; for example in aortic or pulmonary stenosis, and in severe pulmonary hypertension. The cardiac output cannot increase sufficiently to maintain an adequate cerebral circulation and at the same time supply the metabolic needs of the exercising muscles. Recovery occurs when the cerebral circulation is restored as the needs of the muscle decrease.

Syncope in tachycardia

Fainting attacks due to disturbance in cardiac rhythm may result from extreme bradycardia, cardiac standstill or from extreme tachycardia. At the higher rates occurring with abnormal rhythms, the cardiac output begins to fall off because the heart cannot fill during diastole, nor is there enough time for adequate coronary flow which is predominantly diastolic as the capillary bed is closed during systole. When the heart rate is extremely fast (over 250/min) any effort readily causes syncope.

Adams-Stokes attacks: syncope in bradycardia

If the heart stops, unconsciousness follows in about 15 sec and irreversible cerebral changes occur within 4 min. Repeated (Adams-Stokes) attacks of cardiac syncope occur most commonly in patients with complete heart block who often have pulse rates of 28-36/min from a slow escape pacemaker in the ventricles; this bradycardia is not the cause of their syncopal attacks, which are due to episodes of cardiac standstill (pacemaker failure) or ventricular fibrillation. Recovery after cardiac standstill is immediate and complete provided that permanent cerebral damage has not occurred or an epileptic fit does not follow. The sudden onset and rapid recovery help to distinguish a Adams-Stokes attack from a simple faint.

POSTURAL SYNCOPE

Neurological disease or drugs interfering with circulatory reflexes during postural change can produce postural hypotension (e.g. diabetes mellitus with neuropathy, syringomyelia, tabes dorsalis and ganglion blocking drugs). In other situations the filling of the circulation is inadequate (e.g. in the chronic sodium and water depletion of severe adrenocortical insufficiency). Loss of vascular tone after prolonged bed rest may produce a similar situation after major surgical operations, and is one reason for asking the physiotherapist to begin exercising post-operative patients recumbent.

ATHEROMA

Arterial obstruction and consequent damage to vital organs arises from focal degenerative changes in arterial walls. The presence of cholesterol in the lesions has suggested to many that hypercholesterolaemia is responsible. Cigarette smoking certainly accentuates the changes, though whether from hypoxia due to carbon monoxide, or from a vasospastic action of nicotine is not yet clear. The lesions begin as deposits in the endothelium, usually at sites of stress, i.e. at major arterial bifurcations, such as the division of the common iliac or common carotid arteries, or at sites of repeated bending as in the major epicardial coronary arteries or the poplited artery at the knee. Cholesteral deposition in the lesions leads to the build up of a thick layer, the atheromatous plaque, which may ulcerate and discharge its contents into the blood to produce more distal embolic obstruction. Platelet thrombi frequently form on the plaque and may give rise to more distal emboli. The frequency of this process has suggested that the plaque may arise from an organized platelet thrombus with secondary lipid infiltration.

The recent discovery of the part played by prostaglandins has suggested a basis for the individual susceptibility to atheroma. Two prostaglandins synthesized in the body from arachidonic acid play a part in coagulation homeostasis. Prostacyclin (PGI$_2$) is synthesized in the arterial wall and acts as a clearing factor preventing platelet deposisation, while thromboxane (TXA$_2$) synthesised by platelets acts as an aggregating factor, increasing the tendency to platelet thrombi, when released by damage to platelets. Some therapeutic hope is offered by the discovery that drugs such as aspirin, which inhibit prostaglandin synthesis, have long-lasting effects on platelets. As unnucleated cells they cannot reform the damaged enzyme, but synthesis of PGI$_2$ by vessel walls quickly recovers. TXA$_2$ levels do not return till fresh platelets are liberated from the marrow, with a half life of about a week.

PERIPHERAL ISCHAEMIA

Atheroma affecting the major limb vessels produces progressive local arterial obstruction and results in a gradual reduction of the maximum attainable blood flow. The severity of the resulting disability depends on the extent to which a collateral arterial supply opens. During exercise the inflow of blood to the limb is limited and there is a drop in arterial pressure distal to the obstruction. Intermittent claudication represents ischaemic pain arising in muscles distal to the obstruction. Eventually occlusion of major vessels may lead to tissue death in the distal parts of the limb (gangrene).

Raynaud's phenomenon. This is a condition affecting the fingers, and less commonly, the toes. The skin shows paroxysmal colour changes of pallor followed by cyanosis and then redness, accompanied by severe pain. These changes are usually provoked by cold. In Raynaud's 'disease' the primary lesion involves digital arteries, rendering them unduly susceptible to normal constrictor stimuli, and leads eventually to organic narrowing and atrophic changes in the tissues of the digits. The cause of the condition is not known, but in some cases it forms part of the clinical syndrome of a 'collagen' disease, especially scleroderma, systemic lupus erythematosus and rheumatoid arthritis, or is produced by embolic block of digital arteries from more proximal atheroma of large arteries.

DISORDERS OF THE VEINS

Venous obstruction. At most sites in the body there are several veins with free intercommunications draining the tissues. This protects against the development of a high venous pressure and consequent oedema following obstruction of just one vein. Furthermore, moderate elevation of venous and capillary pressures do not cause oedema because the normally negative interstitial pressure provides a safety factor. Oedema cannot appear until interstitial pressure has risen from its mean normal value (-7 mm Hg) to atmospheric pressure. Although complete obstruction of a single vein thus produces no appreciable effects on a tissue, complete and sudden obstruction of all veins draining a tissue causes tissue death, accompanied by engorgement and haemorrhage. Severe but incomplete obstruction causes only oedema.

Venous valvular incompetence. The valves are not essential for venous return, which can still occur passively; but if valves are incompetent the massaging effect of skeletal muscle contraction elevates venous pressure distal to the valves and in the capillary bed. During exercise the venous pressure is higher in the leg than in normal subjects. Persistent elevation of venous pressure results in dilatation of veins and oedema, and has a bad effect on tissue nutrition. Although dilated and tortuous superficial varicose veins often appear in association with raised venous pressure (either general, as in cardiac failure, or local, as in pregnancy) they occasionally occur without apparent cause. In these patients the vein wall is often congenitally defective in elastic tissue.

Loss of venous tone. The capacity of the veins is potentially very great, and their calibre is normally reduced by tonic contraction of muscle in their walls. Failure of this tone in enough veins below heart level causes pooling of the blood with failure of venous return and cardiac output. The role of this mechanism in syncope and other conditions of circulatory failure is not fully known, but it is probably partly responsible for the postural hypotension following sympathectomy or the use of α-sympathetic-blocking drugs.

MYOCARDIAL ISCHAEMIA

Angina is a referred visceral pain which occurs when the heart muscle is short of oxygen. Anything which interferes with the supply of oxygen to the heart muscle can provoke angina. The usual cause is atheromatous change in the coronary arteries, but anoxia or anaemia, by reducing the amount of oxygen carried in the coronary blood, can provoke anginal pain which may disappear when the oxygen content of the coronary blood is restored to normal. Both aortic stenosis and incompetence can interfere with the flow of blood to the coronary vessels in situations where left ventricular hypertrophy requires a greater flow than usual, and syphilitic aortitis may involve the mouths of the coronary arteries and mechanically reduce the coronary flow. Any condition which increases the metabolic needs of the heart muscle will tend to provoke angina. Exercise and emotion are by far the most important precipitating factors, but thytotoxicosis and hypertension also tend to make angina worse. Emotion may precipitate angina by increasing the cardiac output, by producing tachycardia, by raising the blood pressure and by increasing skin and muscle blood flow. The tendency for angina to be worse after a heavy meal is due to the fact that the circulatory needs of the stomach and intestines have also to be met. Nitrates are thought to relieve angina by lowering arterial pressure and hence reducing the work of the heart, but they also produce venous pooling and reduce cardiac output with an added effect on cardiac work. During angina the ischaemic segment of left ventricle fails to relax properly (a partial contracture) and the resulting stiffness may increase filling pressure in a manner resembling left ventricular failure. A corresponding fall in cardiac output may accentuate the ischaemia.

Coronary artery spasm has been clearly established as a cause of some unusual anginal attacks, particularly those unrelated to exercise or emotion. Its role in other situations is under investigation, but it rarely if ever occurs in normal vessels.

Acute myocardial infarction. This results from more severe arterial obstruction, usually coronary thrombosis on an atheromatous plaque. It produces prolonged anginal pain without strict relation to effort which sometimes persists for several hours until necrosis of the affected myocardium destroys the sensory nerve terminals. There is increasing suspicion that in some patients the coronary thrombosis may be secondary to stasis of blood in a region of critically impaired coronary flow initiated by increased demands for myocardial work in the presence of extensive coronary artery disease or by coronary artery spasm. Some cases of cardiac arrest without infarction may have a similar basis and can recover without myocardial damage if treated promptly.

The local myocardial necrosis or infarction acutely interferes with left ventricular function and may produce very low cardiac output (cardiogenic shock) or left ventricular failure; both may be present, but confusion of the two as 'pump failure' is unhelpful as different treatments may be needed.

In the early stages of infarction the findings are often dominated by severe autonomic disturbances, with increased sympathetic tone more frequent in anterior infarction and leading to undue tachycardia which may extend the area of damage by increasing oxygen demand in marginal areas, and setting the scene for cardiac arrest due to ventricular fibrillation; increased vagal tone is more frequent in inferior (diaphragmatic) infarction, due to different distribution of autonomic sensory receptors, and may lead to nausea, bradycardia and a low arterial pressure which can extend damage by reducing coronary blood flow to marginal areas. This transient reflex vagal shock may if untreated extend the infarct and go on to the almost uniformly fatal true cardiogenic shock with severe hypotension, confusion and loss of renal function due to a critically reduced cardiac output.

ARTERIAL HYPERTENSION

Arterial blood pressure is not constant, but is subject to moment-to-moment variations as the circulation adjusts to varying demands for tissue perfusion. The selection of a single measurement as *the* blood pressure induces a false simplicity. Increased activity of the sympatho-adrenal system can raise the blood pressure acutely, as in emotion or exercise. Either cardiac output or peripheral resistance, or both, may increase. Other factors can acutely raise blood pressure (e.g. compression of venous reservoirs by contracting muscle, and increased levels of circulating angiotensin).

Chronic elevation of systemic arterial pressure is a feature of several distinct diseases, but in none is it fully understood. *Phaeochromocytoma* is perhaps the least enigmatic because the hypertension is probably caused by an excessive secretion of both noradrenaline and adrenaline by the adrenal tumour. The attacks of pallor, hypertension and (usually) reflex bradycardia seen when catecholamine release is intermittent may be reproduced by infusing noradrenaline intravenously. While surgical removal of the tumour is commonly followed by a fall in pressure, some degree of hypertension may persist or recur. These clinical failures illustrate a principle common to many secondary hypertensive states—pathological elevation of arterial pressure from whatever cause may eventually induce a self-perpetuating hypertension that is uninfluenced by removal of the factor that initiated the rise of pressure. Possibly damage to the kidney during the initial hypertensive phase gives rise to continued hypertension by a second mechanism.

In *coarctation of the aorta* the vascular obstruction is usually severe enough to limit the escape of blood to the periphery, so leading to a rise both in systolic and diastolic pressures proximally in the upper aorta. Beyond the narrow segment the aorta is often mainly fed by the flow in collateral vessels, e.g. around the scapula. The systolic pressure is reduced, but the diastolic is often above normal, and pulsation is reduced. From the failure of the central hypertension to evoke compensatory splanchnic vasodilatation it may be inferred that the carotid sinus regulating mechanism is probably adapted to a higher than normal blood pressure, as in most other forms of chronic hypertension. The cause of the raised blood pressure is not known, but it seems reasonable that it may have features in common with the hypertension of renal artery stenosis (see below). The hypertension is usually relieved but not always abolished by surgical correction of the anomaly.

The adrenal cortex plays some part in the maintenance of blood pressure, not only by the influence of its mineralocorticoids on salt and water balance, but also because glucocorticoids seem to be required for the maintenance of normal blood pressure. There is at present some dispute about the effects of adrenal cortical hormones on vascular reactivity to nervous and chemical influences. The hypertension of Cushing's syndrome is probably due to the oversecretion of cortisol, and the consistently raised systemic arterial pressure seen in primary hyperaldosteronism is probably due to an excess secretion of salt-retaining adrenal cortical steroids, notably aldosterone. Rarely other unusual steroids may be responsible. The hypertension of pre-eclamptic toxaemia in pregnancy may also involve endocrine changes; though in this disease the cause of the disturbance has not been fully elucidated.

Renal hypertension remains enigmatic. A raised arterial pressure is common in acute and chronic glomerulonephritis, often found in chronic pyelonephritis and renal artery stenosis, and is occasionally seen in congenital, allergic, toxic, neoplastic and degenerative lesions of the kidney. The available clinical and pathological data suggest that a reduction in the blood supply to the kidney is an important factor. This has led to intensive investigation of experimental hypertension in laboratory animals, from which it may be concluded, that the acutely ischaemic kidney liberates a specific secretion (renin) which interacts with a constituent of the plasma to produce a vasoconstrictor substance (angiotensin). The latter is thought to be responsible for the generalised arteriolar constriction that produces the initial rise in blood pressure. The extent to which this sequence contributes to the hypertension of kidney disease in man remains conjectural and attempts to demonstrate pressor substances in the blood in experimental renal hypertension in animals have only been successful in the early stages. However, angiotensin can exert an indirect and slowly developing pressor action when present in concentrations which are not enough to raise blood pressure acutely. It seems likely, therefore, that the renin-angiotensin system plays an important part in the pathogenesis of chronic renal hypertension. Further support for this has come from the therapeutic use of specific inhibitors of the renin-angiotensin system such as inhibitors of converting enzyme. Sodium retention may also contribute to renal hypertension (see below).

In most patients with hypertension no cause, renal or otherwise, can be found, and such patients are therefore described as having *essential hypertension.*

The malignant phase of hypertension. High arterial pressure of any origin may pass into a malignant phase characterized by papilloedema and focal necrotic lesions of the renal arterioles and probably similar-sized arteries elsewhere in the body. This syndrome is at present widely believed to be qualitatively different from 'benign' hypertension, as the renal arteriolar changes provoke the renin response (above) and renal function is rapidly impaired. It seems likely that in most cases malignant hypertension is produced by an excessively high or rapidly rising blood pressure. Patients with malignant hypertension almost always secrete an excess of aldosterone as part of the renin response, though they do not thereby become oedematous. We do not yet know whether this is functionally important; but it helps to explain the low plasma potassium concentrations sometimes seen in such cases.

Theories to explain essential hypertension

The response to sympathetic nervous activation is to increase peripheral resistance and cardiac output, and the (false) association in the popular mind of hypertension with nervousness and anxiety has suggested that increased sympathetic stimulation may be responsible for the initial stages of essential hypertension, going on to a self-perpetuating process as in other forms of secondary hypertension. False hopes in this direction have been raised by the use of inadequate controls and reports of a 'high-output' stage of early hypertension only rarely include similarly investigated normal subjects. Plasma catecholamine levels are not a good measure of sympathetic activity, and have anyway been somewhat equivocal in essential hypertension, though raised adrenaline levels may be found in some patients. In addition, the inbred spontaneously hypertensive Okamoto rat, which provides a close analogue of essential hypertension in man, seems to have a dominant neurogenic component.

The most popular current theory centres round the kidney in respect of sodium and water excretion. A certain renal perfusion pressure is needed to maintain salt (Na) loss from the body. Perhaps on the basis of a congenital lesion the kidneys may require gradually increased perfusion pressure to maintain the excretion of dietary sodium, and hypertension develops, thus preventing blood volume expansion causing an unduly high cardiac output.

Other current theories centre on some defect of cellular cation transport, perhaps associated with a deficient natriuretic ('third') factor—but there is still considerable inconsistency in the measurements of transport defects, as well as considerable differences of opinion about their interpretation.

Principles of observations, tests and measurements

THE CHARACTER OF THE PERIPHERAL PULSE AND ARTERIAL BLOOD PRESSURE

The rate of rise of the pulse wave is influenced by the rate at which blood is entering the aorta from the heart and by the compliance of the arterial tree. In aortic valve stenosis ejection is slowed. The pulse pressure (the volume of the pulse) depends chiefly upon the volume of blood ejected by the left ventricle into the aorta and to a lesser extent upon the force of this ejection and the rigidity of the arterial walls. If the cardiac output is low the pulse is 'small'. If the output is high the pulse is 'full' and may be abnormally prominent in the large arteries in the neck; in contrast to the large pulse of aortic incompetence, the large pulse of a high cardiac output is always accompanied by a tachycardia. The height of the diastolic pressure is an indication of the peripheral resistance. Anything that allows the blood to run off from the arterial system too rapidly during diastole causes a low diastolic pressure and a 'water-hammer' pulse, giving a sharp knock from the radial artery with the arm raised. Examples are onward run off, as in the vasodilatation of hyperthyroidism and anaemia, and backward run off, as in aortic incompetence.

Pulsus paradoxus

The term is a misnomer because the phenomenon is really a gross exaggeration of a normal tendency rather than paradoxical. If the systemic arterial pressure and fullness of the pulse diminish appreciably during inspiration (pulsus paradoxus) it is likely either that the heart muscle itself or the pericardium is abnormally rigid, or that the pericardium is tightly stretched by fluid within it. In severe cases the systemic arterial pulse may disappear altogether during a deep inspiration. The phenomenon is clearly mechanical in origin, because it disappears immediately the pericardial pressure has been reduced by aspiration in cardiac tamponade. A conventional explanation of the phenomenon is that the tug of the diaphragm on a rigid or tight percardium during inspiration reduces the filling of the right side of the heart. However, this explanation fits neither with animal experiments nor with certain clinical observations. Another suggestion is that when there is a limit to the total of right heart volume + left heart volume, as by a rigid or tense pericardial sac, an increase in right heart volume, which is normal during inspiration because of augmented venous return from the reduced intrathoracic pressure, can occur only at the expense of filling of the left side of the heart. The cardiac septum is deflected, the filling of left atrium and ventricle is impaired, and left ventricular stroke output falls. If right atrial inflow of blood is kept constant, experimentally, then pulsus paradoxus cannot be produced by increasing intra-pericardial pressure. Pulsus paradoxus does not occur in lax (low pressure) pericardial effusion, but it may be produced by accentuated intrathoracic pressure swings in massive pulmonary embolism and severe airways obstruction.

VENOUS PRESSURE

The normal venous pulse consists of positive *a* and *v* waves, with *x* and *y* troughs (Fig. 2.2). Clinically the pulse is best seen with the patient reclining with the head supported and the sternomastoid muscle relaxed, at some angle to the horizontal which renders the internal jugular vein visibly pulsatile. The *a* wave is due to atrial contraction and is absent in atrial fibrillation. It is presystolic in timing, comes with the first sound because of slight delay for the pulse to reach the neck, and increases on inspiration. It may be just visible in normal subjects lying at 45° head-up tilt. The *a* wave is increased in conditions in which the right atrium or right ventricle is beating against a raised resistance. The best examples are seen in pulmonary stenosis and pulmonary hypertension when the hypertrophied right ventricle is stiff and ventricular septum intact, and in tricuspid stenosis. In the presence of ventricular septal defect the right ventricular pressure does not exceed that in the left ventricle and the right atrial pressure wave is less marked. A *c* wave synchronous with the arterial pressure peak was observed in early work done with external recording. It is an artefact of underlying carotid pulsation and with catheter tip recordings the *c* wave is unobtrusive and frequently not seen at all.

The *x* descent after the *a* wave results from atrial relaxation plus the pull of the contracting ventricle, which tends to enlarge the atrium. It is followed by the *v* wave, which is caused by atrial filling during ventricular systole, and times with the second heart sound. The *v* wave is augmented by ventricular regurgitation in tricuspid incompetence and is then better called a positive systolic wave. The peak of the wave occurs at the time that the tricuspid valve opens, and is followed by the *y* descent, which represents ventricular filling. In tricuspid stenosis the rate of fall of venous pressure during the *y* descent is slow while in tricuspid incompetence and constrictive pericarditis with a similar high venous pressure it is rapid. In this condition or with a stiff right ventricle the pressure quickly rises from the *y* dip to a diastolic plateau level.

The nature of the cardiac rhythm may also be determined by observing the venous pulse in the neck. In atrial fibrillation the *a* waves are absent as is the *x* descent. The appearance of large 'cannon' waves indicates that the right atrium is contracting at a time when the tricuspid valve is closed. This occurs at irregular intervals in complete heart block where the atria and ventricles contract at different rates, and also occurs regularly in nodal rhythm when atria and ventricles contract simultaneously and also occasionally in ventricular tachycardia when an independent atrial contraction falls on a closed tricuspid valve. In atrial flutter, regular flutter waves due to atrial contraction are occasionally seen at a rate of about 300/min.

The 'central' venous pressure

The mean jugular venous pressure is raised in right heart failure and also in conditions of gross fluid overload even though the heart is normal. If pressure is very high the venous pulse may be difficult to see, the head of the column of blood being above the ears with the patient reclining at 45°. Internal jugular pulsation is looked for first with the patient sitting upright, then at 45° to the horizontal, then supine. An empty external jugular vein which can be made to fill by pressure at the base of the neck is a reliable indication that mean central venous pressure is no higher than the point of pressure. A full external jugular vein may reflect a high pressure in the superior vena cava, but can also be produced by local compression. It is then necessary to rely on the observation of pulsation in the internal jugular vein to estimate central venous pressure. Central venous pressure can be measured directly by means of a thin catheter passed into the superior vena cava, but has technical difficulties for the inexperienced.

Normally inspiration sucks the blood into the chest and lowers the pressure in the superior vena cava: it may be useful to ask a patient to breathe in sharply to demonstrate the upper level of filling of the neck veins in cases in which central venous pressure is excessively high. Gentle pressure over the liver is of great diagnostic value if in doubt, in that venous pulsation is accentuated but arterial pulsation is unaffected. However the anterior position of the artery high in the neck usually distinguishes it from venous pulsation. Even when the venous pressure is very high so that the crest of the pressure wave is not visible the character of the venous pulsation may still be modified by this manoeuvre. In some cases of cardiac tamponade and constrictive pericarditis or severe right heart failure cardiac filling may be impaired by inspiration so that the central venous pressure may rise paradoxically.

Central venous pressure is the cardinal measurement for distinguishing between circulatory failure and low blood pressure due to failure of the heart (venous pressure high) and that due to blood or plasma loss, or to dilatation of veins (venous pressure low), but left heart failure may occur in isolation, with a normal venous pressure, and it may be necessary to measure pulmonary venous pressure with catheter in the pulmonary artery, where pressure distal to a wedged or balloon catheter quickly equilibrates with the pulmonary venous pressure.

THE CARDIAC IMPULSE

Palpation of the cardiac impulse is valuable in deciding whether either ventricle is beating abnormally or against an excessive load. In the presence of left ventricular hypertrophy the apex beat is sustained. Normally the apex beat is completed within the first half of systole, well before the second heart sound. The apex beat may be heaving as well as sustained when the left ventricle is dilated as well as hypertrophic. Right ventricular hypertrophy produces a palpable outwards thrust in the third and fourth intercostal spaces to the left of the sternum; a large pulmonary artery gives a similar thrust in the second space. If both atrium and ventricle on one side are dilated, the site of the apex beat cannot be used to determine which ventricle is involved. Occasionally a double apex beat is palpable, and relates to the sound of forceful atrial contraction (4th heart sound) giving a corresponding presystolic filling thrust.

NORMAL AND ABNORMAL HEART SOUNDS

As a result of recent work physiological explanations can be advanced for most of the abnormal heart sounds heard in various conditions. Some extra sounds may occur in normal subjects, while others always indicate disease. Taking the cardiac cycle in sequence, the first abnormal sound is a presystolic or atrial sound, which gives rise to presystolic triple rhythm of 3 heart sounds per cycle. This fast triple rhythm found in heart failure is often called a 'gallop rhythm'. This sound is always abnormal and is associated with forceful atrial contraction. It may be due to right atrial contraction in pulmonary stenosis, pulmonary hypertension or cor pulmonale; in left ventricular failure it is associated with left atrial contraction. The sound is probably due to blood distending the ventricle, because it occurs immediately after atrial contraction is finished.

Splitting of the *first heart sound* into its two components of tricuspid and mitral valve is commonly heard in health, particularly in a depressed sternum, but is widened when either bundle branch is blocked, so that the ventricles contract asynchronously; i.e. the blocked impulse delays the corresponding ventricle. The loudness of the first heart sound depends on the position of the atrioventricular valves when ventricular systole starts. In health, ventricular filling is complete before the end of diastole and the atrio-ventricular valves tend to close. In mitral or tricuspid stenosis, ventricular filling is prolonged and systole starts when the valve is still wide open. Its sudden closure increases the loudness of the first heart sound. The pliability of the valve also plays a part, because the first heart sound is not loud in patients with a heavily calcified valve. The timing of atrial contraction also influences the loudness of the first heart sound; and patients with a short P–R interval on the electrocardiogram may produce a loud first sound. The soft first sound in acute rheumatic fever is associated with a prolonged P–R interval, so that the valve is starting to close again before ventricular systole.

Systolic clicks occur during the ejection phase of ventricular contraction. They may originate in the pulmonary artery or aorta and are thought to be due to sudden changes in tension in the walls of the vessel concerned. They never occur in healthy hearts, but are loudest in mild pulmonary and aortic valve stenosis when they arise from sudden distension of the fused core of valve tissue, and in pulmonary hypertension or aortic dilatation. The great vessel from which the click originates is usually dilated.

The normal splitting of the *second heart sound* and its variation with respiration has already been described. The timing of either component may be delayed if ventricular contraction is prolonged or delayed. Thus in right bundle branch block, when right ventricular contraction is delayed, the second sound is widely split, but the width of the split still increases on inspiration. In atrial septal defect, the second heart sound is widely split and the width of the split is characteristically uninfluenced by respiration. The probable explanation for this is that inspiration produces such a small percentage increase in the large right ventricular stroke volume and in the increased left ventricular filling through the defect, that pulmonary valve closure is not further delayed. If left ventricular contraction is prolonged in left bundle branch block or aortic stenosis the aortic valve closes after the pulmonary, and the normal relationship between the two components of the second sound is reversed. As a result expiration, by shortening right ventricular contraction, widens the gap between the two sounds. This phenomenon is referred to as reversed or paradoxical splitting of the second sound, and may be difficult to detect in severe aortic stenosis as the aortic second sound is soft. Pulmonary valve closure may be inaudible in severe pulmonary stenosis, when the pulmonary blood flow is small. The second heart sound is then single. In less severe pulmonary stenosis pulmonary valve closure is delayed, but is often audible though faint. If the right and left ventricles are in free communication and have equal pressures as in Fallot's tetralogy and Eisenmenger's complex, the second heart sound is usually single. Ventricular filling starts soon after the second heart sound and if there is mitral or tricuspid stenosis, the sudden opening of the atrioventricular valve often results in a loud, high-pitched sound known as an *opening snap*. It is caused by movement of the aortic cusp of the mitral valve. The opening snap is a distinctive feature of rheumatic mitral valve disease, and if the left atrial pressure is high the mitral valve will open soon after aortic valve closure, so that the opening snap will be early. If the valve is calcified and immobile the opening snap may be lost.

The *third heart sound* occurs during the period of rapid ventricular filling, early in diastole. It is heard in normal children, but in adults is evidence of abnormal filling of either ventricle. It is also heard in conditions in which a large amount of blood enters the ventricle rapidly, as in mitral incompetence, ventricular septal defect, and persistent ductus arteriosus. If the ventricle itself is abnormal, a third sound may occur with a normal flow as in left ventricular failure, constrictive pericarditis or cardiomyopathy ('constrictive myocarditis'). This sound gives rise to diastolic triple rhythm and with rapid heart rates, when diastole is short, the third sound may coincide with an atrial sound and produce a 'summation gallop'. The transient slowing produced by carotid sinus pressure may allow distinction of an atrial or third sound; sometimes both are present. There is a tendency for an atrial sound to be present in small stiff failing ventricles and to give place to a third sound as the ventricle dilates in more severe disease.

HEART MURMURS

Heart murmurs are due to the turbulent flow of blood through normal or abnormal valves or intra-cardiac defects. An increased flow through a normal valve may produce a murmur, the best example being the pulmonary systolic murmur of atrial septal defect. In this condition the right ventricular stroke volume at rest may be twice the normal maximum on severe exercise, and this large flow is thought to be the cause of the murmur. Systolic murmurs are of three types:

(1) Those due to *systolic ejection*, which start after the first sound and finish before the second; these ejection murmurs, often called 'diamond-shaped' because of their phonocardiographic appearance, are due to flow through the aortic or pulmonary orifices, and denote either increased flow through normal orifices or normal flow through stenotic orifices. (2) *Pansystolic murmurs* which fill the whole of systole. These latter occur in mitral and tricuspid incompetence and ventricular septal defect, and fill systole because the pressure gradient, causing the flow responsible for the murmur continues throughout systole, so that flow from high to low pressure chambers can begin with the first heart sound. (3) *Late systolic murmurs*: in some forms of mitral incompetence (e.g. ruptured chordae tendinae and (hypertrophic obstructive cardiomyopathy-HOCM) the murmur occurs in the second half of systole only. These

murmurs are common with lax mitral chordae which allow mitral valve prolapse after mid-systole as the ventricle shrinks. The murmur may be initiated by a sharp mid-systolic click which can occur alone in minor variants. In HOCM the murmur begins with the late onset of subaortic obstruction in mid-systole, but is mainly due to associated mitral incompetence at this time. A similar late murmur from the collateral arteries in coarctation is systolic but delayed in relation to the heart sounds.

Diastolic murmurs are of two types. Those due to incompetent flow from the aorta or pulmonary artery into the ventricle start immediately after the second sound and are high pitched. Diastolic murmurs due to impeded flow from an atrium into a ventricle do not start until isometric relaxation of the ventricle is complete, about 0·1 s after the second sound. These delayed murmurs are low-pitched and their duration and intensity are related to the degree of stenosis and to the flow through the orifice. A similar short murmur may occur with an increased flow through a normal valve, in patients with left-to-right shunts.; from the mitral valve in persistent ductus or ventricular septal defect, and from the tricuspid valve (and accentuated by inspiration) in atrial septal defect. The accentuation of the late diastolic (presystolic murmur) in mitral stenosis is due to atrial systole and is therefore lost in atrial fibrillation.

Continuous murmurs occur when there is a pressure gradient and flow between two parts of the heart throughout the cardiac cycle. The commonest example is persistent ductus arteriosus, where the pressure in the aorta is continuously higher than in the pulmonary artery. If there is no pressure gradient, because of pulmonary hypertension, the continuous murmur is absent.

Inspiration, by increasing the venous return to the right heart, increases the right ventricular output, so that it makes right heart sounds and murmurs louder. Expiration squeezes blood out of the lungs and encourages flow into the left heart and consequently left-sided sounds and murmurs are louder on expiration.

EXERCISE RESPONSES

Cardiac function is most readily tested by exercising the patient. It is important either to observe the patient during exercise, or better to measure the ventilation, oxygen uptake and pulse rate during steady exercise either during a step test or on a treadmill or bicycle ergometer. In general, steady sustained exercise is limited by the maximum oxygen uptake. This is governed by different factors in different diseases but in heart disease it depends on the maximum cardiac output and the maximum extraction of oxygen from the blood by the tissues. By the Fick principle:

$$\text{Maximum oxygen uptake} = \text{Max. cardiac output} \times \text{Max. O}_2 \text{ extraction}$$
$$\text{(in ml/min)} \qquad \text{(in l/min)} \qquad \text{(in ml/l)}$$

The maximum oxygen extraction depends on the level of haemoglobin in the blood and is normally about 140 ml/l. This value is relatively fixed, so that the maximum oxygen uptake in heart disease reflects the maximum cardiac output.

There is evidence in healthy subjects that the heart rate and oxygen consumption increase proportionately with increasing grades of steady exercise, and that the maximum oxygen uptake is reached with a pulse rate of about 195/min. The maximum oxygen consumption can be predicted by extrapolation from oxygen consumption at submaximal levels of exercise (e.g. at a pulse of 140/min), although not, of course, in atrial fibrillation where rate is not strictly related to demand. The cardiac output response varies with posture. In upright subjects stroke volume increases considerably during transition from rest to exercise, while in the supine position stroke volume is larger at rest and so does not increase as much when exercise begins.

VALSALVA'S MANOEUVRE

The measurement of the effect of Valalva's manoeuvre on the arterial blood pressure has been used as a test of ventricular function. In the test, the patient blows up a column of mercury to 40 mm Hg and maintains this pressure for 10 sec. The changes in systemic pressure which occur in normal subjects are shown in Fig. 2.3(a). The arterial pressure rises with the blow, as the increase in intrathoracic pressure is transmitted to the blood in the great vessels. The principal effect of the rise in intrathoracic pressure is to cut off the venous return to the right atrium. As a result, filling of the heart is reduced and the arterial pressure falls as the central reservoir of blood in the heart and lungs is gradually emptied. The fall in arterial pulse pressure and mean pressure, acting through the baroreceptor reflexes, results in a rise in peripheral resist-

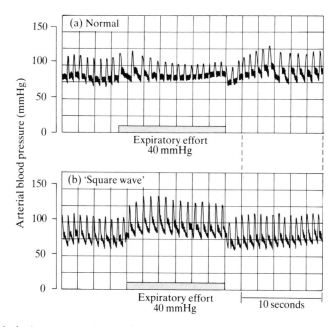

Figure 2.3 *Valsalva's manoeuvre* (see text).

ance, and towards the end of the period of strain is often a slight tachycardia and rise in the mean arterial pressure. When pressure is released, the filling of the heart is suddenly restored and the cardiac output increases. At this stage there is still peripheral vasoconstriction and the increased flow into a constricted arterial bed produces a rise in arterial pressure above the resting level, which is usually referred to as 'overshoot'. This rise in pressure in turn stimulates the baroreceptors and produces cardiac slowing and the pressure returns to normal. Bradycardia after release of Valsalva's manoeuvre is the best clinical sign of a normal response.

In a typical abnormal response, shown in Fig. 2.3(b), the arterial pressure does not fall when the venous return to the right atrium is arrested. This may be either because there is left ventricular failure, with excessively high filling pressure initially, or because there is an increased amoung of blood in the central reservoir of the heart and lungs, or because of a combination of both factors. As the arterial pressure does not fall, there are no changes in rate or pressure during and after the period of strain. The abnormal response may be recognized clinically by the absence of bradycardia after the pressure is released. This abnormal response to Valsalva's manoeuvre has been called a 'failure' response, but it is better to call it a 'square-wave' response, because it also occurs in patients with large left-to-right shunts with large volumes of blood in the lungs, but no evidence of heart failure. ('Square wave' refers to the shape of the systemic arterial pressure tracing during the manoeuvre).

When the autonomic nervous system is disabled (e.g. by diabetic neuropathy, tabes dorsalis, or adrenergic neurone-blocking drugs) no overshoot or bradycardia occur after release of expiratory effort. The arterial pressure gradually regains its previous level.

ELECTROCARDIOGRAPHY

Excitation of cardiac muscle is associated with a reversal of the electrical potential between the inside and outside of the membrane of the individual fibres. The spread of the excitation process from cell to cell through the heart causes a changing electrical field which can be recorded at the surface of the body. The deflections recorded during sequential activation of the heart are labelled arbitrarily from P to U; the P wave represents atrial activation, the PR interval (normally about 0·20 s—one large square on the recording paper) the delay, mainly at the atrio-ventricular node before the complex wave from ventricular activation (QRS) (normally less than 0·10 s—$2\frac{1}{2}$ small squares on the record).

Ventricular recovery setting the background for reactivation is evident as the large slow T wave normally in the same direction as the major wave of the QRS. A smaller U wave may follow as a late part of this process. An adequate account of the theory and practice of electrocardiography is beyond the scope of this book.

The electrocardiograph can give information about:

1   *The site of the pacemaker or the nature of the cardiac rhythm.*
If the excitation process does not arise in the sino-atrial node the P wave becomes abnormal in shape or position or disappears.

2   *Disorders of conduction or of the excitation process.*
Blockage or delay of the spread of the excitation process at any site alters the timing and shape of ventricular potentials, block of conduction in one of these bundle branches widens the QRS complex (bundle branch block-BBB).

3   *The size of the muscle mass in individual chambers of the heart.*
Hypertrophy increases the size of the relevant deflections.

4   *The state of viability or metabolism of the cardiac muscle.*
Death of a portion of the muscle due to obstruction of the artery supplying it can be diagnosed and localized with considerable precision from loss of the R wave and its partial replacement by an initial negative deflection Q.

Experience has also shown that cell damage, electrolyte disturbances, and the action of certain drugs cause characteristic changes in the T waves which may be diagnostically helpful, as in potassium depletion and intoxication, digitalis intoxication, and hypothyroidism. Similar changes are also helpful in the diagnosis of pericarditis and pericardial effusion.

Systolic time-intervals

A simple record of ECG, carotid pulse and phonocardiogram gives good information on the various phases of systole. Left ventricular ejection time (LVET) is obtained from the carotid pulse, and pre-ejection period (PEP) from the time from the Q wave to onset of the pulse.

Exercise testing

A functional assessment of the behaviour of the circulation is best obtained from the response to the natural stress of exercise such as walking on the treadmill or the use of an exercise bicycle. It is possible to measure work load from total body oxygen consumption, either collecting expired air with a mouthpiece and nose-clip, or measuring the changes in oxygen concentration in air flowing through a hood.

The *electrocardiogram obtained during exercise* provides a useful indication of myocardial ischaemia on the basis of local ST segment shift over the affected area. ECG mapping may increase sensitivity, but it is usual to record a single left ventricular lead (say V_5) as screening test. Some modification of the usual limb lead positions is necessary to avoid movement artefact; a site on the trunk near the base of the limb and away from muscles is best, for instance over the scapula. Some ST depression is usual in the tachycardia of exercise, but usually affects only the first part of the ST segment. Horizontal depression of the whole ST segment is the positive response. Such changes often persist when the tachycardia of exercise has subsided, but the old, after-exercise test is probably less sensitive than testing during exercise. In patients with transient arrhythmias diagnostic information can often be obtained from 24 hour tape-recorded ECGs during normal activity, using a small portable commercial tape recorder (Holter monitoring).

RADIOLOGY

The information obtained from radiography of the heart tends to be anatomical in nature and in general relates to the size and position of different chambers of the heart and the great vessels. The assessment of atrial size can best be made by X-ray and, while it is not always possible to distinguish the left ventricle from the right, radiology often gives information about ventricular size. The position and size of the great vessels may be of considerable importance in congenital heart disease, and radiology is particularly helpful in such conditions as coarctation of the aorta, persistent ductus arteriosus, Fallot's tetralogy and transposition of the great vessels. The radiological examination of the lungs may give some indication of left atrial pressure. In pulmonary congestion due to any cause, the presence of Kerley's horizontal interlobular lines at the base of the lungs indicates that the left atrial pressure is raised and has produced interstitial oedema. In full pulmonary oedema, wing-shaped shadows fill most of the lung fields. At an early stage an elevation of left atrial pressure produces Y-shaped shadows due to distention of the upper zone pulmonary veins (normally empty in the erect position). If severe pulmonary hypertension develops the pulmonary artery and its main branches dilate with accentuation of upper lobe blood flow diversion, and tapering of the peripheral lower zone arteries can often be detected. The amount of blood in the lungs can also be assessed radiologically and a distinction can be made between the oligaemic lungs of pulmonary stenosis and the pulmonary plethora in patients with a left-to-right shunt with distended arteries and veins.

The principal use of fluoroscopy is in the detection of valve calcification, which may be of importance in determining the site of a lesion; but it is also helpful in assessing pulsation in various parts of the heart and great vessels, and can give useful information about the volume of blood passing through a particular chamber.

ANGIOCARDIOGRAPHY

The analysis of serial or cinematograph X-rays of radio-opaque dye passing through the heart and great vessels is an important method of outlining the anatomy, and is particularly valuable in the localization of shunts, and in assessing the presence and severity of valvular incompetence.

OXIMETRY

An ear oximeter is designed to measure the arterial oxygen saturation from the output of a photoelectric device recording the light transmitted through the warmed ear. It is thus a measure of the 'blueness' of the patient. Oximetry is particularly useful in detecting the changes in arterial oxygen saturation such as may occur when a patient exercises. The absolute level of saturation is difficult to measure accurately by oximetry. The clinical recognition of cyanosis is difficult until the arterial oxygen saturation falls below 75 per cent. It is important in congenital heart disease to know whether the arterial saturation falls with exercise, and this may be difficult to determine without oximetry. The distinction between cyanosis due to a shunt and cyanosis due to lung disease may be difficult to make on clinical grounds. The recording of the change in arterial saturation by oximetry while the patient breathes oxygen is an important means of determining the cause of the cyanosis. In lung disease without shunting cyanosis is due to failure of the blood to become fully saturated with oxygen during its passage through the lungs. This deficiency is rapidly corrected by breathing 100 per cent oxygen and the saturation rises at the rate of 15 per cent per minute or more. The cyanosis of a shunt results from mixing of venous blood with fully saturated blood coming from the lungs. In this case, when the patient breathes 100 per cent oxygen, the saturation rises slowly because of increased transport of oxygen in solution in the pulmonary blood. The rate of rise is seldom more than 8 per cent per minute and depends on the relative amounts of the pulmonary and shunt flows.

Another use of oximetry is in the detection of atrial septal defect. In patients with small or medium-sized atrial septal defects, the left-to-right shunt is transiently reversed after Valsalva's manoeuvre. This shunt reversal produces a temporary drop in arterial oxygen saturation about 3 sec after the end of the period of strain. The drop in saturation, which is usually about 3 per cent, can be detected by oximetry. With large atrial septal defects, the response to Valsalva's manoeuvre is square wave in type and shunt reversal does not take place.

CARDIAC CATHETERIZATION AND MEASUREMENT OF CARDIAC OUTPUT BY THE FICK PRINCIPLE

Cardiac catheterization sets out to measure the pressure and blood flow in all the accessible chambers of the heart. Right heart catheterization began with insertion of a catheter in a peripheral vein. A great advance was the finding that a catheter wedged in a peripheral pulmonary artery measured pulmonary venous pressure. The same effect is obtained distal to a balloon catheter inflated in the pulmonary artery branch, and can be performed at the bedside by a float-in technique without radiological control. Nowadays the left heart is usually entered by retrograde arterial catheterization across the aortic valve. Intracardiac pressures are measured with an electromanometer connected to the catheter and blood flows are estimated either by indicator (see below) or by using the Fick principle, which involves the analysis of the oxygen content of blood samples take from different parts of the circulation and measurement of the oxygen uptake in the lungs.

$$\text{Cardiac output (in l/min)} = \frac{\text{Oxygen consumption (in ml/min)}}{\text{Arteriovenous oxygen content difference (in ml/l)}}$$

Oxygen is convenient for this purpose, but the Fick principle is, of course, equally applicable to any substance taken up or removed by the lung. The same principle is used to measure the blood flow of individual organs, in which case the flow is usually expressed as the 'clearance' of the indicator.

The determination of the cardiac output in the absence of a shunt entails the collection of blood from a systemic artery and mixed venous blood, usually from the pulmonary artery. These samples are taken during the collection of expired air for the measurement of oxygen uptake, and the method is accurate only if the patient is in a 'steady state'. In patients with shunts, the Fick principle can be applied to measure both the pulmonary and systemic flows. In measuring the pulmonary flow, samples of pulmonary arterial and venous blood are required. A pulmonary venous sample cannot always be obtained and an oxygen content of 97 per cent must then be assumed. In measuring the systemic flow in the presence of a shunt, systemic arterial and mixed systemic venous samples are required. The systemic venous sample is obtained from a site where the blood is not mixed with blood shunted from the left side of the heart. The site of entry of oxygenated blood is of diagnostic value in detecting the level of the shunt.

The measurement of intra-cardiac pressure is important in the diagnosis of pulmonary stenosis and pulmonary hypertension. In pulmonary stenosis there is higher systolic pressure in the right ventricle than in the pulmonary artery; and in pulmonary hypertension both pressures are increased. In assessing the severity of either condition, it is important to consider the pulmonary blood flow. The relationship between pressure and flow is linear in the homogeneous pulmonary vascular bed and is generally expressed as resistance:

$$\text{Resistance (in mm Hg/l/min)} = \frac{\text{Mean pressure difference (in mm Hg)}}{\text{Blood flow (in l/min)}}$$

In this general formula, the mean pressure difference may be between any two points in the circulation. For example, in calculating the pulmonary vascular resistance, the mean pressure difference between the pulmonary artery and the left atrium is used; and for the systemic resistance, the mean right atrial pressure is subtracted from the mean systemic arterial pressure. The normal pulmonary vascular resistance is 1·5 mm Hg/l/-min and the normal systemic resistance about 10 times greater. The systemic vascular resistance is a less useful measurement because of the heterogeneity of the different organs vascular beds in parallel. The pulmonary artery pressure may be raised because of increased blood flow without any rise in resistance.

One of the most important diagnostic aspects of right heart catheterization in congenital heart disease is that the catheter may be passed into the left heart through a defect. The pressure tracing and oxygen content of the blood sampled are important

means of identifying the chamber entered. In rheumatic heart disease right heart catheterization is more concerned with the assessment of the severity of the lesion than
with the anatomical diagnosis. In mitral stenosis, the severity of the obstruction at the
mitral valve can be assessed from the height to which the left atrial pressure is raised at
rest and during exercise, and from the degree to which the cardiac output is reduced.
The pressure in the left atrium can be measured indirectly by impacting the catheter in
a distal branch of the pulmonary artery. This 'wedge' pressure has been shown to
reflect closely the left atrial pressure.

Direct measurement of the pressure gradient across the mitral valve can be
obtained by remote puncture of the inter-atrial septum, and introduction of a catheter
from the right into the left atrium.

Patients with mitral stenosis sometimes develop pulmonary vasoconstriction which
causes further impedance to the circulation. Resistance from this cause can be calculated from the pressure gradient between the main pulmonary artery and the wedge
pressure.

Severity of left ventricular and aortic lesions is assessed by left heart catheterization.
An arterial catheter is advanced upstream into the left ventricle. Angiography can also
be performed in the same way.

INDICATOR DILUTION STUDIES, AND THE MEASUREMENT OF CARDIAC OUTPUT
BY THE STEWART-HAMILTON METHOD

Indicator dilution techniques may be used either for the measurement of the cardiac
output or for the detection of shunts in congenital heart disease. The cardiac output
can be calculated from a record of the concentration of injected dye passing a given
point in the systemic circulation (Fig. 2.4). The technique used is to inject a known
quantity of dye into a vein and record the change in colour due to the passage of dyed
blood, either at the ear or in a systemic artery, with a photo-electric device.

The formula used for the calculation of cardiac output by the dye method is:

$$\text{Cardiac output (in l/min)} = \frac{60.\text{I}}{\text{C.T}}$$

where I is the amount of dye injected in mg, C is the mean concentration of dye during
the first circulation in mg/l and T is the time taken for the first circulation in seconds.
The principal difficulty with the method is that blood containing dye starts to recircu-
late before the first systemic circulation is complete. As a result, the time taken for the
first circulation has to be obtained by extrapolating the exponential decrease in dye
concentration which occurs before recirculation starts. When dye dilution techniques
are used for the detection of shunts during cardiac catheterization, the information
obtained is qualitative. Dye can be injected into any chamber into which the catheter is
passed. The time taken for the injected dye to appear in the systemic circulation will
depend on the lesion. If the dye is injected into a chamber in the right heart from which

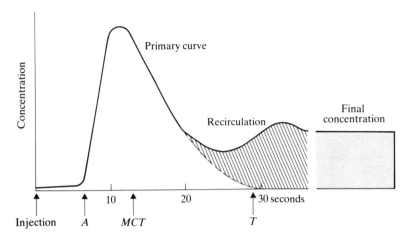

Figure 2.4 *Normal indicator or 'dye' curve.*

The continuous curve shows the concentration of an indicator substance in the blood of a systemic artery following its injection into an arm vein at the time indicated. A is the time of first appearance of the dye at the sampling site. The dark shaded area is the 'primary' curve, i.e. the curve that would be obtained were there no recirculation. Its end, T, has to be drawn by eye, and its position judged by the early part of the descending primary curve. If concentration is on a logarithmic scale the early descending limb is straight. The area under the curve represents the sum of the concentrations at unit intervals of time (CT). Mean circulation time (MCT) is the time taken for half the dye to travel from the site of injection to the sampling site.

Cardiac output can be derived by dividing the quantity of dye injected by the area CT. Since the dimensions of the area CT are concentration (i.e. mass per unit volume) × time, the dimensions of mass (of dye) divided by CT become volume per unit time, which is the cardiac output. The *central blood volume*, i.e. that between the injection and sampling sites, can be calculated by multiplying the cardiac output by the mean circulation time (MCT). The *total blood volume* can be calculated by dividing the mass of dye injected by the final concentration attained after recirculation mixing is complete. If the dye is confined to the plasma, the volume must be divided by the haematocrit.

On-line computing systems have been developed by means of which these calculations can be carried out automatically.

blood is shunted to the left, the dye will appear earlier than if it had passed through the right heart and lungs before reaching the left heart. If there is a left to right shunt, a proportion of the dye injected into the right heart recirculates through the lungs, so that the disappearance of dye from the systemic circulation is delayed. Useful qualitative information about the presence and site of intracardiac shunts can be obtained from the dye curves following dye injections into different chambers of the heart. Simplified techniques using cold as the indicator (it is lost before recirculation) and recording with a thermistor have been adapted to bedside use.

PLETHYSMOGRAPHY

Plethysmography may be employed to obtain an estimate of the volume of blood flowing through an extremity. The method is based on the principle that when the veins are briefly obstructed, the congestive swelling that occurs is due to arterial inflow, and the rate of swelling is for the first few seconds a precise measure of the rate of the arterial inflow. The test involves specially made apparatus and considerable technical experience, and has little place in routine clinical work, except possibly in the diagnosis of arteriovenous fistula of the extremity. In this condition considerable increase in flow occurs even when the lesion is small.

Radioisotope scanning

This approach shows promise as a non-invasive screening test. Using an isotope that behaves as an analogue of potassium, the myocardium can be outlined with a gamma camera. Most work has been done with ^{201}Tl which has a long half-life and can be distributed to hospitals. Other isotopes like ^{15}N as ammonium require an adjacent cyclotron, as the half-life is short.

Giving the radioisotope during exercise shows the myocardium with a cold-spot due to failure to fill the ischaemic area, and a later scan after redistribution will distinguish temporary ischaemia which fills later from the persistent defect of old infarction. Ca-seeking substances such as the labelled ^{99}Tc pyrophosphates are taken up by recent infarction, which accumulates Ca and shows as a hot-spot. Blood-pool labelling with ^{131}I albumin will show the outline and movement of the cardiac chambers and allow screening for ventricular function; the lack of movement of infarcted areas is usually obvious.

Echocardiography

Ultrasonic reflection gives precise measurement of the distance and shape of intracardiac structures. Originally introduced as a fixed beam echo against time (M-mode) through an intercostal space it proved useful in defining mitral valve movement and ventricular diameter. The principles of radar are now being applied to produce two-dimensional sector scans which give a detailed picture of ventricular function or complex cardiac anatomy in congenital heart disease.

CLINICAL OBSERVATIONS, TESTS AND MEASUREMENTS

The arterial pulse

Although it is usual to feel the radial pulse for heart rate measurement, wave form is often best determined from a larger central artery such as the carotid. The form of the arterial pulse is influenced by the volume ejected by the left ventricle (cardiac output) and by changes in the aortic valve (see p. 78).

The venous pulse and pressure

The venous pressure (p. 80) is of great importance in distinguishing cardiogenic from hypovolemic shock, and in identifying congestive heart failure as the cause of a patient's oedema.

Peripheral oedema

In established right heart failure, oedema collects in the dependent parts of the body, such as the ankles and legs in active subjects; and is recognized by temporary deformation by finger pressure (pitting). In the patient confined to bed it is best found over the sacrum and medial sides of the thighs. In infants it usually appears in the soft tissues of the face.

The cardiac impulse

The normal apex beat is largely the result of the recoil of the emptying left ventricle which ejects blood upwards and to the right into the aorta. It is accentuated by left ventricle hypertrophy. The right ventricle is not normally felt but in conditions of right heart load may produce an outwards thrust just to the left of the sternum in the third and fourth intercostal spaces. The electro-cardiogram is useful in confirming degrees of ventricular hypertrophy.

Auscultation

The normal and abnormal heart sounds and the different types of heart murmur have already been discussed (p. 81).

ROUTINE OBSERVATIONS

Non-invasive studies

Great advance has been made in recent years in the introduction of relatively simple methods to measure cardiac performance with little disturbance to the patient. The *phonocardiogram* (p. 83) gives useful information about the timing of heart sounds and murmurs in cases of difficulty. *Electrocardiography* (p. 86) and *exercise testing* are specially valuable in suspected ischaemic heart disease. *Radioisotope scanning* (p. 93) and *echocardiography* (p. 93) are becoming routine.

Special investigations

More elaborate, invasive studies, such as *angiocardiography* (p. 88) and intrinsic electrical mapping are normally confined to surgical referral centres.

References

GUYTON, A.C., JONES, C.E. & COLEMAN, T.G. (1973). *Circulatory Physiology: Cardiac Output and its Regulation.* (2nd ed.). Philadelphia, Saunders.

DICKINSON, C.J. & MARKS, J. (Eds.) (1978). *Developments in Cardiovascular Medicine.* Lancaster, MTP Press.

HAMILTON, W.F. (ed.) (1962–66). The Circulation. *Handbook of Physiology*, Section 2. Washington D.C., Amer. Physiol. Soc.

KAPLAN, N. (1982). *Clinical Hypertension* (3rd ed.). Baltimore, Williams & Wilkins.

SWALES, J.D. (1983). *Clinical Hypertension.* London, Chapman & Hall.

HURST, J.W. & LOGUE, R.B. (1970). *The Heart, Arteries and Veins.* (2nd ed.). New York, McGraw-Hill.

3

Respiration

Normal function

Respiratory physiology is primarily concerned with gas exchange. This is the process by which the lungs transfer oxygen and carbon dioxide between the environment and the blood. The lungs are also involved in certain aspects of the regulation of hydrogen ion concentration and in various non-respiratory functions.

Gas exchange can be conveniently subdivided into four steps:

1 Ventilation: the mass movement of air in and out of the lungs and the distribution of air within the lungs,

2 Gas transfer: the exchange of O_2 and CO_2 between the alveolar air and the pulmonary capillary blood,

3 Pulmonary blood flow: much of this process is more conveniently considered as part of circulatory physiology,

4 Blood gas transport.

This classification not only simplifies the presentation of the subject but is also a convenient and logical system of approach to clinical problems.

VENTILATION

Pulmonary ventilation is produced by the rhythmic contraction of the inspiratory muscles which cause expansion of the thorax and lungs. This rhythmic contraction is initiated in the brain stem. In ventilating the lungs the inspiratory muscles must overcome the elastic forces of the tissues and the resistance of the airways to the flow of air through them. The volume of ventilation in any given period of time is adjusted in response to changes in the partial pressures of the gases in the blood, which in turn depend upon the ratio of ventilation to the metabolic activity of the tissues. The control is thus by a servo or 'feed-back' mechanism.

TOTAL AND ALVEOLAR VENTILATION

A normal subject at rest with a respiratory rate of 15 breaths/min and a tidal volume of 500 ml has a total pulmonary ventilation of 7·5 l/min (i.e. 7·5 l. inspired and 7·5 l expired). However, about 140 ml of each breath are required to flush the conducting airways or anatomical dead-space, so that the ventilation of the alveoli is only about 5·5 l/min. If the depth of breathing is reduced, the dead-space ventilation becomes proportionately greater, so that in theory, if the tidal volume were to be equal to or less than the volume of the dead-space there would be no alveolar ventilation, however great the total ventilation.

If there are parts of the lung which have reduced blood-flow or no blood-flow, the air going to those parts performs little or no exchange of O_2 or CO_2. This portion of the inspired air is therefore like the air in the conducting passages and can be regarded as dead-space or wasted ventilation. The anatomical dead-space plus the volume of this dead-space-like air (which is sometimes called 'alveolar' dead-space to distinguish it from the anatomical or airway dead-space) is usually referred to as the physiological dead-space. Thus, the effective ventilation is the volume of air reaching gas exchanging alveoli each minute. It is called the alveolar ventilation and equals tidal volume minus physiological dead-space and then multiplied by respiratory rate.

In the recumbent normal subject, the inspired air is almost perfectly distributed throughout the lungs and meets the blood in optimal proportions. The physiological dead-space is not appreciably greater than the anatomical dead-space. In the erect position, however, the alveolar and thus physiological dead-space becomes increased because there is little blood-flow and therefore little gas exchange in the apices of the upper lobes.

LUNG VOLUMES

The total volume of air in the lungs can be subdivided into eight different components as illustrated (Fig. 3.1). Of these the most important are (a) *tidal volume*, which is the volume of air inspired or expired during breathing, (b) *vital capacity*, which is the volume of air expelled by maximum voluntary expiration after maximum inspiration, (c) *functional residual capacity*, which is the volume of air remaining in the lungs at the end of normal expiration, (d) *residual volume*, which is the volume of air remaining in the lungs after maximum expiration, and (e) *total lung capacity*, which is the volume of air in the lungs at maximum inspiration.

RESPIRATORY MUSCLES

During quiet breathing in normal subjects inspiration is produced by the action of the diaphragm and intercostal muscles; expiration is produced by the elastic recoil of the lungs. The diaphragm increases the volume of the thorax partly by its descent and partly by raising and everting the costal margin. If the diaphragm is depressed and flat, both of these actions become less efficient, its descent is restricted and the base of the thorax, instead of being expanded, may be contracted. Accessory muscles such as the sternomastoids, which are normally employed only during deep breathing, are then used even at rest. Their action is probably less efficient than that of the diaphragm and they tend to expand mainly the upper thorax where pulmonary blood-flow and gas exchange are least.

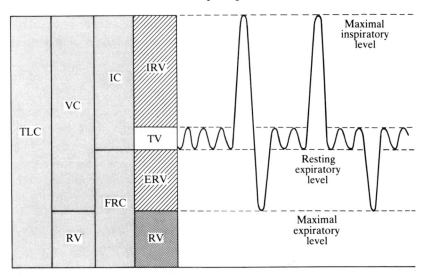

Figure 3.1. *The subdivisions of lung volume.*

There are four capacities and four volumes, viz. total lung capacity (TLC), vital capacity (VC), inspiratory capacity (IC) and functional residual capacity (FRC), and residual volume (RV), inspiratory reserve volume (IRV), tidal volume (TV) and expiratory reserve volume (ERV).

(From Comroe J.H. *et al.* (1962) *The Lung.* Copyright © 1962, Year Book Medical Publishers, Inc., Chicago. By permission.)

In patients who have obstructed breathing, and in normal subjects during deep or forced breathing, expiration may be assisted by the abdominal muscles. These muscles play little, if any, part in normal breathing although they are, of course, employed in coughing and straining.

The neural control of the respiratory muscles is further considered in Chapter 11.

MECHANICS OF BREATHING

Elastic properties of the lungs

Several terms have been used to describe the elastic properties of the lungs, the most popular being 'compliance' which describes the increase in lung volume per unit increase in distending pressure. Compliance, however, varies in a non-linear fashion with changing lung volume. This variation can be taken into account either by relating compliance to the size of FRC and arriving at a single value called specific compliance, or more accurately, by plotting compliance over the whole range of lung volumes and examining the resultant pressure-volume curve. The 'elasticity' of the lungs can also be looked at from the reverse standpoint, as the pressure required to distend them by a given volume. This is usually called the 'elastance'.

The elastic properties of the lungs are functions not only of the elastic tissue proper but also of the structural pattern of the lungs. Furthermore, many other structures within the thorax exert elastic forces and the surface tension of the liquid lining the alveoli provides about half the elastic recoil.

Surface tension

The small size of the alveoli minimizes the distance O_2 and CO_2 have to diffuse and maximizes the area of alveolar surface. But the alveoli have a liquid lining and the pressure generated by surface tension becomes progressively greater the smaller the volume or radius of the alveoli—from Laplace's law applied to a spherical bubble:

$$\text{Pressure} = 2\ \frac{\text{Tension}}{\text{Radius}}$$

However, the surface tension of the air-liquid interface lining the alveoli is reduced many times compared with that of an air-water interface by the properties of the lining fluid, surfactant. This lower surface tension allows the alveoli (and hence the lungs) to be smaller and the transpulmonary pressure distending the lungs to be much less than would be possible if the lining fluid had a high surface tension like water. Apart from its role in helping to maintain alveolar stability, surfactant probably also contributes both to the prevention of transudation of fluid into the alveoli and to certain local defence mechanisms. The composition of surfactant appears to be complex, including lipid, protein and carbohydrate and containing dipalmitoyl phosphatidyl choline as the main individual component. Its synthesis and secretion are by the large alveolar type II cells and the maintenance of the surface film requires periodic large expansion of alveoli, such as by sighs.

The relation of changes in surfactant to disease states is largely uncertain. This is because on the one hand, disease can result in surfactant deficiency and on the other hand, alterations in surfactant could give rise to pulmonary abnormalities. Surfactant deficiency has been well demonstrated in the respiratory distress syndrome of the newborn and may also be involved in the adult respiratory distress syndrome, pulmonary alveolar proteinosis, lung damage from prolonged positive pressure breathing or high inspired oxygen concentrations and some instances of atelectasis, pulmonary oedema and pulmonary vascular obstruction.

Resistive properties of the lungs and airways

The forces opposing inspiration include two major components: firstly, there are the static, elastic forces which we have just considered; secondly, there are the dynamic, non-elastic, flow-dependent or viscous forces which we will now examine.

The initiation and maintenance of airflow through the airways from the mouth to the alveoli or vice versa requires a pressure difference between the mouth and the alveoli. The narrower the airways, the greater their resistance to airflow and the greater must be the pressure difference. If the flow of air were entirely streamlined (laminar) the resistance would be constant and the pressure difference would be proportional to the rate of flow. However, at various sites in the respiratory tract, particularly in the larynx and large upper airways the flow of air is turbulent. When airflow is turbulent, an increase in the rate of airflow demands a disproportionate increase in the pressure difference. For this reason, it is not possible to give a single value for airflow resistance which would be appropriate at all rates of airflow. For practical purposes, resistance is calculated from measurements made during quiet tidal breathing.

The resistance of peripheral airways with a diameter of less than about 2 mm accounts for little of the total airway resistance, especially at high lung volumes. As a corollary, the resistance which arises from tissues sliding over each other must also be negligible, since this component is included in measurements of peripheral resistance.

Mechanics of expiration

In ordinary breathing, expiration is produced by the recoil of the stretched elastic tissues of the lungs. The recoil pressure forces the air up the respiratory passages and maintains the pressure inside them above the surrounding intrapleural pressure. The elastic recoil pressure of the lungs is much greater than is required for quiet expiration in normal subjects and the inspiratory muscles 'pay off' by gradual relaxation. During the deep breathing of muscular exercise the greater distension of the lungs provides greater recoil pressure and the inspiratory muscles 'pay off' more rapidly.

During forced expiration, the intrathoracic (intrapleural) pressure becomes positive because of muscular effort (Fig. 3.2). Its value is less than intra-alveolar pressure (which is intrathoracic plus elastic recoil pressure) but greater than mouth pressure (which is atmospheric pressure or zero). Since there is a progressive pressure drop along the airway from the alveoli to the mouth, there must be a site along the airway where the intraluminal pressure equals the surrounding intrathoracic pressure. This site is referred to as the 'equal pressure point'.

The significance of the equal pressure point is two-fold. Firstly, it divides the airway into upstream and downstream components. While these have no exact, and certainly no constant, anatomical counterparts, the downstream component corresponds to the larger airways and the upstream component to the smaller airways. As we have seen the smaller airways are responsible for little of the total airway resistance, with the result that disease at this site can be clinically silent. Secondly, any further increase in muscular effort and hence in intrathoracic pressure tends to compress further the airway at and below the equal pressure point and not to increase the rate of airflow. The maximum expiratory flow rate (and hence indices such as FEV_1, MMF and flow-volume relations) is thus 'effort-independent' and is determined by the elastic recoil pressure and the resistance of the upstream component. Structural abnormalities of the smaller airway become important during forced expiration: loss of rigidity at or below the equal pressure point can permit the airway to collapse and be converted into a

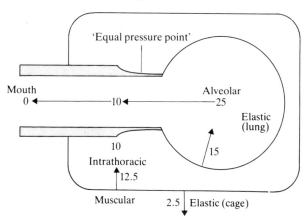

Figure 3.2. *The pressures during forced expiration (1 cmH$_2$O $\simeq$ 0.1 kPa).*
The interrelations between the various pressures during forced expiration are illustrated. Appropriate values are given in cmH$_2$O. The wall of the airway is shown to change from alveolus to mouth but no attempt has been made to show the concomitant change in calibre. During a forced expiration, muscular effort produces a positive intrathoracic (intrapleural) pressure, as soon as muscular contraction is greater than the passive outward elastic recoil of the thoracic cage. Since the pressure gradient down the airway ranges from intra-alveolar values (elastic recoil plus intrathoracic) to mouth values (zero or atmospheric), at some point along the airway a site must exist where the intraluminal pressure is the same as the surrounding intrathoracic pressure. This point is called the 'equal pressure point' and its significance is described in the text.

Starling resistor; any reduction in calibre due to bronchial smooth muscle contraction, mucosal swelling or intraluminal secretions can predispose to premature airway closure and air-trapping.

If the elastic recoil pressure of the lungs is partially lost or if the air-flow resistance of the respiratory passages is increased, the passive mechanism of expiration becomes less effective. There are two possible compensations for this. The first is further distension of the lungs so as to increase recoil pressure and the second is the use of expiratory muscles. The first response could at first sight seem a beneficial result of the hyperinflation that is seen in asthma and emphysema. Unfortunately its effectiveness is limited, firstly, because prolonged stretching of the lungs causes a loss of elastic recoil (increased compliance) and secondly, because the mechanical advantage of the diaphragm and rib cage is reduced at high lung volumes. The second response, the use of expiratory muscles, is also of limited value for although this increases the pressure in terminal airspaces, it also raises the pressure surrounding the airways. Thus, if these muscles contract too forcibly they can cause narrowing or even collapse of the airways, which further increases resistance and limits the rate of airflow. This limitation may be so severe that 'air-trapping' occurs. The narrowing of the airways by forced expiration, while unhelpful to pulmonary ventilation, is valuable in coughing because the linear velocity of the airflow through these narrowed airways is greatly increased, even though the rate of volume flow is reduced. During a vigorous cough all the intrathoracic airways are compressed and the velocity of airflow in the trachea of a normal subject may reach Mach 1 (600 mile/h).

DISTRIBUTION OF VENTILATION

The even distribution of ventilation within the lungs can be disturbed either by regional differences in distending (transpulmonary) pressure or by local differences in mechanical factors (compliance and non-elastic resistance). Regional differences in distending pressure are due to gravity and occur in both healthy and unhealthy subjects. Local differences in mechanical factors are abnormal and are found in many disease states.

Because of the weight of the lung, there is a gradient of intrapleural pressure down the lung, the pressure being about 7·5 cmH_2O less at the top than at the bottom in the erect position (Fig. 3·3). Thus from the bottom to the top of the lung the transpulmonary pressure and recoil pressure becomes progressively greater.

This regional difference in intrapleural pressure gives rise to regional differences in ventilation because the relation between increase in volume per increase in pressure

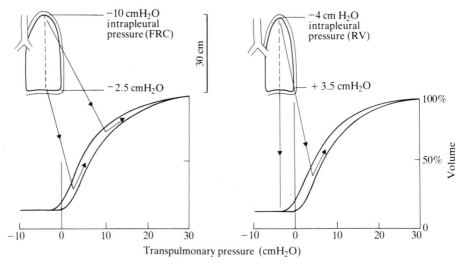

Figure 3.3. *Regional pressure-volume relations at FRC (left) and RV (right) (1 cmH₂O ≏ 0·1 kPa).*
The gradient of intrapleural pressure down the lung is about 7·5 cmH_2O and is due to the weight of the lung. The result is that different parts of the lung operate at different transpulmonary pressures and thus on different parts of the pressure-volume curve of the lung. Since this curve is non-linear, regional volume changes, i.e. ventilation, depend on the lung volume at which the change occurs as well as on the transpulmonary pressure change. Examples and their implications are described in the text.

(From West J., 1977.)

(compliance) is non-linear. During tidal breathing, i.e. at and above FRC, the bottom of the lung receives more ventilation than the top of the lung, because the bottom operates lower down and hence on a steeper part of the pressure-volume curve (Fig. 3.3, left side). At or near residual volume, however, the intrapleural pressure at the base may become positive, so that peripheral airway closure occurs and ventilation is then less at the bottom than at the top of the lung (Fig. 3.3, right side). This effect becomes more marked if the small peripheral airways are diseased or if there is loss of the elastic recoil (as with age or emphysema) which generates the transpulmonary pressure. The lung volume at which airway closure commences is called the 'closing volume'.

The compliance of a lung unit determines the volume of air it receives. The airway resistance to a lung unit determines the rate at which air is received. Thus the product of resistance and compliance, sometimes referred to as the 'time-constant', determines the amount of ventilation received by a unit in a given time (Fig. 3.4). Unequal time-constants of different lung units exposed to the same transpulmonary pressure will result in uneven distribution of ventilation. This uneven distribution of ventilation becomes worse as the respiratory rate increases, since units with longer time-constants receive less and less ventilation. Thus more and more of the lung becomes excluded from ventilation, a phenomenon which is the basis of the test for frequency dependence of compliance.

WORK OF BREATHING

The *mechanical* work of breathing can be estimated by measuring the pressures developed and the volumes of air displaced. The work performed on the lungs can be derived from records of the intrapleural or intra-oesophageal pressure, but the work on the thoracic cage is technically difficult to measure. The total mechanical work of breathing in normal subjects is about 0·6 kilopond metres per minute, of which at least two-thirds is expended on the lungs. In patients with chronic lung or heart disease the mechanical work of breathing may be increased five- or ten-fold.

The *metabolic cost* of breathing can be assessed by estimating the O_2 consumption of the respiratory muscles; this amounts to 1–3 per cent of the total O_2 intake at rest (i.e. about 0·5 ml O_2 per l. ventilation, or 4 ml O_2 per min). At higher levels of ventilation the O_2 consumption of the respiratory muscles increases disproportionately, but in normal subjects it represents only a small portion of the total O_2 uptake even at the highest levels of pulmonary ventilation reached during strenuous physical exertion. In patients with heart or lung disease the O_2 consumption of the respiratory muscles during exercise may reach very high levels. As the maximum rate of O_2 uptake is also limited in these patients, the O_2 consumption of the respiratory muscles may severely restrict the amount of O_2 available to the rest of the body.

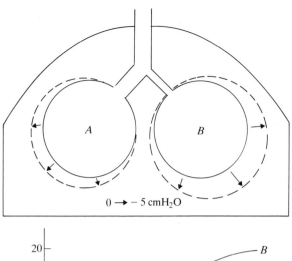

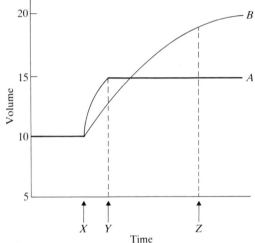

$0 \rightarrow -5\ \mathrm{cmH_2O}$

Figure 3.4. *The mechanical properties of the lungs and the distribution of ventilation.*
(*1 cm H₂O ≃ 0·1 kPa*).

A and B are two 'lung units' (which could be alveoli, lobules or lobes) having the same resting volume (say, 10 mm³). They are enclosed in the thorax. A is less distensible (less compliant) than B so that, when the pressure round them (intrathoracic or intrapleural pressure) is lowered from 0 to -5 cmH₂O, A enlarges only to 15 mm³ whereas B enlarges to 20 mm³. In addition, the airway leading to A is wide and of low resistance to airflow, whereas the airway to B is of high resistance. The effect of this difference in resistance is shown in the tracing of change in volume against time (lower figure). If the surrounding pressure is suddenly lowered at time X, air enters A rapidly and B slowly. If the inspiration stopped at time Y, A would be more ventilated than B, but if the rate of breathing were slower so that inspiration lasted till time Z, B would be more ventilated than A.

Air will flow into A or B until the pressure in A and B is the same as that at the mouth. At time Y airflow into A stops because the pressure in A has become the same as the pressure at the mouth. Air continues to flow into B between times Y and Z because the pressure in B is still lower than mouth pressure. If, at time Y, the intrathoracic pressure were raised (i.e. returned towards O), there would be an interval before the pressure rose sufficiently to stop air flow into B, although A would start to empty immediately. During this period air would flow from A not only to the mouth but also into B. This phenomenon is called 'paradoxical respiration' or *pendelluft* (swinging air).
(Reproduced from the *Postgraduate Medical Journal*, **34**, 30, 1958.)

CONTROL OF VENTILATION AND VENTILATORY PATTERN

The rhythm of breathing is generated in the pons and medulla in groups of neurones customarily referred to as the 'respiratory centre' or 'centres'. In many animal preparations these centres have an inherent rhythmicity but in intact animals and probably in man, the rhythmicity may depend on chemical or other neural inputs. The volume of breathing is chiefly determined by chemical stimuli in the blood and the pattern of breathing by mechanical stimuli relayed in the vagus nerves from the lung. Thus, the chemical composition of the fluid bathing respiratory neurones in the medulla governs the mean inspiratory flow rate ($\dot{V}_I$). The duration of inspiration (T_I) and the intervals between breaths (T_E) are governed by neuronal 'switches', which in animals and perhaps in man, particularly if there is a mechanical disorder of the lungs, are vagally mediated.

The most important chemical stimulus is the arterial P_{CO_2} as sensed by the medullary chemoreceptors. In normal subjects breathing air at sea-level the pulmonary ventilation is adjusted to keep the arterial P_{CO_2} at 40 mmHg (5·3 kPa). There are no cells in the brain which adjust the ventilation in response to changes in the O_2 tension of the blood. However, there are chemoreceptors in the aortic and carotid bodies which respond to hypoxaemia and stimulate the breathing but only when the arterial P_{O_2} falls below about 60 mmHg (8 kPa).

The medullary or central chemoreceptors can have their sensitivity temporarily or chronically re-set in either direction. Firstly, when ventilation is chronically increased, e.g. by the hypoxia of altitude or by a mechanical respirator, the medullary chemoreceptors become adjusted down to the lower P_{CO_2} and will then resist any tendency for it to rise. Thus, patients who are chronically overventilated by a respirator can become distressed if ventilation is acutely reduced, even though the P_{O_2} remains normal and the P_{CO_2} below normal. Secondly, faced with a chronically high P_{CO_2} as occurs in some cases of chronic lung disease, the medullary chemoreceptors can become adjusted up to the higher P_{CO_2} and can lose their sensitivity to change in P_{CO_2}. The ventilation may then be maintained largely by hypoxic drive of the peripheral chemoreceptors. The administration of O_2 will abolish the drive with the result that ventilation can become severely reduced and P_{CO_2} can rise to narcotic levels without the normal stimulation of a central chemoreceptor drive to breathe.

Although the precise mechanisms are unknown, the pattern of breathing is adjusted to minimize the force required. Airway calibre is also subject to a fine adjustment whereby the autonomic nervous system, via its control of bronchial smooth muscle tone, maintains an optimal balance between the resistance and the dead-space of the airways. The final consequence of these mechanisms is that, for any given combination of compliance, airway resistance and dead-space, there is an optimal combination of rate and depth of breathing for each level of alveolar ventilation. The frequency of breathing of normal subjects is close to the optimum and the abnormal ventilatory pattern seen in some disease states, for example in conditions of reduced compliance, also approximates an optimum.

The exchange of O_2 and CO_2 between the alveolar air and pulmonary capillary blood depends on three processes: firstly, the correct distribution of ventilation and blood-flow; secondly, the diffusion of gases through the alveolar air and the various membranes and liquid layers; and thirdly, chemical reactions within the red cells.

DISTRIBUTION OF VENTILATION AND BLOOD-FLOW

The distribution of inspired air to the alveoli may be regarded as a dilution or mixing process, taking as the criteria of perfect distribution that each alveolus shall receive air of the same composition at the same time and in an amount proportional to its volume. Even in normal subjects, these strict criteria are not completely met and some degree of uneven distribution exists. Distribution can be non-uniform on either a temporal or spatial basis or both. Temporal maldistribution occurs when some alveoli are ventilated before others. Although it has been shown using isotopes that certain regions normally do fill or empty before others, the temporal factor has been somewhat difficult to study, involving analysis of the pattern of expired gases or of regional ventilation by external counting. Spatial maldistribution occurs when some alveoli receive more (or less) ventilation than others. This has generally been considered to occur on a parallel basis but it has also been suggested that stratified or series inhomogeneity and collateral ventilation can also be important mechanisms of maldistribution of ventilation. The usual techniques for examining distribution of ventilation, e.g. nitrogen washout methods, largely reflect changes in spatial distribution.

Of greater functional importance than either the distribution of ventilation or the distribution of blood-flow is the distribution of ventilation relative to blood-flow. By *reductio ad absurdum* it will be appreciated that if every alternatve alveolus is ventilated but not perfused and every other alveolus is perfused but not ventilated, then no gas exchange will occur. In this situation the total pulmonary ventilation would behave as dead-space ventilation and the total pulmonary blood-flow would behave as right-to-left shunt. Unless the proportion of ventilation to blood-flow in all alveoli is both uniform and appropriate to the mixed venous and inspired gas composition, the arterial blood gas composition cannot be normal; in other words, ventilation/blood-flow inequality must result in abnormal arterial blood gas values. Although overventilated alveoli will excrete more CO_2 and thus compensate for the CO_2 retention in the underventilated alveoli, they cannot completely compensate for inequality of O_2 uptake because overventilation produces relatively little increase in the quantity of O_2 taken up by the blood.

The only way in which oxygenation of the blood may be corrected in the presence of ventilation/blood-flow inequality is for total ventilation to increase, thus increasing ventilation in the relatively underventilated alveoli and increasing their ventilation/blood-flow ratios. However, since this increased total ventilation must also increase ventilation in those alveoli already relatively overventilated, there must be a reduction in the arterial P_{CO_2}. Thus, the net result of hyperventilation in the face of ventilation/blood-flow inequality is improvement in and possible normalization of arterial P_{O_2} but at the expense of an abnormally low arterial P_{CO_2}.

DIFFUSION

Ventilation moves the gases in bulk up and down the airways but their movements in the depths of the lungs is by molecular diffusion. Oxygen and CO_2 diffuse at almost equal rates in the gas phase. The gases also diffuse through the aqueous layers and membranes between the alveolar air and the interior of the red cells but, because of its greater solubility, CO_2 diffuses through these media much more readily than O_2. Indeed, one of the historical debates of physiology was whether the volume of O_2 uptake could be accounted for by diffusion alone or whether some 'secretion' or active transport was required. We now know that diffusion alone adequately accounts for movement from alveolus to red cell and, furthermore, that there is virtually no O_2 pressure gradient between the alveolar air and the red cell, even during stresses such as exercise or disease. The absence of an alveolar to end-capillary gradient for O_2 should be clearly distinguished from the normal alveolar-arterial difference, which is chiefly due to ventilation/blood-flow inhomogeneity.

CHEMICAL REACTIONS IN THE CELL

Although there is a series of reactions involving O_2 and Hb within the red cell, it is convenient to consider only the overall chemical reaction of O_2 with Hb_4 to form Hb_4O_8. (This polymeric representation is not given elsewhere in this chapter, it being sufficient to distinguish between Hb and HbO_2.) Accepting this simplification: the volume of O_2 that can be taken up by the red cells per unit time is the product of the rate of this chemical reaction and the volume of Hb present in the pulmonary capillaries, although it should be noted that both for O_2 and for CO_2 the reactions may not be complete.

No analogous simplification is possible or justifiable in the case of CO_2. The key chemical reaction is the dehydration of H_2CO_3 to form CO_2 and H_2O. This reaction is catalyzed by carbonic anhydrase in the red cells. But there are a number of other processes to be considered particularly the diffusion of HCO_3^- ions from the plasma into the cells and the other reactions dealt with below (pp. 111–13).

PULMONARY BLOOD-FLOW

As indicated at the beginning of this chapter most aspects of pulmonary blood-flow are discussed as part of circulatory physiology. However, it is appropriate to consider here some general features of the distribution of pulmonary blood-flow and the interdependence of blood vessels and airways in the maintenance of normal ventilation/blood-flow relationships.

The distribution of blood-flow within the lung is chiefly determined by gravity. Thus, in the erect normal subject, blood-flow is much greater at the base than at the apex. However, the gradient of increasing blood-flow down the lung is not linear but depends on the interrelations between pulmonary arterial, pulmonary venous and alveolar pressures (Fig. 3.5). In zone 1, at the top of the lung, arterial pressure is less than alveolar pressure and there is no blood-flow. Zone 1 is small in normal subjects even at rest. In zone 2, in the middle of the lung, arterial pressure is greater than alveolar pressure, which in turn is greater than venous pressure. In this situation, the vessels act like Starling resistors and the driving pressure is the arterial-alveolar differ-

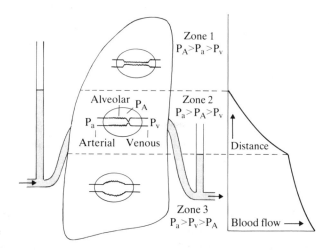

Figure 3.5. *The dependence of local pulmonary blood-flow on the interrelations between alveolar, arterial and venous pressures within the lung.*

There are three combinations of alveolar, arterial and venous pressures that can occur in the lung, and thus there are three zones of different local blood-flow. This pattern is determined by gravity and is illustrated for the upright lung. In zone 1, the alveolar pressure is greater than the arterial pressure, and no blood-flow occurs. In zone 2, arterial is greater than alveolar which in turn is greater than venous pressure. A Starling resistor effect is produced with blood-flow being determined by the arterial-alveolar pressure difference. In zone 3, both arterial and venous pressures are greater than alveolar pressure, and blood-flow is determined by the arterial-venous pressure difference.

(From West J., 1977.)

ence. In zone 3, at the base of the lung, venous pressure is now greater than alveolar pressure and the driving pressure is the arterial-venous difference. At the bottom of the lung, particularly at low lung volumes, blood-flow may decrease. This phenomenon is referred to by some as zone 4 and is due to compression of extra-alveolar vessels when the lung is poorly expanded. Although the factors determining the distribution of blood-flow in the lung have been elucidated by studying the erect subject, the same principles apply in any other position, except that the effect of gravity operates over a smaller vertical gradient.

The normal distribution of blood-flow is appropriately matched to the distribution of ventilation and important homeostatic mechanisms exist to maintain this relation. If local ventilation is impaired, there tends to be a compensatory reduction in blood-flow. It is likely that this mechanism is mediated via local hypoxic vasoconstriction, a mechanism which operates to a varying degree in different individuals. In most subjects, it appears to be very active, while in some it is relatively inactive or absent. One of the clinical implications of the operation of this mechanism is that it can be readily undone by bronchodilator agents, since these are usually vasodilators as well. Since vasodilatation will generally occur before bronchodilation, temporary worsening of ventilation/blood-flow relations can result with some fall in arterial Po_2. A second possible mechanism to reduce local blood-flow can occur in areas which are severely but not completely obstructed. The presence of gross local hyperinflation may cause stretching, flattening and mechanical obstruction of the capillaries in the alveolar walls, so that again reduced local blood-flow accompanies the poor local ventilation.

Conversely, if the local blood supply is impaired, mechanisms exist which can reduce local ventilation. Vascular obstruction results both in alveolar hypocapnia, which can produce bronchoconstriction, and in impaired production of surfactant, which can lead to alveolar collapse. If the vascular obstruction is due to pulmonary embolism, bronchoactive and vasoactive substances such as histamine and serotonin may be released, for example from platelets. Embolism can thus be accompanied by constriction both of large conducting airways (giving increased airway resistance) and of small peripheral airways and alveolar ducts (giving decreased compliance).

There thus exists in the lung a remarkable series of local regulatory mechanisms which are able to maintain the matching of ventilation and perfusion, even in the presence of considerable maldistribution of air or of blood. These mechanisms, however, are never perfect enough to compensate completely for abnormal local ventilation or blood-flow, so that some or even considerable V/Q inhomogeneity always remains. Moreover, those mechanisms are overwhelmed in the presence of severe and widespread disease when there just may not be enough functioning lung tissue to accept the shifted air or blood.

BLOOD GAS TRANSPORT

The main facts about blood gas transport from the standpoint of pulmonary physiology have been established for so long and are so well known that a reiteration is unnecessary. There are, however, certain aspects which are insufficiently appreciated clinically and which will be discussed.

OXYGEN

The relation between the partial pressure of O_2 (Po_2) and the volume of O_2 carried in the blood is given by the dissociation curve. Usually the curve is plotted with percentage saturation along the ordinate. In Fig. 3.6 the actual O_2 content is plotted as well as the percentage saturation because this mode of presentation stresses the differences between the curves of O_2 and CO_2. Anaemia lowers the O_2 curve as plotted from the left-hand ordinate and polycythaemia raises it.

Clinically the 's' shape of the curve is important in several ways. Firstly, a consider-

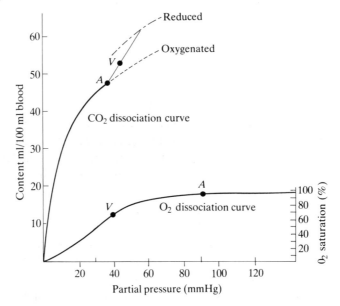

Figure 3.6. *The carbon dioxide and oxygen dissociation curves of blood.* (*1 mm Hg ≏ 0·13 kPa*).

These curves were obtained by equilibrating blood with CO_2 or O_2 at various partial pressures and measuring the quantity of the gas contained by the blood. V and A represent the usual findings in mixed venous (V) and arterial (A) blood in a resting normal subject. The O_2 dissociation curve as plotted on the left-hand ordinate (i.e. ml/100 ml) is that of blood with a normal haemoglobin concentration. The right-hand scale (per cent saturation) is independent of variations in haemoglobin concentration.

The CO_2 dissociation curve is that of blood with normal bicarbonate and haemoglobin concentration. Reduction of haemoglobin allows more CO_2 to be carried at any given partial pressure. Therefore, the simultaneous addition of CO_2 to, and the removal of O_2 from, blood in the tissues produces the solid line $A-V$. The importance of the steeper slope of the CO_2 curve and the 'S' shape of the O_2 curve are described in the text.

(Modified from Riley R.L. & Cournand A., *J. appl. Physiol.* **1**, 825, 1949.)

able reduction in the Po_2 below the normal arterial value does not significantly reduce the oxygenation of the arterial blood. Hence a reduction of arterial saturation below 90 per cent does not occur until arterial Po_2 has fallen to 60 mmHg. (8 kPa) Secondly, the change in shape below 60 mmHg (8 kPa) means that any further reduction in Po_2 causes a disproportionately severe desaturation. Thirdly, this change in slope means that overventilation of parts of the lung cannot compensate for underventilation of other parts of the lung, because further increase in Po_2 above the normal arterial value causes a negligible increase in the volume of O_2 taken up.

The shape of the O_2 dissociation curve is affected by a number of physiological variables, the most important of which are raised Pco_2, increased (H^+) and raised temperature, all of which shift the curve to the right and therefore facilitate the removal of O_2 in the tissues. The opposite changes in Pco_2, (H^+) and temperature shift the curve to the left, as does decreased red cell concentration of certain organic phosphates, notably 2,3-diphosphoglycerate (2,3 DPG). In chronic hypoxia (anoxic and anaemic) the concentration of 2,3 DPG is increased, shifting the curve to the right and thus facilitating the unloading of O_2 in the tissues. In some other conditions such as carbon monoxide poisoning, transfusion of stored blood, metabolic acidosis and some rare red cell enzyme defects, the concentration of 2,3 DPG is decreased, shifting the curve to the left and hindering the unloading of O_2 in the tissues. These shifts of the curve are expressed as changes in 'P 50', the Po_2 at 50 per cent saturation (a shift to the right increases P 50 and vice versa). The normal P 50 is 26–28 mmHg. (3·5–3·7 kPa)

CARBON DIOXIDE

Carbon dioxide is carried in the blood in three main forms (Table 3.1):

Table 3.1. Forms in which CO_2 is present in the blood

	Arterial	Mixed Venous	Difference
Pressure (mmHg)	40	46	6
(kPa)	5·3	6·1	0·8
Content (ml/100 ml blood)	48·5	52·5	4
Solution	2·5	2·8	0·3
HCO_3^-	43	46	3
Carbamino	3·0	3·7	0·7

in simple solution, as bicarbonate, and combined with protein (chiefly haemoglobin) as carbamino compounds. The change in composition as the blood passes along the tissue capillaries are brought about as follows. Dissolved CO_2 diffuses through the plasma into the cells where the enzyme carbonic anhydrase acclerates the formation of carbonic acid

$$CO_2 + H_2O \rightleftharpoons H_2CO_3.$$

The carbonic acid dissociates

$$H_2CO_2 \rightleftharpoons H^+ + HCO_3^-.$$

Most of the H^+ ions are buffered by the haemoglobin. (As reduced Hb^- is a stronger base than $oxyHb^-$ the reduction of the haemoglobin in the tissues simultaneously increases the amount of H_2CO_3 that can be carried at the same P_{CO_2}.)

Most of the HCO_3^- ions diffuse out into the plasma and Cl^- ions enter the cell to restore equilibrium. Reduced haemoglobin has a greater capacity for forming a carbamino compound than oxyhaemoglobin so the removal of O_2 enables more CO_2 to be carried in this form. In the pulmonary capillaries all these processes are reversed.

The importance of the red cells in CO_2 transport should be noted. Although most of the CO_2 is carried in the plasma as HCO_3^-, the red cells are important in four ways:

1   The hydration of CO_2 to form carbonic acid can only occur at sufficient speed in the red cells, there being no carbonic anhydrase in the plasma.

2   The reactions described by the above equations can only proceed fully to the right (converting CO_2 into HCO_3^-) if H^+ ions are removed. Similarly, the reactions can only proceed to the left if H^+ ions are donated. Haemoglobin is the buffer which accepts and donates H^+ ions.

3   Changes in the degree of oxygenation of the haemoglobin alter its affinity for H^+ ions in such a way as to facilitate CO_2 uptake when O_2 is removed and vice versa.

4   The greater the reduction of the haemoglobin in the tissues, the greater is its capacity for forming carbamino-haemoglobin.

The second and third points noted above, coupled with the exchange of HCO_3^- and Cl^- mentioned earlier, account for the difference between 'true' plasma and 'separated' plasma. Thus, if whole blood is exposed to a high CO_2 tension these factors enable the plasma to take up more CO_2 than it does if the cells and plasma were previously separated at a low CO_2 tension and the plasma then exposed to the high CO_2 tension by itself (see also p. 234).

Table 3·1 gives typical values for arterial and mixed venous blood at rest. The relation between total CO_2 content and tension is expressed in the CO_2 dissociation curve (Fig. 3·6). The main curve is that of normal oxygenated or arterial blood. The curve crossing from the oxygenated to the reduced segments is the one obtained if an equivalent amount of O_2 is removed from the blood as CO_2 is added. It is therefore called the 'physiological' dissociation curve.

Fig. 3·6 shows the two very important differences between the dissociation curves of O_2 and CO_2. Firstly, the CO_2 curve is much steeper, implying that much larger volume changes occur for the same change in partial pressure. Secondly, the slope of the curve above and below the arterial values for CO_2 is such that overventilation of parts of the lungs (producing a low P_{CO_2}) can remove CO_2 to compensate for underventilation of other parts of the lungs. Little such compensation for local variations in ventilation can occur for O_2, as discussed above.

Table 3·1 shows that the bulk of the CO_2 in the blood is in the form of HCO_3^-. Any alteration in the HCO_3^- concentration due to non-respiratory changes therefore alters the CO_2 dissociation curve. The concentration of HCO_3^- is increased in metabolic alkalosis and reduced in metabolic acidosis (Chapter 5). Metabolic alkalosis therefore 'raises' the CO_2 dissociation curve and metabolic acidosis 'lowers' it.

BODY GAS STORES

The body of a normal man (excluding the gases in the lungs) contains 'stores' of 1 l. O_2 and 17 l. of CO_2. Practically all the O_2 is in the blood, whereas the bulk of the CO_2 is in the tissue fluids as bicarbonate. (There are even larger amounts of CO_2 in bone, but little of it is exchangeable). These differences in storage capacity and distribution are of considerable clinical importance. A change in the volume of ventilation or of the composition of the inspired air changes the O_2 stores to a new level within 2 minutes, whereas the CO_2 stores take more than 15 minutes to reach a new level. Also, addition or removal of O_2 produces a much greater change in the tension than is produced by the same change in volume of CO_2. A clinical example which illustrates the importance of these differences is provided by a patient who has underventilated while breathing O_2. If he is given air to breathe instead of O_2, the O_2 tension in the blood and tissues will fall rapidly. Increasing the ventilation to normal cannot completely counteract this fall because the high partial pressure of CO_2 in the mixed venous blood maintains the alveolar CO_2 tension high for several minutes, thus diluting the O_2 in the alveolar air. The maintenance of a normal alveolar O_2 concentration during this time requires considerable overventilation, often beyond the ventilatory capacity of such patients. The resultant hypoxaemia can be considerable and may be worse than before O_2 was given in the first place.

RESPIRATORY EXCHANGE RATIO

The respiratory exchange ratio is the ratio of volume of CO_2 expired from the lungs to the volume of O_2 taken in (Vol. CO_2 exp. ÷ Vol. O_2 insp). In a steady state this ratio is the same as the metabolic respiratory quotient (RQ). If ventilation is changed some minutes elapse before the body gas stores adjust to the new steady state. During this unsteady state the respiratory exchange ratio does not reflect the metabolism of the tissues. The dissociation curves of CO_2 and O_2 (Fig. 3·6), as pointed out above, imply that the volume of CO_2 given off for a change in the gas tension of the alveolar air is much greater than the volume of O_2 exchanged for an equal change in tension. Changes in ventilation (which affect the alveolar tensions of O_2 and CO_2 almost equally, but in opposite directions) therefore cause an initially much greater change in CO_2 output than in O_2 uptake. An unsteady state with overventilation therefore causes a high respiratory exchange ratio and an unsteady state with underventilation causes a low respiratory exchange ratio. Once the body gas stores have changed to their new level, i.e. once a new steady state is reached, the respiratory exchange ratio again equals the metabolic RQ.

In the preceding paragraph it has been shown that overventilation causes a high respiratory exchange ratio and underventilation causes a low one. This is true not only of the lungs as a whole in an unsteady state; it is also true of individual regions of the lungs (lobes, alveoli) in the steady state. Those with a high ventilation/perfusion ratio have a high respiratory exchange ratio, those with a low ventilation/perfusion ratio have a low one.

RESPIRATORY AND CIRCULATORY CHANGES IN EXERCISE

The basic changes in respiratory function during exercise are increases in ventilation and blood-flow (cardiac output). The following list records the order of changes in these and other important variables during severe exercise (about 1800 kilopond metres/min) in a young normal subject.

Pulmonary ventilation (minute volume) increases from 8 to 100–120 l/min and pulmonary blood-flow (cardiac output) from 5 to 30 l/min. The O_2 intake and CO_2 output increase from about 0·25 l/min to 4 l/min; the arterial O_2 tension may fall slightly; the H^+ activity rises from 40 to near 50 nM (pH falls from 7·40 to about 7·30); the acidaemia lowers the arterial O_2 saturation 2–3 per cent; the CO_2 tension usually falls. Blood lactate concentration rises from less than 1 to over 10 mmol/l. Fig. 3.7 shows the changes in a normal young man at the levels of physical work used in clinical exercise testing.

The changes in these variables are not necessarily proportional to the intensity of the physical work as judged by the rate of CO_2 output. Ventilation increases in proportion during light and moderate exercise, but increases disproportionately during severe exercise, chiefly because of the development of acidaemia. A corollary of this excess ventilation is that the arterial CO_2 tension is lower during severe exercise than during moderate exercise.

NON-RESPIRATORY FUNCTIONS OF THE LUNG

Although the chief function of the lungs is gas exchange (or 'external respiration'), they are also involved in a number of other functions.

Lung defence

The inhaled air is conditioned by the upper respiratory passages so that it is fully humidified and at body temperature before reaching the depth of the lung. The upper airways function as a heat and water exchanger, recovering most of the heat and water during expiration. Particles greater than 2 or 3 microns are deposited on the mucociliary lining and swept up to the pharynx. Smaller particles which reach the alveoli are

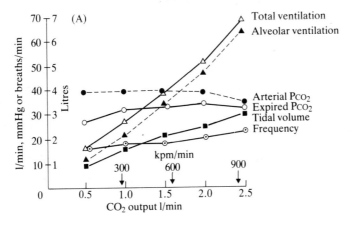

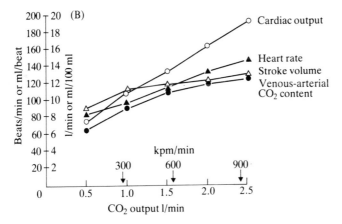

Figure 3.7. *The respiratory (A) and circulatory (B) changes during progressively increasing work.* These are average values obtained in young men. 300 kpm/min is approximately equivalent to walking on the level, 600 to bicycling and 900 to running.

cleared by phagocytosis. The stability of the alveoli depends on the secretion of surfactant by the type 2 cells.

Lungs as a filter

The lungs act as a filter of particulate matter from the systemic venous system and thus protect other organs, such as the brain, from the hazards of embolization. Only a very small percentage of particles larger than red blood cells pass through the pulmonary capillary bed. It is likely that this is a continuously operative function and is not confined to disease states. The lung itself tolerates embolism better than other organs because of its dual blood supply and its efficient disposal mechanisms.

Metabolic function

The lungs also have important roles in metabolism (especially of lipids), in activation and deactivation of many biologically active substances (such as serotonin, histamine, catecholamines, angiotensin, polypeptide hormones), in trapping and release of formed elements and as a potential reservoir of blood to maintain left ventricular filling on a short-term basis. The role of the lungs in the metabolism of angiotensin is of particular importance, since the pulmonary endothelial cells are the main site in the body of 'converting enzyme'. The relatively inactive decapeptide angiotensin I (derived from the action of renin on angiotensinogen) is converted to the highly potent octapeptide angiotensin II in a single pass through the pulmonary circulation.

Disordered function

HYPOXIA

The physiology of hypoxia is so clearly related to the processes of pulmonary and circulatory physiology that a detailed account would be largely a reiteration of much that has already been said. This section therefore merely summarizes the main points.

The following list of causes of hypoxia is logically based upon the stages in the passage of O_2 from the environment to the tissue cells and therefore also helps to indicate the rational treatment.

1 Reduced O_2 tension in the inspired air.
2 Inadequate alveolar ventilation.
3 Impaired pulmonary O_2 uptake.
4 Venous-arterial shunts.
5 Insufficient functioning haemoglobin.
6 Inadequate blood-flow through the tissues.
7 Poisoning or inadequacy of cellular enzymes.

Conditions in groups 1–4 cause 'hypoxic' hypoxia, that is to say, a reduction in the arterial P_{O_2}. Conditions in group 5 cause 'anaemic' hypoxia, that is to say, a reduction in the amount of O_2 in the arterial blood but no reduction in the P_{O_2}. Conditions in group 6 cause 'stagnant' hypoxia, that is to say, normal arterial O_2 values but excessive O_2 extraction from the blood leaving the tissues. Conditions in group 7 cause 'histo-toxic' or cytochemical hypoxia and are associated with normal arterial O_2 values and diminished O_2 extraction from the blood.

Hypoxia is thus a broad term referring to inadequate tissue O_2 utilization for any reason. Hypoxaemia is a narrower term referring only to decreased carriage of O_2 in arterial blood, as in groups 1 to 5.

Conditions in groups 1 to 4 cause central cyanosis (p. 129) if sufficiently severe. Cyanosis does not occur in simple anaemia or carbon monoxide poisoning but does occur in methaemoglobinaemia and sulphaemoglobinaemia. Peripheral cyanosis may occur in group 6. Cyanosis does not occur in group 7.

Although the supply of O_2 available to the tissues is necessarily reduced in hypoxaemia, this reduction causes little damage until the arterial saturation falls to about 50 per cent and the arterial Po_2 falls to about 30 mmHg (4 kPa) (Fig. 3.8). An arterial Po_2 of 20 mmHg (2·6 kPa) seems to be intolerable. Lesser degrees of hypoxaemia cause increases in ventilation, blood-flow, red cell formation and other changes but these are essentially adaptive responses.

Although acute hypoxia causes some disturbance of the breathing and often a slight degree of hyperpnoea it does not usually cause dyspnoea. Chronic hypoxia in the presence of normal ventilatory control causes hyperpnoea and undue breathlessness on exertion.

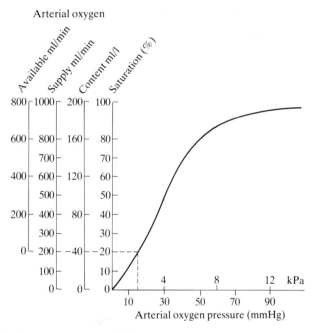

Figure 3.8. *The O_2 dissociation curve and O_2 supply.*
The inner ('saturation') vertical scale is the usual one which expresses (content ÷ capacity) × 100. The next scale ('content') gives the O_2 content per 1. of blood assuming a normal Hb concentration (15 g/100 ml, 150 g/l) and O_2 combining power (1·3 ml O_2/gHb). The next scale ('supply') gives the systemic O_2 flow per min assuming a cardiac output of 5 l/min. The outer scale ('available') allows for the fact that many vital tissues cannot extract the last 20 per cent of O_2 from Hb because they cannot tolerate a capillary Po_2 below about 15 mmHg. Therefore 40 ml/l or 200 ml/min of O_2 are 'unavailable'. As the O_2 requirement of a resting subject is about 200 ml/min, an O_2 saturation of 40 per cent providing an O_2 supply of 400 ml/min is about the tolerable lower limit in uncomplicated hypoxaemia. Evidence of disordered function in the brain, liver and other vital organs appears at saturations below 50 per cent, but may appear at higher levels if there is anaemia or if the circulation is impaired.

Suspicion of hypoxia must depend not upon respiratory but on neurological disturbance. The neurological manifestations of hypoxia are protean, varying from impairment of the highest cerebral functions in mild chronic hypoxia to convulsions and irreversible cerebral damage in a few minutes of acute severe hypoxia.

DYSPNOEA

The term dyspnoea is used to describe either excessive awareness of breathing or discomfort in the act of breathing. There are probably several mechanisms, of which the following are the most important. Firstly, a sensation due to an imbalance or 'inappropriateness' between the volume or flow demanded and the forces required to meet this demand. This is the most important sensation giving rise to dyspnoea in lung or heart disease and is associated with changes in the mechanical state of the lungs or chest wall. Secondly, an increased or excessive ventilation or a high frequency of breathing (sometimes referred to as 'hyperpnoea') can be distressing but these sensations are not strong. They are unimportant in resting subjects and their main importance is in compounding or aggravating the sensation of inappropriateness. These features are seen in patients with anaemia, metabolic acidosis or low cardiac output. Finally, discomfort may arise from irritation or spasm in the major central airways, as in asthma, or perhaps from receptors in pulmonary arteries, as in pulmonary embolism.

Although chemical disturbances in the blood, acting through the chemoreceptors, or mechanical changes in the lungs, acting through the vagi, are important causes of respiratory drive, they are not sources of sensation. The sensations responsible for the symptom of dyspnoea are generated by the motor response to the respiratory drive.

Orthopnoea is probably due to an increase of pulmonary blood volume causing a reduction in compliance, but other factors such as disturbed ventilation/perfusion relationships, the mechanical advantage of the respiratory muscles and reflex mechanisms may be important.

Cheyne-Stokes breathing is due to a hunting behaviour of the respiratory control. The possibility of hunting is intrinsic to any system in which the sensing element (in this case the medullary chemoreceptors) is separated from the effector element (the lungs) so that there may be a time lag between the two (between the medullary P_{CO_2} and the alveolar P_{CO_2}). Hunting normally does not occur because this lag is small, because the respiratory centre is not sufficiently sensitive to CO_2 and because the volume of gas in the lungs damps the change in alveolar gas composition when ventilation changes. In Cheyne-Stokes breathing the most important causal factor appears to be an excessive ventilatory response to CO_2 which overcomes the smoothing properties of the system. The excessive ventilatory response is due to damage to supramedullary nervous pathways which normally inhibit either the respiratory centres or the motor pathways to the respiratory muscles. A prolonged lung-to-brain circulation time or perhaps very low lung volumes could also be important.

LIMITATION OF EXERCISE TOLERANCE

In all exercise lasting more than a minute or so, the rate of work is limited by the capacity to exchange and transport O_2 and CO_2. The effects of a limited capacity in each of the links in the chain or respiratory and circulatory systems connecting the tissues with the air will be considered individually.

Ventilation

Inability to increase ventilation in proportion to CO_2 output and O_2 intake should cause the alveolar and arterial P_{CO_2} to rise and the alveolar and arterial P_{O_2} to fall. Usually dyspnoea caused by the ventilatory stimulus of rising P_{CO_2} causes the subject to stop before significant hypoxaemia develops, provided the initial arterial P_{O_2} is not already reduced.

Pulmonary oxygen transfer

Inability to increase the capacity of the lungs to transfer O_2 in proportion to an increasing O_2 uptake causes an increased alveolar-arterial P_{O_2} difference. The fall in arterial P_{O_2} brings into play two adaptive responses; firstly, an increase in ventilation which, by increasing alveolar P_{O_2}, decreases the hypoxaemia; secondly, an increased cardiac output which compensates for the lessened arterial oxygen content by increasing the total arterial flow.

Cardiac output

Inability to increase cardiac output in proportion to CO_2 output or O_2 intake causes a wide veno-arterial difference for CO_2 and O_2 and a fall in the tissue P_{O_2}. This may cause an increase in anaerobic metabolism (see below).

Anaemia

The effect of anaemia is similar, with respect to O_2, to a reduction in cardiac output, in that the supply of O_2 to the tissues is reduced. The important difference, however, is that in anaemia the cardiac output is often increased, thereby compensating for the reduced O_2 content per unit of blood by an increased total volume of blood-flow.

Peripheral oxygen transfer

Inability of the muscles and their vascular bed to increase their capacity to transfer O_2 in proportion to an increasing O_2 usage, or an enzymatic limitation in the ability of the muscle cells to take up O_2, causes an increased Po_2 difference between the blood and the tissues and may cause an increase in anaerobic metabolism. The difference between this form of limitation and a reduction in cardiac output is that the arteriovenous O_2 difference is not increased.

Metabolism

If the processes discussed above are unable to increase the supply of O_2 to the exercising muscles in proportion to the energy consumption, excessive anaerobic metabolism may occur causing the addition of lactic acid to the tissue fluid. This lactic acid displaces CO_2 from $HCO_3{}^-$. As anaerobic metabolism is relatively inefficient, the total CO_2 evolved through this mechanism is about four times that released by aerobic metabolism for the same work. This excess of CO_2 increases the load on all the above mechanisms and the rise in blood H^+ activity causes an excessive increase of ventilation so that the arterial Pco_2 falls.

What actually stops you?

Excessive demands on local muscle groups may cause local symptoms and in disease, symptoms of local significance, such as angina pectoris, intermittent claudication or wheezing, may signify the limit of tolerance. But in unfitness and in many disease states, such as heart disease, lung disease or anaemia, the message is provided by breathlessness. The above considerations have shown how often limitation at some other site in the chain of respiratory and circulatory processes throws a strain on the breathing either by causing an increased CO_2 load or by increasing the ventilatory drive in other ways.

REDUCED VENTILATORY CAPACITY

The ventilatory capacity may be reduced firstly, by conditions which obstruct the airways and secondly, by conditions which restrict the expansion of the lungs or thoracic cage or cause weakness of the respiratory act.

Airways obstruction

Obstruction of the upper extrathoracic airways, particularly in the larynx, affects inspiration more than expiration; obstruction of the intrathroracic airways affects expiration more than inspiration. Narrowing of the intrathoracic airways may result from processes in the bronchial wall or lumen, such as smooth muscle hypertrophy or spasm, mucosal swelling or the presence of secretions. Narrowing may also result from loss of external elastic support due to destruction of the lung, thus allowing the airways to be compressed by high external pressure during expiration.

Non-obstructive or restrictive defects

These terms describe limitation of expansion of the lungs due to pulmonary causes (such as fibrosis or congestion), to extrapulmonary causes (such as pleural effusion, pneumothorax or kypho-scoliosis) or to disease of the respiratory muscles or of the central or peripheral nervous system.

The use of the term restriction in this context is not ideal and it should be remembered that real restriction to expansion of the lungs, while the commonest, is not the only cause of a 'restrictive' spirometric pattern.

INADEQUATE VENTILATION

If ventilation is inadequate for the metabolic requirements of the body, whether at rest or on exercise, the alveolar Pco_2 rises and Po_2 falls. The arterial Pco_2 also rises and the arterial Po_2 falls but because of the shape of the O_2 dissociation curve, the O_2 saturation does not fall until the degree of underventilation is considerable. The alveolar and arterial Po_2 may fall by half before hypoxaemia becomes manifest as cyanosis.

Inadequate ventilation may be due either to a reduced ventilatory capacity ('can't') or to a reduced central ventilatory drive ('won't'). Although the first is much commoner, it is important to remember that the degree of ventilatory incapacity at which underventilation appears varies greatly between patients (see Fig. 3.13). Many develop inadequate ventilation with hypercapnia when, as judged by assessment of their ventilatory capacity, they could maintain normal alveolar gases. In these patients it appears that there is also some impairment of central respiratory drive. In a small group of patients, inadequate ventilation occurs in the presence of normal ventilatory capacity.

Although the terms 'ventilatory failure' or 'hypoventilation' have been used to describe the state of respiratory function when arterial Pco_2 is increased, the commonest cause of raised arterial Pco_2 is widespread ventilation/blood-flow inhomogeneity combined with an inadequate compensatory increase in ventilation (again either 'can't' or 'won't'). Although in theory hypercapnia could be caused solely by extreme ventilation/blood-flow imbalance, particularly where there is stratified or series maldistribution of ventilation so that one region receives gas from another, in practice hypercapnia tends to occur only when the ventilatory capacity is also impaired. This is because, when ventilatory capacity is normal, total ventilation can be readily increased to maintain a normal alveolar ventilation despite the increased physiological deadspace.

DEFECTIVE GAS TRANSFER

The exchange of O_2 and CO_2 between the alveolar air and the pulmonary capillary blood may be unduly hindered firstly, when the membrane is thickened or reduced in area and secondly, when the volume of the capillary bed is reduced. The uptake of O_2 is more severely affected than the elimination of CO_2 because being less soluble, O_2 diffuses much less rapidly.

The term 'alveolar-capillary block' has been popularly applied to impaired O_2 transfer and equated with a diffusion defect, especially in those conditions in which some morphological basis for this concept has been found. However, in all these conditions, sufficient ventilation/blood-flow inequality occurs to explain any impairment of gas transfer. Moreover, on theoretical and experimental grounds, it is doubtful if severe enough membrane abnormalities can exist in practice to cause a significant diffusion defect. Thus, an alveolar to end-capillary gradient for O_2 does not occur in practice. The term 'alveolar-capillary block' is therefore misleading and it is more reasonable to refer to the functional abnormality as a gas transfer defect.

Faced with impaired ability to exchange gas, the lungs could either maintain normal arterial blood gas composition in the presence of reduced O_2 consumption and CO_2 production, or maintain normal O_2 consumption and CO_2 production at the expense of abnormal arterial blood gas composition. In practice, since the O_2 consumption and the CO_2 exchange are determined by the metabolic needs of the body, the result of impaired ability to exchange gases is abnormal blood gas composition, viz. decreased arterial P_{O_2} and increased arterial P_{CO_2}. When the arterial P_{O_2} falls, the mixed venous P_{O_2} must also fall. However, the fall in the latter may become less marked if cardiac output increases to maintain normal blood gas transport in the face of arterial hypoxaemia. The raised arterial P_{CO_2} stimulates the central chemoreceptors to increase ventilation. The net result of the combination of ventilation/blood-flow inequality and hyperventilation is that the arterial P_{CO_2} becomes normal but that the arterial P_{O_2} is only partially improved. This is because hyperventilation is relatively ineffective in increasing the amount of O_2 taken up by the blood. Hyperventilation can normalize the arterial P_{O_2} only if ventilation/blood-flow inequality is mild and then only with a level of ventilation that results in a low arterial P_{CO_2}. Hyperventilation of this degree clearly could not be caused by a chemoreceptor drive since the P_{O_2} is near normal and the P_{CO_2} is low.

IMPAIRED BLOOD GAS TRANSPORT

In anaemia or in the presence of abnormal compounds of haemoglobin the volume of O_2 that can be taken up by the blood in the lungs is reduced but the P_{O_2} of the arterial blood is normal. As there is no impairment of CO_2 transport or exchange, the arterial P_{CO_2} and CO_2 content, and therefore ventilation, are usually normal. In severe or chronic situations, however, ventilation may be increased by hypoxic stimulation of the peripheral chemoreceptors with consequent lowering of the arterial P_{CO_2}.

Changes in CO_2 transport affect ventilation by altering the arterial P_{CO_2} or H^+ concentration (Chapter 5). Thus in metabolic acidosis the increased (H^+) stimulates ventilation and reduces the arterial P_{CO_2}. For reasons as yet incompletely known the converse is not always true. In metabolic alkalosis, the arterial P_{CO_2} is often normal and not, as would be expected, raised.

DISORDERED FUNCTION IN SPECIFIC DISEASES

Nearly all diseases which affect respiration disturb many aspects of lung function. There are no patterns of disordered function which are pathognomonic of individual diseases. This is not surprising since the lung, like other organs, has only a limited range of responses to a wide variety of possible insults. Functional abnormalities thus cannot be used to make anatomical, pathological or clinical diagnoses. In this chapter, a few common conditions have been chosen to illustrate the interrelationships of the various functional changes that can occur. The functional changes in other conditions can usually be deduced from a knowledge of their pathology.

Paralysis of the respiratory muscles

Paralysis of the respiratory muscles reduces the ventilatory capacity; if ventilatory capacity falls below that required for resting metabolism, the arterial P_{CO_2} rises. However, a common consequence of paralysis of the respiratory muscles is impairment of local ventilation in some regions of the lung due to inability to change position, to take a deep breath or to cough effectively. These processes result in ventilation/blood-flow inequality and hence in a widened alveolar-arterial P_{O_2} difference. Arterial hypoxaemia can thus occur either in the absence of elevation of the arterial P_{CO_2} or disproportionate to any such elevation.

Pneumothorax

The introduction of air into the pleural space, by allowing the thoracic cage to expand and causing the lungs to shrink, raises the intrathoracic pressure. No serious disturbance results from these changes unless, for example as a result of a valvular opening, the intrathoracic pressure rises above atmospheric pressure. The effect of a severe reduction of lung volume or of a rise of intrathoracic pressure above atmospheric is to occlude intrathoracic airways during expiration causing air-trapping. The maintenance of ventilation then requires a forceful contraction of the inspiratory muscles to overcome the elastic forces of the distended thoracic cage. In a severe case, there may be areas of atelectasis in the underlying lung causing a reduction in arterial P_{O_2} but the effect is seldom sufficient to reduce the saturation significantly because blood-flow in the affected lung is also reduced. Only in very severe cases is there ventilatory failure with raised P_{CO_2}.

Lung cysts

Cysts in the lung substance produce virtually the same functional effects as pneumothorax. Only those with a valvular bronchial communication produce an important disturbance of function. Cysts with free ventilation just increase the dead-space.

Pneumonia, congestion, infiltration

These all cause a 'restrictive' reduction in ventilatory capacity and defective gas transfer. The compliance is reduced and there is hypoxaemia without hypercapnia—the Pco_2, in fact, is often low because either hypoxaemia or abnormal stimuli from the lungs or circulation (or both) cause the ventilation to increase.

Adult respiratory distress syndrome (ARDS)

This condition, sometimes called 'shock lung', is the commonest cause of acute respiratory failure in acutely ill medical and surgical patients. Although its aetiology remains unresolved, it has been clearly shown to be a syndrome of acute pulmonary oedema and haemorrhage due to abnormally increased pulmonary capillary permeability. This gives rise to widespread airway closure with atelectasis and to extensive interstitial infiltration, so that lung volumes are reduced, compliance is decreased and there is veno-arterial shunting with resultant hypoxaemia which is often severe. These changes may be dramatically improved by mechanical ventilation and especially by the use of positive end-expiratory pressure (PEEP).

Pulmonary arterial obstruction

Reduction of blood-flow through a part of the lung causes its ventilation to become 'dead-space'. This effect tends to be reduced by compensatory changes in local airway calibre which can reduce ventilation in areas of decreased blood-flow. Compliance may be decreased (due to alveolar duct constriction) and airway resistance increased, sometimes to the point of producing clinical wheeze, but the most characteristic feature of pulmonary vascular obstruction is hyperventilation with hypoxaemia and hypocapnia. The stimulus for the hyperventilation is uncertain but is probably reflex. Hypoxaemia is due to venous admixture or even true shunting but the mechanism of production of areas with low ventilation/blood-flow relationships remains controversial.

Asthma ('reversible' airways obstruction)

There is widespread narrowing of both large and small airways due to one or a combination of bronchial smooth muscle hypertrophy and contraction, mucosal oedema and accumulated secretions. The most fundamental abnormality is unknown but the most basic functional change that can be demonstrated is increased bronchial reactivity. The most prominent secondary change is increased airway resistance and airflow limitation, and hence a greatly increased resistive component to the work of breathing. Air trapping leads to marked hyperinflation.

Although generalized, the airway narrowing is never uniform, so that uneven distribution of ventilation and ventilation/blood-flow inequality follow. Hypoxaemia worsens as the airway obstruction becomes more severe, but at least in mild to moderate asthma, hypocapnia is usual. The stimulus for the increased ventilation is uncertain. In more severe asthma, hyperventilation becomes increasingly difficult due to the great work of breathing it entails. The arterial P_{CO_2} then rises and this should be regarded as a sign of severe decompensation.

Chronic non-specific airways obstruction (chronic bronchitis and emphysema)

Chronic bronchitis is characterized by excessive mucous secretion in the bronchial tree. Emphysema implies an abnormal enlargement of the air spaces distal to the terminal non-respiratory bronchiole, accompanied by destructive changes in their walls. The main feature common to these conditions is irreversibility of the airways obstruction, in that although it may vary somewhat, a major degree of obstruction always remains. Although most patients have features of both conditions, it is possible to recognize extremes of pure emphysema (type A) and of pure chronic bronchitis (type B). The former has also been termed the 'pink puffer' and the latter the 'blue bloater'.

Typically, patients with emphysema (type A) do not underventilate but remain severely dyspnoeic and maintain more normal alveolar gas concentrations until their ventilatory capacity becomes severely reduced. They are not so prone to cardiac enlargement and oedema unless in the terminal stage of their disease. The gas-exchanging surface as assessed by the CO transfer factor is reduced. Compliance is significantly increased and the pressure-volume curve shifted up and to the left. On exercise they develop an increased alveolar-arterial P_{O_2} difference and hypoxaemia.

Typically, patients with chronic bronchitis (type B) develop underventilation with hypercapnia more readily than the reduction of their ventilatory capacity would seem to warrant. This is due to impaired ventilatory drive. They are also more prone to develop cardiac enlargement and oedema at an earlier stage of their disease. The gas-exchanging surface as assessed by the CO transfer factor is fairly well maintained. At rest, they have an increased alveolar-arterial P_{O_2} difference which is due to a large volume of poorly ventilated but well perfused lung. When the total ventilation is increased on exercise, the ventilation of the poorly ventilated regions may increase disproportionately and the alveolar-arterial P_{O_2} difference falls.

Although gas transfer is decreased and compliance increased in emphysema but not in chronic bronchitis, these two conditions are not clearly distinguishable on the basis of simple tests. Indeed their functional similarities tend to be much greater than their differences. Moreover, as noted earlier, most patients do not fall clearly into one or other category but have features common to both conditions. To compound this difficulty, the definition of chronic bronchitis is a clinical one and the definition of emphysema a morphological one. Thus, there are major pitfalls in trying to make a clear and meaningful separation between the two conditions, using presently available criteria.

Small airway disease

It has become recognized in recent years that obstruction of airways less than about 2 mm in diameter is the chief cause of abnormal gas exchange in chronic bronchitis, probably in emphysema, in asthma during remission, in bronchiectasis, in left ventricular failure and in patients with chronic cough and sputum but with normal routine lung function measurements. Many different clinical and morphological entities may thus share the common pathway of small airways obstruction. Small airways abnormalities are typically silent clinically but they may be detected functionally by decreased MMF, increasing closing volume, abnormal flow-volume curve, abnormal frequency dependence of compliance and impaired gas exchange. Although 'small airway disease' may represent an early stage of disease, its long-term significance in patients without other disability remains to be established.

Cor pulmonale

Right heart failure may occur secondarily to any respiratory disease in which there is severe hypoxaemia. The most obvious mechanism is pulmonary vascular obstruction, due either to vasoconstriction from hypoxia or to compression, distortion or obliteration of vessels by the underlying disease process. Hypoxaemia may also impair myocardial function and reduced left ventricular performance can contribute to the consequent cardio-pulmonary derangements. Chronic fluid retention with venous engorgement and oedema usually occurs only when there is CO_2 retention. In this situation, as well as the right heart failure, impaired renal handling of water has been implicated.

CO_2 narcosis

Patients with any chronic lung disease who are persistently hypoxic have part of their stimulus to breathe supplied by the arterial hypoxia. If this is relieved by the administration of excessive O_2, ventilation may fall acutely, causing severe CO_2 retention and acidosis, the effects of which are largely on the nervous system causing unconsciousness, muscular twitching and raised intracranial pressure. Drugs which depress the respiratory centre may produce a similar effect. Such narcotic levels of CO_2 can be obtained only when the inspired air is O_2 enriched. Should CO_2 narcosis develop and the O_2 then be discontinued, the patient is exposed to the risk of very severe hypoxia for a number of reasons: a high P_{CO_2} reduces the effect of any hypoxic drive; the shallow breathing during CO_2 narcosis causes the lungs to deteriorate; and O_2 is washed out of the lungs much more rapidly than the CO_2 accumulated in the body stores.

'Primary' alveolar hypoventilation

This term has been applied to a group of patients who can breathe normally but do not. The characteristic finding is a raised P_{CO_2} in a patient who has a relatively normal ventilatory capacity. Other common features include arterial unsaturation, poly-cythaemia, obesity, oedema, somnolence and neurological damage. The resemblance to Dickens' fat boy has led to the popular synonym of 'Pickwickian' syndrome, but the term is misleading because obesity is not an essential feature.

In fact, it is unlikely that these patients are a homogeneous group. Hypoventilation may be produced by damage at various levels in the brainstem and the other features are very variable.

Sleep apnoea

It has become recently appreciated that some subjects are prone to recurrent, brief episodes of apnoea during sleep. By definition, the diagnosis of the syndrome requires more than 30 such episodes lasting more than 10 seconds during seven hours of nocturnal sleep of both rapid eye movement (REM) and non-REM type. The usual cause of apnoea is upper airway obstruction probably due to loss of muscular tone of the pharyngeal muscles but with persistent diaphragmatic movement. Sometimes there may be a central origin, in which case diaphragmatic movement is absent. Mixed types can also be found. The net result is recurrent acute hypoxaemia, often of considerable severity. Patients usually present with either excessive daytime somnolence due to sleep deprivation or because of loud snoring at night. They may also have mental and personality changes, morning headache, hypertension and cardiac failure; most are over-weight, middle-aged males.

Principles of clinical observations, tests and measurements

EVALUATION OF EXERCISE TOLERANCE BY QUESTIONING

Exercise on the level has to be vigorous to stress the respiration or circulation. A normal man has an aerobic capacity considerably greater than is required for running or bicycling and he can lose three-fifths of his capacity before walking becomes difficult to sustain. So, when trying to assess exercise tolerance, it is usual to supplement enquiry about exercise on the level with questions about hills and stairs. Most people climb stairs at a rate which requires the performance of physical work which is greater than their steady state aerobic working capacity, that is, at a rate they cannot sustain for more than a minute or so. And yet it is possible to exercise for half a minute or so without making any demand on the respiration or circulation. This is why we think nothing of climbing one or two floors but will usually wait for the lift should we want to go up three or more. It also means that the answer to the common type of question 'How many stairs can you climb without getting unduly short of breath?' must lie somewhere between 30 and 120 stairs or 2–6 domestic flights. In answering this question, patients base their estimate not on their steady state aerobic working capacity but on the distress they experience in repaying the oxygen debt and CO_2 accumulation during the unsteady state. The more severe the limitation of exercise tolerance, the more it becomes a matter of 'unsteady state' and the less reliable are changes in the patient's exercise tolerance as judged by questioning in indicating the true state of cardiorespiratory capacity. If important decisions of prognosis or treatment are at stake it is always preferable to supplement an assessment by questioning with an objective evaluation by studying the patient's performance during exercise. The measurements made need not be complex: measurement of the subject's true working capacity on a treadmill or bicycle ergometer together with such simple measurements as ventilation, pulse rate and gas exchange will often be adequate.

PHYSICAL SIGNS OF AIRWAYS OBSTRUCTION

Forced expired time

The best physical sign of diffuse intrathoracic airways obstruction is a prolongation of the time taken to deliver the vital capacity. In a normal subject this is accomplished in 3–4 seconds. Prolongation beyond 6 seconds means that the ratio of forced expired volume in 1 s (FEV_1) to vital capacity (VC) is less than 60 per cent.

Overinflation of the chest

Airways obstruction leads to overinflation of the chest. This is because the airways are narrower during expiration than inspiration and this phasic difference becomes exaggerated by the presence of an obstructive process. If the expiratory airflow is impaired relative to inspiratory airflow, less air is expired than inspired until a new equilibrium is reached at higher lung volumes. The new equilibrium occurs when inspiratory and expiratory airflow are again balanced—the former becoming harder and the latter easier, the nearer tidal breathing is to the maximum inspiratory level. The process may be regarded as analogous to that which occurs when the ventricle is exposed to an acute pressure load and moves up a Starling curve. The early result of acute hyperinflation of the lungs is that the airways are held more open and elastic recoil is increased. Unfortunately prolonged stretching of elastic components results in an increased compliance, so that eventually the benefits of hyperinflation become lost. However, the process is not reversible while the obstruction persists, since deflation would be accompanied by less pull on airways and less elastic recoil—and thus further impairment of expiratory airflow.

Over-inflation of the lungs causes an increase in the areas of resonance on percussion. If the process goes on over a few years the thoracic cage itself becomes deformed: the anterior-posterior diameter of the chest is increased; the sternum becomes elevated so that the suprasternal portion of the trachea is shortened; the diaphragm is flattened and the heart becomes lower and more medial in position. The movements of the chest are also distorted: the expansion ('bucket handle' movement) of the upper ribs is lost and replaced by exaggeration of the upward lift ('pump handle' movement). The depression of the diaphragm causes it to lose its mechanical advantage on the costal margin. The normal diaphragm lifts the costal margin and gives little downward pull on the mediastinum; the depressed diaphragm pulls the costal margin inwards and pulls the mediastinum downwards; this can be felt as a descent of the thyroid cartilage during inspiration.

Auscultation of the lungs

There is still no universally accepted system of terminology for breath sounds and adventitious sounds. Breath sounds transmitted through the chest wall are normally modified as though passing through a low-pass filter with a cut-off frequency greater than 200 Hz. Bronchial breath sounds occur when consolidation bypasses the filter.

Squeaks (musical sounds, continuous sounds, rhonchi) are not due to an 'organ-pipe effect' but to the reed-like action of airways which are narrowed almost to closure and whose walls oscillate during phases of airflow, the pitch being determined by the size and length of the airway involved. Their loudness is little guide to severity; severe airways narrowing may be silent because the rate of airflow is insufficient to generate a sound. Wheeze, an extra-auscultatory sound, is related to rhonchi and is attributed to high velocity of flow and increased turbulence in abnormally narrow proximal intra-thoracic airways.

Crackles (moist sounds, discontinuous sounds, rales, crepitations) are thought to be caused in early inspiration by the passage of a bolus of gas through an intermittently occluded airway at low lung volumes and in late inspiration by the explosive springing open of previously closed peripheral airways. By implication, parts of the lung giving rise to crackles are not ventilated during part of the respiratory cycle.

CYANOSIS

Cyanosis is a blue colour of the skin or mucous membranes usually due to the presence of an excessive amount of reduced haemoglobin (or of methaemoglobin or sulphaemoglobin) in the small blood vessels of the tissues. It is widely taught that there

must be 5 g of reduced Hb per 100 ml blood before cyanosis occurs. This figure is a
very rough approximation whose chief merit is to stress that cyanosis cannot occur in
the presence of severe anaemia and that it may be more readily seen in polycythaemia.

Clinically cyanosis can most conveniently be subdivided into *central*, in which the
arterial blood is unsaturated, and *peripheral*, in which the arterial blood is normally
saturated but the extraction of O_2 from the blood in the tissues is excessive. The
distinction is best made by examining the mucous membranes inside the mouth where
blood flow is always good. If these are blue the cyanosis is of central type. Central
cyanosis can be a difficult sign to be sure of, even under the best conditions of illumi-
nation. It is generally held that central cyanosis can be detected by anyone if the
arterial oxygen saturation is less than 70 per cent and by expert observers up to about
80 per cent. These are almost certainly underestimates, since central cyanosis can often
be detected clinically when the saturation is higher than this. Even this degree of
desaturation denotes considerable hypoxaemia and it should be remembered that 70
per cent is less than the usual saturation of mixed venous blood. Central cyanosis (in
the absence of abnormal haemoglobin) always implies considerable arterial hypoxaemia but gives no indication as to aetiology.

Doubts about the presence or absence of cyanosis due to pulmonary disease can
usually be resolved by the administration of O_2. If a patient who has indefinite
cyanosis thought to be due to pulmonary disease does not become pink on breathing
O_2, then the 'blue colour' is most unlikely to be due to arterial unsaturation of
pulmonary cause.

VITAL CAPACITY

Interpretation of changes in the vc requires an understanding of the factors which limit
the depth of voluntary inspiration and of those which limit the depth of voluntary
expiration.

Maximum inspiration

The total lung capacity at full inspiration is, of course, governed to a considerable
extent by the volume of the thorax and therefore by body size. In addition, to mechanical factors, reflex mechanisms also seem to contribute to the setting of the maximum
inspiratory level (TLC). In normal subjects there is a good correlation with age and
height. Deformities or loss of mobility of the bony cage reduce this volume. Weakness
of the inspiratory muscles or decreased compliance (whether due to loss of functioning
lung tissue or to increased stiffness of the lungs) also reduce the volume that can be
inspired.

Maximum expiration

The residual volume of air in the lungs at full expiration is also partly dependent on the size and integrity of the thoracic cage and the power of the respiratory muscles but to a smaller extent than the total lung capacity. The physiological event which limits expiration, particularly in older subjects and in disease, is closure of the intrathoracic airways. In young subjects, many airways are still open at the end of maximum expiration (RV). The patency of these airways depends on a number of factors: the elastic tension in the lungs tending to hold them open, the strength of their walls and the presence or absence of disease processes tending to narrow them. In a normal subject expiration ceases when the elastic tension of the shrinking lung falls to a low level and the intrathoracic pressure rises sufficiently to collapse the small airways which are without cartilaginous support. Loss of elasticity, such as occurs in emphysema, causes the volume at which this closure occurs to be increased. In bronchitis and particularly in asthma, the airways also tend to close prematurely because of the disease processes in their walls.

The complexity of factors affecting the vital capacity make it almost valueless in the diagnosis of specific diseases and of limited use in distinguishing between different disease processes. Moreover, the range of normal values, even when allowance is made for age, sex and size is large ($\pm$ 10 to 20 per cent) and limits its value as a general screening test or as a standard of comparison between patients. However, in the individual subject, the VC is sufficiently reproducible to make it a useful measurement in assessing the progress of a large number of conditions.

Airway obstruction slows the delivery of the VC so that the FEV_1 and the FEV_1/VC ratio are both reduced (Fig. 3.9). Non-obstructive or 'restrictive' reduction of the ventilatory capacity reduces the VC but does not slow its delivery, so that the FEV_1 is reduced but the FEV_1/VC ratio is normal or even, because of increased elastic recoil, unusually high.

FORCED EXPIRATORY VOLUME

In normal subjects and in patients whose airways are healthy, the cessation of airflow which marks the end of voluntary expiration occurs almost simultaneously throughout the lungs. As a result, about 75 per cent of the VC is expelled smoothly and rapidly in one second and the remaining 25 per cent takes only a further 2 or 3 seconds. However, in conditions causing diffuse airways obstruction (asthma, bronchitis, emphysema), the inequality of the disease process in different airways causes them to close irregularly and progressively as expiration proceeds. The result is that in a severe case, less than 40 per cent of the vital capacity is expelled in the first second and the remainder takes much longer. In fact, airflow continues for as long as the patient can wait before he has to take another breath.

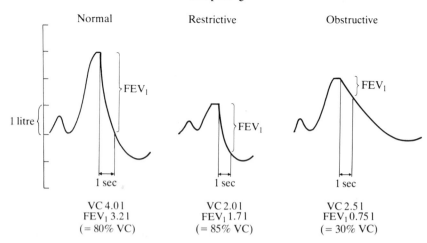

Figure 3.9. *Forced expiratory spirograms.*

The three records are taken from spirometer tracings. Inspiration is up and expiration down, and the records read from left to right. After a deep inspiration the subjects breathed out as rapidly and forcibly as they could. The total volume expired is the (Forced) Vital Capacity (VC) and the volume expired in the first second is the Forced Expired Volume in one second (FEV_1). The 'restrictive' record was obtained from a patient with kypho-scoliosis; the 'obstructive' record from a patient with emphysema. Although the vital capacity is more severely reduced in the patient with kypho-scoliosis than in the patient with emphysema, the proportion of the vital capacity expired in the first second is normal.

A number of methods of quantitatively expressing the rate of delivery of the forced vital capacity have been used. The most popular is to measure the volume expired in the first second (FEV_1) with a spirometer and kymograph or a pneumotachograph and an electronic data processor. This should be normally about 75 per cent of the VC (Fig. 3.9) varying between 80 per cent or more in younger subjects and 70 per cent or less in older subjects. The FEV_1 is reduced by all conditions, obstructive and non-obstructive, which reduce the VC but the FEV_1/VC ratio is reduced only by airways obstruction. The FEV_1 is a much more sensitive index of the severity of airways obstruction than the VC but unfortunately it does not distinguish between the different causes. Thus a patient in acute asthma, whose lungs between attacks are normal, may have the same reduction in FEV_1 as a patient with emphysema. The effect on the FEV_1 of bronchodilator drugs can help in distinguishing between the mechanisms of airway obstruction.

Other methods are also used to express the rate of delivery of the forced vital capacity. Of these the most useful is the maximum mid-expiratory flow rate (MMF) which is the slope of the timed vital capacity measured from 25 per cent to 75 per cent of the expired volume. The MMF correlates highly with the FEV_1 but is more sensitive to mild obstruction and to small airways disease.

Peak expiratory flow

The Wright peak flow meter measures the volume expired in the first tenth of a second of a forced expiration and gives a reading in litres per minute. This simple measurement correlates well with values obtained for the FEV$_1$ and MMF and, like them, is reduced in all conditions, both obstructive and non-obstructive, which reduce the ventilatory capacity.

Maximum breathing capacity

The maximum breathing capacity (MBC), better called the maximum voluntary ventilation (MVV), is measured by asking the patient to breathe as rapidly and as deeply as he can for about 15 seconds. The volume breathed is either collected in a bag or recorded with a spirometer, and the result expressed as l/min. The MBC is little used now because it is no more informative than the FEV$_1$ and is much more difficult for the subject to perform.

    The ventilation which the subject can sustain during exercise is about 30 times the FEV$_1$ but such extrapolation is only approximate. During maximum exercise the ventilation of a normal subject does not reach his MVV. It is not uncommon on the other hand for a patient with severe airways obstruction to attain a ventilation during exercise which is greater than his MVV.

Flow-volume curves

The traditional spirogram records expired volume (on the Y-axis) against time (on the X-axis) (Fig. 3.9). Further information, especially concerning airways obstruction, can be obtained by plotting other, derived parameters during expiration. Thus, volume (on the X-axis) may be recorded against flow (volume divided by time, on the Y-axis). The flow-volume curve shows that maximum expiratory flow gradually decreases over the last part of the vital capacity and various indices have been calculated to quantify this (e.g. flow rate at 50 per cent VC). The slope of the curve cannot be increased by increasing effort; it is thus said to be 'effort independent', a phenomenon due to compression of airways within the chest. In the presence of airways obstruction, the curve is flatter and lower than normal. The change in this curve following helium-oxygen breathing has been used as a means of differentiating the site of airways obstruction. Since helium is of low density, flow improves at sites of turbulence such as large airways but changes little where flow is laminar such as small airways.

ARTERIAL OXYGEN SATURATION AND PRESSURE

The arterial O$_2$ saturation is the percentage saturation of Hb with O$_2$

(100 × content ÷ capacity). It is measured volumetrically or spectrophotometrically on blood drawn from an artery. It can also be measured in the heated ('arterialized') pinna of the ear by oximetry, which is a form of spectrophotometry. Oxygen tension is easily measured by the widely available polarographic electrode systems. Measurements of arterial oxygen tension permit much more accurate study of gas exchange than do measurements of O_2 saturation.

Hypoxaemia is an overall measure of the severity of respiratory disease but it does not discriminate between different mechanisms. It may result from hypoventilation, right-to-left shunting or ventilation/blood-flow inequality. The mechanisms may be separated as follows.

1   With hypoventilation, the arterial P_{CO_2} is raised and P_{O_2} lowered; increasing ventilation so as to normalize P_{CO_2} will also normalize P_{O_2}.

2   Breathing 100 per cent O_2 should result in an arterial P_{O_2} of greater than 550 mmHg (70 kPa). Right-to-left shunting is the only cause of failure to reach this level. The physiological demonstration of right-to-left shunting carries no implication as to its nature or anatomical site.

3   Any hypoxaemia not explained by hypoventilation or right-to-left shunting is due to ventilation/blood-flow inequality. This mechanism is the commonest and most important cause of hypoxaemia.

MEASUREMENT OF ALVEOLAR GAS PRESSURES

The classical (Haldane-Priestley) method of obtaining alveolar air is by forcibly and rapidly expiring down a long tube and analysing the last gas to be expelled. In a trained normal subject the alveolar air obtained by this method is of relatively constant composition and has O_2 and CO_2 tensions close to those of arterial blood (Fig. 3.10a, 'H-P'). An untrained subject may hold his breath or overbreathe just before the forced expiration. This produces an 'unsteady state' in which the alveolar air and arterial blood differ from the 'steady state' values. A way of overcoming this (which is very satisfactory in normal subjects breathing spontaneously) is to analyse the last gas expelled during each expiration while the subject is breathing naturally. This is called 'end-tidal' sampling (Fig. 3.10a, 'E-T'). This gas may be analysed breath-by-breath with a rapid gas analyser or a small proportion of each of several breaths may be collected by a mechanical device and analysed by simpler methods.

In the presence of disturbance of ventilation/blood-flow relationships, which occurs not only in lung disease but in normal subjects ventilated artificially the alveoli do not empty synchronously. Overventilated areas of the lungs tend to empty earlier and underventilated ones empty later. The tracing of CO_2 tension during expiration is therefore much more sloping than in the normal subject (Fig. 3.10b). Therefore neither the Haldane-Priestley (H-P) nor end-tidal (E-T) samples correspond to 'true' alveolar air.

When ventilation/blood-flow relationships are abnormal it becomes very difficult to

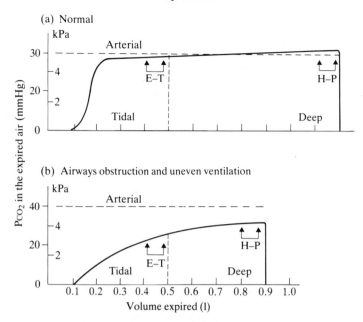

Figure 3.10. *The CO_2 concentration in the expired air during a single expiration.*
    These records show the instantaneous CO_2 tension of the air coming out of the mouth during a resting (tidal) expiration and during a forced rapid deep expiration. There is no CO_2 in the first air to be expelled and then a steep S-shaped curve as the anatomical dead-space is flushed. Following this in the normal subject there is a slightly sloping plateau with a P_{CO_2} close to arterial P_{CO_2} so that either an end-tidal (E-T) or a Haldane-Priestley (H-P) sample provides a good estimate of the P_{CO_2} both of homogeneous alveolar air and of arterial blood. In the lower trace there is no plateau; the record continues to rise fairly steeply as long as the expiration continues. Homogeneous alveolar air therefore cannot be obtained either by an end-tidal or a Haldane-Priestley sample, nor can the arterial P_{CO_2} be estimated by these methods.
    In both normal and abnormal subjects, slowing of the rate of the expiration would cause the slope to become steeper because the mixed venous blood is continuously adding CO_2 to a diminishing volume of air in the lungs.

define 'true' alveolar air, let alone to determine its composition. Theoretically, if the volume and composition of the gas in all the alveoli could be determined, a 'true' mean value for alveolar composition could be calculated. However, this cannot be done, and, moreover, it might give an erroneous idea of the effective alveolar composition—i.e. the composition of the air in the alveoli which are in contact with blood-flow—because many of the alveoli contributing to this 'true' value may be so poorly supplied with blood that their contribution to gaseous exchange is negligible.

'Ideal' alveolar air (Fig. 3.11)

The concept of ideal alveolar air represents an attempt to overcome these difficulties. It makes use of two facts. (1) The respiratory exchange ratio of any average value for

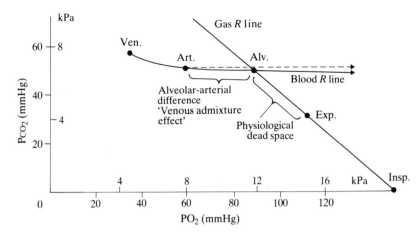

Figure 3.11. *The 'ideal' alveolar air analysis illustrated with the $CO_2 : O_2$ diagram using data obtained on a patient with underventilation and disturbance of ventilation/blood-flow relationship.*

All possible expired and alveolar combinations of Po_2 and Pco_2 which have a respiratory exchange ratio (R) equal to that of the body's RQ must lie on the gas R line. Similarly, all combinations of Po_2 and Pco_2 in blood which satisfy the body RQ must lie on a line (the blood R line) which starts from the mixed venous point (ven.); this line is not straight because the O_2 and CO_2 curves are not straight but it is both nearly straight and flat at Po_2's above 60 because the upper part of the O_2 dissociation curve is flat.

In practice, three points are known: the inspired (insp.), the mixed expired (exp.) and the arterial (art.). The problem is to find the ideal alveolar point (alv.), the physiological dead-space and venous admixture. There are 4 steps. Step 1 says that the gas R line extrapolated from insp. through exp. must pass through alv. Step 2 says that the difference between the Pco_2 of arterial blood and alveolar air (the vertical interval alv.–art.) is very small (the blood R line is nearly flat) so that a horizontal (dashed) line from art. passes close to the alv. which is thus fixed by the intersection of this line with the gas R line. The horizontal difference (alv.–art.) is the alveolar to arterial Po_2 difference. Step 3 says that the ratio of dead-space ventilation to total ventilation is the distance (alv.–exp.) divided by (alv.-insp.). Step 4 says that the venous admixture component is the difference in O_2 content of the blood between alv. and art. divided by the difference in O_2 content between alv. and ven. (the mixed venous point). This fourth step strictly requires that the mixed venous point and the slope of the O_2 dissociation curve be known; in practice both can be estimated with sufficient accuracy for clinical purposes.

(From West J., 1977, slightly modified.)

alveolar air must be the same as that of the body as a whole (i.e. the metabolic respiratory quotient). Unless this is so, the values obtained will not represent a steady state. (2) The differing dissociation curves for O_2 and CO_2 mean that there is only one value of alveolar air which would, if present in all alveoli, cause the volumes of gaseous exchange appropriate to the respiratory quotient.

In practice this 'ideal' value cannot be determined exactly because too much accurate information about ventilation, blood-flow and metabolism is required, but an 'effective' value very close to it is obtained by assuming that the arterial CO_2 tension and the alveolar CO_2 tension are equal. At any site along the alveolar membrane the

P_{CO_2} of the air on one side and the blood leaving the pulmonary capillary on the other side are virtually the same. Hence, the P_{CO_2} of the arterial blood is the mean P_{CO_2} in those alveoli from which the 'average' blood has come. Obviously this estimate of alveolar P_{CO_2} is weighted by the overperfused alveoli in the same way that estimates based on expired air would be weighted by the overventilated alveoli. However, at rest the arterial and mixed venous P_{CO_2} differ only by about 6 mmHg (0·8 kPa) whereas alveolar and inspired air differ by about 40 mmHg (5·3 kPa). Hence the presence of venous or 'shunted' blood (i.e. that from overperfused underventilated alveoli) affects estimates of alveolar P_{CO_2} based on the arterial blood to a smaller extent than the presence of 'dead-space' air (i.e. that from over-ventilated underperfused alveoli) affects any estimate based on expired air.

If the arterial P_{CO_2} is known and is taken to equal the 'effective' alveolar P_{CO_2}, then the 'effective' alveolar P_{O_2} can be calculated using the alveolar air equation.

Physiological dead-space and venous admixture

If all the alveoli were to receive air and blood in the same proportion (e.g. all receive 80 units of air to 100 of blood, or 60 units of air to 100 of blood), the alveolar air expelled from all alveoli would have the same composition and the blood leaving all the pulmonary capillaries would have the same composition. Alveoli which have a higher ventilation in relation to blood-flow than the average contribute air with a lower CO_2 concentration and higher O_2 concentration. This air therefore resembles dead-space air (i.e. air with no CO_2 and 21 per cent O_2). The blood in the capillaries leaving alveoli which have a lower ventilation in relation to blood-flow than the average has a higher CO_2 tension and a lower O_2 tension and therefore resembles mixed venous blood. The problem is to decide what is the true or average alveolar composition. As discussed above the best practical approach is to assume that the arterial CO_2 tension is equal to the 'effective' alveolar CO_2 tension (this assumption becomes increasingly erroneous as ventilation/blood-flow inequality becomes more severe). If this assumption is made, then the overventilated alveoli can be quantified as the increase in physiological dead-space they cause and the underventilated alveoli as an apparent right-to-left shunt or 'venous admixture effect'. This is reflected as an increase in the P_{O_2} difference between the 'effective' alveolar air and the systemic arterial blood (the alveolar-arterial oxygen tension difference) and includes contributions from areas with low ventilation/blood-flow ratios as well as from true right-to-left shunting. This is the basis of the 'ideal' alveolar air analysis which gives the most complete assessment of the distribution of ventilation in relation to pulmonary blood-flow readily available.

More sophisticated analyses of the distribution of ventilation/blood-flow ratios can be made using a recently developed technique based on the elimination from the lungs of a range of injected foreign gases. Six inert gases dissolved in saline are infused intravenously and after a steady state has been reached, arterial and expired gas concentrations are measured by gas chromatography. Since the gases chosen have different solubilities, they partition themselves between blood and gas according to the ventilation/blood-flow ratio. Calculations may then be derived which describe a continuous distribution of V/Q ratios, plotted against ventilation or blood-flow.

Rebreathing methods

As has been shown above, when pulmonary function is disturbed, the determination of effective alveolar gas composition requires the measurement of arterial blood CO_2 tension. The arterial P_{CO_2} may be measured indirectly but bloodlessly by the rebreathing method. This involves the measurement of the mixed venous P_{CO_2} which at rest is about 6 mmHg (0·8 kPa) higher than arterial, with a variation of only a few mmHg even when the cardiac output changes considerably.

The mixed venous CO_2 tension can be relatively simply determined in the following way. A small bag containing CO_2 at a tension estimated to be a few mmHg higher than mixed venous P_{CO_2} is rebreathed. The gas in the bag mixes with that in the lungs (which has P_{CO_2} lower than mixed venous blood) and the resultant mixture then has a P_{CO_2} close to that of mixed venous blood. The mixed venous blood then gives off or takes up CO_2 from the lungs until their tensions are equal. Blood then passes through the lungs without changing its P_{CO_2} and the CO_2 tension of the gas passing backwards

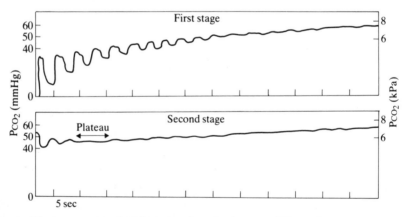

Figure 3.12. *The two-stage method for estimating mixed venous CO_2 tension.*
   This is the record of P_{CO_2} at the mouth of a subject rebreathing from a bag initially containing 100 per cent O_2, its volume being less than twice the subject's tidal volume. During the early part of the first stage the P_{CO_2} rises quickly until the mixed venous P_{CO_2} is reached. The rate of rise then slackens as CO_2 is stored in the body. This rise is allowed to continue until the bag is 'prepared' for the second stage, that is, until the P_{CO_2} is above mixed venous. This first stage in practice usually requires about $1\frac{1}{2}$ min and is followed by a rest of about 3 min during which the subject breathes normally to allow the blood P_{CO_2} to return to normal. During the second stage the subject rebreathes from the bag that he had 'prepared' in the first stage. The CO_2 in the bag, lungs and mixed venous (pulmonary arterial) blood come into equilibrium which is shown as the plateau. The gas in the bag at any time within 20 to 40 sec after the start of the second stage has a P_{CO_2} that of the mixed venous blood.

and forwards between the lungs and the bag does not change. This equilibrium is recognized by recording the P_{CO_2} at the mouth with a rapid CO_2 analyser which shows a 'plateau' of unchanging CO_2 concentration (lower record, Fig. 3.12). The P_{CO_2} recorded during this equilibrium is that of the mixed venous blood. About 20 seconds are available for the attainment and recognition of this equilibrium before blood which has left the lungs unable to give off CO_2 comes back again from the tissues with its P_{CO_2} increased. It is possible to determine mixed venous P_{CO_2} (and thus to estimate arterial P_{CO_2}) using this principle without a rapid CO_2 analyser by getting the subject to prepare the initial CO_2 mixture in the bag by rebreathing O_2. Such is the storage capacity of the body for CO_2 that, once having reached mixed venous P_{CO_2}, the P_{CO_2} in the bag, lungs and blood rises only slowly (upper record, Fig. 3.12). At any time between 75 and 120 seconds after beginning to rebreathe O_2, the P_{CO_2} in the bag is close to the true mixed venous value and is therefore suitable for use as the starting gas in the rebreathing technique proper. It has also been shown that rebreathing a CO_2 mixture so prepared for 20–40 seconds brings the P_{CO_2} in the bag sufficiently close to that of mixed venous blood for most clinical purposes (lower record, Fig. 3.12).

THE INTERPRETATION OF ARTERIAL P_{CO_2}

In interpreting measurements of arterial P_{CO_2} three equations must be borne in mind.

1   *The alveolar ventilation equation.* This relates CO_2 production (which is dependent on metabolic rate), alveolar ventilation (the effective volume of breathing) and arterial P_{CO_2}.

$$\text{Arterial } P_{CO_2} = \frac{CO_2 \text{ Production}}{\text{Alveolar ventilation}} \times 0.86$$

0.86 is an average value at sea level for a factor depending chiefly upon barometric pressure.

This equation can be paraphrased:

$$\text{Arterial } P_{CO_2} \propto \frac{\text{Metabolic rate}}{\text{Effective volume of breathing}}$$

Thus, at any given metabolic rate, inadequate alveolar ventilation causes an increased arterial P_{CO_2}.

An important further step is to decide whether such inadequacy of alveolar ventilation is purely mechanical or whether it reflects impairment of ventilatory control. In other words, is the P_{CO_2} high because the patient cannot breathe enough or because he will not breathe enough? Often both factors contribute. Some assessment of their relative importance may be obtained by comparing the relation between P_{CO_2} and

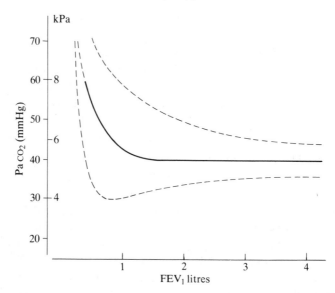

Figure 3.13. *Relation between ventilatory capacity and arterial carbon dioxide tension.*
These curves embrace the range of findings in patients with airways obstruction from a variety of causes. As a general rule the $Paco_2$ remains normal until the FEV$_1$ has fallen to 1 l. The dashed end to the middle curve indicates that such high $Paco_2$ does not occur breathing air. The upper and lower dashed lines show the considerable variation. Some patients have higher $Paco_2$ with lesser impairment of ventilatory capacity; of them it may be said that they are lazy breathers—that they 'won't' as well as 'can't'. On the other hand some patients over-breathe despite more than average loss of ventilatory capacity; these include many asthmatics and the so-called 'pink puffers'.

(Modified from Rebuck A.S. & Read J. *Amer. J. Med.*, **51**, 788, 1971.)

ventilatory capacity in the patient with the relation found in normal subjects (Fig. 3.13). Alternatively, ventilatory drive may be measured directly, as described below.

2 *The alveolar air equation.* This equation depends upon the fact that the sum of the partial pressures of O_2, CO_2, N_2 and H_2O in the alveolar air equals barometric pressure. Nitrogen and H_2O are not exchanged in the lungs, so if the relative volume of O_2 and CO_2 exchanged (i.e. the respiratory exchange ratio or RQ) are known, the alveolar Po_2 can be calculated from the arterial Pco_2.

$$\text{Alveolar } Po_2 = \text{Inspired } Po_2 - \frac{\text{Arterial } Pco_2}{\text{RQ}}$$

(An average RQ of 0·8 may be used in this equation which is a shortened version of the full equation used when greater accuracy is required.)

Thus, for any given inspired Po_2, an increase in arterial Pco_2 is associated with a fall in alveolar Po_2.

3 *The acid : base equation.* The derivation of this equation, of which the following are two versions, is given in Chapter 5.

$$\text{Arterial } H^+ \text{ activity} = 180 \times \frac{\text{Arterial } P_{CO_2} \text{ (kPa)}}{\text{Arterial plasma } (HCO_3^-) \text{ (mmol/l)}}$$
$$\text{(nM)}$$

$$\text{Arterial pH} = 6 \cdot 1 + \log \frac{\text{Arterial plasma } (HCO_3^-) \text{ (mmol/l)}}{0 \cdot 23 \text{ Arterial } P_{CO_2} \text{ (mmHg)}}$$

These equations show that the changes in blood H^+ activity or pH depend upon both the P_{CO_2} and the HCO_3^- concentration.

CONTROL OF VENTILATION

Modifications of older techniques have enabled the ventilatory response to CO_2 to be measured as a simple and standard laboratory procedure. The subject rebreathes from a 4–6 l. bag containing approximately 7 per cent CO_2 in O_2. Ventilation is measured continuously by having the bag enclosed in a bottle and registering the displacement on a spirometer. The CO_2 concentration in the bag, which after equilibration equals alveolar CO_2 concentration, is measured continuously with a rapid CO_2 analyser via a side-line. The CO_2 concentration progressively rises and with it ventilation, until the subject has to stop after a few minutes because of dyspnoea. The progressive increase in ventilation can then be plotted against the increase in CO_2 tension and the slope of the line (V_E/P_{CO_2}) read.

More recently, methods have been described to measure the ventilatory response to hypoxia. Both of these tests assess the sensitivity of the respiratory centre and thus are useful in understanding the variable relation between arterial blood gas tensions and ventilatory capacity (*Fig. 1.13*). However, they do not distinguish between hypercapnic patients who won't breathe and those who can't breathe. Tests have recently been developed which more closely measure the output of the respiratory centre, independently of the mechanical load (i.e. identify patients who won't breathe). These tests are based on the change in pressure at the mouth during transient occlusion of the airway in early inspiration and are virtually pure measurements of ventilatory drive.

THE TRANSFER FACTOR (DIFFUSING CAPACITY)

For many years it was thought that the most important process concerned in the exchange of gases between the alveolar air and the pulmonary capillary blood was diffusion and that loss of or damage to the alveolar-capillary membrane by disease would impair diffusion. Much effort was therefore devoted to the measurement of the 'diffusing capacity' of the lungs. However, it has emerged that these measurements are more dependent on ventilation/blood-flow relationships and chemical reactions than on diffusion and that the less specific term 'transfer factor' would be preferable.

The transfer factor for any gas X is calculated from the following equation:

$$\text{Transfer factor} = \frac{\text{Vol. of } X \text{ taken up}}{\text{Alv. press. of } X - \text{Mean capill. press. of } X}$$

Three measurements are therefore required:

1 The volume of X taken up by the pulmonary capillary blood.
2 The partial pressure (tension) of X in the alveolar air.
3 The mean partial pressure of X in the pulmonary capillary blood.

The volume of gas that will diffuse from the alveolar air to the red cells in the pulmonary capillaries each minute depends upon many factors: the difference between the partial pressure of the gas in the alveolar air and the interior of the red cell; the solubility of the gas in aqueous solutions; its molecular weight; the area and thickness of the following barriers—the alveolar fluid, the alveolar membrane, the capillary membrane, the plasma, the red cell membrane, and the interior of the red cell; and the volume of red cells in the pulmonary capillaries and the speed of certain physico-chemical processes in the red cells. In respect of movement of O_2 and CO_2 across the alveolar-capillary membrane and the other liquid barriers, the much greater solubility of CO_2 outweighs all other factors. For many years, however, there was uncertainty about the ability of the lung to take up O_2 by diffusion in health and/or in disease and much effort was expended in attempts to estimate the diffusing capacity of the lungs for O_2. The major difficulty lies in the estimation of the mean partial pressure of O_2 in the pulmonary capillary blood because the Po_2 changes as Hb takes up the O_2.

The transfer factor for carbon monoxide

Although CO is not a physiological gas, its rate of diffusion throughout the physiological media resembles that of O_2 and the measurements of the transfer factor for CO can give information bearing upon O_2 transfer. The advantage of CO is that it largely removes the difficulty of estimating the mean pulmonary capillary tension. This is because the affinity of haemoglobin for CO is so great that, for all tolerable concentrations, the CO is so completely taken up by the Hb that the partial pressure in the capillary blood is zero. Only the first and second measurements listed above therefore need be made.

Several methods for the estimation of the CO transfer factor have been introduced. They fall into two main groups: the *steady state* methods in which the uptake of CO and the alveolar tension are measured over a period of several minutes; and the *single breath* methods in which the rate of uptake and alveolar tension are calculated from the change in composition of a gas mixture containing CO which is inspired, held in the lungs for about 10 seconds and then expired.

All these methods use low concentrations of CO ($< 0\cdot3$ per cent) in the inspired air. The chief differences between them lie in the estimation of alveolar CO tension. As discussed previously, the determination of the composition of alveolar air is not easy when pulmonary function is abnormal. No method of estimating transfer factor is entirely satisfactory in this respect because disturbance of ventilation/blood-flow relationships cause reductions in the values obtained with them.

Partition of the diffusion pathway into the membrane and intracapillary components

O_2 and CO compete for haemoglobin but do not affect each other's diffusion through the alveolar-capillary membrane. Therefore, by measuring the CO transfer factor at different alveolar O_2 tensions, it can be broken down into a component dependent on diffusion through the membrane and a component dependent on events within the red cell. In normal subjects rather less than half the resistance to CO uptake lies in the membrane. This approach has also permitted estimation of the volume of blood in the pulmonary capillaries exposed to the alveolar air at any instant. In a normal subject, this volume is about 80 ml.

The concept of diffusing capacity is valuable and the transfer factor measurement can provide a quantitative description of the ability of a subject's lungs to transfer O_2 from the air to the blood. Its use to examine the area and permeability of the alveolar-capillary membrane in disease has however, been disappointing for two reasons. Firstly, it has proved difficult to devise a practicable method which is not affected by ventilation/blood-flow inequality. Secondly, changes in the pulmonary blood volume and flow obscure the effect of disease of the membrane. With none of the methods in general use is it possible to maintain that a low value means that the alveolar-capillary membrane is shrunken or thickened. Indeed, it is doubtful if reduction of the diffusing capacity of the alveolar-capillary membrane is ever an important cause of defective O_2 transfer. It is certainly very much less important than disturbance of ventilation/blood-flow ratios.

METHODS OF ASSESSING THE EVENNESS OF DISTRIBUTION OF
THE INSPIRED AIR

Methods of studying the distribution of the inspired air all measure the effects on the expired air composition of a sudden change in the composition of the inspired air. Although they give no direct information about the more important relation of distribution of ventilation to blood-flow, maldistribution of ventilation implies at least some degree of ventilation/blood-flow inequality, since compensatory vascular redistribution is never perfect.

Single breath nitrogen test

In this test, a single vital capacity breath of O_2 is taken. The concentration of nitrogen is then measured at the mouth using a fast analyzer while the subject exhales slowly to residual volume. A typical tracing is shown in Fig. 3.14. In the brief first phase, the

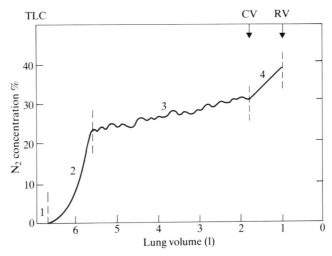

Figure 3.14. *Plot of expired nitrogen concentration against volume in the single breath nitrogen test.*

A normal subject has inspired a vital capacity breath of O_2 and during the subsequent slow maximum expiration, the concentration of nitrogen is plotted against expired volume. There are four distinct phases of the plot, as described in the text. Closing volume (cv) is measured at the transition between phases 3 and 4.

(From West J., 1978.)

concentration of nitrogen remains zero as O_2 is washed out of the upper airways. In the second phase, the concentration rises rapidly as dead-space is progressively washed out by alveolar gas. In the third and longest phase, the alveolar plateau, the concentration is uniform if the O_2 was evenly distributed throughout the lungs in the preceding breath. If ventilation was unevenly distributed, the concentration gradually rises and the slope of the plateau is a measure of the degree of maldistribution. The normal value is 1–1·5 per cent N_2 per litre expired volume. The rise in concentration is probably due to the fact that underventilated areas of the lung receive little of the preceding breath of O_2 and these are also the areas that tend to empty last.

Closing volume

At the end of the single breath nitrogen test, a fourth phase of abruptly rising concentration may be seen (Fig. 3.14). This phase probably reflects the fact that at residual volume, small airways in dependant regions of the lungs are closed and basal alveoli are smaller than apical, although the alveolar nitrogen concentration is uniform throughout the lung. During the breath of O_2, the nitrogen in the initially larger apical alveoli is less diluted than that in the initially smaller basal alveoli. Thus, during the subsequent expiration as basal airways close at the end of the alveolar plateau, expired gas continues to flow from the apical alveoli with their higher concentrations of nitrogen.

Other methods of measuring closing volume involve the injection of a small bolus (e.g. 50 ml) of tracer gas, such as argon, [133]xenon or helium, into the mouthpiece while the subject briefly holds his breath at residual volume. Following maximum inspiration, the concentration of tracer gas is plotted during the following slow expiration and interpreted in the same way as the single breath nitrogen test.

The transition between phases 3 and 4 is taken as the volume at which dependant airways begin to close and is referred to as the closing volume. It is usually expressed as a per cent of vital capacity (the normal value is between 5 and 20 per cent, depending on age), though it may also be expressed as a per cent of total lung capacity, when it is referred to as closing capacity.

The measurement of closing volume appears to be a simple and sensitive test for the detection of early small airway disease, for example in smokers. In more advanced disease with widely distributed airways obstruction, phase 3 is steeper and irregular and no clear transition to phase 4 may be demonstrable. For practical purposes, if the FEV$_1$ is abnormal, the closing volume is usually not measurable.

More detailed indices of uneven ventilation or mixing efficiency may be derived from the tracing of expired nitrogen against time, such as a 'lung clearance index' or calculation of the volume of the 'poorly ventilated space'.

A 'lung clearance index' may be derived by dividing the ventilation required to reach a particular end-point (e.g. 5 per cent expired N$_2$) by the volume of air in the lungs at end-expiration (i.e. FRC, for the measurement of which the nitrogen washout test was primarily applied in the first place).

The volume of the 'poorly ventilated space' may be calculated by plotting the breath-by-breath record of expired nitrogen concentration against time on semilogarithmic paper. It is usually found that the points form a curve with a single inflexion. This curve can thus be analyzed as though it were derived from two separate processes, one representing mixing in a volume of lung which is well or rapidly ventilated and the other representing mixing in a volume which is poorly or slowly ventilated. The poorly ventilated space is either absent or small in normal subjects. In the presence of maldistribution of ventilation, the volume of the poorly ventilated space is increased and its turnover reduced (i.e. the slope of the plot is flatter).

Finally, nitrogen washout may be used to measure lung volumes (FRC and thus RV and TLC). These values are falsely low in the presence of air-trapping for any reason and the difference between TLC measured by this method and by body plethysmography (thoracic gas volume) is a measure of the amount of trapped or non-communicating gas in the chest.

Nitrogen washout

If the inspired gas is changed from air to 100 per cent O$_2$, nearly all of the N$_2$ in the lungs should be washed out in 7 minutes of quiet breathing. A nitrogen concentration in the alveolar air in excess of 2·5 per cent at this time indicates maldistribution. These are generous limits and normal subjects usually wash out nitrogen much faster and more completely than this.

LUNG MECHANICS: COMPLIANCE AND RESISTANCE

The estimation of the lung compliance and airway resistance requires measurement of the transpulmonary pressure, that is, the pressure difference between the alveoli and the surface of the lungs. The pressure at the surface of the lungs—i.e. outside the lungs but inside the thoracic cage—is called the intrathoracic pressure and is usually measured in the oesophagus. The alveolar pressure is equal to mouth pressure when there is no airflow.

The intrathoracic pressure at rest, with no airflow and the respiratory muscles inactive, is normally about 5 cmH$_2$O (0·5 kPa) below atmospheric (Fig. 3.15, point *A*). This pressure is due to the elastic tension or relaxation pressure of the lungs pulling

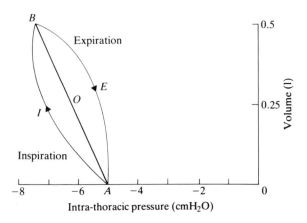

Figure 3.15. *Intrathoracic pressure changes during breathing (1 cmH$_2$O $\simeq$ 1 kPa).*
A record of the relation between lung volume (ordinate) and intrathoracic (pleural or oesophageal) pressure (abscissa). *AIB:* the changes in volume and pressure during a normal inspiration; *BEA:* the changes in volume and pressure during a normal expiration; *AOB:* the changes during a very slow inspiration when the effects of airflow are negligible.

inward on the thoracic cage. (If the thoracic cage were removed at this volume and air prevented from leaving the mouth, the recoil or relaxation of the lungs would cause the pressure in the air spaces to rise to +5 cmH$_2$O.) If the subject breathes in very slowly, so that the rate of airflow is negligibly small, the intrathoracic pressure falls, producing the record *AOB* in Fig. 3.15. The volume inspired (500 ml) divided by the change in oesophageal pressure (−5 to −7·5 = 2·5 cmH$_2$O) gives the compliance of the lungs (200 ml/cmH$_2$O in the example) and the reciprocal is the elastance (5 cmH$_2$O/l in the example). If the subject breathes in at a normal speed the oesophageal pressure produces the record *AIB*. The difference between *I* and *O* is due to the non-elastic resistance offered by the airways to the flow of air and to the viscous resistances of the lung tissues. If a simultaneous record of the rate of airflow is taken, an instantaneous value for $I - O$ divided by the simultaneous rate of airflow is a measure of the non-elastic resistance. A normal expiration produces the record *BEA*, the horizontal distance $E - O$ divided by the rate of airflow again being a measure of non-elastic resistance.

Lung compliance (a static measurement) is derived in practice from the plot of serial volume/transpulmonary pressure relationships measured during the brief plateaux of no flow which are produced when the subject takes a large breath in increments. As previously pointed out, the static compliance, or the volume of lung expansion per unit of transpulmonary pressure increase, depends on a variety of factors which include the amount, arrangement and integrity of the elastic tissue, the surface tension of the alveolar lining fluid and the lung size. Lung size is obviously important since reduction of the amount of ventilated lung tissue reduces compliance, although the elastic properties of the lung tissues may not have changed. A reduced static compliance may thus be due to stiffer lung tissue, higher surface tension of the alveolar lining fluid or decreased lung volume.

If compliance is measured during normal breathing, i.e. dynamically, it will be lower than the static compliance if there is non-uniform distribution of mechanical properties within the lungs (p. 103). Under these circumstances, with increasing frequency of breathing dynamic compliance falls progressively. Frequency dependence of compliance is a sensitive index of small airways disease, although as a test it is difficult to perform reliably.

Measurements of compliance have greatly increased understanding of pulmonary physiology. Generally, however, they are of limited value in the assessment of individual patients, partly because the normal range is wide and partly because changes can be difficult to interpret.

Airway resistance (a dynamic measurement) is usually derived from flow/transpulmonary pressure relationships during tidal breathing. An ingenious method is used to subtract from the total pressure generating flow, the component required to overcome the elastic resistance, thus leaving only the component required to overcome the non-elastic resistance. Other methods may be used to measure airway resistance which do not require an oesophageal balloon. These include the use of a body plethysmograph in which alveolar pressure is measured by applying Boyle's law to the changes in pressure and volume obtained when the subjects pants through a pneumotachograph (which measures flow) or the measurement of pressure/flow relationship at the mouth when the subject is connected to a source of forced air oscillations from a loudspeaker.

The non-elastic, or viscous, resistance arises primarily from the airways (p. 100) with insignificant contributions from the resistance either of tissues sliding over each other or of the tissue fluids. As also pointed out earlier (p. 100), the presence of turbulent flow prevents any single value obtained for non-elastic resistance being appropriate to all values of airflow. For practical purposes, resistance is obtained during tidal breathing and is expressed in cmH_2O per l. per sec. The airway resistance, like compliance, depends partly on lung size. More specifically, the reciprocal of airway resistance (viz. conductance) appears to be linearly related to the thoracic gas volume, for the larger the lung volume at which breathing is occurring, the wider the airways are held open. On the other hand, reduction of the amount of ventilated lung tissues reduces the rate of airflow for a given pressure difference, because there are fewer passages for the air to flow through.

The measurement of airway resistance appears to be of greater potential clinical value than the measurement of compliance, because the normal range is quite narrow, because the resistance of the airways is much the most important component and because the indirect (particularly spirometric) indices which reflect airway obstruction are affected by a range of other factors.

The *spirometric* measurement of the ventilatory capacity provides a rapid, simple and useful assessment of the mechanical properties of the lungs (and thoracic cage). Although it is a gross oversimplification, we may consider that reduced compliance is reflected by a restrictive spirometric defect and increased non-elastic resistance by an obstructive defect.

The *work of breathing* can be calculated from graphs such as Fig. 3.15, in which pressure and volume changes are plotted against each other, because areas on such graphs have the dimension of work, i.e. pressure × change in volume.

The methods described above, based on the measurement of intrathoracic pressure, measure the compliance and non-elastic resistance of structures only within the thorax. The thoracic cage also has elastic and non-elastic properties but these are more difficult to study by available techniques.

Body plethysmography

Lung mechanics can be studied with convenience with the subject seated in an airtight box (plethysmograph). This method has the great added advantage of permitting the rapid, accurate and simultaneous measurement of lung volumes by the application of Boyle's law.

REGIONAL LUNG FUNCTION

Fluoroscopy

On fluoroscopy the intensity of 'lighting-up' of the lungs and the movements of the ribs, the mediastinum and the diaphragm indicate the proportion of ventilation received by individual lobes. Lags in the movement of the diaphragm and the mediastinum indicate unequal obstruction of the airways. A forced voluntary inspiration and expiration magnify regional differences of pressure in the lungs and therefore magnify these regional changes in movement.

Bronchospirometry

This procedure, which involves differential bronchial catheterization and collection of air separately from each lung, is now little used although it has recently been revived using sampling to a mass spectrometer during fiberoptic bronchoscopy.

Radio-isotope techniques

The understanding of regional lung function has been greatly advanced since the introduction of a number of elegant isotopic techniques. The most widely used method is to administer an isotopically labelled gas and to measure the concentration in regions or 'cores' of lung by external counting. ^{133}Xenon, an insoluble gas, has proved to have isotopic characteristics and physical properties that are very suitable for this purpose. In particular, its insolubility ensures that it is confined to the air spaces when inhaled and that when it is injected, it is almost entirely evolved on passage through the lungs. When ^{133}Xe is rebreathed from an appropriate circuit until the counts are stable, its local concentration reflects local lung volume. When a single large breath of ^{133}Xe is taken, the counts reflect local ventilation. When ^{133}Xe is injected as an intravenous bolus, the counts reflect local blood-flow (provided there is some local aeration to receive the evolved gas). Thus, regional ventilation, blood-flow and volume can be quickly and simply assessed. The concentration of expired ^{133}Xe can also be readily measured at the mouth when, like the measurement for example of the expiratory plateau of nitrogen, it gives information about the overall distribution of ventilation.

A less subtle but more widely availabe technique is to scan the radioactivity in the lungs after intravenous injection of macro-aggregates of ^{131}I- or ^{99m}Tc-labelled albumin or other suitable isotopically labelled particulate preparation. These impact in the precapillary pulmonary blood vessels (only a minute number of vessels become temporarily occluded) producing regional radioactivity proportional to the local blood-flow.

With the exception of lung scanning, which is of particular value in the assessment of pulmonary thromboembolism, regional lung function studies have been primarily of physiological interest rather than of clinical application.

Practical assessment

VENTILATORY CAPACITY

Clinical observations

Exercise tolerance and severity of dyspnoea as assessed by questioning; supplemented by observation at rest and when exercising (climbing stairs) as fast as possible.

Note frequency, depth, movements of the chest, action of accessory muscles, wheezing, ability to talk or count between breaths.

Differentiate reduced ventilatory capacity from: overventilation, fatigue, angina or other chest pain, poor morale.

Signs of airways obstruction: forced expired time, overinflation, wheezes.

Signs of conditions likely to hinder expansion: basal congestion, air or fluid in pleura, obesity, deformed thorax.

Signs of muscular weakness.

Routine methods

Spirometry: VC and FEV_1 but not FEV_1/VC ratio or MMF reduced in 'restrictive' disorders; FEV_1, FEV_1/VC ratio, MMF and often VC reduced in obstructive disorders.

Peak expiratory flow.

Effect of bronchodilators.

Ventilatory response to exercise.

Special techniques

Lung mechanics: measurement of airway resistance, static compliance and frequency dependence of (dynamic) compliance.

Flow-volume curves.

Distribution of ventilation.

Lung volumes.

Closing volume.

Arterial P_{CO_2} on exercise.

GAS TRANSFER

Clinical observations

If impaired: hyperventilation and cyanosis; usually inconspicuous at rest but may be provoked by exercise. Cyanosis abolished, ventilation reduced and exercise tolerance improved by O_2.

Routine methods

Spirometry: 'restrictive' defect may be demonstrable.
    Arterial blood gas analysis.
    Transfer factor for CO (diffusing capacity).
    Ventilatory response to exercise.

Special techniques

Ventilation/blood-flow relationships: physiological dead-space and alveolar-arterial P_{O_2} difference.
    Gas exchange on exercise.
    Effects of O_2.

HYPERCAPNIA

Clinical observations

Many signs may be present; none is pathognomonic, all may be absent: mental confusion proceeding to coma, tremor, vasodilation, tachycardia, raised blood pressure (particularly pulse pressure), sweating, papilloedema. Absence of cyanosis does not exclude.

Routine methods

Arterial P_{CO_2} (direct or rebreathing method).
    Arterial pH and HCO_3^-.

Special techniques

Ventilatory response to CO_2 or hypoxia.
    Ventilatory drive.

HYPOXAEMIA

Clinical observations

'Central' cyanosis, i.e. of mucous membranes where blood-flow good. If caused by arterial hypoxaemia due to respiratory failure, the central cyanosis is abolished by breathing O_2. Persistence of cyanosis on O_2 signifies anatomical R–L shunt or abnormal pigment.

Routine methods

Arterial P_{O_2}.

Arterial O_2 saturation (direct or derived from P_{O_2} and pH).
Haemoglobin (and hence arterial oxygen content).

Special techniques

Arterial P_{CO_2}.

Alveolar-arterial P_{O_2} difference.
Arterial P_{O_2} after breathing 100 per cent O_2.
P_{50}.
Abnormal pigments.
Cardiac output.

REGIONAL LUNG FUNCTION

Assessment must always include, and is helped by, assessment of total lung function.

Clinical observations

Physical and radiographic signs of local aeration.

Routine methods

Fluoroscopy.
Lung scanning.

Special techniques

Radioactive gas.
Bronchospirometry.

References

BATES D.V., MACKLEM P.T. & CHRISTIE R.V. (1971) *Respiratory Function in Disease*. Philadelphia, Saunders.
CAMPBELL E.J.M. (1966) Exercise tolerance. *Sci. Basis Med. Ann. Rev.* 128–44.
CAMPBELL E.J.M., AGOSTONI E. & NEWSOM DAVIS J. (1970) *The Respiratory Muscles: Mechanics and Neural Control*. London, Lloyd-Luke.
CHERNIACK R.M., CHERNIACK L. & NAIMARK A. (1972) *Respiration in Health and Disease*. Philadelphia, Saunders.

COMROE J.H., FORSTER R.E., DUBOIS A.B., BRISCOE W.A. & CARLSEN E. (1962) *The Lung: Clinical Physiology and Pulmonary Function Tests*. Chicago, Year Book Medical Publishers.

COTES J.E. (1979) *Lung Function: Assessment and Application in Medicine*, 4th ed. Oxford, Blackwell Scientific Publications.

DEJOURS P. (1966) *Respiration*, Trans. by L.E. Farhi. Oxford University Press.

FENN W.O. & RAHN H. (ed.) (1965) *Respiration. Handbook of Physiology*. Section 3. Vols. I & II. Washington, Amer. Physiol. Soc.

FORGACS P. (1978) *The functional basis of pulmonary sounds*. Chest, 73, 399–405.

HEDLEY-WHITE J., BURGESS G.E., FEELEY T.W. & MILLER M.G. (1976) *Applied Physiology of Respiratory Care*. Boston, Little Brown.

HOWELL J.B.L. & CAMPBELL E.J.M. (ed.) (1965) *Breathlessness*. Oxford, Blackwell Scientific Publications.

JONES N.L. & CAMPBELL E.J.M. (1982) *Clinical Exercise Testing*. Philadelphia, Saunders.

SYKES, M.K., McNICOL, M.W. & CAMPBELL, E.J.M. (1976) *Respiratory Failure*, 2nd ed. Oxford, Blackwell Scientific Publications.

WEST J.B. (1974) *Respiratory Physiology*. Baltimore, Williams and Wilkins.

WEST J.B. (1977) *Ventilation/Blood Flow and Gas Exchange*. Oxford, Blackwell Scientific Publications.

WEST J.B. (1978) *Pulmonary Pathophysiology*. Baltimore, Williams and Wilkins.

4

The Kidney

Normal function

The major function of the kidney is the maintenance of a stable volume and composition of extracellular fluid. This is an essential requirement for the optimal performance of other organs, so that derangements in renal function have widespread consequences within the body. The regulation of extracellular fluid volume (ECV) is achieved by control of sodium and water balance.

The *first step* in volume regulation following the ingestion of water or salt is the stabilization of extracellular fluid tonicity achieved either through changes in water intake or excretion. The urine of many animals, such as the dog, is normally highly concentrated and these animals regulate their water balance by drinking precisely the quantity of water required. Man, however, also drinks for social reasons and regulates his water balance by varying the rate of water excretion by the kidneys. These processes of water intake and excretion respond rapidly to changes in the plasma osmolality. Sodium and attendant anions make the predominant contribution to extracellular fluid osmolality and thus their plasma concentrations are also closely regulated.

The *second step* in volume regulation is the adjustment of the rates of excretion of sodium and anions by the kidneys. A totally isolated kidney perfused with blood promptly increases its rate of sodium excretion if saline is added to the perfusate. Thus one level of control of sodium excretion is exercised by the physical composition of the blood itself acting directly on the kidneys. A fall in plasma protein concentration or haematocrit, or a rise in plasma sodium chloride concentration, both evoke a natriuresis by intrarenal actions and contribute to the prompt restoration of the extracellular fluid volume during volume expansion. An acute increase in renal perfusion pressure also directly increases sodium excretion. At another level, receptors in the low pressure cardiovascular regions, principally atria and pulmonary vessels, can sense small changes of blood volume. Larger changes are sensed by receptors in the ventricles and large arteries. These pressure receptors relay information to the brain via the vagus nerves. Thus the renal excretion of sodium and fluid is modulated via neural and humoral links.

The processes of osmotic and volume control are not entirely independent since the blood volume can regulate the rate of water intake or excretion, and plasma sodium chloride concentration is a powerful determinant of sodium chloride excretion. However, it is only with a severe challenge, as may occur in disease, that the control of volume is sacrificed to the requirements of tonicity or *vice versa*.

The kidney regulates the ECV by maintaining an appropriate balance between the quantity of extracellular fluid filtered by the glomerulus and the quantity reabsorbed by the tubules. These two processes of filtration and reabsorption are themselves interrelated. Thus reabsorption is adjusted to the level of filtration ("glomerulo-tubular balance") and filtration itself is adjusted to the level of reabsorption by the loop of Henle ("tubulo-glomerular feedback"). Thus in the regulation of the ECV the kidney is normally responsive to many stimuli which form a large but well integrated network of controls. These elicit changes in renal excretion by acting on a well-balanced system of filtration and reabsorption. The outcome is a remarkably stable volume and composition of extracellular fluid which varies little despite wide fluctuations in salt and water intake.

STRUCTURE AND FUNCTION OF THE KIDNEY AND ITS COMPONENT PARTS

Each kidney of an adult measures some 12 × 7 × 3 cm and weighs between 120 and 170 g. A section through a kidney shows a well demarcated inner dark area, *the medulla*, with pyramids invaginating the renal pelvis and an outer, pale area, *the cortex*, scattered with red dots, *the glomeruli*. Each kidney at birth contains approximately 1,000,000 nephrons, the number falling progressively during life. A nephron begins with a glomerulus which consists of a tuft of capillaries enclosed within a Bowman's capsule continuous with the proximal tubule. This leads to the loop of Henle, the distal tubule, the collecting duct and thence to the renal pelvis (Fig. 4.1). In the human kidney about 14 per cent of nephrons are "*juxtamedullary*" originating deep within the cortex and sending long loops of Henle down into the medulla. These are of crucial importance for the renal concentrating mechanism. The remainder are "*cortical*" and their shorter loops of Henle do not penetrate far into the outer medulla. This terminology is confusing however, since all glomeruli are found in the cortex (the juxtamedullary cortex contains nephrons of both types) and a rigid differentiation of nephrons into only two classes is invalid on both structural and functional grounds.

Blood vessels and blood flow

The kidneys are usually supplied by a single artery which divides at or just before the renal hilum into about 5 major branches each of which is functionally an end artery, i.e. it does not make a significant anastamosis with neighbouring arteries. Consequently, if one of these vessels is occluded, a segment of the kidney will become infarcted. The *segmental arteries* break up into *inter-lobar* arteries which themselves ramify within the renal tissue to give *arcuate arteries* which follow a curved course in the deep cortex before turning upwards as *interlobular arteries*. Each interlobular artery gives off a large number of branches with thick, well-innervated muscular walls which pass directly to the glomeruli as *afferent arterioles*. The *efferent arterioles* are of two types: those from the cortical glomeruli soon break up into a network of thin-

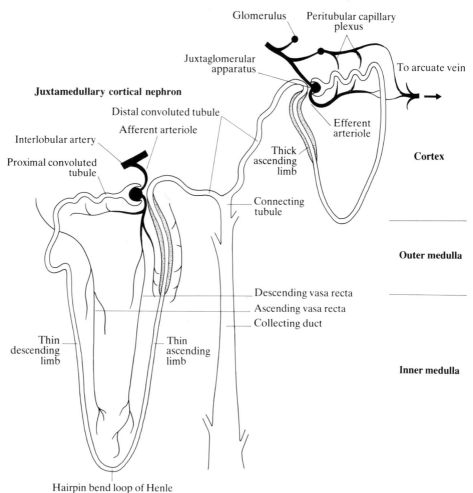

Figure 4.1

Diagrammatic representation of an outer cortical and a juxtamedullary cortical nephron. The blood supply to each comes from an interlobular artery via an afferent arteriole to the glomerulus. The efferent arteriole of the outer cortical nephron breaks up into peritubular capillaries supplying the proximal and distal tubules while that of the juxtamedullary cortical nephron gives an additional branch to the descending vasa recta which supply the capillary plexus in the medulla. Ascending vasa recta and cortical veins originating from the peritubular capillary plexus drain into the arcuate vein. Both types of nephron contain a glomerulus, proximal convoluted tubule, a loop of Henle with a thin descending limb, hairpin bend, thin ascending limb, and thick ascending limb. The loop of Henle of outer cortical nephrons penetrate only to the cortico-medullary border while that of the juxtamedullary cortical nephrons penetrate deep into the inner medulla. The macula densa segment connects the loop of Henle to the distal convoluted tubules at the juxtaglomerular apparatus and the connecting tubule connects the distal tubule to the collecting duct.

walled *peritubular capillaries* surrounding the neighbouring proximal and distal tubules, while those from the juxtamedullary glomeruli are much longer and pass deep down into the medulla where each breaks up into some 30 descending *vasa recta* which supply the capillary plexus of the inner medulla (Fig. 4.1). Some of these juxtamedullary efferent vessels also supply peritubular capillaries in the inner cortex. The juxtamedullary efferent arterioles have a more obvious smooth muscle coat than their cortical equivalents. The kidney has a rich adrenergic sympathetic nerve supply distributed to the vascular smooth muscle, the juxtaglomerular apparatus and the tubules, a parasympathetic nerve supply and a considerable number of afferent nerves whose origins are unclear, but which respond to intra-renal pressure. It has an elaborate system of lymphatic vessels which are distributed mainly to the resistance vessels, the glomeruli and the juxtaglomerular apparati and drain to a few capsular vessels on the surface and a larger number of hilar vessels running between the renal artery and vein.

In a healthy adult, the kidneys receive about 25 per cent of the cardiac output, or 1·3 litres per minute. This flow is well in excess of the metabolic requirements (as shown by the high oxygen content of the renal venous blood), and supports the rapid rate of glomerular filtration (about 120 ml per minute). Stimulation of the renal nerves can cause blood flow to cease, at least temporarily, and sympathetic nerve activity underlies some of the renal vasoconstriction that accompanies anxiety, fear, pain, cold, exercise, hypoxia and haemorrhage. However, in conscious animals, renal blood flow is increased only slightly by denervation, indicating little tonic activity of the nerves on the renal vessels. An isolated perfused kidney can maintain its blood flow and filtration rate during alterations of perfusion pressure within the range of approximately 80–160 mm mercury. However, in the whole animal, a fall in BP produced by haemorrhage activates the sympathetic nervous system and other vasoconstrictor influences which easily override the intrinsic blood flow "autoregulation". Indeed, the kidney is particularly susceptible to ischaemic damage in haemorrhagic hypotension or shock.

Renal vasoconstriction and a fall in GFR, which are independent of the nervous system, occurs during hypercalcaemia, metabolic acidosis and hypocapnia. In healthy subjects, the renal blood flow and GFR are well maintained during dietary salt restriction. However, in elderly subjects or when further sodium depletion occurs as, for example, with the use of diuretic drugs, renal blood flow and GFR can fall considerably. An increase in renal blood flow above the normal range, not usually accompanied by a rise in GFR, can occur during fever, infusion of hypertonic solutions, and as a short-term response to frusemide (but not to thiazide diuretics).

Normal values of RBF and GFR decline quite steeply after the 4th decade of life.

The glomerulus and glomerular filtration

The glomerulus consists of a loop of some 40 capillaries separated from Bowman's space by a series of layers: an endothelium, a basement membrane and an epithelium formed from cells lining Bowman's space. Epithelial cells have large foot processes applied to the basal lamina with spaces termed "slit pores" between them. Mesangial cells are present at the bases of the capillary loops and are continuous with the lacis cells of the juxtaglomerular apparatus (see p. 166). Mesangial cells are phagocytic and contain renin and myofilaments but their function is not yet clear.

The glomerular capillaries filter a fluid virtually free of protein (about 30 mg per litre) but containing crystalloids in the same concentration as plasma water. The first barrier to the filtration of large molecules is the basement membrane. Beyond this, smaller molecules are stopped at the region of the slit pores. The filtration of large negatively-charged molecules such as albumin is further restricted because of fixed negative charges on the filtration barrier. In experimental glomerulonephritis this negative charge is lost, thereby allowing the filtration and excretion of abnormal quantities of albumin.

It is now possible to measure the determinants of glomerular filtration in rats with surface cortical glomeruli by micropuncture methods. The glomerular hydrostatic pressure is low (about 40 per cent of the arterial pressure). It is opposed by a small intratubular pressure and the oncotic pressure of the plasma proteins. This oncotic pressure rises with distance down the glomerular capillaries due to the filtration of a largely protein-free fluid thereby concentrating the capillary plasma proteins progressively. The forces resisting filtration normally rise to equal those favouring filtration just before the end of the glomerular capillaries and at this point filtration ceases. Despite the low mean filtration force, glomerular filtration can proceed rapidly because the water permeability of the glomerular capillaries is 1 or 2 orders of magnitude higher than that of capillaries elsewhere in the body. This model for glomerular filtration is shown diagrammatically in Fig. 4.2. It predicts that an increase in renal plasma flow (RPF) will displace the point at which filtration pressure equilibrium is attained towards the efferent arteriole. A greater fraction of the capillaries are then used for filtration and the GFR rises in parallel with flow until the equilibrium point reaches the end of the capillary (the efferent arteriole). Thereafter, there is filtration pressure disequilibrium and the residual hydrostatic force potentially available for filtration is wasted because of the anatomical constraints of the glomerular capillaries; filtration will no longer rise in parallel with flow and the filtration fraction (GFR/RPF) will fall. The human kidney probably operates normally at, or close to, filtration pressure disequilibrium since increasing plasma flow (whether by reducing the haematocrit or by vasodilatation) does not usually increase the GFR substantially.

An increase in ultrafiltration pressure may occur because of an increase in glomerular capillary hydrostatic pressure, a fall in intratubular hydrostatic pressure or a fall in plasma oncotic pressure (plasma protein concentration). These changes increase the net ultrafiltration pressure at the beginning of the glomerular capillary and increase the rate of fluid filtration but do not greatly alter the point at which filtration pressure equilibrium is attained. Thus an increasing fraction of plasma water is filtered and the filtration fraction (FF) increases.

A fall in GFR will occur with a fall in renal plasma flow, with a fall in glomerular

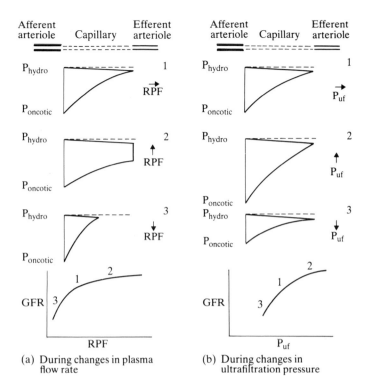

Figure 4.2

Diagrammatic representation of a model of the hydrostatic and oncotic pressure profiles down an outer cortical glomerular capillary from the afferent arteriole (AA) to the efferent arteriole (EA). Under (a) the changes occurring during alterations in renal plasma flow (RPF) are shown and under (b) the changes during alteration in ultrafiltration pressure (P_{uf}). The trans-capillary hydrostatic pressure (P_{hydro}) represents the difference between the glomerular capillary pressure and the intra-tubular pressure. This falls only slightly with distance down the capillary. It is opposed by the oncotic pressure ($P_{oncotic}$) of the plasma proteins and this increases steeply down the capillary due to progressive concentration of the plasma proteins by fluid ultrafiltration. The difference between the hydrostatic and oncotic pressures represents the net filtration pressure. At normal levels of plasma flow (RPF →), the oncotic pressure is shown to rise to equal the hydrostatic pressure by the end of the capillary (filtration pressure equilibrium). At increased levels of plasma flow (RPF↑) equilibrium is no longer attained by the end of the capillary and the full force potentially available for filtration is not utilized. The GFR consequently rises only slightly. At reduced levels of plasma flow (RPF↓), however, the point of filtration pressure equilibrium moves proximally down the capillary and the filtration rate falls with the plasma flow rate. Consequently, glomerular filtration rises steeply with plasma flow only up to a point above which the influence of plasma flow on filtration diminishes (bottom panel). As shown in b, the ultrafiltration pressure determines the rate of fluid filtration and hence the rate of rise of oncotic pressure with distance down the capillary. Thus, at increased levels of ultrafiltration pressure ($P_{uf}↑$), the oncotic pressure rises steeply to approach the hydrostatic pressure by the end of the capillary. The higher levels of ultrafiltration pressure at each point in the capillary lead to a rise in glomerular filtration. Conversely, at reduced levels of ultrafiltration pressure ($P_{uf}↓$), the lower levels of pressure at each point in the capillary lead to a fall in glomerular filtration. Consequently, glomerular filtration rises steeply with ultrafiltration pressure (bottom panel). (For summary of experimental data relating to this model, see Brenner, B.M. and Humes, H.D., *New England Journal of Medicine* **297**, 148–154, 1977).

permeability, with a fall in glomerular hydrostatic pressure, with a rise in the intra-tubular pressure or with a rise in the systemic plasma protein concentration. The fall in GFR in congestive cardiac failure is probably due mainly to a fall in plasma flow, the fall of GFR in glomerulonephritis to a fall in glomerular permeability, the fall of GFR in urinary tract obstruction or osmotic diuresis to a rise in the intra-tubular pressure, while the fall of GFR in hyperproteinaemia (e.g. myeloma) may be due to the rise in plasma oncotic pressure.

In rat kidneys with surface glomeruli, direct measurements have been made of the pressures in the aorta, the glomerular capillaries and the peritubular capillaries to assess the pressure drop across the afferent and efferent arterioles of cortical glomeruli. Relating these to the nephron blood flow rate permits a calculation of the segmental arteriolar resistance. In the anaesthetized rat, of the total arteriolar resistance, about 55 per cent is afferent and 45 per cent efferent. Infusion of noradrenaline increases both afferent and efferent resistance whereas angiotensin has a more pronounced effect on efferent resistance. Glomerular filtration rate is relatively well preserved during angiotension infusion despite a fall in renal blood flow because the glomerular capillary hydrostatic pressure is maintained by the efferent arteriolar vasoconstriction. In response to a fall in perfusion pressure, there is a reduction in both afferent and efferent arteriolar resistances which maintains RBF ("autoregulation"). By use of angiotensin antagonist drugs, it has been shown that maintenance of GFR during reduced perfusion pressure requires an action of angiotensin on the efferent arteriole to counter the fall in its resistance and thus to prevent a fall in the glomerular ultrafiltration pressure. Conversely, the response to a rise in perfusion pressure is an increase in both afferent and efferent arteriolar resistances. By use of drugs that inhibit the formation of prostaglandins, it has been shown that maintenance of GFR during increased perfusion pressure requires an action of prostaglandins on the efferent arteriole to counter the rise in its resistance and thus to prevent a rise in the glomerular ultrafiltration pressure. It remains to be shown to what extent opposing actions of AII and prostaglandins on efferent arteriolar resistances control GFR in other circumstances.

Blood leaving the glomerular capillaries of the "cortical" nephrons enters the efferent arteriole to perfuse the plexus of capillaries surrounding the proximal and distal tubules. The ultrafiltration at the glomerulus of a fluid virtually free of protein and cells concentrates the plasma proteins and red blood cells in the efferent arteriole by a degree determined by the filtration fraction. The viscosity of whole blood is strongly dependent upon the haematocrit (especially at haematocrit levels above 50 per cent) and the viscosity of plasma on the plasma protein concentration. Consequently, the viscosity of the post-glomerular blood will be very high. There is a high level of efferent arteriolar resistance, despite the rather short length and poor muscular coat of the efferent arteriole and this high efferent resistance is essential to maintain the glomerular capillary hydrostatic pressure which permits the high level of glomerular filtration. A high level of efferent arteriolar resistance can be achieved without much active vasoconstriction because of the low intra-vascular pressure and the high blood viscosity at this site.

It is impossible to make direct measurements of vascular pressures and flow rates at juxtamedullary nephrons and thus precise information about afferent and efferent arteriolar resistances are lacking. The efferent arterioles of the juxtamedullary nephrons, however, do have the histological characteristic of resistance vessels. Each supplies about 30 vasa recta which are surrounded by perivascular cells which could be contractile. When rats are dehydrated or treated with vasopressin (ADH) there is an increase in the filtration rate of the juxtamedullary glomeruli and a fall in the pressure in the vasa recta capillaries, with a reduction in blood flow through them, suggesting an increase in efferent arteriolar resistance. These changes might be due to efferent arteriolar constriction or alternatively to the rise in blood viscosity which results from an increase in osmolality of the vasa recta blood during urinary concentration. As the blood flows deep into the medulla, its protein concentration and osmolality increases further and the red cells become crenated. This causes them to lose their normal deformability and raises the viscosity of the blood, perhaps substantially. Passive, rheological changes in the blood may be as important as active vasomotor changes in vascular diameter in determining resistance to blood flow through the vasa recta. Little is known about the overall control of blood flow to this region but blood flow rates to the papilla are very low in anti-diuresis and this is required to maintain the steep solute gradient from the cortex to the papillary tip. All vasa recta terminate in a medullary capillary plexus from which the vessels unite to form the ascending vasa recta which drain into the arcuate veins in the cortex.

The proximal tubule

Each proximal tubule is divided into three structurally and functionally distinct segments. The first two are contained in a long convoluted portion which takes a tortuous course through to the cortex and passes to a shorter "straight" portion (pars recta) which descends through the cortex to end abruptly as the descending limb of the loop of Henle. All segments have a brush border system, an extensive basal labyrinth and numerous mitochondria. The cells are linked to each other at their luminal poles by shallow junctional complexes named tight junctions. Their lateral walls are separated by well-developed intercellular fluid compartments. At their antiluminal borders a basal membrane separates the cells from the interspaces surrounding the peritubular capillaries. The transport properties of the proximal tubule have been studied *in vivo* using samples of tubular fluid withdrawn by micropipettes. Reabsorption is essentially isotonic i.e. the osmolality of fluid in all parts of the proximal tubule closely approximates to that of plasma. Fluid withdrawn from Bowman's space contains ions in the same concentration as a pure ultrafiltrate of plasma. The sodium and potassium concentration of tubular fluid are maintained down the proximal tubule. In the *first segment*, however, there is a rapid decline in tubular fluid bicarbonate concentration with a proportionate rise in the concentration of chloride; the fluid becomes acid. This first 20–25 per cent of the proximal tubule also avidly reabsorbs glucose, phosphate, amino acids and organic acids. There is a small (circa -2 to -5 mV) lumen-negative transepithelial potential difference which opposes sodium reabsorption. In the *second segment*, these concentration gradients are maintained. The *third segment* is characterized by an avid tubular secretory mechanism for para-amino hippuric acid (PAH) and other organic acids. This last segment of the proximal tubule reabsorbs fluid at about one half the rate of convoluted tubules. Active glucose transport is very slight. Here there is a small (circa $+2$ to $+4$ mV) lumen-positive transepithelial potential difference creating a favourable electrochemical gradient for sodium reabsorption. Fluid and electrolyte reabsorption is broadly similar at juxtamedullary proximal tubules but there is less information about their transport characteristics. All proximal tubular segments are capable of active sodium transport as shown by continuing sodium and fluid reabsorption despite the imposition of an unfavourable, uphill, electro-chemical gradient by perfusion of the tubular lumen with low sodium solutions. However, the mechanisms subserving sodium reabsorption differ down the proximal tubule. In the first segment, active reabsorption of sodium is linked to active reabsorption of bicarbonate, glucose and other organic compounds. In the third segment, a fraction of the reabsorption can also be linked to the lumen-positive transepithelial PD favouring passive sodium reabsorption.

Overall, the proximal tubules reabsorb 50–70 per cent of the filtered sodium and fluid, almost all the filtered glucose, amino acids, organic acids, and the small amount of protein which is present, as well as much of the potassium, calcium, phosphate and urea. The quantity of bicarbonate reabsorbed depends critically on the acid-base status (see page 224). It also secretes hydrogen ions, organic acids and bases (including drugs) and ammonia. Thus, by the end of the proximal tubule, the bulk of the filtrate has normally been reabsorbed. Although the tubular fluid remains isotonic with an unchanged sodium concentration, many important modifications to the tubular fluid have already been made that can have a decisive impact on the volume and composition of the final urine.

The cell membrane potential of the proximal convoluted tubule is normally about -60 mV (inside negative) and the intracellular sodium ion activity is low (about 10–20 mmol/l). Consequently, there is a steep electrical and chemical gradient for the passive transport of filtered sodium into the proximal tubular cells (the *first transport step*). Nevertheless, kinetic analysis indicates that this cellular entry step is normally near saturation. A rise in plasma sodium concentration leads to a rise in proximal tubular fluid sodium concentration and reduces the fraction of filtered sodium reabsorbed by the proximal tubule perhaps because this saturable step limits sodium entry from the tubular fluid.

Sodium transport requires an active extrusion of sodium across the basilateral cell membrane, utilizing a membrane-bound Na-K ATPase (the *second transport step*). Inhibition of this Na^+-pump by cardiac glycoside drugs reduces the negative cell membrane potential and raises the intracellular sodium ion activity. There is also reduced sodium transport (although this is not prominent when glycosides are used at therapeutic levels). Acute metabolic (but not respiratory) acidosis also reduces the luminal cell membrane potential, and increases the intracellular sodium ion activity probably by a similar means. This action on active sodium transport could underlie the pronounced reduction in proximal reabsorption of fluid and sodium that occurs in metabolic acidosis.

Sodium (plus attendant ions) transported into the basilateral spaces provides the osmotic force for the reabsorption of fluid. The passive transport of the fluid and ions from the interstitium into the peritubular capillary blood is of paramount importance in the regulation of proximal reabsorption (the *third transport step*). The proximal tubular epithelium has a very high permeability for ions and fluid and a correspondingly low transepithelial electrical resistance, i.e. it is a 'leaky' epithelium. The lateral intercellular spaces and the 'tight' junctions separating them from the tubular lumen are, despite their name, the main leak pathway whereby sodium ions can back-diffuse into the tubular lumen under the influence of a small chemical gradient (aided by an electrical gradient at the first segment of the proximal tubule). Indeed, some experiments have indicated that, on average, to achieve the reabsorption of one sodium ion into the blood-stream, six must be transported into the interstitium since five back-leak into the lumen. This underscores the importance of factors determining capillary fluid uptake in setting the level of proximal fluid reabsorption. These factors are the reverse of those that determine filtration at the glomerulus (see Fig. 4.2). Firstly, the peritubular capillary plasma protein concentration (P_{cap}) which is set by the systemic plasma protein concentration (P_{art}) and the degree to which this is enhanced by ultrafiltration of fluid at the glomerulus (filtration fraction, FF):

$$P_{cap} = P_{art} \times \frac{I}{I - FF}$$

Secondly, the peritubular capillary plasma flow rate which sets the rate of delivery of the oncotically-active particles. Thirdly, the capillary hydrostatic pressure which is set by the balances between the systemic arterial and renal venous pressures and between the resistances upstream and downstream from the peritubular capillary. The oncotic and hydrostatic pressures of the interstitial fluid and the permeability of the capillary epithelium or proximal tubular cell basement membrane are other factors but their practical importance remains to be clearly demonstrated.

The loop of Henle

Only juxta-medullary nephrons give rise to loops of Henle which penetrate deep into the medulla. Both cortical and juxta-medullary nephrons, however, have a descending limb, a thin ascending limb and a thick ascending limb terminating in the macula densa segment at the junction with the distal tubule. The thin limbs have a very simple structure and are not important sites for active sodium reabsorption. The thick ascending limbs resemble the distal tubule with extensive basal infoldings and large mitochondria.

Fluid leaving the loop of Henle is substantially hypotonic and its sodium concentration is correspondingly low (30–60 mmol l^{-1}) even during antidiuresis. This dilution is achieved by reabsorption of salts but not water from the thick ascending limb of the loop of Henle (sometimes called the diluting segment). The transepithelial voltage of this segment is lumen-positive and the reabsorption of chloride is active since it proceeds against an electrical and chemical gradient whereas the reabsorption of sodium and other cations can proceed passively down the electrical gradient. Active transport by isolated, perfused segments of the thick ascending limb of the rabbit is dependent on chloride (but not sodium) being present in the tubular fluid. This chloride transport process generates the lumen-positive transepithelial voltage and is inhibited specifically by frusemide, ethacrynic acid and bumetanide (but not thiazides) acting from within the tubular lumen. In the absence of a positive luminal potential (e.g. absence of luminal chloride or presence of frusemide) reabsorption of sodium and other cations ceases.

As the rate of fluid entering the loop of Henle from the proximal tubule is increased, sodium reabsorption initially increases more than water and the sodium concentration of early distal tubular fluid falls. The capacity to increase sodium reabsorption is, however, limited and at higher rates of fluid flow the sodium concentration of early distal fluid begins to rise. In the physiological range of flow rates, sodium-chloride, but not fluid, reabsorption changes directly with load. The loops of Henle of cortical nephrons reabsorb 25–40 per cent of the filtered sodium but less than 15 per cent of filtered water. Thus, besides being the crucial segment for the development of the osmotic forces required for urinary concentration and dilution, the loop of Henle also makes a substantial contribution to the reabsorption of filtered sodium and chloride.

Distal convoluted tubules

The distal convoluted tubule is located between the macula densa (see below) and the point at which two distal convoluted tubules join to form a cortical collecting tubule. Some 80 per cent of the segment is accessible for micropuncture in the rat and consequently transport processes have been well characterized. Morphologically, the early distal tubule resembles the thick ascending limb of the loop of Henle, while the late segment resembles the cortical collecting tubule. More detailed examination has disclosed four different epithelial cell types but their separate functions are, as yet, unknown.

This segment normally reabsorbs 5–10 per cent of the filtered sodium by an active transport process. Since, at least initially, the tubular fluid sodium concentration is below plasma and a steep lumen-negative transepithelial PD develops down the distal tubule (circa -30 to -60 mV by the late segment) sodium transport must proceed against a steep electrochemical gradient. The net reabsorptive capacity for sodium shows no evidence of saturation over a wide range of loads. At low rates of sodium delivery most of the sodium is reabsorbed in the early segments thereby lowering the sodium concentration of the fluid further down the tubules. Later distal tubular segments therefore must reabsorb sodium against steep chemical gradients and this limits net transport. At high rates of sodium delivery, however, the fraction of the sodium load reabsorbed by the early distal tubule is diminished and the tubular fluid sodium concentration increases as fluid passes down the tubule. This facilitates reabsorption from the later segments and provides a simple means of relating distal tubular sodium reabsorption to load. Both the proximal tubule, the loop of Henle and the distal tubule have this capacity to increase the absolute rate of sodium reabsorption in response to an increase in sodium load and this dampens the potentially enormous effects that changes in GFR or proximal tubular reabsorption can have on sodium excretion (see 'glomerulo-tubular balance', page 172).

Sodium transport from the latter part of the distal tubule (and the collecting ducts) is dependent upon mineralocorticosteroids. Adrenalectomized animals cannot lower the intraluminal sodium concentration or reabsorb sodium normally. This is due to a reduced rate of active sodium transport which can be reversed specifically by mineralocorticosteroids after a latent period of about one hour. Studies on isolated frog skin and subcellular organelles have shed light on the mechanisms of aldosterone action. It involves initially an interaction with specific, high-affinity mineralocorticosteroid receptors and passage of aldosterone into the cell cytoplasm where it binds to a specific protein. This complex is translocated to the nucleus where it binds to chromatin and activates transcription by eliciting messenger RNA synthesis. The postulated aldosterone-induced proteins have yet to be identified, but aldosterone action depends upon formation of new RNA and protein. Evidence is available to suggest that these induced proteins stimulate active transport either by increasing the luminal sodium entry and hence the rate of presentation of sodium ions to the pump in the basilateral membrane ('*permease theory*'), or by increasing the cellular energy sources available for the pump ('*energy theory*'), or by a direct stimulant action on the pump ('*pump theory*'). Regardless of which theories prove correct, each would result in enhanced Na-K exchange across the basilateral cell membrane, leading to aldosterone-stimulated sodium reabsorption. Since the same action would increase potassium accumulation within the cell, it would lead to enhanced K secretion from the cell into the lumen. Finally, since the transepithelial lumen-negative PD in the late distal tubule depends upon active sodium transport, aldosterone would enhance this luminal negativity and facilitate H^+ and K^+ secretion from the cell into the lumen.

The distal tubule can reabsorb bicarbonate, calcium and magnesium and secrete potassium and hydrogen ions and ammonia. It is relatively impermeable to urea and its permeability to water depends critically upon the presence of antidiuretic hormone.

The collecting ducts

The collecting ducts begin as the junction of two cortical connecting tubules. They receive further cortical and medullary connecting tubules and empty into the renal pelvis at the tip of the papilla. Although the collecting ducts normally reabsorb only 2–3 per cent of the glomerular filtrate, they are the site where the final adjustments of urine volume and composition occur, and thus are important in regulating the final urinary composition. Sodium, potassium and hydrogen ion transport are broadly similar to the late distal tubule and are regulated by mineralocorticosteroids. This is the site of the highest transepithelial concentration differences for ions and the epithelium is characterized as "tight" with a high electrical resistance.

RENAL HORMONES

The juxta-glomerular apparatus, renin and angiotensin

The juxta-glomerular apparatus (JGA, see Fig 4.3) consists of the afferent and efferent arterioles adjacent to the glomerulus, a specialized tubular segment at the junction of the loop of Henle and the distal tubule called the macula densa and an intervening region of *lacis* cells (also referred to as the *Polkissen* or the juxta-glomerular cell mass) with which the glomerular mesangium is continuous. As the tubule ascends through the cortex towards its own glomerulus it comes first into prolonged but loose contact with its efferent arteriole. Beyond this, it is related to a conical mass of lacis cells before being briefly but closely related to the afferent arteriole. The JGA has a dense adrenergic neural innervation and a prominent lymphatic drainage. More than 90 per cent of renin is present in myoepithelial cells developed from smooth muscle in the wall of the

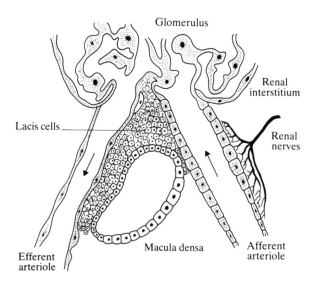

Figure 4.3
   Diagram of the juxtaglomerular apparatus modified from J.O. Davis *et al*, *Circulation Research*, Volume 28 (1974) by permission of the American Heart Association, Inc.

afferent arteriole and containing dense-staining granules in which renin is present in active form. Occasional granular cells are also present in the efferent arteriole and small quantities of renin are present in the lacis and mesangial cells.

Renin is an enzyme that acts on a substrate (angiotensinogen; an α-2 globulin formed by the liver) to cleave a decapeptide angiotensin I which has little or no activity until metabolized by a converting enzyme to the octapeptide angiotensin II (AII). Most renin released from the JGA traverses the interstitium to enter the circulation at the peritubular capillaries. The renin concentration of renal lymph, which drains from this interstitial compartment, is many-fold greater than that of plasma. Indeed, the JGA and renal lymph contain all the components required for AII formation, suggesting an intra-renal as well as a systemic site of action for this hormone. Despite the high lymph renin concentration, the much greater rate of flow of plasma ensures that most renin enters the circulation via the renal vein.

The control of renin release is incompletely understood. A fall in renal perfusion pressure regularly elicits a release of renin even from totally isolated, blood-perfused kidneys. A graded reduction in pressure elicits a graded increase in release. Blood flow initially is maintained, but at a certain perfusion pressure the renal resistance vessels become maximally dilated and the capacity for 'autoregulation' is exhausted. Thereafter flow falls in parallel with pressure. It is interesting that this point of maximum vasodilatation corresponds to the point of maximum renin release. Many other stimuli that elicit an acute release of renin also produce renal vasodilatation. These include beta-adrenoceptor agonist drugs (e.g. isoprenaline, noradrenaline in low dosage) vaso-dilatory drugs (e.g. hydralazine, sodium nitroprusside, dopamine), hormones (e.g. bradykinin, prostaglandin E, prostacyclin) and 'loop' diuretics (e.g. frusemide, bumetanide). Conversely, many stimuli that reduce renin release produce renal vaso-constriction. These include a rise in renal perfusion pressure, alpha-adrenoceptor agonist drugs (e.g. methoxamine, noradrenaline in high dosage) and hormones (e.g. angiotensin II, vasopressin in high dosage). This parallelism between hormone release and smooth muscle tone, together with the common origin of myoepithelial and smooth muscle cells, has prompted the suggestion that similar, calcium-dependent processes may subserve both inhibition of renin secretion and vasoconstriction. In support of this theory are the observations that calcium ionophores which increase the flux of extracellular calcium into cells inhibit renin release while infusions of solutions of magnesium salts which compete with calcium for membrane transport stimulate renin release. Angiotensin II can act in a negative feedback loop to terminate renin release.

Stimulation of renal nerves either directly or reflexly leads to a substantial increase in renin release. This occurs despite renal vasoconstriction and so must involve an additional mechanism. The increase in release is blocked by beta-adrenoceptor antago-nist drugs (e.g. propranolol) whereas beta-adrenoceptor agonist drugs (e.g. isoprenaline) are potent stimuli to renin release. This suggests that renal nerve stimu-lation releases renin through activation of beta-receptors (probably beta-1 class). The renal nerves are important in stimulating renin release during many other circumstances in which sympathetic nervous activity is augmented (e.g. haemorrhage, shock, changing from lying to standing, exercise, fear or the administration of vasodilating drugs). Conversely, patients receiving sympatholytic or beta-adrenoceptor antagonist drugs or those with degeneration of the sympathetic nervous system, have very low levels of renin which respond subnormally to the stimuli of standing or sodium deprivation. Animal experiments show that denervation leads to a loss of renin from the kidneys. Thus nerves may be implicated not only in renin release but also in maintaining renin stores.

The mechanisms subserving the more long-term adjustments in renin release are less completely understood. Dietary sodium chloride restriction or the administration of diuretic drugs regularly increases renin release, whereas sodium loading or expansion of the blood volume with plasma reduces renin release. The inhibition of renin release by sodium chloride is shared by other chloride-containing but not other sodium-containing salts and is blocked by frusemide suggesting that release may be inhibited by chloride reabsorption at the macula densa segment. Moreover, this segment is closely apposed to the renin-containing cells and forms part of the ascending limb of the loop of Henle which is characterized by active chloride reabsorption. However, the mechanism relating renin release from the afferent arteriole to macula densa reabsorption has yet to be elucidated.

The rôle of the renin-angiotensin system has been clarified with the advent of antagonist drugs. Saralasin is an angiotensin analogue (1-sarcosine, 8-alanine angiotensin II) which is a specific competitive antagonist of AII. However, it has an appreciable degree of agonist or angiotensin-like activity and stimulates the sympathetic nervous system. Captopril is an orally-active converting enzyme inhibitor which blocks the formation of AII from AI. However, it also inhibits kininase II which is implicated in bradykinin degradation and potentiates the prostaglandin system as well. Thus results obtained with either class of inhibitors must be interpreted with caution. Nevertheless, their use has suggested that AII is important in maintaining blood pressure during haemorrhage and shock, salt deprivation, diuretic treatment, cardiac failure, cirrhosis of the liver, and adrenal insufficiency. It also maintains high levels of blood pressure in certain patients with hypertension, notably those with high renin levels and renal damage or defective renal perfusion (see page 77). It is an important stimulus to aldosterone secretion and this may be its prime physiological rôle. It stimulates drinking in experimental animals. It is thus involved intimately in blood pressure and fluid volume homeostasis. Most of its effects on tubular function may be explained by its actions on the blood pressure, the renal vessel (see page 160) and on aldosterone secretion (see above) although a direct effect on sodium reabsorption is suggested by some studies. Probably more important is its effect in modulating the filtration fraction through efferent arteriolar vasoconstriction and hence setting the level of plasma protein concentration in the peritubular capillary plasma (see physical factors, page 176). Angiotensin II infusions at sub-pressor doses lead to renal sodium retention by augmenting sodium reabsorption. Angiotensin antagonists increase sodium excretion when used in volume-depleted or salt-deprived animals or some patients with congestive cardiac failure indicating a sodium-conserving rôle for AII in these circumstances.

Renal prostaglandins

Prostaglandins formed within the kidney can modulate renal vascular resistance, GFR, tubular sodium and water reabsorption and renin release in certain physiological and more particularly pathological circumstances. In contrast to earlier work, it is now clear that the cortex, as well as the medulla, is an important site of prostaglandin synthesis. Cortical glomeruli and resistance vessels can synthesize prostacyclin (PGI_2), PGE_2, $PGF_{2\,alpha}$ and thromboxane (TXA_2). However, the major site of formation of PGE_2 and $PGF_{2\,alpha}$ is the medulla where they are formed in the collecting ducts and the medullary interstitial cells. Prostaglandins are not stored in appreciable quantities and thus release represents new synthesis. Phospholipase A acts on phospholipid to produce arachidonic acid which is converted by the cyclo-oxygenase enzyme system to PGG_2 and PGH_2 which are common intermediates for the formation of all the major prostaglandins (Fig. 4.4). $PGF_{2\,alpha}$ can arise by the action of 9-keto-reductase, a cytoplasmic enzyme which acts on PGE_2 or it may arise *de novo* from PGH_2. Enzymes capable of metabolizing PGE_2 and $F_{2\,alpha}$ are present in the cortex and medulla. The biological half-lives of prostacyclin and thromboxane are extremely short since they are unstable in aqueous media. The synthesis of all the prostaglandins can be inhibited non-selectively by non-steroidal, anti-inflammatory drugs such as aspirin and indomethacin while thromboxane synthesis can be inhibited specifically by imidazole and related compounds.

The importance of prostaglandins in the regulation of sodium excretion is controversial. This derives from several causes: in part it is because prostaglandins can have opposing functions e.g. prostacyclin vasodilates whereas thromboxane vasoconstricts; in part it is because they have different distributions within the kidney e.g. those in the cortex are closely related to the glomerulus and vascular pole while those in the medulla are related to the tubules and capillaries; in part it is because, although the production of specific prostaglandins are elicited by specific stimuli, studies of their actions out of their physiological context cannot reveal their true rôle; and in part it is because much of the present knowledge derives from the use of non-selective inhibitors of prostaglandin synthesis. In most, but not all, studies, indomethacin reduces renal sodium excretion and antagonizes the natriuretic action of 'loop' diuretics. This may relate in part to reduced GFR and enhanced filtration fraction and proximal tubular reabsorption. These falls in flow and excretion are particularly prominent in patients with nephrotic syndrome in whom indomethacin reduces the GFR by some 30 per cent. Some but not all experiments have shown also a direct inhibitory effect of prostaglandin E_2 on sodium transport by isolated medullary collecting tubules. As yet, no consistent pattern of prostaglandin release in response to altered sodium intake has emerged

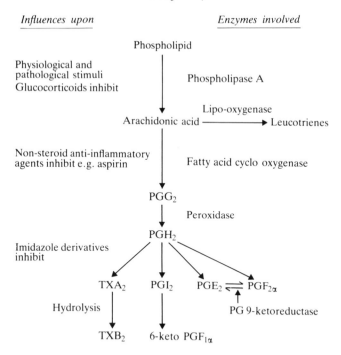

Influences upon   *Enzymes involved*

Phospholipid

Physiological and
pathological stimuli   Phospholipase A
Glucocorticoids inhibit

                                               Lipo-oxygenase
Arachidonic acid ⟶ Leucotrienes

Non-steroid anti-inflammatory
agents inhibit e.g. aspirin   Fatty acid cyclo oxygenase

PGG₂

                                               Peroxidase

PGH₂

Imidazole derivatives
inhibit

TXA₂   PGI₂   PGE₂ ⇌ PGF₂ₐ

Hydrolysis   PG 9-ketoreductase

TXB₂   6-keto PGF₁ₐ

Figure 4.4. *The major pathways for prostaglandin synthesis and metabolism.*
The short-lived endoperoxidases, PGG₂ & PGH₂, are converted to (i) thromboxane (TXA₂) (ii) prostacyclin (PGI₂) and (iii) either prostaglandin E₂(PGE₂) or prostaglandin F₂ₐ. These are interconvertible via prostaglandin-9-ketoreductase. The enzymes involved are not yet fully characterized.

from numerous studies and the part they play in regulating sodium excretion remains to be defined.

The effects of prostaglandins on water excretion are more consistent. PGE₂ can antagonize the increase in water permeability produced by vasopressin (ADH) in the isolated, perfused rabbit collecting tubules (see page 180). Inhibition of prostaglandin synthetase increases urine osmolality and the reabsorption of water and increases the osmolality at the papillary tip. This suggests that prostaglandins, especially E₂, may modulate the action of ADH on the collecting tubules and perhaps the vasa recta capillaries. Moreover, ADH stimulates the formation of PGE₂ and F₂ ₐₗₚₕₐ in the renal medulla and the prostaglandins may therefore act in a negative feedback loop terminating the action of ADH on water reabsorption.

Infusion of AII or noradrenaline or stimulation of the renal nerves leads to the release of prostaglandin from the kidney (PGE₂, F₂ ₐₗₚₕₐ and the prostacyclin metabolites). The vasoconstriction produced is more severe and prolonged after inhibition of prostaglandin synthesis. This suggests that prostaglandins ameliorate the effects of renal vasoconstriction and may underlie the process of tachyphylaxis. PGE₂, and especially· prostacyclin, are potent stimuli to renin release. Indomethacin reduces the release of renin in response to haemorrhage, reduced renal perfusion pressure, sodium restriction and frusemide administration, but not to stimulation of renal beta receptors, suggesting that prostaglandins modulate many, but not all mechanisms for renin release.

Overall, it appears that prostaglandins can exert important effects on renal vascular resistance and tubular function in certain pathological circumstances, such as shock and inflammation, although their rôle in normal physiology is yet far from clear.

Kallikrein-Kinin system

Kallikreins are enzymes that liberàte kinin peptides from kininogen substrates. The kallikrein present in urine is distinct from that found in plasma. It originates from the distal tubules and the collecting ducts. Kinins are inactivated by two kininases, one of which is the same as that which converts angiotensin I to angiotensin II. Bradykinin is a potent renal vasodilator and leads to increases in urine flow and sodium excretion. These diuretic effects may be secondary to its action on renal blood vessels.

Kallikrein excretion is increased by a low sodium diet, probably secondary to an effect of aldosterone. Some studies show a positive correlation between urine flow rate and kallikrein excretion but whether kallikrein contributes to the diuresis or whether the increase in flow simply washes out kallikrein from a tubular compartment is not settled. Recently it has been suggested that renal kallikrein is involved in the conversion of inactive to active renin. But it is not clear where kallikrein, made in tubular cells and released into tubular fluid, meets renin, made in the afferent arteriole and released into the interstitial fluid and blood stream. Further work is required to define what place this system has in normal and abnormal renal physiology.

TUBULO-GLOMERULAR FEEDBACK

Micropuncture experiments have shown that an increase in the rate of flow of fluid through the loop of Henle or an increase in the chloride concentration of fluid injected directly into the macula densa region leads to a reduction in the GFR or blood flow to that nephron. This response can be blocked by frusemide which inhibits active chloride reabsorption. These experiments demonstrate the presence of a mechanism termed "tubulo-glomerular feedback" which relates the GFR and blood flow of a nephron to its rate of chloride reabsorption at its macula densa (or neighbouring) tubular segment. The magnitude of the change in GFR is augmented markedly by prior dietary sodium chloride restriction and is diminished by prior sodium chloride loading. Concurrent saralasin infusion reduces, but does not abolish, the feedback response which appears to implicate AII as a mediator. This presents a paradox, however, since chloride reabsorption inhibits renin release from the kidney (see page 167) yet initiates the 'tubulo-glomerular feedback' process. This problem awaits further clarification. Inhibitors of prostaglandin synthesis also reduce 'tubulo-glomerular feedback' which implicates prostaglandins in addition. The response involves an increase in resistance of the afferent and probably also the efferent arterioles. The rate of sodium chloride reabsorption by the ascending limb of the loop of Henle and the distal tubule is normally closely dependent upon the load presented (see page 164). Thus, the process might restrict glomerular filtration when an excessive load of sodium chloride is presented to the distal nephron. The reabsorption of NaCl and fluid by the proximal tubule accounts for reabsorption of a large but variable fraction of the filtrate and if uncontrolled could produce huge changes in the load presented to the distal nephron. A 'tubulo-glomerular feedback' process would stabilize the distal sodium chloride and fluid load in the face of changes in GFR or proximal reabsorption. The transport of ions (Na, K, NH_4) by the distal tubule and collecting ducts are strongly dependent on the flow of tubular fluid and this feedback process might be required to isolate the subtle distal regulatory processes from changes in function upstream. It is possible that it also contributes to renal "autoregulation". Thus a rise in perfusion pressure would tend to increase blood flow and GFR. A resultant increase in the flow of fluid through the loop of Henle might activate the "tubulo-glomerular feedback process" and restrict blood flow and GFR. However, isolated renal resistance vessels constrict in response to a rise in perfusion pressure and this "myogenic" response provides an alternative explanation for renal blood flow autoregulation.

THE RENAL REGULATION OF SODIUM EXCRETION

An alteration in sodium intake leads to a rapid and precise adjustment in sodium excretion which serves to maintain the extracellular fluid volume remarkably independent of the level of salt intake. Undoubtedly this adjustment is the outcome of many mechanisms working in concert, the relative importance of each depending on the physiological context. For simplicity they will be described individually:

Glomerulo-tubular balance

Of the 180 litres or so of extracellular fluid filtered daily, normally some 178·5 litres (> 99%) are reabsorbed by the renal tubules. Even a very small increase in GFR, if unaccompanied by an increase in reabsorption, would lead rapidly to the excretion of most of the extracellular fluid through the kidneys. The term "glomerulotubular balance" describes the coupling between reabsorption and filtration. It can be seen at the proximal tubule where alterations in fluid filtration normally are matched by proportionate changes in fluid reabsorption so that the same fraction of the filtrate is passed on downstream. Reabsorption is linked to the rate of fluid filtration rather than to the NaCl load. Thus, an increase in sodium chloride load produced by raising the plasma sodium-chloride concentration fails to produce a parallel increase in sodium-chloride reabsorption.

One means of matching reabsorption to filtration is through changes produced in the Starling forces which govern the uptake of reabsorbed fluid into the peritubular capillaries. Where glomerular filtration changes as a consequence of ultrafiltration pressure or glomerular permeability there will be proportionate changes in the filtration fraction and thus in the peritubular capillary plasma protein concentration (see Fig. 4.2). Thus, in this context, a change in GFR will automatically adjust the Starling forces for fluid uptake into the peritubular capillaries which should elicit a balanced change in reabsorption. Where glomerular filtration changes as a consequence of plasma flow rate, the net peritubular oncotic force (the product of the peritubular plasma protein concentration and flow rate) will vary directly with it. Again, the changes in filtration will automatically dictate an appropriate adjustment in the Starling forces to ensure a balanced change in reabsorption. However, increasing fluid flow through the proximal tubule via a micropipette provokes an increase in fluid reabsorption in the absence of changes in the peritubular Starling forces. Probably more than one mechanism is available to the proximal tubule to serve this important process. It is important to realize that where the stimulus used to change GFR also interferes with tubular reabsorption (for example a rapid intravenous saline infusion) then perfect glomerulo-tubular balance no longer applies. As described earlier, increasing tubular fluid flow rates into the loop of Henle, the distal tubule and the collecting ducts increases sodium-chloride reabsorption from each of these segments. The glomerulotubular balance appears to be a general property of the nephron although different segments can utilize different mechanisms.

The kidney has numerous levels of control which serve to stabilize the fluid presented to the distal tubules and the rate of sodium chloride excretion. This is exemplified by its responses to a fall in perfusion pressure. Renin released from the kidneys leads to AII formation which, by its actions on the systemic resistance vessels, helps to restore perfusion pressure and, by its actions on the renal efferent arterioles, helps to restore the glomerular ultrafiltration pressure and filtration rate. Renal autoregulation further minimizes the effects of the fall in perfusion on the GFR. Any fall in the NaCl or fluid load presented to the early distal tubule will restrict 'tubulo-glomerular feedback' and help to restore filtration. Finally glomerulotubular balance, as it affects the different nephron segments, further dampens the effects of any change in fluid load on the excretion of sodium chloride in the final urine. These regulatory processes, although presented as distinct mechanisms, are undoubtedly well integrated. Thus AII formed in response to a fall in renal perfusion pressure may mediate not only the rises in systemic and efferent arteriolar resistance but also may be implicated in "tubular-glomerular feedback" and the coupling of proximal tubular fluid reabsorption to the glomerular filtration rate.

Glomerular filtration rate

Glomerulo-tubular balance renders sodium excretion relatively independent of GFR when the changes are modest or gradual. Indeed, patients with chronic renal failure and very low levels of GFR as often suffer from salt and water depletion as retention.

Extracellular fluid volume expansion with saline can increase sodium excretion even when the GFR is reduced by inflation of a balloon in the aorta above the kidneys. This highlights the predominant importance of tubular reabsorption in the overall regulation of sodium balance.

Aldosterone

Aldosterone can exert dominant effects on tubular sodium reabsorption. Thus patients with Addison's disease or adrenalectomized animals cannot retain sodium and develop serious salt depletion while on a normal salt intake. Administration of aldosterone (or any mineralocorticosteroid) leads to sodium retention and expansion of the extracellular fluid volume. Aldosterone secretion increases regularly during salt deprivation or haemorrhage and decreases during salt loading or blood volume expansion. Thus salt balance could be regulated by the actions of aldosterone on tubular sodium reabsorption. Yet several lines of evidence suggest that the importance of aldosterone in the normal regulation of sodium excretion should not be over-emphasized. Firstly, when the salt content of the diet of normal subjects is abruptly reduced, their rates of sodium excretion fall within 4–8 hours which usually proceeds any detectable changes in plasma or urinary aldosterone levels. Secondly, renal sodium excretion can adjust appropriately to the level of salt intake in adrenalectomized animals maintained on fixed doses of mineralocorticosteroids or in normals given supra-maximal doses. Thirdly, the administration of spironolactone (an aldosterone-antagonist drug) to normal subjects slows the rate at which they reduce their sodium excretion when faced with restricted salt intake but does not affect the final level of excretion achieved. Fourthly, while administration of aldosterone to normal subjects leads initially to salt and water retention, the effect wanes and a new equilibrium is attained when salt and water balance is restored. This process, termed 'mineralocorticosteroid escape' demonstrates that other factors can override the effects of even maximal doses of mineralocorticosteroids. Fifthly, micropuncture experiments have shown that both acute and chronic changes in salt intake are reflected in changes in proximal tubule reabsorption, yet the important actions of aldosterone are confined to the late distal tubule and collecting ducts.

Natriuretic hormones

Failure of aldosterone to account for all the changes in sodium reabsorption that occur in response to changes in ECV prompted a vigorous search for an alternative hormone. Several studies have shown the appearance in plasma or urine of a substance released during periods of ECV expansion and capable of increasing sodium excretion when injected into animals or decreasing proximal tubular sodium reabsorption when injected into the proximal tubule. When a kidney is grafted onto a recipient dog, the rate of sodium excretion of the recipient's kidneys can be set by the previous level of salt intake of the animal donating the kidney, indicating that the kidney itself can release a humoral agent which affects sodium reabsorption. Other experiments however, indicate that the natriuretic material originates from the brain or from the heart. Despite these interesting observations, there is serious doubt about the physiological importance of a specific natriuretic hormone. Thus when saline is added to the blood perfusing an entirely isolated dog's kidney, its rate of sodium excretion increases quantitatively in direct proportion to the saline added. Clearly, this mechanism is independent of the release of any extra-renal hormone. Classical experiments have shown that sodium excretion increases with extracellular volume expansion even when GFR is artificially reduced by inflation of a balloon in the aorta above the kidneys. However, if the balloon is inflated before expanding the ECV the subsequent natriuresis is slight or inconsistent. These results testify to the predominant importance of intra-renal factors in the regulation of sodium excretion during haemodilution. In the course of time numerous hormones will doubtless be shown capable of influencing sodium excretion, but their physiological importance will require critical evaluation.

In some experiments, a bigger increase in sodium excretion has followed the delivery of a sodium load into the gut or the portal circulation than intravenously, prompting the suggestion that the gut or liver contains a sodium-sensing mechanism which initiates rapid changes in sodium excretion. Other experiments, however, have not confirmed the importance of a gut sodium-sensing mechanism and its present status is controversial.

The nervous system

Normal subjects given sympatholytic drugs such as guanethidine or patients suffering from degeneration of the sympathetic nervous system are slow to adjust their rates of renal sodium excretion to changes in input. This is not explicable by subnormal rates of aldosterone secretion since it cannot be reversed fully by administration of mineralocorticosteroid drugs. It is independent of the GFR. It suggests that the sympathetic nervous system can exert an important and direct influence on sodium reabsorption in man.

Micropuncture experiments have shown that enhanced renal sympathetic nervous activity leads to enhanced reabsorption from the proximal and often distal tubules. Tubular cells possess a direct sympathetic neural innervation. Isolated tubular segments contain receptors for catecholamines that regulate the cellular levels of adenylate cyclase. These experiments suggest a direct effect of catecholamines on tubular function. In addition, renal nerve activity has important effects on renal vessels and stimulates renin and prostaglandin release which will also modulate sodium excretion.

The activity in efferent renal nerves is regulated by the pressures in the high and low pressure cardiovascular circuits. Renal nerves also contain many afferent fibres whose activity is increased following saline infusion and decreased following haemorrhage. These afferents respond as if connected to stretch receptors in the post-glomerular vessels or the renal interstitium. A reciprocal relationship between afferent and efferent nerve traffic has been noted which suggests that in the process of volume regulation the kidney could be both a sensor and an effector organ.

Corticomedullary distribution of blood flow

Haemorrhage produces a sharp reduction in blood flow to the outer cortex which is patchy in distribution, while flow through the inner cortex is relatively well preserved. Earlier studies suggested that fluid reabsorption by the deeper cortical nephrons was more complete than the outer cortical nephrons. Consequently, sodium retention should occur if blood flow was directed away from the outer cortex and a greater proportion of the filtrate was distributed to the deeper cortical nephrons. Two lines of evidence, however, argue against this being of prime importance in the regulation of sodium excretion. Firstly, milder degrees of ECV depletion produced by restricting salt intake do not lead to a uniform pattern of changes in intra-renal blood flow distribution. Secondly, a micropuncture evaluation of sodium reabsorption by superficial and deep cortical nephrons during saline infusion showed that, in this context, deeper nephrons reabsorbed less sodium than superficial ones. A true evaluation of the importance of the intrarenal distribution of blood flow and filtration must await the development of better techniques for making these difficult measurements.

Physical factors

Expansion of the extracellular fluid volume (ECV) by salt ingestion or saline infusion dilutes the formed elements of the blood. A fall in haematocrit leads to a fall in filtration fraction (see page 160). This, together with a fall in plasma protein concentration, leads to a marked fall in the peritubular capillary plasma protein concentration. In addition, there is often renal vasodilatation which will increase the hydrostatic pressures in the peritubular capillaries. These changes in the peritubular capillary Starling forces are unfavourable to fluid uptake and thereby reduce proximal tubular fluid reabsorption. The converse sequence of changes follows contraction of the ECV. This "physical factor" hypothesis has been tested directly in micropuncture experiments which have shown that proximal tubular fluid reabsorption can vary directly with the peritubular protein concentration when this has been altered selectively. This model of proximal tubular fluid reabsorption offers an attractive explanation for the relationship between tubular reabsorption and the degree of "fullness" of the blood stream effected simply by changes in the concentration of the components of the blood itself. It offers a ready explanation for how a totally isolated, blood-perfused kidney can excrete quantitatively a saline load added to the blood perfusing it.

Plasma sodium concentration

Experimental studies have shown a direct relationship between plasma sodium concentration and renal sodium excretion which is independent of the ECV. Plasma sodium is normally regulated within very narrow limits due to its substantial effects both on the rate of sodium excretion and (since it is a major osmotic particle of the extracellular fluid) on the intake and excretion of water. However, in clinical situations where an abnormality in plasma sodium concentration is found, there are often drastic changes in ECV whose influence can override those of plasma sodium concentration itself and dominate renal sodium excretion.

Acid-base balance

Changes in acid-base balance can have a decisive effect on renal sodium excretion. This arises partly because of coupled transport between sodium and hydrogen ions, partly because sodium excretion can be dictated by bicarbonate reabsorption and partly because acidosis can have direct effects on tubular function.

An acute metabolic acidosis produced by the ingestion of ammonium chloride produces a diuresis and natriuresis. During more prolonged acidosis ammonium and potassium become the predominant urinary cations. Metabolic acidosis impairs the reabsorption of sodium and fluid by the proximal tubule, apparently by inhibiting the basilateral membrane sodium-potassium ATPase which provides the energy-expending active transport step (page 162). The increased fluid load to the distal tubule is not completely reabsorbed. ECV expansion by salt ingestion or saline infusion leads to a mild metabolic acidosis ('expansion acidosis') which may be one factor conditioning the kidney to excrete sodium and fluid.

Metabolic alkalosis produced by infusion of sodium bicarbonate leads to a large increase in the excretion of sodium and bicarbonate. However, when the alkalosis is due to loss of gastric acid by persistent vomiting, renal sodium and bicarbonate reabsorption are almost complete and glomerular filtration curtailed as a response to extracellular fluid depletion.

Osmotic diuresis

A duiretic is termed osmotic when its action is due to the filtration of significant quantities of poorly-reabsorbable solute. The term is somewhat inappropriate since all diuretics ultimately interfere with the tubular reabsorption of solute. The proximal tubule cannot maintain significant osmotic gradients across it because of its high fluid permeability. Consequently, a solute that is filtered at the glomerulus, yet is not reabsorbed in the proximal tubule will impair the reabsorption of fluid. Active sodium reabsorption from the proximal tubule normally creates the osmotic force for passive water reabsorption, but where this leads to the progressive concentration of a non-reabsorbable solute, the process will become self-limiting. Since active sodium transport continues but water transport is inhibited, a concentration gradient for sodium will develop between the lumen and the blood which will eventually inhibit also the reabsorption of sodium. Normally, more distal nephron segments can increase their rates of sodium and fluid reabsorption when greater quantities remain after incomplete proximal reabsorption. However, in a severe osmotic diuresis where the luminal sodium concentration is diminished, unfavourable gradients for sodium reabsorption will be established along the entire nephron that impairs reabsorption at proximal and distal sites.

Another characteristic of an osmotic diuresis is the excretion of an isotonic urine. Direct measurements of the solute concentration in the medulla and papilla have shown that osmotic diuretics abolish the normally steep gradient, thereby obliterating the force available for concentrating the final urine. This can be attributed both to a reduction in active transport out of the ascending limb of the loop of Henle because of a fall in the tubular fluid sodium chloride concentration and to an increase in the medullary blood flow rate which could wash out the gradient of solute. Thus, although an osmotic diuresis may often produce a considerable fluid depletion, the kidney is rendered insensitive to the action of the antidiuretic hormone (ADH) and this contributes to the diuresis. During a severe osmotic diuresis, the simultaneous inhibition of reabsorption from the proximal and distal tubule and of urinary concentration eliminates many of the defence mechanisms available to the kidneys for the preservation of the ECV. Consequently, a severe and prolonged osmotic diuresis can lead to extremes of ECV depletion. Where the diuresis is less severe or prolonged, the predominant effects are on the elimination of water rather than solutes.

The solute responsible for an osmotic diuresis may be a foreign substance administered specifically to promote a diuresis (e.g. mannitol) or for diagnostic purposes (e.g. radiographic contrast media), or a normal urinary constituent presented in abnormal quantities. Thus, when glucose is filtered at a concentration or load which exceeds the tubular capacity for its reabsorption it will act as an osmotic diuretic. In uncompensated diabetes mellitus not only glucose but also ketone bodies are filtered in excess of the tubules' ability to absorb them. The tubules' ability to reabsorb bicarbonate, sulphate and phosphate is also limited and where these anions are filtered in excess of their reabsorptive capacity they elicit a diuresis. In these cases, there is also necessarily excessive cation excretion, the balance between sodium and potassium in the urine being determined by factors which regulate sodium reabsorption and potassium secretion in the distal nephron. Urea can act as an osmotic diuretic, since distal segments of the nephron are not freely permeable to it (see below). It differs from other osmotic diuretics in that the proximal tubule is relatively permeable to it and this probably accounts for its rather weak effect as an osmotic diuretic. Nevertheless, in chronic renal failure the increased urea load per nephron diminishes fluid and electrolyte reabsorption and this may be important in permitting the damaged kidney to eliminate salts and fluid despite a very low filtration rate.

THE CONTROL OF WATER EXCRETION

The human kidney can vary the osmolality of the urine between wide limits (approximately 70–1200 mosmol/kg). Urinary dilution results firstly from the active reabsorption of solutes from the thick ascending limb of the loop of Henle. In the absence of ADH, the osmotic equilibrium between the dilute tubular fluid and the plasma is prevented by a very low water permeability of this and subsequent nephron segments. Urinary concentration results from the passive reabsorption of water from the distal tubule and collecting ducts into the hypertonic interstitium of the renal medulla. This hypertonicity provides the force for water reabsorption and depends ultimately on active solute reabsorption from the ascending limb of the loop of Henle which adds solute in excess of water to the neighbouring interstitium. This is magnified by the hairpin arrangement of the thin loops of Henle and the vasa recta. The details, particularly of the urinary mechanism, remain controversial, as the process is difficult to investigate experimentally. However, it is now possible to sample tubular fluid and blood from different regions of the medulla and to study the transport characteristics of segments of the loop of Henle using isolated, perfused tubule methods.

In the outer medulla, the vessels and tubules are specially arranged in 'vascular bundles'—whose core contains ascending and descending vasa recta which are surrounded by ascending and descending loops of Henle. Between the bundles lie the thick ascending limbs of the loops of Henle, some descending limbs and the collecting ducts. This arrangement ensures close contact between ascending and descending vasa recta and between ascending and descending loops of Henle which provides an important structural basis for the process of urinary concentration.

During anti-diuresis the osmolality, sodium chloride and urea concentrations in both the tubular and vascular compartments (and probably the interstitium) rises steeply from the corticomedullary junction to the tip of the papilla. Urea appears to play an important rôle since animals depleted of urea by a low protein diet are unable to concentrate their urine maximally until urea is administered. Tubular fluid in the descending limb of the loop of Henle of 'juxta-medullary' nephrons is concentrated as the tubule penetrates into the hypertonic milieu of the medulla by abstraction of water and addition of salt and urea. As it traverses the hairpin bend to enter the thin ascending limb it enters a tubular segment of low permeability to water and urea but high permeability to sodium. The tubule returns towards the cortex through increasingly dilute regions of the medulla and the hypertonicity of its contained fluid is progressively diminished by reabsorption of sodium into the interstitium. At the thick ascending limb, sodium reabsorption is coupled to active chloride transport and again the tubular water permeability is very low. The reabsorption of sodium chloride but not water from the ascending limb into the interstitium and its entry into the descending limb is the so called 'single effect'. This process is multiplied by the tubular counterflow pattern to build up a solute concentration from the cortex to the papilla. In this way the sodium chloride concentration and osmolality of the tubular fluid in the descending limb is always slightly higher than in the ascending limb and the effect is multiplied with distance down the loop. This process is called *counter-current multiplication*. The apposition of the descending and ascending limbs of the vasa recta provides a structural basis for a similar exchange system but based entirely on passive diffusion and this process is called *counter-current exchange*. The effect is to increase the osmolality of tubular fluid and blood flowing down into the medulla at the expense of that flowing out of it, thereby preserving the hypertonic milieu and effectively isolating the medulla from the osmolality of the rest of the body. The collecting ducts pass through increasingly hypertonic medullary regions. In the presence of ADH, their permeability to water and urea is high and water is reabsorbed into the medullary interstitium and thence into the body via the vasa recta. Reabsorbed urea also enters the vasa recta but a portion can penetrate the relatively urea-permeable portions of the loop of Henle to participate in medullary re-cycling of urea (see page 181). ADH may also augment renal concentrating ability by decreasing the vasa recta blood flow, thereby diminishing the wash-out of solutes from the medullary interstitium to the blood stream. On theoretical grounds, even small changes in medullary blood flow should have large effects on the accummulation of solutes within the medulla.

The antidiuretic hormone (officially arginine vasopressin (AVP)) is a nonapeptide synthesized and stored with a carrier protein, neurophysin, in the supra-optic nucleus. Secretory granules migrate down the supraopticohypophyseal tract to the posterior lobe of the pituitary gland from where AVP (and neurophysin) is released into the blood stream. Under normal circumstances the release of AVP, like the sensation of thirst, is controlled primarily by stimuli arising at osmoreceptors located in the hypothalamus. These receptors are very sensitive and plasma levels of AVP are found to double for each 2·5 per cent increase in plasma osmolality. An additional mechanism relates AVP release to blood volume sensed by stretch receptors in the atria and 'low pressure' cardiovascular regions and relayed to the brain via the vagus nerve. The increases in AVP levels which occur during haemorrhage and shock, quiet standing, vasodilatation and positive pressure breathing probably utilize this mechanism. These 'volume-related' changes in AVP can be quantitatively much greater than those achieved by dehydration and at the blood levels achieved, AVP may have systemic vascular actions in addition to actions on the kidney. The release of AVP is also influenced strongly by higher cortical centres. Its release is increased by drugs such as nicotine, morphine, most anaesthetics, clofibrate and chlorpropramide, while alcohol and narcotic antagonist drugs are inhibitors of AVP release.

The actions of AVP on the collecting ducts involve activation of adenylate cyclase, the production of cyclic AMP and an activated protein kinase which phosphorylates a membrane protein leading to an increase in the water permeability of the luminal cell membrane. Calcium, lithium and prostaglandin E all inhibit the rise in adenylate cyclase in response to AVP and inhibit the urinary concentrating mechanism. Potassium depletion inhibits the action of AVP at a step distal to cyclic AMP formation. Chlorpropamide and inhibitors of prostaglandin synthetase (e.g. indomethacin) increase the action of small doses of AVP. Medullary prostaglandins probably modulate the normal physiological actions of AVP on the collecting duct epithelium.

The production of a maximally concentrated urine requires not only the right balance between AVP and prostaglandin action on the collecting ducts but also a delicate balance between solute reabsorption, secretion and cycling within the medulla and medullary blood flow. It is thus not surprising that the renal concentrating ability is one of the first functions to be impaired in progressive renal disease. Renal concentrating ability is also deficient after prolonged over-hydration or osmotic diuresis both of which dissipate the medullary osmotic gradient, by diuretic drugs acting on the loop of Henle (e.g. frusemide), by protein deprivation and by hypercalcaemia and potassium deficiency. In Addison's disease, urinary diluting ability is impaired. Defective urinary dilution in adrenalectomized animals is due to impaired solute reabsorption from the ascending limb of the loop of Henle and excessive antidiuretic hormone concentration. It can be reversed by glucocorticosteroids (e.g. dexamethasone).

THE RENAL EXCRETION OF UREA

Early clearance studies showed that in antidiuretic states some 75 per cent of the filtered urea was reabsorbed but that as urine flow increased the fraction reabsorbed fell to about 50 per cent. Urea clearance remains constant as plasma level is increased by urea infusion. These results were interpreted as showing that urea was reabsorbed passively, largely by the proximal nephron. Subsequent micropuncture studies, however, have shown that urea is both reabsorbed and secreted and is, in fact, recycled within the kidney. The permeability of the proximal tubules to urea is high and about 50 per cent of the filtered urea is indeed reabsorbed there by passive means. Yet by the end of the ascending limb of the loop of Henle more urea is present than in the filtrate. Thus substantial quantities of urea are added to the tubular fluid during passage through the medulla. Since the majority of filtered urea is in fact, eventually reabsorbed, the collecting duct must reabsorb large quantities some of which re-enters the tubular system at the descending limb of the loop of Henle. The ascending limb and the distal tubule are relatively impermeable to urea, at least in the antidiuretic state and provide a pathway to cycle urea between the collecting ducts and descending limbs of the loop of Henle via the medullary interstitium. AVP augments the permeability of the medullary (but not the cortical) collecting ducts to urea (but not other solutes), thereby permitting the deposition of urea in the papilla during antidiuresis and augmenting the concentrating process.

Urea clearance has sometimes been used as an indication of GFR. However, the marked effects of urine flow rate and protein intake on urea excretion and its highly complex tubular handling make it a poor indicator of the GFR.

RENAL MECHANISMS OF ACIDIFICATION AND ALKALINIZATION

The normal adult excretes some 40–80 mmol of acid daily and this can rise to 400 mmol during diabetic ketoacidosis. Acid-base homeostasis requires the lungs to regulate the extracellular CO_2 level and the kidneys to regulate the extracellular bicarbonate level. The kidney normally achieves this end by the conservation of filtered bicarbonate through tubular reabsorption, by the excretion of any excess and by the replenishment of depleted stores by elimination of acid. The urine of meat-eaters is normally acid since protein metabolism produces inorganic acids. Acid elimination entails the elaboration of ammonia and the excretion of titratable acid as hydrogen ions buffered by urinary phosphate (and to a lesser extent creatinine). Only small quantities are eliminated as free hydrogen ions.

Bicarbonate is filtered freely by the glomerulus and reabsorbed actively by the tubules, particularly the initial portion of the proximal tubule. Most bicarbonate reabsorption is secondary to hydrogen ion secretion. This provides tubular fluid H^+ to react with filtered bicarbonate to form carbonic acid which is dehydrated to CO_2 and water. This latter process requires the catalytic action of carbonic anhydrase present on the luminal aspect of the brush border of proximal tubule cells. Carbon dioxide so formed being highly diffusable, enters the cell where, again under the influence of carbonic anhydrase, it is re-hydrated to form carbonic acid. This dissociates to yield hydrogen ions (available for secretion) and bicarbonate ions which pass into the interstitial fluid and are reabsorbed into the blood stream. Several aspects of this process require further description. When carbonic anhydrase is inhibited maximally by drugs some 60 per cent of the filtered bicarbonate appears in the urine. The 40 per cent that is still reabsorbed indicates that some can be transported without conversion to CO_2 in the lumen. Moreover, microperfusion experiments have shown that the maximal rate of hydrogen ion secretion by the proximal tubule is less than the maximal rate of bicarbonate reabsorption. A part of the filtered bicarbonate might be reabsorbed *per se* or alternatively it might be reabsorbed as carbonic acid, presumably by non-ionic diffusion, and dissociate in the tubule cell to hydrogen ions and bicarbonate. This carbonic anhydrase-independent means for bicarbonate reabsorption is probably limited to the proximal tubule since hydrogen ion secretion can account fully for the smaller rates of bicarbonate reabsorption at the distal tubule. Hydrogen ion secretion was believed to be tightly coupled to sodium reabsorption. However, hydrogen ion secretion persists during micro-perfusion of the lumen and peritubular capillaries of the proximal tubule with a solution in which sodium is replaced by choline. Thus, although in several situations sodium reabsorption correlates with acidification the process does not involve a tight coupled counter-transport of sodium and hydrogen ions.

The infusion of sodium bicarbonate increases plasma bicarbonate concentration progressively. Bicarbonate reabsorption rises to a limiting value after which all the excess filtered appears in the urine. This suggests that bicarbonate is reabsorbed by a transport-maximum-limited process $(T_M HCO_3)$. However where experiments have controlled carefully for the inhibitory effects of ECV expansion with $NaHCO_3$ on proximal sodium bicarbonate reabsorption, there is no clear-cut evidence of a $T_M HCO_3$. Nevertheless, although bicarbonate reabsorption may not be a T_M-limited process, renal bicarbonate excretion does normally rise with increasing plasma levels. Conversely, in metabolic acidosis where bicarbonate concentrations are diminished, renal bicarbonate reabsorption is almost complete in the early proximal nephron, allowing steeper pH gradients to develop in the distal nephron so enhancing titratable acid and ammonia excretion. These adjustments provide a renal compensation for metabolic acidosis. An increase in pCO_2 of the blood augments bicarbonate reabsorption presumably by facilitating hydrogen ion formation from carbonic acid within the renal tubular cells and thus promoting H^+ secretion. Conversely, hypocapnia diminishes bicarbonate reabsorption and leads to an alkaline urine. These processes underlie the renal compensation for respiratory disturbances. Perfusion of the peritubular capillaries of the proximal tubule with an acid solution enhances hydrogen ion secretion at any level of CO_2, indicating that acidosis can stimulate proximal hydrogen ion secretion directly. Metabolic acidosis reduces the membrane potential of the proximal tubular cell (inside negative). This would reduce the electrical gradient against which protons must be secreted into the tubular lumen and increase the rate of acidification. During potassium depletion there is a loss of intracellular K^+ and gain of H^+. If this also occurred in renal tubular cells it would explain the enhanced proximal tubular hydrogen ion secretion (and bicarbonate reabsorption). The accompanying metabolic alkalosis is also due to a stimulus to renal glutaminase with a marked rise in ammonia-genesis. Active sodium reabsorption from the distal tubule creates a lumen-negative potential which favours the secretion of hydrogen ions. This may explain the enhanced rates of acidification that occur during increased rates of delivery of fluid to the distal tubule, or increased distal reabsorption produced by aldosterone. Where sodium is delivered with a poorly reabsorbed anion such as sulphate or phosphate the distal tubular lumen electro-negativity is enhanced further. This may underlie augmented renal acidification which accompanies dietary acid loading where the endogenous acids are sulphate and phosphate.

Maximal rates of excretion of titratable acid require not only bicarbonate reabsorption and hydrogen ion secretion but also plentiful urinary buffer. In metabolic acidosis this is normally provided by phosphate, or to a lesser extent creatinine and organic substances. Metabolic acidosis can reduce the reabsorption of phosphate in the proximal tubule and so provide an increase in tubular fluid buffer. All the titratable acid present in the final urine of acidotic rats can be accounted for by that present at the end of the proximal tubule since, although subsequent nephron segments can reduce the urine pH down to lower levels, the reabsorption of buffer reduces the final excretion of titratable acid.

During a metabolic acidosis, renal ammonia production makes an increasingly important contribution to net acid elimination. Renal ammonia is derived largely from deamination of glutamine by glutaminase in the proximal and distal tubules of the renal cortex. The ammonia (NH_3) so formed is highly diffusible and readily enters both blood and tubular fluid. Ammonia is a base and at physiological pH virtually all has reacted with hydrogen ion to form ammonium ion (NH_4^+). The lower the pH the greater the proportion of ammonia present as ammonium. Ammonium ions penetrate cell membranes far less readily than the non-ionized free ammonia. Thus, in acidosis, production of an acid tubular fluid partitions NH_3 as trapped NH_4^+ ions within the tubular fluid compartment. Ammonium excretion is augmented further by enhanced ammoniagenesis due in part to an increased activity of renal glutaminase. In metabolic acidosis the proximal tubule produces substantial quantities of ammonia but, by the early distal tubule, more than half has been lost. The pH of tubular fluid at the tip of the papilla is higher than the proximal tubule and NH_3 may be driven out into the medullary interstitium by the alkalinity of the fluid in the loop of Henle by a reversal of the trapping process. Further ammonia is added to the distal tubular fluid and a large contribution occurs from the collecting duct, most of which probably represents ammonia formed originally in the proximal tubules, expelled by the loop of Henle and trapped in the very acid tubular fluid of the collecting ducts.

Patients with Addison's disease often have a metabolic acidosis and an impaired ability to excrete titratable acid and ammonia whereas those with Conn's or Cushing's syndrome often have an alkalosis due to continuing renal acid elimination. In adrenal-ectomized rats the defects in renal acid elimination require mineralocorticosteroids (for maximal urinary acidification) and glucocorticosteroids (for maximal buffer and ammonia excretion) for their complete reversal. Chronic treatment with excess mineral-ocorticosteroids leads to an alkalosis with a greatly enhanced renal ammonia excretion which is largely explicable by a stimulus to glutaminase activity secondary to pot-assium depletion. The extent to which the adrenal glands contribute to the physiological regulation of renal acid elimination is not yet entirely clear.

RENAL REGULATION OF POTASSIUM EXCRETION

While the colon has the capacity to secrete potassium ions, the renal tubules are quantitatively more important in determining body potassium balance. Some 60–80 per cent of filtered potassium is reabsorbed by the proximal tubule. Superficial, 'cortical' nephrons reabsorb further potassium through the loop of Henle leaving only some 10 per cent of that filtered remaining at the early distal tubule. Importantly, the fraction varies little despite large changes in excretion of potassium in the final urine, which testifies to the dominant importance of potassium secretion by the distal tubule and the collecting duct.

Changes in potassium intake are followed by rapid and accurate changes in renal potassium excretion without measurable changes in plasma potassium concentration. The outcome is to maintain the total body potassium content remarkably stable despite wide variations in dietary potassium intake.

Ordinarily, the distal tubules show net potassium secretion although during severe dietary potassium restriction net potassium reabsorption occurs. The distal tubule cells take up potassium actively at the peritubular cell membrane. This is linked to sodium extrusion by a glycoside-sensitive ATPase. Acute and chronic potassium administration, metabolic alkalosis and mineralocorticosteroids stimulate peritubular potassium uptake whereas metabolic acidosis and potassium depletion inhibit it. Potassium diffuses down an electrochemical gradient from the cell to the lumen. Secretion is stimulated by increasing the cellular K concentration, decreasing the luminal K concentration and increasing the transepithelial lumen-negative electrical potential. Increased sodium reabsorption by the distal tubule stimulates the basilateral exchange of Na for K, augmenting cellular potassium uptake. An increased sodium reabsorption also increases the luminal electronegativity. Moreover, an increased flow of fluid into the distal tubule stimulates Na reabsorption and diminishes the elaboration of a luminal K^+ gradient down the tubule. These three processes ensure a tight linkage between sodium reabsorption and potassium secretion at the distal tubule and collecting ducts. The cortical collecting tubules are also important sites for net potassium secretion (or exceptionally reabsorption) and potassium transport at this site is broadly similar to that at the distal tubule. Acute changes in GFR produce parallel changes in potassium excretion. However, with time, adaptive changes in tubular transport occur and only acute or extreme reductions in GFR (as during acute renal failure) lead to accumulation of potassium in the body. An increase in the rate of flow of fluid into the distal tubule or collecting ducts facilitates the secretion of potassium. This underlies the increase in potassium excretion observed with diuretic drugs which act on the proximal tubule (e.g. acetazolamide) or the loop of Henle (e.g. frusemide, bumetanide or ethacrynic acid) or the distal tubule (e.g. thiazides) and during metabolic acidosis, where proximal tubule fluid reabsorption is also impaired (see page 177). Indeed, the direct effect of metabolic acidosis is to diminish peritubular potassium uptake by the distal tubule cells and diminish potassium secretion. This effect is ordinarily overridden by the marked stimulus of potassium secretion produced by the increase in distal tubular fluid flow rate. A high potassium intake and mineralocorticosteroid drugs augment potassium secretion (see page 174) and during chronic excess of mineralocorticosteroids renal potassium excretion persists despite cellular potassium deficiency.

RENAL REGULATION OF CALCIUM, MAGNESIUM AND PHOSPHORUS EXCRETION

Calcium

Approximately 40 per cent of the plasma calcium is bound to plasma proteins so that the remaining fraction is available for filtration at the glomerulus. An additional fraction (some 7 per cent) is complexed with anions including citrate, phosphate and sulphate, leaving only about half the plasma calcium in the form of free calcium ions. Normally calcium is extensively reabsorbed by the tubules and less than 5 per cent of that filtered is excreted. In the proximal tubule, calcium transport parallels that of sodium and water so that some 50–70 per cent is reabsorbed. Further reabsorption occurs in the loop of Henle and by the early distal tubule, the calcium load is reduced to about 15 per cent of that filtered. Calcium reabsorption by the ascending limb of the loop of Henle is passive and secondary to active chloride transport. It is strongly inhibited by diuretics acting at this site (e.g. frusemide). Overall there is usually a very close correlation between changes in calcium and sodium excretion during alterations in body fluid balance, acid-base load, or the administration of diuretics which act on the proximal tubule (e.g. acetazolamide or osmotic diuretics) or the loop of Henle; they reflect the similar transport processes for Na^+ and Ca^{++} in the proximal nephron. The large increase in calcium excretion during acute metabolic acidosis is due mainly to inhibition of proximal reabsorption (see page 183) but also to an increase in the ultrafilterable fraction of plasma calcium. However, stimuli acting predominantly on distal transport processes can dissociate the excretion of calcium from that of sodium. Thus thiazide and 'potassium-sparing' diuretics (e.g. amiloride and triamterene) reduce the reabsorption of sodium but increase that of calcium. Indeed, thiazide diuretics can be used to reduce calcium excretion in patients with recurrent renal stone formation due to hypercalciuria. Adrenalectomized animals have a reduced rate of reabsorption of sodium but an increased rate for calcium and these defects are reversed by mineralocorticosteroid drugs. Acute administration of parathormone and Vitamin D increase the renal tubular reabsorption of calcium relative to that of sodium, again by actions on the distal tubule and collecting ducts. Prolonged excesses of parathormone or Vitamin D, however, lead to an increase in calcium excretion secondary to a rise in plasma calcium concentration.

Magnesium

Magnesium is an important intracellular ion. Its homeostasis is largely controlled by the kidneys although the details remain obscure. Variations in magnesium intake elicit rapid and precise adjustments in renal magnesium excretion, without detectable changes in the plasma magnesium concentration. The ultrafilterable portion of plasma magnesium is about 70 per cent. Interestingly, magnesium reabsorption by the proximal tubule is restricted and, unlike calcium and most other ions, its concentration increases with length down this segment. There is very extensive magnesium reabsorption by the loop of Henle and this is probably the major site for the control of urinary magnesium excretion. Magnesium reabsorption is increased by magnesium deficiency and decreased by magnesium loading. Magnesium does not normally accumulate in renal failure because of decreased absorption by the gut. There is, as with calcium, a good correlation between sodium and magnesium excretion in acute studies, probably due to parallel changes in absorption in the loop of Henle. Diuretic drugs acting on the proximal nephron decrease magnesium reabsorption, as do mineralocorticosteroids and hypercalcaemia. There is a large increase in magnesium excretion during infusion of magnesium solution due to saturation of a limited reabsorptive process ($T_M Mg$).

Alcohol ingestion, both acute and chronic, inhibits magnesium reabsorption and alcoholism can lead to severe magnesium depletion. Combined depletion of magnesium and potassium can complicate diuretic treatment, malnutrition, congestive cardiac failure and cirrhosis of the liver. In these circumstances, dietary potassium supplements are not retained by the kidneys until the magnesium depletion is reversed. Experimental magnesium depletion provokes a sustained hypersecretion of aldosterone and this may explain the potassium depletion and persistent potassium excretion observed in magnesium depleted subjects.

Phosphate

Phosphate is excreted by glomerular filtration, tubular reabsorption and possibly tubular secretion. Phosphate excretion is normally less than 20 per cent of that filtered. However, acute infusion of phosphate which increases the filtered load increases tubular reabsorption only to a maximum (T_MPO_4) whereafter additional loads are quantitatively excreted in the urine. The T_MPO_4 is reduced by parathormone and dietary phosphate loading. About 85 per cent of plasma phosphate is ultrafilterable and 70–90 per cent is reabsorbed in the proximal tubule. The tubular fluid phosphate concentration falls with distance down the proximal tubule indicating the reabsorption of phosphate in excess of sodium and fluid. The first portion of the tubule is the major site for phosphate reabsorption and most of the important factors which influence renal phosphate excretion act there. Thus phosphate loading reverses the normal proximal tubular profile for phosphate concentration while phosphate depletion produces an exaggerated fall in proximal tubular fluid phosphate concentration. Parathormone and saline loading both diminish proximal phosphate reabsorption whereas parathyroidectomy increases it. Normally, there is probably little further phosphate transport past the end of the proximal tubule. With the development of chronic renal failure the rate of phosphate reabsorption per nephron decreases substantially. This can be due to secondary hyperparathyroidism or to an effect of phosphate retention due to reduced GFR. Renal phosphate excretion during acute changes in the levels of active Vitamin D are mostly secondary to phosphate absorption from the gut, release of phosphate from bone or changes in parathormone levels.

RENAL REGULATION OF GLUCOSE EXCRETION

Normally the renal tubular reabsorption of glucose is effectively complete since the urine contains only trace quantities, undetectable by routine methods. A glucose infusion increases the plasma glucose concentration and hence the filtered glucose load (the product of plasma glucose concentration and GFR). The reabsorbed glucose load initially rises in parallel until a maximal value, after which the extra amount filtered is excreted in the urine. This is not a precise transport maximum limitation (T_M). During glucose infusion a reduction in load produced by a modest restriction of GFR does not eliminate glucose excretion (in fact, the same is true for other substances whose reabsorption has been described as T_M—linked e.g. phosphate, bicarbonate, magnesium). Thus reabsorption responds more closely to changes in glucose concentration than total load. Nevertheless, the renal glucose excretion of a patient with diabetes mellitus can fall if the GFR is reduced and urinary glucose levels cannot be used to predict plasma levels in these circumstances. Glucose reabsorption during increases in the plasma concentration does not show a sharp inflection, but instead exhibits a 'splay' which has been attributed to differences between nephrons in their maximal rate of glucose reabsorption. The degree of 'splay' increases in patients with chronic renal failure and in animals during ECV expansion. In congenital renal glycosuria the 'splay' is so expanded that glycosuria occurs even at a normal level of plasma glucose. The great majority of filtered glucose is usually reabsorbed in the proximal tubule, the site of most active reabsorption being the earliest segment. Glucose transport is a carrier-mediated process which shows saturation kinetics and competitive inhibition by certain other sugars.

In untreated diabetes mellitus, plasma glucose rises well above the renal threshold producing substantial glycosuria. The ensuing osmotic diuresis inhibits the kidney's capacity to regulate the excretion of fluid and ions and can lead to profound volume depletion (see page 177).

RENAL REGULATION OF URIC ACID SECRETION

Uric acid is both reabsorbed and secreted by the renal tubules by a carrier-mediated process. However, most of the filtered urate is reabsorbed and about 85 per cent of that excreted derives from tubular secretion. Uric acid is reabsorbed in the convoluted proximal tubule, secreted in the proximal straight tubule (pars recta) and reabsorbed at some points beyond the proximal tubule (mainly the distal tubule). Organic acids complete with urate for tubular secretion and urate retention is seen with lactic acidosis and with accumulation of organic acids during starvation, severe muscular exercise, diabetic ketoacidosis or ingestion of alcohol. The same carrier for uric acid secretion is probably utilized for the secretion of the 'loop' and thiazide diuretics. Competitive inhibition of uric acid secretion by these drugs may contribute to the hyperuricaemia observed during diuretic treatment. In addition, ECV depletion enhances the reabsorption of uric acid in the convoluted segments and thus the diuretic action itself may lead to a retention of urate. Uricosuric drugs (e.g. probenecid) inhibit the secretion of these diuretic drugs and diminish the renal response to them. Thus hyperuricaemia in patients receiving diuretic drugs is best treated with drugs that inhibit uric acid production. Many uricosuric agents at low dosage reduce urate excretion because they inhibit both tubular secretion and tubular reabsorption.

RENAL ACTION OF DIURETIC DRUGS

The proximal tubule is the major site of action of osmotic diuretics (see page 177) and diuretics which inhibit carbonic anhydrase (e.g. acetazolamide) and a minor site of action of 'loop' and thiazide diuretics, at least in therapeutic doses. Carbonic anhydrase, present both on the luminal surface and within the proximal tubular cells, is required for the reversible catalytic dehydration of carbonic acid to CO_2. This step, proceeding in the tubular lumen, provides one mechanism for bicarbonate reabsorption as CO_2 and proceeding in the tubular cell provides H^+ from CO_2 (via carbonic acid) for secretion into the tubular lumen. The diuretic response to carbonic anhydrase inhibitors follows from inhibition of proximal tubular bicarbonate reabsorption. The ensuing inhibition of sodium and fluid reabsorption increases greatly the bulk flow to more distal segments and overwhelms their reabsorptive capacity resulting in a diuresis and a natriuresis. Potassium secretion is also enhanced. The effect, however, is self-limiting, because the fall in plasma bicarbonate concentration induced by the drug eventually reaches a level at which proximal bicarbonate reabsorption by carbonic-anhydrase-independent mechanisms can compensate fully for the drug's effect. Acute administration of 'loop' diuretics or thiazides can reduce proximal tubular fluid reabsorption. However, the reduction in ECV consequent upon their actions promotes proximal reabsorption and, in the steady state, proximal tubular fluid reabsorption is either unchanged or, more usually, is enhanced.

The most potent diuretics (e.g. frusemide, bumetanide, ethacrynic acid) have their prime sites of action in the thick ascending limb of the loop of Henle ('loop' diuretics) and the early distal tubule where they inhibit active chloride reabsorption and secondarily the passive reabsorption of sodium and other cations. This is also a site of action of mercurial diuretics. When tested *in vitro* on perfused tubular segments, they cause a dose-related fall in the transepithelial, lumen-positive potential and in net ion transport when applied from the tubular lumen. Inhibitors of prostaglandin synthesis (e.g. indomethacin) attenuate the natriuretic action of 'loop' diuretics but the mechanism is not yet clear.

By inhibiting solute reabsorption by the loop of Henle, this class of diuretics will dilute a concentrated urine and concentrate a dilute one. The gradient of solutes in the medulla is dissipated and urinary osmolality approaches that of plasma. This can be utilized in the correction of hyponatraemia due to 'inappropriate ADH secretion' where urinary concentration is inhibited by frusemide and NaCl losses are replaced.

The late distal tubule is a major site of action for thiazide drugs. Since, by this segment, more than 90 per cent of the filtered sodium and fluid normally have been reabsorbed, drugs acting here or downstream necessarily have a somewhat limited potency. Thiazides can inhibit the entry of sodium into the distal tubule cell from the lumen. Continuing sodium extrusion across the basilateral cell membrane lowers the intracellular sodium concentration and increases the transmembrane sodium concentration gradient. Evidence indicates that cellular calcium extrusion across the basi-lateral cell membrane is linked to the passive influx of sodium down its electrochemical gradient. Thus inhibition of luminal sodium entry by thiazides would explain both the inhibition of sodium reabsorption and the stimulus to calcium reabsorption observed in the distal tubule and at the whole kidney level.

The collecting ducts are the major site of action of drugs which inhibit aldo-sterone's action (e.g. spironolactone) and the so-called 'potassium-sparing' drugs (e.g. amiloride and triamterene). Despite different modes of action, the overall effects of these two classes of drug are similar. Aldosterone acts principally on the late distal tubule and collecting ducts (see page 174) and inhibition of its action produces a modest (and variable) natriuresis with a decrease in K^+ and H^+ ion secretion. Spiron-olactone is, of course, most effective when aldosterone levels are highest (primary and secondary aldosteronism, advanced cirrhosis of the liver, nephrotic syndrome) or have been increased by use of other diuretic drugs and is almost ineffective in other circum-stances. Amiloride and triamterene are not aldosterone-antagonist drugs, but like spi-ronolactone, are also most effective where aldosterone levels are high since this stimulates the transport processes of the nephron segments on which these drugs act. Amiloride inhibits the luminal membrane's permeability to sodium and thus acts like thiazides but at a more distal nephron segment.

The secretion of potassium by the late distal tubule and collecting ducts is directly correlated with the tubular fluid flow rate. Consequently, all diuretics that act by inhibiting Na^+ and fluid reabsorption at or before the late distal tubule will promote K^+ secretion and, during prolonged treatment, can lead to potassium depletion. Con-versely, those acting on the collecting ducts will inhibit K^+ secretion and lead to potassium retention (and hyperkalaemia). Thus potassium depletion induced by a "loop" or thiazide diuretic can be treated by the addition of a 'potassium-sparing' diuretic although most patients will not develop sufficient potassium depletion to warrant routine treatment with two drugs.

The pharmacokinetics of diuretics can have important bearings on their actions. Osmotic diuretics are filtered at the glomerulus and cleared at a rate comparable to inulin. Carbonic anhydrase inhibitors, 'loop' diuretics and thiazides are extensively secreted by the organic acid carrier mechanism (which also transports uric acid, PAH and certain other drugs such as penicillin) into the proximal tubular fluid. This process yields such high tubular fluid drug concentrations so that the kidney still responds to 'loop' diuretics even in the presence of a severe fall in GFR. The much higher concentra-tions of these drugs at their sites of action (the luminal aspect of the tubular cell membrane) compared to elsewhere in the body may explain why they have such a high therapeutic ratio (i.e. the dose required to produce unwanted actions compared to dose required to produce wanted actions). Spironolactone is not fully active until metabo-lized, which may explain its slow onset of action.

Abnormal function

ACUTE RENAL FAILURE

Acute renal failure implies an abrupt reduction in renal function. Characteristically, urine flow rate falls below 500 ml per day (oliguria) or rarely may stop completely (anuria) although occasionally, urine flow rates may be maintained (diuretic acute renal failure). The syndrome may be caused by a host of conditions which lead to damage or obstruction of the main renal vessels, the arterioles, the glomeruli, the tubules, or the interstitial tissues which obstruct urinary outflow. However, physiological interest centres around a form frequently encountered in which the kidney failure is secondary to a systemic disturbance often accompanied by hypotension or shock, but structural renal damage is mild or absent (i.e. acute tubular necrosis). No single theory can account for the entire spectrum of abnormalities in acute renal failure due to tubular necrosis. It often occurs in the context of a reduction in the ECF volume following haemorrhage, diuresis with inadequate fluid replacement or fluid loss due to excessive sweating, diarrhoea or burns. During haemorrhage, renal blood flow is severely curtailed and the fraction of cardiac output perfusing the kidneys is reduced. Urinary output drops sharply or ceases altogether. Consequently, clearance studies cannot be undertaken unless urine flow is maintained artificially by an osmotic diuretic such as mannitol when the GFR is seen to be severely impaired. There is a generalized activation of the sympatho-adrenal system. Studies employing renal denervation or sympatholytic drugs indicate that activation of renal alpha-adrenoceptors contributes to the fall in renal blood flow and GFR, whereas activation of beta-adrenoceptors contribute to the release of renin. Haemorrhage releases renin from the kidneys and studies with antagonists of the renin-angiotensin system show that during haemorrhage angiotensin II (AII) helps to maintain the BP but at the expense of the renal circulation. Indeed, during mild degrees of non-hypotensive haemorrhage or ECV depletion, angiotensin antagonists increase renal blood flow and sodium excretion. There is a massive increase in plasma levels of AVP, probably mediated through low pressure intrathoracic volume receptors, arterial baroreceptors and chemoreceptors. AVP, at these high plasma levels, may contribute to renal vasoconstriction and to reduction of the GFR. Release of ACTH provokes glucocorticosteroid secretion and AII provokes aldosterone secretion. Renal prostaglandin synthesis and release is augmented in response to vasoconstriction caused by sympathetic nerve activity and angiotensin. Prostaglandin synthetase inhibitors (e.g. indomethacin) given during haemorrhage reduce renal blood flow further suggesting that local prostaglandin production helps to maintain renal perfusion. As the duration of hypotensive haemorrhage increases, other substances produced in response to tissue damage are released into the circulation (for example, kinins, lysosomal products) and a severe metabolic acidosis develops which causes further renal vasoconstriction. During haemorrhage, the outer cortex, which normally resistance, high-flow vascular compartment, is the region affected most severely by vasoconstriction. This is due to intense afferent arterior vasoconstriction. It is, however, patchy with regions of the outer cortex almost completely deprived of blood supply whilst neighbouring regions are perfused at a relatively normal rate. Probably, regions of the outer cortex oscillate between periods of almost complete cessation of flow followed by short periods of restoration of flow due to the action of local metabolites on resistance vessels. On the other hand, blood flow to the deeper cortex is relatively well preserved. The cut surface of a kidney from an animal killed by haemorrhage shows the contrast between a very pale outer cortex and a deeply-congested inner cortex. The quantitation of the intrarenal distribution of blood flow presents a formidable methodological problem which has yet to be overcome fully. Nevertheless there is a general agreement that the outer cortex takes by far the greater brunt of the fall in renal blood flow during haemorrhage. It is possible that the anatomical differences between 'cortical' and 'juxtamedullary' nephrons (see Fig. 4.1) may underlie the difference in regional blood flow. Thus, the 'juxtamedullary' nephrons with their predominant resistance vessels apparently located distal to the glomerulus may maintain some flow and filtration when flow and filtration to the 'cortical' nephrons with their major resistance vessels located proximal to the glomerulus are severely curtailed.

In the dog, if the haemorrhaged blood is retransfused within a period of one half to 2 hours, the BP usually returns towards normal, the kidneys vasodilate and swell, renal blood flow and GFR increase, albeit not to their previous levels, and urine flow recommences. Initially, tubular function is abnormal with an isotonic urine despite very high levels of ADH and the excretion of an excessive fraction of filtered ions. Nevertheless, histological evidence of tubular damage is usually minimal and normal renal function is restored quickly.

Although the renal tubules normally receive a very high blood supply, the basal oxygen requirement is not large. Thus, during a hypotensive haemorrhage, the severe restriction or curtailment of GFR reduces the oxygen used for active tubular reabsorption and lowers the metabolic requirements of the tubule cells considerably. However, when the hypotensive haemorrhage is more prolonged or severe then signs of tubular damage can occur. There is cellular degeneration and a patchy appearance of gaps in the basement membrane named "tubulorrhexis" which is the hallmark of acute tubular necrosis. These lesions indicate that tubular fluid can leak directly into the interstitium, which indeed becomes oedematous. These lesions affect predominantly the distal tubule. However, it is important to realise that they can be slight or even absent in kidneys showing a severe functional impairment.

The factors that convert a reversible depression in renal blood flow and function into an acute renal failure have been studied extensively in experimental animals. Prior volume depletion or low salt intake worsens the risk of acute renal failure, whereas a high salt intake or the administration of osmotic diuretics (mannitol) at or before the haemorrhage are protective. Other factors contributing to the development of acute renal failure will be discussed below.

Several theories have been advanced to account for the progression of functional vasoconstriction to acute renal failure. The theory of tubular leak implies that filtration persists but fluid leaks back into the interstitium and thence into the blood stream through disrupted tubular cells. Undoubtedly, severe and prolonged renal ischaemia or other serious renal insults, can so damage the tubules as to make them leaky. Nevertheless, in several models of experimental renal failure inulin injected into the proximal tubule can be recovered in full from the final urine indicating the integrity of the tubular system. Another theory holds that tubular obstruction, whether by casts from within the lumen or interstitial oedema raising pressure from without, leads to obstruction of tubular flow. This provides an increase in tubular pressure upstream from the obstruction which opposes filtration. But studies in most experimental models of acute renal failure have not shown a consistent increase in intratubular pressure. However, where acute renal failure is accompanied by the production of abundant casts, as in myoglobinaemic or glycerol-induced renal failure, high intratubular pressures are found regularly, and the induction of a diuresis which prevents the material from blocking the tubules improves renal function. Another theory holds that vascular obstruction by swollen glomerular endothelial cells or by intravascular fibrin deposition leads to mechanical obstruction to blood flow. However, these features are found histologically in only a few cases of experimental and clinical acute renal failure. Acute renal failure is most often the outcome of prolonged renal vasoconstriction. In man, renal blood flow is reduced to about 50 per cent and this fall, if it were due predominantly to afferent arteriolar vasoconstriction, would probably be sufficient to abolish filtration. The sympathetic nervous system and the renin-angiotensin system are certainly implicated in the renal vasoconstriction that precedes acute renal failure. However, they are probably not the prime cause of the renal vasoconstriction, which outlasts any volume deficit or circulatory disturbance, since antagonists of angiotensin or noradrenaline do not improve renal blood flow or function. One theory attributes the renal vasoconstriction during acute renal failure to activation of tubulo-glomerular feedback. This might follow from diminished proximal tubular cell function leading to an increase in sodium chloride load or concentration at the macula densa segment. Against this theory is the observation that frusemide, which blocks 'tubulo-glomerular feedback' responses cannot reverse vasoconstriction in established acute renal failure. At present the cause of the postulated renal vasoconstriction remains an enigma.

Clinical course of acute renal failure

The clinical course of acute renal failure can be divided into 3 phases. The first (or 'oliguric') phase, is characterized by scanty urine formation. A severe curtailment of filtration leads to the retention of products normally excreted. Fluid and electrolyte balance cannot be regulated and continued water ingestion leads to hyponatraemia, water intoxication, and cardiac failure. Potassium retention leads to hyperkalaemia which may be life-threatening and retention of metabolic acids leads to an acidosis.

These factors are worsened by the frequent co-existence of a hypercatabolic state with release of large quantities of intracellular constituents into the extracellular compartment. The rapid development of uraemia (see under chronic renal failure) predisposes to gastrointestinal bleeding, anaemia, impaired wound healing and susceptibility to infection. After about 10 days the 'diuretic' phase commences although a few patients are diuretic from the start. Urine flow usually increases over a few days often reaching supranormal values for a period of about 10 days. Renal blood flow and GFR are severely depressed, but gradually begin to increase. Indeed the blood urea level may not fall initially despite the onset of a diuresis. The patient still has little ability to regulate the ECV or adjust the rate of excretion of important substances despite the return of urine flow since the kidney cannot concentrate the urine and the tubules have little ability to alter the composition of the glomerular filtrate. There follows the 'recovery' phase during which renal blood flow, GFR and function return towards normal. By 6 months most patients have a normal GFR and normal indices of tubular function.

CHRONIC RENAL FAILURE

In the face of a diminishing nephron population, the kidney retains its main homeostatic functions to a remarkable degree, although when challenged the range of responses of the diseased kidney is curtailed. Symptoms of renal failure rarely occur before the loss of some 75 per cent of renal function (as indexed by GFR). In part this is a consequence of the hyperbolic relationship between the plasma level and the clearance of most substances (see Fig. 4.5). Thus a progressive reduction in GFR can produce only a small increase in blood creatinine or urea concentration until GFR falls below about 30 ml min^{-1}. Another factor is the striking adaptation that occurs in the function of the remaining nephrons. The so-called 'intact nephron hypothesis' implies that renal function in chronic renal failure is the outcome of a normal regulatory response in the surviving nephron population which are exposed to an abnormal solute load. The hypothesis provides a helpful simplification of the physiological response of the diseased kidney although certain patients, particularly those whose disease affects predominantly the renal medulla, may have qualitative abnormalities in tubular function superimposed.

The *ability to concentrate the urine* fully is lost early in the course of chronic renal failure although the ability to dilute the urine is preserved until late. The failure to concentrate is due to a failure of the kidney to respond to AVP, probably because of a lack of solute gradient in the medulla. It causes nocturia and thirst. Patients with progressive renal failure do not usually show fluid retention since the remaining nephrons excrete a much greater fraction of the filtered sodium and fluid. However, when challenged with an abnormally high or low salt intake, they show an impaired ability to regulate salt excretion and consequently experience greater changes in ECV. Micropuncture experiments in animals show that, as chronic renal failure develops, there is

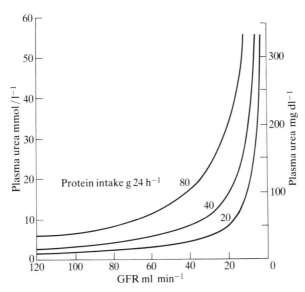

Figure 4.5
The hyperbolic relationship between plasma urea concentration and glomerular filtration rate at different steady-state levels of urea production produced by varying protein intake. Note that a blood urea of 7 mmol l^{-1} corresponds to a GFR of approximately 100 ml min^{-1} at the high rate of urea production, but to a GFR of only 25 ml min^{-1} at the low rate of urea production.

an increase in the filtration rate of surviving nephrons and a significant fall in the fraction of fluid and sodium reabsorbed in the proximal tubules. There is some increase in distal reabsorption but this compensation is incomplete, thereby accounting for increased rates of excretion per nephron. The decreased proximal reabsorption has been related to the filtration of increased amounts of osmotically-active and poorly-reabsorbed solutes that accumulate in the plasma of patients with uraemia. On the other hand, the reduction in reabsorption occurs very quickly after removal of one kidney and must involve additional, but poorly-understood, mechanisms. There is evidence for a natriuretic substance released into the urine of uraemic patients, but its nature and importance are unresolved. Occasional patients have a more profound defect in sodium conservation and can become volume-depleted on a normal salt intake. They frequently have renal disease with prominent structural damage to the medulla.

The *incidence of hypertension* increases progressively with the development of renal failure. Blood pressure can usually be reduced by removal of sodium and fluid at haemodialysis and many patients can be maintained with normal levels of BP by this means alone. BP can often be reduced also by angiotensin antagonist drugs (e.g. saralasin). This suggests that in many patients with chronic renal disease, hypertension is dependent upon an expanded ECV with inappropriately high renin and angiotensin levels.

In chronic renal failure, *potassium homeostasis* is normally little disturbed although dangerous hyperkalaemia can occur during severe acidosis, excessive potassium intake or the use of 'potassium-sparing' diuretics. Usually, however, the serum potassium concentration is a little low because of an increase in potassium secretion by the distal tubules of the surviving nephrons. This might be a response to the increase in fluid and sodium load presented to the distal tubule as a consequence of diminished proximal reabsorption.

Metabolic acidosis is a very frequent finding in chronic renal failure. Two mechanisms have been identified. Firstly, there is decreased reabsorption of bicarbonate in the proximal tubule in experimental animal models. This is normally picked up in the distal tubule but nevertheless bicarbonate appears in the urine at lower than normal plasma levels. Decreased proximal bicarbonate reabsorption has been variously attributed to the osmotic load, the depression of sodium and fluid reabsorption, and hypersecretion of parathormone. Secondly, there is impaired renal ammonia production. It cannot be augmented by glutamine infusion and thus is not due to a failure of renal blood flow to provide precursors for ammonia synthesis. The fall in ammonia excretion parallels the fall in GFR and has been attributed to a decreased functional tubular cell mass. The ability to reduce urine pH is relatively well-preserved in chronic renal failure and titratable acid excretion (determined by acidification and phosphate excretion) is likewise normal.

Renal failure is accompanied by *anaemia*, which is not greatly improved by dialysis. Red cell production is diminished, probably because of diminished erythropoietin production due to diminished renal tissue mass. In addition, the red cell life span is shortened and there is frequently an increase in blood loss because of defective platelet function or gastrointestinal haemorrhage. Fluid retention may expand the plasma volume and further depress the haemoglobin concentration. Hyperlipidaemia is frequent and probably contributes to the increased risk of arteriosclerosis in patients with renal failure.

Glucose homeostasis is abnormal; some patients show enhanced sensitivity to insulin and diabetics often require a reduction in dosage. A significant fraction of circulating insulin is removed by the normal kidney and consequently its biological half-life is prolonged in renal failure. On the other hand, non-diabetic patients can develop impaired glucose tolerance though the pathogenesis of this is unclear.

Gonadal function is deranged in advanced renal failure. Men develop impotence, reduced libido and reduced fertility. Many have testicular atrophy, impaired spermatogenesis with reduced levels of testosterone and increased levels of FSH and LH indicating a primary gonadal defect. In addition, some develop gynaecomastia. Women develop oligomenorrhoea or amenorrhoea and their fertility is reduced sharply.

As renal failure develops, nerve conduction velocities are reduced and some patients develop a frank *neuropathy* which is not reversed by dialysis. There is neuronal degeneration and segmental demyelination but some of the functional impairment can improve after successful transplantation. Impairment of cerebral function is often secondary to fluid and electrolyte disturbances or hypertension. The complex subject of renal osteodystrophy is considered elsewhere (Chapter 9).

The search for a single uraemic toxin that would explain these widespread effects has not so far been successful. Whilst the retention of urea and other nitrogenous compounds does have some predictable effects on the kidney (e.g. osmotic diuresis in remaining nephrons) it cannot account for many of the other uraemic manifestations. Patients treated by regular peritoneal dialysis appear to have a lower incidence of some uraemic complications than those treated by haemodialysis. Artificial membranes effectively restrict the filtration of molecules with a molecular weight which exceeds about 1,500 whereas the peritoneum allows them significant passage. This has led to the suggestion that these "middle molecules" may contain an important uraemic toxin.

Principles in the management of patients with chronic renal failure before dialysis

As the GFR falls below 20 or 30 ml min^{-1} symptoms and signs of uraemia appear. Many uraemic symptoms can be controlled by dietary manipulation. Restriction of protein intake reduces the body load of urea and nitrogenous compounds and diminishes the osmotic diuresis of surviving nephrons. A high calorie intake is essential to prevent negative nitrogen balance, because only a small amount of endogenous nitrogen can be re-synthesized from urea into amino acids and protein. Feeding the carbon skeletons of essential amino acids as ketoacids can promote a positive nitrogen balance and diminish urea excretion and their use in chronic renal failure looks promising. Careful attention to salt and water balance is important since small decrements in salt intake can lead to volume depletion and further worsening of renal function, whereas small increments can precipitate volume overload with hypertension or pulmonary oedema. Effective treatment of hypertension is essential since, particularly in the accelerated phase, it determines the rate of decline of renal function.

OBSTRUCTIVE NEPHROPATHY

Following obstruction to urinary outflow, there is dilatation of the collecting system upstream. The kidney increases in size over the first 10 days. Thereafter, atrophy of renal substance is apparent with loss particularly of juxtamedullary nephrons. Even small degrees of obstruction reduce GFR because of a rise in intra-tubular pressure. In experimental animals, partial outflow obstruction leads to an early rise in renal blood flow, but after a day or so it is reduced and after one week is about 30 per cent of normal. Both the early increase and the later decrease in flow rates are reduced by inhibiting prostaglandin synthesis. This suggests that a vasodilator prostaglandin (e.g. prostacyclin) contributes to the early vasodilatation whereas a vasoconstrictor prostaglandin (e.g. thromboxane) contributes to the later vasoconstriction. Eventually the changes in renal haemodynamics become irreversible. Tubular function is characterized by a decreased ability to concentrate the urine, associated with a decline in the medullary gradient for urea and solutes and an increased fractional excretion of sodium and other ions.

Following relief of an acute obstruction the diuresis and excessive excretion of ions often increases further with the restoration of GFR and can lead to considerable fluid and electrolyte depletion. This is probably explicable by diminished fluid reabsorption from the proximal tubule and loop of Henle as shown in micropuncture studies.

URETEROSIGMOID ANASTAMOSIS

Important metabolic complications may ensue when, because of irreversible bladder disease, the ureters are implanted into the colon. Patients can develop hypokalaemia with a chronic hyperchloraemic metabolic acidosis. Potassium depletion is related to prolonged metabolic acidosis which depletes potassium from the cells and increases its renal excretion and to rectal loss of potassium in colonic secretions. The colon can reabsorb water, urea, ammonia, sodium, hydrogen and chloride ions and this recycling presents an increased load for the kidneys which shows an increase in GFR and blood flow and often in renal size. Later however, reflux of fluid up the ureters from the colon leads to ascending infection which diminishes renal function and severe acidosis may then develop.

RENAL TUBULAR ACIDOSIS

Renal tubular acidosis (RTA) results from a failure of the kidneys to secrete hydrogen ions. The majority can be separated into either proximal (bicarbonate wasting) or distal (acidification defects) types. In the proximal form, there is a partial failure of bicarbonate reabsorption allowing increased loads of bicarbonate to enter the distal tubule which overwhelms its limited reabsorptive capacity. The increase in bicarbonate excretion lowers serum bicarbonate concentration and hence the bicarbonate load for proximal reabsorption. A new equilibrium state is attained where the kidney can reabsorb the reduced filtered load of bicarbonate and excrete acid although a systemic acidosis with a low bicarbonate concentration persists. This is similar to the renal adaptation to carbonic anhydrase inhibition.

Patients with distal renal tubular acidosis have different defects. They show a diminished ability to acidify their urine and to excrete titratable acid and ammonia. This has been attributed to a reduced rate of hydrogen ion generation within the tubular cells.

Both types of RTA lead to a hyperchloraemic metabolic acidosis and potassium depletion. In proximal defects, excessive potassium excretion accompanies excessive bicarbonate excretion and thus occurs early in the development of the condition or during treatment with bicarbonate salts. In the distal form, however, excessive potassium excretion persists and depletion can be severe. It is potentiated by increased aldosterone secretion. Unlike the proximal form it is reduced by sodium bicarbonate therapy because this expands the ECV and reduces aldosterone levels. Increased calcium excretion with consequent osteopenic disease, nephrocalcinosis or renal stone formation can complicate the distal but not usually the proximal type of RTA. Acute and chronic acidosis can stimulate parathormone release which may underlie some of the effects on bone and kidneys.

Proximal RTA is often seen as part of a general proximal tubular dysfunction with excessive excretion of glucose, phosphate, uric acid and amino acids. Distal RTA is seen as a primary defect or as a result of renal transplantation or disease which affects the renal medulla.

Adrenal failure represents a speçial case of renal acidosis in which the combined effects of a reduced GFR, a reduced distal tubular capacity to secrete hydrogen ions and an enhanced phosphate (buffer) reabsorption combine to produce a severe defect in renal acid elimination.

'Potassium sparing' diuretics reduce the maximum rates of acid excretion, but only rarely produce an overt metabolic acidosis.

PROTEINURIA AND NEPHROTIC SYNDROME

Normal human urine is virtually free of protein (less than 50 mg/l). Some normal subjects excrete significant protein but only when standing. They do not usually have underlying renal disease. Studies with synthetic macromolecules have shown that at a molecular weight of 60,000 daltons there is little filtration (0·01 per cent), but that filtration rises steeply with decreasing molecular size and by 20,000 daltons approximately 75 per cent of the molecules are filterable. However, haemoglobin has a molecular weight close to 60,000 daltons yet is excreted rapidly in the urine if it appears in plasma, whereas albumin, which has a comparable molecular weight is excreted in minute quantities. The filtration barrier behaves as if it has fixed negative charges which repel negatively-charged molecules such as albumin and restricts their passage considerably more than would be anticipated from their size alone.

The proximal tubule is able to reabsorb a little protein and two separate transport processes have been defined, one for relatively low, and another for higher molecular weight proteins. Tamm-Horsfall glycoprotein appears in small quantities in normal urine. It is added to tubular fluid by the loop of Henle where it may have a physiological function in regulating tubular permeability. It is a major constituent of urinary casts but its excretion has not yet been shown to vary in any very characteristic way in disease.

Most patients with excessive proteinuria have a glomerular lesion. In 'minimal change' glomerulo-nephritis there is striking proteinuria in the absence of gross changes in the glomerulus, although with electronmicroscopy, changes are seen in the epithelial foot processes. Similar lesions can be produced by injection of puramycin into rats. This produces a loss of the normal charge-restriction of filtration and much of the proteinuria is due to the filtration of molecules such as albumin that would normally be repelled electrostatically by the glomerular barrier. Proteinuria can occur also when low molecular weight proteins are filtered in amounts which overwhelm the tubule's reabsorptive capacity ('tubular proteinuria'). This occurs in Bence-Jones proteinuria when abnormal light immunoglobulin chains appear in the plasma. This also produces an excessive tubular reabsorption of protein which can impair proximal tubular function and increase protein excretion further. Tubular proteinura is seldom gross but can be a useful index of tubular damage.

Patients with 'glomerular' proteinuria excrete significant quantities of proteins the size of albumin and larger. Protein clearance normally falls with increasing molecular size or negative charge. The slope of the line relating the log of the protein clearance to the log of the protein size is used as a guide to the 'selectivity' of the glomerular barrier. Some 'selectivity' persists even with severe glomerular damage. In patients with the nephrotic syndrome, those with the most selective proteinuria have the best response to steroid drugs. Patients with 'tubular' proteinuria retain a normal clearance for albumin (circum 0·02 per cent of GFR) and the excreted proteins are of smaller size (e.g. lysosome and beta$_2$-microglobulin).

Patients with the nephrotic syndrome have massive proteinuria (in excess of 5 g per day), hypoproteinaemia and oedema. The hypoproteinaemia is mitigated by increased hepatic synthesis of albumin and lipoproteins, but this is insufficient to prevent a fall in plasma oncotic pressure. Fluid redistributes from the plasma to the interstitial compartment leading to oedema formation. Glomerular filtration may be normal or even high in the early stages of the disease as the fall in the plasma oncotic pressure favours glomerular fluid filtration. However, the tubular reabsorption of sodium is excessive and this produces fluid retention which potentiates the oedema formation. In most animal models of nephrotic syndrome fluid reabsorption by the proximal tubule is normal or reduced, perhaps reflecting the fall in peritubular protein concentration, but there is an enhanced reabsorption at more distal sites. Likewise, patients with the nephrotic syndrome, in whom distal reabsorption has been inhibited by large doses of thiazides and 'loop' diuretics, also show normal or reduced proximal fluid reabsorption. Patients with the nephrotic syndrome have initially a reduced blood volume (probably because of the reduced plasma oncotic pressure) and high renin and aldosterone levels. Although enhanced distal reabsorption in them may relate to the high aldosterone levels, there are others with chronic nephrotic syndrome who have a normal blood volume and normal levels of renin and aldosterone. The basis for their fluid retention is not yet understood.

THE KIDNEY IN LIVER DISEASE

Abnormalities of renal function complicate many hepatic or biliary diseases. The progress of cirrhosis of the liver is marked by increasingly severe sodium retention and declining renal function. Even in the early phases, there is a failure to excrete salt and water when maximally challenged. In these patients renin and aldosterone levels and

GFR are usually not abnormal, and the cause for the enhanced reabsorption of salt and water is unclear. With progression of cirrhosis, renal blood flow falls steeply and the GFR is diminished despite the reduction in plasma oncotic pressure produced by hypo-albuminaemia. These changes in the renal circulation are unaccompanied by structural alteration in the vessels or glomeruli and can be reversed temporarily by intra-renal arterial injection of vasodilator drugs. This suggests a functional renal vasoconstriction. A rather sudden development of oliguric acute renal failure can complicate advanced cirrhosis, or other forms of severe liver disease and is sometimes referred to as the 'hepatorenal syndrome'. There is intense renal vasoconstriction which, in some patients, is precipitated by a reduction in blood volume often due to vigorous diuretic treatment or gastrointestinal haemorrhage. It has been suggested that the vasoconstriction may be caused by endotoxins reaching the systemic circulation through the damaged liver while others have emphasized the importance of angiotensin. Indomethacin produces a large fall in renal blood flow and GFR in cirrhosis, suggesting that prostaglandins maintain renal blood flow in these patients and thus drugs which inhibit prostaglandin synthesis should be used with great care. The agent(s) primarily responsible for the renal vasoconstriction in cirrhosis of the liver remains controversial.

Renin and aldosterone levels are increased in advanced cirrhosis but even here aldosterone is unlikely to be the sole factor dictating sodium retention. Experimental studies indicate that the enhanced sodium reabsorption is primarily attributable to the distal tubule. Some cirrhotic patients develop water retention which produces a fall in plasma sodium concentration despite sodium retention.

In patients with acute and chronic obstructive jaundice the urine may contain epithelial cell casts and protein. These patients are at increased risk of developing acute renal failure, particularly if they are subjected to surgery, even if they are clinically anicteric. This can occur without evidence of hypovolaemia and although the cause is often not established, endotoxaemia, bacteraemia or intravascular coagulation may all be factors in individual patients. In addition, experimental studies show that an ischaemic kidney develops renal failure more readily after the intravenous infusion of bile salts. The incidence of acute renal failure in patients with obstructive jaundice undergoing surgery can be reduced by pre-treatment with osmotic diuretics such as mannitol.

Thus liver disease has wide-spread effects on renal function, the pathophysiology of which is poorly understood. The management of renal and electrolyte problems are often of paramount importance in determining the outcome.

THE KIDNEY IN HEART FAILURE

In established heart failure, the kidney retains sodium and fluid and, because of the increase in venous pressure, the extra fluid is distributed largely into the interstitial space. There is a fall in renal blood flow in proportion to the severity of the cardiac failure. The GFR, however, is relatively well preserved resulting in a rise in the filtration fraction. The cause for these adjustments is not well established but there is evidence of increased sympathetic nervous activity in cardiac failure (high peripheral resistance, high levels of plasma noradrenaline). If renal sympathetic nerves were activated, this might reduce RBF and increase sodium reabsorption by a combination of direct effects of noradrenaline on vessels and tubules and indirect effects mediated through activation of the renin-angiotensin-aldosterone and the prostaglandin systems. In some animal models of cardiac failure, the increase in renal vascular resistance, filtration fraction and aldosterone levels can be inhibited by drugs that antagonize the renin-angiotensin system (saralasin, captopril). However, reversal of fluid retention does not necessarily follow probably because of a fall in blood pressure. These results highlight the complexity of the interactions between the endocrine, cardiovascular and autonomic nervous systems in cardiac failure.

An increase in filtration fraction is found regularly in patients with cardiac failure. This will increase the concentration of plasma proteins in the peritubular capillary plasma and should augment volume reabsorption from the proximal tubule. Indirect evidence of increased proximal tubule reabsorption in patients with congestive cardiac failure is provided by their lower natriuretic response to diuretics which block distal sodium reabsorption. However, in animal models, enhanced distal tubular reabsorption is also seen. Early in the development of cardiac failure, aldosterone levels are often high and may contribute to sodium retention, but in chronic, stable cardiac failure aldosterone may be normal despite avid salt retention. Thus, as with other common causes of oedema, the pathogenesis of the renal fluid retention cannot be explained simply by excessive aldosterone secretion and remains to be elucidated.

Several mechanisms participate in the enhanced water retention in congestive cardiac failure. Arginine vasopressin levels are high and alcohol (which inhibits AVP release) can induce water diuresis. High AVP levels may be initiated by abnormal signals generated by volume receptors in the "low pressure" chambers. Some patients appear polydipsic, perhaps a consequence of the actions of angiotensin on the drinking mechanism. Finally, enhanced proximal tubular volume reabsorption diminishes the presentation of fluid to the renal diluting segment (ascending limb of the loop of Henle) and diminishes the capacity to generate free water. The combined influences of these mechanisms, all of which tend to conserve water, is to further expand the body fluid volume which in some patients, leads to hyponatraemia.

Where cardiac failure is secondary to chronic lung disease (cor pulmonale) additional factors may contribute to the fluid retention. The proximal tubular reabsorption of sodium with bicarbonate is augmented by an increase in blood P_{CO_2} as occurs in respiratory failure. Polycythaemia developing in response to chronic hypoxaemia is an additional factor that increases the filtration fraction (see below) and promotes sodium and fluid reabsorption. Restoration of a more normal haematocrit sometimes leads to the excretion of retained fluid but care is necessary not to lower the haematocrit below about 50 per cent since this can affect the supply of oxygen to the tissues adversely.

THE KIDNEY IN ANAEMIA AND POLYCYTHAEMIA

The requirement of the kidney for plasma from which to form a glomerular filtrate poses unique problems for the renal circulation in adjusting to changes in blood haematocrit. These adjustments involve active changes of diameter of the resistance vessels (primarily the afferent arteriole) together with passive changes in resistance which are a consequence of the alteration in blood viscosity that depend on the haematocrit. These adjustments normally maintain the level of GFR following acute or chronic changes in haematocrit. A fall in haematocrit produces an initial rise in blood flow due to the lesser viscosity of the blood. This is gradually overcome by renal vasoconstriction which returns blood flow to previous levels and, after some days, below this. Since a greater proportion of blood is now plasma, there is a normal or increased plasma flow and a fall in filtration fraction. Conversely, an acute increase in haematocrit reduces renal blood flow only transiently, and in the steady state renal blood flow is increased despite the high viscosity. Renal plasma flow, however, is reduced and GFR maintained by an increase in glomerular ultrafiltration pressure, probably due to the very high viscosity of the ultrafiltered blood perfusing the efferent arterioles. These remarkable changes in renal haemodynamics however, cannot be achieved by aged or diseased kidneys and in these circumstances the passive effects of blood viscosity on blood flow may predominate. Moreover, in severe anaemia, the changes characteristic of cardiac failure supervene.

THE KIDNEY IN PREGNANCY AND THE EFFECT OF THE CONTRACEPTIVE PILL

By the end of the first trimester of pregnancy, the normal kidney shows a rise in its blood flow and GFR of approximately one third. By the end of pregnancy these values return to normal. Thus there is some fall in blood urea and creatinine concentrations during early pregnancy. There is also a change in tubular function as shown by the lowered renal threshold for glucose, which can lead to renal glycosuria, and salt and water retention. There is an increase in the frequency of orthostatic proteinuria. These changes in renal function cannot be related easily to increases in the plasma levels of either oestrogen or progesterone. Oestrogen administered to non-pregnant human subjects does not increase renal blood flow or GFR, but it does cause a modest degree of sodium retention. The effects of progesterone on renal haemodynamics are controversial. Large doses produce an increase in renal blood flow and GFR but not at the plasma levels found in early pregnancy. Moreover, progesterone is natriuretic in human subjects.

Patients developing toxaemia during the second half of pregnancy have hypertension, oedema and proteinuria. Their glomeruli show some endothelial cell proliferation with swelling of the endothelial and epithelial cells which appear to reduce the luminal diameter of the glomerular capillaries. Proteinuria probably reflects the structural changes in the glomeruli. In toxaemia, renal blood flow is often normal or only mildly reduced, although the GFR is reduced more consistently. In severe eclampsia, however, there is generalized vasoconstriction and a large reduction in renal blood flow and GFR. The oedema and hypertension of toxaemia of pregnancy has not been adequately explained and cannot be attributed to increased aldosterone secretion. In the great majority of patients, these functional changes reverse rapidly following delivery of the baby.

Studies of normal women taking an oestrogen-progestogen contraceptive pill have shown a small reduction in renal blood flow. A significant number develop hypertension. The plasma renin substrate concentration and plasma renin activity are increased and this relates primarily to the oestrogenic component. Whether the rise in blood pressure is related to the renin-angiotensin system or to the fall in renal blood flow or to the sodium retaining effect of oestrogen remains to be elucidated.

MISCELLANEOUS TUBULAR DISORDERS

There are several rare diseases in which renal tubular function is abnormal and only a few will be selected as examples. In one group amino acid transport systems in the kidney and intestine are defective. In *cystinuria* there is a failure to reabsorb the dibasic amino acids, ornithine, lysine, arginine, cystine and homocystine with excessive renal excretion. This condition is inherited by an autosomal recessive pattern. It predisposes to the formation of renal calculi since cystine is relatively insoluble, particularly in acid or concentrated urine. Stone formation can be diminished by hydration and urinary alkalinisation.

More widespread defects in proximal reabsorption may be encountered as inherited conditions in the *Fanconi syndrome* or may be acquired due to *heavy metal poisoning* with lead or cadmium, as a toxic response to certain drugs or as a complication of conditions which damage the proximal tubular cells such as those which lead to Bence-Jones proteinuria. In these conditions there is a failure to reabsorb amino acids, phosphate, bicarbonate, glucose and uric acid with excessive renal excretion and diminished plasma concentrations of these substances.

A number of rare conditions lead to a clinical picture of rickets which fails to respond to usual doses of Vitamin D (*Vitamin D resistant rickets*). In some, there is a primary defect of phosphate reabsorption (hypophosphataemic rickets). In others, there is a deficiency of the active form of Vitamin D, 1-25-dihydroxy cholecalciferol. This can be inherited as an autosomal dominant and the defect corrected either by giving very high doses of Vitamin D, or preferably, by 'physiological' doses of the active metabolite or an analogue of it. The disorder is believed to be due to a failure of the renal hydroxylase which is required for adding the 1-hydroxyl radical to the 25-hydroxy cholecalciferol produced by the liver. An even rarer syndrome has been described which apparently represents end-organ unresponsiveness to active Vitamin D.

Disorders of urine concentration

Several conditions reduce the kidney's ability to concentrate the urine. In nephrogenic diabetes insipidus there is polyuria, thirst and dehydration due to a renal unresponsiveness to AVP. This usually presents in infancy as a sex-linked recessive disorder. Volume-depletion produced by thiazide diuretics alleviates somewhat the polyuria. A failure to concentrate the urine occurs also in potassium depletion and is probably due to enhanced production of prostaglandins which inhibit the action of AVP. Hypercalciuria also inhibits urinary concentration and often leads to severe dehydration. Here the cause is multifactorial, but increased tubular fluid calcium concentration inhibits sodium reabsorption at many sites and may thereby diminish the active transport steps required for elaboration of the medullary solute gradient. Lithium administration can mimic an incomplete form of nephrogenic diabetes insipidus, probably due to diminished sodium transport by the loop of Henle. Finally, an osmotic diuresis or prolonged water ingestion due to an inappropriate water intake can dissipate the medullary solute gradient and render the kidney temporarily insensitive to ADH.

BARTTER'S SYNDROME

This is an uncommon disorder usually presenting in childhood with distinctive metabolic abnormalities. There is hypokalaemia, potassium depletion and metabolic alkalosis. The blood pressure is normal despite a considerable increase in renin and aldosterone secretion; there is resistance to the pressor action of angiotensin II. Hyperplasia of the juxtamedullary apparatus is seen in renal biopsies with increased numbers of renin-containing granules but these features are also found in other situations associated with prolonged hypersecretion of renin, e.g. Addison's disease. There are also reports of hyperplasia of the prostaglandin-containing interstitial cells of the medulla.

Hypokalaemia and evidence of potassium depletion are prime features of Bartter's syndrome and the major cause of symptoms. No doubt they are exacerbated by the observed hypersecretion of aldosterone. However, additional factors are implicated since neither aldosterone antagonist drugs nor adrenalectomy correct fully the potassium depletion. Some patients fail to conserve sodium on a low salt intake despite the high levels of aldosterone, although in the majority sodium conservation is normal. The high renin and aldosterone levels cannot usually be returned to normal by a high salt intake or plasma infusion suggesting that volume depletion is not the only cause for the enhanced secretion rates. The excretion of prostaglandin E_2 is increased. The administration of indomethacin reduces plasma renin and aldosterone levels and also the excessive sodium and potassium excretion. It increases the serum potassium concentration although this still usually remains below the normal range, but it does restore the pressor sensitivity to infused angiotensin II.

The underlying cause of Bartter's syndrome is hotly disputed. One theory attributes this to a defect in potassium transport leading to enhanced renal loss of potassium. Potassium depletion is a constant feature in Bartter's syndrome and in other situations leads to an augmented renal production of prostaglandin E. Excessive prostaglandin production in turn, could underlie the excessive secretion of renin. Vascular unresponsiveness to angiotensin II is probably not a specific defect and is found in most circumstances where angiotensin II levels are high.

Principles of tests and measurements

RENAL CLEARANCE AND GLOMERULAR FUNCTION

Many commonly used tests of renal function depend on clearance methods. The renal clearance of any substance is the volume of plasma required to supply the quantity of that substance appearing in the urine in unit time. Where U is the concentration of substances in the urine, V the urine flow rate and P the concentration in the plasma, then the term UV/P indicates the renal clearance rate. Clearance gives a quantitative measure of the kidney's ability to excrete a substance. Clearance data are readily interpreted where the rate of excretion increases directly with the plasma level, in which case the clearance is a fixed number independent of the plasma concentration (e.g. the clearance of inulin). For other substances such as glucose, the clearance increases beyond a threshold level of plasma concentration, while for others such as urea the clearance is determined by the rate of urine flow.

Clearance measurements are relatively easy to perform. They require little alteration in the normal physiological state, are safe, easily reproducible and can provide quantitative information about whole kidney function in man. However, they cannot distinguish accurately variations among different populations of nephrons, nor localize function to a specific nephron segment and they cannot separate reabsorption or secretion for substances which undergo bidirectional transport (e.g. uric acid).

Clearance can be estimated most accurately by administering the substance as a bolus to fill its volume of distribution and maintaining its plasma level by constant infusion. The plasma concentration (P) refers to renal arterial blood but this is usually assessed from peripheral venous samples. The error involved is slight providing the plasma levels are steady. It is difficult to select the plasma level appropriate for the clearance period when plasma levels are changing rapidly. A plasma sample can be taken at the beginning and end of the urine collection period, the level plotted on semilogarithmic paper and the plasma concentration some 2–4 minutes prior to the mid-point of the urine collection taken as P to allow for the time elapsed between filtration of the marker and its appearance in the bladder. There are other common sources of error. The first clearance following a sudden increase in urine flow is usually spuriously elevated due to the rapid passage of marker concentrated in the tubules and collecting system into the bladder and corresponding errors follow clearances during rapid reductions in urine flow. Variations in the completeness of bladder emptying are a recurring problem. They are minimized by performing clearances at high urine flow rates (e.g. under conditions of water diuresis) since the fraction of urine remaining in the bladder is diminished as the total minute output increases. Further problems arise for substances excreted by non-ionic diffusion where variation in urine pH or flow can exert profound effects on clearance. Drugs may compete for a transport process and thereby affect clearance values (e.g. probenecid can reduce the clearance of para-amino hippuric acid, PAH). Particular care is required to establish an appropriate plasma level with substances whose clearance is determined by a transport-maximum reabsorptive or secretory process.

Several attempts have been made to circumvent some of these problems and to simplify clearance procedures. Markers for the estimation of GFR (e.g. EDTA) or renal plasma flow (e.g. Hippuran) are eliminated almost exclusively by the urine. Consequently, after a bolus intravenous injection, the rate at which their plasma levels decline should relate to their renal clearance (i.e. to GFR or RPF). Such 'single shot' clearance methods obviate the need to obtain urine samples. However, the rate of change of the plasma concentration of the marker will be determined also by its volume of distribution and the rate with which the marker diffuses through this volume. Since measurements are made in the non-steady state, the marker will be in flux, initially mixing within the plasma and then passing out into the extra-vascular spaces and finally, as the plasma level falls, passing back into the plasma. The errors introduced by 'single shot' techniques may outweigh their advantages. They are particularly difficult to interpret where there are abnormalities in the ECV or in the capillary circulation (e.g. cardiac failure). A classical clearance estimation (UV/P) remains the most precise method for getting quantitative information. Even so, the standard error of the mean for clearance values performed under optimum conditions approaches 10 per cent.

Another method with the advantage of not requiring urine collection is that of constant intravenous infusions. Here, the marker is given as a bolus (to fill its volume of distribution) and constant intravenous infusion (to maintain a steady plasma level) by a very accurate injection pump. Eventually, a stable plasma concentration is achieved when the rate of clearance of the marker is equal to its rate of infusion. The latter is calculated from the concentration of the marker infused and the pump speed. There are two advantages: firstly, clearance measurements are made at a time when there is equilibrium between the concentration of the marker in the compartments through which it is distributed (as judged by a constant plasma level) and secondly, the volume of distribution of the marker can be estimated by dividing the quantity retained in the body (i.e. the quantity infused minus the quantity excreted) by the plasma level. Since markers used for estimation of GFR are retained within the extra-cellular compartment the method allows a simultaneous determination of the GFR and ECV which can be very useful when assessing the renal responses to physiological stimuli.

Glomerular filtration rate

The marker used for estimating GFR must be freely filterable at the glomerulus and neither secreted nor reabsorbed by the tubules. Inulin is a standard marker against which all others are compared. It is an artificial fructose polymer of molecular weight 5,200. Direct measurements by micropuncture have confirmed that it is indeed freely filterable at the glomerulus and, when injected directly into the first part of the proximal tubule, is quantitatively recovered in the urine, thereby demonstrating no significant reabsorption. Inulin, however is now rarely used for clinical work since its chemical estimation is time-consuming and it diffuses only slowly through its extracellular volume of distribution. Radio-labelled (^{3}H and ^{14}C) inulins are used in animal experiments but for clinical work it is convenient to use a gamma-emitting radio-label attached to a glomerular filtration marker. Ethylene-diamine-tetracetic acid (EDTA) labelled with ^{51}Cr is used frequently and its clearance corresponds very closely to that of inulin. Other convenient markers include sodium iodothalamate or diatrizoate labelled with radioactive iodine.

There are advantages to the use of an endogenous marker which is continually added to the circulation and maintained at a relatively constant plasma level. This eliminates the need for intravenous infusion and automatically ensures that the substance is at or near its equilibrium value. The clearance of urea bears a rough proportion to the GFR but, as described above (see page 181) is dependent on the urine flow rate. Even at the highest flow rates urea clearance in man is only about 70 per cent of inulin, indicating significant reabsorption. Urea has been largely abandoned as a marker for GFR. In man, the clearance of creatinine is a better indication of GFR. In man and dogs some creatinine is reabsorbed at low urine flows but this fraction is very much less than for urea. More important is the secretion of a small but variable fraction of creatinine by the tubules and creatinine clearance can exceed inulin clearances by up to 30 per cent. For clinical practice, autoanalyser measurements of creatinine in plasma and 24-hour urines provide the best estimate of GFR without recourse to infusion techniques. As a further approximation the GFR can be estimated from the plasma concentrations of urea or creatinine. In severe renal failure, the plasma concentrations of urea and creatinine are certainly increased. Nevertheless, 'normal' values can occur when the GFR is reduced to about 50 ml/min, for two reasons; firstly, their rate of production is not constant between individuals, and urea synthesis varies greatly with protein intake and catabolism. Secondly, there is a curvilinear relationship between the plasma concentration of these substances and the GFR. (Fig. 4.5). With a progressive fall in GFR the plasma concentration increases initially only slightly and this increases excretion so that a new steady state is achieved at a slightly augmented plasma level. As the GFR falls further, greater increments in plasma level are required to provide the same rate of excretion and thus the rate of rise of plasma concentration increases steeply below a GFR of about 20–30 ml/min. Plasma creatinine is independent of dietary protein intake but creatinine production is increased by glucocorticosteroids and trauma, particularly to muscle, and may fall in chronic renal failure. A plot of the reciprocal of the plasma creatinine against time in patients with progressive renal impairment often produces a straight line relationship whose slope can provide an approximate indicator of the rate of decline of GFR.

Accurate measurements of GFR are the cornerstone for assessment of renal function. Even for a substance which is reabsorbed or secreted, its overall renal handling can only be understood by reference to GFR, since its rate of reabsorption or secretion is calculated from the difference between its rate of filtration and its rate of excretion. Where precise quantitative information is required there can be no substitute for repeated estimates of GFR using classical clearance methods with a marker which is neither reabsorbed or secreted. In the damaged kidney, these conditions can never be met with confidence. In acute renal failure in animals, inulin filtered at the glomerulus can leak back into the circulation through the damaged tubules. There is presently no way of circumventing this potential problem in assessing the GFR in acute renal failure in man. It is probably less important in chronic renal failure, but here there is an indication of abnormal renal handling of urea and creatinine and the creatinine clearance often exceeds that of inulin.

Renal blood flow

Renal blood flow (RBF) is conveniently calculated from renal plasma flow (RPF) estimated by clearance methods using the formula: RBF = RPF/(1 − haematocrit). Where a substance is completely cleared from the plasma by a single passage through the kidney and all the cleared material is excreted in the urine, the clearance is equal to the RPF. The renal extraction of PAH, diorast or radioiodine-labelled Hippuran normally exceeds 80 per cent in man and their clearances are used to estimate RPF. These substances are filtered at the glomerulus, secreted by the proximal tubule and are not reabsorbed. The secretory process is saturable and at high plasma levels the clearance of all these compounds falls sharply to approach GFR. They are transported by the weak acid secretory process which can be inhibited by probenecid.

Large errors in the estimation of RPF occur when the renal extraction of the markers is reduced. However, if renal venous samples can be obtained and the clearance calculated using the renal arteriovenous difference for the marker a true estimate of RPF is obtained even where renal extraction varies. Indeed, where renal venous samples are available, the same marker used to estimate GFR can also be used to estimate RPF using the Fick principle. Thus, where the rate of renal excretion and the arterio-venous difference of a substance are known, RPF can be calculated since there is only one rate of plasma flow which could provide the measured concentration difference across the kidney for the measured rate of excretion.

RBF can also be estimated in man from the rate of washout of a radioactively-labelled inert substance such as [131]Xenon. Xenon dissolved in saline is injected directly into the renal artery via a catheter and its rate of disappearance from the kidney monitored by external radioactive counters. Xenon is a lipid-soluble, inert substance which rapidly diffuses into renal tissue. Its rate of clearance approximates to the rate of blood flow to that particular region. The curve produced for the rate of decline of xenon with time is complex and contains at least three superimposed exponential functions, probably representing flow to different regions of the kidney. Quantitatively much the most important compartment is one cleared very rapidly of xenon, represented largely by blood flow to the outer cortex of the kidney. Outer cortical blood flow normally accounts for more than 90 per cent of total renal blood flow and thus this method is an acceptable estimate of the total blood flow in normal circumstances. It has the great advantage over clearance methods in that the marker is removed by the blood stream. It can therefore provide useful information about renal blood flow in conditions associated with anuria or leaky tubules (e.g. acute renal failure) where classical clearance techniques cannot be applied. Nevertheless problems of administering the marker and interpreting the results on diseased kidneys preclude this as a routine clinical method for blood flow assessment.

Finally, in patients undergoing surgery, renal blood flow can be measured from the output of an electromagnetic flow probe encircling the renal artery. Even this method however, is subject to considerable inaccuracy since the output from the probe depends upon its orientation around the vessel, the composition of the blood and the haematocrit.

TUBULAR FUNCTION

Since individual segments of the renal tubule have such strikingly different transport properties no single test can give a reliable measurement of overall function. Assessment of tubular function requires firstly an accurate assessment of GFR (and ideally RPF) followed by quantitative assessment of one or more of a selection of individual tubular functions. The choice of tests depends upon which nephron functions or anatomical segment is suspected of being involved in the disease process.

Maximum rate for reabsorption or secretion (Tm or Ts)

The tubular secretion of PAH or Hippuran is confined to the proximal tubule. Estimation of the maximum rate of secretion of PAH or Hippuran should thus provide a quantitative estimate of function of this part of the nephron. The Tm for PAH requires firstly that glomerular filtration rate be estimated by an independent means. The renal plasma flow can be estimated conveniently in the usual manner from the clearance of PAH and the plasma level of PAH then increased to 20–30 mg 100 ml^{-1} by raising the rate of infusion and the plasma level and rate of PAH excretion re-estimated. The rate of PAH filtration is given by the plasma concentration times the GFR. The rate of excretion is given by the urine concentration times the urine flow rate and the rate of secretion by the difference in these two. Providing that the plasma level is well above the threshold (above 15–20 mg/100 ml) the rate of secretion is assumed to be maximal, and in this range, independent of plasma concentration.

Phosphate is reabsorbed by a Tm limited process, virtually confined to the proximal tubule. Tm PO_4 is estimated as for PAH but with an infusion of a buffered phosphate solution to raise the plasma phosphate concentration well above its threshold value. The Tm PO_4 varies markedly with previous phosphate intake, with the time of day, and with the plasma glucose concentration as well as with factors which influence parathormone levels. The action of parathormone on the kidneys can also be assessed by estimating the rate of cyclic AMP excretion. But, of course, neither test is specific for parathormone action. As research procedures the Tm for other reabsorbed substances (glucose, bicarbonate, magnesium, sulphate) can be estimated using similar principles.

Renal water handling

These tests depend on the measurement of the osmolality of the plasma and urine and the rate of urine flow. The osmolar clearance, (C_{osm}) is calculated as the product of urine osmolality (U_{osm}) and flow rate divided by the plasma osmolality (P_{osm}). Thus, when the urine has the same osmolality of plasma, the osmolar clearance will equal the urine flow rate. Where the urine is more dilute than the plasma, the kidney has, in theory (although not in practice) had to add water free of solutes to the urine. This quantity is called the free water clearance (C_{H_2O}) and is calculated by subtracting the osmolar clearance from the urine flow rate. This is not a true clearance since free water is not added to tubular fluid but is a convenient mathematical calculation. Where the urine is more concentrated than the plasma, the kidney has, in theory, reabsorbed free water from the glomerular filtrate and C_{H_2O} becomes negative. This is either referred to as negative free water clearance or, the negative sign is disregarded and it is referred to as tubular reabsorption of free water $T^c_{H_2O}$.

Maximum urine concentrating ability can be measured by restriction of all fluids to produce a physiological stimulus to AVP release. In practice an overnight deprivation of water for 16 hours does not produce an adequate stimulus in all subjects and more prolonged water deprivation is required to test concentrating ability accurately. The renal response to AVP can be tested conveniently by combining over-night water deprivation with an injection of a supramaximal dose of antidiuretic hormone. Normally, the kidney should concentrate the urine in excess of 750 mosmol/l and values in excess of 1,000 are often found in healthy young subjects. A normal concentrating ability requires a very delicate balance between transport functions, complex counter-current flow rates, tubular permeability, and hormone action and it is hardly surprising that it is one of the most sensitive indices of renal dysfunction. A failure to concentrate the urine fully is found in a wide variety of renal diseases and in hypertension, congestive cardiac failure, protein malnutrition, potassium depletion and hypercalcaemia as well as any situation associated with an osmotic diuresis or excessive and inappropriate water intake. As renal function deteriorates in renal failure, urine osmolality comes progressively to approach that of plasma.

Tests of urinary dilution are performed infrequently since the water load required for the tests can produce a dangerous degree of water retention when urinary dilution is impaired. A failure to excrete a water load adequately is an early feature of Addison's disease.

Quantitation of free water clearance can be used to provide some insight into sodium reabsorption by different nephron segments. Thus, urine is made dilute by the reabsorption of sodium without water in the ascending limb of the loop of Henle. After a maximal water load which inhibits AVP release, there is little further water transport across subsequent tubular segments. Thus, the urine flow rate provides an estimate of the rate of fluid delivery to this segment and urine flow rate divided by GFR gives the fraction of the filtered fluid delivered to the diluting segment. Moreover, since the tubular fluid is diluted by sodium reabsorption, the rate of free water clearance is an index of the rate of sodium reabsorption by the diluting segment and the product of C_{H_2O} and P_{Na} estimates the quantity of sodium reabsorbed at the diluting site. The assumptions implicit in these calculations are considerable and the results obtained thus represent only approximations. Nevertheless, they have been shown useful in indicating the site of action of diuretic drugs in the kidney.

Balance studies

It is a remarkable feature of the normal kidney that when sodium in the diet is abruptly restricted, renal sodium excretion begins to decline in 8 hours and within 2 to 3 days reaches a new steady state very close to intake. The rate of equilibration to a low salt intake is diminished in renal failure, in patients treated with spironolactone, and in patients with defective sympathetic nervous function. Long term balance studies are rarely practicable and where such information is sought, changes in body weight, or total body water and its distribution between bodily compartments, or in the distribution volume of radio-isotopically labelled sodium are required.

Divided renal function

The presence of a functionally important stenosis of one renal artery can often be demonstrated by separate analysis of the function of each kidney. Ureteric cannulae are used to collect urine from each kidney separately. The presence of a mild stenosis is shown by a similar rate of blood flow and filtration yet an enhanced rate of water reabsorption on the affected side. Thus, on that side, the urine flow rate is reduced and the concentration of creatinine and solutes is increased. These patterns of changes are quite specific for renal artery stenosis but the problems inherent in obtaining the samples often outweigh the benefits achieved.

A second method for predicting the outcome of renal arterial surgery in hypertensive patients utilizes a comparison of plasma renin activity in samples of blood drawn from each renal vein. The probability that renal arterial surgery will cure hypertension is increased if the ratio of renin activity in the vein from the 'affected' kidney, compared to the 'unaffected' kidney, exceeds 1·6 to 2·0. A high level of renal venous plasma renin activity on the "affected" side results from an increase in renin release often accompanied by some reduction in renal plasma flow rate. Equally important, however, is the idea that the ratio of renal venous plasma renin activity between the two kidneys can only be increased if the 'unaffected' kidney is not releasing large quantities of renin and has a well maintained rate of plasma flow. Indeed, renin release from a truly 'unaffected' kidney will be suppressed by the high arterial angiotensin levels (and high blood pressure) consequent from renin release from the "affected" kidney and its renal venous plasma renin level will be at, or below, that in arterial plasma.

Acidification

The maximum rate of tubular bicarbonate reabsorption can be estimated as for the Tm PAH (see page 210) during infusion of sodium bicarbonate to raise the plasma level well above the threshold. It is reduced in patients with chronic renal failure and in those with proximal-type renal tubular acidosis and is increased by hypercapnia or potassium depletion.

Tubular hydrogen ion secretion is assessed by giving the patient an acid load and following its renal elimination. Hydrogen ions are most conveniently given in the form of ammonium chloride which is metabolized in the body to urea and hydrochloric acid. For the short test, NH_4Cl (0·1 g kg BWt^{-1}) is given by mouth and the renal acid elimination studied 5 hours later. The urine pH normally falls below 5·3, and the titratable acid excretion should exceed 25 umol min^{-1} and ammonia excretion 35 umol min^{-1}. For the prolonged test, the ammonia chloride is given over 5 days. Urine pH should fall below 5 and the combined outputs of titratable acid and ammonia exceed 120 mmol day^{-1}. Normal acid-base homeostasis is shown by maintenance of plasma bicarbonate within 5 mmol l^{-1} of the initial value.

A failure to reduce urine pH is required in the diagnosis of renal tubular acidosis. In chronic renal failure minimal values of urine pH may be normal but the rate of ammonia production is markedly impaired. Excessive ammonia production at a given level of urine pH is characteristic of potassium deficiency (which complicates many cases of renal tubular acidosis).

Protein excretion

The filtration and excretion of proteins is determined by their size and charge. In patients with 'glomerular' proteinuria, the clearance of a selected group of plasma proteins is related to their molecular size. The slope of a double logarithmic plot of protein clearance against the molecular size indicates the 'selectivity' of the protein-uria. Those with the most 'selective' proteinuria characteristically respond best to corticosteroid treatment. In order of decreasing size, the proteins most often used are orosomucoid, albumin, transferrin, caeruloplasmin, IgG, IgA and alpha$_2$- macro-globulin.

RENOGRAPHY

Useful additional information about renal function or structure can be obtained by following the passage of a radio-labelled tracer through the kidneys by external scintil-lation counting (renography) or imaging (gamma-camera). With a modern computer-ized gamma-camera system it is possible to select an "area of interest" and obtain an activity-versus-time output from that region, thereby combining the advantages of the imaging system with the quantitative assessment of the scintillation counter. There are a wide variety of markers and radiolabels: Hippuran is selectively concentrated by renal tubular secretion (see above): EDTA or diethylenetriaminepentracetate (DTPA) are filtered by the glomerulus and not reabsorbed: dimercaptosuccinate (DMSA) binds to proximal tubule cells and very little appears in the urine. The standard renogram usually employs an injection of ^{131}I Hippuran with scintillation detectors placed over each kidney recording the activity with time. It is very helpful to have a third detector placed over the heart to indicate the changes in blood level. Following an intravenous injection, there is a rapid increase in activity over the kidney as the bolus injection arrives (first phase). This has little diagnostic importance. There follows a more gradual increase as the Hippuran is transferred progressively from the plasma to the proximal tubule cells and thence to the tubular lumen where it is concentrated by water reab-sorption and passes down the nephron. This second phase achieves a peak at about 3 minutes and is followed by the third phase coincident with the appearance of Hippuran in the bladder. During this latter phase Hippuran is still being abstracted from the plasma by proximal tubular cells, concentrated in tubular fluid and passed down the nephron, but there is also removal of the transported Hippuran by the renal collecting system. It is thus a complicated function of plasma levels, secretion, tubular fluid reabsorption, transit and removal. However, normal renograms have almost simulta-neous and quite well delineated peaks. In unilateral renal vascular disease the affected kidney shows a delayed peak and an inpaired third phase. This pattern in a hyper-tensive patient indicates the need for further diagnostic tests in search of a correctable renal arterial lesion. The interpretation of the renogram is difficult in patients with oliguria. Thus in pre-renal and renal causes of acute renal failure there is an impaired third phase but a similar picture can be produced in bilateral outflow obstruction. Obstruction is rarely perfectly symmetrical, however, and a difference between the two kidneys is used as an indication of a post-renal cause, but the use of the gamma-camera improves the diagnostic accuracy in this circumstance. Better definition of the renogram is obtained by the subtraction of the non-renal background which is monitored by the third detector over the heart. It is, however, necessary to equate the blood background curve with the tissue background curve by use of a previous injection of iodinated human serum albumin (which is contained within the vascular system and is not filtered at the kidney). The individual tissue background curves, suitably normalized, can be subtracted from the total output recorded from the left and right scintillation counter to give the left and right kidney Hippuran curves. A comparison of the heights of the curves at the two kidneys between $1\frac{1}{2}$–$2\frac{1}{2}$ minutes following injection provides an approximate estimate of the distribution of renal blood flow between the two kidneys. During this period Hippuran is being actively taken up by the kidneys and none has yet left the nephron to enter the renal pelvis. It should be noted that this method does not provide any quantitative estimate of total renal blood flow nor total flow to each kidney and, to make this calculation, total blood flow must be estimated by some other means. A difference in height of the renogram could also occur because of a difference in the distance between the kidneys and the detectors which is a real source of error. Also, when proximal tubular lesions in one kidney impairs Hippuran uptake from the blood stream, there will be an underestimate of blood flow to that kidney.

Practical assessment

CLINICAL OBSERVATIONS

Acute renal failure

Careful history to detect cause of renal failure and presence of any hypotensive episodes, assessment of fluid balance from duration of oliguria and record of fluid and salt intake and output, changes in body weight, changes in plasma protein and packed cell volume and clinical state. Oedema, raised jugular venous pressure, signs of pulmonary oedema and a high blood pressure suggest fluid overload, whereas flaccid skin, sunken eyeballs, hypotension with an exaggerated postural drop suggest fluid depletion. Urine volume is initially low in most patients (less than 500 ml/24 h) and rises during recovery. Concentrated urine and signs of volume depletion indicate need for trial of intravenous fluid replacement (may require central venous pressure monitoring). Mannitol induces a diuresis in dehydrated patients but there is little or no response in established acute renal failure. In established renal failure, the urine osmolality is equivalent to plasma; proteinuria and glycosuria often present. Monitor carefully serum potassium. Hyperkalaemia and volume overload are urgent indications for dialysis.

Chronic renal failure

Seek history of recurrent urinary infection, analgesic abuse, disorders of micturition, drug intake, previous episodes suggesting nephritis, previous measurements of blood pressure or ureteric colic. Assess duration from the history, degree of anaemia, and duration of nocturnal polyuria. Uraemic symptoms include thirst and nocturia (failure to concentrate), nausea, anorexia and diarrhoea (gastritis and acidosis) impaired level of consciousness and convulsions (hypertensive encephalopathy, water intoxication, hypocalcaemia, drugs), air hunger (acidaemia), cramps (salt and water depletion), tetany (hypocalcaemia) pruritis (raised blood urea), bone pain, tenderness and proximal muscle weakness (renal osteodystrophy), ecchymoses and gastrointestinal haemorrhage (impaired platelet function).

Nephrotic syndrome

History of diabetes, glomerulonephritis, sensitivity reactions, rashes, arthritis, drug intake or malaria. Frothy urine indicates massive proteinuria. Peripheral oedema, pleural and pericardial effusions indicate hypoproteinaemia. Quantitative urinary protein excretion and assessment of clearance of individual proteins to assess "selectivity".

ROUTINE METHODS

Urinary deposit White blood cell excretion exceeding 300,000 per hour in timed urine specimens suggests pyelonephritis, colony count over 10^5 organisms per ml suggests significant bacteriuria, granular casts (debris of tubular cells) indicates glomerular disease, doubly refractile casts which contain lipoprotein are seen in massive proteinuria. A few hyaline casts are not abnormal. In haematuria, erythrocyte casts suggest a renal lesion.

Serum creatine concentration measured by autoanalyser reflects glomerular filtration rate; best combined with 24 hour urine creatinine excretion to quantitate GFR.

Blood·urea concentration depends on protein intake, protein metabolism and GFR.

Plasma sodium concentration is normal or high in Conn's syndrome and water depletion and is normal or low in Addison's disease, sodium depletion, water intoxication or excess ADH action.

Plasma potassium concentration is raised in acute renal failure and sometimes in chronic renal failure. Above 7·5 mmol/l cardiac arrhythmias or cardiac arrhythmia or cardiac arrest may occur. It is raised in Addison's disease and can be high in patients receiving potassium sparing diuretics. Most other diuretics lower serum potassium. It is low in Conn's syndrome, in some renal tubular defects, renal tubular acidosis, loss of gastrointestinal fluids due to vomiting or diarrhoea, liquorice ingestion, diabetic coma or recovery from acute renal failure or renal outflow obstruction. Hypokalaemia predisposes to digoxin toxicity.

Plasma chloride, particularly in relationship to plasma sodium concentration is low in prolonged vomiting and raised in renal tubular acidosis.

Plasma bicarbonate concentration is reduced in renal failure, metabolic acidosis and respiratory alkalosis. A low concentration, without azotaemia and with the urine pH being at 6 or above, suggests renal tubular acidosis. Hypokalaemia is usually associated with raised plasma bicarbonate.

In chronic renal failure if the *plasma calcium concentration* is low, osteomalacia is common. *Plasma phosphate concentration* is low in proximal tubular defects and hyperparathyroidism. It rises in chronic renal failure with a falling GFR.

Plasma albumin and *cholesterol* concentration are usually inversely related in the nephrotic syndrome.

References

BLACK D. & JONES N.F. (1979) *Renal Disease*, 4th ed. Oxford, Blackwell Scientific Publications.

BRENNER B.M. & STEIN J.H. (1978) Sodium and water homeostasis. *Contemporary Issues in Nephrology*. Vol. I. London, Churchill Livingstone.

BRENNER B.M. & STEIN J.H. (1978) Acid base and potassium homeostasis. *Contemporary Issues in Nephrology*. Vol. II.

BRENNER B.M. & STEIN J.H. (1979) Hormonal function and the kidney. *Contemporary Issues in Nephrology*. Vol. IV.

ORLOFF J. & BERLINER R.W. (1974) Renal physiology. *Handbook of Physiology*, Section 8. Washington D.C., American Physiological Society.

GIEBISCH G., TOSTESON D.C. & USSING H.H. (1979) Membrane transport in *Biology*: Vol. IV A, Transport organs. New York, Springer Verlag.

MOFFATT D.B. (1975) *Biological Structure and Function 5: The Mammalian Kidney*. Cambridge, Cambridge University Press.

PETERS G., DIEZI J. & GUIGNARD J.P. (1979). Renal adaptation to nephron loss. *The Yale Journal of Biology and Medicine*, **51**, p. 1–180.

5

Hydrogen Ion (Acid:Base) Regulation

Normal function

INTRODUCTION: TERMINOLOGY AND PHYSICAL CHEMISTRY

Water partially dissociates to form H^+ and OH^- ions:

$$H_2O \rightleftharpoons H^+ + OH^- \tag{1}$$

In pure water and in neutral solutions the concentration of each is equal to 10^{-7} mol/l at 20°C and about $10^{-6.8}$ at body temperature. Their product is a constant (the ionic product of water, 10^{-14} at 20°C) even in solutions containing other ionic material. Therefore the concentration of one ion automatically fixes the concentration of the other.

pH. pH was originally introduced as a mathematical convenience in dealing with the very wide range of H^+ concentrations $[H^+]$ met in chemistry. As conceived, pH would be the negative logarithm to the base 10 of $[H^+]$. Thus pH 7 would represent $[H^+] = 10^{-7}$ mol/l or, in the usual units of chemical pathology, 0·0001 mmol/l. Unfortunately neither pH nor $[H^+]$ as defined in this way can readily be measured (p. 233). pH as measured in practice approximates to the H^+ *activity* rather than the H^+ *concentration.* The activity of a solute is an expression not of how many particles there are but of how many there seem to be and is commonly indicated by round brackets: (H^+). This is equal to $[H^+]$ multiplied by an activity coefficient which is 1 only at infinite dilution and is usually < 1. As most of the quantitative relationships to be quoted are based on pH measurements, (H^+) rather than $[H^+]$ will be used in this chapter. This distinction is biologically important in that it is the activity rather than the concentration of a substance which matters (see Chapter 1, p. 2). Unfortunately many techniques used in chemical pathology measure concentration, but one may note that sodium ion *concentration* can be measured by flame photometry, and sodium ion *activity* by ion selective electrodes. One further point: pH and (H^+) are related to a molal (mass/mass) rather than a molar (mass/volume) scale so that (H^+) should be quoted, for example, in nano moles/kg (nmol/kg) of solvent, or as nano-molality (nM) (the prefix 'nano-' means 10^{-9}).

Acids and bases. In aqueous solution, an acid is a substance which increases (H^+) (lowers pH); a base is a substance which decreases (H^+) (raises pH). A buffer is a system of an acid and base in which the acid is only partly dissociated. The degree of this dissociation varies in such a way as to oppose any change in (H^+).

In the most widely used chemical terminology (that of Bronsted) the term 'acid' is used to describe hydrogen ion (proton) donors:

$$HX \rightleftharpoons H^+ + X \tag{2}$$
Acid

'Base' is used to describe hydrogen ion (proton) acceptors:

$$X + H^+ \rightleftharpoons HX \tag{3}$$
Base

The 'stronger' an acid or base, the more the equilibria in equations 2 and 3 lie towards the right. Most protons are associated with water molecules to form hydronium ions, H_3O^+, but as these react by virtue of the extra proton, no serious inaccuracy results from considering the processes in terms of H^+ alone.

Substances which neither donate nor accept H^+ ions are neither acids nor bases. Acids and bases are not the same as anions and cations. An acid, for example, may be electrically neutral, an anion, or a cation:

e.g.

$$HCl \rightleftharpoons H^+ + Cl^- \tag{4}$$
Electrically
neutral

$$H_2PO_4^- \rightleftharpoons H^+ + HPO_4^{--} \tag{5}$$
Anion

$$NH_4^+ \rightleftharpoons H^+ + NH_3 \tag{6}$$
Cation

Whether a substance behaves as an acid or a base is not necessarily a fixed characteristic. Substances having two or more modes of dissociation may be acids or bases depending on the (H^+) of the solution. For instance, HCO_3^- (a base at physiological (H^+)) is an acid at low (H^+): $HCO_3^- \rightarrow H^+ + CO_3^{--}$; and $H_2PO_4^-$ (an acid at physiological (H^+)) is a base at high (H^+): $H_2PO_4^- + H^+ \rightarrow H_3PO_4$.

Much of the apparent difficulty of acid : base terminology has arisen from the use of terms which would not satisfy the chemical usage set out above. Thus it is incorrect to call sodium ion (a cation which is neither a hydrogen ion donor or acceptor) a 'base'. Terms such as 'fixed acid', 'organic acid' or (more confusing still) 'buffer acid' will not be used in this book. One point of difficulty is the naming of substances such as sodium hydroxide (NaOH) which are not strictly bases (in this case the hydroxyl ion is the true base); it may be helpful to think of them as *alkalis* (hydroxyl ion donors).

PHYSIOLOGICAL ACTIVITY OF HYDROGEN IONS

Extracellular fluid

The normal (H^+) of arterial blood is about 0·00004 m-molal (equivalent to pH 7·4). As the decimal places are unwieldy it is convenient to multiply by a million to yield 40 n-molal (nmol/kg). The normal range is about 36–44 nM (which happens to be 7·44–7·36 pH units—the similarity of these numbers is a logarithmic coincidence—see Fig. 5.1). The (H^+) of the interstitial fluid depends on the local balance between metabolism and blood flow and is probably 3–25 nmol/l higher than in the arterial blood. The range of extracellular (H^+) compatible with life is about 20 to 160 nM (pH 7·7–6·8). These figures show that the 'reaction of the blood' need not be as constant as is often implied.

Intracellular fluid

Estimates of intracellular (H^+) vary from 50 to 1000 nM (pH 7·3–6·8). These estimates have been obtained with a variety of techniques ranging from direct intracellular recording to overall average 'whole body' values based on the distribution of weak acids, such as DMO (5, 5-dimethyl-2, 4-oxazolidinedione). A best estimate of the whole body pH using DMO is about 100 nM (pH 7·0); this must be heavily weighted by the bulk of muscle and its metabolic state at the time of measurement. Even if an average figure were more accurately known either from individual cells or for the body as a whole, it is doubtful what it could mean, because regional variation both between tissues and within individual cells must be very great. Nonetheless, the knowledge of (H^+) within individual tissues (e.g. the myocardium or liver) may prove to be of importance when function is disturbed.

REGULATION OF (H^+)

There is no one receptor or centre which is specifically sensitive to (H^+) and which integrates all aspects of its regulation. Nevertheless, the mechanisms which resist change in (H^+) are efficient and well co-ordinated. They are three in number.
1   Buffer systems within the body.
2   The renal regulation of H^+ excretion.
3   The respiratory regulation of CO_2 excretion.

THE BUFFERING ACTION OF BODY FLUIDS

The first line of defence against a change in (H^+) is provided by the buffer systems of the body fluids. If a strong acid or alkali is added to the extracellular fluid, 30–50 per cent is buffered by systems immediately available in the extracellular fluid (notably bicarbonate-carbonic acid) and the remainder is either buffered in the cells and bone or by the release of buffers from these. This buffering is so effective that the addition of 1 mmol of strong acid per l of body fluid only increases (H^+) by about 5 nmol/l. That is, all but 5 of every million H^+ ions added are buffered. The sites of these buffers govern their availability and the magnitude and rate of the resultant change in (H^+). The buffers of blood and extracellular fluid are immediately available; those of intracellular fluid are less so but are still operative in a few minutes; the immense buffer capacity of bone requires hours or days to become effective. These differences are partly due to blood flow and also to the tissue rates of exchange with fluids.

Bicarbonate, carbonic acid, dissolved CO_2 and hydrogen ion concentrations

Bicarbonate and carbonic acid are quantitatively the most important buffer system in the extracellular fluid.

$$HCO_3^- + H^+ \rightleftharpoons H_2CO_3 \qquad (7)$$

Addition of H^+ ions causes the equilibrium to move from left to right and removal causes it to move from right to left. Chemically this is not a very good buffer system because carbonic acid is quite a strong acid. Its unique physiological virtue lies in the fact that H_2CO_3 is in equilibrium with the dissolved CO_2 of the body fluids:

$$H_2CO_3 \rightleftharpoons CO_2 + H_2O \qquad (8)$$

Hence, by varying the volume of CO_2 excreted from the lungs, the equilibrium in equations 7 and 8 can be adjusted to resist changes in (H^+) (see also p. 233).

The equilibrium conditions for equation 7 can be expressed in the following version of the law of mass action:*

$$(H^+) = K' \frac{[H_2CO_3]}{[HCO_3^-]} \qquad (9)$$

* The Henderson-Hasselbalch equation is derived from the equilibrium constant (K') of the reaction as follows:

$$(H^+) = K' \times \frac{[H_2CO_3]}{[HCO_3^-]}$$

Taking logarithms, and changing the sign of all terms:

$$-\log (H^+) = -\log K' - \log \frac{[H_2CO_3]}{[HCO_3^-]}$$

which can be rewritten:

$$pH = pK' + \log \frac{[HCO_3^-]}{[H_2CO_3]}$$

Carbonic acid forms a small virtually constant portion (about one molecule per 1000) of the CO_2 in solution (equation 8), so equation 9 can be rewritten:

$$(H^+) = K' \frac{[CO_2]k}{[HCO_3^-]} \tag{10}$$

where k is the equilibrium constant of the reaction of CO_2 with water to form H_2CO_3. Usually k is incorporated in K' and Hasselbalch's version of the equation is used:

$$pH = pK' + \log \frac{[HCO_3^-]}{[CO_2]} \tag{11}$$

If $[HCO_3^-]$ and $[CO_2]$ are expressed as mmol/l and (H^+) as nmol/l, K' for plasma is about 800. Using the same units for $[HCO_3^-]$ and $[CO_2]$, pK' is about 6·1.

As the concentration of dissolved CO_2 is proportional to the partial pressure of CO_2 ($[CO_2]$ mmol/l = 0·03 P_{CO_2} mmHg or 0·225 P_{CO_2} kPa), the equations can be further written:

$$(H^+) \text{ nmol/l} = 24 \frac{P_{CO_2} \text{ mmHg}}{[HCO_3^-] \text{ mmol/l}} \tag{12}$$

$$= 180 \frac{P_{CO_2} \text{ kPa}}{[HCO_3^-] \text{ mmol/l}} \tag{13}$$

$$pH = 6·1 + \log \frac{[HCO_3^-] \text{ mmol/l}}{0·03 \, P_{CO_2} \text{ mmHg}} \tag{14}$$

$$= 6·1 + \log \frac{[HCO_3^-] \text{ mmol/l}}{0·225 \, P_{CO_2} \text{ kPa}} \tag{15}$$

Equations 9, 10, 12 and 13 are versions of the Henderson equation; 11, 14 and 15 are versions of the Henderson–Hasselbalch equation.

These equations enable the third variable to be calculated if the other two are known. There are many nomograms and other graphical ways (Fig. 5.1) for solving the equations. Although satisfactory for most clinical purposes, such calculations of the third variable from knowledge of the other two are not entirely accurate, chiefly because the 'constants' K' and pK' vary slightly with temperature, (H^+) and other factors.

Other buffers: haemoglobin

The acid formed in the largest quantity in the body is, of course, carbonic acid itself which is formed from the CO_2 produced in the tissues. The most important buffer for dealing with carbonic acid is haemoglobin:

$$H^+ + Hb^- \rightleftharpoons HHb \tag{16}$$

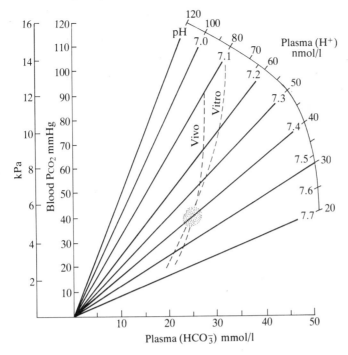

Figure 5.1. *The* $P_{CO_2} : [HCO_3{}^-]$ *diagram.*

This is a graphical representation of equations 12 to 15 in which the acid : base ratio of the $CO_2:HCO_3{}^-$ system is used to indicate (H^+) and pH. The central shaded ellipse represents the normal range found in arterial blood in resting subjects. The 'vitro' curve shows the changes in (H^+) and $[HCO_3{}^-]$ when the P_{CO_2} of blood with a normal Hb concentration is changed *in vitro*; the 'vivo' curve shows the changes in the arterial blood when the P_{CO_2} of the body is changed acutely (over a few minutes or hours). The *in vivo* curve is steeper and the change in (H^+) greater because the buffering effect of Hb, instead of being concentrated in the blood, is shared throughout the extracellular fluid. In this diagram, as in others, such as the Astrup-Siggaard-Andersen log P_{CO_2}:pH diagram and the Davenport $[HCO_3{}^-]$:pH diagram, changes in position of the buffer or dissociation curve are produced by non-respiratory changes in $[HCO_3{}^-]$ while changes in slope of the curve are due to changes in the concentration of the blood buffers, particularly Hb.

This reaction minimizes the rise in blood (H^+) associated with the transport of CO_2 from the tissues to the lungs. Haemoglobin is of particular importance in this respect because its buffering power varies with oxygenation and reduction. Reduced Hb^- is a stronger base (i.e. will bind more H^+ ions) than oxy-Hb^- so that the removal of O_2 from the blood in the capillaries of the tissues simultaneously and inevitably provides a buffer for much of the H_2CO_3 added to it. The situation of Hb within the red cells is of particular importance to the $CO_2/HCO_3{}^-$ buffer system. CO_2 in solution enters red cells freely and carbonic anhydrase, which catalyses equation (8) is also intracellular. Hydrogen ions are thus formed in close proximity to the buffer with which they react.

Haemoglobin is able, of course, not only to buffer carbonic acid but also all acids, and in the blood (as opposed to the interstitial fluid) the concentration of haemoglobin makes it quantitatively as important a buffer as carbonic acid–bicarbonate. The plasma proteins also act as buffers but their capacity in mmol/g is only one-third that of haemoglobin. As the haemoglobin concentration is normally twice that of the plasma proteins, haemoglobin is normally six times as important as the plasma proteins in the total buffer capacity of the blood.

HYDROGEN ION BALANCE AND THE KIDNEY

Hydrogen and hydroxyl ions are taken in with the diet and are also produced by many metabolic processes. Normally there is an excess absorption and production of H^+ ions amounting to about 50 mmol or more daily. Most of these H^+ ions are derived from sulphuric and phosphoric acids formed by the breakdown of protein and other complex substances. The concentration in the body is maintained constant by the excretion of an equal amount in the urine. The two most important mechanisms by which the kidney tubules accomplish this are: (1) by adding H^+ to buffers such as the HPO_4^{--} in the glomerular filtrate, to form $H_2PO_4^-$, and (2) by forming ammonia NH_3, which carries H^+ out as ammonium NH_4^+.

An alkaline diet tends to cause a decrease of (H^+) in the body fluids. In such circumstances the kidney reduces ammonia formation, excretes more HPO_4^{--} and less $H_2PO_4^-$, and may also excrete HCO_3^-.

HYDROGEN ION BALANCE AND THE LUNGS

CO_2 and water are the end products of many metabolic processes. The CO_2 is formed in solution and some of it combines with H_2O to form carbonic acid which in turn partly dissociates

$$CO_2 + H_2O \rightleftharpoons H_2CO_3 \rightleftharpoons H^+ + HCO_3^- \qquad (17)$$

CO_2 is excreted in the lungs and the reactions move from right to left so that the H^+ ions formed in the tissues disappear as H_2O. The concentration of CO_2 in the blood leaving the lungs depends upon the partial pressure of CO_2 in the alveolar air. A change in the alveolar P_{CO_2} therefore alters the equilibria in equation 17. A rise in alveolar P_{CO_2} shifts the reactions to the right, thereby increasing (H^+) in the arterial blood.

It will be appreciated, therefore, that, although H^+ ions are not themselves excreted through the lungs, any alteration in the balance of CO_2 production and elimination alters the (H^+) of the body fluids by altering the P_{CO_2}. The relation between CO_2 production, the volume of the breathing and P_{CO_2} is a simple one, as will be appreciated from the following derivation.

The amount of CO_2 excreted by the lungs each minute equals the amount of CO_2

in each litre of alveolar air multiplied by the number of litres of alveolar air expired. This can be restated as follows: the volume of CO_2 expired equals the alveolar concentration of CO_2 times the alveolar ventilation:

$$\text{Vol. } CO_2 \text{ excreted} = \text{Alv. } CO_2 \text{ concentration} \times \text{Alv. ventilation} \qquad (18)$$

The alveolar CO_2 concentration is related to the same P_{CO_2} as the arterial blood, and in a steady state the CO_2 excreted by the lungs must equal the amount produced by the tissues; so: Vol. CO_2 produced is proportional to Arterial P_{CO_2} × Alv. ventilation or:

$$\text{Arterial } P_{CO_2} \propto \frac{\text{Vol. } CO_2 \text{ produced}}{\text{Alveolar ventilation}} \times \text{Barometric pressure} \qquad (19)$$

It follows, therefore, that any disturbance of the balance between CO_2 production and alveolar ventilation causes a change in P_{CO_2} and therefore in (H^+), not only in the arterial blood but throughout the body fluids. The size of the potential changes involved is very great. A normal subject at rest excretes 15 moles (330 l.) of CO_2 through the lungs each day. This contrasts with the 50 mmol of H^+ excreted by the kidneys normally, and the 600 mmol they can excrete when functioning maximally. An illustration of this contrast is that suppression of urinary formation of H^+ for 30 minutes causes no detectable change in the (H^+) of the extracellular fluid. Suppression of respiratory CO_2 excretion for 30 minutes would cause the (H^+) to exceed 100 nmol/l (pH below 7).

Respiratory response to changes in (H^+)

The preceding paragraph has emphasised the importance of changes in ventilation in altering the (H^+) by changing P_{CO_2}. As described below this mechanism is the cause of respiratory acidosis or alkalosis; it is also the main homeostatic or compensatory response to non-respiratory (metabolic) acid:base changes. This response is mediated by chemoreceptors close to the surface of the medulla which seem to adjust pulmonary ventilation when non-respiratory acids or bases are added to the body so as to maintain the cerebrospinal (H^+) at about 50 nmol/l (pH 7·30).

Effects of raising and lowering P_{CO_2} (Figure 5.1)

The effects of changes in P_{CO_2} upon the acid:base composition of the blood are physiologically important and must be remembered when interpreting clinical problems (p. 236). Three different situations must be considered. The changes will be expressed as the changes in $[HCO_3^-]$ and (H^+) produced by a 1 mmHg change in P_{CO_2}:

$$(\Delta HCO_3^- : \Delta P_{CO_2} \text{ and } \Delta H^+ : \Delta P_{CO_2})$$

The values quoted below are linear approximations which are applicable over the range P_{CO_2} 25–75 mmHg (3–10 kPa).

(*i*) *Blood* in vitro. Although this situation may seem irrelevant it is important for two reasons: first, the changes have been used in several analytical approaches (such as those of Siggaard-Andersen and Astrup); secondly, many older accounts of the body's response to a changing P_{CO_2} are based on studies of blood *in vitro*. *In vitro*, at a normal Hb concentration, the $\Delta HCO_3^- : \Delta P_{CO_2}$ is 0·3 mmol.l^{-1}.mmHg (2·25 mmol.l^{-1}.mmHg^{-1}) and $\Delta H^+ : \Delta P_{CO_2}$ is 0·3 nM/mmHg (2·25 nM/kPa).

(*ii*) *Acute changes in blood* in vivo. If the P_{CO_2} of the body is acutely raised or lowered, the arterial blood changes are $\Delta HCO_3^- : \Delta P_{CO_2}$, 0·1 mmol.$l^{-1}$.mmHg^{-1} (0.75 mmol.$l^{-1}$.kPa^{-1}) and $\Delta H^+ : \Delta P_{CO_2}$, 0·6 nM/mmHg (4.5 nM/kPa). These values are reached within 10 min of the change in P_{CO_2} and last for several hours.

(*iii*) *Chronic changes in blood* in vivo. If the P_{CO_2} of the body is raised or lowered for several days, the arterial blood changes are very similar to those of blood *in vitro*, $\Delta HCO_3^- : \Delta P_{CO_2}$, is 0.3 mmol.$l^{-1}$.mmHg^{-1} (2.25 mmol.$l^{-1}$.kPa^{-1}); $\Delta H^+ : \Delta P_{CO_2}$ is 0.3 nM/mmHg (2.25 nM/kPa).

Explanation for these changes. (i) Blood *in vitro*, with Hb of 15 g/dl is well buffered so that if P_{CO_2} is raised the H^+ ions are mopped up and the equilibrium of the reaction of equation 17 remains over to the right.

(ii) When the P_{CO_2} of the body is acutely raised *in vivo* the bulk of the CO_2 stays in the extracellular fluid (plasma plus interstitial fluid) which is much less well buffered than blood. It is a reasonable approximation to regard these changes as the result of adding the CO_2 to a liquid (blood plus interstitial fluid) whose only buffering is supplied by Hb; i.e. a notional Hb concentration of about 4 g per 100 ml. Consequently the rise in (H^+) is greater and that in $[HCO_3^-]$ is less.

(iii) If the P_{CO_2} is kept raised, the renal tubule cells generate H_2CO_3, excreting the H^+ and passing the HCO_3^- back into the plasma. Thus, whereas (i) and (ii) are entirely physico-chemical matters, (iii) is physiological, and depends for its efficacy on the time course of the renal reactions.

Disordered function

ACIDOSIS AND ALKALOSIS: DIFFICULTIES IN TERMINOLOGY

Much of the difficulty in following accounts of acid:base disturbances hinges on differences in usage. The chief of these is the ambiguity of the meaning of the terms *acidosis* and *alkalosis*. Sometimes they are used in a chemical sense; sometimes in a physiological one. Thus, some people imply that 'an acidosis is an increased blood (H^+)', that is, the disturbance in the chemical composition is the acidosis. Others imply that 'an acidosis is a disturbance of function causing an accumulation of H^+ ions'; i.e. the disturbance of function rather than a change in chemical composition is the acidosis. Although logically undesirable the distinction would not matter very much if the change in function always caused the change in composition; but this is not so. Because of compensatory or opposing disturbances the (H^+) may be normal or low. The physiological usage is to be preferred and in this account the terms *acidosis* and *alkalosis* will be used to imply *abnormal processes or conditions which would cause a deviation of (H^+) if there were no secondary changes*. When describing abnormalities of chemical composition it is preferable to specify the variable in question and to say whether it is increased or decreased. But it is sometimes convenient to use the following terms: *acidaemia* and *alkalaemia* mean a high and low blood (H^+); *hypercapnia* and *hypocapnia* mean a high and low blood P_{CO_2}.

The (H^+) in the intracellular fluid is different from that in the extracellular fluid and may vary independently in disease. The terms intracellular and extracellular acidosis or alkalosis should therefore be used to make it clear which fluid compartment is under discussion. If the site is not specified then it is usual to imply that the extracellular fluid is being considered. This usage is observed throughout the following account and elsewhere in this book.

Intracellular and extracellular changes do not diverge in respiratory disturbances because CO_2 readily crosses cell membranes and can be converted into H_2CO_3 in any body fluid. Divergence in degree and sometimes in direction may occur in non-respiratory disorders because HCO_3^- does not readily cross the membranes of cells other than red cells. Such divergences may be reflected in the acid:base composition of the CSF. An example of this situation is seen following the administration of HCO_3^- to correct a metabolic acidosis; the blood $[HCO_3^-]$ immediately increases but the CSF $[HCO_3^-]$ follows much more slowly leaving the CSF still acid, whereas acidity in the blood has decreased. Hyperventilation thus continues so that arterial P_{CO_2} remains low and blood (H^+) may fall below normal.

RESPIRATORY (GASEOUS) AND METABOLIC (NON-GASEOUS)
ACIDOSIS AND ALKALOSIS

There are many ways in which (H^+) can be increased or decreased. Physiologically and clinically they are best separated into two groups.

1 Disturbances of respiration which affect (H^+) through their effects on CO_2 and H_2CO_3.

2 All other disturbances.

Changes in the first group are usually called 'respiratory' acidosis or alkalosis, and those in the second are called 'metabolic'. These terms are not really satisfactory. For instance, an increase in (H^+) due to increased metabolic CO_2 production has to be called 'respiratory'.

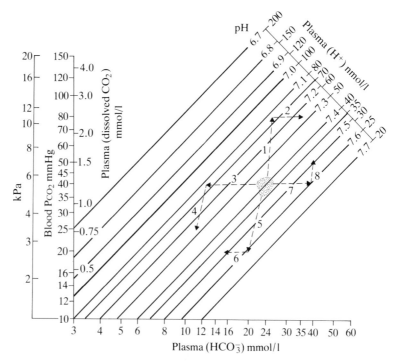

Figure 5.2. *The changes in pH, Pco$_2$ and bicarbonate in the four primary disturbances of hydrogen ion regulation.*

The axes are the same as Fig. 5.1, but they are plotted logarithmically, thus covering a greater range and increasing the discrimination at lower values of Pco$_2$ and [HCO$_3^-$].

$$CO_2 + H_2O \rightleftharpoons H_2CO_3 \rightleftharpoons H^+ + HCO_3^-$$

$$(H^+) = 24 \, \frac{Pco_2 \text{ mmHg}}{[HCO_3^-]} = 180 \, \frac{Pco_2 \text{ kPa}}{[HCO_3^-]}$$

$$pH = 6\cdot1 + \log \frac{[HCO_3^-]}{0\cdot03 Pco_2 \text{ mmHg}} = 6\cdot1 + \log \frac{[HCO_3^-]}{0\cdot225 \, Pco_2 \text{ kPa}}$$

Respiratory acidosis. Pulmonary ventilation is reduced. Alveolar and arterial Pco$_2$, and therefore [H$_2$CO$_3$], increase (arrow 1). (H$^+$) increases (pH falls). The kidney retains HCO$_3^-$ and excretes H$^+$ and therefore tends to restore (H$^+$) towards normal (arrow 2).

Respiratory alkalosis. Pulmonary ventilation is increased (arrow 5). Alveolar and arterial Pco$_2$, and therefore [H$_2$CO$_3$], fall. (H$^+$) decreases (pH rises). The kidney excretes more HCO$_3^-$ and retains H$^+$ ·and therefore tends to restore (H$^+$) towards normal (arrow 6).

Metabolic acidosis. Addition or production of H$^+$ causes a rise in (H$^+$) (fall in pH). Buffering of the rise in (H$^+$) involves the combination of H$^+$ and HCO$_3^-$ to form H$_2$CO$_3$, thus causing a reduction in [HCO$_3^-$] (arrow 3). As a result of this buffer action there should also be increases in [H$_2$CO$_3$], in the concentration of dissolved CO$_2$ and therefore in Pco$_2$. Two things prevent this. Firstly, the excess CO$_2$ would be very easily excreted by the lungs to maintain Pco$_2$ constant. Secondly, the rise in (H$^+$) stimulates the breathing so much that, in fact, more CO$_2$ is washed out, causing Pco$_2$ and [H$_2$CO$_3$] to fall. The (H$^+$) is therefore returned towards normal (arrow 4).

Also, renal causes of acidosis or alkalosis are not well termed 'metabolic'. It might be more satisfactory to use the terms 'gaseous' for those due to changes in CO_2 concentration and 'non-gaseous' for the others. However, these terms have not gained acceptance so the conventional 'respiratory' and 'metabolic' will continue to be used in the following account.

There are four types of disturbance: respiratory acidosis, metabolic acidosis, respiratory alkalosis and metabolic alkalosis. These rarely occur clinically in isolation; there is usually a secondary change tending to restore (H^+) to normal and often a further complicating primary disorder which may tend either to restore or aggravate a change in (H^+). The main effects of the primary disorders of H^+, Pco_2 and HCO_3^- will now be described. It must be emphasized that this account will mention only simple features of the underlying physiology; a fuller description would require an extensive consideration of many other topics notably in respiratory and electrolyte physiology.

A. *Respiratory (gaseous) acidosis*

Respiratory acidosis is present whenever the ratio of alveolar ventilation to CO_2 production falls (see p. 225, equation 19) whether by a rise in CO_2 production or a fall in alveolar ventilation; it is characterized by *hypercapnia,* an increase in arterial Pco_2. *Primary respiratory acidosis* is synonymous with *ventilatory failure* (p. 121). The primary change is a rise in Pco_2 (arrow 1, Fig. 5.2). Given time, the kidney compensates for the rise in (H^+) by excreting more H^+ ions and retaining bicarbonate (causing the secondary change illustrated by arrow 2 in Fig. 5.2). Thus the secondary response to a respiratory acidosis is an increase in $[HCO_3^-]$ occurring over hours or days. The degree of acidaemia depends on the rapidity of development and severity of the acidosis. The changes in blood (H^+) and $[HCO_3^-]$ when Pco_2 is acutely raised (the '*in vivo* dissociation or titration curve of blood') are shown in Fig. 5.2 (p. 228). In acute respiratory acidosis the renal response is negligible. If ventilation ceased at rest, H^+ ions would accumulate in the body at a rate of about 10 mmol per minute. This is over 30 times faster than the rate at which the kidney can excrete H^+ ions.

B. *Metabolic (non-respiratory, non-gaseous) acidosis*

Metabolic acidosis is present if there is an accumulation of acid other than H_2CO_3 in the extracellular fluid, or a loss of base.

Metabolic alkalosis. Loss or neutralization of H^+ ions causes a reduction in (H^+) (increase in pH). Buffering of the reduction in (H^+) is partly effected by dissociation of H_2CO_3 to H^+ and HCO_3^-, thus increasing $[HCO_3^-]$ (arrow 7). This buffer action would tend to cause a fall in $[H_2CO_3]$ and in Pco_2. However, the fall in (H^+) depresses the breathing so that reduced amounts of CO_2 are excreted and Pco_2, and $[H_2CO_3]$, rise. The (H^+) is therefore restored towards normal (arrow 8). (*Lancet,* 1962, ii, 154 modified.)

An increase of acid occurs: if an excess of H^+ ions is ingested (e.g. ingestion of $NH_4Cl \rightarrow H^+ + Cl^- + NH_3$); if an excess of organic acids (e.g. lactic acid or aceto-acetic acid) is produced; or if the kidney fails to excrete sufficient acid.

A decrease of base occurs if there is loss of alkaline alimentary secretions as by diarrhoea or biliary fistula.

The primary change in metabolic acidosis (arrow 3, Fig. 5.2) is a rise in (H^+) and a fall in $[HCO_3^-]$. The rise in (H^+) causes an increase in pulmonary ventilation mediated immediately through peripheral chemoreceptors and eventually through increasing acidity of the CSF and thus a reduction in PCO_2 so the direction of the secondary change is as indicated by arrow 4 in Fig. 5.2. The response to a metabolic acidosis therefore causes hypocapnia, and both primary change and response cause reductions in plasma HCO_3^- concentration.

In minor degrees of metabolic acidosis (e.g. following a short bout of brisk exercise), the respiratory response can soon correct the initial acidaemia, but in severe metabolic acidosis an acidaemia occurs which can only be corrected by excretion of the acid, or (in the case of the organic acids such as lactic and aceto-acetic) by its oxidation to CO_2, which can be excreted by the lungs, and to H_2O. It is important to remember that, although buffering and hypocapnia can prevent acidaemia, the full restitution of normal acid:base composition requires the removal of the offending acid or addition of the missing base.

Unless there is a disorder of renal secretion of H^+, the urine in metabolic acidosis is always acid (the pH usually being less than 5·3).

C. *Respiratory (gaseous) alkalosis (hypocapnia; hypocarbia)*

Respiratory alkalosis exists whenever alveolar PCO_2 is lowered by excessive ventilation. It occurs in artificial or voluntary overbreathing and during excessive stimulation of the breathing by a number of nervous and chemical disorders such as anoxia, various forms of cerebral disease, hepatic failure and salicylate poisoning. Respiratory alkalosis is also a feature of certain disorders of the pulmonary circulation, notably thrombo-embolic pulmonary hypertension, in which respiration is reflexly stimulated.

The primary change is a reduction in PCO_2 (arrow 5, Fig. 5.2). The kidney reacts by excreting more HCO_3^- ions and reabsorbing more H^+ ions (thus excreting an alkaline urine) causing the secondary changes illustrated by arrow 6, Fig. 5.2. Thus the secondary response to a respiratory alkalosis is a decrease in $[HCO_3^-]$ and both the primary change and the response cause reductions in plasma HCO_3^- concentration. Another factor which opposes the fall in (H^+) during an acute respiratory alkalosis is an increase in the blood concentration of lactic and pyruvic acids, due to increased glycolysis by the red cells. During prolonged hyperventilation, lactic acid may also be produced in the respiratory muscles.

As in the case of respiratory acidosis, the presence or absence of an alkalaemia depends chiefly on the acuteness of the development of the alkalosis. If it develops rapidly the kidney is overwhelmed, but in a chronic state severe alkalaemia is unusual.

The urine is initially alkaline but usually becomes acid later.

D. *Metabolic (non-respiratory, non-gaseous) alkalosis*

Metabolic alkalosis is present when there is a loss of acid other than H_2CO_3 from the extracellular fluid or an increase of base.

A loss of acid occurs during vomiting when there is loss of HCl, and a gain in base occurs when absorbable alkalis such as sodium bicarbonate are administered in excess. There are also several conditions associated with potassium depletion in which altered renal function causes a metabolic alkalosis because the kidney excretes too many H^+ ions. Renal loss of Cl^- and retention of HCO_3^- also occurs following administration of thiazide diuretics, frusemide or ethacrynic acid.

It is in the group of disorders causing metabolic alkalosis that the dissociation between extracellular and intracellular (H^+) is best recognized. Potassium depletion commonly causes an extracellular alkalosis at the same time as an intracellular acidosis. This is believed to be caused by H^+ ions taking the place of K^+ inside the cells.

The primary change in metabolic alkalosis (arrow 7) is a fall in (H^+) and rise in $[HCO_3^-]$. The secondary change should be a rise in Pco_2 (arrow 8), but this is often small and may not occur at all. The reason for this lack of respiratory response is not clear but may be connected with the divergence between the intracellular and extracellular change in (H^+).

The lack of respiratory response leads to the development of a significant alkalaemia in most cases of primary metabolic alkalosis.

Both the primary and secondary changes cause an increase in plasma HCO_3^- concentration. The urine is alkaline at some stage in all those conditions due to abnormal loss of H^+ or abnormal absorption of base, because the kidney responds by an increased excretion of basic ions (HPO_4^{--} and HCO_3^-) and an increased reabsorption of H^+. However, in the metabolic alkalosis of potassium depletion the urine is acid because the alkalosis is actually produced by an excessive renal excretion of H^+ ions. The urine is also paradoxically acid when metabolic alkalosis is associated with extreme sodium loss, as after prolonged vomiting.

MIXED DISTURBANCES

Mixed disturbances are very common and may have opposing or additive effects on blood (H^+).

Opposing disturbances

Respiratory alkalosis and metabolic acidosis. This combination occurs, for example, if renal failure is combined with heart failure and pulmonary congestion. A better example is salicylate poisoning in which an initial respiratory alkalosis due to the stimulation of the ventilation by the salicylate radical is complicated by a later metabolic acidosis.

Respiratory acidosis and metabolic alkalosis. This combination commonly occurs in patients with ventilatory failure and oedema ('cor pulmonale') who are overtreated with diuretics causing K^+ and/or Cl^- depletion. This combination is commonly associated with intracellular acidosis, said to be caused by movement of H^+ into cells in exchange for K^+, at the time of the initial acidosis. Loss of K^+, due to the diuretic, makes the return of K^+ to the cells less easy. The urine is acid.

Additive disturbances

These set particularly sinister traps for the unwary partly because they usually occur in patients who are so ill in other ways that they are overlooked and partly because they have opposing effects on the plasma $[HCO_3^-]$ which may thus not be very high or very low. This may mislead those who rely only on a measurement of venous $[HCO_3^-]$ as part of routine electrolyte examination in management.

Respiratory acidosis and metabolic acidosis. This combination is probably the final common pathway of respiratory and circulatory failure and is therefore frequently found. The poor pulmonary circulation and underventilation cause CO_2 retention; the poor peripheral circulation causes stagnant hypoxia and lactic acidosis; and the poor renal circulation reduces the ability of the kidney to excrete H^+ ions.

Respiratory alkalosis and metabolic alkalosis. This combination can occur as a complication of respiratory failure if excessive artificial ventilation is given and K^+ and/or Cl^- depletion is induced by excessive diuretic therapy.

DISTURBANCES OF INTRACELLULAR $[H^+]$

As indicated on page 227 little is known about intracellular (H^+) and it is doubtful to what extent values quoted for it are meaningful. Certain generalizations are, however, important. First, CO_2 appears to cross cell membranes so easily that respiratory (gaseous) disturbances tend to alter the intracellular (H^+) in the same direction as the extracellular. Secondly, this parallelism does not necessarily apply to metabolic (nongaseous) disturbances. Thus in potassium depletion there may be a raised intracellular (H^+) associated with a low extracellular (H^+) (see above). The addition of strong acids or bases to the extracellular fluid may cause an acute paradoxical shift in the intracellular reaction. For example, the addition of HCl (as NH_4Cl) increases the extracellular (H^+) and stimulates the breathing thus lowering intracellular P_{CO_2} and $[H_2CO_3]$. As the added H^+ and Cl^- ions do not enter the cells at once, the intracellular (H^+) may temporarily fall.

Studies of isolated muscle suggest that intracellular (H^+) may change less when extracellular (H^+) is raised (by manipulation of either Pco_2 or $[HCO_3^-]$) than when it is lowered.

Principles of tests and measurements

A full evaluation of the acid:base state requires knowledge of two of the three components of the acid:base equation so that the third can be calculated.

It is important to realize that a description of the chemical composition of the blood, however complete, does not necessarily explain the mechanism causing any disturbance and also to realize that the same change in chemical composition can result from different disturbances.

Knowledge of either the clinical picture or of sequential measurements leading up to that under consideration may be needed before a true interpretation can be made.

pH or (H⁺)

pH as measured approximates to the negative log of H^+ activity. The measurement consists of an electrometric comparison between the unknown solution (blood) and a series of standard buffers to which pH values have been assigned using sound physico-chemical principles. Technically, this comparison is now easy, but some theoretical and practical difficulties remain. The ionic strength (a function of total ion concentration and valency) of the solution to be analysed has an important effect and cannot be determined accurately for solutions containing protein ions. The present position therefore is that changes in pH in a given solution can be measured very accurately (± 0.005 units in the physiological range); the absolute pH value in terms of the agreed scale can be less certainly determined; the interpretation of the 'absolute' pH value in terms of H^+ ionic activity is, in turn, less certain; finally, the relation between the H^+ ionic activity and the H^+ concentration of a solution is uncertain.

THE RESPIRATORY VARIABLE: ARTERIAL Pco_2

Arterial Pco_2 may be measured by means of the Pco_2 electrode (in which pH is the actual variable measured), by other methods based on pH measurement, or by measuring the Pco_2 of alveolar air or by using rebreathing to estimate the Pco_2 of mixed venous blood from which at rest the arterial Pco_2 can be calculated with acceptable accuracy.

The 'electrode' method. The pH electrode is covered with a membrane permeable to CO_2 but not to HCO_3^-. The thin space between the electrode and this membrane is occupied by a matrix such as cellophane holding a dilute bicarbonate solution. When the whole is placed in blood, the CO_2 diffuses through the membrane, forms H_2CO_3 and changes the pH of the solution to a value depending on the Pco_2 of the blood. Efficient manual and automatic systems are available for the performance of this measurement.

The 'interpolation' (Astrup) method. The pH of the blood is measured and then the blood is equilibrated with two gases of known Pco_2 and the pH measured again after each equilibration. This procedure allows the standard bicarbonate concentration to be determined. The original Pco_2 can be calculated from the original pH. In practice these steps can be performed rapidly and semi-automatically.

Blood Pco_2 measurement using a pH meter. The locked relationship between (H^+) (or pH), Pco_2 and $[HCO_3^-]$ means that Pco_2 can be calculated from pH measurements if $[HCO_3^-]$ is known.

The 'indirect' method. Blood Pco_2 can be estimated by measuring pH and bicarbonate or CO_2 concentrations (see above) and calculating Pco_2 making use of the Henderson–Hasselbalch equation or one of its graphical forms.

THE NON-RESPIRATORY (METABOLIC) VARIABLE: BICARBONATE,
ALKALI RESERVE, BASE EXCESS

As a starting point, non-respiratory factors (including renal function) can be said to create an extracellular fluid with $[HCO_3^-]$ of 25 mmol/l and respiratory function to create an extracellular fluid with a Pco_2 of about 40 mmHg (5·3 kPa) having a concentration of dissolved CO_2 of 1·25 mmol/l so that the ratio of 20 : 1 produces a (H^+) of 40 nmol/l or a pH of 7·4. So why not simply measure the $[HCO_3^-]$? Most people would say that this is the best thing to do but, unfortunately, the $[HCO_3^-]$ is not entirely independent of changes in Pco_2. If Pco_2 is raised or lowered it is equivalent to the addition or removal of $[HCO_3^-]$ and the H^+ ions liberated by the dissociation of $[H_2CO_3]$ are buffered by protein, notably Hb:

$$H_2CO_3 + Hb^- \rightleftharpoons HHb + HCO_3^-$$

The effect of this is that the plasma $[HCO_3^-]$ as eventually recorded by whatever analyser is used depends, first, on *in vivo* factors. It is necessary to consider the Pco_2 of the body as a whole and the Pco_2 of the particular sample of blood taken (which in turn depends upon the local blood flow and metabolism of the tissues and the presence or absence of any stagnation in the vessel from which the sample was taken). Secondly, recorded values reflect *in vitro* changes induced as a result of contact with air during sampling, storage, transport, etc. Thus if blood is allowed to lose CO_2 so that, say, its

P_{CO_2} falls to 20 mmHg (2·7 kPa) and the plasma is then separated from the cells, re-equilibration of the plasma with a P_{CO_2} of 40 mm (5·3 kPa) will not restore the $[HCO_3^-]$ concentration to its original value.

In vitro, the sum of $[Hb^-] + [HCO_3^-]$ per unit volume of blood is unaffected by changes in P_{CO_2}. The change in $[HCO_3^-]$ due to a change in P_{CO_2} is accompanied by an equal and opposite change in $[Hb^-]$ (ionized Hb). But this is not true *in vivo* because HCO_3^- ions migrate to the interstitial fluid so that, per unit volume of blood, $[Hb^-] + [HCO_3^-]$ falls. Hence a change in P_{CO_2} occurring *in vivo* may appear to produce a non-respiratory change when compared with what would happen to the same blood if its P_{CO_2} were changed *in vitro*.

Practical solutions

A number of approaches to these problems have been advocated and three are in common use.

1   Use arterial (or arterialized) blood and interpret the $[HCO_3^-]$ in relation to the measured P_{CO_2}. Several types of automatic analyzer are now available, which measure pH, P_{CO_2} and Hb in a sample of arterial blood and then calculate the other acid: base variables.

2   Use venous or capillary blood and equilibrate it with a P_{CO_2} of 40 mmHg (5·3 kPa) and enough oxygen to saturate the haemoglobin so that the respiratory variable is standardized. The $[HCO_3^-]$ of the plasma is then measured. This is the principle of the standard bicarbonate (and the now obsolete alkali reserve).

3   Use venous or capillary blood and titrate the blood to determine its base concentration either in absolute terms or by reference to a given value. The first of these is the principle underlying the buffer base measurement and the second underlies the base excess measurement. The easiest acid with which to titrate the blood is carbonic acid; this is essentially what the Astrup technique does when the relationship between change in P_{CO_2} and change in pH is measured and used to calculate the base excess.

Choice of measurement

For the evaluation of non-respiratory disturbances the best measurement is probably the first, in which the $[HCO_3^-]$ is evaluated by making allowance for the P_{CO_2}. In other words, $[HCO_3^-]$ is compared with the value which is to be expected for the titration curve or dissociation curve of the body as a whole such as Figs. 5·1 or 5·2.

INTERPRETATION OF BLOOD ACID:BASE MEASUREMENTS

The main purpose of what follows is to emphasize that a given set of values may be capable of several interpretations and that graphs, charts, nomograms, etc., which attach physiological or clinical terms to certain combinations of blood values are, at best, oversimplifications.

Table 5.1. *Normal values for arterial blood*

Measurement	Definition	Unit	Normal range
Hydrogen ion activity		mol/kg nM	$36-44 \times 10^{-9}$ 36-44
pH	Neg. $\log_{10}$ H^+ activity		7·44-7·36
Plasma CO_2 content (plasma T_{CO_2})	Vol. CO_2 extractable at existing P_{CO_2}	mmol/l	23-29
Plasma bicarbonate concentration	At existing P_{CO_2}	mmol/l	22-27
Plasma standard bicarbonate concentration	At $P_{CO_2} = 40$ mmHg (5·3 kPa) and O_2 saturation 100%	mmol/l	23-26
CO_2 pressure (P_{CO_2})		mmHg kPa	36-44 4·8-6·0
Base excess	Base concentration as measured by titration to pH 7·4 at P_{CO_2} 40 mmHg, 5·3 kPa	mmol/l whole blood	−3−+3
Whole blood buffer (buffer base) concentration	The sum of concentrations of the buffer anions of whole blood bicarbonate, plasma proteins, haemoglobin	mmol/l whole blood	42-55

LOW P_{CO_2} AND LOW HCO_3^-

A P_{CO_2} below 20 mmHg (2·7 kPa) usually signifies a respiratory alkalosis and there is usually an unequivocal alkalaemia ($H^+ < 30$ nM, pH > 7·5). A [HCO_3^-] below 15 mmol/l usually signifies a metabolic acidosis and there is usually an unequivocal acidaemia ($H^+ > 60$ nM, pH < 7·25). However, lesser degrees of disturbance can be difficult to interpret for a number of reasons: temporary disturbances of the breathing at the time of sampling; analytical scatter; the frequent coincidence of respiratory alkalosis and metabolic acidosis. Interpretation is made easier if either the clinical picture or a previous set of acid:base values is known. The latter may indicate whether the current values have been reached from a more obvious respiratory alkalosis or metabolic acidosis.

HIGH P_{CO_2} AND HIGH [HCO_3^-]

A P_{CO_2} over 55 mmHg (7·5 kPa) almost invariably signifies a respiratory acidosis, but the accompanying acidaemia is usually modest ($H^+ < 50$ nM, pH > 7·30) and may be ble 5.1 *continued*

Changes in disease		Remarks
Increased	Decreased	
cidaemia > 70 severe	Alkalaemia < 30 severe	
lkalaemia > 7·55 severe	Acidaemia < 7·15 severe	
rimarily in metab. alkalosis > 36 severe. Secondarily in respirat. acidosis	Primarily in metab. acidosis < 12 severe. Secondarily in respirat. alkalosis	Venous blood taken without stagnation 1–2 mmol/l greater
esp. acidosis > 80 severe	Resp. alkalosis < 20 severe	If lungs healthy, alveolar P_{CO_2} = arterial P_{CO_2} (p. 137). Rebreathing methods (p. 138) can be used for most clinical purposes
Metab. alkalosis	Metab. acidosis	Normal range and severity of disturbance depend on blood Hb. concentration and saturation. Negative values may be called 'base deficit'

absent. A $[HCO_3{}^-]$ over 36 mmol/l usually signifies a metabolic alkalosis and, in the absence of a coexistent respiratory acidosis (see below) there is usually an unequivocal alkalaemia ($H^+ < 30$ nM, pH $> 7·50$). However, again, lesser degrees of disturbance can be difficult to interpret, particularly in patients with respiratory acidosis. In a patient with an arterial P_{CO_2} of 65–70 mmHg (8·7 to 9·3 kPa), a $[HCO_3{}^-]$ of 30 mmol/l could signify an acute disturbance in which renal retention of $HCO_3{}^-$ had not yet occurred, but it could also signify a coincident metabolic acidosis due, for example, to anoxic lactic acidosis. In a patient with the same P_{CO_2}, a $[HCO_3{}^-]$ of 35 mmol/l could signify a chronic disturbance in which $[HCO_3{}^-]$ retention had occurred ("good compensation"), but it could also be accounted for by coincident metabolic alkalosis, due, for example, to K^+ and Cl^- depletion caused by diuretic therapy.

Practical assessment

CLINICAL OBSERVATIONS

There are four lines of evidence.

I   *Nature of underlying disease*

e.g. Renal disease, diabetes mellitus likely to cause metabolic acidosis; chronic lung disease more likely to cause respiratory acidosis.

2   *Symptoms and signs of abnormal* (H^+) *or* Pco_2

Of limited value, firstly because, with few exceptions, they are not discriminating; secondly because they are often dominated by the manifestations of the underlying condition or associated electrolyte changes; thirdly, (H^+) may vary from 30 to 60 nM (pH 7·5–7·25) and Pco_2 from 30 to 60 mmHg (4 to 8 kPa) without symptoms.

Increased (H^+): twitching, flapping tremor, confusion; proceeding if (H^+) over 80–100 nM (pH 7·1–7·0) to coma and convulsions.

Increased Pco_2: similar, plus tachycardia, hypertension, peripheral vasodilation, sweating, papilloedema; coma (CO_2 narcosis) if Pco_2 over 100 mmHg (13 kPa).

Decreased (H^+) or Pco_2 : paraesthesia, tetany, convulsions.

3   *Altered respiration*

Of most value in metabolic acidosis and respiratory alkalosis when ventilation is increased ('air hunger').

Of less value in metabolic alkalosis and respiratory acidosis because: firstly, normal breathing is so unobtrusive that reduction is difficult to detect; secondly, under-ventilation is uncommon in metabolic alkalosis (p. 231); thirdly, in respiratory acidosis the breathing is usually disturbed by deranged pulmonary function: total ventilation may be normal or increased but much of it is wasted on dead space so that alveolar ventilation is decreased (Chapter 3).

4 *Urinary reaction*

Urine acid (pH less than 7·4, usually less than 6·0) in metabolic and respiratory acidosis, and (paradoxically) in extracellular alkalosis with intracellular acidosis (p. 232).

Urine alkaline (pH greater than 7·4) in acute respiratory alkalosis but may be acid if chronic because urinary reaction then depends chiefly on the diet and metabolism.

Urine may be alkaline in metabolic alkalosis but usually not, because inappropriately acid urine is the cause (e.g. in K^+ depletion).

ROUTINE METHODS

P_{CO_2} by electrode
pH by glass electrode, giving (H^+)
$[HCO_3^-]$ from above, by calculation
$[HCO_3^-]$ in venous plasma
Mixed venous P_{CO_2} by rebreathing method.

Astrup method with arterialized capillary blood gives P_{CO_2} pH, standard bicarbonate, base excess. Automatic electrode systems measure P_{CO_2} and pH, and sometimes Hb and use microprocessor to calculate $[HCO_3^-]$, standard $[HCO_3]$, base excess &c.

Also: Plasma electrolytes for evidence of associated or causative conditions: $[K^+ + Na^+] - [Cl^- + HCO_3^-]$ 'the anion gap'—normally less than 20 mmol/l indicating presence or absence of abnormal ions.

SPECIAL TECHNIQUES

Also: Blood for abnormal radicals: ketones; lactate, salicylate.
Tests for renal capacity to acidify urine.
Respiratory function tests.

References

COHEN J.J. & KASSIRER J.P. (1982) *Acid/Base*. Boston, Little Brown.

COHEN R.D. & ILES R.A. (1977) Lactic acidosis: some physiological and clinical considerations. *Clinical Science and Molecular Medicine*, **5**, 405–10.

DAVENPORT H.W. (1974) *The A.B.C. of Acid-Base Chemistry*, 6th ed. Chicago, University of Chicago Press.

KAUFMAN H.E. & ROSEN S.W. (1965) Clinical Acid-Base Regulations—the Bronsted Schema. *Surgery, Gynecology and Obstetrics*, **103**, 101–12.

NARINS R.G., JONES E.R., STOM M.C. & RUDWICK M.R. (1982) Diagnostic strategies in disorders of fluid, electrolyte and acid-base homeostasis. *American Journal of Medicine*, **72**, 496–520.

SCHWARTZ W.B. & COHEN J.J. (1978) The nature of the renal response to chronic disorders of acid-base equilibrium. *American Journal of Medicine*, **64**, 417–28.

TIZIANELLO A., DE FERRARI G., GURRERI G. & ACQUARONE N. (1977) Effects of metabolic alkalosis, metabolic acidosis and uraemia on whole-body intracellular pH in man. *Clinical Science and Molecular Medicine*, **52**, 125–35.

6

Blood

RED CELLS

Normal Function

The function of red cells is to increase the efficiency of gas exchange between the tissues and the environment. The red cell contains haemoglobin; an oxygen carrying molecule exquisitely adapted to its function. Its enclosure within cells is necessary to provide a high oxygen carrying capacity per unit volume of blood without increasing blood viscosity. The shape of the red cell provides a large surface area for gas exchange.

STRUCTURE

The mature red cell or erythrocyte is a biconcave disc. It is surrounded by a cell membrane; a phospholipid-cholesterol structure in which several glycoproteins are embedded. Some of these function as ion pumps which maintain, through an active process, the electrolyte composition of the cell. Many of these glycoproteins are recognised by the immune response, when the cells of one individual are transfused into another. Because of inherited variations in the way in which sugars are added to the amino-acid backbone of the surface glycoprotein (a process termed glycosylation), these red cell antigens are diverse, but patterns can be seen to follow simple genetic rules. Such blood group antigens are, of course, of great clinical importance, not only in transfusion, but also in the aetiology of several haemolytic anaemias including haemolytic disease of the newborn. The cell membrane is a fluid structure in which glycoproteins can move laterally rather like icebergs in an ocean. Some extend only partially into the membrane, whilst others have hydrophilic portions both on the outside and within the cell. Within the cell is a network of tiny molecular ropes composed of actin, tubulin and other proteins which anchor the inner parts of the surface glycoproteins and maintain the shape of the cell. This requires the expenditure of energy. Also enclosed within the cell membranes is the cytoplasm; a viscous fluid containing haemoglobin.

HAEMOGLOBIN

Haemoglobin consists of a protein (globin) and an iron porphyrin compound, haem. Its total molecular weight is 64,000 and the protein constitutes 97 per cent of this. The protein portion is composed of two separate pairs of amino acid chains, which are held together by non-covalent bonding. There are at least five genes involved in coding for normal haemoglobin components—α, β, γ, δ, ξ. The α genes are on chromosome 16 whereas the others are located on chromosome 11. The genes are transcribed into messenger RNA. These mRNA's can be isolated and translated into their respective globins in cell-free protein synthesis systems. The structural globin gene sequences on RNA and DNA are defined as those required to code for the amino acid sequence of globin. Other sequences within the structural globin genes but not encoded in mature globin mRNA have been found. These intervening sequences or introns are excised and the remaining RNA spliced during the production of active globin mRNA by the red cell precursor. The different types of normal haemoglobin are listed in table 6.1. HbA ($\alpha_2\beta_2$) predominates in the adult whilst HbF ($\alpha_2\gamma_2$) is the most common in the foetus. The mechanism by which these molecular switches seen during the change from embryonic to foetal and subsequently adult haemoglobin, are not understood.

Table 6.1 Normal haemoglobins

Type	Occurrence	Composition
HbA	Adult	$\alpha_2\beta_2$
HbA$_2$	Adult	$\alpha_2\delta_2$
HbF	Foetal	$\alpha_2\gamma_2$
Gower I	Embryonic	$\zeta_2\xi_2$
Gower II	Embryonic	$\alpha_2\xi_2$
Portland	Embryonic	$\zeta_2\xi_2$

Each individual globin chain is associated with one haem group. There are a number of helical sections in the amino acid sequences of each globin sub-unit. The haem component lies in a crypt between two of these helices. The iron in the haem group is in the ferrous (reduced) state. Its oxidation to the ferric form is prevented by the surrounding side chains of the amino-acids on the protein helices. When oxygen is carried by the molecule there is movement between the alpha- and the beta-chains of adult globin. The conformation of the entire molecule thus changes between the deoxy- and oxy-haemoglobin form. The affinity of oxygen for each of the haem residues is dependent on this conformation. As oxygen binds to the haem groups, the conformation changes sequentially giving rise to the classical oxygen dissociation cure for haemoglobin.

A variety of small molecules are also able to affect the overall conformation of haemoglobin and thus its affinity for oxygen. The pH of the surrounding fluid determines the charges on certain specific acid residues in the moledule. For example, oxygen affinity falls as pH falls, so facilitating the release of oxygen in metabolically active tissues where the released CO_2 lowers the pH. In the lungs pH rises as CO_2 is excreted thus increasing the affinity of haemoglobin for oxygen. This relationship of oxygen affinity to pH is termed the Bohr effect. Affinity is also affected by 2,3-

diphosphoglycerate (2,3-DPG). This molecule is an intermediate in the conversion of glucose to pyruvate. An increased red cell 2,3-DPG concentration favours the release of oxygen shifting the oxygen dissociation curve to the right. Red cell, 2,3-DPG is increased as an adaption to hypoxia caused either by a fall in the oxygen tension of inspired air or by cardiopulmonary disease. This at least partially compensates for the hypoxia at the tissue level. Haemoglobin is therefore well adapted to its function as a carrier of oxygen from the lungs to the tissues.

OTHER INTRACELLULAR CONTENTS

The red cell is not a simple delivery package. It requires energy to maintain its shape, activity and electrolyte composition. Red cells are continually being damaged in their journey through the cardiovascular system. A limited number of repair mechanisms, which are also energy dependent, are available to repair damage to the cell membrane. There are no mitochondria in mature red cells. Phosphorylated nucleotides, the source of energy for most biochemical processes, are produced by the pentose phosphate pathway, rather than by oxidative phosphorylation. As red cells age, the enzyme systems become depleted. Defects continue to accumulate in the cell and eventually it is removed from the circulation by the reticuloenclothelial system. The lifetime of the cell is therefore determined by the efficiency of its energy transformation processes.

RED CELL PRODUCTION

Red cells are produced in the bone marrow. There is now evidence that marrow contains pluripotent stem cells, which are continually replicating. Their products differentiate to form erythroid, granulocyte and thrombocyte precursors. These, in turn, are still able to replicate and further differentiate. A major problem in studying the process of differentiation in bone marrow has been the limitation of classical staining techniques. Although haematologists have learnt to recognize cell types within the marrow, little is known about the exact timing of the genetic switches occurring in differentiation. Three processes go on simultaneously in the marrow. The first is replication. The second is differentiation—the switching on and off of gene blocks that will eventually determine the characteristics of the differentiated products; haemoglobin synthesis begins in this phase. Finally, there is maturation during which the products coded for by the switched-on genes, are assembled so allowing the cell to take on its fully differentiated form.

The rate of red cell production is controlled by a glycoprotein (M.W. 39,000) called erythropoietin. The source of erythropoietin is uncertain. Originally thought to be synthesized in the kidney, it is now clear that patients who have had both kidneys removed are still able to produce the hormone. It seems that a variety of tissues are able to produce erythropoietin, although the kidney is probably a major site. Erythropoietin increases the number of erythroid cells in the bone marrow. It does this by stimulating the differentiation of the early red cell precursors. A constant low level of erythropoietin is normally found in plasma.

In most anaemias erythropoietin levels rise, whilst over-transfusion decreases erythropoietin levels. In conditions where the oxygen tension delivered to the blood is reduced, either due to external factors such as high altitude, or from chronic respiratory disease, the level of erythropoietin rises and the rate of red cell production increases. This is a slow process requiring several weeks to become effective. If hypoxia is sudden, there is a sudden release of immature red cells into the peripheral blood from the marrow. This effect is triggered by a second hormone-reticulocyte releasing factor.

The maturation from stem cell to reticulocyte normally takes about four days. About half of the total body content of reticulocytes are in the bone marrow, the remainder gradually maturing in the circulation. The rate of red cell production is 15 to 20 mls per day. This maintains the haematocrit at approximately 45 per cent which provides the most efficient delivery of oxyhaemoglobin to the tissues. Lower values reduce oxygen delivery, because of an inadequate haemoglobin concentration, whilst higher values decrease blood flow because of increased viscosity.

RED CELL DESTRUCTION

Red cell life span is approximately 120 days. The exact mechanism of red cell destruction is not known. The gradual depletion of energy transformation processes results in the inability within the cell to repair defects accumulated in the cell membrane. Damaged cells are recognized and destroyed by phagocytic cells in the spleen, bone marrow and liver. The rate of red cell destruction is therefore predetermined by the constitution of the cell as it leaves the marrow, and is not subject to any external control processes.

REQUIREMENTS FOR RED CELL PRODUCTION

Healthy red cell production requires external supplies of iron, vitamin B12 and folic acid and also a variety of poorly understood local factors within the bone marrow.

Iron metabolism

A normal western adult diet contains about 15 mg of usable iron per day but only 1 mg is absorbed. Because of the large difference between the amount absorbed and the amount ingested, intestinal absorption of iron is critical in our understanding of iron metabolism. Ferrous iron within the intestinal lumen is absorbed in the duodenum and upper jejunum. In the cells it complexes with an iron carrying protein—apoferritin—to produce ferritin. From radio-iodine studies there is evidence that iron uptake is regulated by the apoferritin: ferritin ratio in duodenal cells as they emerge from the crypts of Lieberkuhn. In iron deficiency this ratio is high and increased absorption results. Ferrous iron is released from ferritin, and is coupled to the serum protein transferrin. Normal plasma iron concentration is 12–26 μmol/l. The total binding capacity of plasma is three times this (between 40–70 μmol/l). Serum transport of iron is therefore never a limiting factor in its delivery to the site of use. Transferrin in the serum enters the marrow and surrounds the erythroid precursors. A specific receptor site on such cells binds to transferrin and removes the iron. The unsaturated apotransferrin is released and recycled. Once inside erythroid precursors, iron enters the mitochondria and is incorporated into haem by haem synthetase. This enzyme requires the expenditure of energy and the presence of globin. There is, thus, a feedback control mechanism allowing balanced production of both haem and globin within the immature red cell.

Some of the iron within red cell presursors is taken up by intracellular apoferritin. The resulting ferritin is the cell's own private store of iron to use during its differentiation and maturation. Diseases in which the supply of iron to red cell precursors is plentiful, but the incorporation of iron into haemoglobin is defective, will result in increased ferritin deposits within the cell. These deposits can be seen on light microscopy. Cells containing such deposits are called sideroblasts.

To guard against the consequences of changing availability of exogenous iron, there is a storage mechanism within the body. About 30 per cent of total body iron is stored either as ferritin or haemosiderin, in the liver, bone marrow, spleen and muscle. When iron availability is low this stored iron acts as a buffer to maintain the serum iron bound to transferrin. Iron enters into all cells. It is therefore continually being lost by exfoliation in the gut, skin and hair. In women, menstruation results in the regular loss of an average of 30 mg a month, whilst pregnancy requires 250 mg per child. Lactation also results in significant depletion of iron. For these reasons, particular attention is given to iron intake during ante-natal care. There is no control over iron excretion. Abnormal iron intake, caused for example by repeated blood transfusions will result in accumulation of iron in the body and its deposition within the tissues. This results in haemosiderosis.

Vitamin B$_{12}$

Vitamin B$_{12}$ is synthesized only by micro-organisms. All animals require an external source of the vitamin. The daily requirement in man is about 5 microgrammes. Vitamin B$_{12}$ is a cyanocobalamin of a molecular weight of 1400 daltons. It consists of two components, a planar group resembling a porphyrin ring and a nucleotide which lies at right angles to it. In the centre of the pyrrole ring is a single cobalt atom, essential for the metabolic functions of the vitamin. These are two-fold. Firstly it is responsible for hydrogen transfer in certain intramolecular conversions. The interconversion of the two isomers, methylmalonyl CoA and succinyl CoA, requires the transfer of one hydrogen atom. A second type of hydrogen transfer in which Vitamin B$_{12}$ is involved is the reduction of ribonucleotides to deoxyribonucleotides. This is essential for producing precursors for DNA replication. A third biosynthetic process in which the vitamin is involved is the methylation of methionine, and through this the generation of tetrahydrofolic acid. Active DNA synthesis takes place in red cells precursors during cell replication and maturation. If inadequate B_{12} is available, large abnormal cells called megalocytes appear in the bone marrow. The morphological changes seen in the marrow arise from the inability of nuclear events to keep up with the normal sequence of differentiation.

Vitamin B_{12} is absorbed by a complex process. A glycoprotein called intrinsic factor is normally produced by the human gastric mucosa. This couples to Vitamin B_{12} through the small intestine. In the terminal ileum there are specific mucosal receptors found within the microvilli. The B_{12}-intrinsic factor complex binds to these specific sites and the B_{12} enters the mucosal cells and subsequently the blood.

Folate compounds

The folates are those compounds related to pteroyl glutamic acid (folic acid). Pteroyl glutamic acid consists of pteridine group linked, through para-amino benzoic acid, to glutamic acid. These compounds occur naturally as conjugates in which multiple glutamic acids are attached through peptide linkages to the gamma carboxyl group of the adjacent glutamic acid. Folate compounds are widely distributed in nature; green leaves, such as spinach, lettuce and cabbage being the richest sources. About 50 microgrammes are required daily, although the average diet contains considerably more. Folic acid is absorbed in the proximal jejunum. Absorption is an active process, requiring expenditure of energy by the intestinal mucosa. Absorbed folic acid is distributed to all tissues.

There are three oxidation levels for folic acid: folic acid (F); 7,8-dihydrofolic acid (FH2); and, 5,6,7,8-tetrahydrofolic acid (FH4). It is this latter compound that permits the transfer of one-carbon units in a whole range of biochemical processes in the cell. The conversion of folic acid to tetrahydrofolic acid is carried out by an enzyme, dihydrofolate reductase. This enzyme is readily blocked by folate analogues such as methotrexate and aminopterin. There are a variety of tetrahydrofolic acid one-carbon derivatives which can be used in biosynthetic reactions.

The intermolecular transfer of carbon is an essential process within dividing cells. The conversion of the deoxyuridine to thymidine requires the insertion of a methyl group at the 5 position of the pyrimidine ring. Other sites where one-carbon transfer is important includes the conversion of serine to glycine, methionine synthesis and histidine synthesis. The failure to produce adequate thymidine results in blocking of DNA synthesis. The red cell precursors are sensitive to deficiency in folic acid and megaloblastic anaemia results.

Abnormal function

Two disease states can result from the disordered physiology of red cells; anaemia and polycythaemia. Anaemia refers to the reduction of haemoglobin concentration below the normal level. This results from a reduction in total red cell mass although an increase in plasma volume due to fluid retention can occasionally result in a dilutional anaemia. Polycythaemia is an increase in total red cell mass either due to abnormal erythropoietic stem cell proliferation or secondary to increased erythropoietin production.

ANAEMIA

Anaemia reduces the oxygen carrying capacity of the blood. A variety of compensatory mechanisms reduce the severity of the resultant hypoxia. These include an increase in 2:3-diphosphoglycerate so shifting the haemoglobin dissociation curve to the right and thus yielding more oxygen from the oxyhaemoglobin. In addition the cardiopulmonary system adapts with increased cardiac and respiratory rates so increasing tissue blood flow. The symptoms of fatigue and lassitude reduce the anaemic patients' exercise tolerance so reducing tissue oxygen demands. Anaemia is caused by (a) blood loss, which may be acute or chronic, (b) defective red cell production due to lack of essential nutrients, or congenital or acquired disorders of erythropoiesis, and, (c) impaired red cell survival due to haemolysis.

Blood loss

Acute blood loss does not produce acute anaemia. For this reason a haemoglobin determination is of no value in determining the quantity of blood lost immediately after trauma or a massive gastrointestinal bleed. It is not until the plasma loss in the circulation is corrected that the anaemia becomes apparent. Massive but occult acute blood loss can occur in disorders of haemostasis. Blood may extravasate in the retroperitoneum producing little or no symptoms other than those of acute anaemia.

Chronic blood loss occurs regularly in menstruating women. Up to 100 ml of blood can be lost with each monthly cycle. In addition the nutritional stress of pregnancy and lactation accounts for the frequency of mild anaemia in women of child bearing age. Chronic blood loss can occur in a variety of diseases. Peptic ulceration, gastrointestinal neoplasms and inflammatory bowel disease can result in blood loss into the gut. A variety of renal diseases can result in haematuria with subsequent anaemia. In patients with chronic blood loss the fall in circulating red cell volume triggers erythropoietin release and subsequently increased marrow red cell production. The subsequent increase in demand for iron eventually fails to be met and so anaemia results.

DEFECTIVE RED CELL PRODUCTION

Iron deficiency

Iron deficiency anaemia is one of the commonest diseases in the world; it has been estimated that there are approximately 150 million sufferers. It results from poor diet, malabsorption, excessive loss of blood from the intestinal or genital tracts, or from such physiological events as growth and pregnancy. Because of the considerable iron stores, there may be a long delay between cause and effect as anaemia does not develop until iron stores are exhausted. If the loss of iron is 1 mg/day greater than intake, then it takes approximately 6 years before the haemoglobin concentraiton falls to half the normal value. When iron stores are almost depleted, the plasma iron begins to fall and in severe depletion falls below 2 μmol/l. The total iron binding capacity of the plasma is frequently increased (above 70 μmol/l). At this stage, the marrow is hyperplastic and microscopy shows microblasts and reduced numbers of sideroblasts; there is an absence of stainable iron. The absorption or iron is stimulated by unknown mechanisms and may reach 40–90 per cent of the ingested iron. At the onset of iron deficiency, the marrow delivers normochromic and normocytic cells, but later the cells are hypochromic and microcytic.

B_{12} deficiency

Inadequate absorption of B_{12} results from the failure of the gastric mucosa to secrete enough intrinsic factor (pernicious anaemia, gastrectomy) or to failure of the intestinal mucosa to absorb the B_{12} intrinsic factor complex (extensive small intestinal resection, Crohn's disease). Pernicious anaemia is the result of a failure of the gastric mucosa to secrete sufficient intrinsic factor for the absorption of B_{12}. Gastric atrophy is always found together with a failure to secrete hydrochloric acid. Pernicious anaemia is considered to be an autoimmune disorder, in that it is associated with the presence of antibodies against gastric antigens. Whether these antibodies are causal in relation to the disease or whether they are secondary manifestations is still uncertain. In approximately 55 per cent of patients, IgG antibodies active against intrinsic factor can be found in the plasma. Eighty-five per cent of the patients have plasma antibodies against a cytoplasmic component of gastric parietal cells, although this antibody is also seen in gastritis without pernicious anaemia (40 per cent) and occasionally in apparently normal people (less than 10 per cent). In approximately half of the patients, an IgA anti-intrinsic factor antibody can be found in the gastric secretion and there is evidence that this latter antibody does at least partially inhibit the uptake of B_{12}

After extensive gastrectomy it takes at least 5 years for the stores to become sufficiently depleted to lead to megaloblastic changes in the marrow. The body content of B_{12} must be reduced to 5–10 per cent of the normal amount before abnormalities appear.

An inadequate dietary intake of B_{12} is rare but may occur in strict vegetarians, called vegans; it takes 10 to 20 years to develop.

Diversion of B_{12} away from the intestinal mucosa may also occur as, for example, when a tape-worm (Diphylobothrium latum) or intestinal bacteria utilize B_{12}. The presence of a large tape-worm or of a diverticulum or blind loop giving rise to sepsis in the ileum may result, therefore, in the development of the B_{12} deficiency.

B_{12} deficiency affects all tissues and histological abnormalities are found in many types of epithelial cells. Since marrow tissue probably has the highest turnover rate in the body, the disease is most manifest there. There is a failure in the development of red cell precursors accompanied by the characteristic appearance of megaloblasts.

Megaloblastosis results from the inadequate supply of deoxynucleotides and hence impaired DNA synthesis. Examination of the marrow with DNA stains has shown that there are a number of dividing cells with a DNA content greater than that seen in haploid cells but less than that seen in diploid cells. These cells which have failed to double their DNA content subsequently die.

In pernicious anaemia, the rise in marrow erythrocyte/granulocyte ratio, in faecal urobilinogen output and in plasma iron turnover indicates approximately a three-fold increase in total erythropoietic activity. However, studies on the survival of red cells in the peripheral circulation have shown that not only is the life-span shortened but also that, despite the increased marrow activity, the rate of production of viable cells is only normal or less than normal, depending on the stage of the disease. Thus there is a considerable amount of ineffective erythropoiesis with destruction of red cells before they leave the marrow.

A haematological response to 1–3 μg B_{12} per day is evidence of B_{12} deficiency. Megaloblastic changes of B_{12} deficiency respond temporarily to folic acid administration in doses of 5 mg/day, which is approximately 100 times the minimal daily requirement.

FOLATE DEFICIENCY

The clinical effects of folate deficiency are the result of a failure of the production of DNA. This is chiefly manifested in haemopoietic tissue, but also results in dermatitis, and in gonadal failure. Histological changes in the marrow are similar to those seen in B_{12} deficiency. The peripheral blood shows pancytopenia which macrocytes and hypersegmented neurophils. If iron deficiency is also present, then the morphology of iron deficiency predominates and megaloblastic changes in the marrow may only appear after iron treatment.

The causes of folate deficiency fall into 3 main groups: (a) the diet may be low in folate compounds or there may be malabsorption due to disease of the gut, when folate deficiency is part of a general picture of malabsorption. (b) deficiency occurs when the requirements for folate are increased and exceed that present in the diet; approximately

25 per cent of pregnant women in Britain develop folate deficiency if cytological changes in the marrow are used for evaluation and the incidence if even higher if serum and red cell folate levels are used as criteria for diagnosis. This is the result of increased requirements by the foetus. Megaloblastic anaemia in pregnancy is almost always due to folate deficiency, unless malabsorption is present, when B_{12} deficiency may also occur. The onset of lactation will precipitate deficiency if the dietary intake of folate is only just sufficient for other requirements. Folate deficiency may also develop as a result of increased marrow activity in chronic, acquired haemolytic anaemia, hereditary spherocytosis and sickle-cell anaemia. It may also be the cause of the aplastic crises seen in these diseases. Finally, both hyperthyroidism and infection increase the demand for folates, which can result in megaloblastic changes in a person whose total body content of folate is already almost depleted for some other reason. (c) deficiency also results from interference with folate metabolism by drugs, such as the folic acid antagonists (methotrexate, pyrimethamine), anticonvulsants (diphenylhydantoin, barbiturates and pyrimidone) and ethanol (which depresses formylase activity, an enzyme concerned with transfer of formyl groups from folic acid derivatives); folate deficiency is seen in alcoholics both with and without cirrhosis.

SIDEROBLASTIC ANAEMIAS

Under normal conditions 30–50 per cent of nucleated red cell precursors can be shown to have one or more fine granules of ferritin just visible as very small blue dots diffusely scattered throughout the cytoplasm. In the presence of iron overload there is a small increase in the number of ferritin granules, which are also larger in size than normal.

In other ill-defined diseases in which there is thought to be a disorder of haem synthesis, many iron-containing granules can be demonstrated in a ring around the nucleus. These granules have been identified by electron microscopy as damaged mitochondria. It is thought that when haem synthesis is defective, an excess of iron accumulates on the mitochondria and brings about their damage. It is these diseases with the ring of iron-staining granules which are termed the sideroblastic anaemias. The aetiology is obscure and there are probably several mechanisms involved in producing the abnormality; at least one type is inherited and appears to be a sex-linked recessive condition affecting only males. Drugs such as isoniazid, chloramphenicol, phenacetin and ethanol can also produce anaemia associated with these characteristic ring sideroblasts.

THE ANAEMIA OF CHRONIC DISORDERS

One of the commonest forms of anaemia seen in hospital practice is that associated with chronic disorders such as chronic infections, malignancy, rheumatoid arthritis, renal failure and a large number of rarer disorders. The anaemia is characterized by the low serum iron concentration, a low total iron binding capacity and an increase in total body iron stores. The anaemia is usually mild. Commonly the red cells are unchanged in morphology, being both normocytic and normochromic. The anaemia is usually progressive within the first 2 months of the disease and then stabilizes at a fairly constant level. Although the red cell changes may suggest iron deficiency, the anaemia of chronic disorders will not respond to iron therapy.

True iron deficiency-anaemia and the anaemia of chronic disorders can be distinguished by measuring the total iron binding capacity of the serum and by assessing the bone marrow iron stores histochemically. In the anaemia of chronic disorders the TIBC is low with plentiful marrow stores of iron. This reflects an inability to utilise iron rather than iron deficiency.

Several mechanisms have been defined to explain the anaemia of chronic disorders. First, there is a small decrease in the life span of the red cells, usually to approximately half the normal value. The cause of this early destruction of red cells is unknown. Secondly, the marrow does not respond in the normal way to the anaemia; the rate of production of red cells is approximately normal or slightly increased but not adequate to fully compensate for the anaemia. This lack of response is due to the failure of the anaemia to bring about an increase in the production of erythropoietin, the usual method of maintaining haemoglobin concentration but the marrow itself is capable of responding to erythropoietin when injected intravenously. The cause of this failure to increase erythropoietin production is also unknown. Thirdly, there is a failure to release iron from the reticuloendothelial system, a conclusion drawn from the fact that radioactive iron bound to haemoglobin and injected as non-viable red cells is poorly reutilized. This failure to release iron from stores results in low serum iron levels and fewer sideroblasts. Although the failure to mobilize iron plays a role in the anaemia, the rate-limiting factor in the rate of red cell production is insufficient erythropoietin activity. A few patients with chronic disorders also have a superimposed iron deficiency as well, and therefore show a partial response to iron therapy.

INCREASED RED CELL DESTRUCTION (HAEMOLYTIC ANAEMIA)

INTRINSIC RED CELL DEFECTS

Intrinsic red cell defects which result in a shortening of red cell life span can be due to congenital abnormalities in the red cell membrane, the cytoplasmic enzyme systems or in the haemoglobin. These defects make the mature erythrocytes more susceptible to damage during their passage through the circulation and thus to the development of a haemolytic anaemia. Increased red cell destruction releases large quantities of haem into the cells of the reticuloendothelial system which is subsequently converted into bilirubin. Thus in haemolytic anaemia the concentration of unconjugated bilirubin rises. In addition the bone marrow is stimulated to release more immature erythrocytes to compensate for the anaemia and thus the reticulocyte count rises. The shortened life span of the erythrocytes can be clearly demonstrated by labelling a sample of a patient's red cells with sodium [51]chromate and reinjecting these into the circulation. The subsequent fall off in blood radioactivity is a measure of red cell survival.

Red cell membrane defects. Hereditary spherocytosis is a rare autosomal dominant disease in which there is an abnormality in a structural protein associated with the cell surface. This abnormality results in the loss of normal disc shape of the red cell and an increased susceptibility to damage in the circulation, especially within the spleen. Splenectomy can improve the anaemia in many of these patients. Another similar inherited disorder is congenital elliptocytosis.

Abnormal enzyme systems. **Glucose-6-phosphate dehydrogenase (G-6PD) deficiency** is due to a sensitive mutation leading to a single amino acid change in the enzyme. The inherited defect is of intermediate dominance linked to the X chromosome. Males are hemizygous (X)Y (where (X) is the abnormal chromosome); most females are heterozygous (X)X, but because of the presence of one normal X chromosome, they usually only have a level of about half the normal value, although occasionally they may have very low values, owing to variation in expression of one X-chromosome. Homozygous females (X) (X) are rare but have enzyme levels similar to males. The abnormality is widely distributed, the incidence varying from 60 per cent in Kurdish Jews to less than 1 per cent in Caucasians.

There are many different types of genetic mutations causing deficiency; three of these genetic mutants are common. The normal G-6PD is designated A. A mutant known as A− occurs in African negroes and only give rise to a mild degree of deficiency. A− has an abnormal electrophoretic mobility but a normal functional activity. A more severe form of deficiency occurs in the Mediterranean area, designated G-6PD-Mediterranean. This mutant has a normal electrophoretic mobility but an abnormal functional activity. A third type of mutant is found in patients with non-spherocytic haemolytic anaemia.

When G-6PD concentration is reduced to 10 per cent of the normal level the pentose phosphate pathway becomes defective, resulting in a lowered content of NADPH. As NADPH is the coenzyme for glutathione reductase, there is a consequent reduction in cell reduced glutathione (GSH). When NADPH and GSH are deficient the entry of compounds which have oxidising properties or cause excessive hydrogen peroxide production, causes damage to proteins and consequent cell destruction. Thus haemoglobin is first oxidised to methaemoglobin, followed by oxidation of the globin portion of the molecule. The oxidised haemoglobin may then form a mixed disulphide with glutathione, which precipitates and is visualized by staining as Heinz bodies. Usually G-6PD deficiency is benign, the only abnormality being a 25 per cent reduction in the lifespan of the red cells, although severe jaundice may occur in the newborn. There are over 40 drugs which are known to accentuate G-6PD deficiency. The most important are primaquine and quinine, some sulphonamides, the nitrofurans, aspirin and phenacetin. Certain bacteria and viruses (bacterial pneumonias, infectious hepatitis, viral respiratory infection and infectious mono-nucleosis) and the Fava bean also have the same action. These agents cause a rapid fall in the already depressed levels of NADPH and GSH. Red cells are destroyed and the haemoglobin concentration falls. The older red cells, already depleted of enzyme are mainly destroyed so that after 7–10 days the haemoglobin level ceases to fall and partial recovery may occur despite continuation of the drug. The severity of the anaemia depends on the nature and concentration of the drug and on the extent of the pre-existing G-6PD deficiency.

In **hereditary methaemoglobinaemia** the rate of conversion of methaemoglobin to reduced haemoglobin is inadequate due to abnormalities in methaemoglobin reductase or diaphorase resulting in a shortening of red cell life span.

Haemoglobinopathies. Inherited defects of haemoglobin production fall into two groups. In the first type there is a quantitative change usually resulting from a single amino acid substitution in one of the globin chains. This results in an alteration in the tertiary structure of the globin and a change in its physico-chemical properties. The second type of disorder—the thalassaemias—are caused by quantitative changes in the amount of α- or β-globin produced. Here the globin chains may be normal in structure but the imbalanced production results in functional haemoglobin defects.

Well over 100 abnormal haemoglobins have now been described. These are the result of a mutation in a cistron leading to substitution of an amino acid in either the α- or β-chains of haemoglobin. In the heterozygous state all haemoglobin abnormalities are harmless, with the exception of haemoglobin-s which may give rise to disease following extreme hypoxia. In the homozygous state, Hb-S gives the most severe disease and haemoglobins C and E (β-chain defects) and D (α-chain defect) give rise to only a moderate reduction in haemoglobin.

In **sickle-cell disease** an uncharged valine molecule is substituted for a negatively-charged glutamic acid in the β-chain. This single alteration causes Hb-S molecules to form polymers when oxygen tension is reduced with a resultant change in shape of the red cell. In homozygous SS disease, polymer formation occurs at the oxygen tension of normal venous blood, whereas in the heterozygous SA disease polymer formation does not occur until the haemoglobin oxygen saturation falls to approximately 40 per cent, or lower. The formation of haemoglobin polymers not only leads to sickling but also increases the rigidity of the red cell. The viscosity of whole blood therefore rises and it has been suggested that this change is the cause of the infarcts which characterize the disease. However, patients with Hb-SS disease are always anaemic and the viscosity of Hb-SS blood with a haematocrit of 25 per cent is the same as that of normal blood with a haematocrit of 45 per cent. It is perhaps more likely that the infarcts result from the blocking of small capillaries by rigid sickled cells which obstructs other cells behind them. Delay in passage through the capillaries must also play a part as sickling takes 2–3 min to appear *in vitro* on exposure to low oxygen tensions and passage through the capillaries is normally of the order of 5 sec. Prolonged stagnation of cells certainly occurs in the spleen; studies with ^{51}Cr-labelled cells have shown splenic sequestration and splenic puncture has demonstrated that most red cells in the spleen are sickled, many of them irreversibly.

The red cells are destroyed by phagocytosis. In children this may occur in the spleen, but in adult life the spleen is usually destroyed by infarction, so that destruction is presumably either in the liver or bone marrow. The changes on the surface of the red cells which lead to phagocytosis are not known. A small proportion of cells have a decreased resistance to osmotic lysis and also to mechanical stress *in vitro*, but it is not known whether either of these mechanisms play a part *in vivo*.

The severity of the anaemia is approximately proportional to the amount of Hb-S present. Homozygous SS disease is the most severe, and the Hb-S content is 80–100 per cent, the remainder being Hb-F. The disease does not manifest itself in the immediate post-natal period because there is still a high concentration of Hb-F in the circulation. Hb-S is also seen in association with other abnormal haemoglobins which modify the severity of the anaemia. Thus, sickle-cell-thalassaemia, SF, has 70 per cent Hb-S and is less severe than SS disease; the SC combination produces only 50 per cent Hb-S and is milder still, although there is considerable overlap in extent of the anaemia in the different types. Sickle-cell trait (SA) does not give rise to anaemia unless there is a severe lowering of oxygen tension.

When homozygous abnormalities are found in *haemoglobins C, D, and E,* there is some shortening of the red cell life span. The haemoglobin concentration may also be reduced, but not below 10 g/dl. Target cells are a marked feature of the peripheral blood. The low red cell haemoglobin content results in flattening of the cells. Most examples of these abnormalities are found as a result of the routine screening of red cells for abnormal haemoglobins.

A number of abnormal haemoglobins have been found in which the oxygen carrying capacity is impaired. The haemoglobins M are a group in which the oxidation product, methaemoglobin, is not easily reduced by the normal cell methaemoglobin reductase, thus causing cyanosis. Other haemoglobins have been described where there is an abnormally high or low affinity for oxygen. In the former, polycythaemia is commonly found and in the latter, cyanosis (haemoglobin Kansas).

A number of congenital non-spherocytic haemolytic anaemias of variable severity are due to the presence of abnormal, unstable haemoglobins (haemoglobins Koln, Seattle, Ube-I, Zurich and 'heat unstable haemoglobins') which are easily oxidised and denatured with the cells. The disease is present from the time of changeover from foetal to adult haemoglobin and the instability of the haemoglobin is manifested by the presence of Heinz bodies and by the rapid appearance of methaemoglobin on in vitro incubation without glucose.

Thalassaemias. The α- and β-thalassaemias are caused by decreased or absent α- or β-globin production. The application of the techniques of modern molecular biology to the study of these diseases has provided new insights into the molecular mechanisms involved. The heterozygous state of β-thalassaemia is asymptomatic; the chromosome bearing the normal allelic genes compensates by increased β-chain production. In addition the production of foetal haemoglobin (HF—$\alpha_2\gamma_2$) persists into adult life. Homozygous β-chain thalassaemia—Cooley's anaemia—is usually fatal in early adult life. The anaemia is caused partly by a reduction in functional haemoglobin and partly by the precipitation of excess α-chains within red cells resulting in haemolysis. Owing to the need for frequent blood transfusion, patients with Cooley's anaemia become iron overloaded with resultant liver and cardiac damage. The β-thalassaemias can be further subdivided into β^+ and β^0 diseases. In β^+ thalassaemia small amounts of β-globin are produced whilst in β^0 no β-globin can be detected. A variety of molecular mechanisms for these deficiencies have been identified. In some patients there are base changes in the β-globin structural gene, in others defective RNA is produced due to abnormalities in transcription or in processing, whilst in some mRNA is produced which cannot be translated. In some patients a deficiency in β-globin production is associated with δ-globin abnormalities—$\delta\beta$—thalassaemia and the rare syndrome, hereditary persistence of foetal haemoglobin (HPFH).

The α-thalassaemias are characterised by a decrease in α-globin synthesis. Four possible states of α-thalassaemia exist as there are four α-globin genes in each diploid cell. When only one globin gene is deleted a silent carrier state occurs. When two out of four are deleted the α-thalassaemia trait is present, with morphological changes in the red cell but no significant anaemia. Deletion of three out of four genes results in Haemoglobin H disease. In this syndrome, Haemoglobin H (a tetramer composed of four β-globin chains) occurs and a mild haemolytic anaemia results. When all four α-globin genes are deleted severe anaemia results during foetal development with subsequent death *in utero*. Table 6.2 classifies the different thalassaemias.

Table 6.2 The Thalassaemias

Disease	Anaemia	HbA $\alpha_2\beta_2$	HbA $\alpha_2\delta_2^2$	HbF $\alpha_2\gamma_2$	HbH β_4
β^+ thalassaemia	severe	↓	N	↑	o
β^0 thalassaemia	severe	↓	N	↑	o
$\delta\beta$ thalassaemia	mild	↓	↓	↑↑	o
HPFH	none	↓	↓	↑↑↑	o
α thalassaemia carrier	none	N	N	N	o
α thalassaemia trait	mild	↓	↓	↓	o
HbH disease	moderate	↓	↓	↓	↑↑
Hydrops foetalis α thalassaemia	Intrauterine death				

Haemolytic disease of the newborn. Approximately 17 per cent of all Rh-negative mothers who give birth to an Rh-positive child subsequently produce anti-D if they have a second pregnancy with an Rh-positive child. Immunization is the result of a transplacental haemorrhage of foetal cells, which may occur to a small extent during pregnancy, but which occurs most frequently during birth. The extent of the haemorrhage is usually small, the volume being less than 1 ml of foetal cells in 98 per cent of women. The chance of becoming immunized increases as the volume of the transplacental haemorrhage increases. However, only about 60–70 per cent of all Rh-negative women are capable of producing anti-D; the reason for this is unknown.

Owing to the fact that so few red cells get across the placenta during pregnancy, only about 1 per cent of Rh-negative women are immunized by the time of the birth of the first child, and even then the concentration of anti-D is usually so low that it is very rare for the firstborn to be affected (unless the mother has been previously transfused with Rh-positive blood or has had an abortion). If the primary immunization is the result of the transplacental haemorrhage occurring during birth, then in about half the women, the anti-D reaches a sufficiently high concentration in the plasma during the following six months to be detected by standard agglutination methods. In the remainder, anti-D does not become detectable until they develop a secondary response to the very small number of Rh-positive red cells which cross the placenta during the course of the subsequent pregnancy. The IgG anti-D is then actively transported across the placenta and brings about the destruction of the foetal red cells. The passive injection of anti-D into the Rh-negative mother within 36–72 hours of giving birth to an Rh-positive child reduces the incidence of Rh immunization to about 1–2 per cent in a subsequent pregnancy with an Rh-positive child; that is, passive anti-D injection gives about 90 per cent protection against immunization. The injected anti-D has been metabolized by the time of the next pregnancy. The mechanism of action of injected anti-D is unknown; it has been suggested that it brings about the rapid destruction of the foetal red cell and its associated D-antigen in the phagocytes of the spleen and in this way prevents the D-antigen from stimulating the antibody-producing cells.

Anti-A can also cause haemolytic disease of the newborn in a group O mother giving birth to a group A infant. The disease is almost invariably mild.

Haemolysis may also occur in a variety of disease states as a secondary phenomenon. These include the leukaemias, Hodgkin's disease, non-Hodgkin's lymphomas, systemic lupus erythematosus and acute viral pneumonia. The mechanisms involved are not clear but increased destructive activity of the reticuloendothelial system may well be responsible. A variety of drugs can also result in haemolytic anaemia.

In **autoimmune haemolytic anaemia** the increased red cell destruction is mediated by antibody directed against a red cell surface component. Lysis results either from agglutination of red cells and their subsequent clumping in the circulation or through complement mediated destruction. Why patients with this disease should start producing a red cell auto-antibody is unknown.

Principles of tests and measurements

The two most useful measurements to assess erythrocyte disorders are the estimation of haemoglobin concentration and packed cell volume (PCV). The most useful index obtained from these two measurements is the mean cell haemoglobin concentration (MCHC), which is the ratio of haemoglobin concentration (g/dl) to packed cell volume and is characteristically reduced in patients with any alteration in haemoglobin metabolism, such as iron deficiency and thalassaemia and occasionally in the anaemia of chronic disorders. Size and shape of red cells are usually assessed by examination of a stained blood smear. Electronic methods of measurement also give the mean cell volume (MCV) of the individual red cells, which is reduced in iron deficiency and the haemoglobinopathies and is increased in the macrocytic anaemias. Electronic measurements also give the mean cell haemoglobin (MCH), which is also reduced in the same circumstances as the MCHC.

METHODS FOR ASSESSING RED CELL PRODUCTION AND DESTRUCTION

Moderate to marked degrees of haemolysis can be identified by observing a persistently raised reticulocyte count and serum bilirubin concentration together with a reduced haematocrit. When severe intravascular haemolysis is taking place free haemoglobin and methaemalbumin appear in the plasma and haemoglobin is excreted in the urine. For a more complete assessment and for the provision of semi-quantitative or quantitative data, there are several further tests that can be used for the evaluation of red cell production and destruction.

The erythroid-granulocytic ratio

Marrow hypo- or hyperfunction is directly reflected in the number of red cell precursors that it contains. Unfortunately the total numbers of red cell precursors cannot be estimated directly owing to the difficulty of measuring marrow volume but an indication of marrow activity is given by determining the ratio of erythroid to granulocytic cells, which is normally of the order of 1 : 3. This ratio is valid only if the number of granulocytic cells are normal and the only guide to this is a normal blood white cell count. This test can grade marrow activity into the following categories: hypoactive, normal, moderately hyperactive (approximately 3 times normal) and extremely hyperactive, the latter being approximately 6 times the normal activity.

Plasma iron turnover

There is a continuous turnover of iron bound to transferrin, and most of the iron turnover is the result of the metabolism of haemoglobin. Approximately 30 mg of iron is turned over each day and about 20 mg of this is delivered to the marrow for haemoglobin production. Most of the remainder goes to the parenchymal cells of the liver. Thus about two thirds of the iron turnover depends on the state of activity of the marrow, and this is high enough to allow the measurement to be used as an index of marrow function.

The rate of iron turnover can be estimated by injecting $^{59}FeCl$ intravenously and determining the rate of removal from the plasma. If the plasma iron concentration is measured at the same time, the results of these two estimates can be used to calculate the total iron turnover per day.

Owing to the large variation found in normals, hypoplasia of the marrow cannot be accurately assessed. On the other hand, increased erythropoietic activity can be assessed by this method and a two-fold increase in plasma iron turnover (i.e. 60 mg iron through the plasma per day) is considered significant.

Urobilinogen turnover

When red cells are broken down, either in the marrow (ineffective erythropoiesis) or at the end of their life span, haem is not reutilized but is converted to bilirubin in the reticuloendothelial system by the removal of iron and opening of the tetrapyrrolic ring by splitting of a methene ($=CH-$) bridge with the production of carbon monoxide. Bilirubin is converted to urobilinogen by bacteria in the gut. Estimation of urobilinogen excretion is thus in the first place a measure of red cell destruction, but if the patient is in a state of equilibrium so that the red cell mass is constant, it is also a measure of the rate of production. The theoretical normal rate of excretion of urobilinogen in a 70 kg man is approximately 200 mg a day, but only about 75 per cent of this is found in the faeces on chemical analysis. Failure to obtain complete recovery suggests that there may be alternative pathways of metabolism but none is known at the present time. Alteration of the bacteria flora of the gut by antibiotics may also decrease the extent of the conversion to urobilinogen. Thus although urobilinogen excretion is usually found to be increased when the other indices also indicate increased production of haemoglobin, it is not a reliable or accurate measurement of red cell production. Increased urobilinogen concentration in the gut causes increased absorption and excretion in the urine, where it can be detected by using Ehrlich's aldehyde reagent.

A rise in plasma unconjugated bilirubin is a qualitative indication of increased pigment catabolism, but it cannot be used quantitatively because of the effect of liver function on bilirubin levels. It is not significantly raised until the red cell life span is reduced below 50 days.

Change in total erythropoietic activity of the marrow is thus reflected by the changes in the erythroid-granulocytic ratio, plasma iron turnover and pigment excretion. On the other hand effective erythropoietic activity, i.e. the number of viable cells which are delivered into the circulation, is reflected indirectly by the reticulocyte count and by iron utilization and can also be calculated directly from an estimate of red cell volume and life span.

Reticulocyte count

When red cells are released from the marrow, they still contain some ribonucleic acid (RNA) which disappears over the following 1–2 days. Staining with supravital dyes, such as brilliant cresyl blue, precipitates the RNA as a blue reticulum, the characteristic feature of a reticulocyte. The normal number of reticulocytes is 1·5–2 per cent of the total number of red cells, the value varying with the technique used. In anaemic subjects the observed reticulocyte percentage must be corrected for the patient's haematocrit, otherwise the value will be too high. The reticulocyte count is the most practical way of determining the rate of red cell production but is not an accurate measure of viable red cell production under all conditions for two reasons (a) red cells may remain in the marrow during the reticulocyte stage—this probably occurs in pernicious anaemia (b) red cells may be released at a less mature stage, giving rise to an increase in the number of reticulocytes in the peripheral blood which does not represent an increase in production. These more immature cells have large aggregates of reticulum. The release of immature cells may be suspected if there are nucleated and large polychromatic red cells present; this situation occurs in certain types of haemolytic anaemia.

Red cell utilization of ^{59}Fe

When ^{59}Fe is injected intravenously into normals approximately 80 per cent or more normally appears as ^{59}Fe-haemoglobin in the red cell mass by 10–14 days. A decrease in the percentage of iron appearing in the peripheral circulation occurs when there is decreased or ineffective erythropoiesis.

Red cell life span and site of destruction

Severe haemolysis can be easily demonstrated by reticulocyte counts and by estimation of serum bilirubin. Minor degrees of haemolysis cannot be demonstrated with certainly with these tests and the greatest value in the determination of red cell life span is found when the rate of destruction of red cells is only 2–3 times normal. The life span can be determined using cells labelled with ^{51}Cr. A red cell volume determination can be carried out with the same injection of ^{51}Cr-labelled red cells. With a knowledge of the mean red cell life span and the red cell volume, the daily rate of red cell destruction can be calculated. If production and destruction are in equilibrium, this value also represents the rate of production. Furthermore, surface counting over the liver and spleen following injection of ^{51}Cr-labelled red cells indicates the main site of destruction.

^{51}Cr as Na_2 ^{51}CrO$_4$ is added to the red cells and the ^{51}Cr becomes attached to haemoglobin. The ^{51}Cr-haemoglobin bond is unfortunately not stable and after reinjection of the cells approximately 6 per cent of the ^{51}Cr elutes within the first day. Thereafter the elution rate is about 1 per cent per day. A correction for this elution can be made and the corrected value for survival of radioactivity then represents the survival of the cells. The mean cell life can be obtained graphically by plotting the survival curve, drawing a tangent to the initial slope and extrapolating to the time axis. The point where it meets the time axis is the average life span.

When ^{51}Cr-labelled red cells are destroyed in the liver or spleen, the ^{51}Cr remains within the reticuloendothelial cells of these organs and is only slowly lost at the rate of approximately 3 per cent per day. These ^{51}Cr deposits can be detected by surface counting over the heart, liver and spleen, but the results cannot be interpreted on an exact quantitative base because 50 per cent of the γ-rays produced by ^{51}Cr are absorbed by each 5 cm of path through which they travel in the tissues. Thus changes in the size of the liver and spleen and in the thickness of tissue overlying them alter the amount of radioactivity reaching the surface. Nevertheless, abnormal changes in the rate of accumulation of ^{51}Cr in the liver and spleen can be detected. Accumulation of ^{51}Cr exclusively in the spleen is a reasonably reliable indication that the patient will benefit from splenectomy. In normal individuals, radioactivity detected over the heart closely follows that of the blood; spleen and liver radioactivity is initially (within 30 min of injection) approximately 70 per cent of that over the heart and radioactivity from all areas falls progressively when measured daily. In patients who are destroying red cells in the spleen, radioactivity over the spleen is at first equal to or greater than that over the heart and rises progressively during the following 5–10 days to reach a value approximately 2–5 times as high as the heart radioactivity.

Surface counting can also be of value in detecting the presence of splenunculi after removal of the spleen. If ^{51}Cr-labelled red cells are either heated or coated with anti-D antibody they are removed from the circulation rapidly by splenic tissue, but only at a relatively slow rate by other tissues such as the liver or bone marrow. Thus if cells treated in either of these ways are injected into a patient with a splenunculus, they will accumulate in this organ over a period of 3–6 hours and their presence can be detected by surface counting.

^{51}Cr-labelling of red cells can also be useful in determining whether haemolysis is the result of an intrinsic or extrinsic defect of the red cell. Cells with an intrinsic defect normally also have a reduced survival in a normal recipient, whereas cells damaged by an extrinsic defect usually have a near normal survival in normal recipients.

Plasma haptoglobins

Haemoglobin released into the plasma combines with haptoglobins (MW 100,000); the resultant complex is too large to be filtered through the glomeruli but is rapidly cleared by the reticuloendothelial system. Plasma haptoglobin concentration is measured indirectly by the ability to bind haemoglobin; normal plasma binds 1 g haemoglobin per 1 l plasma, i.e. all the haptoglobin in the plasma will combine with the haemoglobin released from approximately 10 ml of red cells. Haptoglobin levels are reduced even when red cell destruction is extravascular because some haemoglobin leaks into the plasma from engorged phagocytic cells. There is a correlation between plasma haptoglobin levels and mean cell life span. If the mean cell life is less than 30 days, the haptoglobin binding capacity is reduced to a barely detectable level.

TESTS USED IN DETECTION OF DEFICIENCY STATES

B_{12} deficiency

The most convenient and reliable single test for the detection of B_{12} deficiency is the determination of the serum B_{12} concentration. B_{12} levels are determined either micro-biologically or by radioimmunoassay. Normal values usually range from 150–900 pg/ml, but each laboratory should define its own lower limit since slight modifications in technique give different results. Owing to the close interrelationship between B_{12} and folate in metabolic processes, B_{12} levels may be low in folate defi-ciency. Low values are also seen in pregnancy and in iron deficiency. Levels below 80 pg/ml are the result of B_{12} deficiency or a combination of B_{12} and folate deficiency. There is some overlap however between the 'normal' and 'deficient' range and normal values do not exclude deficiency.

High levels of serum B_{12} are found in some patients with myeloproliferative dis-orders due to high levels of a protein which binds B_{12}. In acute and chronic liver disease high levels are also found owing to release of B_{12} from the liver. If these conditions are present together with B_{12} deficiency, the serum level of B_{12} may be brought into the normal range.

B_{12} absorption. B_{12} deficiency is almost always the result of malabsorption. The most satisfactory test of the extent of absorption depends on measuring urinary excretion after giving ^{57}Co-labelled B_{12} (Schilling test). ^{57}Co-labelled B_{12} (1·0 μg) is given by mouth and at the same time a large dose (1 mg) of unlabelled B_{12} is given intramuscu-larly. This large dose of unlabelled B_{12} overburdens the binding capacity of the protein responsible for B_{12} transport in the plasma and this results in the urinary excretion in the 24 hours following administration of about one-third of the ^{57}Co—B_{12} that is absorbed. In normals, 10 per cent or more of the administered dose is excreted. Values below 10 per cent may be taken as evidence of defective absorption; in pernicious anaemia it is usual to find values below 5 per cent.

The advantage of the Schilling test is that it can detect malabsorption of B_{12} even though the patient has been treated with B_{12}. Moreover, if a second test is carried out with the addition of intrinsic factor, a differentiation can be made between pernicious anaemia and small bowel disease (e.g. steatorrhoea), since with the former B_{12} absorp-tion will return to near normal levels whereas in the latter, B_{12} absorption will remain low.

Folate deficiency

Tests currently available for folic acid concentration and metabolism are not always reliable indicators of folate stores. Thus the diagnosis of folate deficiency as the cause of megaloblastic anaemia must rest principally on the exclusion of B_{12} deficiency. Megaloblastic anaemia in the presence of normal plasma B_{12} levels is almost always due to folate deficiency. However, serum B_{12} levels are sometimes low as a result of folate deficiency and exclusion of B_{12} deficiency then rests on finding a normal B_{12} absorption test. The following tests for folate deficiency should therefore only be used as confirmatory evidence.

Serum folate levels are measured microbiologically, using the growth rates of lactobacillus casei which measures the reduced co-enzyme forms as well as pteroylglutamic acid. Normal values for serum folate are 3–16 μg/l. In folate deficiency the majority of patients have values below 3 μg/l although it is possible to have normal values. Serum folate levels may be reduced below normal values soon after the appearance of aetiological factors which lead to folate deficiency, such as diminished intake. Thus serum folate values may be low while stores are still adequate.

Red cell folate levels may also give an indication of folate stores. Normal values are 100–800 μg folate/l red cells, i.e. 40–50 times the concentration in the plasma. Because of the slow turnover rate of red cells, folate levels may lag behind changes in total body folate. Red cell folate levels however will not distinguish between B_{12} and folate deficiency, as they are also low in B_{12} deficiency. In pernicious anaemia red cell folate levels return to normal soon after adequate therapy with B_{12}.

Plasma iron

There is a wide range of plasma iron concentration in normal subjects from 12–26 μmol/l. Plasma iron values show great lability; day-to-day variations are usually $\pm$30 per cent and there is diurnal swing of about 50 per cent, values being highest during the earlier part of the day.

Increased plasma iron levels. Plasma iron levels represent a balance between the rates at which iron enters and leaves the circulation. Raised plasma iron levels may be considered under 4 categories (a) increased red cell destruction brings about the release of iron into the plasma but an increase in plasma iron level is usually only seen when ineffective erythropoiesis is also present, as in pernicious anaemia and thalassaemia or when there is a block in the incorporation of iron into haem as in lead poisoning and pyridoxine deficiency (b) when there is decreased production of red cells as in aplastic anaemia, less iron is removed from the circulation and elevated plasma levels are found (c) when there is an increased mobilization from stores, as after an acute haemorrhage, more iron is mobilized than can be used by the marrow. Plasma levels are also raised (at the same time as the bilirubin) in acute hepatitis; this is the result of the release of iron from damaged liver cells and (d) plasma iron levels are raised when body iron stores are abnormally loaded, as in idiopathic haemochromatosis and in transfusion haemosiderosis.

Decreased plasma iron levels. Plasma iron levels between 8–14 μmol/l are often found in the presence of infection, neoplasm, chronic renal disease, rheumatoid arthritis and after tissue trauma. It is possible that plasma iron levels are controlled by the rate of release of iron from the reticuloendothelial system and this function appears to be impaired in these conditions, resulting in the release of less iron into the circulation.

Iron deficiency anaemia is usually associated with a reduction in plasma iron levels. Iron levels usually do not fall until there is total exhaustion of iron stores and may not fall even when there is a mild degree of anaemia present. When haemoglobin concentration falls below 9 g/dl low plasma iron levels are invariable. It is relatively common for the plasma iron level to fall as soon as iron stores are depleted but before the haemoglobin concentration has fallen producing a state of 'iron deficiency without anaemia'.

Total iron binding capacity (TIBC). Iron is transported in the plasma by transferrin and normally only about 30–40 per cent of the transferrin is saturated with iron. The total amount of transferrin present can be estimated in terms of the total iron binding capacity, which normally falls in the range of 45–70 μmol/l plasma. An increase in TIBC is usually seen in iron deficiency anaemia although this is only a reliable diagnostic indicator when the haemoglobin concentration falls below 9 g/dl. This increase in TIBC in iron deficiency anaemia is to be contrasted with the fall seen in patients with infection, neoplasma and rheumatoid arthritis, and is thus a useful method of distinguishing the latter group from true iron deficiency. Although the TIBC may be high in iron deficiency, the saturation of transferrin with iron falls to values below 16 per cent. A low serum iron and low transferrin saturation together with a haemoglobin concentration within the normal range is the method of diagnosing iron deficiency without anaemia. Another cause of decreased transferrin level is hypoproteinaemia seen in nephrosis, kwashiorkor and chronic liver disease.

TESTS USED FOR DETECTION OF INTRINSIC RED CELL DEFECTS

Abnormal haemoglobins

The presence of abnormal haemoglobins can usually be demonstrated by electrophoresis on paper or starch gel. This results from the fact that the abnormal amino-acid in the molecule alters the total charge. In some cases, there is no alteration in charge but the presence of an abnormal haemoglobin can be demonstrated by heat precipitation ('heat-unstable haemoglobins'). It is necessary to use several buffer systems for electrophoresis as abnormal mobility sometimes only shows up in one particular system.

Heinz body formation

Heinz bodies are darkly staining granules of denatured haemoglobin seen in peripheral red cells. They occur in some of the haemoglobinopathies (because of the increased susceptibility of the abnormal haemoglobin to oxidation) and also in red cell enzyme deficiencies (e.g. G6PD deficiency).

LEUCOCYTES

Normal function

Leucocytes can be divided functionally into three cell series, the granulocytes, lymphocytes and monocytes. Granulocytes are derived from bone marrow stem cells; within the bone marrow they undergo differentiation and maturation in a similar manner to that of the erythroid precursors and the cells mature granules appear in their cytoplasm. The staining pattern of these granules determines their classification as neutrophils, eosinophils or basophils.

The *neutrophil granulocyte's* main function is phagocytosis. It is attracted to areas of immunological activity by a variety of lymphokines—chemical messengers produced by stimulated lymphocytes. In addition neutrophils home on areas of tissue destruction and inflammation drawn by a variety of substances released during the inflammatory response. The mechanisms by which phagocytic cells recognise their targets are not fully understood. Once contact is made the target is surrounded by cytoplasmic extrusions from the granulocyte and subsequently totally engulfed. Enzymes are discharged into the vacuole containing the target by lysosomes from the cytoplasm. In this way the engulfed particle is subjected to considerable hydrolytic and proteolytic enzyme activity. A second mode of enzyme release occurs when neutrophil granulocytes are exposed to bacterial toxins or to complement fixing antibody-antigen complexes. The granulocyte swells, its organelles coalesce and eventually the cell dies releasing its enzyme content. These dead and dying neutrophils contribute to the cellular debris that constitute pus. Such debris collects in tissue spaces following certain infections so forming an abscess.

Little is known about the function of the basophil or eosinophil granulocyte. Both are phagocytic cells. The eosinophil is particularly associated with inflammation caused by allergy and parasitic infections.

The number of granulocytes in the blood is usually maintained within close limits. The time taken for maturation from stem cell to mature granulocyte is about 12 days. Mature cells are released at a steady rate under normal circumstances. Certain infections can result in the release from the bone marrow of more immature cells resulting in a rise in the peripheral white count. Cells normally exist in two functional populations; a circulating and a marginal pool. The marginal pool consists of granulocytes attached to the endothelial cells lining the cardiovascular system. Transfer from marginal to the circulating pool is triggered by a variety of hormones including catacholamines and corticosteroids. Stress, whether psychological or physical can thus increase the circulating white count. The average lifespan of the mature granulocyte is only 12 hours.

The cells of the *monocyte-macrophage series* are also phagocytes. They arise from lymphatic tissue, circulate in the peripheral blood and enter the extracellular fluid of most tissues. Their function is phagocytic: either directly ingesting bacterial and fungal pathogens or mopping up dying granulocytes. In some diseases a symbiosis is reached between macrophage and ingested microorganism. In leprosy and tuberculosis living pathogens can exist within a viable macrophage for several months. This is why these infections require prolonged courses of antibacterial chemotherapy for their treatment.

Lymphocytes are produced in lymph nodes, thymus and extra nodal lymphatic tissue. The cells circulate in the peripheral blood penetrating through the capillary endothelium and entering most tissues. Their physiology is discussed in Chapter 7.

Abnormal function

GRANULOCYTOPENIA

Damage to bone marrow stem cells can occur on exposure to a variety of drugs. Many anti-cancer and immunosuppressive drugs work by inhibiting the rate of cell division. Bone marrow stem cells are rapidly dividing and therefore sensitive to such drugs. Chloramphenicol, phenylbutazone, amidopyine, carbimazole and a variety of other drugs can cause granulocytopenia by switching off stem-cell turnover. This can occur suddenly even if a patient has been maintained for a long period of time on the drug without problems. The granulocytopenia may be irreversible, continuing even after cessation of the drug. Other causes of granulocytopenia include the destruction of cells by an enlarged, diseased spleen or the production of autoantibodies against cell membrane components. When the granulocyte number in the circulating blood falls below 10^9/litre the risk of developing a severe life-threatening infection increases considerably. Such infections are often caused by organisms not normally pathogenic such as the commensal bacteria of the intestinal lumen. It is now possible to maintain a reasonable granulocyte level in patients by granulocyte transfusions. Unfortunately the short lifespan of the mature, active cells means that transfusions have to be given frequently to maintain an active circulating pool. Such transfusions are often necessary when giving aggressive combination chemotherapy to cancer patients.

Increased granulocyte count

Increased numbers of circulating granulocytes can result from shifts between the marginal to circulating pool and from the release from bone marrow of relatively immature cells. Both mechanisms are involved in the neutrophilia seen in infection and in stress. Increased stem cell turnover due to the development of malignant clone results in leukaemia. The leukaemia type and hence its prognosis is determined by the degree of differentiation in the cells of the malignant clone. Stem-cell neoplasms, such as acute undifferentiated myeloblastic leumaemia, carry a very poor prognosis. No differentiated features can be seen in these leukaemic cells which rapidly infiltrate the bone marrow, destroying normal stem cells and their differentiated products. Chronic myeloid leukaemia, on the other hand, results in the production of well differentiated mature granulocytes which are able to carry out phagocytic functions. Death from chronic myeloid leukaemia usually results from the development of a stem-cell type leukaemia after 2–3 years of the chronic phase. The leukaemias represent growth control disorders in which an irreversible change has occurred resulting in malignant transformation. Leukaemoid reactions are another cause of a high white count. Here an abnormal number of relatively immature cells are produced for a limited period of time. Leukaemoid reactions are seen in patients with certain infections and in those with malignant diseases.

PLATELETS

Platelets are circulating cell fragments about 2 μm in diameter. Normal blood contains between 150,000 to 300,000 per cu mm. Platelets are derived from megakaryocytes in the bone marrow. Megakaryocytes are derived from the differentiation and maturation of the pluripotent bone marrow stem cell. The final cell in the differentiation pathway is large and multinucleate. From these megakaryocytes platelets are produced by a process of budding. The platelet contains no nucleus and almost no intracellular machinery. Like that of leucocytes and erythrocytes the number of circulating platelets is held remarkably constant. Production rate appears to be controlled at the level of stem cell by the turnover of the early thrombocyte precursors by a glycoprotein hormone thrombopoietin. The site of production and the feedback mechanisms involved in the control of this hormone remains unknown. The lifespan of a normal platelet is from eight to ten days. After this time they are removed from the circulation either by adhering to the vascular endothelium or by engulfment by phagocytes. Their function, both normal and abnormal, is discussed in chapter 8.

Practical assessment

RED CELLS

Clinical observations

Haemoglobin concentration: Colour of mucous membranes unreliable if Hb 9–17 g/dl. Below 9 g/dl pallor reliable; above 17 g/dl plethora and peripheral cyanosis.

Identification of anaemia: Evidence of iron deficiency: brittle spoon-shaped nails (koilonychia); atrophy of tongue, especially at edges; cracks at corners of mouth; dysphagia (post-cricoid web). Evidence of B_{12} deficiency: atrophy of tongue; neurological damage (sub-acute combined degeneration of cord, peripheral neuropathy, psychosis).

Evidence of haemolysis: Severe: jaundice; haemoglobinuria. Moderate: excess uro-bilinogen in urine.

Routine methods

Haemoglobin: normal values $15\cdot5 \pm 2\cdot5$ g/dl in men; $14\cdot0 \pm 2\cdot5$ in women. Haemato-crit and Hb allows calculation of mean cell haemoglobin concentration (MCHC). Normal: 32–36 g/dl; less than 32 indicates iron deficiency or defective Hb synthesis.

Mean cell volume and mean cell haemoglobin: These can only be measured accurately by electronic methods. The range of the normal values of MCV is 82–92 fl and for MCH is 27–31 pg for each red cell.

Stained smear: shape of red cells; reticulocytes. Red cell count: allows calculation of mean cell volume.

Marrow aspiration: cellularity; erythroid/granulocyte ratio; iron stores; megaloblasts with B_{12} or folate deficiency.

Haemolysis. Increased red cell destruction: plasma Hb and haptoglobin; bilirubin; total faecal pigment. If incompatible cells transfused, can be detected by Coombs' test.

Increased red cell production: reticulocyte count; increased marrow erythroid/myeloid ratio; increased fragility of spherocytes; positive Coombs' test in autoimmune haemo-lysis.

Deficiency states: serum iron and iron binding capacity; serum B_{12} and Schilling test for intrinsic factor; serum folate and red cell folate.

Special tests

Red cell destruction by ^{51}Cr-labelled cell survival; site of destruction can be identified by counting over involved organ.

Red cell formation by studies of ^{59}Fe turnover (ferrokinetics).

Abnormal haemoglobins by electrophoresis; foetal Hb by denaturation resistance.

Red cell enzymes Identification of glucose-6-phosphate dehydrogenase deficiency or glutathione instability. Heinz bodies indicate defect in oxidative metabolic pathway or presence of an unstable haemoglobin.

WHITE CELLS

White cell counts and smear. Counts about 50,000/cmm rarely seen except in leukaemia; acute infections 15,000–40,000. Atypical cells in glandular fever; primitive cells in leukaemias.

Special stains. Peroxidase staining for identifying cell type in acute leukaemia (lymphoblasts fail to take up stain); alkaline phosphatase deficiency may help in distinguishing myeloid leukaemic cells from normal.

Biopsy of marrow or lymph gland usually needed for firm diagnosis of leukaemia or lymphoma.

References

BANK, A., MEARS, G., RAMIREZ, F. (1980). Disorders of Human Haemoglobin. *Science*, **207**, 386–493.

BUNN, H.F., FORGET, B., RANNEY, H.M. (1977). *Human Haemoglobins*. Philadelphia, Saunders.

HUGHES-JONES, N.C. (1979). *Lecture Notes on Haematology*, 3rd ed. Oxford, Blackwell Scientific Publications.

JACOBS, A., WORMWOOD, M. (1974). *Iron Biochemistry and Medicine*. London, Academic Press.

Medicine (3rd series) (1980). *Blood and Reticuloendothelial Diseases*. Vols 27–29.

PROUDFOOD, N.J., BARALLE, F.E. (1979). Molecular cloning of the human ζ-globin gene. *Proc. Natl. Acad. Sci.* **76**, 5435–5439.

WINTROBE, M.M. (1974). *Clinical Haematology*. Philadelphia, Lea and Febiger.

7

Immune Mechanisms

Normal function

Our initial concept of the immune system solely as a protective apparatus against invasion by various micro-organisms must now be reappraised. Advances in cellular and molecular biology in relation to normal and abnormal states of immunity have emphasized the regulatory role of the immune system. Through their main function, the distinction between self and nonself, elements of this system accomplish both a primary defensive role and also one of surveillance. It is upon this background that we will describe the immune response, including the important cooperative functions of its various components, in normal and disease states.

ELEMENTS OF THE IMMUNE RESPONSE

Cellular elements

Lymphocytes: There are at least three functionally distinct populations of lymphocytes: thymus-dependent T-lymphocytes, bone marrow-dependent B-lymphocytes, and other non-T, non-B lymphocytes. The dependence of a lymphocyte refers both to its anatomical distribution and to the source of certain specific influences on maturation of a given subpopulation. T-lymphocytes require circulation through and hormonal influences from the thymus gland early in development in order to achieve functional differentiation. They are distributed in the thymus, the peri-arterial sheaths of the spleen and in the paracortical areas of lymph nodes. The fully mature T-lymphocyte participates in cell-mediated immune reactions such as tuberculin skin hypersensitivity, graft rejections, and in responses to tumours and viruses. Functions of differentiated T-cells include help or suppression of the humoral immune response by B cells (see below) as well as direct or indirect cytotoxicity (killer T cells). T lymphocytes are recognized in the peripheral blood by a peculiar capacity to form conglomerates with sheep red blood cells known as E (erythrocyte)-rosettes. Antibodies, which can be labelled with fluorescein, are available that bind to surface determinants present only on T-cells. It is therefore possible to count mature T-cells in the blood.

While B-lymphocytes are thought to be derived from the same putative bone marrow stem cell as T-lymphocytes, they have their phylogenetic origin in the Bursa of Fabricius in the chicken, the equivalent of which in the human and other mammals is thought to be the bone marrow. B-cells are distributed in B-dependent areas such as

268

the germinal centres and primary follicles in spleen and lymph nodes. These cells characteristically bear surface immunoglobulins and are the direct precursors of active antibody-synthesizing and secreting cells which at their end stage of differentiation are known as plasma cells. B-lymphocytes actively partake in humoral immune responses by making and secreting immunoglobulin molecules. Both T- and B-lymphocytes are induced to differentiate and mature through complex signals brought into play during the cellular processing of antigens. Memory cells of both types can store information for repeat ('second set') encounters with previously recognized antigens, thus hastening the development of a given immune response.

Antibody synthesis by B-cells is distinct from secretion of antibody. The immuno-deficiencies of B-cells are experiments in nature that have lead to the understanding that a cell may be arrested in its development at any of a number of stages: e.g., at the stem cell stage, leading to no capacity to synthesize antibody at all, or at the early B-cell stage in which antibody is synthesized but not secreted. Similarly T-cell deficiency states have pointed out the various stages of differentiation of that subpopulation of lymphocytes. This will be further discussed in the section to follow. Non-T, non-B cells comprise a small but important subpopulation of circulating lymphocytes, which includes natural cytotoxic (natural killer, NK) cells as well as hematopoietic stem cells. The former may function in immune surveillance against neoplastic or virus-infected cells; relative deficiencies of natural cytotoxic cells have been described in several neoplastic and inflammatory conditions.

Close cooperation among macrophages, T- and B-lymphocytes is necessary in order that the immune response proceed normally. Factors (interleukins, IL) derived from macrophages (IL-l) or T-cells (IL-2) act upon other T-cells or upon B-cells in the course of a specific immune response to enhance or suppress cell proliferation. T-cells are capable of regulating both humoral and cell-mediated immune responses, including T-cell effector functions such as the 'killer' effect (Fig. 7.1). Cell surface molecules on macrophages and T-cells which are selected products of immune response and histo-compatibility genes (see below) 'restrict' immune cellular cooperation and function. A cytotoxic T-cell is restricted in its ability to kill another cell, for example a virus infected cell, unless it is first able to recognize the target cell as 'self' with shared major histocompatibility complex determinants displayed on the cell surface. The inability of a particular strain or species to mount effective immune responses (antibody formation, T- and B-cell cooperation, T-cell cytotoxicity) to viral, graft, or tumour antigens can often be explained by specific differences in genetic repertoires. (See genetic control, below).

Subcellular elements

Immunoglobulins: There are five major immunoglobulin classes produced by B-lymphocytes: IgM, IgG, IgA, IgE, IgD. During a primary immune response, B-cell development is marked by a series of rearrangements of the immunoglobulin genes,

IMMUNE RESPONSE

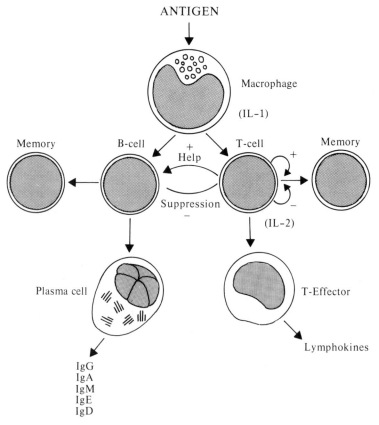

Figure 7.1 *Simplified schema indicating B and T cell involvement in the development of immune responses. See text for explanation.*

allowing for 'switching' to a given class of antibody. This produces specific clones of B-cells having selective memory to produce immunoglobulin molecules of a given class, subclass and idiotype. The idiotypic determinant of an immunoglobulin makes each molecule unique and may be important in immuno-regulation (see below). This means that any B-lymphocyte in the circulation bears a specific antibody molecule with a very definite arrangement in terms of amino-acid sequence and secondary or tertiary structure. Each immunoglobulin molecule of each class and subclass can be shown to contain heavy chains connected to light chains by disulphide linkages with specific regions and domains of the molecule that are repetitive and others that are variable and afford the plasticity of response to specific antigen. Immunoglobulins bind to antigens through two combining sites at the amino-terminus (Fragmen antibody, Fab, where heavy and light chains have combined). The carboxy-terminus of the heavy chains contains the part of the immunoglobulin responsible for biological function (Fragment crystallizable, Fc) and is of major importance in the fixing of complement components (see below) on cell surface antigens prior to their phagocytosis, in the binding of cells to specific targets and in the selective transport of IgG across the placenta.

The J-chain is a specific polypeptide in the immunoglobulin molecule, made by plasma cells, which is necessary for the synthesis of all polymeric immunoglobulins. It is found early in the ontogeny of developing B-cells.

IgM. IgM is the first immunoglobulin to appear in response to specific antigen both in phylogeny and in ontogeny; e.g. it is the only immunoglobulin class in fish. The IgM molecule is a pentamer, molecular weight 1 million daltons, does not cross the placenta, is found within the intravascular compartment, rises rapidly in response to antigen and falls within weeks, is succeeded by the IgG response to the same antigen, and is not characteristically a part of the booster response. IgM is found on the surface of very early B-cells, prior to development of the capacity for antibody synthesis or secretion. IgM fixes complement to cell surfaces.

IgG. IgG is by far the most common class of immunoglobulin and is divided into four subclasses (IgG_1–IgG_4). This molecule has a sedimentation coefficient of 7S, molecular weight 150,000, crosses the placenta, circulates with a half life of approximately twenty-three days, equilibrates between the extravascular and intravascular compartments, and rises gradually and falls slowly (relative to the rapid IgM response) following primary exposure to antigen. It is the antibody class most represented in commercial gammaglobulin preparations, and characterizes the booster (secondary) response in the serum to repeat antigenic stimulation. IgG also fixes complement and enhances phagocytosis (opsonization).

IgA. There are two subclasses of IgA, the immunoglobulin responsible for local immunity at mucosal surfaces. IgA is locally synthesized by plasma cells at these sites and then transported through epithelial cells by undergoing unique changes; these include dimerization and the addition of a moiety derived from mucosal cells known as secretory component. Serum IgA is monomeric, and is synthesized by plasma cells in the bone marrow and peripheral lymphoid tissue. Although the IgA molecule is in relatively low concentration in serum it is the major immunoglobulin in all external secretions such as tears, gastrointestinal secretions or milk. The secretory IgA molecule is highly resistant to proteolysis and prevents colonization by bacteria on epithelial surfaces. IgA does not opsonize or fix complement.

Parenteral immunization usually results in little evidence of IgA antibody in secretions. However, Salk vaccine (killed, parenteral) is as good as Sabin (live, oral) in protection. Presumably this protection is achieved via circulating antibody which prevents a viraemic phase of poliomyelitis; no antibody is detectable in secretions after Salk vaccination whereas secretory IgA is detected following Sabin.

Resistance to mucosal infection is better correlated with local mucosal antibody than with serum antibody response. Inflammation of mucosal tissue causes serum IgA levels to rise.

IgE. This class of immunoglobulin attaches specifically to receptors found on blood basophils and tissue mast cells. It is found in extremely low concentrations in the circulation although it is present in most mucosal tissues where it, like IgA, is locally synthesized by plasma cells. It is found usually in monomeric form. IgE (reaginic or homocytotropic antibody) is responsible for the antigen-specific release of histamine and other mediators of inflammation from basophils and mast cells resulting in typical allergic (immediate hypersensitivity) reactions.

Recently, it has been proposed that IgE antibodies participate in immune reactions which have biological significance over and beyond allergy, since the various mediators released from mast cells or basophils are themselves important regulators of inflammation and the primary immune response. IgE may serve a useful purpose by releasing mediators from sensitized cells thus focussing circulating factors including antibodies and cells at the very site where they may be needed. This is the so called "gatekeeper" theory of the function of IgE.

IgD. Least is known about the function of this immunoglobulin. It appears on the surfaces of B-cells early in their ontogeny together with IgM. Neonatal or cord lymphocytes are found to be coated with IgD; it is possible that IgD functions as a receptor on early B-lymphocytes. Other suggestions are that IgD is actually a vestigial component of the immune response. IgD measurement is not yet of use in clinical practice.

Many immunoglobulins are also known to have regulatory roles in homeostasis. We shall return to this point later in discussing disorders of the immune response.

Lymphokines and monokines

A wide range of biologically active molecules, from very small to relatively high molecular weight substances, are known to be secreted by lymphocytes or monocytes during the course of their participation in immune responses. Some of these components are known to be cell-surface determinants while the source of others is not known. Lymphokines are classically released from antigen-stimulated T-cells; for example, migration inhibition factor (MIF) regulates movement of phagocytes at sites of inflammation, IL-2 (Fig. 7.1) enhances the proliferation of T-cells, and other factors regulate antibody production by B-cells or influence the growth and differentiation of a number of other cell types. Monokines include interferon (a suppressor of viral replication) and B- or T-cell activating factors including IL-1 (Fig. 7.1), which is capable of stimulating T-cells at sites of antigen processing. The various lymphokines and monokines are subjects of active investigation at the present time, since they represent moieties which appear essential for cell-to-cell communication during the course of an immune response.

Complement system

A series of plasma proteins known as complement, mostly present in the form of inactive enzyme precursors, can be sequentially activated in a way analogous to the coagulation cascade (Fig. 7.2). When activated by an immunoglobulin directed against an antigenic determinant on a red cell surface, a series of enzymatic cleavages takes place, generating complement components C_1 through C_9 and resulting in actual membrane damage to that cell. Haemolysis is effected as a consequence of either direct cell lysis (intravascular haemolysis) or phagocytosis of complement coated erythrocytes (extravascular haemolysis).

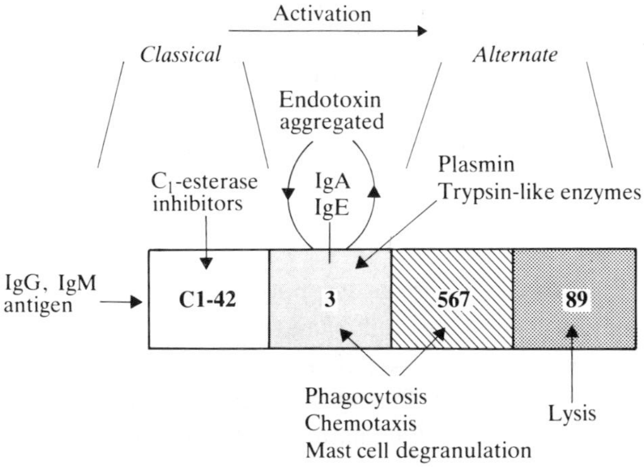

Figure 7.2 *Schematic diagram of the complement activation system.*

Complement is activated classically either by antibody or by antigen-antibody complexes (Figure 7.2). The first component of complement is physically altered upon interaction with a binding site revealed only on the complement-binding (Fc) portion of the antibody after it has interacted with antigen. The cascade then occurs. Each enzymatic step results in the proteolytic cleavage of the next complement component to fix. Many of these fragments are either biologically active themselves or inactivators of other component molecules, so that the complement system amplifies and regulates itself. For example, C_3a and C_5a (active fragments of C_3 and C_5) have chemotactic activities and release histamine from mast cells; C_3b is fixed onto cell surfaces and leads to modulation of cell surface membranes with subsequent alteration or damage; and C_3b-inactivator is an inhibitor specific for active C_3b but not for native or inactive C_3. Activation of C_1 is likewise regulated by a prototypical complement inactivator known as C1-esterase inhibitor. The clinical relevance of this regulator will be discussed in a later section.

Other modes of activation of the complement system are through plasma enzymes such as trypsin and chymotrypsin, or directly by the action of certain bacteria which stimulate the cleavage of C3. The 'alternate' complement pathway can be activated directly by a number of substances without mediation by enzymes or antibody; an example is the action of a complex polysaccharide, endotoxin, which interacts at the C3 stage (bypassing C142) to cause complement activation. This alternate pathway without antibody mediation includes proteins such as properdin and is probably important in resistance to certain bacterial infections.

Other serum proteins

The kallikrein-kinin system, the coagulation system, the fibrinolytic pathways and the complement cascade have complex inter-relationships in terms of their activation in the plasma compartment during the course of an immune response. Activated Hageman factor (factor XII) of the coagulation system appears to be pivotal in "priming" several of these pathways. During the course of an immune response interactions between these systems result in inflammation, tissue alteration and its subsequent repair.

Neutrophils and their products. Circulating polymorphonuclear neutrophils (PMN) are engaged in adherence, chemotaxis and phagocytosis of antigen at tissue sites. They bear specific surface receptors for IgG (Fc-receptor) and complement (C3b-receptor) which aid them in these functions. Complement fragments are important in regulation of PMN numbers and activity in inflammation. Lysosomal enzymes and other proteins derived from activated neutrophils are known to increase vascular permeability, lyse connective tissue substances, and activate fibroblast repair mechanisms. Very often, actual tissue injury in response to antigenic stimulation is effected by neutrophils and other granulocytes. In situations where a particular antigen cannot be ingested, such as at the glomerular basement membrane in immune complex nephritis, neutrophils may release more lysosomal enzymes than usual. Through subsequent complement activation further neutrophil chemotaxis occurs, leading to a vicious circle of inflammation which persists as long as antigen remains present.

Eosinophils, basophils and mast cells. The granulocytes have special roles as effector cells of the immune response. The eosinophilic granule contains substances which can directly kill certain parasites, or participate in allergic states. Basophils and the tissue mast cells both bear a surface receptor for IgE and have distinct metachromatic granules containing vasoactive substances, such as histamine, which alter vascular permeability.

Other cells and substances. Platelets contain vasoactive amines such as serotonin and are known to be important in altering vascular permeability. These cells also bear receptors for immunoglobulins (Fc-receptor). Substances derived from platelets during the course of their activation and aggregation, have now been shown to cause smooth muscle proliferation in the intima of blood vessels. The coagulation cascade plays an important role in determining the tissue response to injury and inflammation. When vasculitis occurs it is common to find immunoglobulins, complement and fibrin deposited together in blood vessels and at injured tissue sites.

Interacting elements

Genetic control of the immune response: Histocompatibility antigens (HL-A). The nature and extent of an immune response to a particular antigen in a given individual are subject to control by a set of genes on the short arm of the 6th chromosome in man. Cell surface components coded for by the genes located on these chromosomes are known as histocompatibility antigens since they allow the discrimination of tissue types of individuals within the same species. Various loci have been defined by serological reactions (HLA antigens) as opposed to those which are defined by cellular reactivity (HLA-B, C and D). A large body of data derived mainly, so far, from murine systems, has shown that immune response (Ir) genes can regulate the cellular and humoral aspects of the immune response. The location of the Ir gene in man is not yet known with precision. Circumstantial evidence points to a close association of these genes with the major histocompatibility complex, especially with the HLA-D antigens. It is thought that the HLA-D system is more closely related to immune response genes than HLA-B; for example, tissue typing for organ transplantation may have more predictive value for graft "take" if HLA-D, rather than HLA-B is assessed.

Another set of antigenic determinants on B-cells, the so-called B-cell allo-antigens or Ia-like antigens (because of the analogy of these to the serologically defined Ia antigens in the mouse which are products of the Ir gene) have now been explored in man. These antigens hold much promise in disease association in that they may directly reflect differences within a species of immune response genes. The products of the major histocompatibility complex and immune response genes as reflected on the surfaces of lymphocytes offer great potential insights into disease patterns and disease susceptibility.

Differences in the amount or type of antibody formed against a given antigen are probably related to Ir genes. Often, the only clues we have in man that a given abnormality of the immune response is genetically linked relate to the finding in a given disease subpopulation of a very high incidence of histocompatibility antigens of a given specificity. This is then taken as presumptive evidence that the Ir gene is linked to the HLA determinants. Examples of these associations will be given in subsequent sections (Table 7.1). Close to 95 per cent of patients with ankylosing spondylitis are found to have HLA-B27 phenotypes and this has rendered the test of diagnostic value in

Table 7.1. Frequency of HLA-B27 phenotype in Caucasian populations

	% HLA-B27	
	Positive	Negative
Normal	5–10	90–95
Ankylosing spondylitis	85–95	5–15
Reiter's syndrome	40–70	30–60

the evaluation of patients with low back pain associated with sacroiliitis. The presence of HLA-B27 makes the individual more than one hundred times as likely to develop such a condition as someone without this membrane marker.

Exactly how HLA associations explain the pathogenesis of diseases is of course unclear. A high or low level of immune response to a given environmental agent may explain such an association. Similarly, if the HLA antigen or the cell surface itself acted as a receptor for a pathogenic agent, the association could be explained. With further refinements in tissue typing methods it may become possible to predict the likelihood of the incidence of certain diseases and even offer accurate assessment of future responses to therapy and prognosis.

Feedback control of the immune response

The 'network theory' of the immune response proposes a direct role for immunoglobulins in the control of immune responses. Simply put, this states that for each specific regulatory T-cell and its corresponding immunoglobulin-producing B-cell, there exists another T-cell/B-cell set which controls synthesis of another immunoglobulin having reactivity against a unique determinant (idiotype) on the first immunoglobulin. The anti-idiotype antibody thus serves as a mechanism of maintaining homeostatic balance. Since each immunoglobulin bears its own idiotype (see immunoglobulins, above) a complex network of T- and B-cell interactions evolves whenever the organism is exposed to antigen. There is increasing evidence to support this theory.

Tolerance

The mechanism whereby the immune system "overlooks" or "tolerates" self tissues is a complex and rarely passive process. Neonatal tolerance can be induced actively in experimental situations by introducing antigens very early in ontogeny. When subsequently exposed to these antigens, the organism is incapable of making a normal immune response. This type of suppressed response may be due to a variety of different mechanisms: (1) deletion of an entire clone of antibody producing B-cells (clonal deletion); (2) the induction of a suppressor cell activity which inhibits antibody responses; (3) a less common mechanism whereby antibodies to the clone of antibody producing B-cells against a specific antigen are made and thus interfere with the immune response to the tolerizing antigen. Consider for example, how the paternal graft (fetus) may survive transplantation rejection by the mother.

Many different protocols have been used experimentally to induce tolerance. The theory of "forbidden" clones which are allowed to proliferate when tolerance is broken (clonal selection theory) is now less widely held as such. It is apparent that the immune system does recognize and react to most antigens that used to be considered "sequestered": for example, thyroglobulin. Those autoimmune situations characterized by autoantibodies are probably therefore, an exaggeration of the norm rather than a bizarre and unusual event. In this context, tolerance is seen as related very closely to immunoregulation. The close cooperation of helper and suppressor cells with antibody producing B-cells is interfered with by manipulation of antigens or their carrier proteins and thus can lead to tolerance. Thus, the tolerant state, in many instances, is but another oversuppressed immune state.

Abnormal function

PATHOGENETIC MECHANISMS

The immune response has been divided into five types, based on Gell and Coombs' original classification. A specific pattern of elements normally involved in each type of response can be shown to be able to give rise to a given abnormal functional state, leading to a clinically recognizable disorder.

Type I: Immediate hypersensitivity or allergy. Attachment of antigen to specific IgE antibodies affixed to basophil and mast cell surfaces, with consequent liberation of potent inflammatory mediators, is the mechanism of type I immunity. Although mainly an IgE-basophil reaction, T-lymphocyte control of IgE antibody responses is also of importance, as well as a genetic control of IgE responsiveness in general to sets of antigens that are ubiquitous.

Type I immunity is exemplified by anaphylactic reactions, angioedema and allergic asthma. The release of histamine and other vasoactive substances leads to increased vascular permeability and broncho-constriction. Eosinophils are attracted to the sites of tissue damage by chemotactic peptides from basophils and are a consistent feature of these responses. Antihistamines, aminophylline, corticosteroids, and blockers of basophil or mast cell degranulation (e.g. disodium cromoglycate) can inhibit this type of immune response. Other disease states such as ulcerative colitis, migraine headaches or skin reactions (urticaria and related phenomena) may have at least in part an involvement of the IgE-basophil system in their pathogenesis.

Type II: Cytotoxicity. Complement fixing antibodies formed by host B-lymphocyte clones against self or altered tissue is the mechanism involved in type II immunity. This response results either from introduction into the organism of foreign material (e.g. drugs), or from subtle but significant alterations of antigenic sites on host tissues. The end result is cytotoxicity with cellular lysis or damage; 'killer' cells may or may not hasten this tissue destruction brought about by antibody (antibody-dependent cellular cytotoxicity or ADCC). For example, viruses may attach to cell membranes and induce cells to synthesize new and altered proteins which can be incorporated into their cell membranes. An immune response may then be generated which attacks cells expressing these antigens. Conceptually, "autoaggression" of a physiologic nature has occurred in such cases, but not autoimmunity. Elimination of the cells expressing these antigens causes restitution of normal function, while incomplete elimination allows continuing attack on "self" components and may lead to clinical abnormality. The expression of hepatitis B-antigen on hepatocytes causes a specific anti-liver cell immune response to occur. Elimination of the virus-infected cells allows for hepatic regeneration, but persistence of virus in the presence of a continuing immune response is one cause of chronic active hepatitis.

Type II immunity is classically seen in such conditions as autoimmune hemolytic anemia with antibodies directed against red cells, either of unknown cause or as a result of alteration of red cell surfaces by antigens such as drugs (penicillin) or antibodies directed against the drugs.

Goodpasture's syndrome is characterized by antiglomerular basement membrane antibodies against kidney and lung tissue which share antigenic determinants. This disorder is directly related to cytotoxicity effected by antitissue antibodies and can be significantly altered by plasmapheresis, which removes damaging antibodies from the circulation.

Type III: Immune complexes. Type III immunity involves the formation of complexes in the circulation or at tissue sites between specific immunoglobulin molecules and the antigens against which the antibodies are directed. During the course of a self-limited bacterial or viral infection immune complexes are regularly formed and cleared by the reticuloendothelial system, the size of the immune complex depending upon the nature of the antigen and the antibody. Soluble immune complexes occur in antigen excess and are therefore a normal early component of any immune response; later large complexes are formed which are readily cleared by the reticuloendothelial system. Problems arise, however, with persisting soluble immune complexes most often as a result of the production of insufficient amounts of antibody, relative to the amount of antigen present. The Arthus reaction, due principally to the localization of immune complexes at tissue sites (where antigen is present in excess) is another instance of type III immunity.

A type III immune response is the pathogenetic mechanism involved in systemic lupus erythematosus (SLE) or hepatitis B-antigen-positive periarteritis nodosa. These chronic illnesses are characterized by the formation of circulating soluble immune complexes and activation of complement. In SLE there is also the presence of anti-tissue antibodies and, specifically, of anti-DNA antibodies of the IgG class. Organ damage due to vasculitis such as glomerulonephritis, is effected by the attraction of neutrophils to the sites of immune complex deposition and complement activation with release of proteolytic enzymes from neutrophils and other granulocytes. In post streptococcal glomerulonephritis or rheumatic fever cross-reactive antigens to streptococci are found to be present in myocardium and within renal tissue and may lead to the development of these specific disease patterns following a type II or III immune response.

The finding of low C3 and C4 complement levels in serum gives indirect evidence of complement activation by circulating immune complexes, and is used to monitor disease activity. Antinuclear antibody (ANA) and rheumatoid factor (RF) are antibodies directed against altered self antigens (nuclear or immunoglobulin determinants) and serve as markers of abnormal self-immunoregulation. While the exact titers of these antibodies do not correlate well with severity of illness, some predictive and diagnostic value in SLE can be attached to antibodies against double stranded DNA, now easily measurable in standard clinical immunology laboratories (see below).

Therapy in immune complex disease is directed against the formation of antibody or at the final effector inflammatory mechanism, by the use of either anti-inflammatory or so-called 'immunosuppressive' drugs (which do not work so much by suppressing immune responses as by altering immune balance). In refractory problems, prolonged plasmapheresis has been used with some success as an adjunct to medication, in an attempt to remove circulating immune complexes and so bring about clinical remission.

Type IV: Delayed hypersensitivity. Type IV immunity is classically defined by skin reactions with a delayed time course. These may be indicative of a disease process or simply measure immune responsiveness to a T-dependent antigen, such as tuberculin. T-lymphocytes, together with macrophages, are involved in this type of immune response and certain viruses, most fungi, and grafts are dealt with by a primarily T-lymphocyte-macrophage interaction.

Type IV immunity is found pathologically in tuberculosis and certain disseminated fungal infections. It is a 'hypersensitivity' reaction of the T lymphocytes and macrophages which release soluble substances and promote tissue damage. Certain conditions, such as Wegener's granulomatosis, are associated with T-cell and macrophage reactions leading to inflammation of blood vessels with life-threatening compromise of major organs. Basophils, T-cells and macrophages can accumulate in delayed hypersensitivity reactions; the complex interaction of all of these cells may be necessary for the complete expression of this type of inflammation.

Type V: Stimulatory hypersensitivity. This category of immune response, proposed by Roitt, refers to the effect of an immunoglobulin molecule on a cell surface receptor in tissues that are usually endocrine in type. Either by changing affinity of cell surface receptors, or by actually altering the functional state of the receptor, immunoglobulins can be shown to significantly affect the binding and action of hormones on target tissues. Hyperthyroidism is a well known example where an IgG molecule (thyroid stimulating immunoglobulin, TSI), acts upon the thyroid stimulating hormone (TSH) receptor on the thyroid glandular cell to produce more hormone, resulting in the clinical state.

Antibodies may also exert negative, suppressive influences upon cellular function, directly causing illness. Examples of these hypoendocrine states include juvenile insulin-resistant diabetes mellitus with islet cell antibodies, or myasthenia gravis, in which an anti-acetyl-choline receptor antibody interferes with normal neuromuscular transmission. In several other endocrinopathies, antibodies to hormonal receptors can be shown to exist; these include adrenal antibodies in Addison's disease, gastric parietal cell antibodies and intrinsic factor binding or blocking antibodies in pernicious anemia, melanocyte antibodies in vitiligo, and parathyroid antibodies in hypoparathyroidism.

SPECIFIC DEFECTS IN IMMUNE RESPONSIVENESS

Immunodeficiency

Combined T- and B-Lymphocyte Deficiency. Severe combined immunodeficiency (SCID) is a condition of early infancy presenting as failure to thrive with susceptibility to overwhelming, often fatal, infections. The fundamental abnormality exemplified by these conditions is incomplete development of both T and B-dependent systems. There is atrophy of the thymus gland and other lymphoid tissue, lymphopenia and hypo-gammaglobulinemia. Unless treated, death supervenes due to uncontrolled infections with normal or opportunistic pathogens—bacterial, viral, fungal or protozoan.

Analysis of these conditions has revealed considerable heterogeneity. While in some cases the total absence of lymphoid tissue, and especially of the immunocompetent areas of the thymus, speak to the absence of a bone marrow stem cell for both T and B cells as well as for macrophages (reticular dysgenesis), there are instances where other stages of abnormalities have been shown; for example, T and B-cell immunodeficiency without absence of macrophages, or T-cell deficiency with normal numbers of B-cells but without immunoglobulin in the serum. Some infants with SCID respond only to bone marrow transplantation while others will respond to transplants of thymic epithelium as well, suggesting an earlier precursor cell defect in the former group. Still others may be helped by hormonal manipulation with fractionated thymic extracts such as thymosin or even by soluble factors derived from cultured thymic epithelium without the need for transplantation of the epithelium itself. Experiments such as these have led to the understanding that an orderly sequence of maturation of T- and B-cells from bone marrow through thymic processing exists, brought to light by abnormal states of lymphocyte differentiation 'blocked' at various points. Follow-up of various treatment protocols can be achieved through analysis of cell surface markers on peripheral blood T- or B-cells or by the selective study of specific lymphocyte functions (helper or suppressor) during therapy.

Autosomal recessive enzyme deficiencies such as adenoside deaminase (ADA) or nucleoside phosphorylase (NP) have been described associated with severe combined immunodeficiency in childhood. It may be that the nucleosides which accumulate are selectively toxic to lymphocytes in the absence of enzymes allowing normal break-down. This may lead to the inability to T- or B-lymphocytes to mature normally and develop an immune response. It is possible to reconstitute immunocompetence in some of these patients by transfusion of red cells containing the absent enzyme.

Primary T-Cell Immunodeficiency. Very infrequently a pure primary deficiency of T-cells exists such as in DiGeorge's syndrome. T-cells are deficient or absent from both the peripheral blood and those lymphoid areas which are 'T-dependent'. The major clinical features are congenital aplasia or hypoplasia of the thymus, peripheral blood lymphopenia, and abnormal or absent T-cell function, but normal gammaglobulin levels and B-cell numbers. A related syndrome associated with hypoparathyroidism may accompany this condition, thought to be a consequence of interference with normal development of the third and fourth pharyngeal pouches in embryonic life. Patients succumb early in infancy to viral, fungal or protozoan infections and may also suffer from hypocalcemia or congenital heart disease. Thymic grafts can restore or cure the immune abnormalities.

Miscellaneous T-lymphocyte Immunodeficiency Syndromes. A number of poorly charac-terized T-cell immunodeficiency syndromes have been described. These include chronic mucocutaneous candidiasis, which is associated with deficient T-cell and macrophage immunity to candida, often in the presence of an endocrinopathy. Nezelof's syndrome, ataxia-telangiectasia and Wiskott-Aldrich syndrome are all disorders in which an immunoglobulin deficiency of one or more subclasses co-exists with variable T-cell immunodeficiencies. A primary immunodeficiency associated with thymic tumours has been known to occur at any age with or without the association of autoimmune disorders. Treatment in many of these cases with infusions of gammaglobulin and/or immunostimulatory drugs can sometimes be of help.

Pure Primary B-Lymphocyte Deficiency. Another type of deficient immune response is lack of immunoglobulins. A congenital sex-linked agammaglobulinaemia described originally by Bruton is often associated with complete absence of B-cells and inability to synthesize any immunoglobulins. Patients succumb to overwhelming bacterial or viral infections usually after the first three months of life when maternally transferred IgG wanes. A late-onset but often genetically transmitted hypogammaglobulinaemia has been described at all ages associated with variable severity of infections and vari-able onset of disease (common variable hypogammaglobulinaemia). In many of the latter patients circulating B-cells are normal in number and in function and the defect resides in an abnormal T-cell subpopulation within the peripheral blood which sup-presses normal B-cell maturation and antibody synthesis and secretion.

Selective primary IgA deficiency is common in the Western world and has an incidence of about 1 : 800 in the general population. Many patients with IgA deficiency suffer from recurring gastrointestinal and upper respiratory tract infections. Since gammaglobulin preparations contain very little, if any, IgA and since this IgA is not transported across the mucosal epithelium, treatment by parenteral injections of gammaglobulin is of little help. Patients with IgA deficiency have an increased incidence of allergies, suggesting that IgA may help limit access, at the mucosal level of potentially harmful antigens, such as viruses or ragweed. This may also account for the increased incidence of autoimmune disorders in such patients and may indicate why breast feeding is associated with a lower incidence of later development of atopic diseases.

A selective deficiency of secretory component has now been described in which the entire IgA molecule is synthesized but the secretory piece is absent. Patients are prone to an increased incidence of infection. A number of different types of IgA deficiency can be defined.

In some instances abnormal suppressor T-cells have been demonstrated selective for IgA synthesis, while in others specific absence of an IgA precursor B-cell is noted. Description of abnormal suppressor mechanisms in some patients with agammaglobulinaemia has led to a deeper understanding of normal T and B cooperative functions.

Deficiencies of Phagocytic Function. Disorders of phagocytes have been described, some of which are associated with a specific inability of neutrophils or monocytes to perform normal bacterial killing. In most of these disorders, a genetically inherited enzymatic defect resulting in impaired peroxide or superoxide production in the phagocyte leads to an increased susceptibility to infection with micro-organisms such as staphylococcus and gram-negative bacilli. The most well-studied of these is an X-linked form known as chronic granulomatous disease, which presents with chronic discharging lesions in lymphoid organs, skin and lungs. Aggressive therapy with antibiotics, supportive treatment and blood cell transfusions have been necessary in these conditions but are limited in usefulness.

Other phagocytic dysfunctions include the Chediak-Higashi syndrome, characterized by recurrent bacterial infections, irregular hair pigmentation and giant cytoplasmic granules in phagocytes and platelets; Job's syndrome which presents with recurrent cold staphylococcal abscesses; lazy leucocyte syndrome with defective chemotaxis of neutrophils in association with neutropenia; and a syndrome of defective chemotaxis in association with elevated IgE levels and recurrent dermal staphylococcal infections.

Complement Deficiency. Specific assays for each of the complement components have enabled delineation of various immunodeficiencies associated with single complement component absence. Deficiencies of the first component of complement are associated with either hypogammaglobulinaemia (Clq deficiency) or with systemic lupus erythematosus (SLE)-like syndromes (Clr and Cls deficiencies). C2 and C4 deficiency have been associated in familial aggregations of sle-like disorders. C3 and C3b inactivator deficiency have been associated with bacterial infections on the basis of inability to generate normal chemotactic factors for neutrophils and other granulocytes. CI esterase inhibitor deficiency presents as an hereditary disorder called angioneurotic edema. Although rare, this particular form of potentially fatal reaction can be specifically treated by androgens, which increase the synthesis of this particular complement component. Deficiencies of the later complement components have in some cases been associated with increased Neisseria infections (C6 and C8 deficiencies especially). Defects of the alternate pathway of complement activation have been described in association with recurrent bacterial infections, especially when splenic function is impaired. Some complement deficiencies can be treated with a plasma infusion at regular intervals.

Secondary immunodeficiency. The commonest causes of secondary immunodeficiency are parasitic infestations and other chronic infections (e.g. malaria, trypanosomiasis, leprosy, tuberculosis and fungal disorders), protein calorie malnutrition, neoplasia and cytotoxic chemotherapy. Acute and chronic viral infection can be associated with abnormalities of immune responsiveness; congenital cytomegalovirus infection is an example of a chronic persistent viral illness associated with elevated immunoglobulins and impaired cell-mediated immune responses.

Malignant disease is often accompanied by variable deficiencies of T- or B-cell immunity. In Hodgkin's disease, for example, it is usual to find skin test anergy and the presence of serum factors which suppress the activation and normal function of circulating T-cells. Increased T-cell suppression of normal B-cell antibody synthesis is found in chronic lymphatic leukemia in a high proportion of cases.

Splenectomy is associated with a deficiency of a phagocytosis-promoting tetrapeptide called 'tuftsin', with subsequent inability to mount normal IgM responses to polysaccharide-encapsulated bacteria. There is an increased risk in these patients of overwhelming infection with such organisms.

Finally, a variety of chronic conditions including alcoholism, uraemia and collagen vascular diseases (see below) may all be associated with inability to mount normal cell-mediated immune responses or with hypergammaglobulinaemia due to T-cell suppressor dysfunction. The normal immune response as a function of aging is only now being studied systematically. There is evidence for T-cell hypo-responsiveness and increased autoantibody formation with advancing age.

EXCESSIVE IMMUNE RESPONSIVENESS

Many rheumatic disorders can be shown to be associated with a tendency to increased B-cell responsiveness. The development of hypergammaglobulinaemia, autoantibodies, circulating immune complexes and variable T-cell abnormalities are all a result of inability to regulate immune responsiveness early in ontogeny. Whether or not a viral stimulus or some other mutagen triggers this excessive responsiveness is not yet clear in humans, but in mice there is clearly a genetic basis for the suppressor T-cell deficit.

The New Zealand black and white mice have a very high incidence of a disorder indistinguishable from human systemic lupus erythematosus. This is accompanied by autoimmune haemolytic anaemia with a positive Coomb's test and glomerulonephritis together with immune complex deposition in a variety of organs. Careful analysis of the original defect within the immune system has led to the delineation of a T-cell differentiation abnormality, leading to inability to suppress normal B-cell responses. These animals can be prevented from developing the disease by regular injections of the soluble immune response suppressor substance (SIRS) prepared from activated normal mouse lymphocytes.

A T-cell suppressor defect has now also been shown in SLE in man as well as in juvenile rheumatoid arthritis where circulating lymphocytotoxic antibodies selective for a T-suppressor cell population are found. The DNA-anti-DNA immune complex of SLE and the rheumatoid factor—altered immunoglobulin complexes—in rheumatoid arthritis are a result of excessive autoantibody production. In treatment of these disorders corticosteroids and immunosuppressives, as well as various immunostimulatory medications (such as levamisole or thymic hormones) have been used. The underlying rationale for the latter is an attempt to re-establish balance in immune responsiveness, rather than simply suppress the immune response nonspecifically.

Excessive antibody production may also be due to benign or neoplastic proliferations of antibody-synthesizing B-cells (for example, benign monoclonal gammopathy, multiple myeloma, or macroglobulinaemia). Interestingly, one important feature of certain malignant plasma cell proliferations is hypogammaglobulinaemia for non-involved subclasses while there is an increased total amount of gammaglobulin due to the monoclonal gammopathy. This leads paradoxically to increased susceptibility to infections despite the high levels of monoclonal immunoglobulin.

DISORDERS INVOLVING ABNORMAL REGULATION

Under this category there are certain disorders where neither deficient nor excessive immune responsiveness is noted. However, certain features of these disorders lead one to believe that an abnormal tissue regulatory function of the immune system has been encountered.

Progressive systemic sclerosis is an example where although there is a paucity of immunological abnormalities, it is possible that abnormal regulation of either fibroblast or endothelial cell growth leads to progressive tissue damage. Lymphocyte-derived factors controlling fibroblast proliferation and serum factors affecting endothelial cells have both been described in scleroderma.

Another example of abnormal regulation is in the area of haematopoietic stem cell abnormalities, principally aplastic anemias and pure red cell aplasia. (see Chapter 6) In these situations, abnormal regulation by T-lymphocytes of haematopoietic cell development is thought to underly the pathogenesis of subsequent marrow failure. It is becoming clear that treatment of some of these disorders can be brought about successfully by therapies which include attempts at abrogating an abnormal T-lymphocyte suppression of haematopoiesis.

Finally, disorders of endocrine organs such as hyperthyroidism in Graves' disease and hypothyroidism or other endocrinopathies are associated with disorders of immunoglobulins and their interactions at cell surfaces (see above).

Principles of tests

Tests of immune function can be categorized in accordance with types of immune responses described above.

Type I immunity (Immediate hypersensitivity or allergy)

Skin testing. A wheal and flare reaction within 15 minutes to intradermally injected antigen (allergen) is used to measure allergy. Basophil or mast cell IgE (on the cell surface) specifically reacts with antigen to release cellular granules containing histamine and other vasoactive substances, leading to the skin reaction.

Degranulation of blood basophils. This test is used as a direct *in vitro* counterpart of allergic skin testing. Peripheral blood or buffy coat basophils are identified and counted in a haemocytometer by staining live cells with toluidine blue at acid pH, rendering the granules metachromic. The putative allergen is then added *in vitro*, and another count of metachromasia performed. The difference between counts before and after addition of allergen is a direct measure of basophil degranulation and indirectly of the amount of specific IgE present on the cell surface.

Radioallergosorbent test (RAST). Antibodies of the IgE class to any given antigen can be precisely measured by a radioimmunoassay. Radio-labelled antigen is reacted with the specific IgE presumed present in serum. Bound and free radiolabelled antigen are then separated, and radioactivity measured in the free phase. Radioactive counts are inversely proportional to the amount of specific IgE originally present.

Type II immunity (Cytotoxicity)

These tests involve the measurement of specific cytotoxic antibodies either in the patient's serum or already bound to target tissues.

Coombs test. Auto-antibodies attached to red cells are measured directly by agglutinating the antibody coated cells with a pure, standardized human anti-gammaglobulin reagent, which bridges immunoglobulin molecules between cells. The indirect Coombs test assesses the presence of red cell antibodies in serum by incubation of test sera with human red cells, followed by agglutination with the anti-gammaglobulin reagent.

Anti-tissue antibodies. Anti-glomerular basement membrane antibodies are assessed by serum binding to heterospecific kidney, lung or placental basement membranes; this is demonstrated through use of a fluorescein labelled anti gammaglobulin reagent, and fluorescence under ultra-violet light.

Chromium release assay. In experimental situations, antibody-dependent cytotoxic effects upon tumour cells or virus infected cells can be assessed using serum which contains the antibody together with an assay for lytic injury to the target cells, which have been coated previously with chromium 51. Radioactivity released into the medium is a direct measure of cell damage, proportional to the amount of antibody present.

Type III immunity (Immune complexes)

Autoantibodies. Antinuclear antibodies and other antibodies against cellular or sub-cellular components (antimitochondrial antibodies, smooth muscle antibodies, anti-thyroid antibodies, parietal cell antibodies) are measured by immunofluorescence techniques, after incubation of animal tissues with patient serum. These antibodies generally do not cause direct tissue damage *in vivo* but are markers of certain immune complex and autoimmune conditions.

DNA-binding antibodies in serum are assessed using either kinetoplasts (rich in double stranded DNA) from a specific protozoan (crithidium) or calf thymus DNA.

Rheumatoid factor is assessed by the coating of latex particles with rabbit or human IgG and then agglutination of the test particles by the serum containing an IgM which agglutinates by its specificity for non-active or altered IgG molecules.

Complement. Components of this system can be measured by inhibition of agglutination of complement coated red cells, using specific antibodies to complement proteins. The total haemolytic complement (CH_{50}) level is a measure of complement mediated haemolysis of red cells by serum containing activated terminal components (C5, 6, 7, 8, 9) of the complement pathway.

Immune complexes. Circulating soluble antigen-antibody complexes can be assayed by several methods. One of these involves incubation of serum with a lymphoblastoid (Raji) cell line which specifically binds activated C3 and thus represents an indirect assay of complement-fixing complexes present.

Type IV immunity (delayed hypersensitivity)

Skin testing. Erythema and induration of the skin 24–48 hours after intradermal injection of viral, mycobacterial and fungal antigens is quantifiable. Mononuclear cell infiltration in response to these antigens is the main histopathological findings. In some situations, basophils are also found in delayed hypersensitivity reactions.

Lymphocyte blast transformation. Antigens such as tuberculin, candida, mumps or trichophytin can cause proliferation of sensitized mononuclear cells *in vitro*. This is the counterpart to delayed hypersensitivity in the skin and can be measured by counting the amount of a radioactive precursor (H^3-thymidine) that is incorporated into DNA over a given time, in the presence of either a specific antigen or nonspecific mitogen such as phytohemagglutinin (PHA).

Contact hypersensitivity. Agents such as dinitrochlorobenzene (DNCB) can be used to elicit a primary T-cell and macrophage mediated immune response, expressed by erythema and induration within two to three weeks of primary contact and then within 48 h on rechallenge.

Quantification of primary elements of the immune system

B-cells and immunoglobulins. Quantification of serum immunoglobulins is performed either by radial diffusion in gels (Mancini technique) into which antibodies to a given class have been incorporated, or by automated nephelometry (using precipitation with anti-immunoglobulin as an end point).

Immunoglobulin responses are assessed by *in vivo* synthesis on rechallenge following specific vaccination (for example, Schick test for diphtheria antitoxins). Specific sensitization with neoantigens can be used to elicit primary antibody responses. The presence or absence of naturally occurring isohemagglutinins (anti-A, anti-B) in patients of the appropriate ABO blood groups can also be used as a measure of B-cell competence.

In vitro stimulation of peripheral blood lymphocytes with pokeweed mitogen leads to B-cell activation and secretion of synthesized immunoglobulin into culture supernatants. These can be directly assayed for content of class specific immunoglobulins.

B-cells. Surface immunoglobulin bearing cells are counted in the peripheral blood by fluorescence with class specific anti-immunoglobulin to which fluoresceinated dye has been conjugated.

T-cells, macrophages and their products. Absolute lymphocyte counts are obtained from differential counting of stained peripheral blood smears. T-cells are enumerated directly by their ability to form rosettes with sheep red blood cells.

Macrophages are not normally counted but their function can be measured by the ability of peripheral blood lymphomononuclear cells to ingest carbonyl iron particles or by adherence to glass or plastic surfaces.

Migration inhibition factor is a lymphokine secreted in response to antigenic stimulation. It is assayed by the ability of lymphocyte culture supernatants to inhibit the egress of macrophages from small capillary tubing in the presence of specific antigen.

Neutrophil function tests. Random migration, chemotaxis to specific stimuli, phagocytosis, and bacteriocidal capacity of peripheral blood granulocytes can be quantified and compared with controls. Specific micro-organisms can be used to assess differential ability in phagocytic and bacteriocidal function.

*Nitroblue tetrazolium test (*NBT*).* This test measures the ability of neutrophils to form hydrogen peroxide, using a change in colour of the dye as an end point.

Practical assessment

History and physical examination. The presence or absence of genetic patterns, the type of abnormality of immune response (excessive or deficient) and the severity of affliction must be assessed before any further intervention.

A complete blood count should be performed including differential and absolute lymphocyte numbers. The presence of lymphopenia may be an important clue to immunodeficiency. Further quantitation of lymphocyte subpopulation by enumeration of T- and B-cells in the peripheral blood may be of help. Histopathological assessment of lymphoid tissue may be needed, as in the diagnosis of immunodeficiencies in the neonate.

T- and B-cell function. Serum or secretory immunoglobulins, isoheamagglutinins, and responses to skin tests assess both directly and indirectly T- and B-lymphocyte function. IgE antibodies are measured by skin tests for immediate hypersensitivity as well as the basophil degranulation test. On rare occasions the RAST is used, but a simple determination of total serum IgE can be of benefit since this is markedly elevated in parasitic and allergic conditions.

Hypergammaglobulinaemia is seen in autoimmune and other excessive immune responses, while hypogammaglobulinaemia is a hallmark of immunodeficiencies, either primary or secondary.

Secondary elements in the immune response. Complement levels, granulocyte numbers and function and other plasma proteins can be important adjunctive assessments of the integrity of the immune system. Hypocomplementaemia is present in many immune complex diseases and in some situations of primary immuno-deficiency. Neutropenia may accompany primary neutrophil disorders or may be secondary to defects in the bone marrow or increased destruction by antibody.

Tests of phagocytic function or chemotaxis are sometimes useful in assessing many secondary immunodeficiencies such as those related to diabetes, chemotherapy and alcoholism. The NBT can be easily performed and quantitated as a screen for chronic granulomatous disease and female carriers of that X-linked recessive condition.

Autoimmune conditions. These can be assessed by a screen for various tissue autoantibodies such as ANA, rheumatoid factor, thyroid and other endocrine specific autoantibodies.

Immune complexes in the serum, together with cryoglobulins, and serial determinations of DNA binding may be of help in assessing and following the severity of immune complex diseases such as SLE.

Lymphocyte function tests. These can be used to specifically diagnose immunodeficiency and to follow their evolution or management. Blast transformation induced by phytohemagglutinin assesses cell mediated immune responses that involve T-lymphocytes and macrophages, while pokeweed mitogen stimulation (above) can be used to assess B-cell function *in vitro*.

Mixed lymphocyte reactions (MLR) from potential donors and recipients of transplanted organs are variants of the lymphocyte blast transformation test and can be used to assess and predict the degree of expected graft compatibility. This is particularly helpful in situations where bone marrow grafting is being contemplated.

Newer tests. Punch biopsy of skin from sun exposed or unexposed areas can be used to supplement information obtained from other autoimmune tests. The 'lupus band' is visualized by immunofluorescent techniques at the dermal-epidermal junction, and is a result of deposition of immunoglobulin and complement at that site. The skin can thus serve as a mirror for systemic illness, such as glomerulonephritis related to SLE.

Certain laboratories now can assess specific T-cell helper or suppressor function for a given antigen. For example, suppressor cell dysfunction, leading to excessive anti-DNA antibody production, has now been documented in SLE. Tests such as these may soon be used routinely in the management of some disorders.

References

BELLANTI J.A. (1978) *Immunology II*. Philadelphia, W.B. Saunders Company.

BENACERRAF B., UNANUE E.R. (1979) *Textbook of Immunology*. Baltimore, Williams and Wilkins.

STITES D.P., STOBO J.D., FUDENBERG H.H., WELLS J.V. (eds) (1982) *Basic and Clinical Immunology*, 4th ed., Los Altos, California, Lange Medical Publications.

PARK B., GOOD R.A. (1974) *Principles of Modern Immunobiology: Basic and Clinical*. Philadelphia, Lea and Febiger.

ROITT I. (1984) *Essential Immunology*, 5th ed., Oxford, Blackwell Scientific Publications.

SAMTER M. (ed.) (1978) *Immunological Diseases*, 3rd ed., Boston, Little Brown and Company.

8

Haemostasis and Thrombosis

Normal function

The normal haemostatic mechanism functions to prevent blood loss from intact vessels and to prevent excessive bleeding from severed vessels. A number of complementary mechanisms have evolved in mammals that limit blood loss from damaged vessels. Together these form the haemostatic system, which includes the platelets, the coagulation factors, and the vessel wall.

PLATELETS

Platelets are cytoplasmic fragments derived from the megakaryocytes in the bone marrow. Platelet production is controlled by a humoral factor called thrombopoietin. Thrombopoietin stimulates megakaryocyte production and maturation. The site of production of thrombopoietin is uncertain, but its production is induced by thrombocytopenia and suppressed by thrombocytosis. Two-thirds of the platelets circulate in the blood stream and they are in equilibrium with the remainder which are in the spleen. Young platelets are larger, more dense, and metabolically and functionally more active than older platelets. Platelets survive for seven to ten days in the circulation.

Platelets contain a number of organelles including electron opaque granules, the dense granules; less dense granules termed the alpha granules; lysomal granules; and mitochondria. A cytoplasmic tubular system continuous with the surface of the platelet greatly increases the effective surface area. Platelets do not contain nuclei, and have a limited capacity to synthesize proteins so that drugs such as aspirin which irreversibly inactivate specific enzymes produce a permanent defect in the platelet.

Platelets contribute to haemostasis directly and indirectly by interacting with the other components of the haemostatic system. Platelets adhere to sub-endothelial surfaces and then bind together to form clumps (aggregation). Platelets contribute to the morphological and functional integrity of endothelial cells. They also provide a phospholipid surface (traditionally termed platelet factor III) on which many of the coagulation reactions occur.

Platelets adhere to a number of surfaces including collagen, polymerizing fibrin and prosthetic vascular devices. The adhesion reaction requires a specific component of the coagulation factor VIII complex which is termed the von Willebrand factor. When platelets are exposed to a specific stimulus they change shape by extending pseudopods and aggregate. Platelet aggregation has been studied extensively *in vitro* using the platelet aggregometer, an instrument that measures the change in light transmission when platelets in suspension are exposed to aggregating agents. A small number of substances have been identified that produce platelet aggregation *in vivo*. These include arachidonic acid, serotonin, adenosine diphosphate (ADP), collagen, noradrenalin and thrombin. Platelet aggregation can be produced by the prostaglandin pathway, by endogenous ADP release and by a third mechanism that is independent of either ADP or the prostaglandins. Most aggregating agents initiate the prostaglandin pathway by activating the enzyme phospholipase A_2. This cleaves arachidonic acid from membrane phospholipids and through a series of oxidative reactions, arachidonic acid is converted to prostaglandin endoperoxides and thromboxanes that produce platelet aggregation. Aspirin and other non steroidal anti-inflammatory drugs block prostaglandin synthesis by inhibiting the enzyme cyclo-oxygenase. Thrombin and collagen cause aggregation through the prostaglandin pathway and also directly by releasing ADP from the dense granules. Thrombin also causes aggregation by at least another as yet unknown mechanism.

When the endothelial lining of vessels is disrupted, collagen is exposed causing platelets to adhere and form an unstable plug which is stabilized by fibrin strands. Thrombin is, therefore, a pivotal enzyme in haemostasis since it both causes platelet aggregation and stabilizes the platelet plug by inducing fibrin formation.

The platelet release reaction is a secretory process characterized by the release of granular contents. ATP, ADP and serotonin are released from dense granules. Platelet factor IV, beta thromboglobulin and platelet mitogenic factors are released from alpha granules and various proteolytic enzymes are released from the lysosomal granules.

The released lysosomal enzymes have the potential to contribute to vascular damage at the site of platelet aggregation. The mitogenic factors stimulate smooth muscle proliferation in the sub-endothelial layers of blood vessels. This is important in initiating repair of damaged vessels but could also contribute to atherosclerosis which has been shown experimentally to follow chronic endothelial damage.

Platelets enhance blood coagulation by providing a surface that allows optimal interactions of the coagulation factors to occur. The vitamin K dependent clotting factors VII, X, IX and II attach to the platelet surface through calcium bridges which bind the γ-carboxyglutamic acid residues on the coagulation protein. In addition factor Xa binds to a specific site on the platelet surface, a reaction that greatly increases Xa activity.

BLOOD COAGULATION

Synthesis and characteristics of coagulation factors

There are ten recognized coagulation factors. All are plasma proteins and with the exception of fibrinogen, all are present in trace amounts (μg/ml). Nine of the ten coagulation factors (factors I, II, V, VII, IX, X, XI, XII and XIII) are synthesized in the liver and the synthesis of four of these (factors II, VII, IX and X) is vitamin K dependent. The site of synthesis of factor VIII is uncertain. The portion of the factor VIII molecule that carries the major antigenic sites (VIIIR : Ag), and also supports ristocetin-induced platelet aggregation (VIII : RCF) is synthesized by endothelial cells and megakaryocytes. The site of synthesis of the coagulant portion of the factor VIII molecule (VIII : C) is unknown.

The turnover of the coagulation factors has been evaluated by isotopic methods and plasma transfusion in patients with factor deficiencies. Factor VII has the shortest half-life (4–5 hours) and factor XIII has the longest half-life (12 days). Fibrinogen also has a relatively long half-life (4–5 days). Most factors can be found on the surface of platelets, (reflecting passive adsorption to the platelet membrane) and factors I, V, VIII, XI and XIII are also located within platelets. After blood clotting, certain coagulation factors (VII, IX, X) remain active and are termed serum factors, while others, factors V, VIII, II, I and XIII are consumed or inactivated during blood coagulation and are termed plasma or consumable factors.

The coagulation proteins also have variable stability on storage. Factors V and VIII are relatively labile and lose activity when stored as bank blood, but the other coagulation factors are stable throughout the period of useful storage of whole blood. The blood platelets are the most labile of the haemostatic factors and are no longer functional in whole blood after 24 hours.

Factors II, VII, IX and X and the inhibitor, protein C, are termed vitamin K dependent factors. In the presence of vitamin K these factors are carboxylated, a reaction which occurs after the factor has been synthesized. The γ-carboxyglutamic acid portion of these coagulation factors forms the site which binds to calcium and phospholipid on the platelet surface. Protein C is a recently discovered inhibitor of blood coagulation that will be discussed subsequently.

The coagulation process

Coagulation proceeds through a series of sequential enzymatic reactions which can be initiated by a variety of stimuli and terminate with the conversion of soluble fibrinogen into insoluble fibrin. Most coagulation factors occur in trace concentrations (1/10,000 the concentration of albumin), yet, for coagulation to occur, each must interact with its specific substrate. These interactions also require a specific tertiary alignment of the coagulation proteins. The coagulation process could not occur if the protein-protein interactions occurred at random unless they were present in much higher concentrations. Platelets provide a surface which allows the coagulation factors to align in a way to optimize their specific interactions. In general, activation is associated with the loss of a portion of the coagulation molecule exposing the active enzymatic site. This, in turn, attacks a specific site on the next coagulation protein.

Blood coagulation may be produced by either activation of the intrinsic pathway

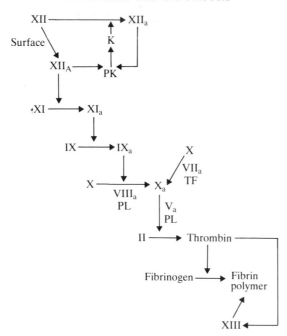

Figure 8.1. *The normal coagulation sequence*
 TF represents tissue factor, PL represents phospholipid, PK represents prekallikrein.

which is relatively slow, or by activation of the extrinsic pathway which is a faster process (Fig. 8.1). While these two pathways are inter-related *in vivo*, it is convenient to consider them separately.

Intrinsic pathway. The intrinsic pathway is initiated by the activation of factor XII when this factor is exposed to a non-endothelial surface. Factor XII undergoes a conformational change without change in molecular weight. It activates factor XI in the presence of high molecular weight kininogen. High molecular weight kininogen is not modified by this interaction, and it is likely that it facilitates the binding of factor XI to a surface. Conformationally changed factor XII, (XIIA) slowly activates prekallikrein to kallikrein (Fletcher factor). Kallikrein participates in a number of processes including inflammation and kinin formation. Kallikrein cleaves factor XII enzymatically to produce factor XIIa that slowly activates factor XI. The physiological importance of factor XII, prekallikrein and high molecular weight kininogen is not clearly understood, since deficiencies of each are not associated with bleeding.

 In the presence of calcium, factor XI activates factor IX. Factor IXa binds to a surface (platelet phospholipid) and activates factor X, a reaction requiring factor VIII. Factor Xa activates prothrombin and this activation occurs on a phospholipid surface and requires calcium and factor V. The activation of prothrombin results in the release of thrombin. Thrombin cleaves two small peptides from fibrinogen (fibrinopeptide A and fibrinopeptide B), and the soluble protein, fibrinogen, polymerizes into insoluble fibrin gel. Thrombin also activates factor XIII which, in the presence of calcium, stabilizes the fibrin polymer by cross-linking fibrin strands.

Both factor VIII and factor V function as co-factors whose role is to bind and align the activated factor on the platelet surface. Their ability to perform these functions is greatly enhanced by their prior exposure to thrombin.

The extrinsic pathway. In vitro, extracts of certain tissues known as tissue thromboplastins have the potential to bypass a number of early steps of the intrinsic pathway. This tissue thromboplastin dependent reaction is known as the extrinsic pathway. Tissue thromboplastin plus calcium activates factor VII which then activates factor X. Blood coagulation then proceeds as was described in the intrinsic pathway, since the steps beyond the activation of factor X are common to both the intrinsic and extrinsic pathway (Figure 8.1).

Interaction between the intrinsic and extrinsic pathway. The separation of the coagulation cascade into intrinsic and extrinsic pathways is arbitrary, since they are linked at a number of steps *in vivo*. Both pathways are necessary for normal haemostasis. Patients with deficiencies of coagulation factors limited to the intrinsic pathway (factor VIII, IX and XI) have haemorrhagic disorders as do patients with a deficiency of factor VII which is limited to the extrinsic pathway.

The coagulation "cascade" does not produce an orderly series of reactions, *in vivo*, since many of the reactions are accelerated by feedback mechanisms which function as "amplification loops". For example, activated Xa and thrombin are both capable of activating factor VIII. Similarly, activated factor X and activated factor VII can activate factor IXa, and thrombin is capable of activating factor V, factor VII and factor X. Factor VII can be activated by factor XIIa and factor IXa.

It has been suggested that the extrinsic pathway functions to rapidly produce small amounts of thrombin which then initiates fibrin formation and also feeds back to accelerate blood clotting by the intrinsic pathway.

Regulation of coagulation. Blood coagulation is limited *in vivo* by a number of mechanisms. These include: (a) the low concentrations of the coagulation proteins, (b) normal blood flow that results in continuous dilution of activated coagulation factors and (c) the rapid removal of activated coagulation factors by the liver and other reticuloendothelial organs, and (d) their inactivation by specific inhibitors.

Specific coagulation inhibitors function to limit coagulation. The most important is anti-thrombin III which binds to the active enzymatic site of activated factors XI, IX, X, VII and thrombin. The rate of this reaction is greatly increased by heparin. Other inhibitors include α_2 macroglobulin (which inactivates thrombin, plasmin and kallikrein), α_2 antitrypsin (XIa and thrombin), Cl inactivator (XIIa, XIa and kallikrein).

The coagulation process is also autoregulated in a number of ways. Thrombin can inactivate factors VIIIa and Va, while Xa can inactivate factor VIIa. The recently described vitamin K dependent protein, protein C, following activation by thrombin, is a potent inhibitor of factor Va + VIIIa.

THE FIBRINOLYTIC SYSTEM

The formation of a fibrin thrombus is an important physiological process that limits the amount of blood loss from injured vessels. Its subsequent lysis and removal is important in the maintenance of unimpeded blood flow. This is accomplished through the fibrinolytic system. The fibrinolytic system also removes extravascular fibrin in wounds and inflammatory exudates. A simplified scheme of the fibrinolytic system is shown in Fig. 8.2. Plasminogen, an α-2 globulin proenzyme, is synthesized by endothelial cells and eosinophils and distributed throughout the extracellular compartment. The active enzyme plasmin is released from the proenzyme by plasminogen activators which are produced by endothelial cells of the blood vessel wall. The amount of plasminogen activator depends upon both the type of blood vessel as well as the organ in which the vessels are located. Capillaries have the greatest amount of plasminogen activator with lesser amounts in the endothelial cells of veins and the least amount in the endothelium of arteries. Organs rich in plasminogen activators include the heart, lungs and adrenals, while the liver, spleen and placenta have considerably less.

The release of vascular plasminogen activators is caused by a variety of stimuli including ischemia, occlusion and vasoactive amines, in particular, serotonin. The granules of platelets are rich in this amine, and serotonin is released during aggregation. Plasminogen can also be activated by urokinase, an antigenically distinct plasminogen activator synthesized in the kidney, and excreted into the urine where it plays a role in the maintenance of the patency of the renal tract. The released plasminogen activator is rapidly inactivated by both circulating inhibitors and the liver.

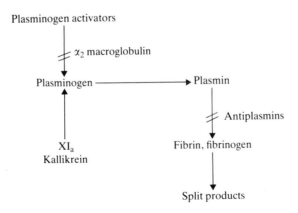

Figure 8.2. *The fibrinolytic system*

The release of plasmin from plasminogen can also be mediated by the intrinsic coagulation system. Both kallikrein and factor XIa can activate plasmin, although they are less potent activators than urokinase. Plasmin degrades fibrin, fibrinogen, factor V and VII.

Circulating plasmin is rapidly inactivated by antiplasmin which binds to the fibrin-binding site of plasmin. Antiplasmin, therefore, has little affinity for plasmin if the latter protein is already bound to fibrin. A number of theories have been proposed to explain why plasmin preferentially digests fibrin and not fibrinogen. These include: (a) plasminogen coprecipitates with fibrin and so is incorporated into the thrombus. Plasminogen activator, which is released from the adjacent endothelial cells, is then able to convert plasminogen to plasmin close to its substrate, fibrin. (b) circulating plasminogen activator may preferentially adsorb to fibrin. This, in turn, converts plasminogen to plasmin. (c) plasmin has a greater affinity for fibrin than it does for antiplasmin. Therefore, fibrin can dissociate plasmin-antiplasmin complexes.

RELATIONSHIP BETWEEN THE COAGULATION SYSTEM, FIBRINOLYSIS AND INFLAMMATION

The coagulation system is linked to the inflammatory response and the fibrinolytic system (Figure 8.3) by factor XII. Factor XIIa converts prekallikrein to kallikrein which in turn mediates the conversion of bradykinin to kinin, increasing vessel per-

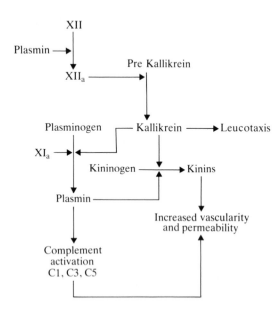

Figure 8.3. *The relationship between blood coagulation, inflammation, fibrinolysis and complement activation*

meability. Plasmin can be activated by XIIa and XIa. Plasmin in turn can activate XIIa, and the complement components C1, C3 and C5.

CONTRIBUTION OF THE BLOOD VESSEL WALL TO HAEMOSTASIS

When a vessel is severed, the amount of blood loss is influenced by: (a) the intravascular pressure, (b) the rapidity of formation and integrity of the haemostatic plug and (c) the pressure exerted by the extravasated blood when the bleeding occurs into the surrounding extravascular compartment. Blood flow continues from the severed vessel until haemostasis is achieved through vessel constriction, the coagulation process or platelet plug formation or until the extravascular pressure equals the intravascular pressure. The mechanism of vessel contraction is poorly understood but thromboxane A_2, a potent vasoconstrictor released by platelets, may participate.

Abnormal function

DEFECTS OF PLATELETS

Quantitative defects

Thrombocytopenia can result from decreased platelet production, increased platelet destruction or sequestration of platelets in the spleen. Decreased platelet production can result from a reduced number of megakaryocytes in the bone marrow or from ineffective platelet production from normal numbers of megakaryocytes. A reduction in the number of megakaryocytes in the marrow is caused by myelosuppressive agents such as radiation, toxic chemicals, drugs and alcohol, infective or inflammatory disorders and by replacement of normal marrow by malignant cells. Ineffective platelet production occurs in megaloblastic anaemia and certain myeloproliferative disorders. Thrombocytopenia associated with increased platelet destruction is usually caused by acquired extra-cellular factors. These include the immune mediated platelet consumptive states of idiopathic thrombocytopenic purpura (ITP), systemic lupus erythematosus, certain lymphomas, as well as septicemia and drug-induced thrombocytopenia. Increased platelet destruction can be caused by non-immune mechanisms such as disseminated intravascular coagulation, diffuse endothelial damage that accompanies burns, vasculitic states and anaphylactic reactions. The bone marrow can compensate by increasing platelet production three to four fold when thrombocytopenia occurs acutely, and up to eight fold in chronic thrombocytopenic states. Thrombocytopenia due to increased splenic pooling is usually associated with marked splenic enlargement.

There is not a clear relationship between the number of circulating platelets and the tendency to bleed. In states of decreased platelet production, the frequency of spontaneous bleeding begins to rise when the platelet count is less than $20,000/\mu l$. Bleeding is more likely to occur at higher platelet counts if there is an associated platelet defect, or another haemostatic abnormality. In contrast, in states of increased platelet turnover, there are a greater proportion of 'young platelets' that are haemostically and metaboli-cally more active, and bleeding may not occur despite severe thrombocytopenia.

Qualitative defects of platelets

Virtually any physiological aspect of platelet function has been associated with a congenital or acquired dysfunction which can lead to bleeding. These include: (1) failure of platelets to adhere, (2) failure of platelets to release their granular contents, (3) failure of the postaglandin pathway, (4) failure of platelets to respond to aggregatory stimuli (receptor defects) and (5) failure of platelets to support coagulation.

Platelet adhesion to the subendothelium requires a specific portion of the coagu-lation factor VIII molecule. This component of VIII is also required for platelet aggre-gation induced by the antibiotic ristocetin. Normal adhesion requires both the presence of the factor VIII molecule, and a specific receptor for this molecule on the platelet membrane. The plasma factor is absent or decreased in von Willebrand's disease, and the platelet receptor (glycoprotein 1), is absent in the congenital platelet disorder known as Bernard-Soulier syndrome.

Fibrinogen is required for platelet aggregation, and the fibrinogen receptor is missing on platelets from patients with Glanzman's thrombasthenia. These platelets do not aggregate and patients with Glanzman's thrombasthenia can have a severe bleed-ing tendency. Defective prostaglandin production occurs with aspirin ingestion or as a congenital platelet defect caused by specific enzyme deficiency. Defective release of platelet nucleotides is a feature of a congenital disorder, known as storage pool defect.

DISORDERS OF BLOOD COAGULATION

Inherited disorders

Inherited deficiencies of each of the coagulation factors have been described. These deficiencies can either be quantitative due to impaired synthesis of the coagulation factor, or qualitative due to the synthesis of a biologically inactive molecule.

There is a rough correlation between the severity of bleeding in patients with an inherited coagulation abnormality and the biological amount of the deficient coagu-lation factor. Patients with less than 1% of normal activity are usually severely affected. Those with 1–5% activity are moderately affected, those with 5–20% are mildly affected, and patients with more than 20% activity rarely have a bleeding disorder. Exceptions to this generalization occur in patients with factor XII deficiency (Hageman's Disease), Fletcher factor deficiency (pre-Kallikrein deficiency) and Fitz-gerald factor deficiency (kininogen). These congenital deficiencies are not associated with significant bleeding. Factor XIII deficiency is also an exception. Patients with levels of 1 to 2% may not have a bleeding diathesis, although lower levels are associ-ated with bleeding.

Acquired disorders

The acquired coagulation disorders are more frequently encountered in clinical practice than the inherited disorders and, unlike the inherited disorders, are almost invariably associated with multiple coagulation factor deficiencies. Acquired coagulation disorders can be caused by impaired synthesis of coagulation factors, circulating inhibitors to coagulation factors (or reactions) or by increased consumption of coagulation factors.

Impaired synthesis. Defective or decreased synthesis of coagulation factors occurs in vitamin K deficiency and in liver disease. Vitamin K deficiency can occur in states of impaired fat absorption and in patients with a poor diet. The physiological function of vitamin K is to catalyze the carboxylation of the coagulation factor protein after its synthesis. Thus, in vitamin K deficiency, there are quantitatively normal amounts of the vitamin K dependent clotting factors (factors II, VII, IX and X) although functionally, their level is greatly decreased. Oral anticoagulants (warfarin) produce a vitamin-K like deficiency state.

The haemostatic defect in liver disease is multifactorial, and can include decreased synthesis of clotting factors, decreased production of the natural inhibitors (antithrombin III), decreased clearance of activated coagulation factors and excessive fibrinolysis. Of these, the impaired synthesis of coagulation factors is usually most important. Decreased coagulation factor synthesis usually occurs sequentially and is related to the severity of hepatic dysfunction. Moderate liver disease is associated with decreased levels of factors XI, XII and the vitamin K dependent factors (II, VII, IX and X). Depression of factor V usually occurs with severe liver disease, and hypofibrinogenemia occurs only if hepatic damage is very severe. Malabsorption of vitamin K due to impaired bile salt section can contribute to the coagulation defect of severe liver disease.

Circulating inhibitors. A number of inhibitors interfere with coagulation. Heparin catalyses the antithrombin III-dependent inactivation of thrombin and factors Xa, IXa, XIIa and XIa. Heparin is used therapeutically in the treatment of venous thrombosis. The pathological production of heparin can occur in systemic mastocytosis.

Non-specific inhibitors interfere with a general coagulation sequence rather than a specific factor. The fibrin split products that are produced by plasmin digestion of fibrin and fibrinogen inhibit the conversion of fibrinogen to fibrin, by interfering with the normal polymerization of fibrin and competing with fibrinogen for thrombin. A non-specific inhibitor originally described in association with systemic lupus erythematosus is termed the lupus inhibitor. It interferes with the interaction of factor Va, and Xa with platelet phospholipid.

Inhibitors directed against specific coagulation factors usually follow repeated transfusions of the specific factor, but can also occur spontaneously. The most common is the factor VIII inhibitor which is associated with multiple factor VIII transfusions in severe haemophiliacs. Inhibitors can also occur spontaneously in autoimmune disorders, post partum and in elderly individuals.

Reduced survival of coagulation factors. The consumption of many coagulation factors occurs in a syndrome termed disseminated intravascular coagulation (DIC). DIC can be considered to result from coagulation activation which overwhelms the normal control mechanisms. The depletion of the coagulation factors and platelets by consumption can lead to bleeding. The widespread deposition of fibrin in the microcirculation is responsible for the other clinical aspect of this syndrome, the generalized small vessel thrombosis that results in pulmonary, renal and cardiac dysfunction.

The most frequent cause of DIC is bacterial septicemia which initiates coagulation by a number of different mechanisms, including the direct activation of factor XII, endothelial damage with resultant activation of factors XI and XII, and by stimulating the release of leukocyte pro-coagulant material that results in activation of factor X. DIC mediated by factor X activation can also occur when tissue factors are released into the blood stream by surgical trauma; obstetrical accidents, such as amniotic fluid embolism, antepartum haemorrhage and abortion; acute leukemia, and certain types of carcinoma.

DIC can also follow extensive endothelial damage that occurs in vasculitis, extensive burns and anaphylactic reactions. The fibrinolytic pathway is also activated with the result that DIC is invariably accompanied by elevated fibrin-fibrinogen degradation products. Other characteristic haemostatic abnormalities include positive paracoagulation assays and thrombocytopenia. The deposition of fibrin strands within the microcirculation forms a physical barrier that damages the circulating red cells and can result in a microangiopathic haemolytic anaemia.

Pathological fibrinolysis. An increase in plasma fibrinolytic activity can produce a haemorrhagic state; however, this is a far less common cause of bleeding than is disseminated intravascular coagulation. Primary pathological fibrinolysis occurs when large amounts of tissue plasminogen activator are released into the blood stream, particularly in patients with hepatic dysfunction in whom there is impaired clearance of plasminogen activator. This situation occasionally occurs in chronic liver disease during or immediately after portal-caval shunt operations. Pathological fibrinolytic states can also complicate extensive trauma, acute leukemia, obstetrical accidents and prostatic surgery. A number of factors contribute to the haemostatic defect in pathological fibrinolysis. These include the dissolution of the fibrin component of the haemostatic plug, depletion of fibrinogen, factor V and factor VIII by circulating plasmin, and interference with normal fibrin clot formation and platelet aggregation by the fibrinogen-fibrin degradation products.

DEFECTS IN THE VESSEL WALL

Defects in the vessel wall itself, or in the perivascular connective tissue are usually characterized by cutaneous purpura and minimal bleeding from mucous membranes and the gastrointestinal tract, but occasionally they can lead to life-threatening haemorrhages. While defects in the vessel wall can be associated with coagulation or platelet disorders, most occur as an isolated haemostatic defect.

Defects in perivascular and sub-endothelial structures

Inherited disorders. The inherited connective tissue disorders are in this group, and the associated haemostatic defects are usually minor. Marfan's syndrome is a connective tissue disorder characterized by arachnodactyly. These patients can have a mild haemorrhagic tendency. Pseudoxanthoma elasticum is an uncommon disorder affecting elastic fibres of the sub-endothelium. Gastro-intestinal haemorrhage can be a major problem in these patients.

Acquired defects. Vascular bleeding due to loss of the supporting extra-vascular connective tissue occurs in patients with rheumatoid arthritis, in patients who have been treated with high doses of corticosteroids and in elderly individuals. The skin in the affected parts tends to be inelastic, smooth and thin. Atrophy in the subcutaneous tissue results in the skin being freely movable over the deeper tissues so that vessels passing from deeper structures of the skin are easily torn. The resultant bleeding occurs into the skin or mucous membranes and is manifest as petechiae or ecchymosis. Muscle haematoma, visceral bleeding and haemarthorosis do not occur.

Amyloidosis is a syndrome associated with the perivascular deposits of amyloid resulting in increased vascular fragility and extensive purpura. The vitamin C deficiency state, scurvy, is characterized by defective collagen and basement membrane synthesis. Associated haemostatic abnormalities include subperiosteal and gingival bleeding and easy bruising with characteristic perifollicular haemorrhages. Rarely the bleeding tendency can be severe and haemarthrosis can occur.

Vascular disorder

Inherited disorders. Hereditary haemorrhagic telangiectasis is a common disorder associated with mucocutaneous telangiectasis. It is frequently characterized by epistaxis and gastrointestinal haemorrhage.

Acquired vascular disorders. Diffuse vascular defects can lead to severe vascular bleeding. These conditions are usually inflammatory and are termed vasculitis. Occasionally the initiating mechanism is understood as exemplified by the vasculitis which complicates serum sickness, and anaphylactic reactions. More frequently, the underlying etiology is unknown. The bleeding occurs at the site of the vessel wall abnormality and tends to spontaneously recur at the site.

THROMBOSIS

It is traditional to consider haemostasis and thrombosis as representing two end of the same spectrum; one physiological and the other pathological. The thrombus and the haemostatic plug have many features in common. Their development is similar and they are both composed of platelets and fibrin but important differences also exist. Haemostatic plugs are largely extravascular and thrombi intravascular. Haemostatic plug formation is initiated by vessel wall damage, whereas the initiating stimulus to thrombosis may be vessel damage, platelet aggregation or activation of blood coagulation.

A thrombus is an intravascular deposit composed of fibrin and formed blood elements. The relative proportions of these elements differ from that in blood because their accumulation is partly selective. In addition, the relative proportion of cells to each other and to fibrin is influenced by haemodynamic factors, and therefore the relative proportion is different in arterial and in venous thrombi. In general, thrombi that form in high flow systems are composed principally of platelets whereas those formed in regions of reduced flow have a much greater red cell and fibrin component.

Normal function

The blood is frequently exposed to thrombogenic stimuli but, in most instances, significant thrombosis is prevented by a number of important protective mechanisms. Thrombosis only occurs if protective mechanisms are impaired or the initiating stimuli overcome them.

The vessel wall

Normal intact vascular endothelium is non-thrombogenic and doesn't react with platelets or coagulation factors. Protection may be due in part to the negative charge on the surface of endothelium; to the production of prostaglandin I_2 by endothelial cells, a powerful inhibitor of platelet aggregation; the presence of heparin-like mucopolysaccharides on endothelial cells; the production of plasminogen activator; and the presence of ADPase on the endothelial surface that may function to inhibit platelet aggregation.

When the vessel wall is damaged, there is endothelial cell loss and exposure of subendothelium which is both reactive to platelets and the blood coagulation system. Experimentally, even mild trauma to the outside of the vessel leads to leukocyte adhesion and migration into the subendothelium and to desquamation of endothelial cells. Platelets and leukocytes rapidly accumulate at sites of endothelial loss and the blood coagulation mechanism is activated.

Blood flow

The normal pattern of blood flow is important in preventing thrombosis because it separates the formed elements of the blood from endothelium. It also facilitates the mixing of activated coagulation factors with their inhibitors and the clearance of activated coagulation factors by the liver and reticuloendothelial system.

Abnormal function

Thrombosis occurs when thrombogenic stimuli to which the body is frequently exposed, overcome or bypass the normal inhibitory and protective mechanisms.

The vessel wall and the coagulation factors

The vessel wall may be damaged by physical factors such as trauma, burns and proteolytic enzymes released from platelets and leukocytes; as a result of interaction with intravascular stimuli; and by damage mediated by bacteria, endotoxin or antibody acting directly upon the endothelium. Thrombi occur at three major sites; veins, arteries and the microcirculation.

Venous thrombi form in areas of slow flow or stasis and often begin as small deposits in valve cusp pockets or venous sinuses most frequently in the deep veins of the leg. The initial deposits are composed of a mixture of platelets and fibrin in varying proportions with interspersed red cells. As the thrombi grow and become occlusive, their composition changes. The fully developed occlusive thrombus is composed of red cells with interspersed fibrin. Venous thrombosis is a common complication in bed-ridden hospitalized patients. It occurs in between 10 and 30% of patients undergoing major surgery and in up to 50% of patients undergoing major orthopaedic surgical procedures. In most patients, the venous thrombi are confined to the calf and are asymptomatic. However, a percentage extend into the proximal veins and these proximal thrombi may be complicated by pulmonary embolism. Most pulmonary emboli are also asymptomatic; however, those arising from larger venous thrombi, particular those in the proximal venous segment, frequently produce clinical manifestations. If the emboli are very large or occur in patients with compromised cardiorespiratory function they can be fatal.

Congenital deficiency of antithrombin III is associated with venous thrombosis. A decrease in the functional antithrombin III activity can also occur postoperatively in patients with liver disease, premature infants, and in females taking the oral contraceptive pill.

Reduced fibrinolysis (caused by either antiplasmin levels or decreased plasminogen activator) can lead to thrombosis. This is a rare cause of postoperative venous thrombosis.

Arterial thrombi usually occur at sites of atheromatous arterial narrowing or ulceration. Flow is disturbed at these sites and platelets can adhere to the exposed subendothelium. Because flow is rapid in arteries, thrombi usually remain localized as mural thrombi and do not have the same tendency as venous thrombi to become totally occlusive. These mural thrombi frequently embolize and may produce transient disturbance or organ function such as cerebral ischaemia. If the degree of arterial stenosis is marked, the arterial thrombus can become totally occlusive and lead to severe ischaemia and infarction of the involved organ or tissue.

Disseminated thrombosis of the microcirculation can be caused by activation of blood coagulation, diffuse endothelial damage and by disseminated platelet aggregation. Patients may present with bleeding because of consumption of coagulation factors and platelets in the thrombotic process or ischaemic organ damage as a result of intravascular thrombosis. Most microvascular thrombi are of the mixed type containing fibrin and platelets but, in certain conditions, for example thrombotic thrombocytopenic purpura, the thrombi are almost entirely composed of platelet aggregates.

Thrombosis of the cardiac chambers complicates diseases of the endocardium, myocardium or valve structures. These include rheumatic heart disease, bacterial endocarditis, myocardial infarction, myocardiopathy and atrial fibrillation from any cause. The most serious complication of cardiac thrombosis is systemic embolization which can produce stroke, mesenteric ischaemia, renal ischaemia or ischaemic gangrene of the limbs.

Principles of clinical observations, tests and measurements

Spontaneous bleeding

Spontaneous bleeding into skin and mucous membranes occurs in vascular and platelet disorders and is caused by extravasation of blood through the intact vessel wall. Spontaneous bleeding into muscles, joints and the peritoneal cavity can also occur with severe coagulation disorders and probably arises from vessels that have been damaged during the course of normal activity.

Petechiae

Petechiae are red spots 1 mm in diameter caused by blood that has extravasated through small blood vessels due to abnormal vessel permeability. These lesions are seen in vascular disorders, thrombocytopenia and platelet function defects. They occur in the latter two conditions because platelets are important for the maintenance of normal vascular integrity.

The tourniquet test

The tourniquet test is performed by applying a blood pressure cuff to the arm and inflating the cuff to midway between systolic and diastolic pressure for 5 minutes and observing the arm distal to the cuff for evidence of petechiae. The test is positive when vessel permeability is increased either because of a vessel wall or platelet defect but can also be positive in haemostically normal individuals. For this reason, this test is seldom used.

The bleeding time

The bleeding time is the time required for bleeding to cease from a small standardized skin incision. The bleeding time reflects the early stages of haemostatic plug formation and is prolonged if there are defects in the vessel wall, thrombocytopenia, or abnormalities of platelet adhesion, release, or aggregation. The bleeding time is usually normal in coagulation disorders, probably because the arrest of bleeding from the small subcutaneous vessel cut by performing the test occurs independently of fibrin formation.

The clotting time

The whole blood clotting time is the length of time required for venous blood to spontaneously clot in a tube at 37° C. The result of this test is influenced by the level of coagulation factors of the intrinsic pathway, by the platelet count and the nature of the surface of the clotting tube. Because of these variables, the clotting time has a wide normal range and is insensitive to coagulation factor deficiencies. Thus, the clotting time may be within the normal range with coagulation factor levels as low as 2–3% of normal.

ACTIVATED PARTIAL THROMBOPLASTIN TIME (APTT)

The APTT is like the whole blood clotting time, but many of the variables of the whole blood clotting time are controlled and therefore this test is much more sensitive and specific for coagulation factor abnormalities. The variables produced by haematocrit, platelet count and the surface of the clotting tube are removed by:
1   performing the test on platelet poor plasma (this excludes the effects of platelets and red cells);
2   by maximally activating Factor XII with agents such as kaolin or celite (this excludes the variable degree of surface activation produced by the glass surface); and
3   by adding a standard amount of phospholipid (platelet Factor III substitute).
    The APTT has a narrow normal range and is sensitive to single coagulation factor deficiencies of 30% or lower. The test is therefore a good screening test of the intrinsic pathway.

ONE STAGE PROTHROMBIN TIME

This test is performed by activating the extrinsic pathway by adding tissue extract to plasma. It is sensitive to deficiencies of Factors VII, X, V, II and fibrinogen. The sensitivity of this test is not the same for each of these factors. It is prolonged when the level of VII, X and V is less than 30% of normal, but is not prolonged unless the level of Factor II (prothrombin) is less than 10%.

THROMBIN CLOTTING TIME

The thrombin clotting time is performed by adding a standard amount of thrombin to plasma. This test is sensitive to concentrations of fibrinogen less than 150 mg%. Depending upon the concentration of thrombin used, the test is also sensitive to heparin, and to the split products of fibrin and fibrinogen.

THE WHOLE BLOOD CLOT LYSIS TIME

The test is performed by allowing blood to clot spontaneously and observing the time it takes for complete lysis to occur. This is an insensitive test and is affected both by the concentration of circulating plasminogen activator and the level of inhibitors to fibrinolysis.

THE EUGLOBULIN LYSIS TIME

The euglobulin lysis time is performed by preparing a euglobulin fraction of plasma, clotting this fraction and observing the time it takes for complete lysis to occur. This test is more sensitive than the whole blood clot lysis time because the euglobulin fraction of plasma contains plasminogen activator, plasminogen and fibrinogen but is relatively free of fibrinolytic inhibitors. This test is sensitive to even mild increases in fibrinolytic activity.

FIBRIN PLATE ASSAY

This test is performed by applying a small volume of plasma or reconstituted euglobulin precipitate to a fibrin plate and measuring the area of lysis. This test is sensitive to relatively low concentrations of plasminogen activator but has the practical limitation that the endpoint is read 18 hours after the sample is applied. This test is sensitive to mild increases in fibrinolytic activity.

PLASMINOGEN ASSAY

This test is performed by converting plasminogen to plasmin by adding a plasminogen activator such as streptokinase or urokinase to plasma and then assaying the proteolytic activity of plasmin using casein as a substrate.

ANTIPLASMIN

The antiplasmin assay is performed by adding plasmin to plasma and after a period of incubation assaying residual plasmin by its proteolytic effect on casein.

TESTS OF FIBRIN SPLIT PRODUCTS

Fibrin/fibrinogen split products are formed by the action of plasmin on fibrin or fibrinogen. These products share antigenic determinants with fibrinogen and are detected in the serum either by a haemagglutination inhibition assay using tanned red cells to which fibrinogen has been absorbed or by a direct agglutination assay using latex particles coated with anti-fibrinogen antibodies. Increased levels of fibrin/ fibrinogen split products are found in serum in disseminated intravascular thrombosis and in pathological fibrinolytic states.

PLATELET AGGREGATION

Platelet aggregation is tested using a turbidometric technique. The test is performed by shining a light beam through platelet rich plasma and recording the change in transmitted light (optical density) as the platelets aggregate. When platelets aggregate more light is transmitted through the plasma. Various aggregating agents are added and the change in optical density produced by platelet aggregation recorded.

^{125}I-LABELLED FIBRINOGEN SCANNING

This technique is based on the fact that isotopically labelled fibrinogen is incorporated into forming thrombi. The radioactive thrombus is then detected by external isotopic scanning. This is a very sensitive technique for detecting small venous thrombi but suffers from the disadvantage that it cannot detect iliac vein thrombi.

IMPEDANCE PLETHYSMOGRAPHY (IPG)

The IPG measures changes in electrical conductance in the leg in relation to venous occlusion caused by inflating a blood pressure cuff. It is insensitive to calf thrombi, but sensitive to the clinically more important proximal vein thrombi.

Practical assessment

CLINICAL OBSERVATIONS

Generalized haemostatic defect. Bleeding from multiple sites, spontaneous bleeding, family history.

Vascular and platelet defects. Skin and mucous membrane bleeding with petechiae and bruises. Postoperative bleeding commences at the time of trauma and once stopped does not usually recur.

Coagulation defects. Haemarthrosis, deep muscle haematomas, retroperitoneal bleeding, haematuria. Postoperative bleeding may be delayed and tends to be recurrent over days.

ROUTINE TESTS

Vascular and platelet abnormality. Long bleeding time, positive normal prothrombin time, normal partial thromboplastin time and normal thrombin clotting time.

Coagulation abnormality. Bleeding time normal, prothrombin time prolonged (deficiencies of the extrinsic pathway), partial thromboplastin time prolonged (deficiencies of the intrinsic pathway), thrombin time prolonged (hypofibrinogenaemia or fibrin split products).

SPECIAL TESTS

Platelet defect. Abnormal aggregation with ADP, collagen or adrenalin.

Coagulation defect. Coagulation factor assays.

Excessive fibrinolysis. Short euglobulin lysis time, low plasminogen level.

References

BENNETT B. (1977). Coagulation pathways: interrelationships and control mechanisms. *Seminars in Haematology*, **14**, 301.

BIGGS R. & RIZZA C.R. (eds.) (1984). *Human Blood Coagulation, Haemostasis and Thrombosis.* 3rd ed. Oxford, Blackwell Scientific Publications.

BOWIE E.J.W. & OWEN C.A. Jr. (1977). Haemostatic failure in clinical medicine. *Seminars in Haematology*, **14**, 341.

HIRSH J. & GENTON E. (1973). Thrombogenesis. In *Physiological Pharmacology*, Vol. V, Root W.S. & Berlin N.I. (eds.) New York, Academic Press Inc.

KELTON J.G. & BLAJCHMAN M.A. (1979). Platelet transfusions. *C.M.A. Journal*, **121**, 1353.

LEWIS J.H., SPERO J.A. & HASIBA U. (1977). *Coagulopathies*. Disease-a-Month, Year Book Medical Publishers, Inc. **23**, 1.

MARCUS A.J. & ZUCKER M.B. (1965). *The Physiology of Blood Platelets*. New York. Grune & Stratton.

MUSTARD J.F. (1968). Haemostasis and thrombosis. *Seminars in Haematology*, **5**, 91.

PENINGTON D., RUSH B. & CASTALDI P. (eds.) (1978). *de Gruchy's Clinical Haematology in Medical Practice*, 4th ed. Oxford, Blackwell Scientific Publications.

POLLER L. (ed.) (1969). *Recent Advances in Blood Coagulation*, London, Churchill.

9

Bone

Normal function

The skeleton makes up one-sixth of the body weight in man and half of this is mineral. Dried de-fatted bone is 25 per cent calcium; there is a kilogramme of calcium in the body and 99 per cent of this is in bone; extra-osseous calcium is only about 10 g. Tracer studies show that there is a rapid exchange of calcium between plasma and bone but the total exchangeable calcium in the body is only a small fraction of the total (about 0·5 g). By weight, there is just over twice as much calcium as phosphate in bone though bone contains nearly 90 per cent of all the phosphate in the body.

Calcium is present mainly as hexagonal plate crystals of apatite, the general formula for which is $3[Ca_3(PO_4)_2]CaX_2$; in bone it is hydroxyapatite that is present $3[Ca_3(PO_4)_2]Ca(OH)_2$ though, especially at the surface, carbonate may enter the crystal, as can fluoride.

Structure

In cortical bone there are well-organized Haversian systems: blood vessels surrounded by lamellae of bone with embedded osteocytes connected by canaliculi to the central canal. In this way the osteocytes are all within 0·1 mm of a capillary. Cortical bone surrounds trabecular bone, which accounts for about one-fifth of the mass of bone. The trabeculae are made up of thin plates of bone separated by marrow. Formation of bone (by mononuclear osteoblasts) and destruction (by multinuclear osteoclasts) occurs at the surface of these plates. The mononuclear osteoblasts occur wherever bone is growing, whether it be under the periosteum or at the surface of trabeculae.

The osteoblasts contain a phosphatase. In blood, the phosphatase arising from bone is an alkaline phosphatase but it is generally believed that to be active in bone formation it would have to be active under acid conditions as well. Collagen and the mucopolysaccharides of bone substance may contribute to the initiation of crystallization by 'epitaxy', i.e. the components of hydroxyapatite are held in the position that they will have in the crystal so that the process of seeding can start. This process of crystallization may be hindered by pyrophosphates. Removal of pyrophosphate may allow crystallization to proceed and phosphatases (which are also pyrophosphatases) may regulate this process.

Organic matrix

The remarkable strength of bone is attributable not only to its mineral content, but to the complex of mineral set in an aqueous phase together with an organic matrix. Collagen is the main constituent of the matrix and accounts for about 90 per cent of it. The remaining 10 per cent is non-collagenous material mainly glycoproteins and proteoglycans.

Collagen is the main extracellular protein of the body and about 60 per cent of it is found in bone. It is made of three polypeptide chains which are folded into a rod-like triple helical molecule about 300 nm long but only 1·5 nm in diameter. The chains are called alpha chains and each of them is coiled into a left-handed helix with about three amino acids per turn. Three helical chains together are then wound around each other into a right-handed triple helix to form the rigid structure of the collagen molecule. Each chain has approximately 1000 amino acids and every third amino acid in the sequence is glycine. The presence of glycine is very imortant for the formation of the collagen molecule as it occupies the restricted space in which the free helical alpha chains come together in the centre of the triple chain. The triple helix requires the presence of proline and hydroxyproline. The molecules of collagen, called tropocollagen, overlap and leave gaps between their ends which are called hole zones. These holes probably provide the sites for the initial deposition of hydroxyapatite crystals in the organic matrix. There are a number of genetically determined, types of collagen in different tissues of the body with different alpha chains: in bone there is a combination of 2 alpha, and 1 alpha$_2$ molecules in the triple helix.

Collagen is first synthesized in the fibroblasts and osteoblasts as a longer molecule, procollagen, (Fig. 9.1), which contains additional polypeptide extensions at both the amino and carboxy-terminal ends of the three chains. As with other proteins, synthesis starts with the translation of the specific mRNA for each chain of collagen on membrane-bound ribosomes. There is now evidence which suggests that the pro-alpha chains are initially synthesized with additional pre-sequences at the amino-terminal ends. The post translation events include hydroxylation of the prolyl and lysyl residues by hydroxylating enzymes in the cisternae. Hydroxylation is followed by glycosylation of hydroxy-lysyl residues and synthesis of interchain disulphide bonds. Helix formation of these modified pro-alpha chains then occurs in the rough endoplasmic reticulum and the procollagen molecule moves to the Golgi complex to be secreted by the cell.

Extracellularly, the polypeptide extensions of procollagen are cleaved by peptidases and collagen molecules, which are formed in this way, are assembled rapidly into fibrils and cross-linking begins. Immature fibrils do not have the strength of the mature ones, until cross-linked by a series of covalent bonds. The organization of collagen initially depends on the charge properties of the molecule so that negatively charged regions of one lie alongside positively charged regions of another. Once this pattern has become established, co-valent cross-links are formed and the strength is established.

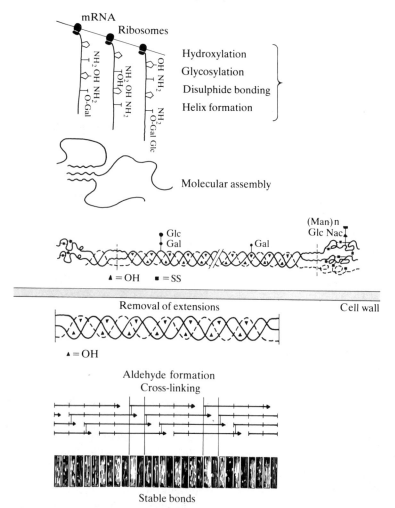

Figure 9.1. *A schematic representation of the synthesis of collagen starting within an osteoblast and continuing outside the cell. Calcification begins in the 'hole zones' between ends of molecules.*
(Reproduced from Smith and Francis in *Recent Advances in Endocrinology and Metabolism* Vol. 2 1982 Edited by O'Riordan, published by Churchill Livingstone).

CALCIUM AND PHOSPHORUS BALANCE

A normal diet supplies about 25 mmol (1 g) of calcium and about 25 mmol (800 mg) of phosphorus daily. The intake of both minerals varies considerably depending on the diet though phosphorus intake, is not so variable as that of calcium because of the presence of phosphorus in all natural foods.

The main site of calcium absorption is the upper small intestine. The duodenum has the greatest ability to absorb calcium per unit of area but most calcium is absorbed by the ileum. In a person with a normal calcium balance, about 12·5 mmol of dietary calcium is absorbed (total absorption) of which about 7·5 mmol is secreted back into the gut in digestive juices. Therefore, most of the ingested calcium is found in the faeces (20 mmol). The net absorption of calcium is related to the intake in a curvilinear manner rising steeply initially as calcium intake rises but after that, less steeply. On average, net calcium absorption is about 20–30 per cent of the intake. The calcium in the blood equilibrates with that of bone and there is a continuous exchange of the mineral between blood and bone. The kidney filters about 250 mmol of calcium daily of which about 245 mmol are reabsorbed and only the remaining 5 mmol is excreted. Calcium balance becomes positive when the need is greater, for example during growth and in pregnancy.

Phosphorus is also absorbed by the upper gastrointestinal tract by the duodenum and jejunum. The net absorption of phosphorus is a linear function of the ingested phosphorus over a wide range of intake. On average it accounts for about 50–60 per cent of the intake. Although phosphorus transport is passive following the active transport of calcium across the intestine, there is also an active transport system which is under hormonal control. The main site of regulation of phosphorus balance is in the kidney and is dependent upon a number of factors as for example, glomerular filtration, parathyroid hormone and sodium excretion rate.

THE PARATHYROID GLANDS AND PARATHYROID HORMONE

There are usually four glands. The two lower (inferior) glands are derived from the endoderm of the third pharyngeal pouch; they migrate downwards with the thymus and so may be anywhere between the angle of the jaw and the superior mediastinum. The two upper parathyroids develop from the fourth pharyngeal pouch; they are usually found in the middle third of the posterior border of the thyroid (usually deep to the plane of the inferior thyroid artery and recurrent laryngeal nerve) whilst the lower glands are usually in front of this plane. The commonest place for the lower glands is about half a centimetre below the lower pole of the thyroid. Their position is much more variable than that of the upper glands. Identification of the inferior thyroid artery helps to localize the glands since this vessel commonly supplies both upper and lower parathyroid glands. Usually, each gland is 6 mm long and 2 mm thick with a weight of 30–40 mg. There are cords of cells often arranged in lobules: the dominant cell (the chief cell) is 5 to 12μ in diameter with a large nucleus and clear cytoplasm: in some cells there are large vacuoles (the water-clear cells). There are also oxyphil (acidophil) cells which are larger than the chief cells. Probably both chief cells and oxyphil cells secrete parathyroid hormone.

PARATHYROID HORMONE

Structure and biosynthesis. Parathyroid hormone is a polypeptide with 84 amino acids arranged in a single chain without any disulphide bridges. Its molecular weight is approximately 9,500. The structure of parathyroid hormone has been established in three species. There are several differences in amino acid sequences between bovine and porcine forms of parathyroid hormone. The sequence of human parathyroid hormone differs from the other two species at 13 amino acid residues. These differences in the sequences affect the immunological properties of the molecules and thus their measurement by radiommunoassay.

Only part of the molecule is essential for biological activity; this is the aminoterminal third of the hormone. Synthetic peptides with the sequence of the first 34 amino acids have been shown to possess all the functions of the intact hormone having both calcium mobilizing and phosphaturic activity.

It is now recognized that parathyroid hormone is synthesized by successive proteolytic cleavages of larger precursor molecules, namely pre-proparathyroid and proparathyroid hormones. By following labelled amino acids in cultures of parathyroid glands it is possible to show the formation of proparathyroid hormone before the hormone itself. The prohormone has 90 amino acids in its sequence, with a hexapeptide extension at the amino-terminal end of the molecule of parathyroid hormone. It was later found, however, that a precursor for proparathyroid hormone also existed and this was named pre-proparathyroid hormone. The existence of pre-proparathyroid hormone was demonstrated in cell-free systems and represents the initial product of the translation of the messenger RNA for parathyroid hormone on membrane bound ribosomes. This can be shown in cell-free systems derived from the wheat germ and ascites tumours. These are capable of synthesizing protein when primed with the appropriate messenger RNA (in this case, prepared from parathyroid glands) without destroying the sequence of the formed peptide, because of their limited proteolytic activity.

Pre-parathyroid hormone has 115 amino acid residues and the elongation exists again, as with proparathyroid hormone, at the amino terminus. In cultures pre-proparathyroid hormone reaches a maximum concentration within 2 minutes after which proparathyroid hormone appears. Parathyroid hormone itself appears after 15 minutes. (Fig. 9.2). The biosynthesis, therefore, of parathyroid hormone follows the same scheme which applies to the synthesis of other secreted proteins, such as insulin or albumin.

Another protein distinct from parathyroid hormone is also secreted by bovine parathyroid glands during *in vitro* studies. This protein, named parathyroid secretory protein, has a molecular weight of about 150,000. Its physiological role is unknown but it has been shown that it is released from the parathyroid glands in response to changes of calcium concentration in the same way as parathyroid hormone.

Secretion. Parathyroid hormone in the circulation is heterogeneous consisting of fragments of the intact molecule. The main secretory product of the parathyroid glands is the 84 amino acid peptide but this is rapidly metabolized in the kidney and the liver. This results in the generation of fragments from the middle and carboxy-terminal regions of the molecule which are biologically inert. There is also evidence that a small

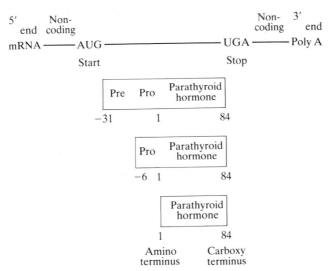

Figure 9.2. *The synthesis of parathyroid hormone and precursor peptides.*
(Reproduced from O'Riordan J.L.H. *et al.* (1982) *Essential Endocrinology,* Blackwell Scientific Publications.

amount of these fragments may be secreted by the glands, thus contributing to the heterogeneity.

The nature of the material secreted may depend on the concentrations of calcium. For example, at low calcium concentrations, more intact hormone is secreted while at high calcium levels the secretion of carboxy-terminal fragments may be increased. The circulating carboxy-terminal fragments are biologically inert and their clearance from the circulation is slow (half disappearance time about 3 hours). This contrasts with the rapid clearance of the intact hormone or of the biologically active amino-terminal region of the molecule, which is 2–5 minutes. The physiological importance of these inert fragments is unknown at present, but awareness of their presence in the circulation is essential for the interpretation of the measurements of circulating parathyroid hormone by radioimmunological techniques.

Control of secretion. Studies of the secretion of parathyroid hormone have depended on the development of immunological assays for the measurement of the hormone in the circulation. The main regulator of parathyroid hormone secretion is the concentration of ionized calcium in the blood. This has been shown in animal studies by direct perfusion of the parathyroids with blood to which different calcium concentrations were added. A fall of serum calcium stimulates parathyroid hormone secretion while a rise of the concentration of calcium suppresses the secretion. However, the change in secretion rate does not bear a simple relationship to the change of serum calcium as in animal experiments. It has been demonstrated that the glands secrete some parathyroid hormone even during hypercalcaemia. Serum phosphate, at least in acute experiments, does not affect the secretion of the hormone. Changes in the concentrations of serum magnesium may affect the secretion of parathyroid hormone in the same way as those of serum calcium, though to a lesser extent. Other modulators of parathyroid hormone secretion include vitamin D metabolites which may affect the secretion of the hormone by a direct action on the parathyroid glands. β-adrenergic agonists stimulate parathyroid hormone release. Blocking agents inhibit parathyroid hormone release induced by β-adrenergic agonist stimulation but do not block the effects of low calcium concentrations, suggesting that the parathyroid cells contain specific β-receptors.

Mode of action. Parathyroid hormone acts primarily on the kidney and on bone. As with many other peptide hormones, its action is mediated by cyclic AMP which behaves as second messenger. The hormone is bound to surface receptors of the cells of its target tissues and the hormone-receptor complex activates the enzyme adenylate cyclase which converts ATP to cyclic AMP. Cyclic AMP in turn modulates cellular processes by activation of a cascade of enzymes, mainly kinases, which result in the final biological response. There is another component of the system within the cell membrane: this links the receptor (on the outer surfaces) to the catalytic unit (on the inner surface). This component is known as the 'G' or N protein, because it requires guanosine triphosphate to be active, and it can bind this nucleotide.

In the kidney parathyroid hormone acts in the proximal convoluted tubules and the early portions of the distal segments of the nephrons and produces phosphaturia by reducing the tubular reabsorption of phosphate. It also increases the reabsorption of calcium. In addition, the hormone affects the renal handling of sodium, magnesium and bicarbonate and in high concentration it can produce generalized aminoaciduria.

In bone parathyroid hormone causes mobilization of calcium by stimulating the osteoclasts. It may also stimulate bone resorption by osteocytic osteolysis. Several biochemical effects have been observed during parathyroid hormone-induced resorption *in vitro*. These include an increase in cyclic AMP, in hyaluronate synthesis, and in lysosomal enzymes, and a decrease in collagen synthesis and in alkaline phosphatase. The action of parathyroid hormone on bone is complex and concentration dependent. Low doses of parathyroid hormone when given to animals have been reported to have an anabolic effect on bone. At normal hormone concentrations there is continuous bone remodelling; at high concentrations the rate of bone resorption is increased and exceeds that of bone formation.

Receptors play a major role in the modulation of the responses of target organs to stimulation by peptide hormones and their number and/or affinity may change in response to changes of the circulating hormone concentrations. This phenomenon is called 'down regulation', 'resistance' or 'desensitization' and appears to be of great importance in the regulation of the action of parathyroid hormone on its target tissues. Prolonged stimulation of the cyclase system with high concentrations of parathyroid hormone leads to an impairment of the production of cyclic AMP both *in vivo* and *in vitro*.

VITAMIN D

Chemistry and metabolism

During the past decade one of the most exciting areas in the endocrinology of calcium metabolism has been the recognition of the hormonal nature of vitamin D. Today, vitamin D is considered as a precursor of a potent hormone which is responsible for the biological effects of the vitamin; the active form is 1,25-dihydroxycholecalciferol (Calcitriol) (Fig. 9.3).

The natural form of the vitamin is D_3 (cholecalciferol) which is formed in the skin by the ultraviolet irradiation of 7-dehydrocholesterol. The irradiation opens the B ring of 7-dehydrocholesterol to form previtamin D_3 and rearrangement of the molecule yields vitamin D_3. In addition, vitamin D_3 can be obtained from foods which contain it naturally such as liver, or the yolk of eggs. Vitamin D_2 (ergocalciferol) is a synthetic compound which is obtained by irradiation of ergosterol from yeasts and fungi. This is used for fortification of foods, for example, milk or dairy products and for the treatment of metabolic bone diseases, and it is further metabolized in the body like vitamin D_3.

Vitamin D_3 produced either in the skin or absorbed from the small intestine (Fig. 9.4), is rapidly accumulated in the liver where the side chain is hydroxylated to form 25-hydroxycholecalciferol. This is the most abundant form of the vitamin in the circulation and its measured concentration is the best index of the vitamin D status in animals and in man. 25-hydroxycholecalciferol circulates bound to a carrier protein (Vitamin D-Binding Protein). The nature of this protein varies with the species. In man, it is an alpha-2-globulin. 25-hydroxycholecalciferol is transported to the kidney where it undergoes further metabolic transformations. The most important takes place

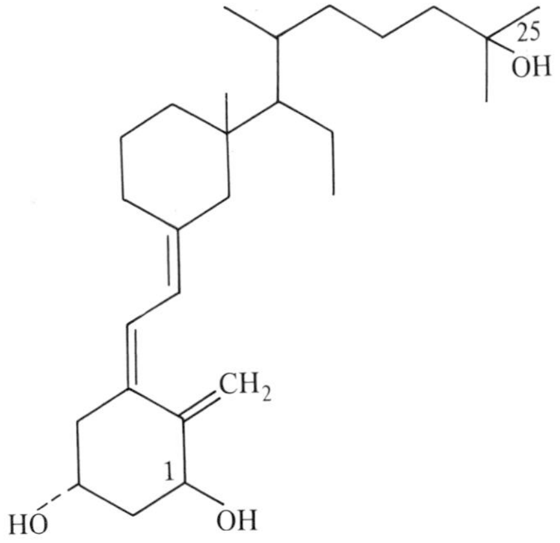

Figure 9.3. *Active vitamin D (calcitriol) structure.*

in the kidney mitochondria, and consists of a specific hydroxylation of the ring structure of the molecule. The metabolite thus produced is 1,25-dihydroxycholecalciferol (Fig. 9.3) which can be considered as a sterol hormone through which vitamin D exerts its actions. 1,25-dihydroxycholecalciferol is produced exclusively in the kidney and acts on its target organs according to the needs of the body.

Alternative metabolic pathways exist resulting in the production of other metabolites more polar than 25-hydroxycholecalciferol. These include 24,25-dihydroxycholecalciferol and 25,26-dihydroxycholecalciferol, which are formed by hydroxylations of the side chain of the molecule. 24-hydroxylation also takes place largely in the kidney.The site of the 26-hydroxylation is not yet known. A trihydroxy metabolite, 1,24,25-trihydroxycholecalciferol, has also been identified (Fig. 9.4). The physiological significance of these compounds is uncertain at present, as is that of other metabolites

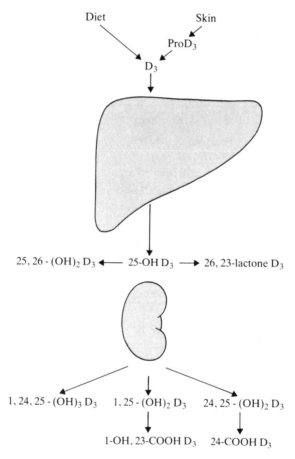

Figure 9.4. *A schematic representation of the formation and further metabolism of vitamin D to a number of hydroxylated metabolites. There are other metabolites, including acidic derivatives and a lactone.*

which include two water soluble side chain carboxylic acids and a lactone that have been identified.

Regulation of metabolism

Dietary intake and endogenous synthesis of vitamin D are both very variable and there is therefore need for regulation of the production of the more active derivatives. This can occur at the stages of the hepatic and renal hydroxylations.

The activity of the liver 25-hydroxylase can be subject to product inhibition by 25-hydroxycholecalciferol itself. This step, however, is not tightly controlled, since any suppression of the enzyme can be overcome when large amounts of substrate (vitamin D) are provided. More, important, therefore, is the regulation of the renal hydroxylases. In the kidney the production of 1,25-dihydroxycholecalciferol is controlled by calcium, phosphate and parathyroid hormone. During hypocalcaemia 25-hydroxycholecalciferol is preferentially converted to 1,25-dihydroxycholecalciferol, whereas in normo- or hypercalcaemia, the main product is 24,25-dihydroxycholecalciferol (Fig. 9.5). This action of serum calcium may be mediated through changes in the secretion of

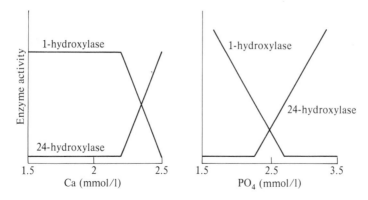

Figure 9.5. *Changes in hydroxylase activity with calcium and phosphate concentration changes (see text).*

parathyroid hormone which increases the activity of the 1-hydroxylase. Phosphate, on the other hand, appears to have a direct effect on the renal enzymes. Phosphate depletion leads to increased production of 1,25-dihydroxycholecalciferol while hyperphosphataemia inhibits the activity of the 1-hydroxylase. 1,25-dihydroxycholecalciferol itself may also control its own production by a negative feed-back mechanism. As the calcium requirements increase in physiological states such as growth, pregnancy and lactation, the hormonal changes which occur during these conditions have been also implicated in the regulation of the production of 1,25-dihydroxycholecalciferol. Oestrogens, prolactin, growth hormone and human placental lactogen can all modulate the synthesis of 1,25-dihydroxycholecalciferol.

Physiological effects of vitamin in D

Vitamin D increases intestinal calcium absorption. 1,25-dihydroxycholecalciferol acts on its target tissues in a hormonal fashion by a receptor mechanism similar to steroid hormones. On entering intestinal cells, 1-25-dihydroxycholecalciferol is bound by a specific cytoplasmic receptor which is a protein of 47,000 mw. The hormone-receptor complex is then rapidly transported to the nucleus where it interacts with the genome resulting in production of specific mRNA. It is thought that nuclear binding is determined by non-histone proteins and this activates specific genes leading to synthesis of mRNA. The mRNA is then translated into proteins such as Calcium Binding Protein which promote calcium transport across the intestine. Specific receptors for 1,25-dihydroxycholecalciferol have also been identified in other tissues such as bone, kidney and the parathyroid glands.

The most important net effect of vitamin D is to ensure that newly formed bone is calcified. Without vitamin D, the bone matrix remains uncalcified leading to the development of rickets in children and osteomalacia in adults. Up to now, it has been difficult to demonstrate a direct effect of vitamin D compounds on bone formation and it may be that calcification is facilitated by maintenance of adequate calcium and phosphate concentrations at the calcification sites. Paradoxically perhaps, 1,25-dihydroxycholecalciferol has been shown to increase bone resorption; this action may be essential during re-modelling to provide adequate minerals from old bone for the newly formed bone.

CALCITONIN

In thyro-parathyroid perfusion studies in dogs, Copp discovered a hypocalcaemic factor which accounted for a reduction in serum calcium that could not be explained simply by parathyroid suppression. This factor was initially thought to be produced by the parathyroids but was subsequently shown to be secreted by the thyroid gland. It was named calcitonin or thyrocalcitonin. Its physiological role in man remains to be established.

Calcitonin has been found in a number of species and its amino acid sequence has been determined. It consists of 32 amino acids with a disulphide bridge between residues 1 and 7. Apart from these similarities, there is considerable variability between the sequence of amino acids in the calcitonins of the different species.

In mammals, the hormone is secreted by the C-cells or parafollicular cells of the thyroid, while in birds and fishes, it is secreted by the ultimobranchial glands. In both cases the origin of these cells is the neural crest and they belong to the so-called APUD system. Calcitonin is synthesized by successive proteolytic cleavages of larger precursors, namely pre-procalcitonin and procalcitonin.

The secretion of calcitonin is regulated by various factors of which the best studied is the concentration of calcium in the blood. An elevation of serum calcium stimulates the secretion of the hormone while a decrease of the concentration of calcium in the blood produces a decrease in calcitonin release. Magnesium may stimulate the secretion of the hormone in normal subjects. Gastrointestinal hormones such as glucagon and pentagastrin can also regulate calcitonin secretion. Intravenous administration of both agents produce a sharp increase in circulating calcitonin in both animals and man. However, large doses are needed and these effects may not be physiologically important. Neuroendocrine factors such as adrenergic agents and somatostatin have been implicated in the control of calcitonin release, but their significance has yet to be established. Females have lower concentrations of circulating calcitonin than males and in females, the concentrations of the hormone may decrease with age.

The complete sequence of calcitonin appears to be essential for biological activity and amino acid deletions from the molecule result in loss of the activity. Calcitonin acts on bone, kidney and the gastrointestinal tract, probably through adenylate cyclase system and the production of cyclic AMP. It decreases bone resorption by inhibiting the osteoclastic activity while it has no effect on new bone formation. In the kidney it increases the urinary excretion of calcium, phosphate, sodium and potassium. In the gastrointestinal tract, calcitonin inhibits gastric secretion and bile flow. Other effects such as promotion of wound and fracture healing, uricosuria and inhibition of lipolysis have also been reported.

Disordered function

PRIMARY HYPERPARATHYROIDISM

The prevalence of primary hyperparathyroidism may be as high as 1 in 1000. The incidence is about two to three times higher in women and it increases with age so that it is commonest in post-menopausal women.

About 80 per cent of cases are due to an adenoma of a single gland. Multiple adenomata occur in 4 per cent of cases and carcinoma in 1 per cent. Hyperplasia of all four glands is found in the remainder. This can affect either the clear or the chief cells of the parathyroids. The lower glands are affected four times more commonly than the upper glands. Histological distinction between an adenomatous and a hyperplastic gland is not always easy unless there is a rim of normal tissue. The recognition of a carcinoma of the parathyroids can also be difficult, as the histological features of malignant cells are usually scarce. The diagnosis is often made on clinical grounds only, e.g. local spread of the tumour, or rarely the presence of distant metastases.

The pattern of clinical presentation of primary hyperparathyroidism is changing over the years; this is due to the greater awareness of physicians of its existence, to the improvement of the techniques for measuring serum calcium and to the wide application of multichannel analysers for biochemical screening. In this way, new cases are being diagnosed at an earlier stage than in the past, and patients who present with symptoms not related to the classical description of the disease are being discovered with increasing frequency. For example, in an analysis of 100 consecutive cases of primary hyperparathyroidism seen at The Middlesex Hospital, the diagnosis was made fortuituously in 17 of them during investigation of an unrelated complaint.

Renal calculi are the commonest presenting problem (45 per cent). The stones usually consist of calcium oxalate and calcium phosphate and they are radio-opaque. About 20 per cent of women with renal calculi have primary hyperparathyroidism, but in men the proportion is much lower. Overall, the disease accounts for about 5 per cent of all patients with renal stones, but in patients with recurrent stone formation, the incidence can be as high as 15 per cent. Quite apart from calculus formation, calcification can occur in the renal parenchyma causing nephrocalcinosis which is usually maximal in the renal papillae. Radiologically visible nephrocalcinosis may or may not be found in association with renal stones.

It is commonly thought that stone formation in patients with primary hyperparathyroidism is due to the hypercalciuric and phosphaturic effects of parathyroid hormone. Stones, however, can be found in the absence of hypercalciuria which is by no means, an invariable finding in primary hyperparathyroidism. It is an oversimplification to think that the formation of stones depends only on the solubility product of calcium × phosphate.

Osteitis fibrosa cystica

Hyperparathyroidism was originally recognized in patients with osteitis fibrosa cystica. This most commonly produces pain, occasionally it leads to deformity and sometimes the bone may fracture. Occasionally, there is a swelling due to a brown tumour, an osteoclastoma, of the jaw, skull, or long bones. Osteitis fibrosa is the clinical presentation in only 10 per cent of patients with primary hyperparathyroidism. Overactivity of the osteoclasts and increased resorptive surfaces are seen; there may be extensive fibrosis and in the most severe form there is also cyst formation; these cysts may be cellular or they may be haemorrhagic. An element of osteomalacia may also be present. In these cases, the histological appearance is similar to that seen in patients who are deficient in vitamin D and have secondary hyperparathyroidism.

In patients with serious hyperparathyroid bone disease, the parathyroid tumours removed are larger than those of patients with other forms of primary hyperparathyroidism. Osteitis fibrosa patients also have on average, a higher serum calcium and a higher immunoassayable parathyroid hormone concentrations.

Thirst and polyuria

These can be symptoms of hypercalcaemia, whatever the cause of the elevation of serum calcium concentration. A high serum calcium is nephrotoxic, leading to reduced net reabsorption of water in the tubules. This, with vomiting can lead to dehydration, which in itself will raise further the serum calcium concentration and set up a vicious circle. Thirst and polyuria are not common unless serum calcium is over 3·0 mmol/l.

Gastrointestinal disturbances

There is said to be an increased incidence of peptic ulceration in patients with hyper-parathyroidism; between 10–25 per cent of patients with primary hyperparathyroidism, give a history of peptic ulcer. This incidence may, however, not be higher than in the general population. If there is an association the basis of it remains to be established, though there is evidence that a high serum calcium increases the secretion of gastrin and thus gastric acid production.

In some patients with hyperparathyroidism, the hypercalcaemia appears to cause acute pancreatitis. Again, the mechanism is not clear, but the elevation of serum calcium concentration may affect the secretion of other gastrointestinal hormones apart from gastrin. In pancreatitis, however, from other causes, hypocalcaemia may be a problem as calcium is sequestered by precipitation of the calcium by fatty acids released from areas of fat necrosis.

Association with other endocrine disorders

Overactivity of one endocrine gland may be accompanied by over-activity of another in the same patient. This occurs in the pluri-glandular syndrome or multiple endocrine neoplasia (MEN). Classically, two types of MEN are described, in both of which primary hyperparathyroidism can occur. Thus, in MEN type I (Wermer's syndrome) the para-thyroids are involved in about 90 per cent of cases and over half of these have hyperplasia of all four glands. About 80 per cent have islet cell tumours, 65 per cent have pituitary tumours. Chromophobe adenomas comprise the majority of these but eosinophilic as well as basophilic tumours may occur. In 40 per cent of the patients, the adrenal cortex is involved showing cortical hyperplasia, single or multiple adenomas or nodular hyperplasia; these, however, rarely cause clinical symptoms. The syndrome is inherited as an autosomal dominant trait. In MEN type II (Sipple's syndrome) hyper-parathyroidism is associated with phaeochromocytoma and medullary carcinoma of the thyroid. Mucosal neuromata can also be found. Although the majority of patients with pluriglandular syndrome present with either the first or the second combination of endocrinopathies, in recent years it has become apparent that any combination of endocrine neoplasms is feasible, and division into Types I and II may be of little value. The pluriglandular syndrome can also be associated with the Zollinger-Ellison syn-drome with overproduction of gastrin.

SECONDARY HYPERPARATHYROIDISM

Secondary hyperparathyroidism is a common and early complication of chronic renal disease. Its aetiology is complex and a number of factors contribute to the elevated circulating parathyroid hormone concentrations. Phosphate retention which occurs early in the course of chronic renal failure may stimulate the parathyroid glands indirectly by depressing serum calcium concentrations. Defective renal production of 1,25-dihydroxycholecalciferol (which leads to impaired intestinal calcium absorption and resistance to the skeletal action of parathyroid hormone) may also contribute to the development of secondary hyperparathyroidism in chronic renal failure. These mechanisms either independently or in combination produce hypocalcaemia which stimulates the parathyroids and leads to hyperplasia of the glands. Although these mechanisms seem to provide a clear explanation for the genesis of hyperparathyroidism in chronic renal failure, there are still some unanswered questions. For example, it is not clear what leads to persistence of the hyperparathyroidism once serum calcium has been restored. Also, examination of the suppressibility of parathyroid hormone secretion by administration of calcium has shown that although in the majority of patients the parathyroids are responsive, in some secretion seems to be autonomous and not suppressible by hypercalcaemia.

Hyperparathyroidism is one major component of renal osteodystrophy and is demonstrable in over 90 per cent of patients with longstanding renal failure. The other component is osteomalacia. Osteoporosis (a term meaning reduction in mass of bone and matrix) can also be a feature, and there may also be osteosclerosis (meaning more dense bone). Hyperparathyroidism usually resolves after renal transplantation but it may persist for years.

Secondary hyperparathyroidism also occurs in chronic vitamin D deficiency, though not all patients with vitamin D deficiency and hypocalcaemia have hyperparathyroidism, indicating that the glands may fail to respond to the hypocalcaemic stimulus. It seems that in this condition, serum calcium is not the only modulator of parathyroid hormone secretion and other factors such as vitamin D metabolites themselves may be involved.

CAUSES OF HYPERCALCAEMIA OTHER THAN HYPERPARATHYROIDISM

In adults, the commonest cause of hypercalcaemia is *malignant disease*. Tumours can raise serum calcium by a number of mechanisms. In the majority of patients, hypercalcaemia is due to bony metastases and hence bone destruction but there are also a number of patients in whom hypercalcaemia can occur in the absence of any metastases to the skeleton. In these cases, production of humoral factors by tumour cells has been implicated in the development of hypercalcaemia. These include prostaglandins, especially prostaglandin E. Indomethacin, an inhibitor of prostaglandin synthesis, can correct the hypercalcaemia of some patients with cancer, perhaps by inhibiting prostaglandin synthesis. Parathyroid hormone-like peptides and osteoclast activation factor (a macromolecular protein which causes bone resorption by activation of osteoclasts) may be important in the genesis of hypercalcaemia of malignancy.

Vitamin D intoxication is another important cause of hypercalcaemia. It occurs readily when large doses of vitamin D are given so that regular supervision is necessary. It can also occur after administration of the new potent 1α-hydroxylated derivatives of vitamin D. Discontinuation of treatment with 1α-hydroxylated forms of vitamin D results in a more rapid fall in serum calcium than if the patient had been treated with vitamin D as the half-life of disappearance of these components from the circulation is much shorter than that of the parent vitamin.

Sarcoidosis can also cause hypercalcaemia and hyper-sensitivity to vitamin D. In hypercalcaemic sarcoid there is abnormally high production of the hormonal form of vitamin D, namely 1,25-dihydroxycholecalciferol, which accounts for the apparent hypersensitivity of such patients to vitamin D.

Other causes of hypercalcaemia include thyrotoxicosis, excessive ingestion of milk and alkali by patients with peptic ulcer, adrenal insufficiency, immobilization of patients with Paget's disease and familial hypocalciuric hypercalcaemia. This familial syndrome is of obscure aetiology and as the name implies, its feature is low urinary calcium excretion in the presence of hypercalcaemia. It is inherited as an autosomal dominant trait and its course is usually benign. Some patients with this condition have had attempted parathyroidectomies which were often unsatisfactory, with either persistence of hypercalcaemia or development of permanent hypoparathyroidism after the operation. In view of the benign course of the disease, surgery may best be avoided if the diagnosis is clear. Nevertheless, as in some patients, grossly abnormal parathyroid tissue is found and the decision on management must depend on clinical findings.

HYPOPARATHYROIDISM

All four parathyroids may be removed accidently during thyroidectomy, or intentionally during total pharyngo-laryngectomy for laryngeal carcinoma. Surgical removal of the parathyroids is the commonest cause of hypoparathyroidism in man. Parathyroidectomy leads to hypocalcaemia and tetany usually within 36 h. This causes first paresthesiae and then carpopedal spasm. Chvostek's sign (contraction of the facial muscles when the facial nerve is tapped) and Trousseau's sign (production of carpopedal spasm when the blood supply to the arm is partially occluded) are both positive. Symptoms are produced when serum calcium falls to about 1·8 mmols/l. Untreated, longstanding hypoparathyroidism may cause cataracts. Tetany due to hypocalcaemia has to be differentiated from that due to hysterical over-breathing (with normocalcaemia) and from hypomagnesaemia and hypokalaemia.

Idiopathic hypoparathyroidism

This condition usually becomes manifest in childhood. It may be sporadic or familial, and results from failure of secretion of parathyroid hormone. It causes hypocalcaemia and hyperphosphataemia and is often manifest as convulsions. Grand mal epilepsy can be produced by hypocalcaemia and measurement of serum calcium is essential in the investigation of epilepsy. Calcification of the basal ganglia may occur. Occasionally, papilloedema is found; this disappears when normocalcaemia has been established. It is possible that there is an immunological basis for primary hypoparathyroidism as it may be associated with Addisonian adrenal insufficiency, pernicious anaemia or muco-cutaneous candidiasis. Antibodies to parathyroid tissue have been reported to occur in such patients but this has been difficult to confirm. Antibodies to other endocrine organs are sometimes present in the serum of patients with primary hypoparathyroid-ism. Occasionally, there is congenital absence of parathyroid glands associated with deficient cellular immunity and cardiac abnormalities. This is the diGeorge syndrome, which is due to failure of development of third and fourth pharyngeal pouches.

Transient hypocalcaemia may occur in the neonatal period as a result of maternal hypercalcaemia due to primary hyperparathyroidism. This usually recovers sponta-neously but it may take sometime (months) or occasionally may become permanent.

Pseudohypoparathyroidism

Pseudohypoparathyroidism is an uncommon inherited condition which was the first demonstration of an endocrine disease where impaired target organ responsiveness to a hormone seemed to be the primary defect. In the classical description, patients with pseudohypoparathyroidism were fat, with round facies, short stature and mental retar-dation. Brachydactyly may be found; shortening of the fourth or fifth metacarpal bone is particularly common. The presenting symptoms can be tetany, grand mal seizures, paraesthesiae, muscle cramps or syncope. Patients may also have cerebral calcification and cataracts. In recent years patients who have all the characteristic biochemical abnormalities but with normal phenotypes have been recognized as well.

In pseudohypoparathyroidism there is failure of the target organs to respond to parathyroid hormone and as a consequence, patients develop hypocalcaemia and hyperphosphataemia. This defect in the receptors may be due to lack of the regulatory component that is the G-protein (also called the N-protein). Parathyroid function is intact and the glands respond to the hypocalcaemic stimulus with increased secretion of parathyroid hormone. As a result, patients have elevated serum parathyroid hormone concentrations. The diagnosis is established by examining the renal and sometimes bone responsiveness to exogenous parathyroid hormone. The extracellular cyclic AMP and phosphaturic responses to parathyroid hormone are minimal or absent, indicating resistance to the renal action of the hormone. There is also skeletal resist-ance to the action of parathyroid hormone and intramuscular administration of the hormone fails to increase serum calcium concentrations. As a result, despite chronically elevated serum parathyroid hormone concentrations, there is generally no radiological evidence of hyperparathyroid bone disease. A few patients with renal resistance but with radiological and histological evidence of hyperparathyroidism have been described and the condition has been variably termed hypo-hyperparathyroidism, pseudohypoparathyroidism with osteitis fibrosa cystica, or pseudohypoparathyroidism with raised alkaline phosphatase. These cases may represent a spectrum of renal and bone resistance to the action of PTH. There may be a variant of this condition, pseudohypoparathyroidism Type II in which cyclic AMP production is normal but further cellular reaction is defective and as a result, no phosphaturia is seen in response to parathyroid hormone administration even though cyclic AMP production has risen. Patients with pseudohypoparathyroidism may also have associated endocrine defects such as hypothyroidism, blunted prolactin response to TRH stimulation, partial resistance to the urine-concentrating ability of ADH and possible resistance to the action of glucagon.

HYPOMAGNESAEMIA

Patients with magnesium deficiency may present with hypocalcaemia which does not respond to treatment with calcium and/or vitamin D but is rapidly corrected by magnesium repletion. Primary hypomagnesaemia is rare and may be familial. Secondary hypomagnesaemia can occur in coeliac disease, in inflammatory bowel disease as well as after intestinal bypass surgery, or resection and after irradiation of the bowel. Prolonged intravenous feeding and chronic alcoholism also lead to hypomagnesaemia. Treatment with gentamycin causes hypomagnesaemia by increasing the urinary loss of magnesium.

The majority of hypomagnesaemic patients have serum parathyroid hormone concentrations which are inappropriately low for the level of serum calcium. Intravenous administration of magnesium salts into such patients causes a rapid increase in circulating parathyroid hormone, which reaches a peak within 2 minutes. This response indicates that in hypomagnesaemia there is impaired release of parathyroid hormone by the glands. In addition there may be resistance to the renal action of parathyroid hormone with phosphaturic and cyclic AMP responses being impaired. This loss of renal responsiveness is reversible by magnesium repletion which corrects the hypomagnesaemia and hypocalcaemia. It should be noted that this phenomenon appears to be specific for the parathyroid hormone-dependent adenylate cyclase since the responsiveness of other adenylate cyclases, (such as the hepatic system responding to glucagon) is normal in the presence of hypomagnesaemia.

RICKETS AND OSTEOMALACIA

In children, the commonest presentation of rickets is with knock-knees or bow legs which develop when the child begins to stand. At the same time, there may be thickening of the wrists and prominence of the costochrondral junctions (giving a rickety rosary) and more rarely, bossing of the skull. These changes are due to a failure of remodelling. In adults, osteomalacia can present in a variety of ways. Pain is the most prominent feature and may be focal or diffuse; it occurs particularly in the legs and there may be difficulty in walking. The patient may also have a waddling gait. Pain in the ribs and back may be severe. Muscular weakness may be a prominent feature and there may be a proximal myopathy though differentiation of the effects of myopathy and those of pain is often difficult. Some causes of rickets and osteomalacia are shown in Table 9.1.

Table 9.1. Conditions leading to rickets or osteomalacia

Vitamin D Deficiency
—Dietary lack
—Lack of UV light
—Malabsorption
    Coeliac disease
    Pancreatic insufficiency
    Blind loop

Disordered metabolism of vitamin D
—Failure of 1ed-alpha hydroxylation

  Inherited
    Vitamin D dependent rickets

  Acquired
    Chronic renal failure
    Tumour induced

—Receptor defect
    End-organ resistance to 1,25-dihydroxycholecalciferol

Hypophosphataemic states
—X-linked hypophosphataemia
—Other Renal Tubular Defects
    Renal tubular acidosis
    Fanconi's syndrome
    Cystinuria

Drug Induced
—Anticonvulsants
—Aluminium hydroxide

Other conditions
—Neurofibromatosis
—Polyostotic fibrous dysplasia
—Wilson's disease
—Hypercalciuric rickets
—Osteopetrosis
—Primary hyperparathyroidism

Vitamin D deficiency

Vitamin D deficiency is still common in many parts of the world. This results from a combination of inadequate synthesis in the skin and inadequate dietary intake. In

North Africa, for example, despite the sunny climate, babies frequently develop rickets because their diet is deficient in Vitamin D and they are not allowed to go out of doors. In Britain, the condition has been eliminated in infants and children by giving vitamin D supplements, but it is a major problem in immigrants of Asian origin from India, Pakistan, East Africa and elderly people living indoors. Measurement of circulating concentrations of 25-hydroxycholecalciferol have established that rickets and osteomalacia existing in these groups are due to deficiency of vitamin D. In symptomatic patients, serum concentrations of 25-hydroxycholecalciferol have been found to be below 3 ng/ml. A number of other explanations have been proposed to explain the development of rickets and osteomalacia in Asians, e.g. that pigmentation of skin reduces vitamin D synthesis. This explanation does not explain why vitamin D deficiency is not seen in West Indian immigrants to Britain who have a darker skin. Another theory, suggests that high phytate content of the chapati flour may bind calcium in the gut and reduce calcium absorption, but this effect is not significant. Supplementation of the chapati flour with vitamin D can correct the deficiency and heal the osteomalacia. It is difficult to establish that the Asian immigrants have lesser exposure to sunlight, and so reduced dietary intake of vitamin D must be important.

In vitamin D deficiency, serum calcium may be low or normal and serum phosphate low, normal or raised. Serum alkaline phosphatase is usually high but can occasionally be normal even in the presence of histological and radiological evidence of osteomalacia. Urinary calcium excretion is low and there is usually secondary hyperparathyroidism. Vitamin D deficiency responds to physiological doses of vitamin D or of 1,25-dihydroxycholecalciferol (calcitriol). The 1-hydroxylase in the kidney is very active and supra-normal concentrations of 1,25-dihydroxycholecalciferol are formed during treatment with vitamin D even when the concentration of the 25-hydroxy precursor has only been raised to normal. The overactivity of the 1-hydroxylase persists for many months until the bone disease has healed.

Vitamin D resistance

Resistance to the action of vitamin D occurs in renal failure because of loss of 1-α-hydroxylase. This may account for the osteomalacic component of renal osteodystrophy.

Deficiency of 1-hydroxylase can also be genetically determined; it occurs in pseudo-vitamin D deficiency, also known as vitamin D-dependent rickets. This is an autosomal recessive syndrome which presents in early life with hypocalcaemia, secondary hyperparathyroidism and rickets which responds to pharmacological doses of vitamin D. Administration, of physiological doses of 1,25-dihdroxycholecalciferol results in healing of rickets, thus suggesting that there is impaired conversion of 25-hydroxycholecalciferol to 1,25-dihydroxycholecalciferol. Furthermore, the circulating concentrations of 1,25-dihydroxycholecalciferol are low in this syndrome, indicating that it is the defective production of the hormone which is responsible for the resistance to vitamin D treatment.

Resistance to the action of 1,25-dihydroxycholecalciferol

True end-organ resistance to the action of 1,25-dihydroxycholecalciferol can occur. There have been only a few cases described and most of them are girls who have total alopecia as a common characteristic somatic feature. There may be rickets and/or hypocalcaemia with abnormally high circulating concentrations of 1,25-dihydroxycholecalciferol in the presence of normal serum 25-hydroxycholecalciferol. Administration of large doses of 1,25-dihydroxycholecalciferol fails to increase serum calcium, thus suggesting resistance of the target organs possible at the receptor level.

X-linked hypophosphataemic rickets

This condition is characterized by rickets associated with normocalcaemia and normal parathyroid activity, and needs large doses of vitamin D to heal it. The characteristic abnormality is hypophosphataemia which is due to a primary renal tubular defect leading to reduced renal tubular re-absorption of phosphate. Aminoaciduria and glycosuria can also be present. The mode of inheritance is X-linked dominant, but autosomal dominant inheritance has also been described. The frequency is about 1 : 25000. The concentrations of 25-hydroxycholecalciferol in serum are normal but there has been some debate about the role of 1,25-dihydroxycholecalciferol in this condition and both normal and low concentrations of the hormone have been reported. It responds to 1,25-dihydroxycholecalciferol in fairly small doses, though phosphate supplements may also be needed.

OSTEOPOROSIS

In osteoporosis there is a reduction of bone mass per unit volume without any change in the nature and mineralization of the bone tissue present. In other words, the bone is qualitatively normal but qualitatively deficient. This is essentially a histological definition. Clinically, the diagnosis is generally made on radiological grounds on the basis of reduced bone density. For careful assessment of this, allowance has to be made for the penetration of the film which may be done if reference standards are radiographed at the same time. Osteoporosis is a disease characterized by normal concentrations of calcium, phosphorus and alkaline phosphatase in the circulation.

In osteoporosis the bones are brittle and break easily. As a result, fractures of the wrist and of the neck of the femur can occur readily following minimal trauma. Since trabecular bone particularly in the vertebral column is especially affected crush fractures can occur, leading to kyphosis and consequent loss of height. A crush fracture of a vertebra usually causes severe pain for a month or so, but between episodes the patient is usually remarkably well. Not infrequently, however, radiographic evidence of a crush fracture is found without there being a history of severe back pain.

The commonest form of osteoporosis is idiopathic and occurs chiefly in post-menopausal women and among the elderly. The condition seems to be insidious in onset and it seems likely that there is slight prolonged negative calcium balance. In animals, a calcium deficient diet can lead to osteoporosis but there is no evidence this is the cause of the clinical syndrome. Nevertheless, patients with osteoporosis are generally given calcium supplements. It has been reported that infusion of calcium over short periods can lead not just to short-term retention of calcium but also to prolonged restoration of positive calcium balance. Castration can cause osteoporosis and so treatment with sex hormones has also been advocated. Treatment with fluoride has also been recommended since fluoride can enter into the hydroxy-apatite crystal lattice and in fluoride intoxication, the bones are radiologically more dense. In such a chronic condition as osteoporosis, however, it is difficult to obtain evidence that any form of therapy is beneficial. When calcitonin was discovered it was hoped that it would be helpful in the treatment of patients with osteoporosis since it was capable of inhibiting bone resorption. This would only be helpful if bone turnover were reasonably rapid but it appears that in osteoporosis bone turnover is normal (even though bone mass is reduced). In fact, calcitonin therapy instead of inducing a positive calcium balance is found to produce a negative calcium balance by increasing the rate of renal excretion of calcium disproportionately.

Secondary osteoporosis can also occur, particularly in Cushing's syndrome and in thyrotoxicosis. In these conditions there is increased protein breakdown and perhaps it is this that leads to a deficiency in bone matrix. Treatment of the primary condition leads to arrest of progress of the bone disease.

IDIOPATHIC HYPERCALCIURIA

The most important cause of renal calculi formation is idiopathic hypercalciuria. As the name implies, in this condition, there is increased urinary excretion of calcium with normal serum concentration. The upper limit of normal of urinary calcium excretion is about 7·5 mmol/24 hrs in persons on a free diet.

The stones that are formed are composed of calcium oxalate and phosphate. Presumably the rise in the calcium phosphate product leads to increased likelihood of deposition of calcium salts. Other factors can be important, such as for example, deficiency of pyrophosphates which inhibit calcium phosphate precipitation or lack of factors which usually reduce the rate of calcium oxalate crystal growth and aggregation.

Dietary hypercalciuria is uncommon but can occur if calcium intake is extremely high (50 mmol/day = 2 g). In *absorptive* hypercalciuria there is increased intestinal absorption of calcium and therefore this type responds to restriction of the dietary calcium with normalization of urinary calcium excretion. The reason for the increased calcium absorption in these patients is not clear, but it has been suggested that it may be vitamin D dependent. Serum 1,25-dihydroxycholecalciferol concentrations may be significantly raised in patients with absorptive hypercalciuria and account for the increased calcium absorption. Whether this represents an abnormality of vitamin D metabolism *per se*, or whether it is secondary to a fall of serum phosphate (which has also been found to occur in absorptive hypercalciuria) remains to be established. A small number of patients with idiopathic hypercalciuria do not fully respond to dietary restriction of calcium and in this case the primary defect may be in the kidney, resulting in renal leak of calcium. This is termed *renal* hypercalciuria and can be treated with thiazides which reduce the urinary excretion rate of calcium.

PAGET'S DISEASE

Paget's disease of bone is quite common, affecting about 3–4 per cent of the population over the age of 40 at least in Germany and England, but is rare in some European countries e.g. Poland. Only about 5 per cent of patients have symptoms while in the rest, the disease is diagnosed by chance (e.g. X-rays taken for some other reason, or finding of elevated serum alkaline phosphatase in biochemical screening) or at autopsy. The condition is characterized by increased bone turnover with both increased bone destruction and formation. The continuous bone destruction which is due to abnormal osteoclastic proliferation, leads to loss of the normal arthitecture of bone. The disease affects the skeleton in a patchy manner, particularly the pelvis, femora, spine (especially the lumbar spine) and skull and can be localized or diffuse.

Pain is the commonest symptom though sometimes it may be difficult to know whether it is due to Paget's disease *per se* or to another cause, for example, secondary osteo-arthritis of the hip. Deformity and fractures may occur as the bone becomes soft. Neurological symptoms due to nerve compression occur occasionally. The aetiology of the disease is obscure, though it has been suggested that it might be of viral origin as nuclear inclusion bodies have been found in the osteoclasts of affected patients. Serum calcium and phosphorous concentrations are normal; hypercalcaemia, however, can rarely occur during periods of immobilization after major fractures. Indices of increased bone turnover, such as serum alkaline phosphatase and urinary hydroxy-proline excretion rate, are elevated.

MEDULLARY CARCINOMA OF THE THYROID

This tumour accounts for 5 to 10 per cent of carcinoma of the thyroid and is a tumour of the parafollicular cells. Between the malignant cells there may be deposits of amyloid. The patients usually present with a goitre but sometimes they present with metastases to lymph glands. In some patients (rather less than 10 per cent) the condition is familial. In most of the families that have been described, there is a high incidence of an association with phaeochromocytomata and in some families there is an association with multiple mucosal neuromata. There is a rare association with hyperparathyroidism. Medullary carcinomata of the thyroid contain large quantities of calcitonin and the plasma concentration of calcitonin is high, being readily detectable by bio-assay and by immunological assay techniques. However, it is difficult to demonstrate any disturbance of calcium homeostasis or bone structure in patients with these tumours.

INHERITED DEFECTS OF COLLAGEN BIOSYNTHESIS

There are a number of conditions in which the biosynthesis of collagen is disturbed. A defect can occur at any of the steps involved in the initial synthesis of the protein, or its post-translational modifications. The intracellular processes may be affected, for example, when the initial gene product (i.e. the messenger RNA) is abnormal or when there is a failure of normal conversion of the pro-alpha chains to pro-collagen. Extracellular processing of the collagen molecule can be disturbed by the failure of formation of cross-linking between molecules, or a disturbance in the formation of fibrils or their assembly into fibres. (Fig. 9.6).

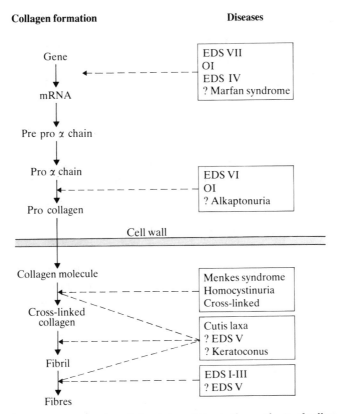

Figure 9.6. *A schematic representation of the abnormalities in the synthesis of collagen occurring within and outside the cell.*
(EDS = Ehler's Danlos syndrome; OI = osteogenesis imperfecta).
(Reproduced from Smith & Francis in Recent Advances in Endocrinology and Metabolism, Vol 2, 1982. Edited by O'Riordan. Published by Churchill Livingstone).

These disorders can be inherited. *Osteogenesis imperfecta* (the 'Brittle Bone syndrome') is a heterogeneous condition, from the clinical and biochemical viewpoints. The main feature of the condition is the fragility of bones and their liability to fracture. This may vary from mild to severe and may be obvious at birth or only become apparent later. The patient may have blue sclerae. The incidence is 1 : 25,000 (similar to that of haemophilia): 80 per cent of cases are dominantly inherited with mild disease and insufficient Type I collagen. Extra-osseous abnormalities of collagen, for example, leading to hypermobility of joints and ruptured tendons and poor wound healing and valvular heart disease can also occur.

The *Ehlers Danlos syndrome* comprises a group of inherited conditions that are characterized by increased elasticity of the skin, which heals poorly and gives paper thin scars. The skeleton is not usually affected but the joints are hyperextensible and dislocate easily and there is a bruising tendency. A number of types are recognized; some are inherited by autosomal dominance while one form is X-linked. In one variant, it has been shown there is a defect of type III collagen formation and these are patients who may rupture arteries.

Homocystinuria occurs in disorders of methionine metabolism. Classical homocystinuria resembles Marfan's syndrome (which may also be due to a collagen disorder, although this has not been fully characterized). Deficiency of cystathionine synthase leads to accumulation of homocyteine which can act as a lathyrogen. Collagen of the skin of patients with homocystinuria is abnormally soluble and the cross links are reduced. This leads to progressively abnormal skeletal growth and scoliosis and lens dislocation, as a result of inhibition of collagen cross-linking by the excessive concentrations of homocysteine present in the tissues.

Disorders of collagen biosynthesis can also be acquired, for example, in deficiency of vitamin C or of vitamin D or as a result of administration of certain drugs. Scurvy serves as a classical example of a disorder of the post translational stage of collagen biosynthesis, since ascorbate is required for the hydroxylation of proline. When there is deficiency of vitamin C, the chains formed contain reduced amounts of hydroxyproline and are not able to form a triple helix at body temperature. This accounts for the clinical features of scurvy and particularly the fact that scorbutic wounds fail to heal. Lack of vitamin D has an effect on bone collagen that in some ways is similar to that of osteogenesis imperfecta, quite apart from the defective mineralization which results from deficiency of vitamin D.

The disorders can also be drug induced; for example, administration of penicillamine causes chelation of the lysine and hydroxy-lysine derived aldehydes, and this interferes with cross linking. In animals, aminonitriles such as β-aminoproprionitrile, are lathyrogens, that is, produce lathyrism, by binding irreversibly to lysyl oxidase and blocking the formation of lysyl derived cross links in newly synthesized collagen and elastin. Subsequent fragility of the connective tissue leads to arterial rupture, bone deformity and scoliosis. Deficiency of metals such as copper, and zinc can also cause disorders of collagen biosynthesis since, they are essential cofactors for cross-linking in collagen and elastin.

Principles of tests and measurements

SERUM CALCIUM CONCENTRATION

This is the single most useful measurement in assessing calcium homeostasis and bone metabolism. The normal range by most modern methods is 2·20 to 2·55 mmol/l. Undue stasis during collection will increase serum calcium by causing haemoconcentration, which raises the concentration of albumin which has calcium bound to it. The normal serum albumin is taken as 41 g/l and for an increase of 6 g/l in albumin the observed serum calcium is reduced by 0·1 mmol/l to give the corrected value. Conversely, it is proportionately increased if serum albumin is low. Blood for estimation of calcium should be taken fasting if it is suspected that the patient has marginal hypercalcaemia since after meals serum calcium may rise by 0·12 mmol/l.

Cortisone suppression test

This test is useful in distinguishing hypercalcaemia due to hyperparathyroidism from that due to other causes, particularly multiple myeloma, sarcoidosis and vitamin D intoxication. In these conditions, glucocorticoid (given as hydrocortisone 40 mg three times a day for ten days by mouth) will reduce the serum concentration to normal but it will not do so in patients with primary hyperparathyroidism. The test can however, occasionally fail to reduce serum calcium when the hypercalcaemia is due to metastases to bone (but then these are usually visible radiologically), or to ectopic production of parathyroid hormone from a neoplasm of non-parathyroid tissue, such as carcinoma of the lung, ovary or kidney.

SERUM PHOSPHATE CONCENTRATION

The normal range (0·7–1·2 mmol/l) usually quoted is for the fasting state in adults, but it is higher in children. Serum phosphate is raised in uraemia and hypoparathyroidism and is low in patients with phosphaturic rickets.

PLASMA PROTEIN CONCENTRATIONS

Interpretation of changes in serum calcium needs knowledge of plasma proteins. The effects of changes in serum albumin have already been discussed. Changes in serum globulins may indicate the cause for hypercalcaemia, e.g. the sharp band of a monoclonal gammopathy, detected on electrophoresis of serum will indicate multiple myeloma.

ALKALINE PHOSPHATASE

This is a useful estimate of bone turnover because serum alkaline phosphatase is raised whenever bone turnover is increased, whether this is due to Paget's disease, hyperparathyroidism or bone metastases. The normal value varies according to the exact method and substrate used so care has to be taken in interpreting results unless the local normal range is known.

Elevation of serum alkaline phosphatase also occurs in a variety of liver diseases and if there is doubt as to the cause of a raised serum alkaline phosphatase, then serum 5-nucleotidase or γ-glutamyl transpeptidase activities can be measured in plasma. If the elevated serum alkaline phosphatase is of hepatic origin, there is usually accompanying elevation of the serum 5-nucleotidase or γ-GT activity.

Low levels of serum alkaline phosphatase occur in the rare condition hypophosphatasia. The clinical condition that results resembles rickets and the patients exrete increased amounts of phosphoethanolamine in the urine.

NEPHROGENOUS CYCLIC AMP ESTIMATION

Estimation of the urinary excretion of cyclic AMP may be of some assistance in the diagnosis of primary hyperparathyroidism. There is, however, a considerable overlap between values in normal subjects and hyperparathyroid patients. Cyclic AMP in the urine is derived from two sources, glomerular filtration and renal synthesis of the nucleotide. This latter component is termed nephrogenous cyclic AMP and about 90 per cent of it is parathyroid hormone-dependent. Estimation, therefore, of its excretion provides a good index of parathyroid activity. The normal range for nephrogenous cyclic AMP is between 0·34 and 2·70 nmol/100 ml GFR. The excretion is generally high in patients with primary hyperparathyroidism and low in patients with hypoparathyroidism and hypercalcaemia of nonparathyroid origin.

RADIOLOGY

The classical radiological sign of hyperparathyroidism is subperiostal erosion of the phalanges, best seen in the middle phalanges. In severe cases, the terminal tufts may also be eroded. X-rays of the skull of a patient with hyperparathyroidism may have the 'salt and pepper' appearance—a mottled appearance best seen on the lateral view. Increase in the size of the pituitary fossa may occur in patients who have hyperparathyroidism as a part of the pleuriglandular syndrome. In severe osteitis fibrosa cystica, there are lytic lesions of varying size and texture, but they usually have a crisp margin. Differentiation from lesions due to myedoma or secondary carcinoma can be difficult.

Nephrocalcinosis may be present in the kidney in patients with hyperparathyroidism; this is usually in the renal papillae. Nephrocalcinosis can occur in a variety of other conditions including renal tubular acidosis and medullary sponge kidney. In the latter case, the calcification is usually of a feathery pattern, occurring in the renal medulla; in patients with medullary sponge kidney, small cystic areas can be demonstrated on an intravenous pyelogram. These cysts are not filled on retrograde pyelography.

The appearance of Looser's zones or 'pseudo fractures' is the one reliable radiological sign of osteomalacia—this diagnosis cannot be based on abnormalities of texture or density. Looser's zones are commonly seen from the inferior rami of the pubis, the medial border below the neck of the femur, the tibia and fibula on the medial border of the scapula, and in the ribs. In rickets, the changes are most obvious at the wrists and knees. The epiphyseal plate becomes broader and cupped and there is delayed fusion. There may also be deformity, such as bow legs, or knock knees or distortion of the pelvis.

In Paget's disease, radiologically, the combination of lytic areas and osteosclerosis will be seen side by side often adjacent to a joint or in the skull. In osteoporosis there is diffuse loss of bone density as seen radiologically. Assessment of this is generally subjective. It is best seen in the vertebrae on a lateral view where there is loss of the normal trabecular pattern but the upper and lower borders of the vertebrae may however, appear more dense, as a sharp white line, and the vertebrae may be partially collapsed in a wedged pattern.

ASSAYS OF PARATHYROID HORMONE

Although a number of bioassays for parathyroid hormone have been described, up until recently, none was sensitive enough for direct measurement of the hormone in blood. More recently, the cytochemical type of bioassay was developed and applied to the measurement of parathyroid hormone. This method is based on the principle that a hormone causes a specific biochemical effect on its target tissue which can be quantitated by microdensitometry. In the case of parathyroid hormone, the activity of glucose-6-phosphate dehydrogenase in the distal convoluted tubule of guinea pigs is measured. In this way, very low concentrations of hormone can be detected. However, this method has the disadvantage that it is technically difficult and time consuming, and therefore not suitable for routine investigations. Now radioimmunoassays are generally used.

The first radioimmunoassay for parathyroid hormone was described in 1963 and this was followed by a number of different assay systems. There were certain problems with these assays which made the interpretation of results difficult. These included the presence of fragments of the hormone in the circulation and the lack of complete discrimination between normal subjects and patients with hyperparathyroidism. These assays were based on antisera raised against parathyroid hormones of animal origin. As there are species differences in the sequence of parathyroid hormone, the human hormone was of lower reactivity. Recently, human parathyroid hormone became available in sufficient quantities for raising antisera and in this way, homologous radioimmunoassays were developed. A further improvement was the development of region-specific assays. Because antisera raised against the intact hormone contain a heterogeneous population of antibodies, assays for defined regions of the molecule provide a better characterization of the measured immunoreactivity. A problem with the use of homologous human radioimmunoassay systems has been the difficulty in labelling intact human parathyroid hormone. To overcome this, the immunoradiometric type of assay was used. In this assay, the antibody instead of the hormone is labelled.

In this way assays against either the amino- or the carboxy-terminal end of the human hormone were developed and applied to the study of the secretion and metabolism of parathyroid hormone in man. The normal range for the homologous amino-terminal specific assay is from undetectable to 120 pg/ml while for the carboxy-terminal is betwen 0·1 and 0·8 ng/ml.

ASSAYS OF VITAMIN D

Assays of all the major metabolites of vitamin D have been developed and these are either competitive protein binding assays or radioimmunoassays. An important prerequisite for that were the improvements in extraction and purification techniques and the synthesis of labelled vitamin D compounds of high specific activity. Usually an initial lipid extraction is necessary followed by chromatographic separation with the use of high pressure liquid chromatography.

Measurement of 25-hydroxycholecalciferol in serum provides a good index of the vitamin D status. Normal concentrations vary in different parts of the world and show a seasonal variation with a nadir in late winter and a peak in late summer. The normal concentration of 25-hydroxycholecalciferol in adults in the U.K. ranged between 3·5–30 ng/ml with a mean of about 12 ng/ml. Assays of this metabolite have been valuable in identifying the cause of rickets and osteomalacia. Concentrations below 3 ng/ml indicate vitamin D deficiency.

The mean concentrations of 24,25-dihydroxycholecalciferol in the U.K. population is about one tenth of the concentration of 25-hydroxycholecalciferol.

Measurement of circulating 1,25-dihydroxycholecalciferol has been particularly difficult because of its low concentrations. The normal range is 22–60 pg/ml and the concentrations are age related, children having higher and elderly people having lower concentrations than adults, and is independent of the concentration of its precursor.

BONE BIOPSY

This can be valuable in establishing a diagnosis of osteomalacia if this is in doubt. The biopsy should be taken from the iliac crest and undecalcified sections should be examined. The presence of uncalcified osteoid is the hallmark of osteomalacia. For this reason, the biopsy must not be decalcified in acid but must instead be embedded in hard plastic and cut with a special microtome.

Practical assessment

Assessment of only five situations will be considered; two of these are clinical entities—renal calculi and bone pain; three are biochemical disturbances—hypercalcaemia, hypocalcaemia and raised serum alkaline phosphatase. Radiology and biochemical tests are essential to the practical assessment of these conditions. Important facets are summarized in Tables 9.2 and 9.3

Table 9.2. Radiologically detectable disorders of bone

X-Rays	Chiefly valuable in assessment or detection of
Lateral skull	Multiple myeloma, Paget's disease, post menopausal osteoporosis (skull unaffected).
Chest PA	Malignant deposits, sarcoidosis, Looser's zones in ribs.
Both hands and wrists	Hyperparathyroidism, sarcoidosis, rickets
Straight abdomen	Nephrocalcinosis, radio-opaque renal stones, kidney size.
Lateral lumbar spine	Severity of bone rarefaction, vertebral collapse metastatic deposits, renal osteodystrophy, osteoporosis, osteomalacia.
Pelvis and hips	Chronic osteomalacia (tri-radiate deformities) Osteomalacia (Looser's zones), Paget's disease.
Both knees AP	Rickets and osteomalacia

Table 9.3. Disturbances of serum and urine in disorders of calcium homeostasis and bone (N = normal, ↑ = increased, ↓ = decreased)

	Plasma Ca mM/L	Plasma inorganic P(mM/L)	Plasma HCO_3^- (mEq/l)	Alkaline phosphatase units/L	Urine Ca mM/24 h	Blood urea mM/L
Normal adult	2·2–2·55	0·6–1·2	20–25	50–280	3–7	3–7
Osteoporosis	N	N	N	N	N	N
Primary hyperpara-thyroidism	↑	N or ↓	N	N or ↑	N or ↑	N
Secondary hyperpara-thyroidism with renal failure	N or ↓	↑	↓	N or ↑	↓	↑
Osteomalacia and rickets	N or ↓	N or ↓	N or ↓	↑	↓	N
Hypopara-thyroidism	↓	↑	N	N	↓	N

RENAL STONE

Clinical observations, tests and measurements

History of renal pain or colic; gout. Urinary symptoms-of haematuria, of passing stones of gravel and history of renal tract infection.

Dietary assessment; calcium intake (including calcium in hard water); oxalate intake; total fluid intake.

Family History (gout; hyperparathyroidism).

Gouty tophi; band keratopathy (ectopic calcification).

Routine tests

Plain X-ray of abdomen; radio-opaque stones-calcium containing or cystine. Non-opaque stones-urate (see as filling defects on intravenous pyelogram). Ureteric hold-up seen on intravenous pyelography.

Analysis of stone; testing urine for cystine.

Serum calcium normal and urine calcium high in idiopathic hypercalciuria.

Serum calcium high in hyperparathyroidism.

X-rays to exclude other causes of hypercalcaemia; radiological changes of hyperparathyroidism.

Cortisone suppression test.

Assay of parathyroid hormone.

Special techniques

Locate parathyroid tumour pre-operatively.

BONE PAIN

Routine observations, tests and measurements

Age; history of onset-gradual or insidious; distribution-focal or diffuse; associated symptoms.

Dietary assessment-vitamin D intake.

Clinical signs-evidence of neoplasm, lymphadenopathy; lump in breast; enlarged glands; splenomegaly. Bone tenderness (in osteomalacia); deformity; thickening of wrists, bow legs, knock knees.

Routine tests

Serum calcium, phosphate, alkaline phosphatase, plasma proteins, electrolytes and blood urea, urine calcium blood count.

Radiological changes of: osteoporosis (lateral X-ray of spine); secondary deposits in bone; Paget's disease; hyperparathyroidism with osteitis fibrosa cystica; osteomalacia and rickets.

In osteoporosis serum chemistry normal. In disseminated malignant disease hypercalcaemia and raised alkaline phosphatase. In Paget's disease, normal serum calcium, raised alkaline phosphatase. In osteomalacia normal or low serum calcium, raised alkaline phosphatase, low urine calcium, pseudofractures. Renal-glomerular failure, raised blood urea. Tubular disorders, low serum phosphate, phosphaturia, glycosuria, aminoaciduria, acidosis.

Special investigations

Assays of vitamin D metabolites.

HYPERCALCAEMIA

Clinical observations

History of peptic ulcer, pancreatitis, Addison's disease, thyrotoxicosis. Thirst and polyuria; vomiting; dehydration; drowsiness; depression. Pluriglandular syndrome; evidence of malignant disease with or without metastases to bone.

Routine investigations

Alkaline phosphatase, plasma proteins, including electrophoresis (to show monoclonal gammopathy); electrophoresis of urine protein (to show Bence-Jones protein).

Lymphadenopathy in sarcoid, especially in hilar regions seen on chest X-ray. Kveim test; iritis, impaired renal function; raised ESR. Evidence of vitamin D intoxication; response to hydrocortisone suppression test. Blood urea raised in milk alkali syndrome and hypercalcaemic sarcoidosis.

Radiological evidence of hyperparathyroidism or of other causes of hypercalcaemia.

HYPOCALCAEMIA

Clinical observations

History of renal disease leading to uraemia. History of malabsorption syndrome. Vitamin D intake, treatment with anticonvulsant drugs.

Age of onset; history of thyroid surgery; family history.

Facial appearance; short metacarpals (in pseudohypoparathyroidism); evidence of Addison's disease and fungal infections (in idiopathic hypoparathyroidism).

Routine investigations

Electrolytes and blood urea, creatinine clearance, tests for malabsorption syndrome. Barium meal and follow through (post-gastrectomy osteomalacia, blind loop syndrome). Intestinal biopsy (Gluten-sensitive enteropathy). Serum magnesium and phosphate.

Special investigations

Measure parathyroid hormone (low in hypoparathyroidism; raised in pseudo-hypoparathyroidism). Test response to parathyroid hormone (measure urinary cyclic AMP after administration of parathyroid hormone).

RAISED ALKALINE PHOSPHATASE

Clinical observations

Evidence of liver disease; evidence of malignant disease-carcinomatosis, evidence of Paget's disease, hyperparathyroidism or osteomalacia.

Routine tests

Serum calcium and phosphate; urinary calcium. Radiological evidence of malignant disease-primary and secondary; Paget's disease. 5 nucleotidase or γ-glutamyl transpeptidase raised if elevated serum alkaline phosphatase due to liver disease.

Reference

Nordin, B.E.C. (ed.) (1984) *Metabolic Bone and Stone Disease*, 3rd ed. Edinburgh, Churchill Livingstone.

10

Skeletal Muscle

Normal function

STRUCTURAL ORGANIZATION OF STRIATED MUSCLE

Depending on its size, a muscle is made up of between a thousand and a million fibres 40–80 μm in diameter. They tend to run the whole length of the muscle and therefore vary from less than 1 cm to over 20 cm in length. Each of these highly specialized cells is bounded by a plasma membrane, a proteolipid bilayer which is responsible for the electrical properties of the fibre. Interspersed in the membrane are various enzymes and transport proteins with further specialization occurring at the endplate where the fibre is excited by the transmitter acetylcholine (ACh) released from the nerve terminal. Outside the plasma membrane is a basement membrane, a protein and polysaccharide layer akin to the collagen of the endomysial connective tissue. The connective tissue matrix contains the blood vessels and nerve fibres which supply the muscle. Some refer to the plasma membrane as the sarcolemma but the latter term should strictly include both plasma and basement membranes. The muscle spindles, specialized sensory end-organs containing intrafusal muscle fibres, are described in Chapter 11.

Between basement and plasma membranes are found myosatellite cells which are of importance for the rapid regeneration of muscle following injury. Beneath the plasma membrane are the many nuclei of the muscle cell. The sarcoplasm contains organelles common to all cells such as mitochondria, ribosomes and Golgi apparatus together with energy substrates (glycogen and lipid), and the enzymes concerned with glycogen synthesis, glycogenolysis and glycolysis. In man, mitochondria are found beneath the plasma membrane and between the myofibrils on each side of the Z-disc.

Each muscle fibre consists of several hundred myofibrils, contractile elements about 1 μm in diameter, which run along the axis of the fibre. In polarized light they have a banded appearance and, because the light and dark bands of myofibrils are in phase, the fibre shows the characteristic striated appearance. Electron microscopy shows that the pattern of cross-striation results from ordered arrays of thick and thin filaments overlapping each other to a variable extent depending on the state of contraction of the fibre (Fig. 10.5 p. 351). The optically dense 'A' bands (anisotropic, or doubly refractile to polarized light) are made up of thick myosin filaments about 15 nm in diameter and 1·6 μm in length. The thin actin filaments which make up the 'I' (isotropic) bands are about 1·3 μm long and 5 nm in diameter. They extend from each side of the 'Z' disc to interdigitate with the thick myosin filaments of the A bands. In the region of overlap, cross-bridges between actin and myosin filaments are seen which are the sites of cyclical interaction. The macromolecular basis of contraction will be considered later. The 'M' line at the centre of the A band is formed by transverse connections between the myosin filaments and the 'H' zone is that part of the A band which is free of overlapping actin filaments (Fig. 10.1). The Z discs divide each myofibril into sarcomeres about 2·0–2·5 μm long at rest.

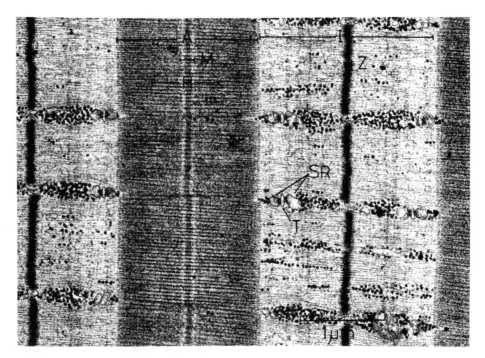

Figure 10.1. *Electron micrograph of normal human muscle showing arrangement of the myofilaments.*
A: A-band; I: I-band; Z: Z disc; H: H zone; G: glycogen granules; T: transverse tubule; SR: sarcoplasmic reticulum. (see also Fig. 10.5).
(Kindly supplied by Dr. D.N. Landon)

The transverse tubules which cross the myofibrils at the junctions of A and I bands are in continuity with the extracellular space. They may be regarded as invaginations of the plasma membrane. Each myofibril is surrounded by a complex longitudinally arranged tubular system, the sarcoplasmic reticulum. As the longitudinal tubules approach the T-tubules, they dilate to form the lateral sacs or cisternae (see Fig. 10.5). Although the lumens of the two tubular systems are not in direct continuity they are joined by 'foot processes'. The T-tubule with its paired adjacent terminal cisterns is known as a triad. As will be discussed these structures serve to link the excitation of the membrane of the fibre with contraction.

INNERVATION OF MUSCLE FIBRES

The motor nerves supplying skeletal muscles are large myelinated fibres having their cell bodies in the ventral grey matter of the spinal cord. Soon after entry into the muscle they divide into numerous terminal branches each supplying a single extrafusal fibre. The motor unit, the functional unit of skeletal muscle, consists of those muscle fibres supplied by a single α-motor neurone. The size of individual motor units varies widely. In general, muscles concerned with fine movement such as external ocular muscles and the intrinsic hand muscles consist of small units. Those of large limb muscles may contain over a thousand fibres.

Muscle fibres of a single motor unit are scattered singly over a wide cross-sectional area of the muscle. The muscle fibres which go to make up a single motor unit have similar histochemical and mechanical characteristics and these parameters are largely determined by the α-motor neurone.

THE NEUROMUSCULAR JUNCTION

The neuromuscular junction is a specialized region where the motor nerve terminal and plasma membrane are separated by a gap of about 50 nm (Fig. 10.2). As the motor nerve approaches the muscle fibre it loses its myelin sheath. The nerve terminal is packed with membrane-lined vesicles containing ACh, and mitochondria which generate energy for ACh synthesis.

At the neuromuscular junction the plasma membrane of the primary synaptic cleft is thrown into folds, the secondary synaptic folds, giving it a large surface area. The primary cleft and its invaginations are rich in acetylcholine esterase (AChE).

The postsynaptic membrane contains the acetylcholine receptors (AChR), glycoprotein oligomers of apparent mol.wt 250,000–300,000 daltons, which traverse the membrane. The ACh binding subunit has an apparent mol.wt of about 42,000 daltons. The ACh ionophore, or sodium channel, is probably carried on a separate subunit. The most widely accepted hypothesis is that the AChR macromolecule possesses at least two conformational states, an "active" one associated with an open ionophore and an "inactive" closed form.

The human motor endplate contains 1–4×10^7 AChR sites which are localized at the apices of the secondary synaptic folds. Recent advances in the understanding of AChR have come from the use of radioactively labelled elapid snake toxins, such as I^{125} α-bungarotoxin which binds specifically and irreversibly to junctional AChR. Using gamma- or scintillation counting or autoradiography, labelled toxins have been used to define the number and distribution of AChR and also to investigate their synthesis and degradation.

The junctional AChR are in a constant state of turnover, with a half-life in the order of one week. They are synthesized in the region of the Golgi apparatus and then

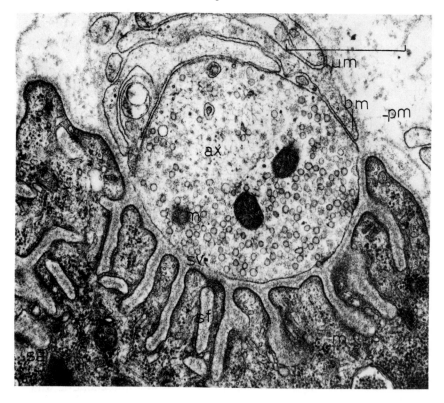

Figure 10.2. *Electron micrograph of rat motor endplate.*
ax: axoplasm; m: mitochondria; sv: synaptic vesicle; sa: sarcoplasm; cf: collagen fibres; sf: synaptic folds; bm: basement membrane; pm: plasma membrane.
(Kindly supplied by Dr. D.N. Landon)

transported to their site of insertion in the postsynaptic membrane. They are degraded by lysosomal enzymes following endocytosis. Factors determining the formation and maintenance of the endplate are complex but they include the presence of an intact and functionally active motor nerve. Following denervation, there is a rapid appearance of newly-synthesized extrajunctional AChR along the whole plasma membrane while the number of junctional AChR begins to decline only after several weeks.

Neuromuscular transmission is mediated by ACh which is held within the synaptic vesicles. Each vesicle contains a 'quantum' of 4000–9000 ACh molecules. In resting muscle the slow random release of ACh quanta generates postsynaptic miniature endplate potentials (mepps) 1–2 mV in amplitude. Arrival of a nerve impulse opens calcium channels in the presynaptic membrane which leads to synchronous release of 50–300 quanta into the synaptic cleft by fusion of the vesicles with the presynaptic membrane. There is in addition to the ACh in vesicles a pool of axoplasmic ACh which is not immediately available for release.

In the resting state a potential difference of about 90 mV, negative with respect to the outside, is maintained across the postsynaptic membrane. This largely results from the difference between intracellular and extracellular concentrations of potassium, which is a more freely permeable cation than sodium. The inward movement of potassium is linked to the active extrusion of sodium by the ATP-dependent sodium pump. The interaction of ACh with AChR opens chemically gated sodium channels (ionophores) and causes a brief increase in sodium conductance. This leads to depolarization of the postsynaptic membrane. The small potential changes of the mepps are insufficient to reach threshold for the generation of a propagated action potential. The much larger endplate potential produced by the synchronous release of many ACh quanta will normally trigger an action potential in the surrounding membrane.

The action potential is generated by influx of sodium ions through electrically gated sodium channels. These channels are opened by depolarization. The sodium influx causes further depolarization and further increase in sodium conductance producing a positive feedback effect and making the inside of the fibre positive with respect to the outside. The rise in potential itself closes the sodium channels. Repolarization of the membrane is largely due to a delayed outward potassium current and the potential returns to its resting level in 2–3 msec. The ionic basis of the action potential in mammalian muscle appears to be broadly similar to that of the more extensively investigated frog muscle.

The opening time of the AChR ionophore is about one msec. The AChE at the endplate rapidly hydrolyses the ACh to choline and acetic acid. This process is complete within a few msec, well before the arrival of the next nerve impulse. Choline is actively taken up into the nerve terminal. By the action of choline acetyltransferase, it combines with acetyl CoA to form ACh. Part of the ACh is enclosed into vesicles, the other fraction remaining in the axoplasm. During neural activity the rate of synthesis and mobilization of ACh is substantially increased, an effect outlasting the duration of stimulation. Control of synthesis is poorly understood but may be partly dependent on intracellular calcium concentration which increases in the nerve terminal following repetitive stimulation.

There are about 10 AChR sites and 10 molecules of AChE for every molecule of ACh released. In man one quantum of ACh opens about 1500 sodium channels each of which conduct approximately 5×10^4 sodium ions. It has been calculated that opening of only 4 per cent of the channels would produce 90 per cent of the maximum depolarization at the endplate. There is, therefore, a large "margin of safety" in neuromuscular transmission. This is supported by experimental observations that the twitch response is unchanged until 70 per cent of the AChR are blocked and is only abolished when 90 per cent of the receptors are occluded.

There are several sites, presynaptic and postsynaptic, where these events can be influenced both by drugs and by pathological changes. The release of ACh is calcium-dependent and is inhibited by raising the extracellular magnesium level. The postsynaptic membrane potential is influenced by relative alterations in extracellular and intracellular potassium. Changes in potassium levels are clinically important *per se.* They also affect the actions of neuromuscular blocking agents.

The macromolecular basis of contraction involves a change in the orientation of the crossbridges between myosin and actin filaments. The cycle of attachment, crossbridge movement, and detachment depends on ATP hydrolysis and is activated by calcium ions released into the sarcoplasm from the sarcoplasmic reticulum.

The myosin molecules of the thick filaments are arranged with their 'tails' longitudinally orientated and their globular 'heads' projecting laterally to form the crossbridges (Fig. 10.3). The thin filaments consist of a double chain of actin molecules with tropomyosin occupying the grooves between them. Near the end of each tropomyosin molecule is a globular troponin complex which consists of three subunits. One of these, troponin-T, binds the troponin complex to tropomyosin. One troponin complex is attached to every seventh actin monomer.

The action potential is propagated from the endplate along the plasma membrane at 3–4 meters per second. The wave of depolarization also spreads internally down the T-tubular system. It has been proposed that the charge is transferred to the lateral sacs (cisternae) by means of the communicating 'feet' and causes the release of calcium ions from the lateral sacs into the aqueous sarcoplasm (Fig. 10.5).

The activation of contraction by calcium is mediated by removal of inhibition present in the relaxed state. In relaxation, the concentration of free calcium in the sarcoplasm is less than 10^{-8} M and the troponin complex inhibits the interactions between actin and the myosin heads. Calcium release raises the concentration to about 10^{-5} M, the calcium ions binding to troponin-C, the calcium binding subunit, to produce a conformational change in the complex. This interaction removes the restraint exerted by the inhibitory subunit, troponin-I. During this process movement of the tropomyosin molecule occurs. This may uncover reaction sites on the actin molecules, the so-called "steric hypothesis". Alternatively, there may occur direct transfer of conformational change from troponin to actin which is then transmitted to adjacent actin molecules. During relaxation the calcium is actively reaccumulated by the sarcoplasmic reticulum and the activation process is reversed.

The exact nature of actin-myosin interaction is not at present known and the role of calcium in activation is probably more complex than indicated. The mechanism of crossbridge movement is not fully understood and is the subject of several different hypotheses. The following sequence, first proposed by Lymn and Taylor, provides a useful working model (Fig. 10.4). In the resting state, the actin-myosin crossbridge is dissociated and the myosin 'head' bears the reaction products of ATP hydrolysis, ADP and P_i. In this state ATPase activity is low. Activation of actin by calcium release allows interaction between actin and myosin 'heads' which is associated with rapid release of ADP $+$ P_i and an increase in ATPase activity. Simultaneously there takes place a conformational change in the myosin head altering its orientation and moving the filaments 5–10 nm with respect to each other. Finally an ATP molecule is attached to the head as detachment of the cross-bridge occurs. The ATP is hydrolysed and the system returns to the resting state.

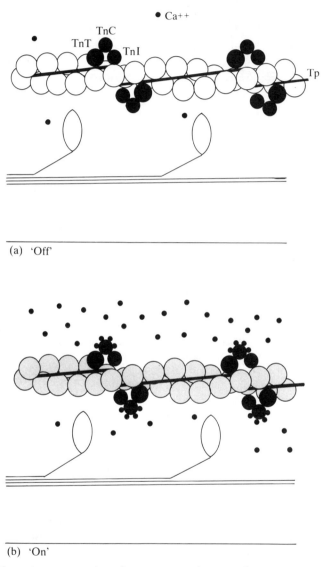

(a) 'Off'

(b) 'On'

Figure 10.3. *Schematic representation of arrangement of contractile proteins.*

 a. The I (actin) filament is a double helix of actin molecules to which are attached: Tp (tropomyosin) and the three subunits of troponin: TnT (tropomyosin-binding), TnC (calcium-binding) and TnI (inhibitory subunit). The myosin heads of the A (myosin) filament are also shown.

 b. On activation, the local calcium concentration rises and calcium ions bind to Troponin C.

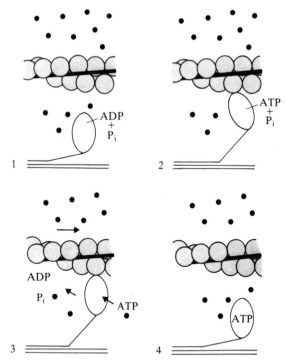

Figure 10.4. *Schematic representation of actin–myosin interaction.*

1. At rest, myosin heads are separated from the actin molecules and bear the reaction products of ATP hydrolysis (ADP and P_i). ATPase activity is low.

2, 3. Rapid release of ADP and P_i occurs in association with attachment and a conformational change in the myosin head causing the actin and myosin filaments to move about 10 nm relative to each other. ATPase activity is high.

4. The attachment of a further molecule of ATP is associated with detachment of the myosin head. The ATP undergoes hydrolysis and the cycle is repeated.

(modified from Perry, 1979, see references).

CONTRACTILE PROPERTIES OF MUSCLE

A. V. Hill proposed a mechanical model of contracting muscle consisting of a contractile component and an elastic component in series. Although some elasticity resides in the tendons, the major part is suggested to be due to elasticity of the cross-bridges. At extended muscle lengths, a third factor is required to explain the contractile properties, a parallel elastic component which is partly due to connective tissue but muscle cell components may contribute.

The tension developed by an active muscle fibre depends on the number of cross-bridges formed and hence on the degree of overlap of the myosin and actin filaments (Fig. 10.5). With whole muscle, the force of contraction is determined by the number of motor units firing. The force of contraction is increased by 'recruitment' of more units and also by increasing the firing frequency of the units. In limb muscles discharge

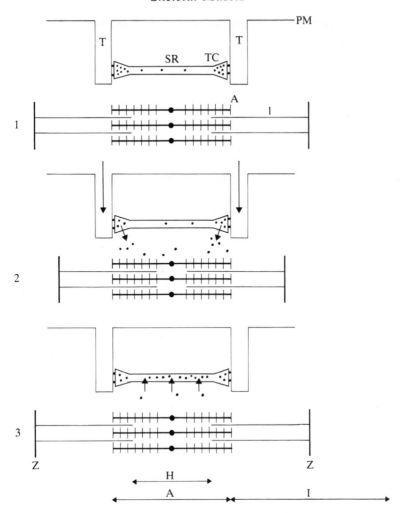

Figure 10.5. *Schematic representation of activation and contraction.*

PM: plasma membrane; T: transverse tubular system; SR: sarcoplasmic reticulum with terminal cisterns (TC) containing calcium ions. Compare arrangements of myofilaments with Fig. 10.1.

1. At rest.

2. Spread of depolarization down T-tubules leads to release of calcium ions and shortening with overlap of A and I filaments. (see Fig. 10.3b)

3. Reuptake of calcium ions into SR causes relaxation.

frequencies of 5–50 Hz are usually found but some units are capable of responding at rates of over 100 Hz for brief periods. Rates of up to 200 Hz have been recorded in the extraocular muscles.

The tension developed by a muscle in a single maximal twitch is less than that achieved during a fused tetanus. The initial force is partly taken up by the elastic component and relaxation has already begun before full tension can develop. In mammalian muscle the ratio of tension produced by a single contraction to that during tetanus (the twitch : tetanus ratio) is about 1 : 5.

Energetics of contraction

Muscle contraction is classified as isotonic or isometric depending on whether the muscle shortens in contraction, although much muscle activity is mixed or occurs during lengthening. In isometric contraction the muscle performs no external work and because of the rise in intramuscular pressure the blood supply may be impaired. In isometric tetanus it is possible to measure the production of an 'activation heat', starting at a high level soon after stimulation and persisting throughout the tetanus. Afterwards a 'relaxation heat' is produced which may be due to the combination of thermoelastic effects and the reaccumulation of calcium by the sarcoplasmic reticulum.

In isotonic contraction, in which the muscle shortens and performs external work, there is an additional 'heat of shortening'. If a contracting muscle is stretched, its heat production is reduced, suggesting an absorption of energy during the stretch. Finally, all these processes are followed by a prolonged 'recovery heat' corresponding to the oxidative resynthesis of the high energy compounds degraded in contraction.

ATP provides the energy for contraction. Changes in the levels of ATP and creatine phosphate during ischaemic contraction can be measured and the theoretical energy liberation can be compared with the measured energy output of the contracting muscle (work plus heat). Results with amphibian muscle suggest that the energy released is in excess of that which can be attributed to ATP breakdown. However, more limited data in the human suggests ATP breakdown is adequate to explain energy production.

MUSCLE FIBRE TYPE CHARACTERISTICS

Adult human muscle consists of different muscle fibre types which may be characterized by their histochemical properties. In the most convenient and widely used classification, type I fibres have low contents of phosphorylase and glycogen and low myosin ATPase activity but high concentrations of oxidative enzymes. They derive their energy largely from aerobic metabolism. In contrast type II fibres are rich in phosphorylase and glycogen, have high myosin ATPase activity and are more adapted to anaerobic metabolism. Type II fibres may be subdivided on the basis of the ATPase reaction into types IIa and IIb, the latter having a lower oxidative capacity. Type I fibres contract more slowly than type II and, by virtue of their higher oxidative enzyme content, type I and IIa fibres are more fatigue-resistant.

In lower animals motor units may be divided into fast twitch and slow twitch units which consist of type II and type I fibres respectively. The former achieve peak tension 10–40 msec after the onset of contraction and the latter 60–120 msec although units of intermediate properties are also found. Fast twitch units therefore require a more rapid rate of stimulation to achieve fused tetanus. The histochemical characteristics of individual fibres belonging to a single motor unit are uniform so that the contractile properties of the whole muscle will depend on the relative proportions of slow and fast units. In mammals such as cats and rodents certain muscles are found to contain predominantly or either slow twitch or fast twitch units. The paler colour of 'fast' (or 'white') muscles is due to their lower capillary density and myoglobin content.

Various differences between slow and fast muscle are recognized. The sarcoplasmic reticulum of fast muscle shows a greater rate of calcium uptake and higher ATPase activity. The troponin of fast muscles has a greater affinity for calcium and the calcium-activated myosin ATPase activity is higher. Differences have also been found in the myosin light chains and in tropomyosin.

In man segregation of fast and slow units into separate muscles does not occur. Although muscles vary in their contraction rates they contain a mixture of motor unit types. Although data is more limited, type I fibres appear to have slow twitch, and type II fast twitch properties as in other mammals. The type I and II fibres are arranged in a mosaic (Fig. 10.6), although their relative proportions differ from muscle to muscle. There is also evidence, particularly from the study of athletes, that there may be considerable variation between individuals in fibre type composition. In animals it appears that fast twitch units are concerned with rapid movement while slow twitch units, capable of sustained activity, subserve a largely postural role. Although there is less direct physiological evidence in man it is reasonable to assume similar roles.

It appears that the contractile and histochemical properties of muscle fibres are controlled by their motor nerves. It is possible, as cross-innervation experiments have shown, to alter the properties of a mature mammalian muscle. For example, if the motor nerve to the slow soleus in the cat or rat is allowed to re-innervate the fast twitch extensor digitorum longus muscle, the latter acquires the contractile and biochemical characteristics of a slow muscle. The nature of this neural control is unknown but may be mediated by the temporal pattern of nerve impulses and by muscle activity. For example tenotomy of the slowly contracting soleus muscle (which largely abolishes activity in the muscle) speeds up the contractile response. This change is prevented by stimulation of the motor nerve at 5–10 Hz for 8 hours per day but not by stimulation at 20–40 Hz.

Muscle fatigue

Fatigue (failure to maintain a target force or workload) is not adequately explained. In isolated nerve muscle preparations, repetitive high frequency stimulation will cause failure of the contractile response. At this time neuromuscular transmission, the muscle action potential and the contractile response to direct stimulation by caffeine may be preserved. This suggests failure of excitation-contraction coupling. In fatigued muscle slowing of relaxation occurs and there is a decrease in the energy required to maintain a given force, but the underlying mechanisms are not clear.

Figure 10.6. *Transverse section of human muscle.* The section has been stained by the myosin-ATPase reaction after preincubation at pH 4·35 and shows type I (stained darkly) and types IIa and IIb fibres.

In the intact body, a number of complicated intracellular and extracellular metabolic changes are likely to occur and, in particular, it is possible that the ischaemia which accompanies strong isometric contraction *may* impair neuromuscular transmission.

In the human, fatigue after severe exercise largely recovers within minutes but full recovery may take several hours. A slight fall in ATP and a marked fall in creatine phosphate concentrations occur in ischaemic exercise but afterwards both rapidly regain resting levels. The prolonged fatigue is pronounced on stimulation of the muscle at 20 Hz but is not prominent at 80 Hz. The force generated at the lower frequencies may be reduced by 50 per cent, implying that the production of a given force will require either recruitment of more units or an increase in firing rates under conditions of fatigue. This may explain the subjective sensation of fatigue.

Although fatigue is a common symptom of muscle disease, there is likewise no clear explanation. There is for example no direct evidence that the depletion of high energy phosphates in ischaemic exercise is greater than in normal muscle.

MUSCLE PAIN

Muscle fibres themselves are insensitive to pain but sensory nerve endings are situated round blood vessels and in fascia. Muscle pain due to spontaneous cramp or to vigorous exercise is a normal phenomenon and is thought to be due to the production of "pain substances" by the active muscle although their nature is unknown.

Tissue hypoxia *per se* does not accelerate the onset of pain and the ischaemia occurring in isometric contraction is presumably associated with pain because of impeded clearance of the putative pain substances from the muscle.

MUSCLE TONE

There is no motor unit activity in relaxed normal muscle. Muscle tone partly depends on the elasticity of the muscle but in states of hypertonus mainly upon contraction of muscle fibres excited reflexly through increased fusimotor activity during stretch. (See Chapter 11).

MUSCLE METABOLISM

Muscle is dependent on two main energy sources, carbohydrate (glucose and glycogen) and free fatty acids (FFA). The fibre is able to synthesize and store glycogen. Skeletal muscle, which accounts for about 40 per cent of body weight contains 60 per cent of the total body glycogen.

FFA are stored as triglycerides. A description of the principal energy-generating pathways is given in Chapter 14. The net result of all these processes is the generation of ATP and creatine phosphate. The function and ultimately the structure of the fibre is dependent on maintaining relatively fixed intracellular concentrations of ATP of about 7 mM.

Resting muscle utilizes mainly FFA for its energy requirements (about 90 per cent) the remainder being provided by glucose metabolism. Glycogen is the obligate energy substrate for short periods of high intensity exercise greater than 90 per cent VO_2 max (maximal oxygen uptake), and is also the major fuel during the initial stages of sub-maximal exercise (i.e. 50 per cent VO_2 max). The breakdown of glycogen produces pyruvate which under aerobic conditions undergoes mitochondrial oxidation. Under anaerobic conditions pyruvate is converted to lactate and alanine which are trans-ported to the liver where they are metabolized. As exercise continues, endogenous muscle glycogen becomes depleted and is supplemented by blood-borne glucose and by FFA mobilized from adipose tissue by lipolysis. During endurance exercise about 70 per cent of the energy is derived from oxidation of FFA and about 30 per cent from glucose oxidation. Oxidation of branched chain aminoacids may contribute to energy require-ments during prolonged vigorous exercise.

Knowledge of an outline of muscle metabolism is necessary not only to understand normal function but also because a number of muscle diseases are identified as having a specific biochemical basis.

Disordered function

This account will be largely restricted to consideration of muscle weakness and of muscle hyperexcitability. Disorders of tone and the control of movement are discussed in Chapter 11.

Weakness may arise from a number of different functional defects:

(a)   Disorders of the anterior horn cell or motor nerve
(b)   Disorders of neuromuscular transmission
(c)   Disorders of the muscle membrane
(d)   Disorders of excitation-contraction coupling
(e)   Disorders of the contractile mechanism
(f)   Disorders of the energy processes.

DISORDERS OF ANTERIOR HORN CELL OR MOTOR NERVE

Acute denervation

When the continuity of a motor nerve is interrupted, conduction ceases at the level of the lesion. Distal to the lesion, nerve fibres remain excitable and are capable of con-ducting impulses for 3 to 4 days until Wallerian degeneration prevents this. Following denervation of muscle fibres there is a generalized appearance within a few days of newly synthesized extrajunctional AChR (and corresponding ACh sensitivity) which disappear when re-innervation is established. Loss of neuromuscular transmission and resultant muscle inactivity are important in regulating the appearance of extra-junctional AChR, but other factors are clearly involved. In contrast to the rapid appearance of extrajunctional AChR, the junctional AChR are relatively stable and no decrease in numbers occurs for several weeks. ACh sensitivity at the denervated end-plate is maintained for several months. On denervation, progressive muscle fibre atrophy occurs and it is likely that neuromuscular transmission and muscle activity are necessary to maintain normal protein synthesis.

Recovery of function, which is determined by the ability of the motor nerve fibres to regenerate and re-innervate the muscles, depends on the nature of the lesion and the distance over which regeneration has to occur. When the connective tissue of the nerve remains intact (axonotmesis), regeneration may proceed at the rate of 2–3 mm per day and complete functional recovery may occur. However, when the continuity of the whole nerve is interrupted (neurotmesis), regeneration may be delayed by the forma-tion of a neuroma and recovery is usually incomplete. Reasonable recovery is possible for periods of up to one year but beyond this the outlook becomes progressively poor.

Chronic partial denervation

Gradual fall-out of motor nerve fibres occurs in chronic denervating disorders such as motor neurone disease and the chronic peripheral neuropathies. In these conditions the terminal branches of surviving motor nerve fibres develop collateral sprouts which reinnervate adjacent muscle fibres which have lost their nerve supply. This can be demonstrated experimentally in muscle partially denervated by section of one spinal root of a motor nerve. In rats, collateral sprouting is considerably reduced if the partially denervated muscle is subjected to regular periods of direct stimulation. Sprouting also occurs when conduction of the motor axon is blocked by tetrodotoxin or when presynaptic blockade is produced by botulinum toxin. Under these conditions terminal sprouting is inhibited by direct stimulation of the muscle or, with botulinum toxin, by cross-innervation with a foreign nerve. There is disagreement about whether post-synaptic blockade by α-bungarotoxin causes terminal sprouting and one report indicates that it may actually *inhibit* sprouting caused by botulinum toxin. It does however appear likely that muscle activity regulates sprouting. Thus in addition to the control of the motor nerve over muscle characteristics, the muscle appears capable of influencing nerve growth. An interesting finding made in amphibians is that dener-vation on one side may induce terminal sprouting in contralateral motor nerves at the same segmental level, suggesting a central influence on peripheral nerve growth which is carried transneuronally in the spinal cord.

In man, the process of collateral reinnervation results in a gradual increase in the size of the surviving motor units, which may be detected electromyographically some-times even before the onset of clinical weakness. The spatial arrangement of the motor units is altered in chronic partial denervation. Instead of being scattered over a rela-tively wide cross-sectional area of the muscle, the individual fibres of enlarged units become arranged in groups which can be demonstrated histochemically as 'type group-ing'.

Pre-synaptic disorders of neuromuscular transmission

Deficient synthesis or mobilization of ACh are possible causes of failure of neuro-muscular transmission.

Transmission failure is caused by the drug hemicholinium which is thought to block the active reuptake of choline by the nerve terminal. Release of ACh quanta can be inhibited by local increase in magnesium or decrease in calcium ion concentration. Certain antibiotics particularly of the aminoglycoside group, e.g. Gentamicin, may interfere with neuromuscular transmission, probably by impairing the release of ACh. Tetanus and botulinum toxins block ACh release. This is also seen with certain rep-tilian and amphibian toxins, e.g. β-bungarotoxin.

Hypothermia below 30–32° C in man causes impairment of neuromuscular trans-mission, mainly at high frequencies of nerve stimulation, which is probably a presynap-tic effect. Paradoxically, lowering the temperature may improve transmission in the Eaton-Lambert syndrome and myasthenia gravis (see below), and also in botulism and experimental hypermagnesaemia.

In presynaptic disorders, e.g. botulism, transmission may be improved by calcium infusion and by guanidine and 4-aminopyridine, which increase ACh release. The clinical value of these drugs is limited by possible toxic effects.

The myasthenic syndrome (Eaton-Lambert syndrome)

This rare disorder of neuromuscular transmission occurs alone or in association with malignancy, particularly oat cell carcinoma of the lung. Clinically it is characterized by weakness, fatiguability and depression of tendon reflexes, all of which typically improve after sustained voluntary contraction. This phenomenon, post-tetanic facilitation, can be demonstrated electromyographically. The compound muscle action potential increases with repetitive nerve stimulation and after voluntary contraction. The dis-order is due to a reduction in the number of quanta of ACh released.

Post synaptic disorders of neuromuscular transmission

Non-depolarizing (competitive) block. D-tubocurarine, gallamine and pancuronium which are used extensively as muscle relaxants are examples of drugs causing non-depolarizing block. The simplest explanation of their action is one of competition with ACh for the AChR sites. Their effect can be partially overcome by increasing the concentration of ACh at the endplate. Such an effect is produced by anti-cholinesterase drugs, for example neostigmine, which inhibit the normal hydrolysis of ACh although they may also have an additional presynaptic effect.

Depolarization block. If a reduction in potential across the post-synaptic membrane is maintained, this region becomes refractory to the generation of an action potential. This effect is produced by decamethonium and succinylcholine (suxamethonium) and probably represents their principal mode of action as muscle relaxants. A similar situation results from the high levels of ACh at the endplate in overdosage with anti-cholinesterase drugs ("cholinergic crisis") and certain organophosphorus compounds which block AChE activity irreversibly. This type of block is potentiated by anti-cholinesterase drugs. Non-depolarizing blocking agents tend to antagonize the effects of depolarizing agents, presumably by reducing the number of available AChR.

Some depolarizing blocking agents may also have presynaptic effects. Antidromic stimulation of the motor unit is probably responsible for the fasciculation seen following intravenous succinylcholine.

Rare individuals lack, or possess an abnormal, plasma pseudocholinesterase, the enzyme which largely destroys succinylcholine. Here the drug will have a dangerously prolonged action.

It should be emphasized that great species differences are found in the action of neuromuscular blocking drugs.

Myasthenia gravis. Myasthenia gravis is characterized by variable muscle weakness which is typically increased by exertion. Although often generalized, the weakness is commonly more obvious in external ocular and bulbar muscles.

Advances in the understanding of neuromuscular transmission defects have come from the use of radioactively labelled snake α-neurotoxins, particularly α-bungarotoxin from the Formosan krait, Bungarus multicinctus. In myasthenia gravis there is a reduction in the number of junctional AChR. There is strong evidence, both in the human disease and in experimentally induced myasthenia, that circulating anti-AChR antibodies increase the rate of AChR degradation. Antibodies have been shown to cross-link the AChR which accelerates the normal process of internalization and destruction. The effect of the antibodies on blocking of neuromuscular transmission, although this has been shown experimentally, is of uncertain importance. Such antibodies are found in about 90 per cent of cases of generalized myasthenia gravis. They are a heterogenous group and this may explain why overall it is found that the antibody titres correlate poorly with the severity of the illness. However, in *individual* cases, the antibody titre does show some correlation with the clinical state. A temporary clinical improvement can be achieved after an interval of a few days by the removal of the antibody by plasmaphoresis.

10–20 per cent of cases are associated with a thymic tumour and the majority of the rest, especially in younger age groups, show thymic hyperplasia. The exact role of the thymus is still not clear. Its cells bear antigenic similarities to striated muscle and it is

possible to grow cells identical to myotubes from thymic tissue culture. A high proportion of cases with thymoma have anti-striated muscle antibody in the serum, making
this antibody a useful diagnostic marker for thymoma. It is possible that the thymus
initiates and maintains a stimulus to anti-AChR antibody formation although it is not
itself the principal site of antibody formation. After thymectomy over half of myasthenic patients eventually improve or remit.

In addition to circulating anti-AChR antibodies a wide range of immunological
disturbances have been reported in myasthenia gravis. The peripheral blood lymphocytes will undergo blast transformation on exposure to electric eel AChR and there
have been several reports of impaired T-cell function. However, lymphocyte invasion of
the endplate region is only a rare finding and there is no strong evidence for an
important role for disturbed cell-mediated immunity. Lastly, there are known associations of myasthenia with certain HLA types, the most widely reported being that of
HLA-B8 with younger patients with no thymoma. Individual susceptibility is likely to
be important but there is so far no evidence as to the cause of the initial breakdown in
immunological tolerance leading to antibody formation. EMG shows a decrement of the
muscle action potential to repetitive supramaximal nerve stimulation at lower frequencies. The defect in neuromuscular transmission can also be demonstrated with
single fibre EMG recording.

Both the symptoms and the electrophysiological disturbance are partially reversed
by anticholinesterase drugs which increase both the concentration and the life span of
ACh at the post-synaptic membrane. Excessive doses will, however, produce depolarization block and clinical deterioration. It is important to note that additive effects are
produced by any of the factors previously mentioned which adversely affect neuromuscular transmission, e.g. gentamicin. A rare syndrome clinically identical to myasthenia gravis may occur in patients on long-term treatment with D-penicillamine. Most
have anti-AChR antibodies in the serum and recovery usually occurs following withdrawal of the drug.

Other forms of myasthenia are rare. In transient neonatal myasthenia, which may
affect infants born to myasthenic mothers, the maternal anti-AChR antibody crosses
the placenta and clinical effects persist until the maternal antibody is degraded and
new AChR are synthesized by the infant. Congenital myasthenia is not antibody mediated. In this group of disorders a variety of apparently specific defects, presynaptic as
well as postsynaptic, have been identified in individual cases. These include congenital
absence of AChE, defective synthesis of ACh and abnormality of the AChR macromolecule.

DISORDERS OF THE MUSCLE MEMBRANE

Myotonic disorders

Myotonia is characterized by repetitive action potentials in the muscle fibre following stimulation. This results in prolonged contraction following strong voluntary contraction and a distinctive EMG record. A blow to the muscle belly with a patellar hammer may elicit "percussion myotonia" where the stimulation to contraction is a sudden deformation of the muscle membrane.

Myotonia is still observed after nerve section or curarisation and the contractile mechanism is probably normal except in paramyotonia (see below). The best understood of the several syndromes of myotonia is myotonia congenita (Thomsen's disease) which has an excellent animal model in the form of a certain strain of myotonic goat. The defect is probably a reduced chloride conductance in the plasma membrane. With muscle activity the potassium concentration rises in the T-tubular system but this does not normally affect the surface membrane potential because chloride ions are relatively freely 'shunted' to counteract the charge of the potassium ions. When chloride conductance is reduced, this cannot occur and the membrane potential becomes influenced by the ratio of the external potassium concentration to tubular potassium concentration, resulting in depolarization. A similar effect on chloride conductance can be produced experimentally by diazacholesterol and by certain aromatic carboxylic acids e.g. anthracene-9-carboxylic acid, which are known to produce myotonia. It is not clear whether the defect postulated in the "chloride hypothesis" is the sole abnormality.

There is no direct evidence that the same mechanism occurs in myotonic dystrophy. Here the decrease in chloride conductance is less marked. Some workers have reported a decreased resting membrane potential and an increased intracellular sodium concentration. A possible explanation for this is an altered stoichiometry of the muscle sodium pump. Red cell membranes from affected individuals have been reported to exchange two sodium for two potassium ions instead of the normal 3 : 2 ratio. Others have reported an increased sodium conductance in the resting sarcolemmal membrane.

Myotonia also occurs in the periodic paralyses (see below), and occasionally in hypothyroidism although here a more common finding is "pseudo-myotonia".

In paramyotonia congenita, which appears to overlap with hyperkalaemic periodic paralysis, the myotonia is characteristically provoked by cold. Because contraction may be maintained in the absence of electrical activity, there may be a defect of excitation-contraction coupling.

Quinine, procainamide and phenytoin may all improve myotonia. All have a "local anaesthetic" effect on the muscle membrane, probably reducing the number of voltage-dependent sodium channels and hence reducing the excitability of the membrane.

POTASSIUM DISTURBANCES AND THE PERIODIC PARALYSES

It is not known whether the weakness associated with hypokalaemia is due to primary effects on nerve or muscle. It will be recalled that the resting membrane potential is largely dependent on the ratio of extracellular to intracellular potassium ion concentration. Pure hypokalaemia would be predicted to cause hyperpolarization of excitable membranes, however extracellular potassium loss will to a varying extent affect intracellular potassium levels and the actual electrophysiological disturbance is not known. Hypokalaemia will tend to increase sensitivity to non-depolarizing, and decrease the sensitivity to depolarizing, neuromuscular blocking drugs.

The periodic paralyses are rare but dramatic disorders. Although changes in serum potassium and distribution were the first abnormalities to be identified they cannot alone explain the electrophysiological disturbance. The disorders are usually autosomal dominant in inheritance although sporadic cases are also found.

Hypokalaemic periodic paralysis

Attacks of weakness, which may be profound, last several hours or even days, and tend to be precipitated by heavy meals or by prolonged rest after vigorous exercise. They may be provoked by a carbohydrate load and, particularly, with glucose/insulin.

The weakness associated with hypokalaemia may occur with potassium levels of 3 mmol/l and become severe at 2·5 mmol/l. The hypokalaemia is mainly due to potassium entry into the muscle. Such levels of potassium would not be expected to cause weakness in normal subjects. Although usually clinically normal between attacks, some patients later develop a proximal vacuolar myopathy.

During an attack the muscle membrane is unable to propagate an action potential, while the contractile mechanism responds normally to directly applied calcium. *In vivo* studies have given conflicting information about the occurrence of depolarization of the resting membrane potential in attacks. However, study of intercostal muscle *in vitro* has shown substantial depolarization even at normal extracellular potassium concentrations and further depolarization on adding insulin. Intracellular sodium concentration was increased and the depolarization was reversed by replacing 90 per cent of the external sodium with choline. It was therefore suggested that the abnormality was an increase in sodium permeability. An alternative theory, also explaining why depolarization rather than hyperpolarization is found in association with the hypokalaemia, is that the internal potassium is sequestered in the dilated vacuoles which arise from the sarcoplasmic reticulum during attacks.

Attacks can be treated by the administration of potassium, usually orally, and may be prevented by a high potassium and low carbohydrate diet. Chronic treatment with carbonic anhydrase inhibitors, particularly acetazolamide, may also prevent attacks, a beneficial effect which appears to depend on the production of a metabolic acidosis. A sporadic form of hypokalaemic periodic paralysis is occasionally found in association with thyrotoxicosis, particularly in orientals.

Hyperkalaemic periodic paralysis (potassium sensitive
periodic paralysis, adynamia episodica hereditaria)

Attacks tend to be briefer than in the hypokalaemic form, lasting several hours and may be precipitated by rest after exercise and can be provoked by oral potassium administration. Myotonia, particularly of the eyelids, is found between attacks.

Attacks of paralysis may occur with serum potassium levels at the higher limit of normal and may be severe at 7 mmol/l, the rise of serum potassium being mainly due to loss from muscle. Again, these serum levels would not cause weakness in normal subjects. During an attack, the muscle potassium is low and the sodium high and between attacks the muscle potassium is either normal or low. Microelectrode studies have generally shown a decreased resting potential with further depolarization to between -40 mV and -50 mV during paralysis, at which time the membrane is inexcitable. The changes in potassium will not alone explain the findings and the abnormalities in ionic distribution are probably complex. It is likely that increased sodium permeability occurs but unclear why the expected consequent stimulation of sodium/potassium exchange does not occur. Intracellular calcium disturbance, and defective excitation-contraction coupling have also been postulated.

Treatment with chlorothiazide and carbonic anhydrase inhibitors may prevent attacks. Attacks usually respond to high carbohydrate foods. Rarely, intravenous glucose/insulin is needed, which stimulates the re-entry of potassium into the muscle.

Normokalaemic periodic paralysis

A third variety of familial periodic paralysis is found which resembles hyperkalaemic paralysis clinically but in which no change in serum potassium is found. The resting membrane potential may be low between attacks. No explanation is at present available for the condition.

ABNORMALITIES OF EXCITATION-CONTRACTION COUPLING AND OF THE CONTRACTILE MECHANISM

Endocrine myopathies

In certain myopathies associated with endocrine disturbance, particularly those of osteomalacia and of hyper- and hypo-thyroidism, pathological changes in the muscle fibres may be very slight in the face of significant weakness. In other cases atrophy of type II fibres is found. For convenience these disorders are considered together.

In hypothyroidism, the rate of relaxation is decreased, a change unlikely to be due to alteration in the proportions of type I and II fibres. The rate of ATP turnover has been found to be reduced, probably due to decreased utilization, for a given amount of force generation in submaximal contraction. Conversely, in hyperthyroidism ATP turnover is increased. Here, however, the weakness may be due to fibre atrophy. These changes in ATP turnover may reflect alterations in myosin ATPase activity with disordered thyroid function. In hypophysectiomised rats there is a failure of both growth and, specifically, of muscle protein synthesis. Maximal stimulation of muscle protein synthesis is found to occur with physiological levels of thyroxine replacement. With much higher levels of thyroxine (a hundred times the physiological dose) protein synthesis remains unchanged but protein degradation is much increased. This suggests increased protein catabolism may be a factor responsible for the muscle wasting of thyrotoxicosis.

Similar experiments in rats to those with thyroxine indicate that decreased synthesis rather than increased catabolism may be responsible for the muscle wasting caused by corticosteroids. In the myopathy of Cushing's syndrome and of corticosteroid treatment, the principal histochemical abnormality is type II fibre atrophy. Why type I fibres are relatively spared is not known. Microelectrode studies of corticosteroid-treated mice have shown inexcitability of the extensor digitorum longus (predominantly type II) muscle membrane in which excitability was restored by hyperpolarization of the membrane by microelectrodes. No abnormality was found in the fibres of soleus (type I fibres). A reduction in intracellular potassium in type II fibres was suggested to be responsible.

In osteomalacia the weakness is difficult to assess because of the presence of skeletal pain but appears to be unrelated to plasma calcium levels. 25,hydroxycholecalciferol increases the muscle ATP concentration and the rate of protein synthesis in vitamin-D deficient rats in which physiological concentrations of 1,25-dihydroxycholecalciferol have no such effect. This may be relevant to the pathogenesis of the human myopathy.

Malignant hyperthermia

In this rare but important condition, generalized muscle rigidity and hypermetabolism, often fatal, occur following the administration of succinylcholine and certain volatile anaesthetic agents, particularly halothane. *In vitro* study of affected muscle has shown that these agents promote calcium release from the sarcoplasmic reticulum and that excessive calcium release also follows direct application of caffeine, both resulting in continuous contraction and uncontrolled glycolytic and oxidative phosphorylation. Affected individuals (and certain of their relatives) probably suffer from a subclinical myopathy. Some cases follow an autosomal dominant inheritance but the condition is unlikely to represent a single disease entity. The syndrome has, for example, been noted in central core disease and myotonia congenita.

CONDITIONS CAUSING WEAKNESS IN WHICH THE
PHYSIOLOGICAL DISTURBANCE IS EITHER UNKNOWN
OR DUE TO GENERAL FIBRE DESTRUCTION

In the progressive muscular dystrophies (Duchenne, limb-girdle and facioscapulohumeral) weakness is mainly due to widespread muscle fibre destruction. Hypertrophied fibres are frequently found and may represent work hypertrophy by functionally relatively preserved fibres. Disorganization of the myofibrillar pattern is found in a high proportion of fibres.

The most intensive research has centred on Duchenne dystrophy. Earlier suggestions that the fault was one of defective innervation or of functional ischaemia are not now generally accepted. A number of biochemical disorders have been reported but an outstanding problem is the identification of primary pathological processes in a disorder with such widespread structural change.

Recent work has suggested a primary abnormality of the plasma membrane and an associated defect of sarcoplasmic calcium regulation. Structural defects in the plasma membrane have been shown by electron microscopy and freeze-fracture techniques in fibres which appear otherwise normal or which show only mild degenerative changes. Likewise an abnormal permeability of the membrane has been shown which may be relevant to the high concentration of sarcoplasmic enzymes found in the serum which is an early and consistent abnormality. On the basis of abnormal physical and biochemical properties of the erythrocyte membrane, it has also been suggested that there may be a generalized membrane abnormality in Duchenne dystrophy. This hypothesis, however, remains controversial.

Histochemical and X-ray microanalytical methods have shown that the sarcoplasmic calcium concentration is raised at an early stage. Dilatation of the sarcoplasmic reticulum is also often noted supporting the possibility of disturbed calcium regulation. Elevated calcium levels could cause cell damage by two suggested means. First, the calcium might produce sustained myofibrillar contraction and secondly might activate calcium-activated neutral protease (CANP) an enzyme which will degrade certain of the myofibrillar proteins. It is not however, clear whether increased calcium levels result from redistribution of internal calcium ions or from entry of extracellular calcium. This hypothesis of a membrane abnormality leading to intracellular calcium accumulation is the most convincing of current hypotheses.

In polymyositis and dermatomyositis necrosis of muscle fibres may be accompanied by involvement of the terminal branches of the motor nerves. There is evidence for disturbance of cell-mediated immunity in some cases and it is also possible to demonstrate deposition of immunoglobulin in the membranes of damaged fibres. The weakness may be out of proportion to the degree of fibre damage suggesting that this is not the sole mechanism.

Certain congenital myopathies such as central core disease and nemaline myopathy show greatly reduced proportions of type II fibres or histochemical uniformity. Because normal histochemical differentiation is determined by the lower motor neurone, it is possible that these conditions are caused by a disturbance of trophic neuronal function during myogenesis.

DISORDERS OF ENERGY PROCESSES

Although uncommon, these disorders are of considerable theoretical interest. Many specific biochemical defects are now known and this is a certain area of growth of knowledge.

The effect of a block in a metabolic pathway tends to lead to a build-up of substrates which may be either retained in the sarcoplasm (e.g. glycogen, lipid) or removed by the circulation (e.g. lactate). The severity of clinical expression of certain defects may partly depend on whether an enzyme is partially deficient or absent. If no alternative energy pathway is available, exercising muscle may reach an "energy crisis" in which the supply of ATP is thought to fail. Structural damage may result and an important consequence is the release into the circulation of myoglobin, which may cause acute renal tubular damage.

Glycogen metabolism

Five forms of glycogenosis (glycogen storage disease) directly involve striated muscle (types II, III, IV, V and VII). These disorders result from a deficiency of one of the enzymes catalyzing the breakdown of stored glycogen or blood borne glucose, resulting in the accumulation of normal or abnormal glycogen in the sarcoplasm which is readily identified histochemically.

Acid maltase deficiency (II)

This is a lysosomal storage disease resulting in three types of clinical disorder. First, a generalized progressive disease fatal in infancy; secondly, a childhood form with slowly progressive limb-girdle and truncal weakness; thirdly, an adult form often with prominent respiratory muscle involvement. This lysosomal enzyme deficiency, except in the adult form, is generalized. The enzyme defect is detectable in leukocytes and fibroblast culture. The importance of this pathway of breakdown in normal muscle is uncertain and the precise mechanism for the myopathy is unclear.

Debrancher enzyme deficiency (III)

This is a generalized but benign and remitting disease of childhood. Clinical myopathy may present in adult life long after the resolution of visceral manifestations. The deficiency is detectable in erythrocytes and leukocytes.

Brancher enzyme deficiency (IV)

This is a rare generalized rapidly progressive disease in which muscle hypotonia and wasting may be a relatively minor feature.

Myophosphorylase deficiency (V) (McArdle's disease)

This is a disorder characterized by attacks of muscle pain, stiffness and weakness during exertion. During vigorous exercise muscles may develop localized contractures which, unlike true cramps, are electrically silent. Severe attacks of muscle pain may be associated with myoglobinuria. Onset is usually in childhood or adolescence but adult-onset cases have been described. Both absence of the muscle isoenzyme of phosphorylase and the presence of an inactive enzyme have been found. Isoenzymes in other tissue appear to be normal.

The 'second wind' phenomenon, in which decreased activity at the onset of symptoms may permit more prolonged and strenuous activity, may in part represent a switch to blood-borne energy substrates (glucose, FFA and alanine).

Muscle biopsy shows increased glycogen. Phosphorylase activity is absent in muscle fibres but present in the smooth muscle of blood vessels. The diagnosis is supported by the failure of the venous lactate level to rise after ischaemic exercise, although this will also be found in debrancher and phosphofructokinase deficiencies.

Phosphofructokinase deficiency (VII)

This gives a similar clinical picture to myophosphorylase deficiency but in addition there is a haemolytic tendency due to partial PFK deficiency in erythrocytes in which one of the two enzyme subunits (the muscle subunit) is lacking. Diagnosis can be confirmed by histochemical or biochemical analysis of muscle biopsy material.

Several other enzyme deficiencies including phosphoglucomutase and phospho-hexoisomerase have been suspected in individual cases but not proven by direct biochemical measurement.

FATTY ACID METABOLISM

Carnitine deficiency

Carnitine is the obligatory carrier of long chain fatty acids across the mitochondrial membrane (see Fig. 10.7). Body carnitine is partly of dietary origin and is partly synthesized from lysine in a three stage process involving both kidney and liver. Two disorders of carnitine deficiency are recognized although there may also be intermediate types.

In the 'myopathic' form, a proximal myopathy develops in childhood or teens. Muscle carnitine levels are 5–20 per cent of normal with normal or slightly low serum levels. Because the entry of fatty acids into the mitochondria is reduced, they tend to accumulate in the sarcoplasm as lipid droplets, which is the principal abnormality seen with histochemical stains. Carnitine deficiency may be due to a partial failure of carnitine transport into the muscle which is an active process. Several cases have responded to oral carnitine although, strangely, the muscle carnitine may remain low. Response is also seen with corticosteroid treatment.

In the 'systemic' form carnitine levels are low in plasma, liver and muscle suggesting a generalized defect in carnitine transport across cell membranes. There is fluctuating

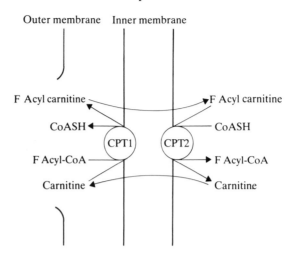

Figure 10.7. *Fatty acid transport across the mitochondrial membrane.* Long chain fatty acids in the sarcoplasm in the form of fatty acyl-CoA combine with carnitine, a process catalyzed by carnitine palmityl transferase I (CPTI) to form fatty acyl carnitine which crosses the inner mitochondrial membrane. CPT2 then catalyzes the dissociation of fatty acyl carnitine to fatty acyl CoA (which then undergoes β-oxidation) and carnitine (which is returned to the sarcoplasm).

muscle weakness, often from early childhood with recurrent episodes of hepatic encephalopathy. Partial deficiency of carnitine is found in skeletal muscle, serum and liver.

Similar low muscle carnitine levels have also been described in patients with chronic renal failure undergoing haemodialysis in whom there was no similar clinical myopathy.

Carnitine palmityl transferase deficiency

There are two CPT isoenzymes, CPT I and CPT II, which are involved in the transfer of long chain fatty acids across the inner mitochondrial membrane (see Fig. 10.7). CPT deficiency, in which enzymes levels less than 20 per cent normal are found, is characterized by recurrent episodes of muscle pain and myoglobinuria. These usually occur after fasting or exertion, conditions under which FFA are most required for mitochondrial metabolism.

The patients are clinically normal between attacks, as is the muscle morphology, but the latter may show lipid accumulation during or after attacks. It is not clear why CPT deficiency should produce a clinical picture so different from the superficially similar carnitine deficiency.

The treatment is of avoidance of precipitating conditions and the taking of a high carbohydrate diet.

Several other 'lipid storage' myopathies have been described in which carnitine and CPT levels are normal and it is likely that other biochemical defects will soon be identified.

MITOCHONDRIAL MYOPATHIES

This group consists of (a) myopathies with an identified or suspected defect in mitochondrial metabolism with or without morphological mitochondrial changes, (b) myopathies without known biochemical defects in which morphological abnormalities predominantly affect the mitochondria. The principal biochemical technique used in their study is the estimation of the respiration of isolated muscle mitochondria in the presence and absence of the normal substrates of oxidative metabolism. CPT deficiency has already been mentioned.

The clinical picture is very varied but in general there is found weakness, exercise intolerance and excessive lactate production while more severely affected cases present as an often fatal lactic acidosis. The increased lactate production reflects a block in the further metabolism of pyruvate by the mitochondrion (see Chapter 14). Pyruvate is then reduced to lactate or transaminated to alanine.

The first such disorder recognized was the extremely rare Luft's disease which results from uncontrolled mitochondrial respiration independent of phosphorylation and of the energy needs of the cell. This "loose coupling" of mitochondrial respiration causes wastage of energy as heat, producing a picture of hypermetabolism with fever, weight loss and tachypnoea.

Specific defects of the pyruvate dehydrogenase complex results in lactic acidosis. In the respiratory chain (Fig. 14.5) several specific deficiencies of individual cytochromes have been identified, producing conditions varying from rapidly fatal neonatal disease to a myopathy compatible with only mild limitation of normal activities.

Striking abnormalities of mitochondrial structure are found in many of these conditions. They may also occur in certain patients with chronic progressive external ophthalmoplegia (CPEO) and this is invariable in those with CPEO associated with neurological and cardiac involvement (Kearns-Sayre syndrome).

Principles of tests and measurements

ELECTROMYOGRAPHY

Electromyography provides a method of recording and analyzing the electrical activity of skeletal muscle. It forms an important element of electrodiagnosis where it may be combined with measurement of motor and sensory nerve conduction and studies of neuromuscular transmission. With these techniques it is usually possible to determine the site of disordered neuromuscular function (anterior horn cell, peripheral nerve, neuromuscular junction or muscle) and sometimes an indication of its pathology.

The principles of muscle physiology and anatomy on which EMG are based have been described (pp. 343–348). The electrical activity set up by excitation of a motor unit reflects the summation of the action potentials of its many individual muscle fibres. Recorded activity will depend on the position, size and characteristics of the recording electrode, on the number and anatomical arrangement of the muscle fibres of the unit and on the degree of synchrony of their excitation. In general, the amplitude of a recorded potential falls exponentially with the distance from the source.

Three main types of electrodes are used. A surface electrode, applied to the skin over the muscle belly, records the summated potentials from a large volume of muscle. The concentric needle electrode (CNE) samples a more restricted area. In "single fibre" EMG, a very fine needle electrode, carefully positioned, is used to sample activity from one or two individual muscle fibres. As well as being displayed on an oscilloscope screen, muscle activity is recorded on film and is also usually amplified and fed into a loudspeaker.

Normal EMG *(with* CNE*)*

In a normal subject the muscle is electrically silent at rest. Insertion or movement of the needle may induce brief discharges, 'insertion activity', due to mechanical stimulation and local injury by the needle.

On gentle voluntary contraction, potentials arising from individual motor units are seen (Fig. 10.8a). These are usually bi- or tri-phasic, about 0·5–2 mV in amplitude and 5–15 ms in duration. They are smaller and briefer in facial and extraocular muscles. The highest voltage components are contributed by fibres nearest the electrode and the early and late components probably arise from activity in more distant fibres of the same unit.

On increasing the force of contraction, progressively more units are recruited and

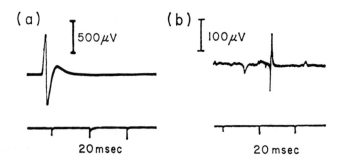

Figure 10.8a & b.
   (a) Normal motor potential from human muscle.
   (b) Fibrillation potential from denervated muscle. Records obtained by a concentric needle electrode. Note the smaller size and shorter duration of the fibrillation potential.

their discharge frequency increases. The potentials on maximal contraction become confluent in what is termed the 'interference pattern'. The first units to appear are of smaller amplitude than those recruited with more vigorous contraction.

The activity of single muscle fibres is rarely seen in normal muscle but may be recorded with fine electrodes (see below).

EMG *in disease*

Any process leading to loss of motor nerve fibres results in a reduction in the number of motor unit potentials and an incompleteness of the interference pattern. The degree of impairment of the interference pattern reflects the severity of the denervation but it is clearly partly dependent on the degree of voluntary effort, making assessment difficult where there is pain or lack of motivation by the subject.

In primary muscle disease there is a random loss of muscle fibres and this causes changes in the appearance of motor unit potentials without significantly altering their numbers until the disease is advanced (Fig. 10.9).

Figure 10.9. *Electromyographic recording obtained by a concentric needle electrode from a patient with polymyositis showing low voltage, short duration polyphasic motor unit potentials.*

Alteration in amplitude of motor unit potentials. Loss of muscle fibres in a unit, such as occurs in primary muscle disease, will reduce the amplitude of the motor unit potential. Conversely potentials will increase in amplitude if the number of constituent fibres increases. In chronic partial denervation collateral sprouting with reinnervation of neighbouring muscle fibres increases the size and compactness of motor units. This process may give rise to giant motor unit potentials up to 20 mV in amplitude. It is shown histochemically as "type grouping" in which clustering of fibres of the same histochemical type occurs in contrast to the normal mosaic.

Alteration of duration and form of motor unit potentials. If temporal dispersion in the spread of excitation through the motor nerve terminals were to occur, this would result in prolongation of the motor unit potential. It is likely that such a process accounts for the prolonged units seen in re-innervation (where the nerve sprouts are initially unmyelinated). Additionally, (in diseases such as the dystrophies), there may be some slowing in the propagation of the muscle action potential which is normally too rapid to influence unit duration. This asynchrony of excitation together with the fall-out of units probably accounts for the polyphasic units found in both primary muscle disease and in chronic partial denervation (Fig. 10.9).

Potentials of shorter duration occur in primary muscle disease when the number of fibres in a unit is sufficiently reduced.

Single fibre recording and disturbances of neuromuscular transmission. In this technique a fine needle electrode is sited so as to record the activity of a pair of fibres from the same unit which will fire within a few milliseconds of each other. Under conditions of steady but gentle voluntary contraction, it is arranged that the oscilloscope trace is triggered by the potential of the first fibre of the pair to be activated so as to display the potential of the second (Fig. 10.10a). In normal muscle, the time interval between the two potentials is only slightly variable, the variation being termed "jitter". In conditions with disturbed neuromuscular transmission, especially myasthenia gravis, where transmission may be on the verge of failure, the degree of jitter increases and, if transmission intermittently fails "blocking" of the second potential occurs (Fig. 10.10b). Abnormal jitter can however be found in other neuromuscular diseases, e.g. polymyositis.

Another useful diagnostic test representing the same phenomenon can be demonstrated with surface electrode. In response to slow repetitive nerve stimulation, usually at 3 Hz, the muscle action potential becomes reduced in amplitude after the first few impulses, usually by 15–30 per cent. This "decremental response" represents the progressive fall-out of some fibres as neuromuscular transmission fails (Fig. 10.11). Both phenomena can be reversed with intravenous edrophonium.

The post-tetanic facilitation seen in the Eaton-Lambert syndrome has already been mentioned.

Changes in excitability

Denervated fibres are abnormally sensitive to many forms of stimulation. This is reflected both by increased insertion activity and also by the presence of "positive sharp waves", rapid positive potentials with slow decline thought to arise from damaged regions of fibres.

Myotonia. The prolonged trains of potentials recorded in myotonia following needle movement or voluntary activity are initially of high frequency, 100–150 Hz, but the frequency rapidly declines over a few seconds producing the so-called "dive bomber" effect when amplified and fed into a loudspeaker.

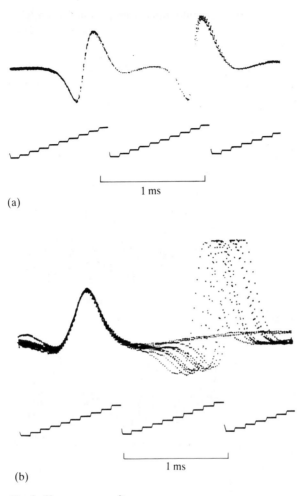

Figure 10.10a & b. *Single fibre* EMG *recording.*

(a) Normal record. The two potential pairs are shown. Multiple recordings are superimposed and little variation occurs in the time interval between pairs.

(b) Increased 'jitter' in a patient with a defect of neuromuscular transmission (a myopathy associated with non-autoimmune myasthenia). Occasional failure of neuromuscular transmission causes intermittent non-appearance of the second potential or 'blocking'. This abnormality is also characteristic of myasthenia gravis.

(Kindly supplied by Dr. N.M.F. Murray).

Fibrillation. Spontaneous fibrillation potentials of 20–300 μV usually lasting less than 2 ms with a rather constant frequency of 2–10 Hz arise from individual fibres (Fig. 10.8b). They are a feature of denervation, occurring 10–20 days after the interruption of the nerve supply, but are also seen in primary muscle disease, particularly polymyositis. They are not simply a feature of extrajunctional AChR formation and denervation hypersensitivity, but are probably due to altered membrane properties producing instability of the resting potential. Despite their stimulation by anticholinesterase drugs, they are not abolished by curarisation. Fibrillation is not visible except in the tongue.

Fasciculation. Fasciculation is due to the spontaneous firing of motor units and is visible through the skin as a fine rippling movement in relaxed muscle. It may arise from the anterior horn cells or more peripherally from the motor nerve terminals,

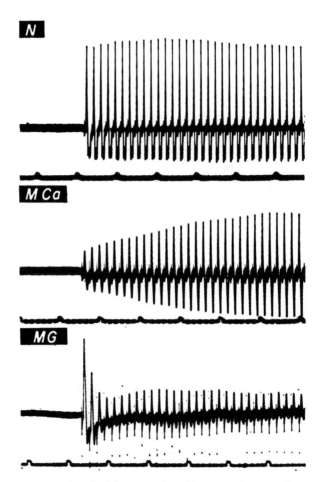

Figure 10.11. *Action potential evoked from muscles of hypothenar eminence by supramaximal stimulation of ulnar nerve at wrist at a rate of 54/sec.*
Action potentials were recorded by electrodes on skin surface over belly and tendon of abductor digiti quinti muscle. Amplification was the same in each record. Time signals are 0·1 sec. N is normal. MCa is myasthenic syndrome with small cell bronchogenic carcinoma. The initial response is low, but successive responses increase rapidly. MG is moderately severe myasthenia gravis. The initial response is normal, but successive responses decrease rapidly.
(From Lambert E.H. & Rooke E.D. (1965) Myasthenic State and Lung Cancer. In Remote Effects of Cancer on the Nervous System, pp. 67–80. Ed. Lord Brain & Norris F.H.)

exciting the rest of the unit antidromically. It has already been mentioned in connection with anticholinesterase drugs and depolarizing muscle relaxants. It is a sign of denervation. However, a coarse fasciculation is occasionally seen in normal subjects usually in the calves and after exertion. This 'benign fasciculation' is not of pathological significance.

It is often useful, as in the diagnosis of motor neurone disease, to detect fasciculation in multiple sites by the use of surface electrodes placed over different muscles. The fasciculation can be recorded on the multiple channels of an EEG.

Myokimia is a somewhat similar but regular undulating spontaneous muscle activity. It is associated with repetitive bursts of identical motor unit potentials at about 50 Hz. Their origin is probably in the spinal cord or brain stem. Myokimia may occur in several conditions including multiple sclerosis, uraemia and in spinal cord injury.

Tetany arises from hyperexcitability of the peripheral nerve and is not a primary muscle disorder. It is usually due to a reduction in extracellular ionized calcium either in hypocalcaemia or in metabolic or respiratory alkalosis. It may occasionally be associated with hypokalaemia or hypomagnesaemia.

Ordinary *muscle cramps* are thought to be of neural origin and are associated electromyographically with fluctuating high voltage, high frequency and irregular bursts of motor unit activity.

In *tetanus* muscle spasms result from uninhibited discharges from the spinal cord. Hyperexcitability at a spinal level is also thought to occur in the *stiff man syndrome* in which there is found sustained involuntary contractions of limb and trunk muscles. Another rare disorder *continuous muscle fibre activity* (*Isaacs' syndrome*) is characterized by progressive generalized fasciculation and muscle rigidity and weakness, the EMG showing a rapid rhythmic discharge. The disorder may be due to hyperexcitability of the motor nerve terminals with antidromic spread. It should be noted that in any condition with uncontrolled muscle spasms, e.g. status epilepticus and tetanus, there may occur secondary muscle damage and myoglobinuria.

SERUM ENZYMES

Many sarcoplasmic enzymes may appear in the serum in muscle damage. The most frequently measured is creatine kinase (CK or CPK) which is found to be most elevated in conditions with widespread muscle fibre destruction particularly Duchenne dystrophy and polymyositis, but also in attacks of myoglobinuria in metabolic myopathies (e.g. myophosphorylase and CPT deficiency). Serial measurements are of value in assessing response to treatment. Mild elevation may be found in chronic denervating conditions. It should be noted that minor trauma or vigorous activity can raise the CK level in normal individuals. It is elevated in 70 per cent of female carriers of Duchenne muscular dystrophy. It may be useful in detecting subclinical myopathies such as the malignant hyperpyrexia trait.

CREATINE AND CREATININE

Creatine is actively taken up by muscle where about 98 per cent of the total body creatine is present as creatine phosphate. In muscle disease both decreased uptake and increased release into the serum may occur, exceeding the renal threshold to produce creatinuria.

Creatinine, the anhydride of creatine, is a muscle metabolite and the daily excretion roughly parallels total muscle mass being approximately constant for any individual (hence its use as an index of renal function). Its excretion falls in muscle wasting conditions.

Neither substance is of much clinical importance in myology.

MUSCLE BIOPSY

Biopsy usually performed under local anaesthesia, should be taken from a moderately affected muscle. The material of choice is fresh tissue rapidly frozen in isopentane cooled in liquid nitrogen. Samples may be stored for many years at −70° C.

In addition to the study of microscopic morphology, the total number of enzyme systems in muscle which can be studied with modern histochemical techniques now exceeds 100. Those enzymes which are relevant to muscle disease are myophosphorylase, PFK, certain oxidative enzymes and myofibrillar ATPase. The ATPase reaction is important in identifying muscle fibre types and their differential involvement in disease.

Biochemical analysis is sometimes indicated where enzyme defects are identified histochemically. These may also sometimes be confirmed in other tissues, e.g. red blood cells and leucocytes in debrancher deficiency.

The use of needle biopsy has shown the feasibility of analyzing relatively small muscle specimens histochemically and biochemically. The technique is principally used to sample quadriceps and is less suitable for other muscles. Although giving somewhat less satisfactory specimens than open biopsy the technique allows serial biopsies to be performed. It is well suited for studying the biochemical changes of muscle in exercise.

Practical assessment

CLINICAL OBSERVATIONS

It is important to enquire about the strength of foetal movements, neonatal respiratory or feeding difficulties, delay in motor milestones, and childhood failure to be able to compete physically with peers. These may be features of congenital myopathies or spinal muscular atrophy. A positive family history should be sought. That a large number of drugs can cause muscle disease should be borne in mind.

Most myopathies are painless. Pain is sometimes found in the myotonic disorders and in denervation. However, pain normally implies active fibre destruction, involvement of intramuscular blood vessels or connective tissue, or defects in energy processes.

Associated clinical features may indicate the cause of a myopathy. Examples are skin lesions or Raynaud's phenomenon (a collagen vascular disease); eye signs tremor and tachycardia (thyrotoxicosis); bone pain and steatorrhea (osteomalacia); truncal obesity and hypertension (Cushing's syndrome); weight loss (carcinomatous myopathy).

Fasciculation occurs with lesions of the anterior horn cells, peripheral nerves and nerve roots. If widespread and accompanied by wasting and weakness, it suggests motor neurone disease.

Slowness of relaxation after forcible muscle contraction and on percussion of the muscle belly indicate myotonia. Associated dystrophic features are present in myotonic dystrophy. Myotonia also occurs in hyperkalaemic periodic paralysis.

Attacks of muscular spasm in the extremities with associated parathesiae suggest tetany. In latent tetany it may be possible to demonstrate positive Trousseau's and Chvostek's signs, both of which depend on increased excitability of peripheral nerves.

True hypertrophy (muscle enlargement without weakness) occurs in myotonia congenita and in the early stages of muscular dystrophy. It is also seen in athletic training. Pseudohypertrophy (enlargement with weakness) is characteristic of Duchenne dystrophy but is sometimes seen in other dystrophies, polymyositis and spinal muscular atrophy.

Muscle wasting usually implies either fall-out of fibres or atrophy of individual fibres due to primary muscle disease or denervation. However, wasting also occurs in upper motor neurone lesions, e.g. parietal wasting, and is common in prolonged disuse and in debilitating diseases e.g. TB or malignancy. Quite rapid wasting, often without detectable weakness, is seen in the quadriceps following knee injury. In neuromuscular diseases wasting is invariably accompanied by some weakness, although the reverse is not always true. Weakness without wasting may be seen in myasthenia gravis, periodic paralysis and in some of the metabolic myopathies.

Weakness and wasting may have a characteristic distribution. Selective, symmetrical involvement of certain proximal muscles or parts of muscles is found in muscular dystrophy. Diffuse proximal weakness is more suggestive of an acquired myopathy or a spinal muscular atrophy. Distal weakness occurs in peripheral neuropathies, myotonic dystrophy and in the rare distal myopathy. Asymmetrical weakness and wasting involving both proximal and distal muscles suggests a neurogenic lesion e.g. motor neurone disease.

Muscles supplied by the cranial nerves may be involved early in myotonic dystrophy (facial and sternomastoids), facioscapulohumeral dystrophy (face and sternomastoids), motor neurone disease (bulbar muscles), ocular myopathy and myasthenia gravis (external ocular muscles).

Worsening of weakness by exercise is characteristic of myasthenia gravis but may also occur in polymyositis and even in non-muscular conditions such as multiple sclerosis. Precipitation of weakness by rest following a heavy meal or by rest following exercise would suggest one of the periodic paralyses.

Tendon reflexes are usually depressed or absent in muscular dystrophy due to involvement of the muscle spindles. Loss is also seen in the congenital myopathies. In acquired myopathies, e.g. thyrotoxic myopathy and polymyositis, reflexes are preserved or even exaggerated. Hyperreflexia in the presence of gross wasting and weakness is characteristic of motor neurone disease, due to the presence of an upper motor neurone lesion.

ROUTINE INVESTIGATIONS

Blood count and ESR are usually abnormal in polymyositis or collagen vascular disease. Disturbed serum K^+ levels may be found in periodic paralysis. Plasma Ca, HCO_3^-, Mg^{++} and K^+ may be abnormal in tetany and the arterial pH may be high. Serum CPK has been discussed. Endocrine assays will be relevant where an endocrine myopathy is suspected. ECG should be performed in those muscle diseases with possible cardiac involvement, e.g. dystrophia myotonica. Anti-AChR antibodies should be sought in generalized myasthenia gravis. Simple pulmonary function tests such as FVC are important where respiratory muscles may be involved. Electromyography is a most useful technique in confirming the presence of myopathy, denervation, a disturbance of neuromuscular transmission or myotonia.

SPECIAL INVESTIGATION

Muscle biopsy

The application of biochemical, histochemical and ultrastructural techniques to the study of muscle biopsy material has greatly increased the value of this investigation in the diagnosis of neuromuscular disorders. Ideally it should always be undertaken when there is doubt about the diagnosis on other grounds.

Biochemical analysis of other tissues

In the enzyme deficiencies (e.g. glycogen storage disease except type V) the defect can be identified in erythrocytes, leukocytes or fibroblast culture.

Edrophonium (tensilon) test

This may be diagnostic in myasthenia gravis. It is important to attempt to record some objective improvement. A partial response may occasionally be found in other conditions when there is abnormality in the region of the neuromuscular junction such as polymyositis (involvement of nerve terminals) and motor neurone disease (reinnervation).

Lactate levels and the ischaemic exercise lactate test

Lactate usually is measured in the forearm venous blood after occlusion of the arterial supply by an inflated sphignomanometer cuff. In disorders with a block in glycolysis, e.g. myophosphorylase or PFK deficiency, the lactate levels fail to rise after ischaemic exercise. Conversely, in conditions with defective oxidative metabolism, resting lactate levels may be high and rise excessively after moderate exercise. It may be possible to show a reduced oxygen consumption in these individuals.

Provocation tests in periodic paralysis

Attacks of hypokalaemic paralysis may be provoked by carbohydrate load or glucose/ insulin and, in the hyperkalaemic variety, by oral K^+. ECG monitoring is essential.

Antenatal diagnosis

Cultured cells from amniotic fluid have led to prenatal diagnosis in acid maltase deficiency. Raised foetal serum CPK is found in Duchenne dystrophy but this test is not at present felt to be entirely reliable. Another approach in Duchenne dystrophy is to offer, after foetal sexing, abortion to carriers who have a high risk of having affected male children.

References

Physiological

AIDLEY D.J. (1978) *The Physiology of Excitable Cells.* 2nd ed. Cambridge University Press.
FELIG P. and WAHREN J. (1975) Fuel homeostasis in exercise. *New England Journal of Medicine,* **293,** 1078–1084.
HUXLEY A.F. (1974) Muscular contraction. *Journal of Physiology,* **243,** 1–43.
KATZ, SIR BERNARD (1966). *Nerve, Muscle and Synapse.* New York. McGraw-Hill.
PERRY, S.V. (1979) The regulation of contractile activity in muscle. *Biochemical Society Transactions,* **7,** 593–617.
VRBOVA G., GORDON T. and JONES R. (1978). *Nerve–muscle Interactions.* London. Chapman-Hall.

Clinical

British Medical Bulletin. Volume 36, number 2. The muscular dystrophies.
DUBOWITZ V. and BROOKE M.H. (1973) *Muscle Biopsy: a modern approach.* London. Saunders.
EDWARDS R.H.T., YOUNG A. and WILES C.M. (1980) Needle biopsy or skeletal muscle in the diagnosis of myopathy and the clinical study of muscle function and repair. *New England Journal of Medicine,* **302,** 261–271.

GOODGOLD J. and EBERSTEIN A. (1977) *Electrodiagnosis of Neuromuscular Disease.* 2nd ed. Baltimore. Williams and Wilkins.

HUIJING F. (1975) Glycogen metabolism and glycogen-storage diseases. *Physiological Reviews,* **55**, 609–658.

Journal of Neurology, Neurosurgery and Psychiatry (1980), **43**, number 7. (Whole issue devoted to myasthenia gravis).

LANE R.J.H. and MASTAGLIA F.L. (1978) Drug-induced myopathies in man. *Lancet* **ii**, 562–565.

LAYZER R.B. and ROWLAND L.P. (1971) Cramps. *New England Journal of Medicine.* **285**, 31–40.

MORGAN-HUGHES J.A. (1980) Diseases of Muscle. *Medicine,* **36** (second series).

MORGAN-HUGHES, J.A. (1981) Defects of the Energy Pathways of Skeletal Muscle. In *Recent Advances in Clinical Neurology.* Vol. 3. Editors, Matthews, W.B. and Glaser, G.H. Edinburgh and London. Churchill-Livingstone.

VINKEN P.J. and BRUYN G.W. (editors) (1979). Disease of muscle. Volumes 40 and 41, *Handbook of Neurology.* North Holland. Amsterdam.

WALTON, SIR JOHN (editor) (1981) *Disorders of Voluntary Muscle.* 3rd ed. Edinburgh and London. Churchill-Livingstone.

11

The Nervous System
in the Control of Movement

Normal function

The enormous amount of experimental work carried out on reflex action over the last century has naturally been associated with attempts to explain movement in terms of increasingly complex reflex actions. But in the last few years the role of "central patterning" in movement has been attracting more attention. Bizzi and Evarts have nicely summarized the generally accepted features of a reflex action, such an action being

> "unlearned (based on inherited neural circuits), predictable from the inputs, uniform and adjustive or protective in purpose. To these two characteristics two other attributes, heavily charged with philosophical implications have been added: first, that the reflex is "involuntary", second that it is not dependent on consciousness."

Well known examples of reflex actions in man are the tendon reflex and the pupillary light response.

A centrally-patterned movement, in contrast to a reflex action, lacks this tight "input-output coupling". It is predominantly determined by the existing neural connections. Sensory input may serve to trigger the movement, but once initiated it runs its course in a manner analogous to a "motor-tape". During the execution of the movement, sensory input could play a variable role. In lower animals there are examples of repetitive movement in which peripheral sensory feedback simply exerts a tonic influence, and others where it provides phasic reinforcement or timing cues. For more complex centrally-patterned movements, Bullock has stated that sensory input is of decisive importance in

> "creating the permissive steady state centrally (making the frog want to jump), in directing action adaptively (aiming his jump), and in perfecting details during the action in some cases . . . But central patterning is the necessary and often the sufficient condition for determining the main characteristic features of almost all actions, whether stimulus triggered or spontaneous."

Centrally-patterned movement can be initiated automatically as well as triggered. Breathing is perhaps an example of the former, coughing probably an example of the latter.

While it may be accepted that central-patterning is a feature of certain movements in man, the nature and extent of feedback control remains critical. This will be discussed in a later section, but here the point must be made that feedback may be "internal" as well as "peripheral" or "external". By internal feedback is meant feedback arising from within the central nervous system rather than from external structures (e.g. muscle spindles) affected by the movement. Internal feedback is potentially important since it provides a means of monitoring the motor action at all stages of its execution including the period before movement has occurred, and thus allows an earlier error correction than with peripheral feedback alone.

SEGMENTAL MECHANISMS IN SPINAL CORD

The motor unit

The alpha motoneurone (anterior horn cell) is the final common pathway in the execution of a movement. Its axon and branches, and the muscle fibres it innervates, constitute a motor unit. Muscles carrying out fine movements tend to have motor units containing relatively few muscle fibres; for example, the external rectus of the eye has an average of 10 fibres per unit. In contrast, some motor units in large limb muscles may have more than 1,000 fibres. The numbers of motor units in some human muscles can now be estimated physiologically; an intrinsic muscle of the hand contains about 100 units whereas the soleus, a much larger muscle, has approximately 1,000. In a healthy subject the number of units in a muscle remains constant until the end of the sixth decade, at which time a progressive loss of units commences.

Within the same muscle the motor units show considerable variation in size and in other features; three types of units are now recognized. One type (slow-twitch-oxidative or type I) has a relatively small number of muscle fibres and these contract slowly; the fibres contain many mitochondria and have a good capillary blood supply. These fibres can be recognized histochemically because they stain weakly for alkaline myofibrillar ATPase. These motor units are used for weak contraction, both voluntary and reflex; they are well suited for continuous activity and are therefore employed in the maintenance of posture. Another type of unit (fast-twitch-glycolytic or type IIB) is reserved for strong but brief efforts; these units are composed of comparatively large numbers of fibres which contract more rapidly, contain few mitochondria, and stain strongly for alkaline myofibrillar ATPase. A third type of unit (fast-twitch-oxidative-glycolytic or type IIA) has some features of the other two types—fast twitch, many mitochondria, and strong alkaline myofibrillar ATPase activity; as in the type IIB fibres, the glycolytic enzyme systems are well developed. These units participate in contractions of intermediate or greater intensities.

Within any one motor unit the muscle fibres are homogeneous since their biochemical characteristics are dictated by the motoneurone; this controlling action is achieved partly by the pattern of impulses reaching the muscle fibres and partly by special messenger substances transported along the axons to the muscle fibres (neurotrophic action). In a normal muscle the motor units overlap with each other; because of differences in their staining reactions, the contrasting fibres present a checkerboard appearance. Alterations in this arrangement are of great significance for the recognition of denervation (see Chapter 10).

If a muscle contraction is to become stronger, the discharge frequencies of the motoneurones increase and additional motor units are called into activity. In a maximal contraction the discharge frequency may be initially as high as 100 impulses/ sec before declining to a steady rate of about 20 impulses/sec. In partially denervated muscles, some compensation is achieved through the use of higher discharge rates which, in turn, generate more tension; in addition, the surviving motor units are larger (through collateral reinnervation, see Chapter 10). Due to the loss of motor units however, the smoothness of contraction is lost; this is well seen in a patient with advanced motor neurone disease.

Muscle spindles

Muscle spindles are length-sensing mechanoreceptors arranged in parallel with the main muscle. Stretching of the main muscle increases the discharge from spindle sensory endings while shortening has the converse effect. The sensory endings are of two types. Primary endings are served by fast conducting group Ia afferent fibres which make monosynaptic connections with their own and with synergistic motoneurones (Fig. 11.1) and inhibitory connections (via inhibitory interneurones) with antagonistic motoneurones. Spindle secondary endings are served by group II, slower conducting fibres which make disynaptic neuronal connections. Group I and group II fibres, via a relay in the spinal cord, project to the cerebellum and group Ia afferents also project to sensory cortex. Primary endings are more sensitive than secondary endings to the dynamic component of the stimulus, i.e. to the rate of change of length.

The sensory endings in the spindle lie on "intrafusal" muscle fibres which are of two types, nuclear bag and nuclear chain fibres. These intrafusal muscle fibres are innervated by gamma motoneurons ("fusimotor neurones") which lie in the anterior horn along with alpha motoneurones. Gamma ("fusimotor") discharge causes intrafusal muscle fibre contraction and activates the spindle sensory endings as if stretching of the main muscle had occurred. Gamma motoneurones can be functionally identified as "static" or "dynamic" by their effects on the spindle primary afferent discharge, stimulation of dynamic gamma fibres increasing the dynamic response of primary endings during stretch. Thus the CNS can control the dynamic sensitivity of primary endings by adjusting the balance of excitation to static and dynamic gamma fibres.

The *stretch reflex* depends upon spindle primary endings and is served by a monosynaptic reflex arc. Sherrington carried out his original work on the stretch reflex in the decerebrate animal (i.e. in an animal showing decerebrate rigidity) and distinguished both a phasic and a tonic component in the reflex response, the former occurring

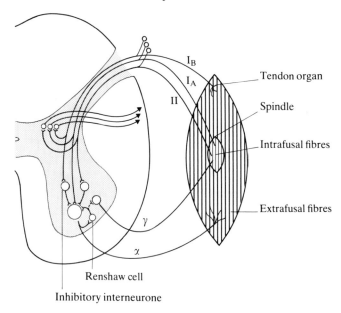

Figure 11.1. *Diagram showing the principal features of the segmental proprioceptive mechanism.*
  The spinal cord is shown in transverse section. Neurones with an inhibitory action are shown as filled circles. α, alpha motoneurone; γ, gamma motoneurone; Ia, group Ia afferent fibre from spindle primary ending; Ib, afferent fibre from Golgi tendon organ; II, afferent fibre from spindle secondary ending. Collateral branches from the proprioceptive afferent fibres are shown projecting to supraspinal structures via a relay at segmental level.

during the period of actual stretching and the latter during the period of constant maintained stretch. The response to muscle stretch in intact man is not quite the same. A brief stretch elicits the tendon reflex but in the relaxed limb no sustained stretch reflex can be obtained. However, in a contracting muscle, for example the biceps, the initial response to stretch is again the tendon jerk followed, after a latent period of inactivity, by a sustained response. Both of these responses depend on spindle primary endings (in the decerebrate animal, spindle secondary endings also contribute to the tonic reflex). They are usually also referred to respectively as the "phasic" and "tonic" responses, but they are not directly comparable to those seen in the decerebrate animal.

Tendon organs

These mechanoreceptors are innervated by group Ib (fast conducting) afferent fibres and via an interneurone make inhibitory connections ("autogenetic inhibition") with motoneurones. As in the case of spindle afferents, they also project to the cerebellum. Tendon organs have recently been shown to have a low threshold for activity during active muscle contraction, whereas formerly they were thought only to be active at very high tensions. The receptors would therefore be discharging whenever the muscle was active, their discharge rate increasing in parallel with muscle tension.

Renshaw cells

The alpha motoneurones give off collateral branches from the intramedullary parts of their axons Fig. 11.1). These collaterals form synapses with the Renshaw cells which are also found in the anterior horn. The Renshaw cell axons then pass to form inhibitory synapses on the surfaces of the alpha motoneurones of the same motoneurone pool (recent evidence suggests that they may also make inhibitory connections with fusimotor neurones). Thus a nerve impulse generated by an alpha motoneurone passes via the motor nerve fibre to the muscle, but also passes via the recurrent Renshaw loop back to the same motoneurone pool, where it produces an inhibitory or damping effect. By this means all but the initial impulses arriving at a motoneurone pool during a maintained discharge from the spindle endings will find the anterior horn cells partially inhibited, and will therefore produce a smaller reflex motor discharge than the initial excitatory volley. It seems likely that in the tendon reflex the motor volley leaves the spinal cord before the inhibitory Renshaw loop has had time to come into action.

DESCENDING CONTROL OF THE SPINAL PROPRIOCEPTIVE MECHANISM

We must next consider how the nervous system controls segmental reflex effects, for the unmodified stretch reflex would clearly be an embarrassment in the execution of a movement. Here a further concept needs to be introduced, namely that through its descending projections, the CNS can control transmission of neural impulses through spinal reflex arcs. Thus a number of descending pathways are now known to be able strongly to influence segmental reflex events. This can be shown experimentally by examining the effects of electrical stimulation of the descending pathway under study upon a test spinal reflex set up in alpha motoneurones by selective stimulation of, for example, group Ia afferents. There are two mechanisms by which descending projections may control transmission through segmental reflex arcs: first, by their action on interneurones placed in the spinal reflex arc as shown schematically for the descending pathway identified as *A* in Fig. 11.2; second, by their inhibitory action on the terminal part of afferent fibres just before their synapse with the next neurone (presynaptic inhibition; pathway *B* in Fig. 11.2). These mechanisms are important not only because they provide the means of controlling segmental reflex effects—by the same means, the CNS can exert a descending control of afferent transmission to central structures. In this context it is interesting that the pyramidal tract should have been shown to be one of the most effective pathways by which the CNS influences reflex and afferent transmission. This aspect of pyramidal tract function is not one to which clinicians usually pay much attention but several properties traditionally attributed to this pathway now have to be abandoned.

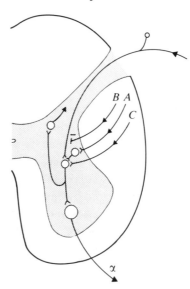

Figure 11.2. *Diagram of connections concerned in the supraspinal control of the segmental neural mechanism.*

Inhibitory interneurone is shown as filled circle. α, alpha motoneurone. The segmental reflex can be inhibited through the descending pathway *A* by means of an interposed inhibitory interneurone, or through *B* by presynaptic inhibition. The motor pathway *C* acts on the alpha motoneurone through an interneurone which is shown giving off a collateral branch projecting to supraspinal structures via a relay at segmental level, thus providing 'internal feedback'.

So far we have been concerned with how the CNS could modify and control segmental proprioceptive input within the spinal cord by its action on the reflex arc. There is a further specific mechanism for altering the proprioceptive input from the muscle spindle. Spindle discharge is determined not only by the degree of main muscle stretch but also by intrafusal muscle fibre contraction which is under the control of gamma motoneurones (Fig. 11.3). Gamma discharge causes reflex excitation of alpha motoneurones through the neural circuit which is known as the "gamma loop" (gamma motoneurone-intrafusal muscle fibre-spindle primary ending—group Ia afferent fibre—alpha motoneurone). Moreover, the CNS can adjust the dynamic response of the spindle to the mechanical stimulus by adjusting the balance between static and dynamic fusimotor excitation.

The gamma motor system therefore provides an alternative route to the "direct" descending excitation of alpha motoneurones. This "indirect" route is via the gamma loop (pathway *B*, Fig. 11.3.). The potential of the latter mechanism for functioning as a servo loop is considered separately below.

Descending motor pathways usually excite their target motoneurone via spinal interneurones which may also form part of the spinal reflex arc (Fig. 11.2). These interneurones are thus intermediary for descending inputs and for segmental inputs to

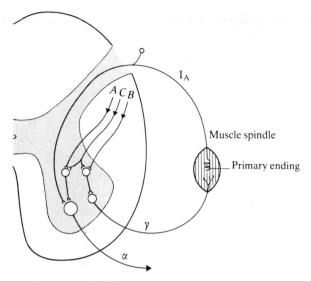

Figure 11.3. *Diagram of the 'gamma loop'.*
A, 'direct' route; *B,* 'indirect' route. *C* shows a descending pathway coactivating alpha and gamma motoneurones.

the motoneurone. But in addition, via an axon collateral, they may project to central structures through a neurone relay at spinal level (Fig. 11.2). This afferent pathway would be in a position to signal information about the state of excitation of the interneuronal pool and could thus provide "internal feedback".

Reciprocal effects on alpha motoneurones to antagonist muscles can be exerted by a single descending pathway through the interposing of inhibitory interneurones as shown in Fig. 11.4. Such an arrangement probably contributes to reciprocal activity of inspiratory and expiratory muscles, excitation of the descending "inspiratory" pathway causing excitation of inspiratory motoneurones and coincident inhibition of expiratory motoneurones.

The gamma loop as a servo system

The way in which the gamma loop could operate as a servo system can readily be appreciated. A command signal for muscle shortening, for example, directed to the gamma motoneurone would cause a length misalignment between intrafusal (spindle muscle fibres) and extrafusal (main muscle) fibres, and give rise to an error signal through spindle 1a fibres. These afferents, being excitatory to their own alpha moto-neurones, would lead to reflex contraction of the main muscle so that extrafusal muscle length tend to "follow" intrafusal muscle length. The system could thus operate as a "follow-up length servo".

While there is evidence that a follow-up length servo may operate in respiration and in the execution of certain reflex movements in anesthetized animals, there is equally strong evidence that a mechanism of this kind is not involved in voluntary movements in man. The critical observations refuting the role of the gamma loop were made with the aid of fine tungsten wire electrodes which were inserted through the skin into peripheral nerves. By careful positioning of these microelectrodes it has been possible to record impulse activity in 1a fibres from muscle spindles and to relate the discharge to the onset of motor unit activity in the muscle. In this way one can indirectly compare the timing of the discharges in gamma and alpha motoneurones. Such experiments have clearly shown that for all types of finger movement—weak or strong, rapid or slow, isometric or isotonic—the discharge of gamma motoneurones does not precede that of the alpha motoneurones but occurs at about the same time. Although gamma discharge can no longer be considered to have such a dominant role in the genesis of movement, it is important nevertheless. By causing the intrafusal fibres to contract, the sensitivity of the spindle sensory endings is maintained during contraction of the extrafusal (main muscle) fibres. Thus, the spindle can still signal information about length despite the unloading effect occasioned by the shortening of the muscle belly. Further, if a contracting muscle is momentarily stretched by an increase in load, the spindles remain able to respond and, by the stretch reflex loop, produce an appropriate response from the alpha motoneurones. Finally, the discharge from the 1a fibres provides the motoneurones with potent background facilitation so that any further descending inputs required for the continuation of the contraction are made corre-

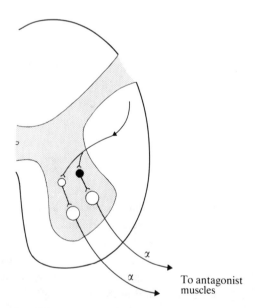

Figure 11.4. *Diagram to show how a descending pathway can exert reciprocal effects on antagonistic muscles.*
Inhibitory neurone shown by filled circle.

spondingly more effective. This facilitatory action can be demonstrated by the production of a "silent period" in the EMG when, during a static (postural) contraction, the muscle is suddenly shortened either by stimulation of the motor nerve or by removal of the load against which it was acting. The silent period develops too soon for this to be a voluntary response but it can be explained by the sudden removal of spindle 1a excitation to the alpha motoneurones as a result of main muscle shortening.

DESCENDING MOTOR PATHWAYS

Corticospinal pathway (pyramidal tract)

There is a mistaken tendency, particularly in clinical circles, to simplify the understanding of motor control by regarding the pyramidal tract as the only descending voluntary motor pathway. This tract derives its name from the fact that its constituent nerve fibres pass through the medullary pyramids on their way to the spinal cord. In man, each tract comprises about 1 million fibres, mostly small myelinated ones but a few (2 per cent) being much larger and fast-conducting. These last fibres arise from the large neurones (Betz cells) in the deeper layers of the motor cortex. Important though the pyramidal tract is for movement (see below), it is not the only descending motor pathway nor are its functions exclusively motor. Thus, the pyramidal tract gains as many fibres from other regions of the cerebral hemisphere, particularly the parietal lobe, as from the motor cortex. Further, many of the fibres are destined for cell groups in the dorsal horn of the spinal cord and in the dorsal column nuclei, both of which process information from incoming somatosensory pathways. These corticofugal "sensory" fibres can control the amount of information being passed on to the brain; in doing so, it is probably that they instruct the sensory nuclei that a movement is about to take place.

In addition to its "sensory" projections, the pyramidal tract also supplies a number of cell groups in the brainstem, including the caudate nucleus, thalamus, red nucleus, pontine nuclei, olive and lateral reticular nucleus (Fig. 11.5); the significance of these connections will be considered later.

Turning now to the "motor" fibres in the pyramidal tract, we find that these are mostly distributed to internuncial neurones lying in the intermediate part of the gray matter of the spinal cord, immediately behind the motoneurones. Some of these interneurones are facilitatory and others inhibitory to particular groups of motoneurones; it is the interneurones which are able to coordinate the motoneurones so that appropriate numbers of motor units in agonist and synergistic muscles are called into activity, while motoneurones of antagonist muscles are inhibited. Most of the projection is, of course, to the side of the spinal cord opposite to the motor cortex, but about 10 per cent of fibres do not participate in the decussation of the medullary pyramids, and terminate ipsilaterally. These uncrossed fibres may well be important in the preservation of some residual movement in patients with hemiparesis following stroke. In

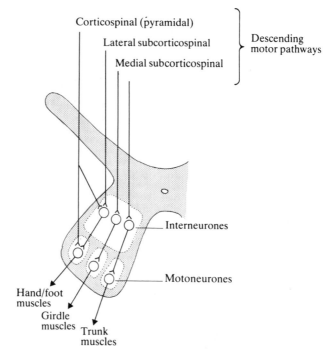

Figure 11.5. *Descending motor pathways and their terminations in the spinal cord.*
See text.

man and in other primates, however, the most important function of the pyramidal tract is the control of finger movements and the tract is therefore indispensable for the performance of such skilled tasks as writing, typing, tying knots, fastening buttons and playing musical instruments. This special responsibility of the pyramidal tract for digital skills is reflected in the fact that the alpha motoneurones for finger and wrist movements, unlike other motoneurones, receive direct excitatory synaptic connections from pyramidal tract fibres. These connections also account for the long-standing observations that hand movements have a low threshold to electrical stimulation of the cerebral cortex and that, following a destructive cortical lesion in man, it is the fine movements of the hand which show greatest impairment.

Subcorticospinal pathways

The subcorticospinal pathways, in distinction to the pyramidal tract, arise from neur-ones in the brainstem. On the basis of their terminal distribution within the spinal cord, the subcorticospinal pathways have been grouped into medial and lateral systems.

The *medial group* of fibres originates in the superior colliculus (tectospinal tract), the interstitial nucleus of Cajal (medial longitudinal bundle), the vestibular complex (vestibulospinal tract) and the pontine and medullary reticular formations (reticulospinal tract). Most of the fibres in these descending tracts terminate on neurones within the ventromedial part of the internuncial zone (Fig. 11.5); unlike the pyramidal tract, very few subcorticospinal fibres project directly to motoneurones.

The *lateral group* of subcorticospinal fibres is less well defined but certainly includes the axons originating in the red nucleus (rubrospinal tract). These fibres also end on interneurones, but only in the dorsilateral part of the internuncial zone. Lawrence and Kuypers (1968), on the basis of ablation experiments in monkeys, have summarized the attributes of the different descending motor pathways in the following way. "The medial brainstem pathways function as the basic system by which the brain exerts control over movement. This control is especially concerned with maintenance of erect posture, integrated movements of body and limbs and with directing the course of progression. The lateral brainstem pathways, at least in regard to the extremities, superimpose upon the above control the capacity for the independent use of the extremity, particularly of the hand. The corticospinal (*pyramidal tract*) connections mediate a control similar to that of the brainstem system but, in addition, provide the capacity for further fractionation of movement as exemplified by individual finger movements".

Two points bear repetition. First, the system of internuncial neurones lying posterior to the motoneurones has a very important role in further processing descending messages and thereby ensuring that the appropriate pattern of motoneurones is recruited while antagonist neurones are inhibited. Secondly, in all movements there is coactivation of the alpha and and gamma motoneurones so that the sensitivity of the muscle spindles is maintained during contraction of the main muscle mass.

Having dealt with the descending motor pathways, it is now necessary to attend to the cell groups in the brain from which these fibres arise. Although, for clarity, each structure will be considered separately, it must be understood that they combine their activities whenever movement is to be attempted.

MOTOR CORTEX

The strip of cortex lying immediately in front of the central fissure, (precentral gyrus), has long been recognized as important for movement. Clinical observations made on patients with focal damage to this area continually remind us that the motor cortex is concerned with activities of the contralateral limbs and that the "map" of movements is inverted on the precentral gyrus. Electrical stimulation of this region, either in conscious patients undergoing surgery or in lightly anaesthetized primates, can evoke contractions on the opposite side of the body. Although the extent of the movements depend on such factors as the depth of anaesthesia, the starting position of the limb, and the characteristics of the stimulus, it has been consistently noted that contractions of the thumb and mouth are most easily elicited. This observation is related to the relatively large areas of motor cortex devoted to the thumb, fingers and face; this spatial dominance is, of course, a reflection of the importance of the digits in tactile exploration and fine movement and of the lips in speech and facial expression.

Within the motor cortex the cells that give rise to the descending axons of the pyramidal tract are found in the deeper layers of the cortex, particularly in layer V. The larger cells with fast conducting axons are silent at rest but give a brisk discharge if movement is to be made. In contrast, most of the smaller cells, from which the slower-conducting pyramidal tract axons arise, are continually active irrespective of the intended movement.

The output from the motor cortex to spinal motoneurones has already been considered in relation to the pyramidal tract. The motor cortex also makes three other types of connection which are important for movement. The first of these is to two brainstem nuclei which themselves give rise to descending motor fibres. These are the red nucleus and the lateral reticular nucleus. Secondly, the motor cortex is connected to the cerebellum and, in return, receives information back from that structure. Figure 11.6 shows that the input to the cerebellum is relayed through the pontine nuclei; new fibres arise which cross the midline to end as mossy fibres within the cerebellar cortex. The pathway from the cerebellum to the motor cortex is also interrupted by a synaptic relay in the brainstem, this time in the ventralis lateralis (VL) nucleus of thalamus. The third projection from the motor cortex is to the caudate nucleus and putamen; like the cerebellum, these nuclei feed information back to the cortex and use the globus pallidus and VL nucleus of the thalamus to do so (Fig. 11.6). The connections of the motor cortex which have been described serve two purposes. Through the projection to the spinal cord the cortex calls motoneurones into activity, allowing the movement to begin. Secondly, through the pathways to the cerebellum and basal ganglia, the cortex informs key regions of the brain of the way in which the movement has been planned. In turn these structures, through their respective feedback loops to the motor cortex, can control the further course of the movement.

BASAL GANGLIA (Fig. 11.6)

Despite the large sizes of the nuclei which make up the basal ganglia, relatively little is known about their function. Until recently, the only evidence that one of their roles is the control of movement came from the study of patients with such disorders as Parkinsonism, Huntington's chorea and hemiballismus (see below). The most prominent parts of the basal ganglia are the caudate and lentiform nuclei which, together with the claustrum, comprise the corpus striatum. The division of the lentiform nucleus into the putamen and globus pallidus is justified on morpholigical and neurophysiological grounds. Thus, while the putamen resembles the caudate nucleus in its fibre connections and in having predominantly small neurones, the globus pallidus differs from both in having many large cells and contrasting projection. Whereas the caudate nucleus and putamen receive information from all regions of the cerebrum,

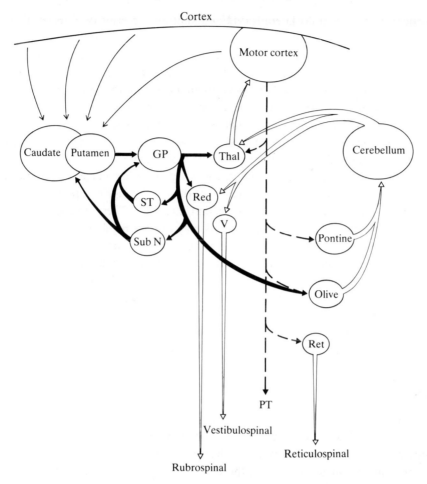

Figure 11.6. *Main neural pathways involved in the preparation and execution of voluntary movements.*

GP, globus pallidus; *PT,* pyramidal tract; *Ret,* mesencephalic reticular formation; *ST,* subthalamic nucleus; *Sub N,* substantia nigra; *V,* vestibular nuclei.

See text.

including the motor cortex, they send almost their entire output to the globus pallidus. It is now known that there is an orderly map of projections from the cerebral cortex upon the caudate and putamen; it is the latter nucleus which receives the fibres from the motor cortex. Although the cortex provides the major input to the basal ganglia, there is also an important projection from the substantia nigra; this last pathway is a dopaminergic one and is affected in patients with Parkinsonism.

The output from the basal ganglia is channelled through the globus pallidus and part of the substantia nigra; these two nuclei exhibit a number of similarities and are now regarded as comprising one functional unit. Both structures send their main projections to the ventralis lateralis (VL) and ventralis anterior (VA) nuclei of the thalamus. In turn, the VL nucleus makes powerful excitatory synaptic connections with cells of the motor cortex while VA projects to the premotor area. The VL thalamo-cortical projection thus completes the following feedback loop with the motor cortex: motor → putamen → globus pallidus → VL (thalamus) → motor cortex. The globus pallidus also projects to the subthalamic nucleus, red nucleus, mesencephalic reticular formation, inferior olive and hypothalamus. Through the connection to the red nucleus, and thence to the rubrospinal tract, the basal ganglia are able to influence motoneurone excitability without involving the motor cortex.

The proximity of the subthalamic nucleus to the other nuclei considered above suggests that it should be considered as one of the basal ganglia; it receives fibres from the globus pallidus and projects back to that structure. These anatomical considerations indicate that the subthalamic nucleus may exert a regulatory function over the output nuclei (globus pallidus and substantia nigra) of the basal ganglia. A lesion of the subthalamic nucleus, either in man or in monkeys, produces gross involuntary movements of the contralateral limbs termed hemiballismus.

It is still clinical custom to class disorders involving the basal ganglia as "extra-pyramidal". While this term has a certain convenience, its rigid application is precluded by the close functional interrelationship now recognised to exist between the motor cortex and pyramidal tract, on the one hand, and the basal ganglia, on the other.

CEREBELLUM

Many anatomical and physiological details of cerebellar circuitry have been worked out in the past few years and have invited hypothesis about the function of the cerebellum in the control of movement. The cerebellum is extraordinarily uniform in its anatomical arrangement and can be thought of as being composed of a large number of identical units of the type schematically shown in Fig. 11.7. The only output is inhibitory through the Purkinje cell which is central to each unit. The inhibitory action of Purkinje cells is exerted on the cerebellar nuclei and the vestibular nuclei. There are two possible routes of input (1) the climbing fibres whose cells of origin lie in the inferior olivary nucleus in the medulla and which synapse with the Purkinje cell; (2) the mossy fibres which synapse with the granule cells in cerebellar cortex. The axons of the granule cells from "parallel fibres" which are the chief interconnections of individual Purkinje cells. The parallel fibres also synapse with the Golgi cells and stellate cells, both of which are inhibitory, the former acting on granule cells and the latter on Purkinje cells.

The main sources of cerebellar input (Fig. 11.6 and 11.7) are cerebral cortex, spinal interneurones and peripheral receptors. The pyramidal tract neurones, through collateral branches to the inferior olivary and pontine nuclei, project to the cerebellum via both climbing fibre and mossy fibre inputs. This circuit arrangement would clearly allow the cerebellum to provide internal feedback for output from cerebral cortex, and

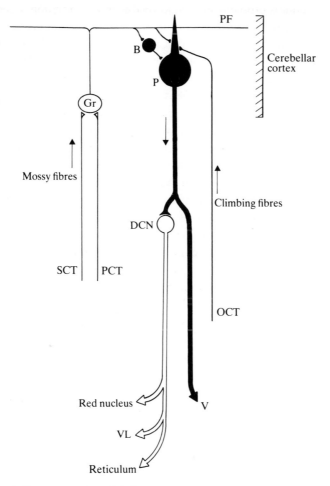

Figure 11.7. *Cerebellar circuitry (simplified).*
  B, basket cell; *OCN*, deep cerebellar nuclei; *Gr*, granule cell; *OCT*, olivocerebellar tract; *P*, Purkinje cell; *PCT*, pontocerebellar tract; *PF*, parallel fibres; *Ret*, reticular formation; *SCT*, spinocerebellar tracts; *V*, vestibular nuclei; *VL*, ventralis lateralis nucleus of thalamus.
  See text.

thus to correct errors in motor activity elicited from cerebral cortex and carried out by command signals through the pyramidal tracts and subcorticospinal tracts. The cerebellum could act as a comparator of information derived from three sources, namely "command" signal from the motor cortex, signals from spinal interneurones which are receiving both segmental and descending inputs and information from peripheral receptors. From these comparisons, the cerebellum could compute and initiate appropriate corrections through its output connections.

  As well as acting to correct errors in motor activity elicited from the cerebral cortex, the cerebellum is probably also concerned with the initiation of movement. This hypothesis was originally put forward to provide an explanation for the disturbance in initiating movement which is noticeable in patients with cerebellar lesions. It receives some support from recent studies in patients with Parkinsonism undergoing stereotaxic operations. Some neurones recorded from the ventralis lateralis nucleus of thalamus are found to be activated by voluntary movement but not by passive movements of the limbs or by muscle palpation; this suggests that they might be concerned in the initiation of the movement rather than with the sensory response to the movement itself. Other studies (see below) have shown that the cerebellum plays a crucial role in setting the times for the agonist and antagonist motoneurones to discharge when rapid movements are made. Also relevant are the findings from animal investigations; these have demonstrated that Purkinje cell discharge precedes movement, but the timing in relation to thalamic and pyramidal tract neurones has not yet been established. In this connection, too, it may be significant that neurones of the ventralis lateralis nucleus of the thalamus exert a very powerful excitatory effect on pyramidal tract neurones.

The cerebellum is also thought to be concerned in the learning of skilled movements. The problem of how this is done has led control and communication theorists to develop models based on the details of cerebellar circuitry established by neurophysiologists. Such a liaison is probably essential for solving the complex problems in this field, and because these models are open to experimental testing, they provide a further stimulus for experiment. The following recently proposed model is an example. In this model, all movements it is assumed, can be broken down into "elemental movements" so that to learn to carry out an action it is necessary to learn to copy the correct sequence of elemental movement. Two main theoretical possibilities then exist for the execution of a movement; at the appropriate instant, either the required "elemental movement" can be turned on or all the unrequired "elemental movements" can be turned off. Cerebellar design is thought to make the latter alternative the more attractive. The model proposes that large (deep) pyramidal tract neurones can drive elemental movements and that the set of situations in which these cells respond includes those where they are needed. These neurons receive positive feedback (which would be necessary for the initiation and maintenance of movement) through the loop involving pontine and cerebellar nuclei and ventralis lateralis nucleus of thalamus (see Fig. 11.6). This loop has a suitably short reflex time, perhaps as little as 6 msec. The recognition of which elemental movements should be turned off is the task of the small (superficial) pyramidal tract neurones, which can potentially prevent firing of the large pyramidal tract neurones through their action on Purkinje cells. How the superficial pyramidal tract neurones recognize what corrections should be made is likely to depend on a complicated analysis for "it involves ideas about what the animal is trying to do".

But cerebellar circuitry is such that it could "learn" to link the results of this analysis (i.e. which corrections should be made) with the specific context in which the movement might occur, and to store the resulting information. Central to this process of learning is the prediction, as yet unproven, that the synaptic connection between the Purkinje cells and the parallel fibres (Fig. 11.7) is facilitated when the parallel fibre is active at about the same time as the climbing fibre to that Purkinje cell. By this means, the Purkinje cell could learn to recognize such contexts, contextual information being provided by the massive afferent inflow to mossy fibres.

CLASSES OF MOVEMENT

Respiratory movement

Respiration provides an example of an automatically initiated movement, which has features of central-patterning. More is known at present about the role of peripheral feedback than internal feedback in the regulation of the central respiratory drive, but the latter may prove to be at least equally important.

The rhythmic central respiratory drive originates in chemo-sensitive brainstem structures, whose detailed workings need not concern us here. The nature of the central respiratory drive is such that excitation is delivered to one set of respiratory moto-neurones (e.g. inspiratory) and coincident inhibition to the antagonist set. These reciprocal effects are probably exerted at spinal levels through inhibitory interneurones as shown on page 388. A tight alpha-gamma linkage, i.e. a close co-activation or co-inhibition of alpha motoneurones and fusimotor neurones, characterizes the central respiratory drive. Thus excitation in a descending "inspiratory" pathway distributes via interneurones excitation both to alpha and gamma motoneurones innervating inspira-tory muscles, and inhibition to expiratory alpha and gamma motoneurones. Reflex excitation round the gamma loop (servo-drive) is unquestionably significant in the control of intercostal muscle activity but not essential because weakened movement continues when the loop is opened. In the case of the diaphragm, alpha drive probably predominates. In both muscles, proprioceptive input from tendon organs is important, especially so for the diaphragm.

Respiratory muscles are active in other non-respiratory movements such as those of posture and volition. The drives serving these functions are in part at least carried by independent descending pathways and thus interact with the respiratory drive and with each other at spinal level as well as centrally. Intercostal muscles in particular have an important postural role. One way in which the cerebellum can control their activity is by adjusting the balance of the alpha-gamma linkage and some intercostal gamma mononeurones are in fact dominated by their cerebellar input, receiving no effective respiratory modulation.

Viscerally-evoked movement

This group of movements includes acts such as coughing, swallowing, sneezing, hiccup and vomiting. The muscles of the abdominal wall, the intercostal muscles, the diaphragm, laryngeal and pharyngeal muscles, and the tongue are the principal somatic muscles involved. Although these acts are sometimes described as "reflex" this term is probably inappropriate when used as defined earlier. They are better described as centrally-patterned movements of the triggered type for it is characteristic of these movements that features of the input are not easily identifiable in the pattern of the output. A brief mechanical stimulus either to larynx or trachea, for example, evokes a complex cough response involving partial co-activation of inspiratory and expiratory muscles and carefully timed control over laryngeal opening. Feedback may not be very significant in this kind of movement; certainly experimental interruption of potential feedback loops has little effect on the act of swallowing.

Postural movement

In a sense, postural movements are voluntary in that the organism wills the postural goal (to stand, to sit, etc.). But, once learned, the movements whereby this goal is achieved and maintained do not have to be individually willed. Reflex mechanisms involving the spinal proprioceptive apparatus are of particular importance, but they require supraspinal control to be effective. The chronic spinal animal cannot stand and even the decerebrate animal (low mid-brain transection) with its exaggerated stretch reflex and well-developed crossed extensor reflex, can only stand when appropriately placed and is easily toppled over. Removal of cerebral cortex alone, however, leaves the animal able to stand and walk. Structures particularly concerned with exerting the descending control of the spinal proprioceptive mechanism during the maintenance of posture are the vestibular mechanism, which has close connections with cerebellum, brainstem nuclei and the basal ganglia. (One of the main deficits in patients with Parkinsonism is the disturbance in postural control.) Feedback plays an important part in the automatic corrections necessary in the maintenance of posture. In man, for example, the gamma loop is known to be active in a postural contraction, and removal of the cerebellum, which is involved in internal feedback, leads to severe disorders of postural control.

Walking

Walking is of interest in demonstrating the importance of supraspinal structures in facilitating quite complex motor programmes organized within the spinal cord. Animals which have received transections of the neuraxis at the level of the spinal cord or midbrain are still capable of walking if supported by a harness and placed on a treadmill. However, the spinal mechanisms for this coordinated activity can only become effective if the transected descending motor pathways are activated, either by electrical stimulation or by the injection of L-dopa.

The interpretation of these findings is that, within the mesencephalic reticular formation, there is a group of monoaminergic neurones which are concerned with the control of certain spinal reflex mechanisms including walking. The axons of these cells project to interneurones within the cord and release norepinephine; possibly serotoninergic neurones are also involved. These descending pathways raise the excitability of the walking-interneurones; additional excitation is received from afferent fibres of the limbs, particularly those from muscle spindles and Golgi tendon organs. Once the excitability of the cord is high enough, an automatic sequence of neuronal discharges takes place. As one population of alpha and gamma motoneurones fires the local interneurones ensure that antagonist motoneurones are simultaneously inhibited and that preparations are made for exciting the motoneurones responsible for the next phase in the stepping motion. The interneurones must therefore control not only extensor and flexor motoneurones supplying muscles acting on a single joint, but also those producing movement at other joints in the same limb; further, at each phase of the stepping cycle there must be appropriate excitation or inhibition of motoneurones serving the other limbs. It seems remarkable that so much detailed programming should have been built into the spinal cord; it is chastening that some of the key observations and speculations on the phenomenon of automatic walking were made 70 years ago; since that time the nature of the neural control circuits has remained unknown although several attractive models have been proposed.

Voluntary movement

Voluntary movements are superimposed upon other motor demands such as those of posture or breathing, but are not, of course, always dominant. During breathholding, for instance, a "willed" expiratory movement is soon involuntarily subject to an increasing rhythmic respiratory modulation. The interactions between these separate and often conflicting demands will be occurring at the spinal level as well as centrally. Voluntary movements differ from respiratory, postural and viscerally-elicited acts in requiring the cerebral cortex for their normal execution. One could expect that both internal and external feedback would be particularly important in this type of movement which demands the greatest accuracy in performance.

Although we do not understand how the central nervous system decides upon and executes a voluntary movement, a number of clues are available to suggest some of the mechanisms involved. In the first place, it must be recognised that the discharge from the motor cortex into the pyramidal tract only occurs late in the proceedings at a time when other regions of the brain have completed much of the planning. Even in reaction time experiments, in which a subject must respond in the shortest possible time to an external stimulus, approximately a tenth of a second is available for neuronal transactions to be conducted within the brain. This time, although brief, would be sufficient for neural circuits involving up to 50 synapses to be operating; probably only the last few connections in the pathway are within the motor cortex. If not in the motor cortex, in which part of the brain, then, might the desire to move arise? Perhaps the most likely answer would be one of the association areas, since it is the evolutionary development of these regions in the primate brain which parallels the acquisition of thought processing. Recent microelectrode studies in unanaesthetized monkeys have, in fact, revealed a population of neurones in the posterior parietal area which discharge well in advance of a movement. Of particular significance is that the movements appear to be spontaneous; similar movements taking place in response to external stimuli are not associated with activity of these neurones. Studies of patients with apraxia suggest that following the formulation of the desire to move, the first steps in planning the movement may also be conducted in the parietal lobe. In these patients it is impossible to perform certain purposive movements, the nature of which is understood, despite the absence of paralysis, ataxia or sensory loss. In right-handed persons the lesions are invariably located in the left parietal lobe. As the planning of the movement proceeds other parts of the brain, including the basal ganglia and cerebellum, are called into activity; this preparation for movement can be detected electrophysiologically in the form of a gradually increasing negativity (readiness potential) which is largest over the vertex. The role of the basal ganglia in voluntary movement is least well understood. Clinical observations on the akinesia of Parkinsonian patients suggests that the substantia nigra has some sort of priming action. It can be shown that, despite the absence of willed movement, the appropriate motor programmes are intact in the Parkinsonian brain. Thus, a patient unable to move may catch a ball thrown unexpectedly in his direction and may also walk if a pattern of stripes, or a series of minor obstacles, are put in his path. More dramatic still is the effect of L-dopa in permitting the resumption of normal motor activities.

If the movement is to be a sudden (ballistic) one, the pattern of motoneurone involvement falls into three parts. First, there is excitation of agonist motoneurones so as to move the limb in the appropriate direction. Next follows excitation of antagonist muscles so as to slow the limb as it nears its target; lastly, there is another burst of activity in agonist muscles as the final adjustment of the limb is made. In each phase of the movement there is co-activation of alpha and gamma motoneurones. Not only are these three components programmed in the brain but the duration of the first agonist contraction is fixed and independent of the distance to be moved. As a consequence a large movement is achieved by the excitation of additional motoneurones, thereby enabling more force to be developed within the same interval of time; hence a correspondingly greater propulsion of the limb is achieved. Although the three phases of movement are not independent of spinal influences, they still occur if the subject is denied information from the periphery, as in the patient with a severe peripheral neuropathy. From studies of patients it would appear that one of the roles of the cerebellum is to fix the lengths of the phases of excitation in agonist and antagonist motoneurones. In patients with cerebellar disease the initial phase is sometimes lengthened and the movement correspondingly prolonged, thereby causing overshoot; on other occasions the inhibitory phase is lengthened, producing undershoot.

The neural system underlying slow movements differs from that considered above, for it is often only possible to detect activity in agonist muscles without any checking being exerted by antagonist muscles. Animal studies suggest that the putamen is specially concerned with movement of this type, whereas it has little interest in ballistic tasks; the cerebellum is active in both types of movement.

Once a movement is underway the controlling mechanisms become operational at several levels. The cortical neurones receive positive feedback during the execution of the movement through the collateral pyramidal tract loop involving the cerebellum and ventralis lateralis nucleus of thalamus. The cerebellum can also act to correct errors during the execution of the movement on the basis of internal feedback from spinal interneurones, and on peripheral feedback from muscle, joint, and skin receptors. The alpha-gamma linkage in the motor drive allows the gamma loop, acting as a servo mechanism, to stabilize the movement against expected loads, but this servo can be interrupted by autogenetic inhibition in the event of unexpected load changes.

Disordered function

LOWER MOTONEURONES

In seeking the cause of muscle weakness it is still convenient to follow traditional neurological practice by considering whether the lesion is more likely to be of the upper or of the lower motoneurone type. For this purpose the term "lower motoneurone" is used synonymously with "motor unit" to encompass not only the motoneurones in the brainstem and spinal cord but their axons, the neuromuscular junctions and the muscle fibres. In most patients the distinction between upper and lower motoneurone lesions can be readily made. In a patient with a lower motoneurone lesion there is gross muscle wasting, caused partly by disuse but mainly by the loss of trophic factors normally supplied to the muscle fibres by the motoneurones. The tendon reflexes are depressed or absent, partly because of the smaller main muscle mass available to respond in the reflexes, and also because the intrafusal muscle fibres are usually involved in the disease process. Fasciculations may also be seen in patients with lower motoneurone disturbances and are contractions of single motor units; the impulses responsible for the contractions arise spontaneously in the motoneurones or in the distal parts of their axons, probably at the point where the axons begin to divide within the muscle belly; occasionally they result from irritation of ventral root fibres.

UPPER MOTONEURONES

The term "upper motoneurone" is preferable to "pyramidal tract" because the signs traditionally described under the latter heading are not those of a pure pyramidal tract lesion but rather one involving both the pyramidal tract and the subcorticospinal pathways. In using the term "upper motoneurone" to describe the long motor tracts, one should remember that most of the fibres of this projection, in their course from cortex to alpha and gamma motoneurones, are interrupted by relays with neurones of the basal ganglia and brainstem nuclei, and with interneurones at spinal level. A single "upper motoneurone" with its cell body in the cerebral cortex and its axon making direct synaptic contact with a lower motoneurone is uncommon, but such a mono-synaptic corticospinal pathway does exist to motoneurones of muscles moving the fingers. This accounts for the clinical observation that individual movements of the fingers are those most affected by lesions directly involving cerebral cortex.

The features of an upper motoneurone lesion are weakness, spasticity, increased tendon reflexes and extensor plantar responses. Spasticity and increased reflexes may be absent in the intitial period of shock following an acute lesion. Studies in primates indicate that the depression of tendon reflexes during this early phase is largely a consequence of reduced gamma motoneurone excitation. The weakness is most marked in the extensor muscles in the upper limb and in the flexor muscles in the lower limb. The reason for this is the inhibitory bias on the antigravity muscles exerted by the long motor tracts, so that their interruption reveals a relative strength in these muscles. Individual movements of the fingers are usually the last to recover. Habitually bilateral movements of the trunk and upper face are relatively spared the effects of lesions of upper motoneurones. Recordings from paretic muscles reveal that motoneurones are recruited with difficulty and, once activated, may continue to discharge beyond the intended conclusion of a movement; this overflow of excitation aggravates the diffi-culty in changing the direction of a movement.

Spasticity is characterized by an increased resistance to passive stretch which becomes more pronounced as the speed of the stretch is increased. In some muscles, the increase in resistance is followed by sudden relaxation, the clasp-knife effect. This is usually easily demonstrable in biceps and quadriceps muscles but does not occur, for example, in hamstrings. The clasp-knife effect in the quadriceps is probably due to the inhibition exerted on the alpha motoneurones mainly by Golgi tendon organ afferents. In the biceps and triceps muscles, on the other hand, the clasp-knife effect probably has a different cause, being a consequence of the velocity-dependent nature of the stretch reflex. The slowing of the movement brought about by the reflex contraction causes in turn a reduction in the reflex response and therefore a decrease in the resistance. Spasticity indicates damage to the subcorticospinal pathways and does not occur with a pure pyramidal tract lesion. It is most prominent in the flexors of the upper limbs and in the extensors of the lower limbs, and is a manifestation of hyperactive stretch reflexes. This hyperactivity is a consequence of an increased excitability of both alpha and gamma motoneurones. It is possible to reduce the contribution of the gamma neurones to spasticity by blocking impulse conduction in the gamma axons with local anaesthetic. In most patients spasticity is greatly diminished and in some a surprisingly large residuum of voluntary movement is revealed.

Increased tendon reflexes have a similar cause to spasticity. These are "release" effects following destruction of neural tissue, i.e. they occur as a result of interruption of an inhibitory process active in the intact animal. This leaves unopposed a tonic facilitation originating from lower structures, in particular from the vestibular system.

The extensor plantar response is confusingly named because the response is really part of a protective flexion withdrawal reflex. This is present in the new-born but is soon suppressed by the descending motor control system in the process of learning to stand and walk. Interference with this descending pathway, as in a lesion of the upper motoneurones, releases the more primitive flexion reflex.

EXTRAPYRAMIDAL SYSTEM

Degeneration of the nigro-striatal pathway is the most constant pathological finding in idiopathic and post-encephalitic Parkinsonism, and appears to be at the centre of the functional disturbance. The neurones forming this pathway use dopamine as their neural transmitter and it was the observation that dopamine was deficient in the corpus striatum in patients with Parkinsonism which led to the therapeutic use of L-dopa. The clinical features of Parkinsonism are rigidity, disturbance in postural control, hypokinesia and tremor. There is some weakness but the reflexes are not consistently altered. The rigidity of Parkinsonism is characterized by smooth resistance to passive displacement (lead-pipe or plastic rigidity). In many patients this rigidity is regularly interrupted, giving rise to the "cog-wheel" effect. The latter is not due to a superimposition of tremor because the frequencies are different.

The rigidity of Parkinsonism resembles the spasticity which follows an upper moto-neurone lesion in reflecting overactivity of both alpha and gamma motoneurones. Like spasticity it can be reduced by blocking impulse conduction in gamma motor axons, thereby indirectly diminishing the background reflex excitation of alpha motoneurones by muscle spindle afferent fibres.

Tremor of Parkinsonism has a frequency range of 3–7 c/s, and does not occur synchronously throughout the body. It is absent during sleep, is most prominent at rest and is diminished by voluntary movement. Unlike the rigidity of Parkinsonism, the tremor is largely independent of background facilitation of alpha motoneurones by muscle spindle afferents since it persists after the gamma axons have been blocked by local anaesthesia.

Microelectrode recordings made in patients undergoing stereotactic surgery for Parkinsonism indicate that the pacemaker for tremor is situated in the brainstem and involves the ventralis lateralis nucleus of the thalamus. In this nucleus cells are found to discharge at the frequency of the tremor in one of the contralateral limbs. The fact that the same neurones are unaffected by passive manipulation of the tremulous part and that their discharges precede the resumption of tremor after a period of quiescence suggests that their activity is involved in the genesis of the tremor rather than being a consequence of the latter. In experimentally induced Parkinsonism in primates, neurones of the motor cortex can also be observed to discharge in a similar manner. This oscillation in the discharge pattern of certain thalamic neurones appears to come about as the result of interference with the corpus striatum. When one remembers that the ventralis lateralis nucleus of thalamus projects strongly to cortical neurones, forming an important pathway in the control of motor activity, it is not surprising that periodic discharge of these neurones leads to a tremor discharge in alpha motoneurones. Stereotaxic lesions placed in the ventralis lateralis nucleus of thalamus can abolish tremor in the contralateral limbs.

Hypokinesia means slowness in initiation of movement, poverty in its range and difficulty in its rapid repetition. Posture and balance are also defective, the patient being unable to compensate for imposed postural displacement. The gait is slow and shuffling, or sometimes festinant, and arm swinging is impaired.

From questioning Parkinsonian patients, it appears that the difficulty in initiating movement is not due to any loss of will or to an inability to decide what must be done; rather, it results from a failure of those motor mechanisms, operating at a subconscious level, which would normally implement a decision. A patient with Parkinsonism finds, to his consternation, that he has to will each part of a movement which would normally have been undertaken in a smooth and automatic manner. Unbroken motor sequences reappear after L-dopa administration and may also do so if the patient is suddenly taken unawares by a novel stimulus. Other strategies may also temporarily restore movement—for example, bouncing a ball may enable a previously immobile Parkinsonian patient to walk, as may the placement of small obstacles in his path. This last observation suggests that visual information is of unusual significance in the disordered motor transactions of the Parkinsonian brain.

Ballism is the term used to describe sudden involuntary flinging movements of a limb. Most commonly it is one arm which is affected but sometimes the ipsilateral leg is also involved, the condition then being termed *hemiballism* or *hemiballismus*. It is possible to deduce some of the underlying pathophysiology by considering the nature of the movement; it must clearly result from a sudden synchronous discharge of motoneurones innervating synergistic groups of limbgirdle muscles. Both the globus pallidus and the ventralis lateralis nucleus of the thalamus appear to be necessary for these movements to occur on the opposite side of the body. In most, but not all, cases the pathological lesion involves a small part of the subthalamic nucleus; curiously, a larger lesion, produced stereotactically, may be effective in abolishing the abnormal movements.

Chorea is another disorder of the basal ganglia. Thus, in Huntington's chorea there is a striking loss of small neurones in the caudate nucleus while in Sydenham's chorea there are necrotic areas in this nucleus, with the lentiform nucleus and surrounding areas of brain also being commonly involved. Undoubtedly, the most common cause of chorea at the present time is L-dopa therapy for Parkinsonism and neuroleptic drugs for the management of psychotic patients (see below). It is characteristic of choreic movements that they are brief and jerky and, at first sight, may appear to have been intended by the patient. They may include facial grimaces, rapid protrusion of the tongue and gesturing with the arms; sometimes the patient may simply appear to be fidgeting excessively. Although brief, the movements possess a cohesion which places them beyond such elemental disorders as tremor and ballism.

Athetosis is another sign of extrapyramidal disturbance and is closely related to chorea. Athetoid movements are rather slower than those of chorea, however, and, in the case of the hands, typically take on a sinuous writhing quality. The hands are held pronated with the wrists flexed and the fingers hyperextended. The feet may also be involved, as may the jaw, tongue and lips. Not only do the athetoid movements prevent the patient from maintaining a fixed posture but they are precipitated by attempts to make voluntary movement. In most of the brains examined at autopsy the lesions have involved the putamen or globus pallidus. From what is known of the different terminations of the descending motor tracts within the spinal cord, it is probable that athetosis results from abnormal excitatory discharges in the later subcorticospinal pathways (see Fig. 11.5).

Torsion dystonia is the name given to slow twisting or turning movements of the head, neck, trunk or limbs. They can occur spontaneously or may follow the onset of voluntary movement. The torsion movements are very powerful and must therefore involve most of the neurones in the respective motoneurone pools; the distribution of the involved musculature suggest that the abnormal descending discharges utilize the medial subcortiospinal motor pathways (Fig. 11.5). Electromyographic recordings indicate that there is co-contraction of agonist and antagonist muscles. No structure has yet been identified as being uniquely responsible for the genesis of dystonia; instead, autopsy examinations have incriminated the caudate nucleus, putamen, deep cerebellar nuclei and centromedian nucleus of the thalamus. Large bilateral stereotactic lesions of the ventralis lateralis nuclei may be effective in abolishing the disorder, either by interrupting the dystonic motor pathway or by reducing background facilitation.

Spasmodic torticollis, although often regarded as an hysterical trait, is probably nearly always a variant of torsion dystonia. In this condition the head may be bent over to one shoulder, turned to one side, or flexed backwards. Similar postural abnormalities can be produced in animals by lesions placed in the brachium conjunctivum, and there is other evidence to suggest that the cerebellum, and particularly the interstitial nuclei, may subserve this condition. In keeping with this interpretation is the observation that torticollis can be improved following interruption of the cerebellar output to the motor cortex, by a lesion placed in the ventralis lateralis nucleus of the thalamus.

In many of the patients suffering from one of the movement disorders described above, it is difficult to identify a single nucleus or tract which appears to have undergone degeneration from one cause or another. Perhaps this difficulty is not surprising, for experience in drug therapy over the past two decades has shown that movement disorders can be readily produced in patients with psychiatric disorders by manipulating the potency of the dopaminergic and cholinergic pathways in the brainstem. The possibility therefore exists that in patients with spontaneously occurring disorders there may be relatively subtle aberrations involving such mechanisms as the defective synthesis or release of neurotransmitters, or reductions in available receptors. In considering these drug actions, it is now recognised that not only L-dopa but all the major antipsychotic agents are capable of inducing abnormalities of muscle tone or movement. For example, reserpine, tricylic antidepressants and the phenothiazines may all produce a Parkinsonian syndrome in which rigidity and hypokinesia are usually more marked than tremor. In these patients it appears that the balance between dopaminergic and cholinergic systems is altered in favour of the latter. In contrast, treatment with L-dopa, phenothiazine, or butyrophenone may produce a variety of abnormal movement syndromes including facial tics, chorea, athetosis, spasms of individual muscle groups, torsion dystonia and generalized motor restlessness (akathisia). The facial, tongue, neck and back muscles are especially liable to be affected by these movements. When the movements occur as a late consequence of drug therapy, they are collectively termed tardive dyskinesias; they appear to result from dominance of the cholingeric brainstem pathways by the dopaminergic ones.

CEREBELLAR DISORDERS

The signs of cerebellar dysfunction are hypotonia, ataxia of gait, nystagmus, intention tremor, past-pointing, impairment of rapid alternating movements and slowness in initiating and arresting movements. Speech is dysarthric with a scanning quality. A mid-line cerebellar lesion affecting the part of the cerebellum that has the densest connection with the vestibular nuclei causes ataxia of stance and gait, not always with nystagmus and often with relatively normal coordination in the limbs when the subject is at rest. More laterally placed cerebellar lesions cause ipsilateral hypotonia and intention tremor with past-pointing and difficulty in initiating and repeating movements. A postural tremor, present for example when the arms are outstretched, occurs with lesions involving the connection between the deep cerebellar nuclei and the red nucleus.

All these signs of cerebellar dysfunction develop because the cerebellum is no longer able to function satisfactorily as a 'comparator'; that is, it can no longer make effective comparisons of the neural signals indicating the motor task to be undertaken and the impulse messages returning from peripheral sense-organs, monitoring the current progress of the movement. To some extent this difficulty arises because the incoming information is delayed or lost as it traverses damaged fibre tracts in the depth of the cerebellum; the same lesions would, of course, interfere with the output from the cortex via the deep cerebellar nuclei. Alternatively, the information, having arrived at the cerebellar cortex, will be prevented from being processed properly by any local cortical damage. Whereas timing of arriving impulse volleys is relatively unimportant in the primary receiving areas of the cerebral cortex, it is probable that errors of only a few milliseconds are sufficient to disrupt the cerebellar controlling mechanisms. Consequently, the cerebellum is no longer able to order the neural corrections necessary to ensure accuracy and smoothness in the execution of movement. Corrections when made are too large, too late and too prolonged—hence such phenomena as ataxia, intention tremor, dysdiachokinesis and post-pointing. Even for very brief (ballistic) movements, where there is insufficient time for correction, the cerebellum plays a crucial role in deciding beforehand the timing of the phases of contraction in agonist and antagonist muscles (see above); the incorrect timing sequences which result are responsible for dysmetria.

PHYSIOLOGICAL TREMOR

The immediate cause of the tremor is a tendency for motor unit discharge to be grouped at 9–10 c/s. The tremor does not occur synchronously in the two arms. The amplitude is influenced by the contractile properties of the muscle, an increase in the speed of contraction such as is caused by adrenaline or thyrotoxicosis increasing the amplitude of the tremor, and a decrease in the contractile speed having the opposite effect. Why tremor should occur at all remains uncertain. Delay occurring around the stretch reflex arc has been thought to contribute to the oscillation in motoneurone discharge but the persistence of tremor after limb deafferentation excludes this mechanism as the essential cause. Ballistocardiac effects have also been invoked but these make only a 10 per cent contribution to the total tremor. One can say no more at present than that the genesis of physiological tremor seems to depend on spinal or supraspinal mechanisms.

SPINAL LESIONS

It is natural to think of the effects of spinal cord lesions on motor activity only in terms of interference with descending pathways. But as we have seen, during the execution of a movement afferent neural traffic provides both internal and external feedback. One might therefore expect interruption of these afferent pathways to contribute to the functional disturbance, although the extent of this effect is not yet known.

Spinal lesions which interfere with the long motor tracts cause ipsilateral upper motoneurone signs; direct involvement of the alpha motoneurone gives rise to a lower motoneurone lesion at that level. Cervical cord lesions involving the vento-lateral quadrant of the cord will interfere with the descending pathway carrying the central respiratory drive, and if the level of the lesion is above the phrenic outflow (C3–5), all the major respiratory muscles will be affected. This has occasionally occurred during high cervical cordotomy carried out for the relief of pain. In these patients, spontaneous respiratory movements may be severely impaired, sometimes ceasing in sleep, while voluntary movement is relatively spared. The converse may also occur when all voluntary movement is lost, while spontaneous rhythmic respiratory activity continues. These observations emphasize the independence of the descending system concerned with spontaneous and voluntary breathing.

Not only do spinal lesions interfere with voluntary motor pathways but they also interrupt the modulating influence of these and other descending pathways on the excitability of the internuncial neurones within the grey matter of the spinal cord. In turn, the level of excitability of the internuncial neurones will affect the responsiveness of the alpha and gamma motoneurones. For example, after a cord lesion there is an instant loss of excitability in gamma motoneurones such that the tendon jerks are lost and muscle tone, also dependent on the stretch reflex loop, is diminished. At the same time, however, the cutaneous reflexes are changed and abnormal flexor withdrawal patterns are established; at first the only sign of this is the Babinski response, in which there is dorsiflexion of the great toe and fanning out of the lesser toes following stimulation of the outer border of the sole of the foot. Within a few days the net level of excitability rises in the separated segments of cord. The reason for this delay is not known but it seems unlikely to be due to the time taken for degeneration of descending axons since the period of spinal shock lasts only a few hours in "lower" mammals. The hyperexcitability of the alpha and gamma motoneurones causes overactive stretch reflexes and hence the phenomena of abnormally brisk tendon jerks, increased muscle tone and clonus. The flexor withdrawal reflexes also become progressively stronger and can now be initiated by weak mechanical stimuli over a large area of the leg. In their most extreme form there is, in addition to the Babinski response, dorsiflexion of the ankle, flexion of the hip and knee, and emptying of the bladder and rectum. Electromyographic recordings have shown that it may take as long as a few seconds for the flexor withdrawal reflex to commence following the first application of the stimulus. It would therefore appear that very complicated polysynaptic pathways may be involved in amplifying the sensory signal until it reaches the threshold for discharging the populations of flexor motoneurones. Once the withdrawal reflex has been elicited, the reflex pathways enter a period of depressed excitability lasting tens of seconds, during which time it may be impossible to evoke anything more than the Babinski sign.

The excitability of the spinal cord can be enhanced in other ways. For example, strychnine and tetanus toxin both interfere with postsynaptic inhibitory mechanisms; the motoneurones, now freed from a tonic restraining influence, discharge spontaneously or following the least stimulus; massive and severe spasms result and may prove fatal.

Principles of observations and tests

Wasting

Skeletal muscle fibres atrophy when deprived of nerve supply. The death of an alpha motoneurone or interruption of its axon therefore leads to muscle wasting, although the capability of other nevre fibres to branch distally and reinnervate denervated muscle fibres can reduce this effect.

Fasciculation

Spontaneous discharges may arise in alpha motoneurones which are diseased. Discharges may arise proximally so that the whole motor unit is activated, or in a distal branch of the nerve causing activity in only part of the motor unit. Fasciculation is visible through the skin and is characterized by irregular firing rates and its occurrence in the fully relaxed subject.

Contraction fasciculation

Abnormally large motor units may occur in denervated muscle because of distal branching in those nerve fibres which are still intact. In such cases motor unit discharges may be clearly visible in the skin. This contraction fasciculation can be distinguished from ordinary fasciculation by its voluntary initiation and low threshold for activity, and by the regular discharge frequency.

Fibrillation

This is an abnormality found on electromyographic sampling of partially denervated muscle (see below), namely, spontaneous discharge of individual muscle fibres rather than of whole motor units. In the tongue, in contrast to other muscles, fibrillation may be visible to the naked eye.

Tone

This term describes the resistance the examiner perceives when passively stretching a muscle. A component of this resistance will always be due to the physical characteristics of the muscle and tendon for both tissues possess inherent elasticity and offer a degree of resistance to stretch. The remainder of the resistance felt by the examiner is due to a weak tonic stretch reflex which, in a healthy subject, involves only a very small fraction of the motoneurones.

In patients with damage to the corticospinal pathways, stretching evokes a more vigorous tonic stretch reflex. This increased resistance (spasticity) may be felt to increase initially during the course of the movement and also becomes more prominent with increasing velocity of stretch. In some muscles, notably the quadriceps, an initial increase in tone is followed by a sudden reduction, the "clasp-knife" phenomenon. In these muscles, the sudden melting of resistance seems to be caused mainly by the inhibitory effects of Golgi tendon organ afferents. The hyperexcitability of the tonic stretch reflex is a consequence of increased excitability of both gamma and alpha motoneurones.

Resistance to passive muscle stretching which remains constant throughout the range of movement is a feature of patients with extrapyramidal disease. The neuro-physiological basis for this alteration in the stretch reflex, and for the commonly superimposed cog-wheel effect, is not certain, but increased excitability of alpha and gamma motoneurones probably both play a part.

Tendon reflexes

These are phasic stretch reflexes and thus depend on the monosynaptic reflex arc. The afferent limb of the reflex is served by the spindle primary endings and their group 1a afferent fibres.

In some healthy subjects, the set of the reflex is such that the tendon jerk is absent, but it can be made to appear by "reinforcement" in which the subject voluntarily contracts a muscle elsewhere. Reinforcement acts by increasing excitability of alpha and gamma motoneurones. Increased excitability of either alpha or gamma moto-neurones leads to an increase in the tendon jerk, the former acting directly, and the latter indirectly by increasing the sensitivity of the spindle primary endings. Both mechanisms are probably operating in the hyperactive tendon jerks seen in patients with upper motoneurone lesions.

Tendon reflexes, of course, will be reduced or abolished by lesions involving the lower motoneurone. Changes in nerve conduction in the afferent limb of the reflex, as occurs in demyelinating polyneuropathies, can also reduce reflex excitability. This is because the wave of excitation set up peripherally in the 1a afferent fibres is dispersed by the time it reaches the spinal cord as a consequence of differential slowing of afferent nerve conduction. The resulting asynchronous excitatory input is insufficient to excite the motoneurone pool and the tendon reflex fails to appear. In neuropathies of the axonal (dying-back) type, the tendon reflexes will be dispersed because there are too few 1a fibres to excite the motoneurones adequately.

Clonus

When the phasic stretch reflex is hyperexcitable, sustained stretch can lead to repeated reflex contraction, characterized as clonus. This can be elicited most readily in the calf muscles on dorsiflexing the ankle. Anxious healthy subjects may occasionally show clonus. It is a common feature in patients with upper motoneurone lesions in whom the stretch reflexes are hyperactive. The mechanism of clonus is that, as the muscle belly relaxes after its first reflex contraction, the lengthening of the muscle spindles activates the annulospinal endings of the 1a fibres. The activation comes about because the endings are made hyperexcitable by the increased tonic gamma motoneurone discharge which occurs after an upper motoneurone lesion. The volley of impulses set up in the 1a fibres during relaxation then excites the alpha motoneurones and initiates a second reflex contraction; the cycle is then repeated, producing the phenomenon of clonus.

Power

Because of the relative bias exerted against antigravity muscles by the long motor tracts, their interruption usually results in greater weakness of the flexors of the lower limbs and of the extensors of the upper limbs. Lesions of motor cortex particularly interfere with fine movements of the hand. One reason for this is the close (monosynaptic) relationship between some pyramidal tract neurones and the alpha and gamma motoneurones innervating muscles moving the fingers.

In testing power clinically, one is examining the integrity of the neural mechanism concerned with voluntary activation of muscle. It is especially necessary to remember this when assessing the state of the respiratory muscles by the "vital capacity" measurement, for this manoeuvre is a voluntary one and gives no information about the integrity of the neural mechanism concerned with generating spontaneous breathing. For this reason it may falsely indicate to the clinician that no respiratory disturbance is present. So far as the limbs are concerned, it is important to recognise that, in patients with upper motoneurone lesions, the functional disability during such natural tasks as walking and eating may be much greater than clinical testing of power would indicate. This is because there is, in each phase of the voluntary task, insufficient time for the damaged descending motor pathways to build up excitation in the alpha motoneurone pool. In the clinical tests, however, strength is assessed in a group of synergistic muscles over a period of several seconds, and there is time for isometric tension to increase to the maximum level permissible.

In chronic lower motoneurone lesions the severity of the neurological deficit may be masked by two compensatory mechanisms, axonal sprouting and increased motoneurone firing rates. Axonal sprouting enlarges the motor unit territories of surviving motoneurones and enables these units to generate greater than normal tensions. The higher discharge rates of the remaining motoneurones permit greater tetanic tensions to be developed by their colonies of muscle fibres.

Plantar responses

In new-born infants, a cutaneous stimulus to the sole of the foot may evoke a flexion withdrawal reflex characterized by extension of the hallux, fanning of the toes and dorsiflexion at the ankle; if the stimulus is adequate, flexion at knee and hip also occurs. This primitive protective reflex is replaced by the plantar flexion response during early childhood as the child learns to walk, the area from which the withdrawal reflex can be elicited progressively contracting. This change in the reflex pattern must depend upon the developing connections in the motor system. With their damage later, as in patients with upper motor neurone lesions, the reflex reverts to the infantile pattern.

Coordination

The performance of the "finger/nose" and "heel/knee/shin" manoeuvres, and the ability to carry out rapid alternating movements (e.g. supination and pronation of the hand), test a subject's coordination. "Intention tremor" and the disorganization of rapid alternating movements usually indicate a disturbance of the cerebellum or its connections, for the cerebellum is directly involved in the initiation and control of movement. Mid-line structures of the cerebellum are concerned particularly with controlling movements of the trunk, and their damage·leads to an ataxic gait, often without much evidence of incoordination in the limbs tested individually.

ELECTROMYOGRAPHY

The discharge of motor units in a muscle can be recorded from a needle electrode inserted into it. One can identify discharges of individual motor units if muscle contraction is slight, but when contraction is maximal the "interference pattern" which develops in healthy subjects prevents such identification.

The amplitude of a motor unit potential is related to its size (i.e. to the number of its component muscle fibres). When individual alpha motoneurones die, the muscle fibres they innervate will atrophy unless reinnevervated by the distal branching of a surviving neurone. Thus surviving neurones may come to serve a greater number of motor fibres and give rise to larger and more polyphasic action potentials. The presence of action potentials of this type thus implies denervation, as does a reduction in the number of units ("reduced interference pattern").

Fibrillation potentials also occur in partially denervated muscle. These are brief, low amplitude potentials which arise from spontaneous oscillations of the membrane potential at the site of the former neuromuscular junction.

The firing rate of an individual motor unit varies with the muscle but does not usually exceed 50/sec in a steady contraction. In upper motor neurone lesions the firing rates are reduced and poorly sustained.

Conduction velocity in nerve fibres may be measured by stimulating a nerve at two points and finding the difference in the conduction time to the onset of the muscle action potential. If the distance between the two points is known, conduction velocity over this length can be calculated; this value pertains only to the *fastest* fibres in the nerve. Slight slowing of conduction velocity may occur with neurone death, as in motor neurone disease, and implies fall-out of the fastest-conducting fibres. Severe slowing of nerve conduction is characteristic of segmental demyelination. Conduction velocity in sensory nerves can be measured by direct electrical stimulation of the nerve (e.g. through ring electrodes on the finger to stimulate the digital nerves) and recording the nerve action potential at appropriate sites, such as at the wrist.

"H" reflex

This is the monosynaptic reflex discharge evoked in a muscle by electrical stimulation of its muscle spindle afferents. The reflex is thus similar to the tendon jerk but by-passes the muscle spindle by stimulating its group Ia afferent fibres directly. It provides a means of testing the excitability of the motoneurone pool, but it cannot be elicited in all muscles. It is not widely used as a diagnostic technique.

Practical assessment

In clinical practice, disorders of the motor system are usually diagnosed from the history and examination. Indeed, only a minority of the neurological investigations in regular use give direct information about disordered function of the motor system. Plain x-rays, CAT scans (computerized axial tomography), and contrast studies (air-encephalography, angiography, brain scan and myelogram), for example, are primarily concerned with demonstrating anatomical changes, and examination of the cerebro-spinal fluid with providing pathological information. Electrophysiological methods, on the other hand, can give direct evidence of disordered neuronal function in some diseases.

The choice of investigation in patients with motor disturbance depends not so much on the type of motor disorder as on the pathological process which is thought to underlie it. It is not appropriate to outline here the investigative procedure in patients with neurological disease who have as one manifestation of their illness disordered motor function. The summary which follows is concerned only with indicating the clinical and investigative features which may be helpful in establishing the nature of the motor disorder.

DISORDERS OF LOWER MOTONEURONES

Symptoms. Weakness, muscle thinning, muscle twitching.

Signs. Wasting, weakness, fasciculation, depressed or absent tendon reflexes.

Investigations. Electromyography. This may show (a) spontaneous fibrillation, (b) reduced number of motor units active, (c) abnormally large and polyphasic motor units (because surviving neurones branch distally to innervate muscle fibres whose original innervation has been lost. The surviving neurones thus serve greater numbers of motor fibres and give rise to larger and more polyphasic action potentials), (d) reduced motor conduction velocity, most marked in lesions which cause segmental demyelination of the nerve fibres.

DISORDERS OF UPPER MOTONEURONES

Symptoms. Stiffness, heaviness or weakness in limbs; loss of hand dexterity; focal motor epilepsy (Jacksonian attack); alteration of speech.

Signs. Spasticity, clonus, weakness predominantly of extensors in the arms and flexors in the legs, impairment of individual movements of fingers and toes, increased tendon reflexes, extensor plantar response. With pseudobulbar palsy, slurring dysarthria, dysphagia, impaired tongue movements, facial weakness.

Investigations. Electromyography. This may show relatively low rates of firing of motor units, which are poorly sustained.

CEREBELLAR DISORDERS

Symptoms. Clumsiness or unsteadiness of the limbs, unsteadiness of gait, alteration of speech.

Signs. Intention tremor, dymetria, hypotonia, impaired rapid alternating movements, excessive rebound, nystagmus, titubation, scanning dysarthria, wide-based ataxic gait.

Investigations. Neuro-otology. (a) Positional nystagmus of central type. (b) Exaggerated caloric responses. (c) Ocular dysmetria and rebound nystagmus. (d) Ocular square waves or flutter.

EXTRAPYRAMIDAL DISORDER

Symptoms. Stiffness, tremor, slowness, tendency for writing to become smaller, difficulty in turning over in bed, deterioration in walking.

Signs. Expressionless face and infrequent blinking, impaired ocular convergence, blepharoclonus, rigidity, tremor at rest (3–7 c/s) diminished during movement, paucity of movement, festinant gait, speech monotonous in pitch and intensity.

References

BRODAL, A. (1981). *Neurological Anatomy in relation to Clinical Medicine* 3rd ed. London, Oxford University Press.

BUCHTHAL, F., and SCHMALBRUCH, H. (1980) Motor unit of mammalian muscle. *Physiological Reviews* 60, 90–142.

EVARTS, E.V., BIZZI, E., BURKE, R.E., DE LONG, M. and THACH, W.T. (1971) Central control of movement. *Neurosciences Res. Prog. Bull* 9 (1), 1–170.

GRANIT, R. (1970) *The Basis of Motor Control.* London, Academic Press.

Other selected references

BRODAL, A. (1973) Self-observations and neuro-anatomical considerations after a stroke. *Brain,* 96, 675–694.

CALNE, D.B. (1970) *Parkinsonism: physiology, pharmacology and treatment.* London, Arnold.

COOPER, I.S. (1969) *Involuntary Movement Disorders* New York, Hoeber.

ECCLES, J.C., ITO, M. and SZENTAGOTHAL, J. (1967) *The Cerebellum as a Neuronal Machine.* Berlin, Springer-Verlag.

HAMMOND, P.H., MERTON, P.A. and SUTTON, G.G. (1956) Nervous gradation of muscular contraction *Brit. Med. Bull.* 12, 214–18.

LANCE, J.W., McLEOD, J.G. (1975) *A Physiological Approach to Clinical Neurology* London, Butterworth.

LAWRENCE, D.G. and KUYPERS, H.G.J.M. (1968) The functional organisation of the motor system in the monkey. I. The effects of bilateral pyramidal lesions. *Brain,* 91, 1–13.

MATTHEWS, P.B.C. (1972) *Mammalian Muscle Receptors and their Central Actions.* London, Arnold.

PHILLIPS, C.G. and PORTER, R. (1977) *Corticospinal Neurones: their role in movement.* London, Academic Press.

VALLBO, A.B. (1971) Muscle spindle response at the onset of isometric voluntary contractions in man. Time difference between fusimotor and skeletomotor effects. *Journal of Physiology,* 318, 405–431.

The Gut

GASTROINTESTINAL HORMONES

Normal function

Over the past decade a wealth of knowledge has been accumulated regarding the identification, chemical structure and functional interrelationships of the hormones of the gut endocrine system. With this knowledge it has become apparent that these hormones may have a major influence on the normal and abnormal physiology of the gastrointestinal tract.

The cells responsible for the production of these polypeptide hormones are known as APUD cells. This is based on the initial letters of the most important of their common cytochemical and functional characteristics (Amine content and Amine Precursor Uptake and Decarboxylation).

The possibility that the cells of the APUD system are derived embryologically from a single region of neuroectoderm remains conjectural.

Whilst the original concept of the APUD system referred to classical endocrine polypeptide producing cells it has now become apparent that the peptides produced by this system have a spectrum of activity ranging from neurotransmitter to true hormone and that not all cells exhibiting cytochemical APUD characteristics are part of the APUD series (and vice versa).

The cells constituting this diffuse neuroendocrine system may have different functions according to their location; their products being secreted directly into the blood stream (endocrine), into intercellular fluid to affect only neighbouring cells (paracrine) or into a synapse with another neurone (neurocrine).

Many of these peptides are widely distributed through the gastrointestinal tract, pancreas and brain. For some their exact function in both normal and pathological states remains to be elucidated, the ability to identify and characterize them having outstripped investigation of their mode of action. Table 12.1 lists those hormonal peptides found in the gut.

ESTABLISHED GASTROINTESTINAL HORMONES

Gastrin

Gastrin is synthesized and released from cells (G cells) located in the mucosa of the

Table 12.1. Hormonal peptides found in the gut and classified by probable mode of action.

Circulating hormones	Paracrine peptides	Peripheral neural peptides
Gastrin	Somatostatin	Vasoactive intestinal polypeptide (VIP)
Secretin	Vasoactive intestinal polypeptide (VIP)	Somatostatin
Cholecystokinin (CCK)	Substance P	Substance P
Pancreatic polypeptide		Bombesin
Motilin		Enkephalin
Gastric inhibitory polypeptide (GIP)		
Neurotensin		
Enteroglucagon		

gastric antrum and to a lesser extent the upper small intestine. It exists in two main forms; little gastrin, a 17 amino acid molecule (G 17) and big gastrin which contains 34 amino acids (G 34). Another gastrin macromolecule known as big-big gastrin has also been described but whether this is an additional form of gastrin or an artefact is uncertain.

The full range of biological activity is present in the four C-terminal amino acid residues of the gastrin molecule. G 34 is twice as abundant in serum as G 17 although the latter is six times more potent for acid secretion. However, because of differences in their half lives infusions of both G 17 and G 34 produce comparable rates of acid secretion.

Although the major short term action of gastrin is stimulation of gastric acid secretion a number of other physiological responses can be detected following exogenous administration. These include secretion of pepsinogen, increase in gastric blood flow and contraction of stomach circular muscle. In addition the hormone appears to have a long-term trophic action, stimulating cell proliferation in the stomach, colon and small intestine.

Both acetylcholine and histamine are necessary for the maximum effect of gastrin on gastric acid secretion although the exact inter-relationships between the parietal cell receptor mechanisms for these three substances remain unknown.

The regulation of circulating gastrin levels is based on a negative feedback loop which relates antral gastrin release inversely to intragastric pH. Superimposed upon this are a number of other factors which promote gastrin release following a meal. These include increased vagal activity, peptides, amino acids and calcium ions and local cholinergic oxynto-pyloric reflexes induced by antral distension.

Secretin

Secretin is composed of 27 amino acids and the whole polypeptide is necessary for biological activity. Secretin (S) cells are present throughout the mucosa of the duodenum and upper jejunum but predominate in the duodenal bulb. Endogenous release of secretin ocurs exclusively in response to hydrogen ions delivered into the duodenum.

The amount released is proportional to the amount of acid entering the duodenum up to a pH of 3·0. Above this, release falls off until pH 4·5–5·0 when it ceases.

The role customarily ascribed to secretin is stimulation of the secretion of bicarbonate-containing juice by the pancreas and liver with other factors being necessary to potentiate this effect, most notably cholecystokinin and vagal activity. However recently several observations have led to the suggestion that secretin may not be physiologically important; the threshold pH for secretin release is pH 4·5 although duodenal pH rarely goes below this. Other hormones e.g. VIP and CCK are capable of stimulating pancreatic bicarbonate secretion; these and the vagal influence after a meal may be more important than previously realized.

Cholecystokinin

In 1928 it was postulated that the presence of fat in the intestine released a hormone which caused gall bladder contraction. This hormone was named cholecystokinin after it's primary action. In 1943 a hormone was described which was released from the small intestine and caused stimulation of pancreatic enzyme secretion. Accordingly it was named pancreozymin. In 1968 following purification of each substance it became apparent that both hormones were the same, their activities residing in a single peptide. For convenience the name cholecystokinin (CCK) was internationally accepted.

Although cholecystokinin consists of 33 amino acid residues the entire range of biological activities is possessed by the C-terminal octapeptide and is therefore used as a substitute for CCK in experimental studies.

Both gastrin and CCK possess the same 5 C-terminal amino acid residues and are therefore capable of the same pharmacological actions, differing only quantitatively.

Cholecystokinin is released in response to the naturally occurring L-isomers of essential amino acids. D-isomers and non-essential amino acids have no effect. Other potent releasers of the hormone are calcium ions in physiological concentrations and fatty acids, the longer chain fatty acids being more effective than medium and short chain ones.

A large number of biological activities affecting the secretory, absorptive and motor functions of the gastrointestinal tract have been ascribed to cholecystokinin. It's most potent actions are the promotion of gall bladder contraction and pancreatic enzyme secretion. However it also stimulates gastric and intestinal motility and inhibits the contraction of both the lower oesophageal sphincter and the sphincter of Oddi. Further actions include a trophic one (since animal experiments demonstrate pancreatic hypertrophy and hyperplasia following chronic administration of the hormone) and a possible role in the central regulation of appetite.

OTHER GASTROINTESTINAL HORMONES

Although the localization, characterization and pharmacological effects of many of the hormones listed in Table 12.1 are known, their physiological role remains unclear. The amino acid sequences of many of these peptides are similar and several hormones have the same effect in varying degree. The important actions of any gastrointestinal hormone must be those that occur physiologically i.e. such actions should occur with serum levels of the hormones that might be expected following an endogenous stimulus such as a meal. Difficulties arise however, when the peptide is thought to exert it's effect locally e.g. VIP.

Until more evidence is obtained these polypeptides remain candidate hormones. The actions of those for whom the evidence is strongest are listed in Table 12.2.

Table 12.2. Biological actions of some candidate gastrointestinal hormones

Hormone	Action
Pancreatic polypeptide	Decreased pancreatic enzyme output
	Relaxation of gallbladder
Motilin	Stimulation of antral and duodenal motility
	Contraction of lower oesophageal sphincter
Gastric inhibitory polypeptide (GIP)	Glucose-dependent insulin release
Vasoactive intestinal polypeptide (VIP)	Intestinal vasodilatation
	Regulation of gastrointestinal motor tone and motility
	Stimulation of pancreatic bicarbonate secretion
	Inhibition of gastric acid secretion
Somatostatin	Inhibition of gastric and pancreatic secretion
	Inhibition of gastrointestinal motility

Disordered function

The routine measurement of serum gastrin by radioimmuno-assay is now readily available. A variety of conditions may be associated with high circulating levels of the hormone although the only clearly defined pathological roles for gastrin are in the gastrinoma and retained antrum syndromes. Any breakdown in the feedback loop between gastric acid secretion and gastrin release will lead to increased amounts of gastrin.

Gastrin levels rise with age presumably due to a reduction in gastric acid production and patients with hypochlorhydria or achlorhydria (e.g. pernicious anaemia) may have very high levels, even into the gastrinoma range. A similar situation arises after vagotomy where the high gastrin levels are probably secondary to hypochlorhydria, although abolition of cholinergic inhibition may play a role.

Duodenal ulceration

Patients with duodenal ulceration tend to have an increased gastric acid production. Whilst their fasting gastrin levels do not differ from those of normal subjects their post prandial serum gastrin responses are often greater than normal. Studies on gastrin concentrations and G-cell numbers in the antral mucosa demonstrate wide variations in the amounts of gastrin and numbers of G-cells present. However, there are some duodenal ulcer patients who appear to have either true G-cell hyperplasia or G-cell hyperfunction. Since it is now accepted that duodenal ulcer disease is a heterogeneous group of disorders such subgroups may prove to be of importance. In addition patients with renal failure who are known to have an increased incidence of duodenal ulceration also have high circulating gastrin levels. This probably reflects reduced renal clearance of gastrin although there is also a direct correlation between the presence of secondary hyperparathyroidism and G-cell density in these patients. Patients with severe burns who develop stress ulcers (Curling's ulcer) have also been shown to have hypergastrinaemia.

Hypercalcaemia

Patients with hyperparathyroidism have increased serum calcium levels. These patients frequently have duodenal ulceration and hypergastrinaemia. Although calcium ions are known to exert a direct effect on antral G-cells causing gastrin release the concentration required locally to achieve this effect is higher than might be expected in hyperparathyroidism. Many of these patients may therefore have Type I *Multiple Endocrine Adenomatosis* (MEA I) with an occult associated gastrinoma which is known to release gastrin in response to slight increases in the level of serum calcium. Gastrin levels frequently fall in these patients following parathyroidectomy although a high percentage of these patients will go on to develop the Zollinger-Ellison syndrome several years later.

The Zollinger-Ellison syndrome

The clinical features of this syndrome as originally presented by Zollinger and Ellison in 1955 were a diagnostic triad of hypersecretion of gastric acid, rapidly recurring ulceration despite adequate therapy and a non-beta islet cell tumour of the pancreas. With the introduction of assays for gastrin it became apparent that the syndrome was caused by the excess production of gastrin from the islet cell tumour or gastrinoma.

The tumours are frequently multiple and small. At least 60 per cent are malignant and 50 per cent will have metastasized to the liver by the time of diagnosis although they may still be surgically undetectable. In some patients a primary pancreatic tumour cannot be found either because it is so small or because of an ectopic location e.g. the duodenal wall. Generally the tumours grow slowly and it is possible to live for several years with multiple secondaries, providing control of acid hypersecretion can be achieved, usually by the use of a total gastrectomy or H_2—receptor antagonists. Peptic ulcer patients with multiple ulcers or ulcers at unusual sites, and patients with hyper-parathyroidism or a family history suggesting MEA I are candidates for a gastrinoma.

Isolated retained antrum

Hypergastrinaemia may occur following a Polya gastrectomy as a result of retention of antral mucosa in the duodenal stump. The isolated G-cells are free from acid inhibition and the resulting hypergastrinaemia leads to acid hypersecretion and recurrent ulceration.

The Verner-Morrison syndrome

In 1958 a syndrome of refractory watery diarrhoea and hypokalaemia in association with an islet cell tumour of the pancreas was described. An eponym, WDHA, has been given to the syndrome based on the cardinal clinical features of *W*atery *D*iarrhoea (6 litres or more of cholera-like stools per 24 hours) *H*ypokalaemia which is often profound and *H*ypochlorhydria or *A*chlorhydria. Because of the diarrhoea and the associated pancreatic tumour the syndrome has also acquired the name 'pancreatic cholera'.

The association between this syndrome and raised plasma levels of vasoactive intestinal polypeptide (VIP) was first described in 1973 and whilst several gut hormone tumours are capable of giving rise to a similar clinical picture, it has become accepted that the classical Verner-Morrison syndrome is the result of a VIPoma. VIP levels are usually undetectable or very low in normal individuals probably because it is acting as a local paracrine or neurocrine peptide. However, patients with the syndrome have elevated plasma levels of the hormone. Further evidence of its role is that high concentrations of VIP and VIP-producing cells are present in tumour tissue, there is a fall in circulating VIP levels following resection of the tumour or treatment with streptozotocin, and chronic VIP infusion in pigs mimics the clinical picture.

VIP may be produced by both pancreatic and neural tumours hence some patients have ganglioneuromas or ganglioneuroblastomas. The effects of VIP are probably mediated via cyclic AMP in a manner analogous to the effects of cholera toxin or prostaglandins. Whilst the syndrome is rare, early diagnosis is important. At present the average time from onset of symptoms to diagnosis is three years. By this time 50 per cent of tumours have metastasized and a third of patients with resectable tumours die post-operatively because of the profound electrolytic disturbances which occur.

Principles of tests and measurements

Whilst the development of radioimmunoassay has led to the measurement of many gastrointestinal hormones their exact role is often unclear and routine estimations of serum levels of those hormones listed in Table 12.1 are therefore unwarranted. The exceptions are in those patients with symptoms suggestive of either the Zollinger-Ellison or Verner-Morrison syndromes.

For patients with a gastrinoma the establishment of the diagnosis nowadays depends less on acid secretory testing than on measurement of a fasting serum gastrin level and, if necessary, dynamic testing of serum gastrin responses to injections of secretin and/or calcium. However, tests of gastric secretion are simple to perform, an answer is usually rapidly available and useful information may be obtained particularly in difficult cases, e.g. postoperatively. Various criteria for the diagnosis of gastrinoma have been suggested based on acid output and a variety of secretory ratios. Before the measurement of serum gastrins became routine a gastrinoma was rarely detected until the full blown picture of the Zollinger-Ellison syndrome was apparent and hence gastrin levels were greatly elevated. Nowadays greater awareness of situations where a gastrinoma is a possibility has led to the correct diagnosis when the fasting serum gastrin levels is only marginally elevated or occasionally normal. Similarly high levels may be the result of one of the other causes of hypergastrinaemia. To evaluate such cases stimulation tests may be required.

Secretin and calcium stimulation tests have gained the widest use. Although the gastrin responses to secretin and calcium are similar calcium infusion appears to be less discriminatory. In normal subjects injection of secretin tends to cause a fall in serum gastrin whilst in patients with a gastrinoma there is an early and short-lived rise. An increase in serum gastrin of more than 200 pg/ml over basal in response to a bolus injection of GIH secretin will identify patients with a gastrinoma who have borderline elevations of the hormone.

Secretin/Pancreozymin test. Whilst individual measurements of secretin and pancreozymin/CCK are not at present clinically useful, a combination of both hormones given intravenously as a bolus or infusion with assay of duodenal contents for bicarbonate and enzyme concentrations is used as a pancreatic function test (p. 462).

Cholecystokinin-cholecystogram. There is no reliable assay for cholecystokinin but the synthetic octapeptide of CCK is commercially available. Since it causes gallbladder contraction it can be used as a refinement of the standard oral cholecystogram in those patients who may have abnormalities of gallbladder function without cholelithiasis.

Practical assessment

Clinical observations

Recurrent peptic ulceration, ulceration at an early age or unusual site, watery diarrhoea, previous parathyroid surgery, family history suggestive of MEA I.

Routine methods

Serum calcium, acid secreation tests (p. 461). Anatomical localization; isotope scanning, ultrasound, arteriography.

Special techniques

Fasting serum gastrin or plasma VIP level, secretin stimulation test or calcium infusion. Transhepatic portal venous sampling for tumour localization.

MOTILITY

Normal function

PROPERTIES OF INTESTINAL SMOOTH MUSCLE

With the exception of the pharynx and anus the motor function of the alimentary canal is performed by smooth muscle. The importance of smooth muscle in governing this function lies in its mechanical properties which differ from those of striated muscle. Viable intestinal smooth muscle cells maintain a degree of tension but they have no characteristic resting length. Thus the bowel wall has an innate degree of rigidity or tone allowing sustained and often powerful contractions to occur with a very small expenditure of energy. After death, elongation of intestinal smooth muscle cells takes place with the result that the antemortem length of the bowel (250 cm) is considerably less than its postmortem length (650 cm).

Smooth muscle is capable of complex variations in tone even after denervation and these responses to stretch and release of stretch allow viscera to accommodate to changes in volume with minimal changes in transmural pressure. If the muscle is held at constant length its tension falls or if held at constant tension its length slowly increases. It is these properties which explain how the stomach and colon are capable of accommodating large amounts of material with no increase in intraluminal pressure.

Anatomical considerations

Beneath the glandular mucosa the muscularis mucosae consists of an inner circular layer and an outer longitudinal layer of muscle fibres with thin connecting muscle strands responsible for interaction between the layers. Whether the circular layer is helical or formed by closed rings and the longitudinal layer bundles twisting in an open elongated helix or bands of muscle with axes parallel to the gut remains in dispute. These muscle layers vary in thickness in different parts of the gut being particularly prominent in the stomach. Contractions of the circular muscle throughout the intestine act to occlude the lumen whilst contractions of the longitudinal muscle is associated with receptive relaxation of an adjoining segment of bowel. The sphincters of the digestive tract are specialized areas of smooth muscle with their own innervation e.g. the pylorus. The cricopharyngeal sphincter however is purely striated muscle whilst the anal sphincter is striated muscle with additional smooth muscle fibres.

Although five nerve plexuses are described, the two main plexuses are Auerbach's plexus which is predominantly motor in function and lies between the inner circular and outer longitudinal muscle layers and Meissner's plexus which is sensory and lies deep to the muscularis mucosae.

Control of gastrointestinal smooth muscle activity

Throughout the gut there is a myogenic control mechanism manifest by slow spontaneous fluctuations in membrane potential which are propagated via the longitudinal muscle layer as a wave of depolarization and which possess a definite rhythmicity characteristic of the tissue in which they occur. This conducted wave of depolarization may trigger bursts of action potentials which are local and followed by muscular contraction. The conducted waves of depolarization are called slow waves, basic electrical rhythm, pacemaker potentials or electrical control activity whilst the action potentials are sometimes called spike bursts or electrical response activity. Not every slow wave triggers an action potential and a variety of neural and humoral components are capable of initiating, modifying and co-ordinating the motor patterns dictated by this myogenic control system.

Autonomic innervation of intestinal smooth muscle is via both sympathetic and parasympathetic systems. Postganglionic sympathetic nerve fibres arise from the coeliac, superior mesenteric and hypogastric plexuses and terminate in contact with the cell bodies of the intramural nerve plexuses. Parasympathetic innervation via the vagus extends as far as the transverse colon whilst the distal colon is supplied from the pelvic nerves via the hypogastric plexuses.

The chief function of sympathetic innervation is inhibition of intestinal smooth muscle. This is mediated by noradrenaline which blocks release of the excitatory transmitter acetylcholine. In addition sympathetic fibres are excitatory for some sphincters and influence mucosal gland secretion. The effect of the parasympathetic system on gastrointestinal motility is however far greater. The fibres are all preganglionic and excitatory, the effects being mediated by acetylcholine. Section of both vagi causes dilation, diminished tone, slowed peristalsis and delayed gastric emptying; hence the need for a drainage operation in conjunction with a truncal vagotomy.

Whilst the autonomic nervous system is classically divided into adrenergic and cholinergic efferent nerves, morphological studies have suggested the existence of a further peptidergic nervous system with several biologically active peptides occurring both within the nerves of the gut wall and the brain (p-type nerves).

The exact role of these neurocine peptides (Table 12.1) within the peptidergic neuronal system has still to be fully elucidated. However, experimental evidence strongly suggests that they act as neurotransmitters and that the peptidergic neurones constitute the major short neuronal system of the gut. It is probable therefore that this peptidergic nervous system acts as a major modifying influence of adrenergic and cholinergic impulses and mediates local reflex mechanisms. These local reflex arcs serve to co-ordinate smooth muscle activity, the potential effective stimuli of which are: chemical composition of gastrointestinal contents, pH, osmotic pressure and digestion products of protein and fat.

Long intestinal reflexes—The activity of one part of the digestive tract may influence another distant part. An increase in ileal activity and a fall in ileocaecal sphincter pressure accompanies gastric secretion and emptying. This is mediated by physiological quantities of gastrin. Emptying of the ileum through the sphincter is therefore associated with eating and is known as *the gastroileal reflex*.

Humoral influences on smooth muscle include neurohumoral transmitters, gastrointestinal hormones, prostaglandins and drugs. In addition contraction of a smooth muscle cell follows changes in membrane potential which in turn depend on electrolyte distribution, hence alterations in the concentration of potassium, sodium, chloride and calcium ions may be important e.g. hypokalaemia may cause a paralytic ileus.

TYPES OF GUT MOVEMENT

Whilst the term motility is vague and encompasses a variety of features of intestinal movements some other terms describe activities of the smooth muscle of the gastrointestinal tract which can be defined.

Transit is the term applied to the progression of gut contents along the intestinal tract.

Segmentation consists of radial contractions of the circular muscle of the intestine thus dividing a length of bowel and its contents into segments. Activity tends to occur in cycles becoming stronger after eating and being augmented by vagal stimulation or anticholinesterases. Segmentation results in a to and fro movement although slight progression aborally occurs since the frequency of contraction is greater in the upper part of the intestine. Inhibition of segmenting contractions is probably the major mechanism facilitating intestinal transit.

Short propulsive movements are short weak movements travelling at a mean velocity of 1·2 cm per second for an average distance of 15 cm. They move the gastrointestinal contents slowly aborally.

Peristalsis is a progressive wave of contraction preceded by relaxation which moves rhythmically along the gut. Depolarization passes along the longitudinal muscle and spreads to the underlying circular muscle. It is the normal activity of the oesophagus, pyloric antrum, terminal ileum and rectum. In addition it occurs throughout the gut in response to obstruction.

Swallowing

Chewing is not necessary for normal digestion and absorption providing food can be swallowed without difficulty. Swallowing is initiated voluntarily by elevation and retraction of the tongue against the hard palate which propels the bolus into the pharynx. As the bolus enters the oropharynx contraction of the superior constrictors and soft palate closes the nasopharynx, respiration is inhibited and the glottis is closed by elevation and forward displacement of the larynx. All this takes less than one second. Although initiated voluntarily, swallowing proceeds as a co-ordinated involuntary reflex controlled by the 'swallowing centre' in the medulla. Since there is a negative intraoesophageal pressure, there are sphincters at both ends of the organ to withstand entry of air and gastric contents. As the distal pharyngeal muscles contract the cricopharyngeus sphincter relaxes, resuming its resting tone once the bolus has passed. A primary peristaltic wave, which is a continuation of the oral and pharyngeal phase, then propels the bolus to the distal oesophagus. Relaxation of the lower oesophageal sphicter commences 1·3 secs after the commencement of swallowing (as oesophageal peristalsis is commencing) although the oesophageal phase of swallowing takes about 8–10 seconds.

The competence of the cardia

At rest the intraoesophageal pressure is 10 mmHg lower than the intragastric pressure, this difference being considerably accentuated by position and straining. Two main factors probably work together to prevent the reflux of gastric contents.

The lower oesophageal sphincter. Under resting conditions the lower 2–5 cm of oesophagus which lie above and below the diaphragm constitute a physiological high pressure zone. Anatomically there may be a corresponding area of asymmetrical muscle thickening. Using continually perfused open-tipped recording catheters the resting pressure in this segment is 12–30 mmHg above the intraabdominal pressure. A variety of factors both physiological and pharmacological are known to cause alterations in lower oesophageal pressure and some may be of importance in the pathogenesis of oesophageal reflux (Table 12.3), however both the hormonal and nervous control of lower oesophageal sphincter pressure are still the subject of much debate.

Table 12.3. Factors influencing lower oesophageal sphincter (LOS) pressure.

	Increase	Decrease
Foods	Protein meal	Fat meal
		Chocolate
Drugs	Cholinergics	Anticholinergics
	Metoclopramide	Ethanol
	Domperidone	Nicotine (smoking)
		Caffeine (coffee)
		Antihistamines
Hormones	Gastrin	Cholecystokinin
	Motilin	Secretin
	Pancreatic polypeptide	Glucagon
		Progesterone
Other agents	Histamine	Cyclic AMP
	Prostaglandin $F_2\alpha$	Prostaglandin E_2

Mechanical factors. The lower oesophageal sphincter alone is not sufficient to prevent gastro-oesophageal reflux. Since the lower 2 cm of the oesophagus lies below the diaphragm any rise in intra-abdominal pressure compresses this segment as well as raising intragastric pressure. Hence the pressure in the sphincter remains higher than in the stomach. In addition any rise in intra-abdominal pressure leads to an increase in lower oesophageal sphincter pressure although the mechanism by which this reflex is mediated is unknown. Several other mechanical factors have also been proposed as important for the anti-reflux barrier. These include valve like mucosal folds at the gastro-oesophageal junction, the acute angle of entry of the oesophagus into the stomach and contraction of oblique sling like smooth muscle fibres at the cardia. These mechanisms are probably unimportant in man.

Gastric motility and emptying

Gastric slow wave activity originates from a site high on the greater curve which acts as the pacemaker for the stomach. When associated with electrical response activity a peristaltic contraction appears in the gastric body and moves towards the pylorus. These peristaltic waves occur at 3–4 per minute and increase in both depth and velocity as they approach the gastroduodenal junction. They tend to be more vigorous during hunger and diminish in force immediately after the stomach is filled. As the contraction approaches the gastroduodenal junction some of the gastric contents enter the duodenal bulb but the peristaltic wave tends to overtake the bulk of the gastric contents and, as it does so, it propels it back into the main body of the stomach thus ensuring adequate mixing. This is known as retropulsion.

The regulation of gastric motility depends on many mechanisms. There are reflexes mediated by the sympathetic and parasympathetic nerves such as vagal activity and there are local reflexes triggered, for example, by duodenal distension. In addition, the gastrointestinal hormones gastrin and motilin almost certainly play a part.

Vagotomy temporarily diminishes antral electrical activity and distension of the duodenum or jejunum also causes slowing and weakening of antral contractions. Gastrin appears to increase the speed and force of the contractile wave but delays emptying whilst physiological amounts of motilin enhance gastric emptying.

Intragastric aqueous, solid and oil phases each have a characteristic pattern of emptying with fluids leaving the stomach quickest. The rate of emptying is governed by the chemical and physical properties of the chyme entering the duodenum. Fat, low pH and hypertonic solutions have a decreasing order of potency.

These regulatory mechanisms ensure an orderly transfer of contents from the stomach to small intestine which allows an optimum time for digestion and absorption.

Small intestinal motility

The most important movement of the small intestine is segmentation, controlled by the basic electrical rhythm which probably originates from a pacemaker near the ampulla of Vater. Aboral propulsion is accomplished by short propulsive movements and to a lesser extent by segmentation. These movements allow thorough mixing of contents with digestive juices and their exposure to the mucosa for absorption. Although motility of the small intestine is increased following feeding, the propulsive activity is reduced due to increased contraction thus leading to increased resistance to flow. The rate of passage of chyme is fairly rapid through the duodenum but slows as it progresses aborally.

The ileocaecal sphincter is a zone 4 cm long with a resting pressure 20 mmHg greater than colonic pressure. The sphincter relaxes each time a propulsive wave passes along the terminal ileum thus allowing a squirt of chyme to enter the caecum at regular intervals. The *gastroileal reflex* also relaxes the sphincter following eating.

Large intestinal motility

The majority of colonic contractions are of the segmentation type with durations of 12–60 seconds. With contractions occurring in adjacent areas the contents of the large bowel are slowly moved back and forth exposing them to the mucosa for absorption of water and electrolytes.

Propulsion of contents takes place by mass movements consisting of a ring of contraction arising at any point in the colon and moving for a variable distance towards the anus at a rate of 2–5 cm per minute. These movements occur 2–4 times per day in healthy individuals, and appear to be stimulated by the ingestion of food, psychogenic factors and somatic activity; they disappear during sleep. The contents of the descending and sigmoid colon are semisolid. Segmenting contractions in the upper rectum are more frequent than in the sigmoid and proximal colon so that they retard the flow of contents into the rectum which normally remains empty until defaecation is imminent.

The conscious desire to defaecate is initiated by rectal distension. As the rectum is distended it contracts and the internal anal sphincter, which is part of the circular muscle of the gut, relaxes owing to the rectosphincteric reflex. This reflex can be overcome by voluntary contractions of the external anal sphincter if defaecation is not expedient. Relaxation of the internal sphincter is transient and the sensation disappears until the passage of more contents into the rectum.

Defaecation is a combination of involuntary and voluntary procedures. Internal and external sphincters relax and intra-abdominal pressure is increased by contraction of the diaphragm, closure of the glottis and tensing of the anterior abdominal wall muscles; meanwhile the muscles of the pelvic floor relax.

The highest pressures can be generated by adopting a squatting position and the passage of a large stool requires less effort to evacuate it than a small stool. The extent of bowel cleared at defaecation varies from the rectum alone to the whole of the descending colon.

Disordered function

DISTURBANCES OF OESOPHAGEAL FUNCTION

Dysphagia is a sensation of difficulty in swallowing and usually results from oral lesions, neuromuscular dysfunction, mechanical obstruction or a motor disorder of the oesophagus.

Neuromuscular dysfunction

This can cause dysphagia by affecting the three striated muscle groups necessary for orderly swallowing: the pharyngeal constrictors and the tongue, the upper oesophageal sphincter and the upper third of the oesophagus. The most important primary muscle disorder is myasthenia gravis which can affect all the muscles of deglutition.

Lesions affecting the brain stem or the cranial nerves e.g. poliomyelitis, motor neurone disease and cerebrovascular accidents tend to cause paralysis of the palatal muscles which impair closure of the nasopharynx during swallowing and allow nasal regurgitation. These patients are only able to swallow by allowing the bolus to flow slowly off the back of the tongue and down the oesophagus under the influence of gravity. They are, therefore, prone to bronchial aspiration of food and saliva and so the development of pneumonia.

Mechanical obstruction

This is usually caused by neoplastic lesions of the pharynx or oesophagus, by benign stricture formation following reflux oesophagitis, by swallowing caustic material or by extrinsic compression Occasionally cervical oesophageal webs composed of squamous mucosa can cause dysphagia. These may be associated with an iron deficiency anaemia (Plummer-Vinson syndrome). Such patients often have oropharyngeal leukoplakia and are prone to develop hypopharyngeal carcinomas.

Motor disorders of the oesophagus

Achalasia or cardiospasm is characterized by a double defect in oesophageal function. The lower oesophageal sphincter fails to relax normally on swallowing thus impeding the flow of liquid and solid material into the stomach and there is a failure of normal peristalsis in the lower two thirds or the smooth muscle portion of the oesophagus. The normal peristaltic wave is replaced by irregular, unco-ordinated and non-propulsive tertiary contractions. The aetiology of the disorder is unknown but infection with *Trypanosoma cruzi* (Chaga's disease) produces a similar condition. Histologically there is degeneration and loss of the ganglion cells of Auerbach's plexus; those still present are often surrounded by a chronic inflammatory infiltrate.

The motor disturbances can be explained by the effects of denervation on the smooth muscle of the oesophagus and lower sphincter. The effects of denervation lead also to increased sensitivity of the smooth muscle to cholinergic drugs. Although the lower oesophageal sphincter has been shown to be hypersensitive to gastrin, plasma gastrin levels in patients with achalasia are within the normal range.

Diffuse oesophageal spasm is characterized by non-progressive contractions and substernal pain which may be severe and resemble cardiac ischaemia. Dysphagia does not necessarily accompany the chest pain and the lower oesophageal sphincter relaxes normally. Radiologically, a tortuous and 'corkscrew' barium outline occurs due to 'tertiary' non-propulsive contractions forming pseudodiverticula. The aetiology of the condition is unknown although diffuse changes in the vagus nerve have been reported. A similar picture may be seen in patients with reflux oesophagitis.

Motor disorders can occur with collagen vascular diseases, particularly systemic sclerosis, where there is aperistalsis in the lower two thirds of the oesophagus and lower oesophageal sphincter insufficiency which leads to severe reflux. There is both fibrosis of the oesophageal wall and abnormalities in the nervous control of the lower sphincter.

Gastro-oesophageal reflux

Intermittent incompetence of the lower oesophageal sphincter with short episodes of reflux occurs in the majority of healthy individuals after food. Symptomatic gastro-oesophageal reflux however is a common disorder in the middle aged and elderly and is usually due to the presence of a hiatus hernia. The absence of the intra-abdominal portion of the oesophagus allows any increase in intra-abdominal pressure to result in reflux of gastric contents since the lower oesophageal sphincter alone is unable to maintain competence of the cardia. Obesity and changes in posture aggravate the symptoms. Reflux can, of course, occur in the absence of a hiatus hernia and a decrease in resting lower oesophageal sphincter pressure can usually be demonstrated in these patients. The heartburn of pregnancy is an example of this, possibly mediated by the effect of sex hormones on the sphincter. Persistent reflux usually leads to oesophagitis which may result in benign stricture formation.

DISTURBANCES OF GASTRIC MOTILITY AND EMPTYING

Delay in gastric emptying is usually due to outflow obstruction caused by an antral carcinoma or duodenal ulcer scarring. Following truncal or selective vagotomy gastric atony occurs for a variable period of time and the operation must be combined with a drainage procedure such as a pyloroplasty or gastrojejunostomy. This is unnecessary if a highly selective vagotomy is carried out which preserves the vagal fibres to the pylorus and antrum (the nerves of Laterjet).

Gastroparesis may also occur in advanced diabetes, diffuse neuromuscular disorders and in otherwise healthy individuals following an apparent viral infection. Similarly ileal distension is associated with inhibition of gastric motility, the *ileogastric reflex*.

Rapid gastric emptying can occur whenever a surgical stoma is created between the stomach and small intestine. This may give rise to the unpleasant postprandial symptom complex referred to as 'the dumping syndrome' which can be conveniently divided into an early phase associated with hyperglycaemia and a late phase associated with hypoglycaemia.

'Early' dumping occurs during or soon after a meal and consists of weakness and faintness associated with abdominal discomfort and borborygmi. The early passage of hypertonic gastric contents into the small bowel causes distension, an influx of fluid into the lumen to restore jejunal isosmolarity with a concomitant drop in blood volume and a rise in blood glucose. Since symptoms may precede these measurable changes, humoral factors have been postulated and there is evidence to suggest that both 5-hydroxytryptamine and a bradykinin-like polypeptide are released into the circulation in increased amounts. It is a reflection on our lack of precise knowledge that glucagon, GIP, VIP, gastrin, secretin, cholecystokinin, neurotensin and substance P have all been implicated as well.

'Late' dumping produces similar symptoms 2–3 hours after a meal. The rapid passage of carbohydrate into the upper small bowel causes release of enteroglucagon. This sensitizes the pancreatic islets to produce increased amounts of insulin leading to symptomatic hypoglycaemia.

Vomiting occurs when afferent impulses activate the vomiting centre in the medulla. Stimuli include a rise in intracranial pressure, tactile stimulation of the throat, distension or injury to the bladder, uterus or testes, rotation or unequal stimulation of the labyrinths, chemical changes in body fluids, gastric or duodenal distension and drugs or noxious substances activating duodenal receptors. Vomiting usually begins with retching which defeats the normal anti-reflux mechanism by creating a sliding hiatus hernia. The antrum contracts shifting gastric contents into the upper part of the stomach and following a deep inspiration ejection of the contents is achieved by strong contractions of the abdominal wall musculature and diaphragm against a closed glottis. Acute pressure changes across the gastro-oesophageal junction may result in tears of the lower end of the oesophagus (Mallory-Weiss syndrome).

DISTURBANCES OF INTESTINAL MOTILITY

Disorders of intestinal movement, particularly of the colon, are common but their pathophysiology is poorly understood. Distension of an intestinal segment, rough handling of the bowel at surgery or peritoneal irritation can lead to an adynamic ileus known as the *intestino-intestinal reflex*; the resulting gaseous distension tends to exacerbate and prolong the condition.

Pain

Pain of colonic origin is due either to colonic distension or traction on the mesentery. The most frequent cause is the spastic colon variety of irritable bowel syndrome, a disorder of unknown aetiology which may be associated with constipation, or alternating constipation and diarrhoea. The pain is usually localized to the left iliac fossa but can occur anywhere in the abdomen. It is frequently related to meals or defaecation. When pain is present, pressure recordings from the colon show increased segmentation and anticholinergic drugs which inhibit colonic contractions cause relief. However, segmental pressures of the same amplitude and frequency occur in subjects without pain. The exact role of "stress" in modifying these responses and the threshold to pain in an individual patient with the irritable bowel syndrome remains unknown. Recently the effect of naloxone on lowering an experimentally induced irritability threshold in these patients has led to the proposal that endorphins may be important. Similarly recordings of myoelectrical activity in response to cholecystokinin and pentagastrin reveal differences in basal electrical activity between controls and patients suggesting that gastrointestinal peptides may also play a role.

In diverticular disease pronounced muscle thickening occurs and high intraluminal pressures have been reported but pain in uncomplicated disease is uncommon. The symptoms of both the irritable bowel syndrome and diverticular disease are usually improved by the taking of a high fibre diet and subsequent increase in faecal bulk although the role of low fibre diets in the aetiology of both conditions remains conjectural.

Constipation and diarrhoea

Constipation implies the difficult passage of hard stools, frequency being relatively unimportant since between 3 stools per day to 1 soft stool per week is within the normal range. The urge to defaecate occurs when the rectum is distended to a certain critical volume, usually 100–150 ml. This urge can be resisted by voluntary contraction of the external anal sphincter.

If this voluntary inhibition persists over a long period sphincteric relaxation will not occur in response to the normal stimulus of a full rectum. The subsequent use of purgatives may perpetuate the problem by leading to damage and loss of intrinsic innervation and thus to a failure of gut motility. The result is *dyschezia* or rectal constipation. Colonic constipation is due to slowing of colonic transit and tends to occur with inactivity (page 429). Constipation may also occur when neuromuscular disorders; for example, destruction of the lumbosacral cord leads to an inability to appreciate rectal distension. The exact physiological mechanism is complex.

Hirschsprung's disease or aganglionic megacolon is an interesting congenital abnormality characterized by an absence of ganglion cells in a variable length segment of distal bowel. The colon proximal to this segment becomes dilated and hypertrophied in a manner analogous to achalasia of the cardia.

A variety of drugs cause constipation, particularly anticholinergic drugs and ganglion-blocking drugs which inhibit motility, and also opiate derivatives which increase smooth muscle tone and raise intraluminal pressure.

Diarrhoea is the passage of excessive amounts of soft or fluid stools; it is not generally the result of a colonic motility disorder although segmentation is frequently decreased. Thus, in the non-haustrated colon of chronic ulcerative colitis peripheral resistance is low due to diminished segmenting contractions; a free flow of colonic contents occurs which contributes to the diarrhoea. Similarly patients with the painless diarrhoea variety of the irritable bowel syndrome show a marked reduction in wave-like motility. But conversely, there is an increase in colonic motility in anxiety, thyrotoxicosis and the carcinoid syndrome all of which may be associated with diarrhoea.

Principles of tests and measurements

Oesophageal motor disorders

Radiological methods remain the most common and easily available investigation. A barium swallow combined with cineradiography is particularly valuable since frame by frame analysis of co-ordination of the action of the tongue, pharyngeal musculature and upper sphincter is possible.

Manometry is useful for measuring pressures in the body and lower oesophageal sphincter both at rest and on swallowing. The most commonly employed method involves using catheters with lateral orifices that are perfused continuously with water.

The use of several tubes recording pressures at different levels enables orderly peristalsis and abnormal 'tertiary' contractions to be detected. More recently miniature silicon pressure transducers have been used which are especially suited for studies involving the upper sphincter and pharynx.

Tests to demonstrate reflux and/or oesophagitis

Radiology and endosocopy with biopsy are readily available but only patients with a severely incompetent sphincter tend to have free reflux on a barium swallow. The use of a pH probe combined with manometry may be helpful, a fall in pH below 4 or 5 indicating a reflux episode. Alkaline bile reflux, however, may also produce severe oesophagitis, the test is, therefore, of limited use following gastrectomy. The use of a pH probe can be combined with an acid clearing test in which the number of swallows necessary to clear a known amount of acid is measured.

Scintiscanning of radiolabelled saline instilled into the stomach is an elegant new technique. Intra-abdominal pressure is increased by use of an external binder and an episode of reflux represented by radioactivity occurring above the level of the lower sphincter.

The acid perfusion test (Bernstein test) is helpful in differentiating cardiac pain from that of oesophageal reflux. 0·1 N HCl is instilled into the lower oesophagus in a blind fashion and the patients symptoms are noted.

Gastric emptying

The three recognized methods of assessing gastric emptying are by radiology, intubation tests and the use of radioisotopes.

None of these is ideal and most depend on a marker with which to follow the emptying pattern. It is important to realize that gastric emptying patterns for fluids and solids are markedly different (p. 428) hence the use of contrast media is not truly representative and only measures total gastric emptying time.

Intubation tests require aspiration of a nonabsorbable marker e.g. phenol red, polyethylene glycol or chromium-51 and hence have the disadvantages of being invasive and requiring a liquid meal. A double aspiration technique with duodenal sampling can be employed since the duodenal contents are sufficiently liquid to be aspirated. Radioisotope tests utilizing solid meals labelled with a variety of isotopes have been used. The rate of emptying can then be measured by external scintiscanning or a gamma camera. The disadvantages are the expense of the equipment required and changes in the binding of the isotope to the test meal.

Gastric smooth muscle electrical activity can also be assessed using suction electrodes.

Colonic pressure and motility measurements

Pressures in the left side of the colon can be measured using either small air or fluid filled balloons, perfused polyethylene tubes or pressure sensitive radio-telemetering capsules. Whilst balloons are sensitive, false pressure changes can occur on direct contact with moving stool.

Colonic motor activity is assessed by tracing the progress of telemetering capsules or of substances impregnated with radioactive material. If telemetering capsules are combined with a radioactive marker e.g. chromium-51, propulsive activity can be recorded at the same time as pressure studies.

The use of small intraluminal suction electrodes, which can be attached via a standard sigmoidoscope, allows rectosigmoid myoelectrical activity to be measured.

Overall gut transit time can be measured by the use of a non-absorbable marker. The use of coloured dye, e.g. carmine red, is a crude test and only allows measurement of the time of first appearance or disappearance of the marker. Chromium-51 has the added advantage of allowing quantitation of the time for passage of the maximum amount of marker. The ingestion of a number of small barium pellets is probably the best method. The stools are collected and X-rayed and the number of pellets present counted and plotted against time. In addition plain X-rays of the abdomen allow the position of the pellets to be followed.

Small bowel transit time may be assessed by means of the lactulose breath test. Lactulose, a non-absorbable disaccharide, is broken down by colonic bacteria to give hydrogen which is excreted via the lungs. The time taken from ingestion to the appearance of H_2 in the breath is a measure of the mouth to caecum transit time.

These methods are useful research tools for understanding and comparing the effects of drugs and diets on gastrointestinal motility.

Practical Assessment

SWALLOWING AND GASTRO-OESOPHAGEAL REFLUX

Clinical observations

Pharyngeal dysphagia, coughing, aspiration or nasal regurgitation suggests weakness or anaesthesia of tongue, palate or larynx. Dysphagia for both liquids and solids and worse with cold liquids than hot suggests a motor disorder. Indicated site of obstruction only of very limited help in location of lesion.

Incompetence of cardia; postural retrosternal pain, acid regurgitation.

Routine methods

Edrophonium test for myasthenia gravis. Barium swallow; obstruction, incoordination, loss of peristalsis, reflux. Cineradiography. Endoscopy and biopsy; hiatus hernia, reflux, macro- and microscopic evidence of oesophagitis. Bernstein test.

Special techniques

Oesophageal manometry, pH probes and acid clearing test.

GASTRIC EMPTYING

Clinical observations

Gastric outlet obstruction; vomiting of stale food, distended stomach with visible peristalsis, succussion splash.

Routine methods

Gastric aspiration; greater than 50 ml after overnight fast suggests gastric outlet obstruction. Barium meal; delayed emptying of contrast.

Special techniques

Intubation and radioisotope test meals.

INTESTINAL MOTILITY

Clinical observation

Site of pain, relieving and aggravating factors in irritable bowel syndrome. Pain may be reproduced by rectal examination or insufflation of air at sigmoidoscopy. Acute onset of colicky pain suggests obstruction. Absent bowel sounds on auscultation.

Stools: small and stringy or 'rabbity' in irritable bowel syndrome. Blood and mucus suggests inflammatory bowel disease.

Routine methods

Sigmoidoscopy–mucosal inflammatory lesions, melanosis coli suggesting laxative abuse; confirm by rectal biopsy. Plain abdominal X-ray–obstructive pattern of dilated gas filled loops. Barium meal and "follow-through". Barium enema; evidence of obstruction, mucosal disease, absence of haustra. Colonoscopy and biopsy; obstructive or inflammatory lesions.

Special techniques

Colonic pressure and motility studies, myoelectrical activity.

DIGESTION AND ABSORPTION

Normal function

Digestion is the chemical breakdown of food into simpler molecules which can then be absorbed.

Absorption is the transfer of these simpler molecules across the small bowel mucosa into the portal venous blood or lymphatics. Since it is the algebraic sum of transport from lumen to blood and blood to lumen, the term 'net absorption' is more correct.

Both digestion and absorption are able to proceed in the absence of neural or hormonal influences. Digestion of food is carried out by *an intraluminal phase* which occurs extracellularly under the influence of enzymes secreted by the salivary glands, the chief cells of the stomach and the pancreas and by *a membrane phase* under the influence of hydrolytic enzymes associated with the columnar epithelial cells of the villus. Despite the salivary and gastric components little if any digestion occurs until the duodenal and jejunal lumen are reached, where the major products of digestion are absorbed across the small intestinal mucosa. For most substances the site of maximum absorptive capacity is the jejunum although bile salts and vitamin B_{12}-intrinsic factor are absorbed almost exclusively in the distal ileum.

ANATOMICAL CONSIDERATIONS

The surface of the small intestinal mucosa is thrown into folds covered by finger-like villi which protrude into the lumen. These villi occur at a density of 10–40 per mm^2 and are 0·5–1·5 mm in length. Their surface is covered by a layer of columnar epithelial cells (enterocytes), the free borders of which are composed of hundreds of minute microvilli. These microvilli greatly increase the absorptive area of the small intestine and on light microscopy constitute the brush border.

The brush border is the barrier which nutrients, water and electrolytes must traverse before entering blood or lymph. Ultrastructurally, this barrier consists of a luminal unstirred layer of fluid up to 1 mm thick, a layer of acidic polysaccharides (the glycocalyx) covering the microvilli, the cell membrane and cytoplasm of the enterocyte, the basal cell membrane, the intercellular space, the basement membrane and the capillary or lymph vessel wall. The brush border membrane is composed of a bimolecular layer of lipids (primarily sterols and phospholipids) whose polar groups are orientated outwards and covered with protein. This arrangement facilitates the transport of lipid-soluble (non-polar) molecules across the membrane.

TRANSPORT PROCESSES

The enterocyte membrane utilizes a variety of physico-chemical processes to regulate the flux of solutes and fluid across the intestinal mucosa.

Passive diffusion allows particles to traverse the cell layer through intercellular spaces along osmotic gradients. The rate of diffusion is dependent on the molecular size, lipid solubility and net electrical charge of the particles. Highly lipid-soluble (non-polar) compounds are able to dissolve in the brush border membrane and diffuse through rapidly, and in the case of partly ionized compounds only the unionized lipid-soluble molecule is absorbed. Water soluble (polar) compounds however do not readily traverse cell membranes although if small enough (MW < 100) they will diffuse through the aqueous pores.

Facilitated diffusion is the movement of particles from a region of high concentration to one of low concentration or down an electrical gradient until equilibrium is reached.

Active transport is highly specific and allows solute movement to occur against an electrochemical gradient. Unlike passive and facilitated diffusion the process is energy dependent. The active transport of many nutrients such as glucose, galactose and the majority of free aminoacids is based on pumps and carriers which are probably complex proteins located in the cell membrane although their exact nature is not understood.

CARBOHYDRATE DIGESTION AND ABSORPTION (see Fig. 12.1)

An average daily Western diet contains 400 gm of carbohydrate comprising 60 per cent starch, 30 per cent sucrose and 10 per cent lactose. Starch is a mixture of amylose and amylopectin both of which are glucose polymers whilst sucrose and lactose are disaccharides of glucose and fructose, and glucose and galactose, respectively. The

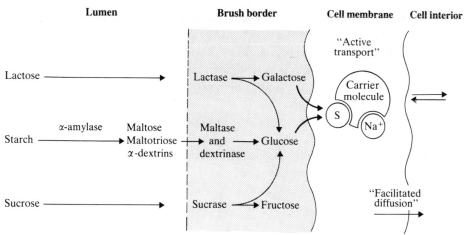

Figure 12.1. *Diagrammatic representation of carbohydrate digestion and absorption.*

intraluminal phase of starch digestion is under the influence of both salivary and pancreatic α-amylases although the oral component is less significant due to the short exposure time. Amylose is split into maltose and maltotriose whilst amylopectin additionally yields α-limit dextrins which are branched saccharides containing an average of eight glucose units.

The membrane phase is completed by brush border enzymes which hydrolyse the oligo- and di-saccharides to their constituent hexoses (Fig. 12.1). These enzymes are; lactase (lactose to glucose and galactose), sucrase (sucrose to glucose and fructose), maltase (maltose to glucose) and α-dextrinase (branched dextrins to glucose and maltose). The resultant released hexoses are of two types. The aldohexoses, glucose and galactose are actively transported across the intestine against a concentration gradient by a shared sodium dependent carrier molecule which uses energy derived from ATP hydrolysis. Sodium enters the cell carrying with it the concomitant co-transported water soluble substrate e.g. aldohexoses, aminoacids, etc. A sodium independent system of glucose absorption which may account for up to 50 per cent of the amount absorbed has also been described.

Fructose, the ketohexose derived from the hydrolysis of sucrose, is absorbed by facilitated diffusion.

PROTEIN DIGESTION AND ABSORPTION

Virtually all ingested protein is completely broken down and absorbed as aminoacids and small polypeptides. Additionally, protein derived from digestive juices and desquamated intestinal cells is also digested and absorbed.

Although protein digestion begins in the stomach with denaturation by gastric acid and hydrolysis of internal peptide bonds by pepsin, the major part of protein digestion takes place within the lumen of the duodenum and upper jejunum. This intraluminal phase is under the influence of the pancreatic proteases (trypsinogen, chymotrypsinogen, proelastase and procarboxypeptidase) all of which are inactive until they reach the bowel lumen where trypsinogen is activated to trypsin by the brush border enzyme, enterokinase. Trypsin in turn acts autocatalytically to continue to activate trypsinogen and the other protease precursors.

The endopeptidases trypsin, chymotrypsin and elastase break down the internal protein peptide bonds to produce peptides which are subsequently acted upon by carboxypeptidase A and B to produce neutral and basic amino acids respectively and small oligopeptides of 2–4 amino acid residues.

The membrane phase of peptide digestion is complex. Whilst there are brush border peptidases capable of breaking down oligopeptides into dipeptides and individual amino acids the majority of cellular peptidase activity resides in the surface epithelial cell cytosol fraction. It appears that tri-, tetra-, and di-peptides with bulky side chains are hydrolyzed at the intestinal cell surface whilst other dipeptides are absorbed intact and hydrolyzed within the cell itself.

A sodium-dependent carrier mechanism analogous to that for glucose is responsible for the absorption of individual amino-acids and their transport across the cell where they are taken up by the portal venous system.

FAT DIGESTION AND ABSORPTION

Dietary fat is composed mainly of triglycerides which are long chain fatty acids linked to glycerol. A small percentage is in the form of medium chain triglycerides. Since triglycerides are not water soluble, dietary fat must first be emulsified for absorption to take place. Although some emulsification takes place as a result of the churning action of the gastric antrum and segmental contractions of the small intestine, the major factor influencing emulsification is the presence of bile salts within the duodenum. Above a certain critical concentration, which is lower for conjugated bile salts than for unconjugated ones, bile salts form mixed micelles. These are large molecular aggregates which possess hydrophilic (polar) groups facing outwards and lipophilic (hydrophobic) groups orientated inwards. Co-lipase secreted by the pancreas attaches to the surface of an emulsified lipid droplet creating a binding site for pancreatic lipase. Lipase acting at the oil-water interface of the lipid droplet hydrolyzes triglyceride ester bonds to form free fatty acids and/or monoglycerides which are in turn incorporated into the bile salt micelle and the process continues (Fig. 12.2). Other lipids such as lysolecithin, cholesterol and fat soluble vitamins also get taken up into the micelle. The end result is water soluble micellar fat. The micelles thus act as labile storage depots transporting lipids across the unstirred layer of the brush border where they are released to traverse the cell membrane either by diffusion or by utilizing a carrier protein. This carrier protein transfers long chain fatty acids to the endoplasmic reticulum where they are reconstituted with monoglyceride to form triglyceride. The bile salts pass on down the intestine

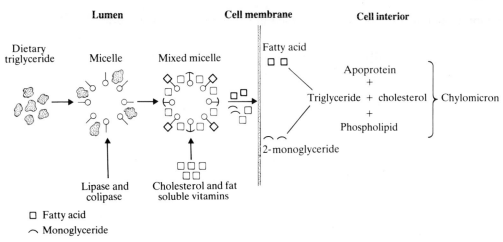

Figure 12.2. *Diagram of the major steps in the digestion and absorption of fat.*

to be reabsorbed in the terminal ileum as a continuation of the enterohepatic circulation.

The final stage in the process of fat absorption is for the reconstituted triglyceride to be coated with apoprotein, phospholipid and unesterified cholesterol to form chylomicrons which enter the central lacteal of the villus from where they reach the circulation via the thoracic duct.

Medium chain triglycerides appear to be absorbed intact and hydrolyzed within the mucosal cell by intestinal lipase. The constituent glycerol and fatty acids then pass directly into the portal blood without re-esterification. This improved efficiency in the handling of medium chain triglycerides allows them to be substituted for dietary fat, particularly in patients with steatorrhoea due to maldigestion.

VITAMIN ABSORPTION

The fat soluble vitamins A, D, E and K are associated with dietary lipid and become incorporated into the mixed micelles which enable them to come into contact with the intestinal surface epithelium. Their absorption is therefore dependent upon the presence of bile salts and an intact small bowel mucosa but not lipase.

The water soluble vitamins are absorbed in most cases by simple passive diffusion. However, vitamin C (L-ascorbic acid) appears to be absorbed in the ileum by a sodium dependent active process and vitamin B_1 (thiamine) by an active process at low luminal concentrations and by diffusion as the concentration increases.

Folic acid in the diet exists predominantly in the form of polyglutamate conjugate. These conjugates are hydrolyzed by a specific brush border peptidase to give active monoglutamic folate which is absorbed in the upper small bowel probably utilizing a specific, active carrier process. Within the mucosal cell monoglutamic folate is converted to reduced methyl folate which is the principal biologically active form circulating in the blood.

Vitamin B_{12} is preferentially bound by a salivary cobalamin-binding protein (R protein) at low pH in the stomach. A combination of partial degradation of this protein by pancreatic proteases and a rise in pH allows dissociated vitamin B_{12} to combine with intrinsic factor, a mucoprotein secreted by the gastric parietal cells. The vitamin B_{12}-intrinsic factor complex then passes down the small intestine to be absorbed intact by a specific active transport system in the distal ileum.

MINERAL ABSORPTION

Calcium is absorbed in the duodenum and upper jejunum by an active transport mechanism involving a specific calcium binding protein (CaBP) and probably other microvillar proteins in addition.

The synthesis of CaBP present within the intestinal mucosal cell is dependent upon the physiologically active form of vitamin D, 1α, 25 dihydroxy cholecalciferol (1α, 25 $(OH)_2 D$) under the influence of circulating parathyroid hormone. The amount of CaBP available is inversely proportional to the dietary intake of calcium.

Iron. The amount of iron absorbed is inversely related to body stores and mucosal cell iron content and occurs as a two stage process of uptake followed by serosal transfer. Iron uptake occurs throughout the length of the small intestine at low intraluminal concentrations by means of an energy dependent active transport process. Following uptake into the mucosal cell iron is either incorporated into the iron binding protein ferritin or serosal transfer occurs. It is this latter process which is localized to the duodenum and which is the rate limiting step in iron absorption.

FLUID AND ELECTROLYTE TRANSPORT

Fluid movement. Seven to ten litres of fluid enter the upper small bowel each day consisting of 1·5 to 2·5 litres of ingested fluid, 1·0 to 1·5 litres of saliva, 2 to 3 litres of gastric juice and 2 to 3 litres of pancreatic, biliary and small intestinal secretions. Of this only about 1·5 litres reaches the colon most of which is absorbed leaving the average stool on a Western diet containing 100–200 ml of water (Table 12.4).

The small intestinal mucosa is freely permeable to water such that the maximum rate of absorption is theoretically at least greater than 1 litre per hour. In practice the gastric mucosa is relatively impermeable to water and the amount of fluid delivered to the small bowel is controlled by the rate of gastric emptying under the influence of osmoreceptors present in the duodenum. Anisotonic fluids therefore enter the small

Table 12.4. Colonic fluid and electrolyte turnover

	Terminal Ileum		Stool		
	Amount ml or mmol	Conc^n	Amount ml or mmol	Conc^n	Amount absorbed ml or mmol
Water	1500	—	100	—	1400
Na	210	140	4	40	206
K	9	6	9	90	—
Cl	90	60	1	10	89
HCO$_3$	75	50	3	30	—

bowel slowly, to be rendered isotonic by the movement of water across the freely permeable mucosa.

This movement of water across the mucosa occurs as a passive response to osmotic pressure gradients. Although this movement may be bidirectional the net effect is absorption as a result of the active absorption of solutes such as glucose and amino acids. The permeability of the small intestinal mucosa to water and solutes is greatest in the jejunum and decreases down the bowel, the colon being the least permeable.

Electrolyte absorption. In the jejunum, where intestinal permeability is high, sodium and potassium are predominantly absorbed by passive movement down concentration gradients. However, two other mechanisms are available. The active transport of glucose and amino acids induces an osmotic flow of water which in turn 'carries' sodium and potassium ions through the large pores, a process known as 'solvent drag'. In addition there appears to be a glucose-dependent brush border carrier molecule which transports sodium into the cell (p. 439). The maintenance of a low intracellular sodium concentration which enables the carrier mechanism to operate is provided by energy derived from ATP hydrolysis under the influence of sodium and potassium ATPase.

Bicarbonate, largely derived from pancreatico-biliary secretions is removed from the jejunal lumen by combination with actively secreted H^+ to form CO_2 and water.

Since the presence of luminal bicarbonate also appears to stimulate sodium absorption it has been postulated that a Na^+/H^+ exchange process exists.

Within the ileal lumen there appears to be a double ion-exchange process such that chloride ion is exchanged for bicarbonate and sodium ion is exchanged for hydrogen ion. The net result is the absorption of sodium and chloride by specific active absorptive processes against electrical and concentration gradients.

Transport mechanisms in the colon are similar to those in the small intestine although quantitatively different. Semi-liquid ileal discharge is transformed into semi-solid stool. Sodium is actively absorbed across the colonic mucosa creating a luminal potential difference 30–40 mV negative to that of the serosa whilst potassium accumulates within the lumen as a result of passive diffusion down the electrical gradient, active secretion in exchange for absorbed sodium and loss in mucus and desquamated epithelial cells. Although stool concentrations of potassium are high the actual amount lost in stool under normal conditions remains relatively small due to the small volume of stool water excreted. Despite net movement into the lumen, stool bicarbonate concentration remains low because of its reaction with organic acids to form carbon dioxide and water (Table 12.4).

Disordered function

MALABSORPTION

In principle, failure to absorb a single substance e.g. Vitamin B_{12} or excessive absorption of a substance such as iron (as in haemochromatosis) should be covered by this term but in practice a malabsorption syndrome implies steatorrhoea with deficiencies in the absorption of fat soluble vitamins, protein, carbohydrate and minerals occurring to a varying degree depending on the underlying disorder. Nevertheless steatorrhoea is far from invariable and patients with a malabsorption syndrome may present with any or one of the complications listed in Table 12.5. These secondary manifestations often overshadow the intestinal complaints particularly in severe cases.

CAUSES OF THE MALABSORPTION SYNDROME

A classification of the major causes of malabsorption is given in Table 12.6.

Table 12.5. Malabsorption and its sequelae.

	Substance malabsorbed	Complication
Water soluble vitamins	Vitamin B complex	Peripheral neurophathy, glossitis, cheilitis and dermatitis.
	Vitamin B12	Neurological sequelae and anaemia
	Folic acid	Anaemia
	Iron	Anaemia
Fat soluble vitamins	Vitamin K	Prolonged prothrombin time
	Vitamin A	Xerophthalmia and keratomalacia
	Vitamin D $\}$ Calcium	Tetany and osteomalacia
	Protein	Hypoalbuminaemia and osteoporosis
	Fat	Steatorrhoea
	Carbohydrate (disaccharides)	Abdominal distension and flatulence
Water and electrolytes	Na^+ K^+ H_2O $\}$	Hypotension, muscle cramps and weakness

Table 12.6. Classification and causes of the malabsorption syndrome

1. Disturbances of digestion.
   Pancreatic enzyme deficiency
   Bile salt deficiency

2. Intestinal mucosal lesions.
   Gluten enteropathy
   Tropical sprue
   Dermatitis herpetiformis

3. Structural lesions of the small intestine.
   Crohn's disease
   Lymphoma
   Amyloid
   Radiation

4. Surgery.
   Small intestinal resection
   Gastric surgery

5. Infections.
   Parasites e.g. Giardiasis
   Small bowel bacterial overgrowth
   Tuberculosis
   Whipple's disease

6. Specific mucosal cell disorders.
   Disaccharidase deficiency
   Vitamin B12 malabsorption
   Cystinuria and Hartnup disease

7. Pharmacological.
   Drugs e.g. cholestyramine
   Alcohol

8. Miscellaneous.
   Endocrine disorders
   Carcinoid syndrome
   Systemic mastocytosis

Disturbances of digestion. Pancreatic enzyme deficiency is usually due to chronic pancreatitis in adults and cystic fibrosis in childhood. In adults malabsorption of fat and protein occurs with weight loss as the predominant feature whilst in children pancreatic insufficiency may lead to failure to thrive, hypoalbuminaemic oedema and vitamin K deficiency in addition to the respiratory complications of the disease. Vitamin D deficiency and metabolic bone disease is not a feature of these disorders. Conjugated bile salt deficiency is caused by severe cholestatic liver disease, prolonged extrahepatic biliary tract obstruction, a diminished bile salt pool or bacterial overgrowth leading to deconjugation of bile salts. The resulting failure to bring fat and fat soluble vitamins into a water soluble phase results in steatorrhoea, bleeding tendencies and eventually osteomalacia.

Intestinal mucosal lesions. Gluten enteropathy and the enteropathy associated with dermatitis herpetiformis primarily affect the upper small bowel causing villous atrophy and a consequent loss of absorptive area. Whilst any of the deficiencies noted in Table 12.5 can occur, folic acid and vitamin D deficiency are particularly common. Withdrawal of gluten from the diet leads to remission of the mucosal lesion in both disorders. The aetiology of tropical sprue is unknown and mucosal abnormalities occur throughout the small intestine. Folic acid deficiency is invariable and the disorder usually responds to a combination of antibiotics and folic acid.

Structural lesions of the small intestine. Diseases in this category give rise to malabsorption due to a combination of mucosal damage, infiltration of the lamina propria and lymphatic obstruction. Patients with Crohn's disease may also have small bowel bacterial overgrowth and/or bile salt malabsorption as a result of ileal disease or resection.

Surgery. Limited resection of the small bowel is tolerated well since it has a large reserve and is capable of increasing its absorptive capacity. However ileal resection leads to an inability to absorb the vitamin B_{12}-intrinsic factor complex and to bile salt deficiency due to interruption of the entero-hepatic circulation. Loss of the ileo-caecal valve may also contribute to bacterial contamination of the small intestine. Patients with ileal resection are also liable to hyperoxaluria and the development of urinary calculi. This appears to be due to a combination of increased colonic absorption of oxalate caused by the presence of bile salts and to the formation of calcium soaps by excess luminal fatty acids, thus reducing the amount of calcium available to bind oxalate and render it non-absorbable.

Malabsorption is a relatively frequent occurrence following gastric surgery and whilst more common after gastrectomy it may also develop after a vagotomy and drainage procedure. A number of factors contribute to the malabsorption, including inadequate mixing of contents with bile and pancreatic enzymes, loss of the reservoir function of the stomach with a decreased transit time, bacterial overgrowth in stagnant loops and abnormalities of pancreatic function following vagotomy. Iron, folic acid, vitamin B_{12} and vitamin D deficiency may also all occur particularly with a Polya gastrectomy. Occasionally latent disaccharidase deficiency or gluten enteropathy is unmasked by gastrectomy.

Infection. Infestation with the parasite Giardia lamblia may infrequently give rise to a malabsorption syndrome. Such infection tends to be particularly common in patients with a gastrointestinal immunodeficiency syndrome.

In Whipples disease the small bowel lamina propria is infiltrated with PAS positive macrophages and electronmicroscopy reveals the presence of bacteria (the nature of which are unknown). Prolonged antibiotic therapy results in remission.

The normal upper small bowel contains approximately 10^4 organisms per gram of contents; however the numbers increase down the length of the bowel and a marked transition occurs across the ileo-caecal valve where caecal bacterial counts of 10^9-10^{12} organisms per gram of contents are usual. Bacterial proliferation in the upper small bowel occurs in a variety of disorders including achlorhydria, post gastrectomy, blind loops, fistulous disease, jejunal diverticula and stasis e.g. in systemic sclerosis. The result is an overgrowth of bacteria, primarily anaerobes, capable of deconjugating bile salts and of combining with vitamin B_{12}, either bound or unbound, rendering it unavailable for absorption.

Specific mucosal cell disorders. Hypolactasia is the commonest disaccharidase deficiency. This may rarely be due to congenital alactasia but is usually acquired in later life. In the primary form intestinal lactase levels are high at birth but decrease in childhood or adolescence and the small intestine is otherwise structurally and functionally normal. Approximately 10 per cent of Caucasians and the majority of Asians and Blacks have low intestinal lactase levels. The secondary form occurs as a result of small intestinal mucosal damage e.g. following gastroenteritis. The presence within the colonic lumen of non-absorbed disaccharide acts as an osmotic load (particularly since bacterial degradation in the colon yields two solute molecules for every one of lactose). In addition bacterial fermentation of lactose results in the production of organic acids with an irritant effect on the colon.

Cystinuria and Hartnup disease are rare congenital abnormalities in the transport of specific amino acids.

ABNORMALITIES OF FLUID AND ELECTROLYTE TRANSPORT

Gastrointestinal Fistulae

Pancreatic and biliary fistulae usually occur post-operatively and fluid loss from either is rarely excessive although it may be up to 1·5 litres per day. Since both secretions contain large amounts of bicarbonate a metabolic acidosis can ensue.

Small intestinal fistulae are nearly always the result of Crohn's disease. The fluid lost is virtually isotonic with plasma hence thirst may not be a feature.

Diarrhoea

The four major mechanisms of diarrhoea are: Disturbances of intestinal motility (p. 433), increased intestinal secretion (p. 456), the presence within the gut lumen of osmotically active substances and inhibition of normal active ion absorption.

Osmotic diarrhoea occurs as a result of the presence within the intestinal lumen of osmotically active solutes. These lead to retention of water and a consequent increase in faecal bulk. This may be the result of maldigestion of ingested food e.g. in lactase

deficiency or because of the ingestion of poorly absorbable osmotically active substances e.g. magnesium sulphate used as a laxative. Clinically such diarrhoea is characterized by the fact that it ceases with fasting.

Inhibition of normal active ion absorption in the colon results in the accumulation of
electrolytes within the colonic lumen and subsequent diarrhoea. In the rare disorder of
congenital chloridorrhoea active chloride ion absorption is defective. Bicarbonate ion
is therefore not secreted into the lumen in exchange for the chloride ion and the stools
become acidotic and the patients alkalotic. A similar mechanism causing inhibition of
active sodium ion absorption may contribute to the diarrhoea associated with defective
bile salt and fatty acid absorption or with mucosal inflammatory disorders.

Principles of tests and measurements

Tests used in the investigation of a patient suspected of having a malabsorption syndrome on clinical grounds fall into two groups, those used as screening tests and those
used to identify a specific cause.

SCREENING TESTS

Quantitative faecal fat estimation

The total daily faecal fat output remains the most useful routine test for documenting
the presence of abnormal digestion or absorption.

The human small intestine has an enormous reserve capacity to absorb fat and
even on a high fat diet normal individuals excrete less than 5 g per day. Of this 1–2 g is
endogenous; desquamated epithelial cells, bile salts and bacteria. Since the output of
fat in the faeces varies from day to day stool collections over a number of days are
required. The use of high fat diets, equilibration periods, long collecting times and inert
markers to ensure accurate stool collections have all been used to improve the reliabil-
ity of the test. In practice, however estimation of the amount of fat present in a
three-day stool collection taken whilst the patient is on an ordinary ward diet remains
the standard method.

Indirect tests of fat absorption

Serum carotene. Carotene is the precursor of Vitamin A and therefore depends upon
normal fat transport for absorption. If the amount of dietary carotene is adequate then
a low level is a useful screening test of fat malabsorption.

Radioisotope fat studies. The concept of assessing the absorption of labelled fat by
measuring the amount of the label present in blood or urine following oral ingestion is

attractive and obviates the need to handle faeces. However, in practice the method appears to be a poor indicator of mild steatorrhoea and has an unacceptably high incidence of false positive and negative results. More recently a method of assessing fat absorption using a ^{14}C labelled fatty acid has been described. Following ingestion of ^{14}C-triolein, $^{14}CO_2$ production in the breath is measured. The use of the shorter acting, more stable ^{13}C may ultimately prove more suitable although it has the disadvantage of requiring costly apparatus.

D-Xylose absorption test

D-Xylose is a monosaccharide which is absorbed in the proximal small bowel and only metabolized to a small degree. Its absorption is not influenced by pancreatic or biliary secretion and it is therefore useful as an indicator of upper small bowel mucosal disease. Following a 5 g oral dose at least 1·2 g of D-Xylose should appear in the urine within five hours. Recently the value of a single one-hour blood xylose estimation has been emphasized. However, as a screening test xylose absorption is of little practical value because of the large number of false negative results which occur in malabsorption syndromes not associated with mucosal disease.

Stool osmolality and electrolytes

The assessment of stool osmolality, sodium and potassium concentrations may be used to differentiate osmotic diarrhoea from secretory diarrhoea (p. 459). Normally $(Na + K) \times 2$ should approximately equal the measured osmolality. A discrepancy suggests that other non-electrolyte solutes are present and contributing to the diarrhoea.

IDENTIFICATION OF SPECIFIC CAUSES OF MALABSORPTION

The assessment of pancreatic function is described on p. 462.

Small bowel bacterial overgrowth.

Urinary indican estimation remains the simplest of the indirect tests for small bowel bacterial colonization. Although dietary tryptophan is normally completely absorbed in the small intestine certain bacteria are capable of metabolizing this to indicans which are excreted in the urine. However there is only a very poor correlation between jejunal bacterial counts and urinary indicans and any disorder causing malabsorption of tryptophan will lead to its conversion to indicans by colonic bacteria.

Breath tests. The ^{14}C glycocholate breath test relies on the deconjugation of ^{14}C-labelled bile salts by small bowel bacteria with the production of $^{14}CO_2$ which is then excreted in the breath. False positive results may occur with ileal resection or disease since colonic bacteria are able to deconjugate the malabsorbed bile salt.

The production of hydrogen from indigestible substrate e.g. lactulose by intestinal bacteria is the basis of the H_2 breath test. An early 'peak' of H_2 in the breath following an oral dose of lactulose suggests small bowel bacterial overgrowth. False positives may occur with rapid intestinal transit (p. 435).

Disaccharidase deficiency

A number of methods are available for determining disaccharidase deficiency. The lactose tolerance test is carried out in a similar manner to a standard glucose tolerance test using 100 g of lactose (= 50 g glucose + 50 g galactose). Lactose malabsorption results in a flat curve and symptoms may be reproduced. Recently, lactose breath tests have proved a simple screening test for disaccharidase deficiency although less accurate than the assay of enzymes in small intestinal mucosal biopsies. In hypolactasic individuals an oral dose of lactose will pass unabsorbed into the colon where it is fermented by colonic bacteria to produce hydrogen which is exhaled.

Tests to assess absorption of individual substances

Folic acid absorption tests whilst described, are rarely required in practice.

Vitamin B_{12}. The standard test of vitamin B_{12} absorption is the Schilling test with and without intrinsic factor, although whole body counting of $^{58}CoB_{12}$ following an oral dose is now regarded as the most sensitive and reliable method.

Calcium. The twenty-four urinary excretion of calcium is a useful screening test of calcium deficiency. Whole body counting of $^{47}CaCl_2$ following a test meal has been described but is rarely used in practice.

Iron deficiency in the presence of a normal diet and in the absence of blood loss suggests malabsorption. Stool or whole body counting of ^{59}Fe following an oral dose of labelled ferric chloride may be useful in difficult cases.

Direct sampling of intestinal contents

Although not used routinely, intubation studies provide information about the net and bidirectional transport of water and ions in the intestine and the absorption of sugars, amino acids and peptides. Using a double-lumen tube a short length of intestine is perfused with the test substance and a non-absorbable marker. The disappearance of the test substance is then measured by sampling the distal part of the perfused segment and applying a correction factor based on the amount of marker retrieved.

Practical assessment

Clinical observations. Diarrhoea, nutritional deficiencies (Table 12.5); onset in child-hood (coeliac disease, cystic fibrosis); abdominal pain, diabetes (chronic pancreatitis); surgery (gastric and small bowel).

Routine methods. Assess deficiencies; blood film, serum B_{12} and folate, iron and total iron binding capacity, calcium and alkaline phosphatase, liver function tests and pro-teins, prothrombin time.

Three-day faecal fat excretion, xylose absorption test and serum carotene. Twenty-four hour urinary calcium excretion, lactose tolerance test, bone X-rays. Exclude ana-tomical abnormality; small bowel biopsy and radiology.

Special techniques. Pancreatic function tests (p. 462), endoscopic retrograde cholangio-pancreatography (ERCP). ^{14}C-glycocholate, lactose and lactulose breath tests. Disaccha-ridase assay of small bowel biopsies. Isotope absorption tests of folate, B_{12}, calcium and iron. Perfusion studies.

GASTROINTESTINAL SECRETION

Normal function

SALIVARY GLANDS

The salivary glands secrete 1–2 litres of saliva per day and are capable or producing up to 4 ml per minute on maximal stimulation. The parotid glands are 'serous' glands producing a watery fluid devoid of mucin whilst the submandibular and submaxillary glands are mixed 'mucous' and 'serous' which produce a more viscid saliva. Per unit weight of tissue salivary glands are able to secrete at a rate several times higher than any other gastrointestinal organ. In addition to its digestive function saliva is necessary for lubrication of food prior to swallowing, facilitating speech, and the maintenance of oral hygiene.

Composition

The major constituents of saliva are water, electrolytes, enzymes and mucroproteins. Although acinar secretion resembles an ultrafiltrate of plasma the final product is hypotonic and contains less sodium but added potassium.

Salivary enzymes include alpha-amylase, capable of breaking up the glycosidic bonds of glucose polymers such as occur in starch and lysozyme which attacks micro-bial cell walls thus contributing to the prevention of dental caries.

Saliva also contains a cobalamin-binding protein known as 'R' protein. Inability to degrade 'R' protein may lead to Vitamin B12 malabsorption. (p. 442)

Control of secretion

Although aldosterone and antidiuretic hormone alter the sodium and potassium concentrations of saliva, initiation and maintenance of secretion is almost exclusively dependent on parasympathetic and sympathetic nerves. The parasympathetic system is the stronger stimulus for secretion and since its nerves are cholinergic, drugs with an anticholinergic action often produce a dry mouth as an unwanted side effect. The role of neurocrine peptides such as substance P which is known to stimulate salivary secretion remains to be elucidated.

THE STOMACH

Structure

The mucosa of the stomach has a surface area of 800 cm^2 and contains several different types of secretory cell. The body of the stomach, particularly the fundic area contains glands which are long and tortuous with the lumina of three to seven glands opening into a depression on the mucosal surface known as gastric pits. Each gland contains several types of cell of which the chief (zymogen) cells are the most numerous and are responsible for the production of pepsinogens. The parietal (oxyntic) cells are interspersed among the chief cells particularly in the mid region of the gland and are responsible for the production of gastric acid.

Mucous cells are situated in the neck region of the gland and together with the surface epithelial mucous cells produce a layer of viscid alkaline mucus which protects the gastric mucosa against the irritant effects of acid, pepsin, bile and pancreatic proteolytic enzymes.

The distal fifth of the stomach (i.e. the antrum but extending for a variable length up the lesser curve) contains glands which do not possess either parietal or chief cells. Interspersed amongst the mucous cells of these glands are the gastrin producing G cells.

Secretion

The parietal cells produce a solution containing approximately 150 mmol of hydrochloric acid per litre and a Vitamin B12–R factor binding macromolecule, the intrinsic factor. The chief cells secrete a small volume of juice whose electrolyte composition approximates to that of extracellular fluid. This juice contains two types of immunologically distinct pepsinogens, group I (PGI) and group II pepsinogens (PGII) which are converted to the active protease pepsin. Pepsin is a heterogeneous group of enzyme proteins capable of hydrolyzing protein and polypeptides in an acidic medium. The surface epithelial cells produce mucus which is a viscid gel composed of various macromolecules and lysozyme whilst the neck chief cells secrete soluble mucus which is chemically and physically distinct. Gastric juice is a mixture of secretions from these four types of gastric secretory cell with in addition contamination by food, saliva and duodenal juice. It contains both an acidic and an alkaline component, the H^+ concentration of the juice rising as the rate of secretion increases.

Some individuals secrete ABH (O) blood group antigens in their gastric juice; this process is under genetic control. Blood group and secretor status may have a bearing on the future development of peptic ulceration.

Control

It is customary to divide postprandial gastric secretion into cephalic, gastric and intestinal phases according to the site at which the receptors are activated. In practice these phases begin almost simultaneously and overlap in time. Tasting, smelling, chewing and swallowing of food, particularly when appetizing, stimulates parietal cell secretion via the vagus, the mediator appears to be acetylcholine and the entire response can be blocked by vagotomy. Hypoglycaemia also activates hypothalamic centres that stimulate secretion via the vagus. The gastric phase is predominantly hormonal (p. 417). Amino acids and peptides are the major releasers of gastrin (predominantly G17) although vagal stimulation during the cephalic phase and antral distension are also important. The latter effect is mediated by local reflexes. All stimulants for gastrin release are inhibited by acid acting directly on antral G cells. In man distension of the intact stomach also causes acid secretion without a concomitant rise in serum gastrin. Whilst protein digestion products in the duodenum are known to stimulate acid secretion this effect does not appear to be mediated by gastrin.

The exact relationship between histamine, acetylcholine and gastrin at the parietal cell level is still a matter of controversy despite the recent demonstration of histamine (H_2) receptors and the develoment of compounds showing specific competitive antagonism to these receptors.

The 'mediator hypothesis' (figure 12.3) suggests that all the secretagogues such as gastrin and acetylcholine stimulate the parietal cell via histamine receptors whilst the 'permissive hypothesis' proposes that histamine continually sensitizes the parietal cell to the effects of acetylcholine and gastrin and that when this action is blocked acid secretion in response to these mediators is diminished.

Studies on isolated parietal cells indicate that histamine, acetylcholine and gastrin all have independent actions although in combination they interact and augment stimulation of the cells.

Chapter 12

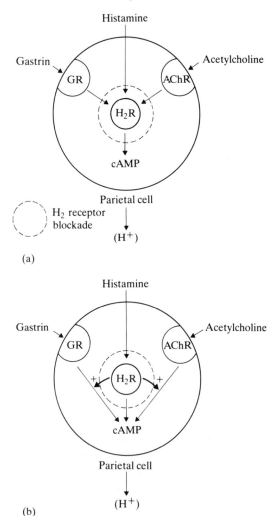

Figure 12.3. *Mechanism of gastric acid secretion.*
   A 'mediator hypothesis'—histamine acting as final common pathway.
   B 'permissive hypothesis'—histamine sensitizes parietal cell to the effects of gastrin and acetylcholine
   (H_2R = Histamine receptor, GR = Gastrin receptor, AChR = acetylcholine receptor)

THE PANCREAS

Structure

The bulk of the pancreas is composed of acinar cells arranged in lobules which produce the digestive enzymes forming its exocrine secretion. The acini are arranged around a branching system of ductules which eventually join in 'herring bone' fashion the main pancreatic duct. Cells of the intralobular and immediate extralobular ducts secrete an isotonic fluid rich in bicarbonate. Interspersed among the secretory acini are isolated clumps of endocrine cells called the islets of Langherhans. These islets are highly vascular and several different types of granular cell can be distinguished using a variety of techniques. Alpha (A) cells comprise approximately 20 per cent of the islet tissue and secrete glucagon, beta (B) cells make up approximately 75 per cent of the islets and produce insulin, delta (D) cells manufacture somatostatin and hPP cells secrete pancreatic polypeptide.

The pancreas is supplied by both sympathetic and vagal parasympathetic fibres which are distributed around the intrapancreatic blood vessels, secretory acini and ducts.

Exocrine secretion

The pancreas secretes up to one litre of juice containing water, electrolytes and enzymes daily. The juice secreted by the acini and duct cells is originally hypertonic but the final juice flowing from the main pancreatic duct at the ampulla of Vater is isosmotic with plasma. This fluid is rich in bicarbonate 95 per cent of which is probably derived from the plasma and the concentration of which rises with the rate of secretion. The acinar cells secrete digestive enzymes from inactive precursors, the zymogen granules. These enzymes play an important part in intraluminal digestion, hydrolyzing food particles into a readily absorbable form. Pancreatic alpha–amylase is secreted in an active state and like all amylases hydrolyzes the glycosidic linkages of natural glucose polymers such as occur in starch. Trypsin, chymotrypsin and carboxy-peptidase are necessary for the breakdown of polypeptide chains to small polypeptides and amino acids. They are formed from their precursors trypsinogen, chymotrypsino-gen and pro-carboxypeptidase. The intraluminal enzyme enterokinase is necessary to initially convert trypsinogen to trypsin, subsequently the reaction proceeds in an auto-catalytic fashion at pH 7·0–9·0. Pancreatic lipase in cooperation with bile salts acting as surface-active agents (p. 440) hydrolyzes triglycerides to produce monoglycerides and free fatty acids.

Control

Pancreatic secretion is controlled by co-ordinated neural and hormonal mechanisms. As with the stomach there is a cephalic and gastric phase of secretion. Visual, olfactory and taste stimulation increases the volume, bicarbonate and lipase output, an effect abolished by vagotomy, whilst antral distension increases pancreatic amylase output, probably due to the effect of gastrin release. Intraluminal contents in the duodenum stimulate pancreatic secretion even when the pancreas is denervated. This intestinal phase is mediated by pancreozymin, secretin and probably VIP. Calcium ions are also necessary since enzyme secretion fails to be maintained by the stimulant in its absence.

Endocrine secretion (see chapter 14)

THE SMALL AND LARGE INTESTINE (see also Fluid and Electrolyte Transport p.442).

The exocrine portion of small intestinal secretion is derived from Brunner's glands in the duodenum which produce a small amount of highly viscid juice the amount of which increases with feeding. During digestion and absorption there is a rapid bidirectional exchange of water and electrolytes across the intestinal mucosa although there is little or no intestinal secretion during the interdigestive phase. Water diffuses passively across the intestinal mucosa along osmotic pressure gradients and can occur in either direction thus maintaining intestinal contents isosmotic with plasma. Since absorption is the net result of the bidirectional movement of salt and water, factors influencing intestinal secretory processes will also affect absorption. Three intestinal secretory mechanisms have been described; those which involve the production of cyclic adenosine monophosphate (CAMP), those occurring as a result of cholinergic stimulation or those which occur because of changes in mucosal hydrostatic pressure.

Cyclic AMP appears to induce intestinal secretion of both salt and water. Since cAMP is derived from adenosine triphosphate (ATP) under the influence of adenyl cyclase a variety of agents capable of activating this enzyme cause increased intestinal secretion. These include VIP and prostaglandins which are both present in the normal bowel and toxins from pathogenic bacteria. *In vitro*, secretion of chloride ions appears to be stimulated by acetylcholine and inhibited by atropine by a mechanism independent of cAMP. This suggests that a parasympathetic control mechanism for normal intestinal secretion exists. Slight increases in pressure on the serosal surface widen intercellular spaces and tight junctions which leads to an increase in mucosal permeability, so causing secretion.

Protein loss in gastrointestinal secretions

Native protein derived from digestive juices and desquamated cells are digested and absorbed in the same way as is ingested protein. Quantitatively 5–15 per cent of the normal turnover of albumin and gamma globulin is accounted for by enteric protein loss into the stool.

Intestinal gas

The composition of gas in the stomach is similar to that of the atmosphere. It is derived from swallowing or the ingestion of food either as part of the foodstuffs structure or by processing (e.g. souffles, bread etc). Further down the bowel, however, intestinal gas is composed of five major gases—nitrogen, oxygen, carbon dioxide, hydrogen and methane. Carbon dioxide is largely derived from neutralization of gastric acid or fatty acids by bicarbonate whilst hydrogen is mainly formed in the colon by the action of colonic bacteria on a non-absorbed carbohydrate and protein (e.g. lactose). Methane is also produced by the action of colonic bacteria although only one third of the adult population appear to harbour the appropriate organisms. The tendency of some stools to float is related to their gas content and not the presence of fat. Normal individuals pass flatus per rectum an average 13·6 times per day producing a total of 200–2,000 ml of gas. The ingestion of non-absorbable legumes (e.g. beans) significantly increases this amount.

Disordered function

SALIVARY SECRETION

Dry mouth (xerostomia) is most commonly temporary and due to evaporation of saliva caused by mouth breathing, anxiety or fear stimulating the sympathetic nervous system, or by drugs possessing an anticholinergic action. More rarely the cause may be destruction of glandular tissue e.g. by irradiation or Sjogren's syndrome, a symptom complex consisting of keratoconjunctivitis sicca, xerostomia and recurrent salivary gland swelling usually in association with a connective tissue disorder. Excessive salivation (ptyalism) may be the result of local disease e.g. stomatits, Parkinson's disease or drugs e.g. iodides and anticholinesterases. In some patients with oesophageal reflux or gastric ulceration the mouth may suddenly fill with a large quantity of salivary secretion; this is known as 'waterbrash'.

GASTRIC SECRETION

Achlorhydria and hyposecretion

Abnormal gastric mucosal states include superficial gastritis, atrophic gastritis and gastric atrophy. Hyposecretion of gastric acid tends to correlate with the histological lesion and the extent of gastritis noted endoscopically, whilst achlorhydria or complete failure of gastric acid secretion occurs with gastric atrophy. A further classification subdivides gastritis into two major groups although not all patients fit neatly into these subdivisions.

Type A gastritis is associated with positive parietal cell antibodies, diffuse involvement of the body mucosa with sparing of the antrum, little or no gastric acid secretion and a familial occurrence. This type is associated with pernicious anaemia and other auto-immune disorders. Since the antral mucosa is spared, hypergastrinaemia occurs (p. 420).

Type B gastritis is four times as common as type A, parietal cell antibodies are absent and the gastric mucosa including the antrum may be involved in focal inflammation; acid secretion is only moderately impaired.

Acid hyposecretion either occurring as a result of gastritis or iatrogenically following the use of H_2 receptor blocking drugs can lead to bacterial overgrowth. Some of these bacteria may be capable of forming nitrosamines which have been implicated in the pathogenesis of gastric carcinoma, the incidence of which is increased in patients with pernicious anaemia.

Peptic ulceration

Benign peptic ulceration cannot occur in the absence of acid although gastric ulcers may occasionally develop in the presence of acid hyposecretion. Whilst patients with duodenal ulceration tend to have an increased acid output this is by no means invariable and other factors such as mucosal resistance, gastrin responses and anatomical site may also be important. Smoking which is known to retard healing and maintain chronicity of peptic ulcers is commoner among patients with duodenal ulceration and recent work has suggested that the tendency of these patients towards acid hypersecretion may be related to their smoking history.

The treatment of peptic ulceration has been greatly modified by the use of H_2 receptor antagonists which give rapid symptomatic relief and significantly increase the rate of ulcer healing when compared with placebo. However unrelated drugs which increase gastric mucosal resistance or coat the ulcer thus 'preventing' acid digestion also appear to give similar healing rates.

Surgery for peptic ulceration is reserved for patients who have complications, or who do not respond to medical management. Most surgical procedures involve either a truncal, selective gastric or highly selective 'proximal' vagotomy thus effectively reducing the level of acid secretion albeit at the expense of a degree of morbidity and recurrence in a minority of patients.

Pyloric stenosis

Pyloric obstruction usually occurs as a result of either duodenal ulceration or an antral carcinoma. Recurrent severe vomiting and the loss of hydrochloric acid as gastric juice gives rise to a hypochloraemic acidosis. Loss of potassium in the vomit can cause hypokalaemia which may be exacerbated by the development of secondary hyperaldosteronism due to sodium loss by the same route.

PANCREATIC SECRETION

A decrease in the output of pancreatic enzymes may occur in chronic pancreatits, after pancreatectomy or in cystic fibrosis. Steatorrhoea due to a reduction in pancreatic lipase only occurs if 90 per cent or more of the pancreas is non-functioning.

Acute pancreatitis. Pancreatic tissue, juice and serum normally contains inhibitors which are capable of inactivating prematurely activated proteases. However under certain pathological conditions pancreatic enzymes may become activated within the gland and lead to autodigestion. Experimentally, pancreatic duct outlet obstruction with reflux of bile up the pancreatic duct gives rise to acute pancreatitis. An analogous situation may occur clinically in pancreatitis associated with biliary tract calculi since a large number of such patients have been shown to pass gall stones in their faeces within 10 days of their attack. Although alcoholic pancreatitis very frequently gives rise to an acute episode it is always on the background of a chronically damaged gland. Other rarer causes of acute pancreatitis include trauma, drugs and hypercalcaemia.

Calcium soap formation due to combination with fatty acids released by extra pancreatic fat necrosis may be the cause of the hypocalcaemia frequently observed during an acute attack of pancreatitis. However other theories include raised glucagon and calcitonin levels and decreased parathormone secretion. Despite the possibility that other intraperitoneal events can give rise to moderate elevations, a rise in serum amylase remains the most valuable aid to diagnosis. It may however, be normal or only slightly elevated in some cases where an acute attack is superimposed upon an already chronically damaged gland.

Chronic pancreatitis. Alcohol is the commonest cause of chronic pancreatitis worldwide although other causes include biliary tract disease, congenital anomalies, protein deficiency and slowly obstructing tumours of the pancreatic duct. Most patients present with pain although if damage is severe enough steatorrhoea with or without diabetes mellitus may occur.

INTESTINAL SECRETION

Secretory diarrhoeas of intestinal origin may occur either as a result of passive secretion caused by an increased hydrostatic and tissue pressure or due to active ion secretion.

An increase in hydrostatic pressure occurs as a result of an elevation in mesenteric venous pressure or a lymphatic blockage. Nevertheless patients with chronic conditions such as severe portal hypertension or constrictive pericarditis do not develop secretory diarrhoea presumably as a result of the development of adaptive mechanisms. Active ion secretion by the small intestine is the most important cause of secretory diarrhoea and in many cases appears to be mediated by high intracellular concentrations of cAMP. The toxins of *Vibrio cholerae*, some enteropathogenic *E coli* and salmonellae all bind to mucosal cell membrane receptors stimulating adenylcyclase. Other activators of adenylcyclase include prostaglandins E_2 and $F_2\alpha$, VIP, bile salts, fatty acids and some laxatives. Theophylline derivatives also raise cAMP levels by inhibiting the normal degradation enzyme, phosphodiesterase. In secretory diarrhoeal states mediated by cAMP, chloride and bicarbonate are actively secreted whilst sodium and potassium enter the lumen passively. Active sodium and chloride absorption is simultaneously inhibited and water moves into the lumen along the osmotic pressure gradient. Normal stool concentrations of sodium are low whilst those of potassium are high (see Table 12.4). However as stool volume increases to over one litre sodium and potassium concentrations come to resemble those in plasma and electrolyte solutions isotonic with plasma are required for replacement therapy. Cholera is an extreme example of a secretory diarrhoea, up to 20 litres of stool water isotonic with plasma being secreted within 24 hours.

Metabolic consequences. Potassium depletion as a result of faecal and urinary losses (due to secondary aldosteronism) occurs in prolonged or severe diarrhoea whilst loss of faecal bicarbonate leads to a metabolic acidosis. Occasionally however, hypokalaemia may be severe enough to cause an increased renal hydrogen ion excretion and a metabolic alkalosis.

GASTROINTESTINAL PROTEIN LOSS

A number of conditions lead to an increased loss of protein into the gastrointestinal tract with resulting hypoalbuminaemia. These include inflammatory bowel disease, giant rugal hypertrophy and intestinal carcinomas. The increase in the synthesis of albumin by the liver to twice the normal rate may be insufficient to counteract severe protein loss resulting in a protein deficiency state.

INTESTINAL GAS

Excessive belching. Repetitive aeructation is preceded by a swallowing movement causing air to enter the oesophagus. Although this movement may sometimes alleviate abdominal discomfort in many cases the condition appears to be functional.

Excessive flatus. In the majority of instances this is due to the consumption of non-absorbable carbohydrates (legumes, some fruits and whole grains) but it may be necessary to consider malabsorption or lactose intolerance as a cause in some patients.

Principles of tests and measurements

GASTRIC SECRETION

Modern diagnostic methods have diminished the usefulness of routine gastric secretion studies. They continue to be of value, however, in the assessment of recurrent ulceration, post-surgical syndromes and hypersecretory states.

Tests are best performed in the morning after an overnight fast. For all tests the sampling tube should be radiopaque, of adequate bore and adjusted under radiological control so that its tip lies in the most dependent portion of the stomach. Mechanical suction is more efficient than hand aspiration. Although a specimen of gastric aspirate contains acid parietal cell secretion, alkaline non-parietal cell secretion, refluxed intestinal, biliary and pancreatic secretions and saliva, in practice tests of gastric secretion are confined to measuring the total amount of secretion of the concentration of acid in small volumes of the aspirated juice.

Hydrogen ion activity is usually measured with a pH meter and titrateable acidity by titration with 0·1 M NaOH to neutrality either electrometrically or using phenol red as an indicator. Titration to physicochemical neutrality (pH 7·0) is probably most appropriate although some authorities prefer pH 3·5. In practice there is probably little difference in the outcome since with hypersecretion the amount of alkali needed to reach the lower titration end point is sufficiently large to establish the diagnosis whilst with hyposecretion the pH provides the relevant information. Although it is possible to use indicators to compensate for pyloric losses this factor is usually ignored in clinical tests, as opposed to research studies.

Basal secretion

After aspiration and measurement of the gastric residue four 15 minute samples are continually aspirated. Basal secretion represents that proportion of the parietal cell mass being stimulated by nervous and hormonal activity under 'basal' conditions which can vary widely. The normal basal acid output (BAO = volume in litres × titrateable acidity in mmol/l) is less than 5 mmol/hour.

Maximally stimulated secretion tests

Histamine and pentagastrin. The subcutaneous injection of histamine (and its synthetic analogue Histalog) are rarely used nowadays to provide maximal acid secretion. Antihistamines which do not affect H_2 receptors are necessary to antagonize the unpleasant peripheral side-effects of histamine on H_1 receptors. Pentagastrin is a synthetic analogue of the physiologically active C-terminal tetrapeptide sequence of gastrin. Since it can be given by subcutaneous, intramuscular and intravenous routes and has relatively few side-effects, it has replaced histamine for routine stimulated secretion tests.

The indices of stimulated secretion in general use are: Maximal acid output (MAO) used to denote the acid output in the hour after maximal stimulation with either histamine or pentagastrin (this is, in fact, not maximal but probably 75–90 per cent of calculated maximal secretory capacity). Peak acid output (PAO) defined as the sum of the two highest consecutive 15 minute outputs after stimulation with histamine or pentagastrin (multiplied by 2).

Augmented histamine or pentagastrin response (AHR or APR) refers to the output during the 15–45 minutes following injection (multiplied by 2).

Insulin test. Insulin induced hypoglycaemia stimulates gastric secretion via a vagally mediated central mechanism. The adequacy of a vagotomy may therefore be tested by this means. The blood glucose level should drop below 2 mmol/l for the test to be valid.

Radioimmunoassay of serum group I pepsinogens

This test whilst not widely available is useful in epidemiological studies for, whilst indirect it seems to correlate very well with stimulated acid secretion.

Tests for intrinsic factor secretion

May be either indirect from its effect on Vitamin B12 absorption as in the Schilling test and the whole body counting method (p. 450) or by measuring it directly in gastric juice which remains a research procedure. Normal intrinsic factor secretion exceeds minimal daily requirements by about 100 times.

Dynamic tests of gastrin secretion (see p. 422)

PANCREATIC SECRETION

Most tests of pancreatic secretion only tend to be helpful when severe disease with steatorrhoea is already present.

Direct tests

Although the secretin-pancreozymin stimulation test remains the standard direct method by which pancreatic exocrine function is assessed it is not widely used due to the care required in its execution and interpretation if meaningful results are to be obtained. The response of the pancreas to intravenous secretin and pancreozymin is assessed by measuring the rate of secretion of juice, secretion of bicarbonate and secretion of enzymes. Juice being collected either by intubation of the duodenum or more recently directly at endoscopic retrograde pancreatography (ERP).

Indirect tests

The Lundh test—involves the collection of duodenal juice for two hours following a test meal of protein, fat and carbohydrate. This juice is then analyzed for its tryptic activity. It provides a simple and helpful if somewhat crude assessment of pancreatic function particularly in the diagnosis of chronic pancreatitis associated with steatorrhoea.

N-benzoyl-L-tyrosyl-p-amino benzoic acid (Bz-Ty-PABA) test—is a recent, simple non-invasive test of pancreatic function which correlates well with the Lundh test. Chymotrypsin cleaves the synthetic peptide to Bz-Ty and PABA which is then excreted in the urine. The amount of urinary PABA excreted in a set time period being proportional to the amount of chymotrypsin available. The test depends, however, on normal small bowel mucosal absorptive capacity, liver conjugating ability and renal function for excretion of the cleaved PABA.

Pancreatic polypeptide secretion tests. Ingestion of a standard meal results in markedly diminished secretion of pancreatic polypeptide in patients with chronic pancreatitis.

Radioimmunoassay of trypsin and lactoferrin. A commercial radioimmunoassay kit is now available for the estimation of trypsin in serum, urine, duodenal and pure pancreatic juice. Preliminary results suggest that the test may become a useful method of assessing the degree of pancreatic exocrine deficiency. Pancreatic juice lactoferrin can also be measured by radioimmunoassay. This glycoprotein is often markedly raised in chronic calcific pancreatitis but normal in acute pancreatitis and pancreatic cancer.

GASTROINTESTINAL PROTEIN LOSS

The available techniques make use of a number of radio-isotopically labelled macromolecules. Originally iodine labelled polyvinylpyrrolidine or albumin was used, however iodine is not an ideal label since it is reabsorbed from the bowel after leaking into the lumen and has therefore been replaced by Chromium-51 labelled albumin. Stools are collected for 4 days following the intravenous injection of the isotope. The normal excretion is 0·1–0·7 per cent of the dose; patients with protein losing enteropathy excrete 2–40 per cent of the ingested dose into the stool.

Practical assessment

SALIVARY SECRETION

Clinical observation. Xerostomia and ptyalism usually clinically apparent.

Routine methods. Plain X-ray; sialogram to delineate duct anatomy.

Special techniques. Suction cup applied to orifice and juice collected for volume (Curry test); normal 3–13 ml over 5 minutes.

Gastric secretion

Clinical observation. Other autoimmune disorders particularly thyroid disease and vitiligo suggestive of pernicious anaemia.

Routine methods. Gastric residue greater than 250 ml indicates some degree of outlet obstruction. Basal acid output (BAO), peak acid output (PAO) and maximal acid output (MAO) after pentagastrin. PAO less than 15 mmol/hour is strong evidence against a duodenal ulcer; PAO greater than 50 mmol/hour strong evidence for duodenal ulcer. MAO greater than 15 mmol/hour supports diagnosis of stomal ulcer. BAO/MAO ratio greater than 60 per cent suggests Zollinger-Ellison syndrome (p. 420). Serum gastrin concentration.

Insulin test for assessing completeness of vagotomy positive if acidity increases by more than 20 mmol/l above basal in any 15 minute specimen.

Parietal cell and intrinsic factor antibodies in pernicious anaemia and gastritis.

Schilling test or whole body counting of radiolabelled B12 to determine cause of low serum B12.

Endoscopy and biopsy to assess gastric atrophy, gastritis and ulceration. Multiple or unusually situated ulcers suggestive of Zollinger-Ellison syndrome.

Barium meal may show bald appearance in gastric atrophy.

Special techniques. Secretin test and calcium infusion test for Zollinger-Ellison syndrome (p. 422). Radioimmunoassay for group I pepsinogens in serum. Intrinsic factor content of gastric juice.

Pancreatic secretion

Clinical observations. Abdominal pain. Diabetes and steatorrhoea if secretion very low. Usually occurs in alcoholic patients.

Routine methods. 3 day faecal fat greater than 60 mmol/day suggestive of pancreatic steatorrhoea. Serum amylase. Diabetic glucose tolerance curve. Plain X-ray abdomen–pancreatic calcification strongly suggestive of alcoholic chronic pancreatitis. Lundh test meal.

Ultrasound is helpful in diagnosing chronic pancreatitis and in differentiating from pancreatic carcinoma. The usefulness of the procedure is heavily dependent on the expertise of the operator.

Isotope scanning. [75]Selenomethionine which accumulates in areas that take up amino acids is the isotope most frequently used. The test gives few false positives but a large number of false negatives. This is largely due to the masking of isotope activity by other organs.

Special techniques. Secretin-pancreozymin test, Bz-Ty-PABA test, radioimmunoassay of serum trypsins and lactoferrin. All useful in establishing pancreatic disease when severe.

Poor discriminatory value between chronic pancreatitis and pancreatic cancer. Endoscopic retrograde cholangio-pancreatography (ERCP) delineates pancreatic duct anatomy. Pure juice can be collected.

Intestinal secretion

Clinical observations. Volume, electrolyte concentration of losses (vomit, aspiration, ileostomy effluent, faeces). Stool electroytes account for all (or nearly all) stool osmolality; diarrhoea persisting on fasting suggests secretory diarrhoea. Gastrointestinal protein loss; oedema and increased susceptibility to infection; features of underlying disease.

Special techniques. Four day excretion of radioactivity following intravenous ^{51}Cr albumin. Small intestinal perfusion with multilumen tubes and occluding balloons for studying relative water and electrolyte changes in disease states.

References

General

CODE C.F. (Ed) *Handbook of Physiology*. Section 6. Alimentary Canal. Vol. 2. Secretion (1967); Vol. 3: Intestinal Absorption (1968); Vol. 4: Motility (1968) Washington, American Physiological Society.

DAVENPORT H.W. (1977) *Physiology of the Digestive Tract*. 4th ed. Chicago, Year Book Med Publ. Inc.

DUTHIE H.L. & WORMSLEY K.G. (Eds) (1979) *Scientific Basis of Gastroenterology*. Edinburgh, Churchill Livingstone.

SIRCUS W. & SMITH A.N. (Eds) (1980) *Scientific Foundations of Gastroenterology*. London, William Heinemann Medical Books Ltd.

SLEISENGER M.H. & FORDTRAN J.S. (Eds) (1978) *Gastrointestinal Disease; Pathophysiology, Diagnosis, Management*, 2nd ed. Philadelphia, Saunders.

Gastrointestinal hormones

BLOOM S.R. (Ed) (1978) *Gut Hormones*. Edinburgh, Churchill Livingstone.

CREUTZFELDT W. (Ed) (1980) Gastrointestinal Hormones, *Clinics in Gastroenterology*, Vol. 9, Philadelphia, Saunders.

Motility

ATKINSON M. (Ed) (1976) Disorders of oesophageal motility, *Clinics in Gastroenterology*, Vol. 5, Philadelphia, Saunders.

BORTOFF A. (1976) Myogenic control of intestinal motility. *Phys. Reviews* **56**, 418–434.

DUTHIE H.L. (Ed) (1978) *Gastrointestinal Motility in Health and Disease*. Lancaster, England, MTP Press Ltd.

Digestion and absorption

Losowsky M.S., Walker B.E. & Kelleher J. (1974) *Malabsorption in Clinical Practice.* Edinburgh, Churchill Livingstone.

Gray G.M. & Cooper H.L. (1971) Protein digestion and absorption. *Gastroenterology* **61,** 535–544.

Gray G.M. (1975) Carbohydrate digestion and absorption: role of the small intestine. *N. Engl. J. Med.* **292,** 1225–1230.

Rommel K. & Goebell H. (Eds) (1976) *Lipid Absorption: Biochemical and Clinical Aspects.* Lancaster, England, MTP Press Ltd.

Secretion

Baron J.H. (1978) *Clinical Tests of Gastric Secretion.* London, Macmillan.

Field M., Fordtran J.S. & Schultz S.G. (Eds) (1980) *Secretory Diarrhoea.* Bethesda Maryland, American Physiological Society.

Janowitz H.D. & Sachar D.B. (Eds) (1979) *Frontiers of Knowledge in the Diarrhoeal Diseases.* New Jersey, Projects in Health Inc.

13

The Liver

Normal function

ANATOMY

The liver, the largest of the solid viscera, weighs from 1·4–1·6 kg in the adult. Blood drains from the splanchnic area via the portal circulation to provide two-thirds of the blood reaching the liver. The remaining one-third is more highly oxygenated hepatic arterial blood. The liver stores carbohydrate as glycogen, regulates the composition of blood traversing the organ, and secretes bile. The bile is both a route for elimination via the gut and a fluid essential for the digestion and absorption of fat.

Embryology, connections and functional units

Embryologically, the liver arises from the foregut as a diverticulum of entoderm which invades the septum transversum. Here it acquires mesenchymal elements to provide vascular and fibrous components; the mononuclear phagocyte (Kupffer) cells are derived later from the bone marrow. Three sets of vessels and the common bile duct enter or leave the liver at the hilum and posteriorly. Conceptual problems concern the manner in which portal vein, hepatic artery, hepatic vein and bile duct communicate with each functioning liver parenchymal cell and how liver cells are organised into functional units. This area is confused by the existence of two systems of nomenclature: One views the terminal branches of the hepatic vein as the "centre" of a functional unit, the lobule and the other regards the portal triad of vein, artery and bile duct as the centre of the live acinus. (Figure 13.1).

Functionally, the acinus is a more logical unit; this chapter will explain this view and also use the concept. However, the ideas are hard to visualize in three dimensions. Extrapolated from sectioned tissue, the misleading simplicity of a concept based upon the central position of branches of a hepatic vein has led to extensive reference to the liver lobule as the functional unit. A principle which has been ignored is that of microcirculatory unity, in that functional units of tissue should be supplied with blood by a single afferent vessel. There is adequate evidence now to support this view and the functional units of the liver, the acini, may thus be seen as a bunch of grapes, with hepatic sinusoids connecting to the hepatic veins which invest and drain the outer aspect of the 'grapes'. Such units may be seen under the microscope, and they are apparent in specimens of liver corroded after filling the vessels with plastic.

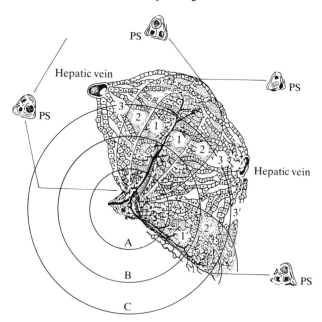

Figure 13.1.

From original publication of Rappaport resurrecting the 'acinar' concept.

Reference: RAPPAPORT, A. M., BOROWY, Z.J. and LOTTO, W. N. *Anat. Rec.* 1954, **119**, 11.

In that view which regards the unit of hepatic function as the group of cells nourished by a single portal triad, the ACINUS is centred on this structure with its arteriole, branch of the portal vein and bile duct. "1" (Zone 1) indicates cells proximate to the triad. Blood from an acinus drains into various branches of the hepatic vein (C.V.) and cells in this position (further from nutrition) comprise Zone 3 ("3"). Zone 2 is intermediate but not anatomically defined. In the old terminology still widely used, the liver is described as made up of lobules. These structures only exist in two-dimensional sections. Lobules are roughly hexagonal around a CENRAL VEIN (C.V.) which draws blood from a number of portal triads (P.S.).

Endothelial cells line the sinusoids where Kupffer (mononuclear macrophage system) cells are also found making up 5–10 per cent of the liver cell mass. Where adjacent hepatocytes are in contact, and on a side of these cuboidal cells away from the sinusoidal aspect, specialized regions of membrane develop forming bile canaliculi. At this point the biliary tree commences: as the portal triad is approached, a number of hepatocytes will bound the canaliculus before transition to bile ductular (epithelial) cells occurs with acquisition of a basement membrane. (Figure 13.2) The anatomy of this junction is uncertain.

Terminal (capillary) branches of the hepatic artery supply highly oxygenated blood and enter sinusoids not more than one-third of the distance between terminal portal veins and terminal hepatic veins, so that oxygen tension is greatest close to the portal tracts. This, together with the composition of blood entering the liver leads to functional differences among liver parenchymal, and perhaps also Kupffer, cells depending

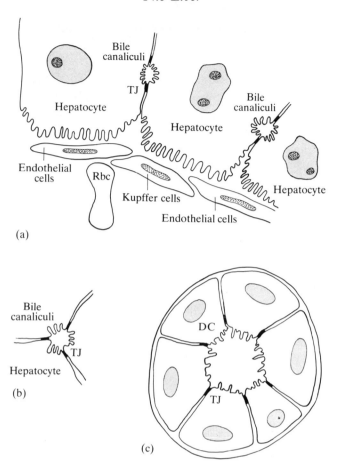

Figure 13.2.

(a) Three hepatocytes (HC) face a sinusoid being separated from the lumen by endothelial (EC) and Kupffer (KC) cells. The sinusoidal aspect of the hepatocytes has a microvillous surface. Where the plasma membrane of adjacent hepatocytes is contiguous, bile canaliculi (BC) are seen bounded by two (Fig. 2a) or more (Fig. 2b) hepatocytes. As bile canaliculi approach the portal tract the number of lining cells increases and the structure is referred to as a bile-duct (Fig. 2c), when it acquires a basement membrane.

upon their position in the acinus (or lobule). (Figure 1) Zone 1 cells close to the portal triad are in the region of presumed highest oxygen tension; these cells are also the first to be exposed to portal blood carrying materials absorbed from the gut as well as gut-derived hormones. (Zone 1 cells are peri-lobular in the old terminology). Zone 3 cells are in the immediate vicinity of the terminal branches of the hepatic vein (THV) where opposite conditions exist, i.e. low oxygen tension, lower concentration of nutrients, etc. These would be centri-lobular cells in the old terminology. Cells in an intermediate position form Zone 2.

The hepatic and portal circulation

Portal blood is 80–90 per cent saturated with oxygen. Total hepatic blood flow is about $1·5$ l min^{-1} but subject to fluctuation as splanchic blood flow varies. The liver abstracts 55 ml min^{-1} of oxygen which decreases oxygen saturation in the hepatic vein to 60–65 per cent. These figures suggest a gradient in oxygen tension of from 60 mmHg in Zone I of the acinus to 30mmHg in Zone III.

Excepting for the muscular hepatic artery, the vessels entering and leaving the liver are distensible vessels of high capacitance and low resistance—in fact, the liver accounts for only 1–2 per cent of total splanchnic resistance. In consequence, any increase of pressure in the portal circulation above the normal range of 5–13 mmHg signifies a very large increase in portal vascular resistance. Small changes in the total cross sectional area of the hepatic circulatory bed will not appreciably raise portal venous pressure.

Lymphatics drain the liver at the hilum. Flow in these is about 0·75 ml/min and the protein content of hepatic lymph is nearly equal to that of plasma. This implies that one fifth of the total albumin pool passes daily through the hepatic lymphatics. Lymphatic channels accompany the venous drainage and anastomose with lymphatics draining the pleural cavity; lymphatic flow between these systems may occur in either direction.

Nerve supply

Both adrenergic and cholinergic nerves appear in the portal triads and it is reasonable to assume that the smooth muscle of the hepatic artery and its branches, the biliary duct system and possibly also the portal vein and its branches receive efferent nerves from both branches of the autonomic nervous system. Afferent nerves certainly innervate the liver capsule and probably also the larger radicals of the biliary duct system. Unfortunately there is too little proven data regarding the function of these nerves for this topic to warrant more space. However, visceral pain may be appreciated when the liver capsule is stretched or the larger biliary ducts distended. Various experiments have suggested that liver cells possess either or both of α and β-adrenergic receptors, but physiologically, hormonal control of carbohydrate homeostasis is probably of more importance than control by the autonomous nervous system.

Growth and regeneration

The liver of both humans and laboratory animals possesses remarkable powers of regeneration. This is seen after surgical removal for major trauma or tumors in man, after partial hepatectomy in laboratory animals and as a continuing process associated with disease states which destroy liver cells. The phenomenon of regeneration has been extensively studied in the rat. In this animal, within one week of removing 60–70 per cent of the organ the original weight is regained; partial hepatectomy may be repeated monthly for many months without the response failing. The architecture of the regenerated liver remains relatively normal, although the initial regeneration results in restoration of mass through diffuse hyperplasia rather than true regeneration as, for example, occurs when a salamander regrows an amputated leg from a newly-formed limb bud. However, the new tissue retains normal vascular and biliary connections. The liver appears to "know" when to regenerate and when to stop regenerating, which is equivalent to saying that the process is subject to homeostatic control. Cross-circulation experiments suggest that humoral factors are responsible for this control with insulin and glucagon playing some role. Portal blood coming from the gastro-, duodeno-, and pancreatico-vessels is necessary for normal growth and function of the liver as well as for rapid regeneration. Conversely, if portal blood is diverted to bypass the liver and enter the systemic circulation directly, important disturbances of hepatic growth result.

THE ROLE OF THE LIVER IN INTERMEDIARY METABOLISM

It is convenient to consider in turn storage, interconversion and regulation of metabolic fuels. The anatomical position of the liver is suited to permit control of energy-yielding substrates: portal blood delivers nutrients directly from the gut which must either be taken up by the liver or allowed to circulate. A 600 kcal meal comprised equally of carbohydrate, protein and fat will deliver about 40 g, 50 g and 22 g of these respective nutrients to the circulation. Apart from chylomicrons which are absorbed via lymphatics, "circulation" means the portal circulation and thus the liver. Blood from the pancreas also enters the portal circulation so that the liver is exposed to concentrations of insulin 3–$10 \times$ higher than other organs. This also applies to other gut-derived polypeptide hormones. The liver is able to store about 390 mmoles (70 g) of glucose as glycogen which, here and in muscle, comprises the total body store of carbohydrate fuel. Removal of the liver, or its destruction by disease, may lead to severe, even fatal, hypoglycaemia. Neither fat nor protein is stored significantly in the liver but this organ does play some role in regulating stores of these fuels elsewhere in the body. The interrelation between protein metabolism in liver and muscle is reviewed on p. 474 and lipoproteins on p. 477.

The major metabolic pathways which permit the utilization and interconversion of the metabolic fuels are described in Chapter 14. Only enzymes of the ornithine cycle which lead to the synthesis of urea, are relatively specific to the liver. Urea is synthesized from ammonia, itself formed when amino acids are deaminated to keto-acids and very toxic in the free form. This intramitochondrial urea synthesizing cycle is closely related to the citric acid cycle.

The urea cycle

While small amounts of unwanted nitrogen are excreted as creatinine or urate, urea is by far the most important route for its elimination; urea levels in blood and urine fall in liver failure even though blood levels of most amino acids rise. Kidney tissue can synthesize small amounts of urea, and the required enzymes have been detected in other tissues, but quantitatively the liver predominates.

Urea synthesis can be described in terms of three cycles. Ammonia (derived from glutamine, asparagine, glutamate, aspartate, adenine and guanine) enters via the α-ketoglutarate-glutamine cycle (Fig. 13.3) which is coupled to formation of aspartate by transamination of oxaloacetate. The reactions numbered 1–5 in the Figure constitute the urea cycle proper where one mole of NH_3 (entering via glutamine) and a second mole (via aspartate) are condensed with carbon (as bicarbonate) to yield urea.

The reactions in the figure have been drawn to emphasize the close interdependence of the tricarboxylic acid (Krebs) cycle and the pathways for disposal of nitrogen as urea. In the rat, and this seems to be true also in man, the overall activity of urea

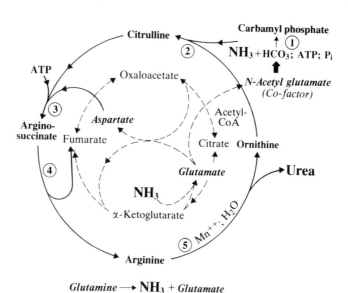

Glutamine $\longrightarrow NH_3$ + Glutamate

Figure 13.3. *Intermediary metabolism directed to synthesis of urea.*
The outer circle (solid lines) comprises the specific reactions of the urea cycle which synthesize arginine from which urea is split off by hydrolysis (5). The inner circle (broken lines) provides an outline of the citric acid cycle to show its close relationship to the urea cycle. Ammonia enters these cycles via glutamate and thence aspartate and N-acetylglutamate. The overall reaction is $2NH_3 + CO_2 = (NH_2)_2CO + H_2O$. Ammonia is transported to the liver either free or as glutamine.

synthesis is related to protein turnover. Thus enzyme activity decreases in starvation and increases when a high protein diet is taken. Activity also increases when glucocorticoid hormones accelerate protein catabolism. Unfortunately little else definite can be said about regulation at a molecular level.

The interconversion of metabolic fuels

Disposal of excess amino acid nitrogen as urea permits protein to provide keto acids which enter either carbohydrate metabolism (the 13/23 glucogenic amino acids) or fat metabolism (the ketogenic amino acids). Glucose can pass through acetyl CoA to be synthesized to fatty acids; these in turn, may be esterified with glycerol derived from the glycolytic pathway to form triglyceride. Cholesterol is also synthesized from acetyl CoA (See Section 5-2) but unsaturated fatty acids, essential for membrane synthesis, must be derived from the diet. Glucose circulates free in the plasma, "free" fatty acids and keto-acids circulate largely bound to albumin. Triglycerides and cholesterol, both free and esterified, circulate bound to carrier proteins as the lipoprotein fraction of plasma. The liver thus interconverts the major metabolic fuels and also provides carrier molecules for those with limited solubility in water.

Effects of feeding and fasting

After an overnight fast, glucose uptake by insulin-dependent organs will have almost ceased, and plasma insulin levels will have decreased to 10–20 μ units/ml. However, insulin-independent tissues will continue to use about 50–60 mmole of glucose per hour. Clearly the 70 g ($\simeq$ 390 mmoles glucose) of hepatic glycogen could not supply this for more than ± 8 hours; thus the human liver becomes almost devoid of glycogen after a 24-hour fast. Actually, glycogenolysis provides only about 75 per cent of glucose released by the liver with the remaining 25 per cent being provided by gluconeogenesis. Most body proteins are being cyclically broken down and resynthesized. The amino acid building blocks circulate, and will be taken up (cleared from blood) by the liver as required for gluconeogenesis. When this occurs in excess of dietary intake, catabolism of tissue protein must exceed anabolism and the muscle mass should be regarded as the principal reservoir of available amino acids. It is worth noting that in the rat some glucose formed from gluconeogenesis is converted to glycogen which is not absent even from the liver of a fasted animal. This pathway appears to be a minor one in humans.

An important mechanism to conserve tissue proteins involving the regulatory role of ketones on gluconeogenesis becomes apparent when fasting continues beyond 2–3 days. Keto acids (or "ketone bodies"), β-(OH) butyrate and acetoacetate, gradually replace glucose as the principal brain fuel, but also inhibit hepatic gluconeogenesis from amino acid precursors. Plasma keto-acid concentration rises in starvation to reach levels roughly equal to the level of plasma glucose (± 5 mmoles/l); unlike glucose this, like the unesterified fatty acids, is transported bound to serum albumin.

Regulation of metabolic energy (also see Chapter 14)

The role of glucagon and insulin are concerned with regulation of the processes outlined above. Insulin permits liver, muscle and adipose tissue to utilize blood glucose. Levels of insulin fall during fasting and rise after feeding a carbohydrate or protein meal. Insulin serves to regulate blood glucose and amino acid levels by altering clearance principally by the liver, but also by muscle and adipose tissue. The molecular mechanisms involved are still not clear. Of practical importance as well as theoretical interest is the large flux of K^+ ions into liver cells when insulin stimulates glycogen synthesis. Intravenous glucose given with insulin can rapidly lower plasma K^+.

The role of glucagon is less clear. As inhibition of glucagon release by somatostatin causes a fall in blood glucose, it appears that glucagon secretion during fasting does help to maintain adequate circulating glucose, but it is not essential. When glycogen is converted to glucose there is efflux of K^+ from the liver, an effect which can be marked after injection of glucagon. If portal blood is shunted away from the liver leaving largely or only an arterial supply, important disturbances of hepatic growth result which are not due only to a loss of perfusing blood volume. Insulin and glucagon have been extensively studied as possible trophic factors. However, substituting only these hormones does not completely substitute for portal blood and control of the growth and mass of the liver appears to be multifactorial.

The liver is the site of insulin degradation by proteolysis. A fragment of the insulin molecule—'C' peptide—is split off and persists in plasma for a relatively long time. Assay of circulating C-peptide can thus be used to determine average levels of plasma insulin activity over periods of several hours.

THE SYNTHESIS OF PROTEIN FOR EXPORT

All tissues synthesize protein, but when the regulation of amino acid and protein metabolism is considered, attention is usually focused on the liver and muscle which are quantitatively so important. Daily albumin synthesis in liver amounts to about 12 g but more than ten times this amount of protein is made daily in muscle. Certain constraints on the protein economy of the body must be borne in mind. First, amino acids can only be stored after synthesis into tissue proteins and muscle proteins constitute the major store. Second, tissue protein can only be synthesized from a correct balance of essential and non-essential amino acids. Third, after protein intake enzymes for protein deamination and ureagenesis increase in activity and dispose of unused amino acid nitrogen as urea. Fourth—and this emphasizes the interdependence of liver and muscle in regulating synthesis and breakdown of protein—three essential amino acids, valine, leucine and iso-leucine referred to as the branched chain amino acids (BCAA), are poorly taken up and oxidized by liver but selectively taken up by muscle. Fifth, and this is of great clinical significance, amino acids can only be stored as protein if controls unrelated to food intake favour anabolism. Trauma and disease tends to be associated with a catabolic state which increased intake may not reverse.

Alanine is the major circulating amino acid transporting amino acid nitrogen from muscle to liver. When insulin levels are low, and glucagon levels high, alanine is cleared from the circulation by the liver and converted to glucose and urea, i.e. gluconeogenesis, at the expense of muscle protein. Against this background of protein turnover in the body as a whole we can examine the 15 g or so of protein synthesized daily by the liver.

The plasma proteins (60–80 g/l) comprise albumin which is a homogeneous protein of low molecular weight (68,000 daltons) and the heterogeneous globulin fraction. Globulins include coagulation factors, immunoglobulins, the complement system, circulating enzymes, and proteins with specific transport functions. In health, the concentration of all plasma proteins lies within fairly narrow limits suggesting a measure of regulation. The concentration of any particular protein reflects the balance between synthesis and degradation and there is relatively little known about the site or determinants of degradation. With the exception of immunoglobulins synthesized by cells of the immune system, plasma proteins are synthesized by the liver and all are glycoproteins, excepting for albumin. The attached carbohydrate residues, which permit recognition of proteins by cell surface receptors and may determine differential tissue uptake, represent post-translational modifications made within vesicles of endoplasmic reticulum and prior to export from hepatocytes.

Albumin

Albumin has three main functions: (1) it contributes two-thirds of the plasma colloid osmotic (oncotic) pressure, (2) it provides binding sites for transport in the plasma of hydrophobic material, and (3) it constitutes a small store of circulating protein—300 g in a healthy adult. Circulating colloid is essential to maintain fluid within the vascular space. One millimole of protein (68 g albumin) per litre will provide 18 mmHg (2·45 kPa) of oncotic pressure and the concentration of albumin in health lies in the range 35–55 g/l. Analbuminaemia (the congenital almost complete absence of this protein) is compatible with life only because there is an increase in other circulating low molecular weight proteins and a compensatory hypertrophy of the lymphatic system.

A 70 kg adult synthesizes 12 g of albumin daily (175 mg/kg day). Since the total pool of albumin is around 300 g, 0·04 of this must be degraded daily if the circulating level is to stay constant. This corresponds to a half life of $\log_e 2/0·04$, or 17, days. Synthesis of albumin occurs only in the liver but degradation is widespread. Regulation of these processes is poorly understood but both synthesis and degradation are sensitive to nutrition, and starvation decreases degradation as well as synthesis. Hepatic albumin synthesis only has a limited capacity to respond to increased demands. Thus when protein loss through damaged kidneys (nephrosis) or skin (the burned patient) exceeds about 40 grams daily, hypoalbuminemia must result.

While the Starling hypothesis describing the role of oncotic pressure in fluid movement in and out of capillaries is qualitatively correct, its original quantitative formulation underestimated the importance of albumin in interstitial fluid in opposing fluid movement back to the capillaries and the role of lymphatics in preventing fluid accumulation in the interstitial space (oedema). It also overestimated mean effective capillary pressure. It is interesting to note that the greater part of the albumin pool (180 g of 300 g) is located in the extravascular space, particularly in the skin.

Coagulation factors (See also Chapter 8)

Blood coagulation requires the conversion of fibrinogen to fibrin by the proteolytic enzyme thrombin. Activation of prothrombin to thrombin occurs at the end of a chain of events which may involve only circulating factors (the Intrinsic pathway), or tissue thromboplastin (the Extrinsic pathway). Both pathways involve the sequential activation of a series of proteolytic enzymes. Four steps in the intrinsic pathway utilise proteins synthesized in the liver with a Vitamin K dependent post-translational modification to the amino acid chain. These four proteins are factor II (prothrombin), factor VII, factor IX (Christmas factor) and factor X (Stuart factor). Vitamin K is an essential cofactor in the conversion of glutamyl residues in the inactive pro-factors to their γ-carboxy derivatives. This process precedes, and is distinct from, the proteolytic cleavage of part of these molecules which constitute activation, and must still precede coagulation.

Fibrinogen is also synthesized in the liver and accounts for approximately 3 g/l of plasma protein. Clinically significant deficiency of fibrinogen almost only occurs due to rapid consumption in the coagulation process and synthesis is, as implied above, independent of Vitamin K. Much less prothrombin circulates (0.1 g/l) and only one-tenth this concentration of factors IX and X. It is partly the short half life, reckoned in hours, of the Vitamin K-dependent coagulation factors (II, VII, IX, X) which underlies their practical importance in liver function. This means that bleeding may be an early sign of failure of protein synthesis in the liver, and this will also be a consequence of a deficiency of Vitamin K.

Other circulating proteins

Important transport proteins are produced by the liver. Transferrin serves to carry iron which is an essential nutrient for all tissues. Each molecule of this glycoprotein has two similar ferric-iron binding sites and the very great affinity for ferric iron is evident from the K_D of 10^{-31} M. The high affinity of this protein for iron also serves to make iron unavailable to pathogenic bacteria for which, as for all living cells, the metal is an essential growth factor, and to prevent loss of iron in urine.

Globulins binding steroids (cortisol, sex-steroids) provide specific sites for the transport of these water-insoluble hormones. Similarly there is a thyroxine binding globulin and a binding protein has been identified for Vitamin A.

An important group of proteins originating in liver are called collectively the acute phase reactant proteins (APRP), which increase in concentration in plasma after tissue injury. Included in this group are acid glycoprotein, α_1-antichymotrypsin and α_1-antitrypsin, ceruloplasmin, haptoglobin and C-reactive protein. Fibrinogen, which has been dealt with above, is also included. Aseptic tissue necrosis, inflammation (whether infective or not) and infiltration by neoplasm will all stimulate an increase in synthesis of APRP's. The response is fast (starting in 4–6 hours and maximal after 48 hours) for acid glycoprotein, α_1-antichymotrypsin and C-reactive protein. The others respond approximately half as quickly. While ceruloplasmin (copper) and haptoglobin (haemoglobin) have a clearly defined carrier function, and two APRP's are inhibitors of proteolytic enzymes which could do harm circulating in plasma, the function of α_1 acid glycoprotein and C-reactive protein is uncertain though appears related to regulation of immune function. α_1-antitrypsin is a protease inhibitor: different forms of this glycoprotein are inherited and some phenotypes are associated with pulmonary or hepatic disease. The link between disease and altered circulating proteolytic activity is however, unclear.

Other proteins of hepatic origin may be detected in the plasma even though they have no known function there. Included in this category are various enzymes which are assumed to leak into the body fluids or be spilled out in the course of liver cell damage: these are considered later.

The lipoproteins

The lipoproteins are those plasma proteins which circulate associated with cholesterol, cholesterol ester and triglyceride. Ratios of protein : lipid are variable and the apoproteins can be seen as solubilisers or transporters of lipid. However, this view would oversimplify their role. They originate in both the gut wall and in the liver and are exported across cell membranes in these organs already loaded with fat. Lipoproteins then undergo reactions in plasma, lose some of their fat and are taken up at sites including the liver, but also skin and fibroblasts.

At least eight lipoprotein-apoproteins have been identified and these occur either in high density (HDL) or low density and very low density (LDL and VLDL) lipoproteins. These have much in common with cell membrane and in fact the lipoproteins originate in smooth endoplasmic reticulum (SER) where enzymes for lipogenesis are located. HDL particles possess a lipid bilayer structure; low density particles have variable structures depending on the amount of lipid they are carrying.

Two classes of enzyme alter lipoprotein particles: the lipoprotein lipases originate in the liver and are subsequently attached to capillary endothelium; these are activated by heparin and hydrolyse triglyceride on chylomicrons, VLDL and LDL. Lecithin-cholesterol-acyl transferase (LCAT) acts at the surface of HDL to exchange β-acyl fatty acids of lecithin with the 3β-hydroxyl of cholesterol and form insoluble cholesterol esters. This is mentioned to explain the origin of cholesterol ester in HDL but neither the function of the reaction, nor the fate of esterified cholesterol is fully understood.

Chylomicrons (the largest of the low density lipoprotein structures and the form in which much of the fat from the gut reaches the liver) are cleared rapidly from plasma with a half-life of about 15 min. They lose their triglyceride and form LDL which persists in the plasma for some hours. The ultimate fate of LDL and HDL, together with its esterified sterol, is not clear but both skin and fibroblasts have receptors which permit uptake of these particles.

Also unclear at this time is the functional significance of the complex protein composition of the different lipoproteins. However, the activity of both lipoprotein lipase and the LCAT enzyme is influenced by this. Uptake by particular organs may depend upon the lipoprotein composition of LDL and HDL particles.

THE SECRETION OF BILE

A system of tubes conveys secretions from the bile canaliculi, via ductules lined with bile duct cells, to the ducts which are hollow muscular structures. Right and left bile ducts converge, are joined by the cystic duct communicating with the gall bladder, and the common bile duct ends at the ampulla (of Vater) in the 3rd part of the duodenum.

Bile is a greenish-gold fluid isosmolar with ECF and with a similar electrolyte composition except for additional HCO_3^-. Bile contains bile salts, phospholipids, and cholesterol in colloidal solution as well as proteins, bile pigments and other organic solutes. About 0·5 ml/min is formed, or 0·72 l/day. Bile is essential for digestion and absorption of fat, and it is a route for excretion via the liver and the faeces of various foreign substances including drugs as well as the bile pigments.

Mechanism for the secretion of bile

Two mechanisms have been identified for the formation of bile; the first produces bile at a rate related to the flux of bile salts and this bile flow certainly occurs across the canalicular membrane of the liver cell. The second is independent of bile salts, is stimulated by the polypeptide hormone secretin and occurs at a site distal to the regulatory site for bile salt dependent secretion. Bile salt independent secretion may therefore occur in the smallest bile ducts and possibly also in the larger branches of the biliary tree. The bile salts may be concentrated up to 1000 × in bile relative to blood. A fluorescent marker, injected into the portal circulation can be visualized reaching the bile ductules within seconds. These facts suggest that the rate of bile secretion is much larger than bile flow in the common duct and that reabsorption of water and electrolytes occurs after secretion by parenchymal cells and perhaps throughout the biliary tree.

Absorption of salts and water is a major function of the gall bladder which receives one-third to one-half of bile draining from the liver. The cystic duct leading from the common bile duct to the gall bladder is devoid of smooth muscle and filling of the bladder is dependent upon the duodenal ampulla being closed. This structure appears to relax during intestinal peristalsis so that when the duodenum is quiescent the gall bladder will tend to fill with bile and relatively little of this fluid will enter the gut. After a meal, particularly one containing fat, the duodenal hormone cholecystokinin will contract the gall bladder, open the ampullary sphincter, and deliver concentrated bile to the duodenum. Cholinergic motor nerves blockable by atropine (muscarinic) supply the biliary tract but their importance relative to various gut polypeptide hormones is unclear. The duodenal ampulla is sensitive to enkephalins (and the opiate drugs) and evidence is emerging for a complex situation in which this activity may also be modulated by gut-derived polypeptide hormones (e.g. VIP-*vasoactive intestinal polypeptide*).

The gall bladder epithelium abstracts from the bile a sodium-chloride-rich fluid, effectively a fluid isosmotic with plasma, thus concentrating the organic constituents of bile.

The bile salts

The structural formulae for a few important bile acids are included together with that for cholesterol to show their relationship, because without this, the nomenclature becomes very confusing. It is worth noting that the COOH group of the bile 'acids' shown forms an amide bond with either glycine or taurine. The pk of the resulting conjugates is lower, and unconjugated bile acids are relatively insoluble at acid pH. Either free or unconjugated bile acids may ionise and form 'salts' with cations.

Bile acids are formed by the cleavage and oxidation to a carboxylic acid of the side chain on cholesterol and by hydroxylation of the steroid ring at up to 3 of positions 3, 7 and 12. All hydroxyl groups are 'Alpha' thus on the same side of the molecule and the polar OH groups, together with the COO$^-$ groups of amino acids (glycine or taurine) with which bile acids are conjugated by amide bonds, provide a water soluble

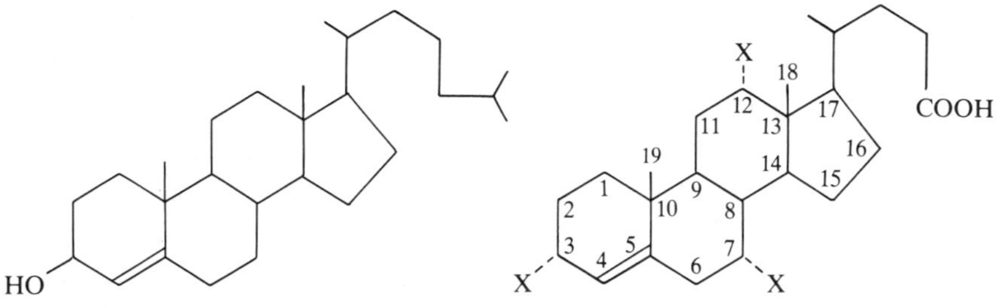

	3α	7α	12α
Cholic	OH	OH	OH
Chenodeoxycholic	OH	OH	H
Deoxycholic	OH	H	OH
Lithocholic	OH	H	H

Figure 13.4. *Structural relationships of cholesterol and bile salts.*

aspect to an otherwise lipophilic molecule. It is the dual solubility (amphiphilic) property that is associated with the physiological importance of bile acids and more so with their conjugates with glycine or taurine.

The diverse chemical structures of the bile salts are further complicated by two important sets of modifications which occur after initial synthesis in the liver of the primary bile acids; firstly these may be converted to "secondary" bile salts by bacteria in the gut. The chief reaction is dehydroxylation e.g. at position 7, giving lithocholic acid. Secondly, bile acids returned from the gut to the liver by the enterohepatic route may be further modified in the liver to give "tertiary" bile acids Ursodeoxy_cholic acid, an epimer of chenodeoxy_cholic (chenic) acid is such a tertiary bile acid.

Bile salts dissolve in water but, as detergents, they associate into micelles or aggregates when their critical micellar concentration (CMC) is exceeded. Such micelles will have a hydrophilic exterior but a hydrophobic interior and, depending upon the effectiveness of the detergent, the micelles can dissolve a variable amount of water insoluble material while remaining in colloidal solution. Phospholipids (e.g. lecithin) are also amphiphilic though much less effective detergents than bile acids and less soluble in water. However, lecithin in an aqueous environment will swell and, given the superior detergent property of the bile acids, actually increase the amount of cholesterol carried in micellar solution.

Bile salts and phospholipids are thus present in bile as micelles which also contain cholesterol. It is possible that secretion across the canalicular membrane occurs as micelles or by the exocytosis of membrane fragments. This is a major excretory route for cholesterol from the body but, as will be described in Part II, the capacity of this system may be exceeded.

In the intestine, the detergent properties of the bile salts solubilise fats, increase access of lipases, and facilitate absorption. Bile salts (conjugates) are then reabsorbed in the terminal ileum and delivered back to the liver via the portal circulation. So efficient is this enterohepatic circulation that each day the small pool of bile salts (2–4 g) is cycled 6–10 times through intestine and liver with a daily loss of only 0·6 g which is replenished by hepatic synthesis. The terminal ileum plays a crucial role in the reabsorption of conjugated bile salts but bile salts that have been deconjugated by bacterial action compete poorly for this carrier mediated process and enter the enterohepatic circulation by simple diffusion. The size of the bile salt pool is regulated within the liver by negative feedback which operates on at least two steps in bile acid synthesis. The first of these, hydroxymethylglutaryl CoA reductase, is a rate-limiting step in *cholesterol* synthesis while the second regulated step is the 7-α hydroxylation of cholesterol, a step which commits a cholesterol molecule to become a bile acid. Thus hepatic rates of both cholesterol and bile acid synthesis are regulated by the bile acid pool. Bile salt micelles carry cholesterol away from the liver in the bile and, in the intestine, cholesterol is not absorbed in the absence of bile salts.

The intracellular localisation of cholesterol and bile acid synthesis is of some interest because the oxidations and hydroxylations which modify the steroid molecule—

mixed function oxygenase (MFO) reactions—are located in the smooth endoplasmic reticulum and are involved in other catabolic activities of the liver. These processes are dealt with in the following section.

CATABOLIC FUNCTIONS OF THE LIVER

Recent years have seen an upsurge of interest in the functions of the prolific smooth endoplasmic reticulum (SER) of hepatic parenchymal cells. Biochemically, this "micro-somal" fraction (as the SER is called after ultra centrifugation) is extremely reactive. The major classes of reaction of interest are conjugations (with glucuronate, glutathione, other amino acids, sulphate) and oxidations. Oxidations are mostly mixed function oxygenations, (MFO's) where the substrate, S, undergoes reactions of the following general type:

$$H^+ + SH + O_2 + NADPH \rightarrow SOH + H_2O + NADP^+$$

The term MFO arises because oxidation and reduction proceed together and the reactions are oxygenases because molecular oxygen is incorporated into the substrates.

The MFO pathways—of which there are several—have low substrate specificity and metabolise steroids, bile acids, foreign chemicals from the gut and numerous drugs. Different substrates may compete for metabolism. Further, the activity of these systems is variable; they are 'inducible', i.e. more (or less) enzyme may be synthesized in response to diverse hormonal and environmental stimuli. When such induction is produced by exogenous materials such as drugs, physiological systems may be perturbed.

Bilirubin metabolism

Haemoglobin, myoglobin, b-type cytochromes and catalase have iron-protoporphyrin IX (haem) as prosthetic groups. Generally, the prosthetic groups are not re-utilized when the proteins are re-synthesized and the haem is converted to bile pigment. Given a haemoglobin concentration of 12 g/dl blood, a blood volume of 5/l, and a mean red cell life span of 120 days, the steady state requires that 600/120 g or 5 g, of haemoglobin is degraded daily. With 4 moles of haem per mole of haemoglobin and respective molecular weights of 594 and 68,000 daltons, this yields 175 mg daily of haem to be excreted as bile pigment from effete red cells. There is a further contribution (± 60 mg) from other haemoproteins, mostly arising within the liver itself.

An enzyme (haem oxygenase) in the endoplasmic reticulum of most—perhaps all—cells converts haem to biliverdin in a reaction which opens the porphyrin ring at the α-methene carbon, yielding carbon monoxide in the process. Biliverdin is then reduced to bilirubin: the specific isomer formed (IXα) is uniquely insoluble in aqueous media due to stabilization of a cyclic and hydrophobic configuration by hydrogen bonding. Bilirubin must therefore be transported to the liver bound to a carrier because only minute amounts can exist free in solution. Albumin functions as the principal bilirubin carrier.

As blood traverses the hepatic sinusoids binding sites within hepatic parenchymal cells compete with albumin for the pigment, and on each pass through the liver about 1/30th of the bilirubin diffuses into hepatic parenchymal cells. An intrahepatic binding site which may be important is on the protein ligandin. This is present in relatively high concentration in the cytosol of liver cells and cells of other tissues and, apart from acting as an intracellular ligand for organic anions, possesses enzymatic activity as a GSH-transferase thus emphasizing a wider role in detoxification processes.

Within liver cells bilirubin is made more water soluble through conjugation of one or both propionyl side chains with glucuronate. The enzyme responsible is one of the uridine diphosphate glucuronyl transferases (UDPGT's) which, as do the mixed function oxygenases, form part of the smooth endoplasmic reticulum (SER) membrane. Mono- and diglucuronyl conjugates of bilirubin then pass across the canalicular membrane into bile by mechanisms unknown.

Bile pigment, like other organic solutes, is concentrated in the bile by the abstraction of isotonic water and electrolyte to give bile its golden-greenish hue. Bacteria from the large gut further alter bilirubin; here it may be deconjugated and/or metabolized to stercobilin imparting a brown colour to the stool or further reduced to colourless derivatives, the stercobilinogens. Some unconjugated bilirubin and stercobilinogens may be reabsorbed in the large bowel and returned via the enterohepatic circulation to the liver. If the liver does not re-excrete stercobilinogen into the bile it may spill into the urine where it is confusingly referred to as urobilinogen.

All of the above facts have clinical relevance in disorders of bilirubin metabolism associated with jaundice. The process can be broken down into discrete steps: (a) haem degradation (b) plasma transport and hepatic uptake (c) conjugation (d) excretion into bile (e) metabolism in the gut and (f) enterohepatic recirculation. It should be noted that nowhere is there evidence for active transport (or facilitated diffusion) of bilirubin. Movement of the pigment is entirely accounted for by movement between binding sites of differing affinity and altered solubility consequent on conjugation. Only the mechanism for movement into the bile canaliculus remains to be fully explained.

Degradation of hormones

The oxidation of cholesterol to bile acid by enzymes in the SER has already been mentioned; so also has conjugation of the resulting bile acids. These oxidation reactions belong to the class of mixed function oxygenations (MFO-reactions) now of such importance. Steroid hormones, like cholesterol, are also susceptible to MFO-hydroxylations and the liver is the principal site for inactivation of estrogen, androgens, progestogens, glucocorticoids and mineralocorticoids by either MFO hydroxylations or conjugations.

The conjugations of steroid hormones differ from those of the bile acids (which involve amide linkage with the amino acids glycine or taurine) and utilize glucuronic acid via UDPG-T like bilirubin, or, unlike bilirubin, sulphuric acid via sulfotransferases which yield steroid sulphates. Whether with glucuronate or sulphate, the effect is to increase water solubility of the steroid thus favouring partition away from cell membrane components and into aqueous phases. Both glucuronyl- and sulfo-transferases are families of enzymes and substrate specificities probably account for the route followed by a particular steroid prior to its excretion in bile or urine.

Metabolism and elimination of xenobiotics

Attention has been drawn to various enzymes which oxydize or conjugate steroids or bilirubin thus increasing water solubility of these molecules and favouring their excretion. These pathways tend to exhibit limited substrate specificity and are also able to accelerate the elimination of foreign lipophilic molecules (xenobiotics). This term embraces many drugs in common use so that the liver plays a vital role in terminating drug action. Drugs for which these enzyme systems have high affinity and capacity are rapidly cleared from portal blood and exhibit a short duration of action unless they happen to be sequestered within cells of other organs. Some therapeutically useful steroids have been chemically modified to slow hepatic oxidation and conjugation. The 17α substitution of the female sex steroid, estradiol, is a case in point and 17α-ethinyl estradiol has made oral contraception possible because of the relatively slow rate of degradation of this semi-synthetic steroid by the oxidative enzymes of the liver SER.

Metabolism of drugs and hormone by hepatic oxidative, conjugating and other pathways is not necessarily associated with subsequent secretion into bile as happens to bilirubin. In general, compounds of higher molecular weight (over 400 daltons) are more likely to be excreted in bile but specific transport processes also appear to operate for drugs such as the penicillins. Study of these processes is less easy and has been less intensively pursued than transport across the renal tubules.

SOME VITAMINS WITH WHICH THE LIVER IS PARTICULARLY CONCERNED

Vitamin B_{12} and the fat soluble vitamins A, D, E and K are either stored in the liver or play significant roles in hepatic function to justify consideration here. Although the water soluble vitamins (the B-group, and C) are important in biochemical processes occurring in the liver, of this group only Vitamin B_{12} is stored in this organ which may contain sufficient B_{12} to last some years. Four forms of the vitamin are known: adenosyl, methyl, hydroxy and cyano derivatives of cobalamin. The hydroxy or cyano derivatives are used therapeutically but adenosyl cobalamin is the major intrahepatic derivative. The liver synthesizes two transport proteins (transcobalamins I and II) which account for circulating levels of B_{12}. Methyl cobalamin is active in the demethylation of 5-methyl-tetrahydrofolate to form tetrahydrofolate, the active form of folic acid so that one consequence of B_{12} deficiency is a lack of the active form of this vitamin.

Vitamin A is available from the diet as the ester (retinyl palmitate) or as carotene, split by the intestinal mucosa to give two moles of the aldehyde (retinal). Both forms are absorbed with fat and stored largely in the liver (0·35 μmol/g) as ester. Retinol, released by hydrolysis of the ester, circulates bound to retinol binding globulin. Deficiency of this factor leads to night blindness and abnormal keratinisation of epithelial surfaces.

Vitamin D is not extensively stored in the liver but this organ plays an important role in metabolism of the vitamin. To be physiologically active Vitamin D must undergo two hydroxylations at the 25 and 1 positions; the first of these occurs in the liver SER, the second in the kidney. Hepatic hydroxylation of Vitamin D is another example of a mixed function oxygenation and if hepatic MFO reactions are abnormally active (as occurs in patients treated chronically with anticonvulsant drugs such as phenobarbitone which induce hepatic MFO activity) it appears that metabolism of the vitamin occurs to polar products other than 25-OH Vitamin D leading to a relative deficiency of the vitamin and osteomalacia.

Like Vitamin D, *Vitamin K* is not stored in the liver and daily requirements must be met from the diet for adequate synthesis of clotting factors II, V, VII and IX. The Vitamin is converted to an epoxide (probably by a MFO reaction) required for the post translational modification of 10 glutamates at the amino terminal end of prothrombin to γ-carboxy glutamate residues. Similar modifications probably activate the other Vitamin K dependent clotting factors. Various forms of Vitamin K (K's 1–3) are recognized which originate from bacteria or plants, or are synthetic; differences affect their route of administration for therapeutic purposes. The coumarin anticoagulant drugs inhibit the formation of Vitamin K epoxide thus preventing the carboxylation of glutamyl residues in "pro"-proteins and interfering with the formation of clotting factors.

Vitamin E (α-tocopherol) has an uncertain status. Tocopherols do inhibit spoilage of unsaturated fats by peroxidation in the same way as butylated hydroxy toluene or p-hydroxy anisole used by the food industry. However Vitamin E probably plays a more specific role protecting intracellular membrane systems from oxidative damage. Apart from a syndrome of haemolysis in newborn children clearly associated with a deficiency of α-tocopherol, the clinical importance of Vitamin E remains uncertain.

Disordered function

Disordered function secondary to disease processes such as damage to parenchymal cells by alcohol (alcoholic hepatitis), or viral disease of the liver or even obstruction to biliary outflow by stones, will present very differently at first but come to overlap more as time passes. Each may end with cirrhosis, portal hypertension and the same stigmata of chronic liver failure. As a single disease process progresses, so a range of different pathophysiologic mechanisms may manifest sequentially.

There are numerous biochemical pathways active primarily in the liver which may be compromised by *inborn errors of metabolism*. Such disorders, including the lipoid and glycogen storage diseases, porphyrias and specific protein deficiencies cannot be considered in detail here. *The rate at which the liver is damaged* will dictate the degree of functional disturbance. At one extreme (fulminant hepatic failure) there is rapid and complete loss of function because of toxins, infectious disease, trauma, or, rarely, vascular disease. At the other extreme, destruction with scarring slowly leading to a cirrhotic liver, may pass unnoticed until some vital aspect of liver function is compromised. This is quite common. *The liver has tremendous powers of regeneration*; destruction of the organ by a disease process will manifest with both the effects of the destructive process, and the accompanying regeneration. If regeneration is accompanied by excessive collagen deposition, a cirrhotic liver results in which the architecture is permanently deranged and the regenerating cells are clustered in nodules which lack normal or adequate connections to the vessels entering and leaving the liver. It is not known why certain pathologic processes tend to be associated with collagen deposition.

CIRCULATORY DISTURBANCES

Contrary to earlier views the hepatic artery may be ligated during surgery (e.g. to control haemorrhage) without liver failure necessarily occurring, even though some part of the organ will infarct. If portal blood flow is maintained liver function will be disturbed for a brief period and then recover. The portal vein can become obstructed early in life.

At birth, circulation through the liver changes as the umbilical veins close; umbilical sepsis may lead to thrombosis of the portal vein so that at the start of extra-uterine life the liver is denied a normal supply of splanchnic blood. Normal development will not occur and the liver will be small, tends to deposit iron, and may fibrose. This underlines the important trophic effect of splanchnic blood draining to the liver.

Obstruction to the circulation through the hepatic sinusoids commonly develops secondarily to hepatic parenchymal or biliary disease when regeneration and scarring lead eventually to cirrhosis. Portal hypertension is the consequence of a marked increase in resistance to blood flow in the portal circulatory bed. It will be recalled (Pt. I) that arterial blood joins portal blood in Zone 1 of the liver acini. Obstruction between this point and hepatic venous outflow will cause a further increase in pressure in the portal vein.

Portal hypertension

With time, collateral veins develop, bypassing the portal circulation through the liver and linking the splanchnic and systemic venous circulations. Anastamoses are found at the lower end of the oesophagus, in the cardia of the stomach, arising from the splenic vein, traversing the diaphragm, around the umbilicus and may also involve the haemorrhoidal veins. Depending upon the severity and duration of portal hypertension, these collateral channels may be huge. Rupture, particularly of oesophageal varices, can cause catastrophic blood loss and is unfortunately a common problem.

The spleen increases in size in portal hypertension with an increase also in its arterial supply. Such large spleens may be associated with accelerated breakdown of formed elements in the blood (hypersplenism) apart from contributing an increased blood flow to the already congested portal circulation.

Shunting of blood through collateral channels from the portal to the systemic circulation has both good and bad features. Good, because the portal pressure is reduced and thus the risk of a collateral vein rupturing but bad because functional liver tissue is being bypassed by the portal blood. Porto-systemic shunting is associated with altered blood levels of amino acids, ammonia and glutamine; porto-systemic encephalopathy is a disorder of brain function attributed to this aspect of compromised hepatic function.

Porto-systemic encephalopathy

Porto-systemic—or hepatic—encephalopathy (PSE) is a term used to describe the disturbed cerebral function which occurs when systemic blood supplying the brain is inadequately cleared by the liver of gut-derived nitrogenous materials. This may occur because of reduced hepatic parenchymal cell function or because portal blood enters the systemic circulation without traversing hepatic sinusoids, or both. The breath of the patient may have a characteristic musty odour (foetor hepaticus). Higher cerebral function is disturbed even to the extent of deep coma, and abnormal motor function manifests as a coarse, typically flapping, tremor. The electroencephalogram is abnormal. At its most severe in acute hepatic failure, PSE may be associated with catastrophic increases in intracranial pressure associated with cerebral oedema.

Just what N-containing compounds cause PSE is not clear. Some correlation exists between disturbance of brain function and circulating levels of ammonia and glutamine; certain toxic amines are formed by gut bacteria. These may be significant or altered levels of circulating amino acids may disturb cerebral neurotransmitter metabolism.

It is often possible to reduce portal venous pressure surgically by establishing a communication between the portal vein and the inferior vena cava (side-to-side shunt) or by implanting the portal vein into the inferior vena cava and tying off the portal vein entering the liver, or by using other operations involving the splenic and left renal veins. Unfortunately, the risk of porto-systemic encephalopathy is high when portal blood is deliberately encouraged to bypass the liver and there must be a good reserve of hepatic function for these operations not to lead to a disastrous deterioration of cerebral function.

Obstruction to the hepatic circulation can also occur distal to the sinusoids. In hepatic veno-occlusive disease obstruction is by thrombosis of hepatic veins within the liver due to toxins, associated with estrogen containing oral contraceptives, or without there being any identifiable cause. Portal hypertension develops very rapidly, to produce ascites which is severe and difficult to treat. The increased pressure in the hepatic veins due to right sided heart failure or constrictive pericarditis is not usually associated with severe adverse effects on hepatic function, however, exceptions to this statement occur and stasis in the hepatic veins in these situations can be associated with considerable damage to parenchymal cells in Zone 3 of the acinus and jaundice.

DISORDERED INTERMEDIARY METABOLISM

The liver is the storage site and regulator of circulating carbohydrate fuels; it also is exposed to high concentrations of regulatory hormones (insulin and glucagon) in portal blood. It is perhaps surprising, therefore, that abnormalities of carbohydrate metabolism are not more prominent clinically in patients with liver disease, but this is not the case. It is true that hypoglycaemia may occur in fulminant hepatic failure, and it can complicate starvation in patients with cirrhosis, but this is rare unless alcohol is involved (see below). A brief description of fulminant hepatic failure is included at this point.

Fulminant hepatic failure

The rapid loss of all categories of liver function is not common, but carries a high mortality when it does occur. Viral infection and paracetamol overdose are the commonest causes in Britain; it is a rare complication of late pregnancy. In children it occurs as part of Reye's syndrome. Encephalopathy will progress rapidly to coma and jaundice will deepen steadily. Bleeding occurs due to decreased circulating levels of clotting factors (p. 491). Hypoglycaemia may occur due to failed gluconeogenesis. Urea production will decrease drastically while high levels of amino acids, glutamine and ammonia circulate. To add to the problems facing such a desperately ill patient acute renal tubular necrosis with renal failure is a common, if poorly understood, complication of fulminant hepatic failure.

Carbohydrate

Carbohydrate homeostasis is seldom normal when liver disease is advanced. Oral glucose tolerance tests (GTT) on cirrhotic patients resemble those in diabetics, and such patients may spill glucose in their urine after a meal. In contrast, the intravenous GTT is usually normal and the abnormality seen after oral loading probably reflects the porto-systemic shunting of both glucose from the gut, and insulin from the pancreas.

Circulating levels of insulin tend to be high in cirrhotic patients; the reason(s) for this are not generally agreed. Possibly with insulin not reaching the liver directly in portal blood, higher than normal levels must circulate systemically to achieve carbohydrate homeostasis, which must depend largely on insulin reaching the liver via arterial blood. The liver is the major site of insulin degradation and clearance of insulin from blood is reduced in liver disease. 'C-peptide', a fraction of proinsulin released in equimolar amount with insulin from the pancreas, is, in contrast, cleared by the kidney. In liver disease, insulin levels increase but 'C-peptide' levels do not.

As already inferred, alcoholic liver disease is associated with significant abnormalities of carbohydrate homeostasis. Alcohol is a liver poison, but in addition, the metabolism of alcohol via acetaldehyde to acetate occurs at the expense of NAD^+ being reduced to $NADH$. Only availability of NAD^+ controls the rate of alcohol oxidation, and the ratio $NADH : NAD^+$ increases; this in turn inhibits gluconeogenesis via the central glycolytic pathway. This abnormality, coupled with the already reduced reserve capacity for gluconeogenesis of a damaged liver, causes hypoglycaemia and may thus cause or exacerbate coma.

The existence of inherited biochemical abnormalities was referred to in the introduction to this section. A number involve carbohydrate metabolism and include: (a) inability to metabolize galactose (galactosaemia) and (b) various enzyme defects preventing the breakdown of glycogen. In the former, circulating galactose is toxic to various tissues while in the latter, failure to maintain blood glucose during starvation is coupled with damage to the liver by the huge amounts of glycogen accumulated.

Fat

Accumulation of droplets of fat within parenchymal cells is seen in the reaction of the liver to both normal and abnormal stimuli. Direct hepatotoxins such as alcohol cause this, and the change also occurs in diabetes mellitus or simply as an accompaniment to obesity. Fat metabolism may be affected in various ways to cause this abnormality. One possibility is that the synthesis of proteins required for export of fat from the liver (lipoprotein apoproteins) is inhibited. Alternatively, there may be decreased levels of ATP in the cells and thus a failure of exporting mechanisms, or there may be overproduction of lipid. Alcohol is thought to cause all three effects. Fat may double the weight of the liver in alcoholic hepatitis; when the toxin is removed and normal nutrition substituted this fat then leaves the liver with a consequent temporary increase in circulating lipid (Hyperlipaemia). Accumulation of droplets of fat in liver also occurs with exposure to other toxins which inhibit protein synthesis (e.g. methionine) or toxins which lead to peroxidative damage to membrane lipids (e.g. carbon tetrachloride).

Amino acids

As would be predicted, urea synthesis is compromised in severe parenchymal liver disease. Blood urea nitrogen falls while levels of amino acids and also of ammonia, increase. Ammonia is toxic to the brain and these events may be accompanied by abnormal brain function—hepatic (or porto-systemic) encephalopathy. Both the suggestion that elevated amino acids might interfere with neurotransmitter synthesis, or that elevated levels of ammonia disturb brain function have been advanced to explain this process but the relative roles of these are still debated.

As already stated hepatic encephalopathy is more likely in those patients in whom a high percentage of splanchnic blood bypasses the liver and in such patients measures are employed which are designed to reduce absorption of amino acids and possibly also to reduce the absorption of toxic products of bacterial metabolism from the large gut. Dietary protein is restricted, antibiotics are given to reduce colonic flora, colon pH is reduced to reduce absorption of amines and intestinal transit time is shortened using lactulose or saline purgatives. The kidney will synthesize urea from circulating ammonia to a limited extent, but a major problem in patients with severely compromised parenchymal function is their inability to tolerate dietary protein while at the same time needing an adequate protein intake to counteract accelerated catabolism of body proteins. It has been proposed that an altered ratio of aromatic amino acids (tyrosine and phenylalanine) to branched chain amino acids (valine, leucine and isoleucine) might be significant but testing of this hypothesis in a clinical setting has not supported it.

DISORDERED PROTEIN SYNTHESIS

Abnormalities of protein synthesis feature prominently in the clinical presentation of liver disease. Albumin, the coagulation factors I, II, VII, and IX to XII, and various other transport proteins are all affected, but albumin and the coagulation factors are presently of most clinical interest. The half-life of albumin is relatively much longer than that of the coagulation factors. A figure of 17 days was given for the albumin $T\frac{1}{2}$ in Part I although rates of synthesis and catabolism vary with the intake of protein and calories, and also with body temperature, thyroid status and the response of patients to disease. Trauma, including surgery, for example, leads to a rapid acceleration in breakdown of body protein. Starvation is associated with slower turnover until body proteins are degraded to provide energy. Accepting these qualifications for statements regarding half-lives of proteins, the half-life of albumin is still 5–10 × longer than that of the coagulation factors. Consequently, if sudden and severe damage to the liver occurs, hypoalbuminaemia will not occur for a week or longer although bleeding may occur within 24–48 hours. On the other hand, in chronic liver disease hypoalbuminaemia and the attendant changes in body fluids and electrolytes, may dominate the clinical picture while reduced levels of the coagulation factors may be apparent only from laboratory tests. Altered levels of transport proteins do occur in liver disease and may distort results from clinical laboratories reporting total (free and bound) concentrations of circulating substances, (this point should be borne in mind when such tests are interpreted. Assays for thyroid hormones, various steroids as well as protein bound drugs are affected by this source of error). This is a suitable point at which to consider the consequences for fluid and electrolyte balance of hypoalbuminaemia.

Hypoalbuminaemia and the development of ascites

Normally albumin provides two-thirds of the plasma oncotic pressure. The problem of hypoalbuminaemia will be discussed from the point of view of ascites which, occurring in advanced liver disease, provides the most frequent clinical problem associated with this state.

Normally the peritoneal cavity contains little free fluid but in advanced liver disease it can accumulate quantities of ascitic fluid in excess of 20 litres causing great discomfort to the patient. There are two reasons for this: (1) a decrease in serum albumin, usually below 30 grams/litre due to failing hepatic protein synthesis and (2), portal hypertension. Neither of these alone provides a sufficient explanation. Decreased plasma albumin will cause the ECF space to expand with increased plasma renin activity and increased circulating aldosterone levels both representing the predictable responses of physiological control of the effective intravascular volume. Local factors tend to determine where the excess extravascular fluid accumulates and portal hypertension will encourage this to localize in the peritoneal cavity. Although this fluid is not an inflammatory exudate it does contain considerable amounts of protein; 10 g/l is typical when the serum albumin may be only twice this figure. There is evidence that part of the newly synthesized albumin from a cirrhotic liver enters ascitic fluid without ever passing through the systemic circulation.

Ignoring the role of lymphatics, the Starling forces which determine formation or resorption of ascitic fluid are:

Where P refers to hydrostatic, and π to osmotic, forces:

$$\Delta P = (^{P}\text{portal vein} - {}^{P}\text{intra-abdominal}) - (^{\pi}\text{plasma oncotic} - {}^{\pi}\text{ascites oncotic})$$

If ΔP is positive, ascites will tend to form. This relationship has been tested in liver disease in patients with and without ascites, and is roughly valid. However, significant amounts of protein-rich fluid can enter the peritoneal cavity from hepatic lymphatics and, presumably, visceral peritoneal lymphatics play a role in removing protein and fluid.

Superimposed on this simple argument for the presence or absence of ascites based on hydrostatic pressures are those events secondary to a reduced effective intravascular fluid volume which will expand the extracellular fluid space, will tend to increase portal venous pressure and will alter the renal handling of salt and water. Plasma renin activity, angiotensin II, aldosterone, vasopressin and possibly an additional postulated hormone currently referred to as natriuretic factor, are all effectors for homeostatic (feedback) systems which maintain extracellular fluid volume, sodium content and osmolality. Some of this fluid is, of course, intravascular where it is effective in maintaining normal circulatory function and plasma protein must be regarded as playing a critical role in maintaining the relative amounts in the extracellular fluid inside and outside the vascular bed.

While protein concentration distinguishes intravascular and extravascular ECF, protein does slowly exchange between these spaces. Maintenance of the gradient is an active process in which the lymphatic circulation and extravascular protein catabolism are involved but knowledge here is incomplete. Patients accumulating ascites will be actively retaining sodium which may almost disappear from their urine. The value of spironolactone, the aldosterone antagonist, in managing this problem emphasizes that hyperaldosteronism contributes to the accumulation of ascitic fluid.

Disordered synthesis of blood coagulation factors

Most of the circulating proteins required for blood clotting are synthesized in the liver and these may be divided into these which do, and those which do not, depend upon Vitamin K for their biosynthesis. Vitamin K may be deficient; this will be associated with an inadequate capacity to complete the synthesis of Vitamin K-dependent clotting factors. Dietary lack (the vitamin is available in fresh plant products), poor absorption, which occurs whenever absorption of fat is interfered with, and inadequate transfer from mother to foetus prior to birth are clinically important problems. Also, the group of anticoagulant drugs which are inactive analogs of Vitamin K (coumarin-type anticoagulants) cause a picture very similar to Vitamin K deficiency.

Acute, severe liver damage may be accompanied by diffuse intravascular coagulation—a hypercoagulability problem—but failure of blood clotting in liver disease is more common. If the primary disease process involves malabsorption of fat and the fat soluble vitamins, parenteral Vitamin K (e.g. phytomenadione) will improve or normalize the disordered clotting process. However, if the available mass of healthy parenchymal liver tissue is inadequate to synthesize enough of the required protein factors, this cannot happen. In liver disease, fibrinogen (Factor I) synthesis is seldom reduced sufficiently for hypofibrinogenaemia to constitute a clinical problem, and it is primarily through deficiency of the Vitamin K-dependent factors (prothrombin or Factor II, Factors V, VII and IX) that liver disease leads to defective blood clotting. These factors are not usually measured singly, but as the prothrombin time (PT) or partial thromboplastin time (PTT) (see Chapter 8).

The most rapidly turning over of the Vitamin K-dependent coagulation factors is Factor VII ($T\frac{1}{2}$ = 3–6 hours) so in the event of massive liver cell damage of rapid onset, 24 hours may suffice to lower Factor VII levels to 1/5 to 1/250 of normal. In chronic liver disease it is common to find prolongation of the prothrombin time and partial thromboplastin time. This is due to inadequate synthesis of the Vitamin K dependent factors all of which will be reduced to a similar proportion of normal in a stable patient, despite their different half-lives. Obviously the disorder cannot be corrected by injecting the vitamin.

Lipoprotein disorders in liver disease

With the liver and gut mucosa being the main sites of lipoprotein synthesis it is perhaps surprising that disorders of lipoprotein function have been relatively neglected in the study of liver disease. Hyperlipidaemia has long been recognized as a complication of both obstructive and some forms of parenchymal liver disease, but important questions regarding mechanisms remain unanswered.

As mentioned in connection with possible causes of alcoholic fatty liver, failure of hepatic protein synthesis could compromise the supply of lipoprotein apoprotein from liver (both VLDL and HDL) (see p. 477 for abbreviations) but not, of course, from the gut (possibly only VLDL). The enzyme LCAT, responsible for forming cholesterol ester by trans-acylation, originates in the liver and may be deficient in parenchymal cell damage. LCAT deficiency reduces HDL and is also associated with abnormal LDL—one of the reasons for finding an atypical LDL, "lipoprotein X", in liver disease.

When bile outflow is obstructed cholesterol synthesis increases leading to hyper-cholesterolaemia and, for a reason not clear, a matching hypertriglyceridaemia. Part of this may be due to reflux of lecithin from the biliary tract back to the blood stream. It has been shown that high circulating levels of triglycerides and sterol in liver disease result in red cell membrane abnormalities (expansion of the membrane, and increased rigidity) giving spur cells and target cells on blood smears. Membranes of other cell types are possibly also affected although changes observed in red cells may reflect the non-renewable nature of components of this cell. Finally, it is interesting to note that hypolipidaemic drugs which lower circulating lipid levels by mechanisms largely unknown in patients without liver disease, tend to have an opposite effect in hyperlipidaemia when this is secondary to liver disease.

DISORDERS OF BILE FORMATION AND FLOW

If the mass of functioning liver tissue is reduced, disorders of bile formation and flow are inevitable. However, these tend not to dominate the clinical presentation so that even in advanced cirrhosis, for example, jaundice may be mild or absent. The mechanisms involved when disordered secretion of bile is significant may be examined at various levels: (1) metabolic processes within liver parenchymal cells, (2) the secretion of bile across the canalicular membrane, (3) the composition and detergent properties of bile, (4) the major problems that arise due to mechanical obstruction of bile flow and (5) interference with the enterohepatic circulation. Clinically, the more obvious disturbances are jaundice due to an increase in circulating bile pigment, pain, due to obstruction of the gall bladder or the bile ducts, infection, and abnormal gastrointestinal function when bile ceases to enter the small intestine.

Routine clinical tests for altered bile pigment metabolism are better developed than tests to investigate the metabolism of other constituents of the bile. However, progress is being made with the physiological chemistry of the bile salts and new tests as well as new forms of treatment are emerging in this area.

While much attention is paid to the treatment of patients who develop biliary calculi, it is worth noting that these manifestations are secondary to the disorders already listed. Unfortunately prevention lags behind surgical or medical management but this is another area where progress is being made.

Retention of bilirubin, and jaundice, will be dealt with mainly in the following section because this will permit widening of the discussion to include other substances (e.g. drugs) handled by similar hepatic mechanisms. This section will deal with the pathophysiology of the bile salts and other general aspects of cholestasis.

Metabolic processes within the liver cell

As the rate of secretion of bile is, in part, dependent upon the rate of secretion of bile acids, a variety of factors which affect bile acid metabolism influence bile flow. The enterohepatic circulation is crucial and requires that excessive amounts of bile salts are not degraded in the gut, that the terminal ileum has not been resected and that rapid intestinal transit does not prevent reabsorption of bile salts.

Secretion across the bile canilicular membrane

When the secretion of bile is obstructed (cholestasis—see below) at the level of the parenchymal cell, abnormalities of the peri-canalicular region may sometimes be seen under the electron microscope. It has been postulated that these may be due to bile salts accumulating to reach concentrations sufficient to disrupt the canalicular membrane. This remains plausible but unproven. The secondary bile acid—lithocholic acid—has been held responsible and cholestasis can be produced in laboratory animals by feeding this substance.

Composition and detergent properties

The micellar solution that constitutes bile is only stable within a limited range of concentrations of the three major components of the micelles—bile salts, lipid and cholesterol. However, when supersaturation occurs as it frequently does, in the gall bladder of apparently normal individuals, cholesterol will not necessarily precipitate to form stones. An additional factor (or factors) is required, such as a nidus (which could be a clump of bacteria) for crystal growth. Nevertheless, lithogenicity of bile defined by purely physico-chemical criteria applied to aspirated bile does correlate with stone formation. With the demonstration that chenodeoxycholic acid or ursodeoxycholic acid taken as dietary supplements will increase the bile acid pool size, and re-dissolve smaller stones in patients with a functioning gall bladder this is now an area of practical importance. The mechanism of this effect is incompletely understood and does not seem to depend solely on a change in the physical properties of the bile; the rate of secretion of cholesterol is also reduced. It is worth adding that there are not major differences among the various dihydroxy and trihydroxy bile acids in their capacity to solubilize cholesterol and thus prevent stone formation.

At present the size of the pool of hepatic, biliary, gut and circulatory bile salts is considered the major determinant of stone formation, with the rate of secretion of cholesterol the second factor to consider. It will be recalled that pool size depends largely on the inhibitory effects on synthesis of bile salts returned from the gut to the liver by the enterohepatic circulation. Factors which reduce this include resection of the terminal ileum, the administration of cholestyramine ion-exchange resin which binds bile salts in the gut lumen and increased bacterial degradation of bile salts within the upper small intestine. Such bacterial activity increases secondary and tertiary bile salts. Diet may also play a role and it is claimed that increased dietary fibre prevents absorption of secondary bile salts from the colon.

Hormones, particularly estrogens, influence hepatic bile salt metabolism. Women during the reproductive phase of life have a smaller bile acid pool. This estrogen effect is more marked in users of oral contraceptives in whom bile is abnormally lithogenic and pregnancy, with high levels of estrogen production, is also associated with a reduced bile salt pool and an increased tendency to form cholesterol gall stones.

Surprisingly, considering the interest in cholesterol metabolism in heart disease, less is known about determinants of cholesterol secretion rates. It appears that cholesterol, newly synthesised in the liver, is favoured for export into bile. Obesity, diet, estrogen therapy and certain disease states also increase cholesterol over the one mmol, approximately, secreted daily.

Cholestasis

This term implies failure of bile to flow and does not distinguish between blockage of the biliary system (extrahepatic cholestasis) and a failure of bile secretion across the biliary canaliculus (intrahepatic cholestasis). (Obstruction to major bile ducts within the liver would still be regarded as extrahepatic). Initially the extra- and intrahepatic varieties will be clearly different. In the former a gland (the liver) has had its secretory duct obstructed while in the latter the secretory machinery is defective. However, with passage of time changes within the liver secondary to the duct obstruction will blunt these differences.

Extrahepatic cholestasis

Stones, tumours, scar tissue, oedema due to inflammation, and parasites are all possible causes of blockage of biliary drainage. Bile pigments cease to be excreted into the gut and jaundice results. The stool becomes pale and the urine darkens. Bile salts suffer a similar fate: digestion and absorption of fat are impaired and the absorption of fat soluble vitamins is impaired. Stools will be not only pale, but also greasy (steatorrhoea) and the level of circulating cholesterol may rise. Deposition of bile salts in the skin is associated with itching (pruritis) and it is possible that the high concentrations of bile salt which occur locally in the liver may cause damage to the biliary canaliculus and intracellular membrane systems.

Intrahepatic cholestasis

In intrahepatic cholestasis the mechanism of obstruction is often ill-defined. However, the bile ducts should be demonstrably patent. Alcoholic liver disease and viral hepatitis can both present with cholestasis although it is uncommon for this consequence to dominate the presentation. The best example of intrahepatic cholestasis is seen as an idiosyncratic response to phenothiazine drugs which do appear, in some individuals, to exert a specific toxic effect on the canalicular membrane with attendant ultrastructural abnormalities.

In any event the consequences are similar. The secretion of bile salts and the bile salt-dependent component of biliary secretion decreases or ceases. As the elimination of bile pigment, cholesterol and lipid are a part of this component, their excretion is also effectively obstructed. However, intrahepatic cholestasis is seldom as complete as can occur when a calculus blocks the common bile duct.

Having made the distinction between intrahepatic and extrahepatic cholestasis it should be pointed out that as smaller biliary radicals are involved by an 'extrahepatic' process—one distal to hepatic parenchymal cells and biliary canaliculi—the overlap will be greater. Cholangitis may affect all levels of the biliary system and the smallest biliary ducts are involved in primary biliary cirrhosis, a condition in which progressive fibrosis of the portal tracts occurs.

Some further effects of cholestasis on bile pigment metabolism are dealt with in the following sections. These will emphasise that when bile pigment and bile salts accumulate in the liver rather than reaching the gut their metabolic fate changes; diversion to the urine is partly a consequence of this altered metabolism.

Quantitative aspects of the accumulation of bilirubin in the circulation are of interest. Normally, about 300 mg of unconjugated bilirubin (M.W. 528)—568 μmoles—is produced daily. This circulates with a volume of distribution slightly greater than that of plasma, or 3·13 l. If disposal of bilirubin were suddenly and completely stopped the level of circulating bilirubin would rise at a rate not exceeding 180 μmol/l/day. Conjugation will increase the volume of distribution, and losses in the urine will also reduce the rate at which jaundice would increase. These facts, together with the fact that the obstruction to the flow of bile is not usually complete, will decrease the rate at which the circulating bilirubin level increases.

Jaundice due to acute obstruction of the common bile duct by a calculus

Gallstones more commonly make their presence known by the associated cholecystitis and when stones do obstruct the common bile duct this is often against a background of previous disease of the gall bladder or ducts; furthermore the obstruction is often incomplete. However, for the sake of illustration let us again consider sudden, fairly complete, obstruction of the common bile duct.

Severe colicky pain will occur when a stone lodges in the muscular common bile duct. Jaundice will develop rapidly (p. 495) due to an increase in circulating conjugated bilirubin. The stool will become clay-coloured due to absence of stercobilin and urobilinogen will be absent from the urine which, however, will be dark because of renal elimination of conjugated bile pigment. Pruritis (itching) may develop as bile salts reflux from the liver into the systemic circulation and deposit in the skin and unusual conjugated bile salts may accumulate in the liver and be excreted by the kidney.

Laboratory tests will find no evidence of haemolysis (unless these are pigment stones *due* to haemolysis), the elevated bilirubin in plasma will be largely conjugated (direct reacting), activity in plasma of alkaline phosphatase and $5'$ nucleotidase are likely to be rising. Other tests of liver functions—protein synthesis in particular—should be normal.

Unless the obstruction is relieved, succeeding days will see additional changes as hepatic parenchymal function is adversely affected by accumulated secretions.

DISORDERED CATABOLIC FUNCTIONS OF THE LIVER

Bilirubin

The physiology of bilirubin was described in Section I as divisible into six discrete processes from heme degradation to enterohepatic circulation. Each step may be disordered. Heme degradation may be accelerated by hemolysis, by ineffective erythropoiesis, or by degradation of extravascular accumulations of blood. These processes will be associated with an increase in circulating unconjugated (indirect reacting) bilirubin but clinical jaundice, which requires circulating bilirubin to exceed about 70 μmoles/l is usually absent or mild.

The capacity for plasma transport of unconjugated bilirubin on albumin may be exceeded by rapid red cell breakdown particularly if this is associated with reduced hepatic clearance. This may occur in the neonatal period when a hemolytic process such as is caused by anti-Rh(D) antibodies circulating in the infant is combined with immaturity of the liver. The lipophilic, unconjugated bilirubin IX α tends to leave the circulation and to deposit in brain tissue leading to kernicterus, or bile pigment staining of the basal ganglia.

The commonest cause of an increase in circulating unconjugated bilirubin is

Gilbert's syndrome, a term used to describe people otherwise perfectly normal, whose plasma bilirubin lies in the range 20–100 μmoles/l. This reflects reduced clearance of bilirubin and is exaggerated by starvation or a fat-free diet. Hepatic uptake of other organic anions such as ICG is also defective but the mechanism is not further understood.

Conjugation of bilirubin is reduced in the neonatal liver and particularly in the livers of premature neonates as already mentioned. There is also an inborn error of metabolism, or group of such errors, referred to as the Criggler-Najar syndrome, in which activity of the enzyme, UDPG-T is reduced so that, as in Gilbert's syndrome, patients have unconjugated hyperbilirubinaemia. However, the Criggler-Najar syndromes are more severe as well as possessing a demonstrable enzymatic defect in bilirubin metabolism.

In intrahepatic cholestasis, as described above, the transfer of conjugated bilirubin into the bile canaliculus will decrease or cease being tied to the bile salt dependent component of bile flow. When this happens bilirubin conjugates (mono- and diglucuronide) will diffuse back to the plasma and undergo limited clearance by the kidney so that the urine darkens.

The fate of bile salts is similar to that of bile pigments in cholestasis. The bile salts undergo further reactions in the liver including oxidation, hydroxylation and conjugation with sulphate or glucuronide and these more water soluble compounds may then be cleared by the kidney. It is worth stressing the general principle that when lipophilic materials are exposed to the mixed function oxygenases and conjugating enzymes of the hepatic endoplasmic reticulum, they are likely to undergo some of a range of reactions which generally tend to increase their water solubility. Although this occurs in cholestasis the retained bile salts are not adequately dealt with by these novel pathways of excretion and the accumulation of the bile salts may potentiate liver cell damage.

Degradation of hormones, and other lipophilic compounds by hepatic MFO reactions

Chronic liver disease, particularly chronic alcoholic liver disease is accompanied by stigmata which have, in men, been attributed to feminisation. These include palmar erythemia, gynaecomastia, and testicular atrophy with impotence. Spider naevi about the arms, face and trunk also occur but as these do not occur in men treated therapeutically with oestrogens their appearance must require additional factors. Testosterone levels are low in men with alcoholic liver disease and it is possible that the abnormal liver tissue present (hyperplastic nodules in a cirrhotic liver) has either accelerated the degradation of testosterone or its conversion to oestrogen by aromatisation of the steroid 'A' ring. Despite considerable and careful experimental work a definite explanation for the feminisation so commonly noted in liver disease is unavailable.

The final point to be made under this heading relates to the fate of bile salts in cholestasis. It was noted that when obstruction lead to retention of bile salt and bile pigments, these underwent various novel reactions in the endoplasmic reticulum of hepatic parenchymal cells. These rather non-specific degradative reactions—oxidations, hydroxylations and conjugations—also serve to terminate the actions of steroid hormones, to degrade Vitamin D and to accelerate the disposal of many drugs and other foreign chemicals. The activity of these pathways depends very much upon hormonal and other environmental effects upon liver cells and some drugs such as phenobarbitone also exert a marked effect on the amount of smooth endoplasmic reticulum in liver cells as well as on the activity of associated oxidative and conjugating pathways. These facts have important pharmacological implications which will not be dealt with here. However the inducing effect of phenobarbitone on these non-specific catabolic functions of the liver has therapeutic implications. Phenobarbitone treatment may, by increasing conversion of bile salts to soluble glucuronides that can be cleared by the kidney, reduce pruritis associated with cholestasis. Such treatment may also reduce jaundice in some patients with hereditary unconjugated hyperbilirubinaemia. There is also faster metabolism of steroid hormones; however, as endogenously secreted hormone is subject to homeostatic control sensitive to circulating levels, this effect will be compensated for.

PRINCIPLES OF CLINICAL OBSERVATION

Tests and measurements

Rapid advances in diagnostic procedures now makes it possible to monitor most aspects of the structure and function of the liver in the course of normal clinical practice and without undue discomfort or risk to patients. In this section, the terms 'clinically' or 'clinically useful' are used to indicate investigations that influence patient management. Tests that are not so described may still provide important information in an appropriate research setting.

GROSS ANATOMY

An ordinary x-ray film of the abdomen yields little information about the liver; at best it may provide a rough guide to its size unless gas shadows, calcification or radio opaque calculi are seen. In contrast, computerized axial tomography (C.A.T.-scan) permits the reconstruction of "slices" through the trunk and provides a very complete image of the liver. Contrast between different types of tissue varies but tumor deposits may be differentiated. A cheaper procedure is the ultrasonic scan which relies on the reflection of sound waves from any interface between regions of different density. Dilated duct systems and dense bodies such as calculi give strong signals. "Ultrasound" is thus an excellent tool with which to investigate the calibre of the bile duct system and to seek biliary calculi. Ultrasound gives poor results in the presence of much ascitic fluid.

The various techniques for imaging* radiation from the liver after injecting radioactive material also provides information about the anatomy of the organ but, as this relies first on uptake and reflects specific functions at a cellular level, it is considered under headings appropriate to mechanisms of uptake.

Blood vessels entering and leaving the liver

Although we speak of portal hypertension due to obstruction of the portal circulation, the term is used when a collateral circulation is found without portal venous pressure necessarily exceeding its average normal value of 8·8 mm Hg (12 cm H₂O) (1·2 kPa)—although it often will. The point is that the important investigations of these vessels are not so much studies of pressure or flow, but radiologic examination after contrast material has been injected to demonstrate normal and abnormal vessels. This is generally the approach used to examine the 3 major sets of blood vessels—hepatic artery, portal, and hepatic veins. Setting angiographic data apart, studies of pressure and flow are of little diagnostic or prognostic value, nor is this information necessarily required even if surgery on the portal system is contemplated.

The portal vein and the porto-systemic collateral circulation

Contrast material may be introduced into the portal vein by at least five routes: (1) a needle in the splenic pulp permits pressure recordings which reflect pressure in the portal vein or contrast medium may also be injected using this route to produce a splenic venogram. (2) The left branch of the portal vein can sometimes be catheterized via the rudimentary umbilical vein. (3) During oesophagoscopy, in patients with a dilated porto-systemic collateral circulation, a varicose vein may be entered and a catheter passed retrogradely to the portal vein. (4) During surgery, the portal vein may be catheterized directly. (5) Trans-hepatic portal venography in which a thin flexible needle is advanced under radiographic control until a branch of the portal vein is encountered.

All these approaches permit pressure measurements although these are not, as pointed out, currently regarded as having clinical relevance. Angiography defines the anatomy of the portal vein and its collaterals (if any) while cine-angiography can be used to indicate the direction and speed of blood flow. A simple barium swallow is usually sufficient to show the presence of oesophageal varices—dilated collateral veins in the submucosa.

---

* The word 'imaging' is used deliberately to embrace scanning techniques where an image is reconstructed after registering count rates on an x–y grid, and the 'gamma-camera' techniques which reconstruct a 2-dimensional image from the output of multiple collimated radiation counters.

Hepatic artery and hepatic vein

The hepatic artery may be catheterized via the coeliac axis; but studies of flow or pressure in this vessel are again not routinely made. The hepatic vein is accessible via the vena cava; if a catheter is wedged into a branch of one of the hepatic veins, it will reflect sinusoidal pressure which is usually similar to portal vein pressure. If evidence is sought for an obstructed venous outflow, either at the level of the terminal hepatic venule (THV), or in larger branches of the two hepatic veins, wedged and unwedged hepatic vein pressures can be measured and compared with portal vein pressure— usually measured using the transplenic route. To identify blockage of the larger hepatic veins (Budd-Chiari syndrome) angiograms may be sufficient. Elevated pressure in the inferior vena cava particularly from cardiac causes, should be borne in mind as a cause for increased pressure recorded in an hepatic vein.

HEPATIC BLOOD FLOW

Blood flow through hepatic artery or portal vein can be measured directly using an electromagnetic flow meter during surgery, but overall blood flow to the liver must usually be determined indirectly using the Fick principle. Ideally, this requires a steady state level of an indicator in the afferent vessels (artery AND portal vein) while blood is sampled from efferent vessels (there are at least two hepatic veins) unless complete uptake can be assumed and indicator concentration in efferent vessels may be taken as zero.

ICG (indocyanine green) may be given by constant infusion and the 85 per cent or greater uptake of dye during a single pass through the liver is accepted as complete. This is a reasonable compromise to avoid simultaneous efferent vessel sampling. ^{125}I-colloidal albumin is even more efficiently extracted from blood by the liver than is ICG.

For a suitable indicator:

$$\frac{\dot{Q}_{BL}}{C_{BL}} = \dot{V}_{LIV}$$

where $\dot{Q}_{BL}$ = quantity of indicator infused per unit time into the circulating blood
         C_{BL} = concentration of indicator in (afferent) blood at steady state
         $\dot{V}_{LIV}$ = total hepatic blood flow per unit time.

At present, measurement of hepatic blood flow is not part of routine clinical studies.

INTERMEDIARY METABOLISM

Carbohydrate

Coma occurs in acute liver failure and may reflect hypoglycaemia, consequent on the acute failure of carbohydrate homeostasis, or encephalopathy thought to reflect abnormal nitrogen metabolism, or both. *Blood glucose* may be measured to test the former possibility but, reflecting our ignorance regarding the cause of hepatic encephalopathy and coma, measurement of blood ammonia or glutamine is not particularly useful. Electrical activity of the brain reflected by the *electroencephalogram* (EEG) will be abnormal when consciousness is depressed and some correlation exists between the increase in amplitude and slowing of the EEG record, and the severity of encephalopathy due to liver failure.

The *fasting blood glucose* is usually normal in chronic liver disease although severe hypoglycaemia may be caused by alcohol ingestion. However *glycosuria* is common and reflects poor glucose homeostasis. Glycosuria requires that fasting glucose level be determined but is not a reason to test endocrine regulation with an oral *glucose tolerance test* (GTT). 550 mmoles of glucose in 500 ml H_2O given orally to a fasted subject is followed by regular sampling of blood and urine for determination of glucose. This may be abnormal (a) because glucose from the gut bypasses the liver through porta-systemic shunts entering the systemic circulation (b) pancreatic blood is also shunted so that surviving liver cells are exposed to lower insulin levels and (c) higher levels of insulin circulate accelerating pheripheral utilisation of glucose. An *intravenous* GTT (2·8 mmoles glucose/kg) by I.V. infusion of a 25 per cent solution over 5–15 minutes) avoids (a) but will not provide additional data of clinical value in patients with a normal fasting blood glucose level.

If inherited abnormalities of hepatic carbohydrate metabolism are suspected, special tests will be required. In galactosaemia an unusual reducing sugar (galactose) will appear in the urine and this is not detected if glucose oxidase is used to measure glucose. In glycogen storage disease, hypoglycaemia occurs. Possibly, the abnormal enzymes can be measured in biopsy material. Both fructose and galactose tolerance tests have been used but both can produce severe symptoms of hypoglycaemia in patients and are dangerous.

Before considering tests of intermediary metabolism involving lipid and protein, it is worth emphasizing that a clinical nutritional survey is a valuable manoeuvre as it reflects (*inter alia*) on the capacity of the liver to process and interconvert carbohydrate, fat and protein. This survey should estimate body fat and lean body mass (skeletal muscle). "Visceral" protein synthesis—e.g. albumin and transferrin—should be examined at least by recording the amounts of these proteins circulating, and function of the immune system should be examined at least with a white blood count, measurement of immunoglobulins and testing for delayed hypersensitivity (cell-mediated immunity) as these functions are compromised in protein-calorie malnutrition.

Protein synthesis

As pointed out in Sections I and II it is important to consider tissue protein and hepatic secreted protein separately. While it is routine now to consider levels of plasma protein circulating in liver disease, the growing appreciation of difficulties with protein nutrition in patients with liver disease will broaden the range of investigations likely to be considered 'routine'.

Measurement of total plasma protein includes albumin, the *coagulation factors*, transferrin (measured as the iron-binding capacity), the lipoproteins, various other minor constituents also of hepatic origin and the *immunoglobulins* which derive from lymphocytes and are subject to different controls. Total plasma proteins may be well maintained even if hepatic protein synthesis is compromised. *Albumin* is routinely measured separately, and levels below 35 g/l suggest decreased synthesis or increased loss. As pointed out in Section II, defective blood coagulation can develop in hours when protein synthesis in the liver is suddenly interrupted. Vitamin K (10 mg i.m.) should be administered to ensure that deficiency of the vitamin is not responsible for the prolonged PT or PTT (See Chapter 8) characteristic of severe parenchymal liver disease. Assay of *individual clotting factors* is possible in specialized laboratories but is not part of routine clinical investigation.

Studies of *protein synthesis and turnover* are also not routine in clinical practice but are an important topic in current research into deranged protein metabolism. A pure protein e.g. albumin, may be labelled with ^{125}I, injected, and the disappearance of radioactivity from plasma measured. The observed decrement will relate to destruction of serum albumin once distribution of the labelled material in the total pool of albumin in the body is complete. Study of synthesis, on the other hand, requires a constant source of labelled carbon precursor and measurement of the rate at which this label appears in a highly purified protein. The label may be ^{13}C (stable, but costly), or ^{14}C (radioactive). The source may be HCO_3^- or an amino acid such as tryptophan, given by constant infusion; however, there will be a difference between tryptophan radioactivity in plasma and within liver cells so that calculation of rates of incorporation, using this technique requires assumptions which can make the data obtained difficult to interpret.

Finally, it should be stressed that muscle synthesises $3 \times$ as much protein as liver (per gm. body weight) with nitrogen metabolism in these two sites differently controlled. Myo-fibrillar protein in skeletal muscle is characterised by methylated histidine residues. This modified amino acid is not reutilized but appears in the urine so that the amount of *methyl histidine* in urine may be used as an index of skeletal muscle breakdown. As more is learnt of these events clinical tests will provide a more complete picture of protein metabolism in liver disease.

Steroid hormones, thyroid hormone, some vitamins, bilirubin, fatty acids, keto acids and many drugs are transported bound to various plasma proteins. The total amount circulating—though not necessarily the activity—may be altered when hepatic protein synthesis is disturbed and this should be borne in mind when tests are interpreted.

Lipids and lipoproteins

Tests of carbohydrate function depended largely on our detailed understanding of blood glucose and on the hormonal control of circulating glucose as well as glycogen stored in liver and muscle. Tests of protein synthesis largely ignored circulating amino acid levels but measured concentrations of circulating proteins for comparison with well-defined normal limits. Neither approach is as profitable in the clinical study of lipid and lipoprotein metabolism. Keto acids, free fatty acids, triglycerides and cholesterol can all be measured in peripheral blood but large fluctuations occur between the fed and fasted states and particularly during the absorptive phase. Hypercholesterolaemia (with an accompanying increase in triglycerides) occurs in liver disease and is more pronounced in obstructive liver disease. However, the wide range of normal levels even in fasted individuals detracts from the diagnostic significance of blood lipid measurements. Reduced LCAT levels occur in chronic liver disease and are associated with a reduction of esterified cholesterol and HDL. Patients with persistent hypercholesterolaemia deposit the sterol in skin, (e.g. just above the inner canthus of the eye—xantholasma) and in tendons, particularly the Achilles tendon, and regions of skin subject to pressure. Clinical examination should seek xanthomas in these sites.

Lipoproteins are separable into classes by ultracentrifugation, column chromatography and electrophoresis. These time-consuming tests permit conclusions about the relative amounts of up to eight apoprotein fractions, the ratio of triglycerides, phosphatides and sterols (esterified and not) and the pattern of fatty acids present in the glycerides and phosphatides. In the future such analyses may provide information of practical value as more is learnt of the control and interconversion of different lipid fractions.

EXCRETORY AND CATABOLIC FUNCTIONS

Despite the importance of bile salts in liver function, there are not yet satisfactory clinical tests for the metabolic pathways involved, for the enterohepatic circulation of bile salts, for bacterial degradation of bile salts or for the separable bile salt dependent and independent components of bile formation. An earlier section considered the processes whereby bilirubin was (a) taken up by liver cells from plasma (b) conjugated with a resulting increase in water solubility and (c) excreted in bile. This sequence provides appropriate headings to subdivide different aspects of hepatic function.

Uptake and excretion of bilirubin

When increased levels of bilirubin circulate, the skin, sclerae and mucous membranes are stained and, with levels above ± 60 μmoles/l, jaundice can be appreciated clinically. If inadequate uptake is responsible, the circulating bilirubin will be unconjugated making the distinction between conjugated and unconjugated hyperbilirubinaemia a test of clinical value. Unfortunately, because of limitations in available methods, routine clinical chemistry laboratories report instead figures for 'indirect'- and 'direct'-reacting bilirubin with these fractions only roughly equivalent to unconjugated and conjugated pigments. Measurement of direct-reacting pigment is less accurate, leading to larger errors when conjugated hyperbilirubinaemia is assessed. The commonest cause of increased unconjugated bilirubin is a harmless inherited abnormality of hepatic parenchymal function, Gilberts' syndrome. Less common are the more severe Crigler-Najjar syndromes. Overproduction of bilirubin by accelerated breakdown of red cells must, however, be excluded when unconjugated hyperbilirubinaemia is encountered.

In the research laboratory bilirubin may be labelled with ^{14}C and used to determine fractional clearance from blood perfusing the liver. This is neither feasible, nor necessarily useful, in clinical practice. Measurement of clearance from the blood of the dye, bromosulphthalein (BSP), used to be a popular clinical investigation in suspected liver disease until it came to be agreed that this result added little or nothing to measurement of direct- and indirect-reacting bilirubin. In this test circulating BSP is taken up by liver cells by a process partly dependent upon the high affinity for BSP of ligandin. The dye is then conjugated with glutathione and excreted as the conjugate into bile. During cholestasis BSP-GSH conjugate refluxes into the blood. These processes are clearly analogous to those concerned with excretion of bilirubin.

When conjugated and unconjugated bilirubin are measured, the test should be seen as a clearance test in which the rate of production is constant (600 μmoles/day), and circulating unconjugated bilirubin is a function of fraction clearance (0·015 min^{-1}) and volume of distribution ($\approx 3\cdot 3$ l), with conjugated bilirubin (accepting limitations in laboratory procedures) reflecting the diffusion back into plasma of this metabolite because of defective secretion into bile.

Because conjugated bilirubin does not circulate bound to albumin it is ultrafiltered at the glomerulus and enters the urine to which it imparts an orange-green hue. Here it may be detected from the purple colour obtained with diazo reagent available as a clinical test in tablet form.

It should not be forgotten that cholestasis does not persist for many days without hepatocellular parenchymal function also becoming abnormal. When bile flow is first obstructed, the increase in circulating bilirubin will involve only the conjugated fractions, but as parenchymal cell function deteriorates unconjugated bilirubin will also increase.

When the cause of hyperbilirubinaemia is increased production due to haemolysis or ineffective erythropoiesis, unconjugated bilirubin is the fraction that is increased; this finding diverts attention from the liver to the haemopoietic system but, as already mentioned, the commonest cause for this finding is Gilberts' syndrome.

The metabolic fate of conjugated bilirubin in the gut has been described. *Pale stools* indicate that the pigment is not reaching the bowel; green stools may indicate intestinal hurry, or altered bacterial activity (antibiotics) interfering with stercobilin forma-tion and increased *urobilinogen* (detected in the urine from the red colour formed with p-dimethyl amino-benzaldehyde in 6N HCl or 'Ehrlich's reagent') indicate firstly, that stercobilin is present in the gut, and secondly, that the liver is failing to clear uro-bilinogen adequately from portal blood. This capacity may be overloaded in haemo-lytic disease and also in mild hepatic parenchymal cell dysfunction. However, with the ready availability of quantitative measures of bilirubin circulating in the blood the clinical importance of qualitative tests such as urine urobilinogen is waning.

Ehrlich's reagent also give a red colour when added to urine containing porphobil-inogen, a monopyrrole precurser of haem. The coloured complex formed with uro-bilinogen is soluble in chloroform, while that formed with the monopyrrole pigment is not. Porphobilinogen must be present in urine to sustain a diagnosis of acute porphy-ria (although the converse does not hold) and some patients carrying the gene for acute intermittent porphyria will always excrete urine giving a positive test for porphobilino-gen with Ehrlich's reagent.

The secretion of bile

Clinical hallmarks of obstructed bile flow, apart from jaundice, are itching due to circulating bile salts, damaged skin due to scratching in an attempt to relieve the itch, disordered digestion of fat, malabsorption of fat soluble vitamins and calcium, and hyperlipidaemia. The actual rate of bile flow can only be measured when bile is draining externally after bile duct surgery.

Quantitative tests on bile are possible when the fluid has been aspirated from the duodenum but a complete collection is not possible. Bile can also be collected quanti-tatively for brief periods during the ERCP procedure.

The pattern of bile acids secreted can be modified by bacterial activity. Deconjuga-tion of bile acids for example, arises when the small intestine is colonized with bacteria able to split bile salts. This may be detected by *thin layer chromatography* of aspirated bile. Enterohepatic circulation of bile salts may be roughly estimated from the amount of the "secondary" bile acids deoxycholate and lithocholate which are formed by 7α de-hydroxylation of cholic- and chenodeoxy cholic- acids in the gut. A better method to assess bacterial overgrowth in the small intestine would have appeared to be the ^{14}C-glycocholate breath test in which an oral dose of this tracer is followed by moni-toring the exhalation of $^{14}CO_2$ derived from oxidation of the glycine moiety after reabsorption of glycine subsequent to deconjugation. However, the precision of this test appears to be poor.

Hyperlipidaemia associated with biliary obstruction may be accompanied by xanthomas and/or xantholasma; investigation requires that serum cholesterol and tri-glycerides be measured when the subject is fasting.

The observations described above and relevant to bile salt metabolism comprise routine clinical practice. In addition, the clearance of erythritol or mannitol by the liver may be used to measure bile salt dependent bile flow. About 15 μmoles of bile acid is secreted each minute and for each μmole there is an obligatory volume of 14 μl water—i.e. 0·2 ml/min of bile salt dependent flow. The bile salt independent fraction is measured by difference.

Measurement of circulating bile salts or free acids has not been possible in routine clinical practice as it is technically difficult. Gas chromatographic and radioimmunoassay methods are altering this, and it is reasonable to predict that measurements of bile salt metabolism will add to our knowledge of liver disease in relation to bowel function much as knowledge of bilirubin metabolism now permits us to understand jaundice.

THE LIVER 'SCAN'; CONTRAST MATERIALS IN THE BILIARY TREE

As mentioned earlier, images may be constructed from radioactivity localized in the liver following i.v. administration of compounds taken up by this organ. Liver scans thus have 2 components: (1) an anatomical component with the scan showing the position, shape, size, regularity and homogeneity of the organ, and (2) a functional component where the amount of radioactivity localizing in the liver (as opposed, for example, to spleen, bone or kidney) provides information on its ability to abstract something from blood.

Various isotopes are currently available which, being differently handled by parenchymal and/or Kupffer cells, reflect different aspects of hepatic circulation and function in their uptake. To begin by stating the obvious, if there is no circulation to a region it cannot take up an isotope. Colloidal sulphur labelled with 99 M-Technetium ($^{99}Tc^m$) is taken up by phagocytic elements and therefore, localises in Kupffer cells. Since hepatocellular adenomas and some nodules of hyperplastic tissue may lack Kupffer cells, sulphur colloid is not taken up in these regions which thus appear to be "cold". A pedunculated structure of this type will not only be cold but it will not even contrast with surrounding normal tissue and so be invisible using this type of scan. The dye, rose bengal, labelled with ^{131}I, is taken up by well differentiated parenchymal cells using the same processes as are involved in bilirubin uptake; excretion into the bile follows. An adenoma may take up Rose-Bengal but, lacking bile canaliculi, be unable to excrete it with normal rapidity and thus retain activity for longer. The isotope 67-gallium (^{67}Ga) is also used to visualize the liver although the basis for this is less clear: it appears to be taken up by cells actively synthesizing protein and with increased lysosomal activity. Adenomata, compared with carcinoma, show poor uptake of ^{67}Ga Iodinated (^{125}I)-methionine is handled as is the non-iodinated amino acid and incorporated into protein so that scanning with this compound also tends to highlight tissues which are actively synthesizing protein. This includes rapidly growing tumors.

Finally a relatively new series of compounds has become available which, because they can be labelled with the short-lived isotope, $^{99}Tc^m$, permit a high temporary radiation level to be associated with only a small total radiation dose to the patient.

^{99}Tc^m- labelled hydroxyindole diacetic acid (^{99}Tc^m-HIDA), defines parenchymal cell function as does rose bengal but permits radioactivity to be followed through the biliary system and into the intestine even when this excretory function is severely compromised i.e. in jaundice.

No routine clinical tests permit the secretion across the bile canalicular membrane to be analyzed. However, a wide variety of procedures enables the structure and function of the remainder of the biliary tree to be assessed. Most of these tests are concerned with the patency of the bile duct system including cystic duct, gall bladder and pancreatic duct, and with detection of calculi in any of these structures. Most also require x-ray imaging after iodinated material has been introduced into the biliary tree.

The simplest test is the oral cholecystogram. About 8 p.m., and after a supper low in fat, the patient ingests an iodinated dye which is absorbed from the gut and excreted by the same mechanisms as ICG. Twelve hours later x-rays are taken which should show the gall bladder, cystic and common bile ducts filled with contrast material. A fatty meal at this stage stimulates cholecystokinin release from intestinal mucosa which, perhaps together with vagal efferents, empties the gall bladder. This test will fail unless dye is absorbed from the gut, taken up by the liver and concentrated by the gall bladder overnight. When these conditions are met, calculi may be outlined in the gall bladder. Risk to the patient is negligible but failure of the test is frequent; it is almost inevitable in the presence of jaundice. In a very similar test, absorption of dye and reliance upon the approximately 10-fold concentration that occurs in the gall bladder is bypassed using an intravenous injection. This may also fail in a patient with jaundice and/or liver disease. Failure of these tests does not exclude success with ^{99}Tc^m-HIDA.

Information regarding the biliary tree is often important in just those patients in whom the simpler tests described above fail. Three methods remain to fill the biliary tree with radio-opaque dye. (a) a thin needle may be advanced under x-ray control through the chest wall towards the hilum of the liver until a bile duct is intersected when contrast can be injected (the percutaneous cholangiogram). (b) a side viewing fibreoptic endoscope can be swallowed and then advanced to the duodenum where the duodenal ampulla is catheterized and contrast material introduced into the biliary or pancreatic duct systems. (Endoscopic retrograde choledocho-pancreato-gram, or 'ERCP'). (c) if all the preceding approaches fail a surgeon may open the abdomen, cannulate the biliary tree and introduce contrast material directly (operative cholangiogram).

ASSESSMENT OF OTHER CATABOLIC FUNCTIONS OF THE LIVER

Considerable stress has been laid on enzyme systems located in the endoplasmic reticulum of liver parenchymal cells and catalyzing oxygenations (including hydroxylations) and conjugations. These reactions are concerned with the metabolism of steroids (cholesterol, bile acids, hormones), bile pigments and xenobiotics (drugs and other foreign chemical substances). As the activity of these pathways varies in response to inducing stimuli, (the drug phenobarbitone is a familiar example) and could be altered in liver disease, clinical tests of these functions would seem to be desirable. Generally, however, these are unavailable.

Antipyrine clearance by the liver

Antipyrine (phenazone) is an obsolete analgesic and antipyretic drug which is extracted from blood by the liver and slowly metabolized. Clearance of antipyrine from blood depends upon hepatic metabolism, not blood flow. This in turn depends upon the activity of hepatic MFO reactions. The subject to be studied ingests 2 g of drug before sleeping; blood samples are taken on rising and 8–10 hours later for determination of antipyrine. Plasma decay of the drug may be plotted using these accurately timed samples, and the half-life determined. Clearance (ml/min) can also be calculated if certain assumptions are made regarding the volume of distribution of the administered drug.

This test will demonstrate the shortening of half-life observed when subjects are exposed to inducing drugs such as phenobarbitone or rifampicin. Half-life tends to be prolonged in the elderly but a consistently impaired clearance of drug is only associated with severe parenchymal cell disease. An alternative approach, the ^{14}C-antipyrine breath test, measures radioactive carbon dioxide exhaled when a subject metabolizes formaldehyde split off from antipyrine during mixed function oxygenation of a labelled methyl group.

Gamma glutamyl transpeptidase (γ-GT)

This enzyme assay is discussed elsewhere (p. 510). It is sometimes referred to incorrectly as a 'test of enzyme induction'. There is no convincing evidence that changes in the activity of γ-GT are associated with altered metabolism of drugs or steroids by the endoplasmic reticulum.

Steroid metabolism

The question may be asked, why, if catabolic functions of the liver associated with the activity of the SER are involved in degradation of steroid hormones, the metabolism by the body of these substances is not used to assess the activity of such pathways. In animal experimentation, particularly in vitro work, this is done with patterns of testosterone metabolism being used to indicate the relative activities of different microsomal hydroxylations. Thus far such studies have not been adapted to studies in man.

CIRCULATING ENZYMES AND CELL-SURFACE PROTEINS ORIGINATING IN THE LIVER

Various proteins derived from liver and biliary tissue enter the circulation and are measurable in peripheral blood. Assay depends upon either the enzymatic properties of these proteins or upon immunoassay. The precise physiological significance of a particular protein need not be known. Information correlating altered concentration in peripheral blood with specific disease states, or pathologic findings, gives diagnostic importance to these measurements. The different categories we can consider include enzymes that leak from damaged or dead cells, enzymes synthesised by liver or biliary tissue under abnormal conditions, and surface marker proteins which indicate a change in the pattern of gene products expressed by liver cells. Such altered gene expression may reflect altered growth characteristics of these cells.

The kinetics of circulating proteins of diagnostic value are significant; when tests are made for a given event, the interval between the suspected event and the test should not be large relative to the half-life in plasma of the substance tested for. Unfortunately data is not always available to determine whether this condition is met.

Enzymes released by damaged cells

Aspartate transaminase (glutamic oxaloacetic transaminase; AST; SGOT) ($T\frac{1}{2} = 47 \pm 5$ hrs) glutamic pyruvic transaminase (SGPT) ($T\frac{1}{2} = 47 \pm 10$ hrs) lactic dehydrogenase (LDH) and iso-citrate dehydrogenase (ICD) are enzymes of intermediary metabolism which increase in the peripheral blood when the liver parenchyma is damaged. Normal levels of AST are under 40 I.U./l; during acute viral hepatitis AST activity may increase to 1000–10,000 I.U./l. However, these enzymes are not unique to liver tissue and increases also occur after myocardial infarction. As distinction between intra-abdominal and intrathoracic disease can be important, more specific markers of liver cell damage have been sought. Because enzymes required for urea synthesis are rather specific to liver tissue (only kidney shares this activity) ornithine transcarbamylase (OTC) has been tested as a more specific marker for liver damage. However, what OTC gains in specificity, it loses in sensitivity, as well as being harder to measure. In consequence, AST is as useful as an index of hepatic parenchymal cell damage as any more specifically 'hepatic' enzyme and supplementary investigation such as the assay of circulating creatine phosphokinase (CPK) (which does not occur in liver) may permit cardiac or skeletal muscle damage to be differentiated.

Enzymes synthesized in response to disease
akaline phosphatase (AP) and 5' nucleotidase (5-NT)

Unlike the circulating enzymes from damaged cells referred to in the preceding paragraphs, alkaline phosphatase and 5' nucleotidase are not indicative of cell damage. However, similar problems with specificity can occur. Alkaline phosphatase may be derived from bone, placenta, kidney, gut or liver and the different electrophoretic mobility of these isozymes may be required to confirm one or other source.

Experimentally, if the bile duct of a rat is ligated, there will be an acute rise in alkaline phosphatase. This association between biliary obstruction and increased alkaline phosphatase has lead to the mistaken view that increased circulating enzyme is synonymous with obstruction. It is not; synthesis of the phosphatase by biliary duct tissue is stimulated by diverse pathologic processes involving the bile ducts and need not be accompanied by mechanical obstruction. In experimental animals this increase may be blocked by inhibiting new protein synthesis with cycloheximide.

Although an ubiquitous enzyme, 5′ nucleotidase activity is seldom increased in plasma in the absence of liver disease. However, if compared with that electrophoretic component of AP derived from the bile duct (and distinct from liver parenchymal cells) 5-NT is less helpful in distinguishing obstructive and non-obstructive disease of the liver.

Gamma glutamyl transpeptidase (γ-GT)

Although routinely measured in most clinical chemistry laboratories the precise significance of γ-GT is even less clear than AP. This enzyme is one of a sequence of reactions referred to collectively as the gamma glutamyl cycle which are located in the plasma membrane of liver cells and are concerned with transfer of amino acids from surrounding fluid into cells. Increased growth or protein synthesis may therefore, be associated with increased activity of γ-GT. There have been claims that the enzyme is induced by drugs that induce their own metabolism in the liver (such as the anticonvulsant drugs), or that it is induced by alcohol but neither claim is substantiated by published accounts of work in humans or animals. However there is a strong association between alcoholic liver *disease* and γ-GT so that the enzyme is empirically useful in following the course of liver disease in alcoholic subjects.

It has been shown when cancer of the liver is initiated by chemicals in experimental animals that foci of neoplastic cells may be identified using a histochemical stain for γ-GT activity and the enzyme may be associated with hepatic parenchymal cells having altered growth characteristics.

Other marker proteins, originating in liver and identifiable in blood

α-fetoprotein is a circulating glycoprotein produced by some liver tumors. Normal levels do not exclude a tumor, but increased levels provide a useful index to monitor therapeutic responses. Other possible markers either have been, or are being, tested.

Practical assessment

HEPATIC AND PORTAL CIRCULATION

Portal hypertension and collateral circulation

Clinical observation: gastro-oesophageal bleeding; splenic enlargement; abdominal wall collateral veins; venous hum; ascites.

Routine methods: study of oesophagus and gastric fundus by both barium (with air contrast) and endoscopy.

Special techniques: intrasplenic pressure; trans-splenic or transhepatic venography; hepatic vein catheterization (wedge pressure). To study flow patterns; cine-radiography coupled with splenic venography, arteriovenography, wedged hepatic venography, trans-umbilical portal venography.

Hepatic blood flow

Clinical observations and routine methods: a small liver suggests low total blood flow. No other observations or routine methods are available. Evidence of porto-systemic encephalopathy suggests diversion of blood from the hepatic circulation.

Special techniques: By Fick principle using plasma clearance of indocyanine green dye. Hepatic vein catheterization desirable but not essential; radioactive particle uptake by the liver and scanning; direct measurement at operation.

Portal-systemic shunting (not easily distinguished from impaired hepatocellular function).

Clinical observation: mental changes; flapping tremor; hepatic foetor. Portal hypertension and venous collateral circulation. Hepatocellular dysfunction.

Routine tests: assessment of portal venous collateral circulation. Assessment of hepatocellular function.

Special tests: EEG; detailed study of portal and collateral circulation (above); blood ammonia; glutamine.

HEPATOCELLULAR FUNCTION

Clinical observations: General nutrition; fluid retention, particularly ascites due to hypo-albuminaemia. Haemorrhage due to reduced synthesis of coagulation factors. Spider naevi, gynaecomastia, testicular atrophy, impotence and hyperaemic palms suggest altered sex hormone metabolism. Mental changes, flapping tremor and hepatic foetor. These suggest severe dysfunction. Ascites suggests portal hypertension coupled with hypo-albuminaemia. Jaundice is more likely to reflect hemolysis or obstruction to bile flow.

Routine methods. Plasma proteins; coagulation tests (PT and PTT) reflect hepatic synthesis of short-lived plasma proteins. Bilirubin: distinguish conjugated and unconjucated fractions. Serum transminases are elevated when hepatocellular damage is occurring. γ-glutamyl transaminase (GGT) may reflect regeneration. Blood urea may fall; blood sugar may be low in hepatic coma. Thrombin clotting time.

Special techniques. Blood ammonia and glutamine levels; urinary amino acids in acute hepatic failure. Serum B_{12} level. Serum iron binding capacity (TIBC) reflects transferrin, the iron transport protein synthesized in the liver. Measurement of albumin synthesis. Scanning of liver tissue following uptake of 99-Tc^m sulphur colloid, ^{125}i-Rose-bengal dye, ^{67}Ga or ^{125}I-methionine. Antipyrine clearance to assess drug metabolizing enzymes. Measurement of individual blood-coagulation factors.

BILE PIGMENT METABOLISM: JAUNDICE

Clinical observations. Colour of skin; pale lemon in prehepatic jaundice; orange-yellow in hepatocellular jaundice; green-yellow in obstructive jaundice.

Colour of urine: dark in all forms of jaundice, less dark if bilirubin unconjugated; yellow and frothy if bile salts present in urine.

Colour of stools: dark in prehaptic jaundice; pale in other forms. Pruritis and evidence of scratching in presence of cholestasis or biliary obstruction.

Routine methods: Plasma bilirubin, conjugated and total. Assessment of haemolysis. Plasma alkaline phosphatases, 5-nucleotidase in some centres and γ-glutamyl transpeptidase.

Special techniques. Bilirubin clearance using ^{14}C-bilirubin. Imaging of biliary tree with ^{99}Tc^m-HIDA. Percutaneous cholangiogram. Endoscopic retro-choledocho-pancreatogram (ERCP). Studies of bile salts by chromatography of duodenal aspirate or ^{14}C-glycocholate test.

References

NEWSHOLME, E.A. The role of the liver in the integration of fat and carbohydrate metabolism and clinical implications in patients with liver disease. In POPPER, H. & SCHAFFNER, F. (eds.) *Progress in Liver Disease*, Vol. 5. New York, Grune & Stratton.

PORTER, R. & WHELAN, J. (eds.) (1978) *Hepatrophic Factors*, Ciba Symposium No. 55. Amsterdam, Elsevier.

STANBURY, J.B., WYNGAARDEN, J.B. & FREDRICKSON, D.S. (eds.) (1978) *The Metabolic Basis of Inherited Disease*, 4th edn. New York, McGraw-Hill.

WEINBREN, K. (1978) The liver: In SYMMERS, W. ST. C. (ed.) *Systematic Pathology*, 2nd edn, pp. 1199–1301. Edinburgh, Churchill Livingstone.

WRIGHT, R., ALBERTI, K.G.M.M., KARRAN, S. & MILLWARD-SADLER, G.H. (eds.) (1979) *Liver and Biliory Disease: Pathophysiology, Diagnosis and Management.* Eastbourne, W.B. Saunders.

14

Energy Sources and Utilization

Abbreviations

NAD$^+$	nicotinamide adenine dinucleotide	DPN$^+$	NAD$^+$
		DPNH	NADH
NADP$^+$	nicotinamide adenine dinucleotide phosphate	FFA	free fatty acids
		p	phosphate
NADH	reduced nicotinamide adenine dinucleotide	K_m	Michaelis constant
		CoA	coenzyme A
NADPH	reduced nicotinamide adenine dinucleotide phosphate	FADH$_2$	reduced flavin adenine dinucleotide
		GTP	guanosinetriphosphate
ATP	adenosinetriphosphate	RQ	respiratory quotient
ADP	adenosinediphosphate	VLDL	very low density lipoprotein
AMP	adenosinemonophosphate	cyt b	cytochrome b
TCA	tricarboxylic acid	cyt b red	reduced cytochrome b

Normal function

The expenditure of energy in the body is required for maintenance of ionic gradients across membranes, locomotion, synthesis of new molecules, secretory processes, detoxification and heat generation. The central driving force which is the basis of all energy-requiring biological processes is high-energy phosphate in the form of ATP. The generation of sufficient ATP to meet the energy requirements of the body involves control of many different metabolic processes. ATP can be generated by the oxidation of carbohydrate, triglyceride and protein. These three major sources of energy can be oxidized at different rates and in different proportions to each other, depending on the body's demand for energy and on the availability of these energy fuels to the body.

The generation of ATP occurs almost exclusively in mitochondria where the energy released by the oxidation of reducing equivalents in the form of NADH (DPNH) by O$_2$ is coupled to the formation of ATP (oxidative phosphorylation). Though NADH is generated during partial oxidation of glucose and amino acids, most of the NADH generated by the body is in the TCA (tricarboxylic acid) cycle. ATP is also formed anaerobically during glycolysis, though the amount formed in this way is quantitatively insignificant for total body energy production.

NORMAL ENERGY GENERATION

Formation of TCA *intermediates from different food constituents*

The fuels for energy generation may be carbohydrate (glucose), triglyceride (FFA) or protein (amino acids). Though the initial metabolism of glucose, FFA and amino acids is different, the final oxidative pathway is common for all metabolic fuels once they enter the TCA cycle (Fig. 14.1).

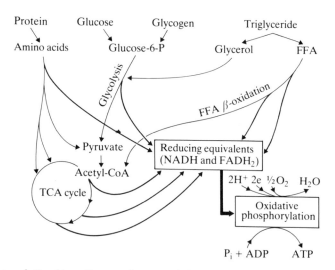

Figure 14.1. *Interrelationships of intermediary metabolism.*
    The thin arrows indicate the metabolic fate of the carbon moieties of the metabolites. For details on the metabolic fate of the carbon moieties of amino acid oxidation see Fig. 14.2. The wide arrows indicate the metabolic sources of the generation of reducing equivalents. For details on the quantitative source of reducing equivalents see Table 14.1.

Amino acid oxidation

Each of the twenty amino acids has a separate oxidative pathway. However, the common intermediates of the oxidation of these amino acids are oxaloacetate, pyruvate, α-ketogluturate, acetyl CoA and acetoacetate (Fig. 14.2). The amino acids that are metabolized to oxaloacetate (aspartate, threonine, valine, isoleucine, methionine), pyruvate (alanine, serine, cysteine, cystine, trytophan) and α-ketogluturate (glutamate, proline, arginine) are *glucogenic* because oxaloacetate, pyruvate and α-ketogluturate may be converted to glucose in liver (in kidney to a small degree) via the gluconeogenic pathway. Those amino acids which form acetoacetate or acetyl CoA are referred to as *ketogenic* (leucine, phenylalanine, tyrosine, lysine, isoleucine and tryptophan) because

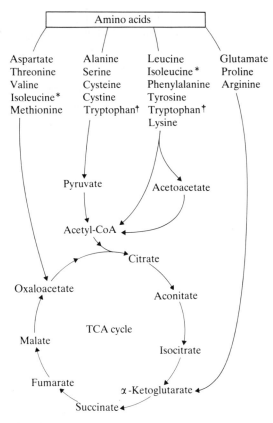

Figure 14.2. *Amino acid oxidation.*
After partial oxidation most of the carbon moieties end up as intermediates of the TCA cycle.

* isoleucine is partially oxidized to acetyl CoA and oxaloacetate.
† tryptophan is partially oxidized to pyruvate and acetyl CoA.

these intermediates cannot be converted to glucose (see p. 000). Most of the NADH from amino acid oxidation is generated by oxidation in the TCA cycle. Thus, the energy generation from amino acid oxidation is largely from the formation of reducing equivalents during oxidation in the TCA cycle.

Glycogenolysis

Liver. Glucose is stored as glycogen, a branched-chain polymer of glucose. In the unbranched part of the glycogen molecule the glucose moieties are linked by 1,4-glycosidic linkage and by 1,6-glycosidic linkages at the branch points. Glycogen degradation (Fig. 14.3) involves the hydrolysis of these 1,4- and 1,6-glycosidic linkages. Phosphorylase hydrolyzes the 1,4-glycosidic bonds to form glucose-1-P. Glucose-6-P is formed from glucose-1-P by P-glucomutase. In liver the major fate of glucose-6-P, derived from glycogenolysis, is the formation of blood glucose by hydrolysis of glucose-6-P. Though glucose-6-P derived from glycogen in liver may be metabolized via glycolysis or the pentose phosphate pathway, it is probably not an important quantitative fate during the post absorptive phase when hepatic glycogenolysis normally occurs. The branch points of the glycogen molecule, representing about 8 per cent of the glycogen, are hydrolyzed by the debrancher enzyme to form glucose directly.

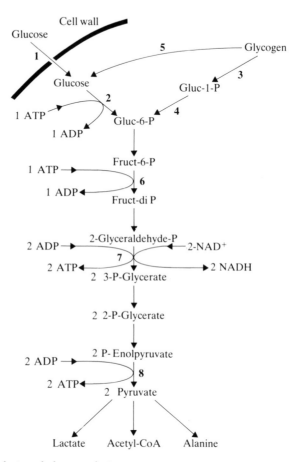

Figure 14.3. *Glycolysis and glycogenolysis.*
    The formation of pyruvate from glucose results in the net formation of 2 ATP per glucose moiety; 3 ATP are formed per glucose moiety from glycogen. Glycogen is degraded to glucose-6-P via glycogen phosphorylase (reaction **3**) and phosphoglucomutase (reaction **4**). A small proportion of free glucose is formed directly by the debrancher enzyme (reaction **5**). Free glucose cannot be formed from glycogen via glucose-6-P unless glucose-6-phosphatase is present (present only in liver, kidney and gut). NADH is generated by glyceraldehyde-P dehydrogenase (reaction **7**). Glycolysis from glucose or glycogen may be regulated at nonequilibrium reactions. This could be glycogen phosphorylase (reaction **3**), P-fructokinase (reaction **6**), hexokinase (reaction **2**) or pyruvate kinase (reaction **8**). Cell membrane translocation of glucose (reaction **1**) is important in muscle and adipose tissue.

Muscle. The fate of the glycogen degradation in muscle is different from liver. Because muscle does not have the enzyme, glucose-6-phosphatase, no free glucose is formed. (About 8 per cent of the glycogen is hydrolyzed by the debrancher enzyme and thus a small amount of free glucose may be formed. This probably explains why muscle glucose production is observed initially with acute exercise). The glucose-6-P is exclusively metabolized via glycolysis since muscle has virtually no pentose phosphate pathway activity. Therefore, muscle glycogen degradation does not contribute directly to an increase in the body glucose pool. However, it can, indirectly, particularly during exercise. The lactate and pyruvate (largely lactate) formed during glycolysis enter into the blood and are chiefly converted to glucose in liver by gluconeogenesis. The glucose may then be reutilized by muscle. This recycling of glucose is referred to as the Cori cycle.

Glycolysis

Glycolysis refers to the partial oxidation of glucose to pyruvate (Fig. 14.3). Its important features are that it does not require oxygen and that it generates H^+. The terminology of aerobic and anaerobic glycolysis is misleading since all glycolysis is actually anaerobic. In sense, the lactate and pyruvate formed in excess of what is oxidized by the TCA cycle is generally referred to as anaerobic glycolysis. This can be an important fate of glucose in exercising muscle and a number of organs if O_2 supply is inadequate.

Glycolysis results in ATP synthesis anaerobically. Two ATP molecules are formed per glucose molecule from glucose and 3 ATP/glucose from glycogen. Thus, some energy as ATP can be generated without oxygen consumption, but, as will be seen later, this is not too important quantitatively, except in mature erythrocytes and in muscle during short bursts of severe exercise.

The first step of glycolysis in muscle and adipose tissue is transfer of glucose across the cell membrane. Unless insulin is present glycolysis from glucose in these tissues is very slow. Thus, the rate-determining step for glucose utilization is probably the entry of glucose into the cell. Liver, kidney, brain and erythrocytes are relatively permeable to glucose and control of glycolysis in these organs does not reside at the cell membrane.

The initial step in the metabolism of glucose is the formation of glucose-6-P catalyzed by hexokinase. Liver has an additional 'hexokinase' referred to as glucokinase which has a high K_m for glucose. The activity of liver glucokinase essentially controls the rate of glucose utilization in liver. Starvation results in a rapid decrease of the glucokinase concentration, thus sharply decreasing the liver's capacity to utilize glucose. Increased glucokinase synthesis, mediated through increased insulin levels, occurs with a carbohydrate rich meal and is associated with increased hepatic glucose utilization.

The other glycolytic reaction important in controlling the rate of glycolysis is phosphofructokinase, which catalyzes the formation of fructose diphosphate from fructose-6-P. The rate of this reaction is controlled by several metabolites and its regulation is probably important in controlling glycolysis in most tissues.

FFA *oxidation*

FFA oxidation (Fig. 14.1) is initiated by the formation of the CoA esters of the fatty acids in the cytosol. The fatty acyl CoA must enter the mitochondria before oxidation can occur. The fatty acyl CoA is converted to fatty acyl carnitine in the cytosol to facilitate transport of the fatty acid moiety into the mitochondria, since the fatty acyl CoA itself cannot cross the mitochondrial membrane. In the mitochondria fatty acyl CoA is reformed. The long chain fatty acyl CoA is then sequentially oxidized. The quantitatively important fatty acid oxidized in humans is palmitic acid, containing 16 carbon atoms. By sequential oxidation and formation of acetyl CoA (2 carbon atoms), one mole of palmitic acid results in the formation of 8 moles of acetyl CoA. The formation of each mole of acetyl CoA is associated with the formation of one mole of FADH$_2$ and NADH. These reducing equivalents are subsequently oxidized in the mito-chondria in the presence of O$_2$, ADP and inorganic phosphate to form H$_2$O and ATP. Since this oxidation is linked to the formation of high energy phosphate (ATP) it is referred to as oxidative phosphorylation.

Oxidation in the TCA *cycle*

Oxidation in the TCA cycle (Fig. 14.4) essentially involves the cyclical oxidation of acetyl CoA. The initial reaction is the condensation of oxaloacetate with acetyl CoA to form citrate. Following a series of oxidations, oxaloacetate is reformed. The important feature of the TCA cycle is the formation of 3 moles of NADH, 1 mole of FADH and 2 moles of CO$_2$ for each mole of acetyl CoA oxidized. There is also formation of 1 mole of high energy phosphate (GTP). The important quantitative role of the TCA cycle is the formation of reducing equivalents (NADH, FADH$_2$) which are the substrates utilized by the electron transport chain to form ATP.

The important quantitative role of the TCA cycle is the generation of reducing equivalents and not the generation of high energy phosphate compounds. Further-more, the TCA cycle involves the cyclical oxidation of acetyl CoA. Thus, there is no net formation of any of the TCA intermediates, which explains why fatty acids cannot be converted to glucose.

Oxidative phosphorylation

Oxidative phosphorylation (Fig. 14.5) to the body is like the engine to a car. It takes specially prepared fuel (NADH and FADH$_2$), oxidizes it in the presence of O$_2$ and

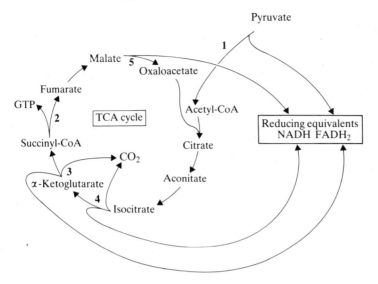

Figure 14.4. *The tricarboxylic acid cycle.*
The TCA cycle involves the cyclical oxidation of acetyl CoA. Two CO_2 (reactions **3** and **4**), one GTP, 2 NADH and 1 $FADH_2$ are formed per acetyl CoA molecule. The oxidation of pyruvate by pyruvate dehydrogenase complex (reaction **1**) to acetyl CoA also results in the formation of 1 NADH and 1 CO_2 per pyruvate molecule.
② succinyl CoA synthetase
③ α ketoglutarate dehydrogenase
④ isocitrate dehydrogenase
⑤ malate dehydrogenase

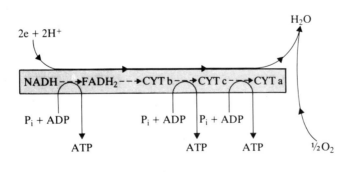

Net reaction (from NADH):
$$2\,e + 2H^+ + 3P_i + 3ADP + \tfrac{1}{2}O_2 \longrightarrow H_2O + 3\,ATP$$

Figure 14.5. *Oxidative phosphorylation.*
The energy released by the oxidation of NADH is coupled to the formation of ATP at 3 possible sites: between NADH and $FADH_2$, between cytochrome b and c, and between cytochrome c and a.

develops power (ATP). Although ATP is also generated at the substrate level in glycolysis and the TCA cycle, these amounts are insignificant compared to the amounts generated in the mitochondria during oxidative phosphorylation.

Oxidative phosphorylation involves the oxidation of reducing equivalents. These multiple steps are mediated via a series of electron carriers, each of which can be reduced by accepting electrons from the preceding carrier and can again become oxidized by transferring electrons to the next carrier. The multiple steps in the electron transport chain facilitate a maximum yield for ATP. Three molecules of ATP are formed from the oxidation of one molecule of NADH while only 2 are formed from FADH$_2$ (Fig. 14.5).

To summarize, the oxidation of fuels, be they glucose, amino acids, free fatty acids or ketone bodies, is linked to the formation of reducing equivalents. The subsequent oxidation of the reducing equivalents in the electron transport chain generates virtually all the energy required for life. Some ATP can be generated directly at the substrate levels, but the capacity is so small that it represents only a small fraction of the energy derived from oxidative phosphorylation. A summary of the quantitative energy relationships from different food sources is shown in Table 14.1.

Storage of metabolic fuels

Metabolic fuels are stored as triglyceride or glycogen. Amino acids in the form of protein can also be a fuel reserve, although proteins, unlike triglyceride or glycogen, always have other biological functions as well. Some energy is also stored as intracellular ATP or creatine-P in muscle, but the amount is so small that almost immediate regeneration is required once depleted. Similarly the body pool of free glucose, FFA and amino acids is very small. For example, in a 70 kg human, basal metabolism at 2000 kcal/day could be maintained by the free body glucose pool for about 1·2 hr, for about 3 min by the FFA pool and for about 1 hour by the free amino acid pool. The intracellular pools of ATP and creatine-P could sustain metabolism for even less time. The body supply of ATP would last about 2 minutes, if organ differences in metabolic rate are ignored.

Thus it becomes clear that the primary fuels for long term storage are triglyceride, glycogen and protein. The relative quantitative importance of these fuels can also be assessed by calculating the length of time basal metabolism could be maintained from the normal body supply for each of these fuels. Assuming 20 per cent body fat, basal metabolism from fat could be maintained for about 70 days. Glycogen in liver (75 g) and muscle (250 g) could sustain basal metabolism for about 15 hours. Assuming that a third of body protein could be lost without severe functional damage, the protein stores could support basal metabolism for about 4 days. Although these calculations are somewhat artificial, nevertheless they do outline the importance of triglyceride as an energy store and the limited energy stores in the form of glycogen and protein.

Table 14.1. Energy generation from major food constituents

	Generation of High-Energy Intermediates													
	Oxidation to TCA Cycle (Acetyl CoA)					Oxidation in TCA Cycle				Oxidation via Oxidative Phosphorylation			Total ATP* Generated
	ATP	NADH	FADH$_2$	CO$_2$	ATP†	NADH	FADH$_2$	CO$_2$	ATP†	NADH	FADH$_2$	O$_2$	
Carbohydrate (glucose)	+2*	+4	0	−2	+2‡	+6	+2	−4	+34	−10	−2	−6	38
Fat (palmitic acid)	0	+7	+7	0	+8	+24	+8	−16	+123	−31	−15	−23	131
Protein (alanine)	0	+1	0	−1	+1	+3	+1	−2	+14	−4	−1	2·5	15

	Energy Balance (kcal/mole)			
	Energy Yield by Bomb Calorimetry	Energy Yield as ATP§	Efficiency	Respiratory Quotient‖
Carbohydrate (glucose)	686	327	47%	1·0
Fat (palmitic acid)	2400	1126	47%	0·695
Protein (alanine)	386	129	33%	0·83

* moles/mole original substrate

† 3 ATP generated from NADH, 2 from FADH$_2$ (see Fig. 14.5)

‡ actually generated as GTP

§ assuming 8·6 kcal/mole ATP

‖ ratio of CO$_2$ produced : O$_2$ utilized

TRANSPORT OF METABOLIC FUELS

Triglycerides

Ingestion of food in excess of what is required for immediate energy metabolism, glycogen storage and protein synthesis is converted to triglyceride, largely in the liver. The newly formed triglyceride is then transported from liver, primarily to adipose tissue as a very low density lipoprotein (VLDL). At the adipose tissue the triglyceride moiety is hydrolyzed to FFA and glycerol by lipoprotein lipase. The FFA is taken up by adipose tissue cells, re-esterified and stored as triglyceride. Free glycerol is metabolized largely in liver. In fat cells from some animal species there is active FFA synthesis and this probably accounts for a substantial fraction of total body lipogenesis. While some FFA synthesis occurs in human fat cells, it remains to be established that this is quantitatively important.

When the adipose tissue triglyceride is required for energy metabolism, it is hydrolyzed to form FFA and glycerol by the hormone-sensitive lipase, an enzyme different from lipoprotein lipase. The FFA and glycerol are released into the blood stream available as fuel for other organs.

Thus, the transport form for stored lipid is triglyceride as very low density lipoproteins. However, when triglyceride is mobilized from adipose tissue it is transported as FFA and glycerol following hydrolysis of stored triglyceride.

Carbohydrates

The predominant transport form is glucose. It may also be transported as lactate or pyruvate, but only during special situations such as exercise where the Cori cycle (p. 000) is operative in shuttling three carbon moieties (predominantly lactate) from muscle to liver for reconversion to glucose.

Proteins

Protein, as a metabolic fuel is transported in plasma as amino acids. Most of the transport of amino acids from skeletal muscle is via glutamine and alanine, probably representing the partial muscle oxidation of those amino acids which can be converted to pyruvate and α-ketogluturate (Fig. 14.2). The plasma has a high protein concentration but it does not serve a significant function in the transport of protein from organ to organ.

Ketone bodies

Ketone bodies (β-hydroxybutyrate, acetoacetate, and acetone) are partial oxidation products of FFA. Acetone is produced by spontaneous decarboxylation of acetoacetate and may reach significant plasma concentrations when the acetoacetate levels are very high. The ketone bodies are not an important transport form of fat in the fed and post-absorptive states obtained in normal day to day situations. However, with starvation, diabetic ketoacidosis and certain metabolic disorders (glycogen storage diseases and others) the ketone body concentrations are high and then may be a major metabolic fuel for muscle, heart and brain. The metabolic advantage of ketone bodies over FFA is unclear, except perhaps that ketone bodies, but not FFA, can be utilized by brain for energy metabolism.

NORMAL ENERGY REQUIREMENTS

Energy is required to maintain basal body metabolism which consists of cardiac contraction, breathing, urine production, maintenance of cellular membrane gradients, heat generation, detoxifications and synthetic reactions. During basal metabolism O_2 consumption is a convenient measure of aerobic metabolic requirements. More than 90 per cent of total body O_2 consumption is accounted for by the major organs; brain consumes approximately 20 per cent, the abdominal organs about 25 per cent, skeletal muscle about 30 per cent and heart about 11 per cent. During exercise the energy requirements in skeletal muscle increase ten to fifteenfold, while the demand in heart increases only about fourfold. Since the metabolic requirements of other organs do not change significantly during exercise, the muscle energy requirements during exercise exceed all other energy requirements in the body by more than tenfold.

Similarly, other bodily functions may demand increased energy. For example, ingestion of a meal will increase O_2 consumption, largely in the abdominal organs. Shivering will increase O_2 consumption, largely in skeletal muscle. Other metabolic functions that can significantly increase energy requirements are heat production and synthesis of new tissue, particularly in the foetus and growing infant.

The total body O_2 consumption is a useful measure of total body energy requirements, although it is not particularly helpful in sorting out the specific fuels utilized in different organs. One method that has been used to assess the proportion of total energy derived from fat and carbohydrate oxidation is the respiratory quotient (RQ). It is the ratio of CO_2 production to O_2 consumption. Fat oxidation produces an RQ of 0·7 while carbohydrate oxidation gives a value of 1·0 (see Table 14.1). The rate of protein oxidation may be assessed by measuring the urinary nitrogen excretion. However, the RQ is valid only if it is assumed that degradative processes are much greater than synthetic processes, and that amino acid metabolism is low compared to carbohydrate and fat. Though these conditions are usually met under closely controlled conditions, the RQ only provides information on the different fuels oxidized in the whole body, not in individual organs. This can be accurately assessed only by determining the extraction of specific fuels across the organ's vascular bed or by determining the oxidation of labelled fuels, if it is known that the metabolic fuel in question is metabolized only in one organ.

REGULATION OF ENERGY METABOLISM

In general the regulation of body energy metabolism occurs at three levels.

Food intake is controlled by appetite and controls total body intake of energy.

Hormones. Generally, these control flux of metabolic fuels from organ to organ, resulting in a redistribution of energy sources.

Intracellular control. This involves control of pathways within an organ and is usually mediated by various metabolic intermediates of a particular metabolic pathway.

Food intake

Food intake is regulated by the appetite and hunger centres. From a physiological viewpoint it would be convenient if this was all that was involved. However, there are many psychological factors that result in excess or inadequate food intake which appears to be quite unrelated to control by the appetite and hunger centres.

Hormones

The complex interrelationships of hormonal effects on energy metabolism control the mobilization and utilization of metabolic fuels. The most important hormone is insulin.

Insulin is a relatively small protein molecule, consisting of 2 peptide chains of 21 (A-chain) and 30 (B-chain) amino acids joined by 2 disulphide bridges. It is synthesized in the islets of Langerhans as proinsulin, a single chain polypeptide with a 'connecting peptide' joining the A and B chains. The proinsulin is hydrolyzed in the β-cells to form insulin which is then stored in the β-granules. About 200 units of insulin are stored in the pancreas while the average daily requirements are 15–50 units. Insulin is released into the portal blood stream from the β-granules, though some insulin may be released directly after synthesis without initial storage in the β-granules.

Hormones, metabolic fuels and the autonomic nervous system have effects on insulin secretion. The insulin secretion is greater when glucose is administered orally than intravenously, suggesting that there is a gut factor stimulating insulin release. Several gut hormones, such as secretin, pancreozymin, gastrin, gut glucagon and gastric inhibitory polypeptide (GIP) have been proposed as the gut factor. GIP is considered the likely hormonal gut factor although definitive proof for this is lacking. Vagal stimulation also stimulates insulin release, though this autonomic effect is thought not to be of critical importance in regulating insulin secretion since denervation or transplantation of the pancreas has no significant effects on acute glucose homeostasis. Adrenaline inhibits insulin release, and is probably of physiological importance during acute stress and exercise. Perhaps the most important physiological controls of insulin secretion in humans are the stimulatory effects of amino acids, particularly arginine and leucine, and of glucose. In humans insulin secretion is not stimulated significantly by fatty acids or ketone bodies.

The cellular mechanisms mediating the stimulation of insulin secretion are unclear, although it appears to be mediated by effects of glucose and amino acid mediators on islet cyclic AMP concentration. The increased cyclic AMP concentration presumably, then, initiates a series of, as yet unknown, events leading to increased insulin release and/or synthesis.

Insulin secretion varies considerably from moment to moment in the same person and also varies widely between individuals. Wide and rapid changes in plasma insulin concentration are possible because it is rapidly degraded by the liver and kidney. The wide variation in insulin requirements between individuals is due to many factors. For example, obesity usually increases the insulin requirements substantially because of increased insulin resistance, thought to be due to decreased biological effectiveness of insulin at the insulin receptor. In lean aerobically trained subjects, insulin requirements and peak plasma insulin levels are low probably because insulin is more effective at the insulin receptor. Obesity is known to decrease the number of cell membrane insulin receptors while the receptor number is increased by weight loss and aerobic training.

METABOLIC EFFECTS OF INSULIN

Adipose tissue

Adipose tissue metabolism is largely concerned with the storage of metabolic fuels as fat during periods of abundant fuel supply and release of fuels during periods of fuel deprivation. The effects of insulin are associated with the process of increased storage of triglycerides. Insulin increases the glucose entry into the adipose tissue, increasing glycolysis, resulting in increased formation of glycerol-P required for triglyceride synthesis. The increased glucose entry results in increased oxidation of glucose through the hexose monophosphate pathway (HMP shunt) increasing NADPH formation required for FFA synthesis. The increased glucose oxidation may also generate acyl CoA equivalents, the final intermediates in FFA synthesis from glucose. Insulin also inhibits the rate of lipolysis by indirectly inhibiting the hormone-sensitive lipase activity, and it increases the activity of lipoprotein lipase, permitting increased FFA uptake from the very low density lipoprotein. Although all these effects of insulin apparently are present in human adipose tissue, it appears that the effects of insulin on stimulating FFA synthesis from glucose are probably of minor importance quantitatively in human adipose tissue. Most of it occurs in liver. Thus, the important effects of insulin probably are the inhibition of lipolysis, the increased generation of glycerol-P and the stimulation of lipoprotein lipase. These effects are modulated by insulin at the cell membrane. Some of these effects are produced by decreasing cell cyclic AMP levels but other mechanisms of insulin action may be of considerable importance. Control of cell glucose entry is not mediated by cyclic AMP.

Muscle

The net effect of insulin in muscle is to increase glucose utilization, glycogen synthesis and protein synthesis. Insulin stimulates glucose entry into muscle by increasing the rate of glucose translocation. It stimulates glycogen synthesis by a mechanism independent from its effect on cell membrane permeability of glucose. It also stimulates protein synthesis by increasing amino acid transport across the cellular membrane and by additional effects on protein synthesis at the level of the polysomes. It also inhibits proteolysis.

Liver

The overall effect of insulin is to increase disposal of blood glucose. This is achieved by insulin stimulating the synthesis of glucokinase, the rate-limiting enzyme for glycolysis in liver. It does not affect glucose permeability of liver cells. It also stimulates lipogenesis by a mechanism independent of its effects in glucose utilization. It stimulates hexose monophosphate shunt activity to generate the additional NADPH required for lipogenesis.

Insulin switches the liver from a glucose-producing to a glucose-utilizing system. This is achieved by a coordinated effect on glycolysis, glycogenolysis, glycogen synthesis and gluconeogenesis. Glycolysis is stimulated by the indirect effects of insulin on glucokinase synthesis. Gluconeogenesis is inhibited by the effects of insulin on the first few steps of the gluconeogenic pathway. The mechanism of this inhibition is not clear, although it appears to be mediated through the effect of insulin on decreasing the hepatic cyclic AMP concentration. Insulin stimulates glycogen synthesis and inhibits glycogenolysis which also appears to be mediated by changes in the cellular cyclic AMP concentration.

Glucagon

Glucagon has a variety of metabolic effects, particularly in liver. It stimulates hepatic lipolysis, glycogenolysis, gluconeogenesis and proteolysis. All these hepatic effects are suppressed by insulin. Glucagon also has some metabolic effects peripherally but the physiological significance of these effects is poorly understood. Indeed, the data to date on glucagon suggest that it probably has a very restricted role *in vivo*. It may be important in preventing hypoglycaemia during the ingestion of protein meals. Digestion of protein increases plasma amino acid levels, some of which stimulate insulin release. The increase in insulin levels would produce hypoglycaemia unless counteracted. This is achieved by the increased plasma levels of amino acids stimulating glucagon secretion. It may be that the glucagon then acts on the liver to stimulate glycogenolysis and possibly gluconeogenesis which then prevents the development of hypoglycaemia by insulin.

OTHER HORMONES

Growth hormone stimulates lipolysis, but its effects appear to be much slower than the catecholamines. Growth hormone secretion is stimulated by hypoglycaemia and suppressed by hyperglycaemia. Though these effects suggest that growth hormone may be important in the regulation of metabolism of energy fuels, there is very little data that suggests that growth hormone has a physiological role in the control of energy metabolism *in vivo*.

Catecholamines, particularly adrenaline, stimulate glycogenolysis in muscle and liver. Catecholamines (both adrenaline and noradrenaline) also stimulate adipose lipolysis. The net effect is an increase in blood glucose, largely from effects on liver glycogenolysis. The plasma FFA and glycerol levels are increased due to effects in lipolysis. The beneficial metabolic effects of catecholamines are the increase in the concentration of glucose and FFA, the major metabolic fuels. Catecholamines are probably important physiologically for regulation of energy metabolism only during acute stressful situations.

ACTH and β-lipotrophin at high concentration stimulate lipolysis *in vitro*, but *in vivo* the effects of ACTH are probably all mediated by stimulating cortisol secretion. Cortisol increases proteolysis, particularly in skeletal muscle. It also increases the hepatic capacity for gluconeogenesis by stimulating increased synthesis of the 4 key gluconeogenic enzymes. Increased cortisol levels thus increase gluconeogenesis by increasing the supply of gluconeogenic precursors by stimulating proteolysis and increasing the hepatic capacity for gluconeogenesis. The role of cortisol in regulation of energy metabolism physiologically is not clear, except that it appears to play a 'permissive' role in the maintenance of normal glucose homeostasis. Thus, the fasting hypoglycaemia observed in adrenal cortical insufficiency is probably related to the lack of normal physiological control of release of gluconeogenic amino acids from the periphery.

INTRACELLULAR REGULATION

The effects of the hormones in metabolism are eventually manifested by changes in intracellular metabolic events. However, the control by hormones is initiated usually by binding of hormone to cell membrane or intracellular receptors. Intracellular regulation, as it is used here, refers to regulation of energy metabolism which is initiated by intracellular events.

The intracellular regulation of energy metabolism is essentially the control of oxidative phosphorylation and those pathways which generate NADH, the main substrate for oxidative phosphorylation. Quantitatively, the most important pathways are glycolysis, FFA oxidation, TCA cycle oxidation and ketone body metabolism. Although amino acid oxidation generates some NADH, it is quantitatively not very significant.

The control of glycolysis limits the rate of glucose oxidation to pyruvate and also the supply of pyruvate for oxidation by the TCA cycle. The important control of glycolysis in *muscle* and *adipose tissue* is the transfer of glucose across the cellular membrane. This is largely controlled by insulin and represents control by extracellular events. In *erythrocytes, brain, kidney* and *liver* cell membrane translocation of glucose is insulin independent and the cell membrane probably is not rate limiting for glycolysis. In these tissues, the control of glycolysis probably is at phosphofructokinase, which converts fructose-6-P to fructose diphosphate. An increase in tissue ATP concentration, which would occur when sufficient ATP is available for metabolic reactions, inhibits phosphofructokinase, thus decreasing glycolytic flux and decreasing the generation of fuel (NADH) for oxidative phosphorylation. In contrast, when the ATP supply is limited, the ATP concentration decreases, removing the inhibition at phosphofructokinase. Glycolysis then increases and generates more NADH and pyruvate. The subsequent oxidation of pyruvate by the TCA cycle similarly generates additional NADH. The AMP concentration, which changes in the opposite direction of ATP, also has effects on phosphofructokinase opposite to ATP.

Some ATP is generated during glycolysis (Fig. 14.3) which may be of some significance in muscle energy generation during acute exercise. It is the exclusive source of ATP generation in mature erythrocytes.

The intracellular regulation of the TCA cycle is determined, to a large degree, by the concentration of NADH, one of its major products. As the NADH concentration (more correctly, the free NADH/NAD$^+$ ratio) increases, it indirectly decreases the oxaloacetate concentration (for discussion of mechanism see p. 518). Since the mitochondrial oxaloacetate concentration is probably rate limiting for citrate synthase and for the TCA cycle any changes in oxaloacetate levels will alter TCA cycle activity.

An additional control of the TCA cycle is the ATP/ADP ratio. At high ATP/ADP ratios (adequate energy generation by oxidative phosphorylation), isocitrate dehydrogenase, an enzyme of the TCA cycle, is inhibited. As the ratio decreases the inhibition is removed. The relative importance of NADH/NAD$^+$ and ATP/ADP ratios in controlling the TCA cycles has not yet been firmly established. However, it appears that the NADH/NAD$^+$ ratio is probably the important one. The subsequent oxidation of NADH by oxidative phosphorylation is similarly controlled. As sufficient ATP is generated, the ATP/ADP ratio increases, decreasing the ADP concentration. This decreases the supply of ADP for oxidative phosphorylation and decreases respiration.

Glycolysis, the TCA cycle and oxidative phosphorylation should be viewed as a functional unit controlling energy production. Any changes in the NADH/NAD$^+$ ratio controls the TCA cycle while changes in the ATP/ADP ratio controls glycolysis, the TCA cycle and oxidative phosphorylation.

Disordered function

DIABETES MELLITUS

Diabetes mellitus is a syndrome characterized by relative or absolute insulin deficiency. Diabetes mellitus is diagnosed by demonstrating glucose intolerance under carefully controlled metabolic conditions (p. 541). When the disease is clinically apparent glucose tolerance testing is unnecessary; a plasma glucose level well above normal is virtually diagnostic. Generally the syndrome may be subdivided into one of three groups.

Juvenile onset diabetes develops chiefly in children and young people. It is character-ized by a complete lack of endogenous insulin secretion. Ketoacidosis develops readily and control of blood glucose concentration may be difficult. Unless treated with insulin these patients rapidly develop ketoacidosis and die.

Maturity onset diabetes characteristically develops insidiously in later life. Patients are frequently obese and generally do not have a tendency to develop ketoacidosis. In the majority of these patients the disease can be managed without either insulin or oral hypoglycaemic agents, by reducing body weight to its ideal value by appropriate dietary management. Insulin secretion is present but has decreased biological effec-tiveness which is reversed by weight reduction and exercise.

Diabetes secondary to other diseases. The development of glucose intolerance is associ-ated with acromegaly (p. 555), phaeochromocytoma (p. 590), Cushing's syndrome (p. 577), pregnancy (p. 667), haemochromatosis (p. 261) and other conditions. There is evidence suggesting that diabetes developing secondary to these conditions is actually 'unmasking' the diabetic trait already present.

Causes of diabetes

The tendency to develop diabetes is increased with a strong family history. Maturity onset diabetes has a strong familial tendency while this is quite low with juvenile onset diabetes. Juvenile onset diabetes is strongly associated with the presence of several cell surface antigens. Plasma islet cell antibodies are present in the majority of patients with juvenile onset diabetes but not with maturity onset diabetes. Juvenile onset dia-betes develops at an increased frequency at the time of year when flu-like illnesses occur most frequently.

The specific cause of diabetes is unknown. In the juvenile-type diabetes it is clear that in the vast majority of cases there is an absolute failure of insulin secretion by the pancreatic β-cells. However, the reason for the absence of insulin secretion is unknown. In the maturity onset diabetes and in the diabetes developing secondary to other conditions, there is relative insulin deficiency. In many cases the plasma insulin concentration during fasting and with a glucose tolerance test may be normal. However, the plasma glucose: insulin ratio is abnormally high.

Although the development of diabetes mellitus has an important genetic component other factors affect its development. Obesity, dietary indiscretion and repeated pregnancy are important components in the development of adult onset diabetes.

A number of metabolic abnormalities are associated with diabetes. Plasma FFA levels generally are higher in diabetics but this probably reflects the relative insulin insufficiency rather than an etiological factor in the development of diabetes. Plasma growth hormone concentration is higher in diabetics but no conclusive evidence is available suggesting that this abnormality precedes the development of diabetes. Maturity onset diabetes is generally associated with insulin resistance and a decrease in the number of insulin receptors. This is reversed by weight reduction and by aerobic exercise training with a corresponding increase in the number of insulin receptors.

The development of diabetic ketoacidosis

Diabetic ketoacidosis represents the extreme end point of insulin lack. The insulin lack produces major aberrations in carbohydrate, protein and fat metabolism. The development of ketoacidosis should be viewed as a continuum from normal insulin secretion to overt diabetic ketoacidosis.

One of the first metabolic derangements produced by insulin lack is an inhibition of glucose utilization in muscle and adipose tissue. However, the other major body sources of glucose utilization, largely brain and erythrocytes, are not affected since they are not insulin dependent. In addition, insulin lack stimulates hepatic glycogenolysis and gluconeogenesis, producing additional body glucose. The first detectable blood abnormality is hyperglycaemia, particularly after carbohydrate meals. Later, fasting hyperglycaemia becomes apparent when the insulin lack is severe enough to stimulate hepatic glycogenolysis and gluconeogenesis. Initially the relatively mild insulin lack will produce glycosuria only after meals. With more severe insulin lack the glycosuria will become continuous and particularly aggravated by carbohydrate rich meals. Since glycosuria is always associated with appropriate water loss, excessive glycosuria is invariably associated with polyuria. This results in dehydration, increased plasma osmolality, increased thirst and polydipsia.

An integral and inseparable part of the derangement of glucose homeostasis is the effect of insulin lack on protein metabolism. Insulin lack inhibits protein synthesis and increases protein degradation. In view of the large body muscle mass these derangements are, to a large extent, produced by skeletal muscle. The increased protein degradation results in a large increase in the body's supply of amino acids. Most of these are metabolized by the liver. Since a third to one half of the increased amino acid load is converted to glucose via gluconeogenesis, this further contributes to the hyperglycaemia. The amino acid nitrogen is converted to urea and excreted in the urine. Negative nitrogen balance results which, at times, can be massive. If this state continues for a considerable period of time sufficient protein degradation occurs to produce muscular atrophy. Though this was frequently seen in the preinsulin era, it is relatively uncommon today.

The most sensitive biological effect of insulin is on fat metabolism. Therefore, gross aberrations in fat metabolism are usually not observed until late in the development of diabetic ketoacidosis. The most important effects of insulin lack on lipid metabolism are in adipose tissue and liver.

In adipose tissue insulin lack results in decreased triglyceride synthesis and increased triglyceride breakdown (lipolysis). Triglyceride synthesis is decreased because insulin dependent glucose entry into the cell is decreased. This results in decreased glycolysis and decreased generation of α-glycerol phosphate, which is probably rate-determining for triglyceride synthesis. In addition, insulin lack probably inhibits triglyceride synthesis by an effect independent of glucose entry into the cell. Lipolysis is probably increased due to the stimulatory effect of insulin lack on adipose tissue lipase activity. The net effect of insulin on adipose tissue is a massive increase in the release of FFA. This increases the plasma FFA levels resulting in other alterations in liver function.

The liver, faced with increased FFA levels increases FFA uptake almost as a linear function of its plasma concentration. It may be oxidized or reconverted to triglycerides and secreted into blood as very low density lipoprotein (VLDL). It cannot be converted to glucose and only a small fraction can be disposed of in other ways, such as cholesterol or phospholipid synthesis. The oxidation of FFA in liver is largely to the ketone bodies, with only a very small fraction proceeding to complete oxidation. Thus, normally the quantitative effect of insulin lack in liver is most likely related to its effects on lipogenesis and ketogenesis. Since lipogenesis is inhibited, the large FFA uptake is oxidized largely to ketone bodies. Although insulin lack does decrease the TCA cycle activity indirectly, the increased diversion of acetyl CoA moieties away from the TCA cycle to ketogenesis is quantitatively much less important than the inhibition of lipogenesis is in generating acetyl CoA for ketogenesis.

The increase in ketogenesis results in increased hepatic production of β-hydroxybutyric and acetoacetic acid. Their blood concentration increases, producing a metabolic acidosis. The blood pH decreases, respiration is stimulated and the P_{CO_2} and plasma HCO_3^- concentrations decrease. Because peripheral oxidation and urinary excretion of ketone bodies is limited, the continued production of ketone bodies results in a progressive worsening of the metabolic acidosis. Eventually the respiratory system can no longer compensate for the severe ketoacidosis and the picture is characterized by hyperventilation, low blood P_{CO_2}, low $[HCO_3^-]$, low pH and a high ketone body concentration (15–20 mM).

The elevation of blood β-hydroxybutyric and acetoacetic acid concentration produces additional problems for the kidney, already troubled by the osmotic diuresis produced by the hyperglycaemia. If the kidney had unlimited capacity to excrete acids the additional load of ketone bodies would merely aggravate the diuresis and rid the body of the excess acids. However, the capacity of the kidney to concentrate H^+ is limited. Thus, β-hydroxybutyric and acetoacetic acid are excreted, to a certain degree, as the free acids, but most H^+ has to be excreted with other cations, largely as the Na^+, K^+, and some as the NH_4^+ salts. Thus, the body gets rid of some of the ketone bodies but largely at the expense of body Na^+ and K^+ loss. This aggravates the dehydration and further decreases the blood volume. The blood pressure decreases and the glomerular filtration rate is diminished. Thus, the body's ability to maintain at least a marginal steady state by urinary loss of glucose and ketone bodies is severely compromised. The development of nausea, mental confusion and diabetic coma at some stage of the acidosis prevents the person from taking appropriate remedial action, such as intake of water and electrolytes. The blood levels of glucose and ketone bodies then again rise precipitously. This rapidly leads to severe acidosis, shock and death.

Hyperosmotic non-ketotic hyperglycaemia

Hyperosmotic non-ketotic hyperglycaemia generally develops in elderly people who may or may not have diabetes. Clinically, it is characterized by dehydration, which may be associated with a semi-stuporous state, coma, or convulsions. The blood glucose concentration is very high (frequently greater than 55 mmol/l or 1000 mg/ 100 ml), the plasma osmolality is raised but the blood ketone body concentration is not significantly increased. Metabolic acidosis is usually not present, though lactic acidosis can be present in severe cases.

Though the cause and pathogenesis of this syndrome is not totally clarified, it probably develops as follows. Hyperglycaemia develops due to relative insulin lack (p. 531). Polyuria follows. The frequently association nausea and vomiting prevents voluntary rehydration. Dehydration rapidly develops and the glomerular filtration rate decreases. The blood glucose concentration now rises rapidly and the hyperosmolar state, with its associated coma, develops. The lactic acidosis which occasionally develops is usually associated with an increased lactate : pyruvate ratio reflecting cellular hypoxia, probably secondary to inadequate oxygenation of tissues. The lack of ketoacidosis though not completely understood, is probably because enough insulin is present to prevent the development of the abnormalities of lipid metabolism (p. 531).

Prediabetes

If diabetes mellitus were simply a recessively inherited disease then all those homozygous for the trait would, by definition, be 'prediabetic'. However, since the inheritance of diabetes is considerably more complex the definition of prediabetes as a genetic entity is difficult. Generally speaking, there is very little disagreement among clinicians in classifying the offspring of two juvenile diabetic parents as 'prediabetic', even though the tests in these children may still be negative.

A variety of tests have been devised to detect 'prediabetes'. The one most frequently used is the oral glucose tolerance test after administration of glucocorticoids. Although this procedure detects abnormal glucose tolerance curves in a certain proportion of patients, its use as a definitive procedure in detecting the prediabetic state is controversial.

Complications of diabetes mellitus

The most serious long-term complications of diabetes are nephropathy, retinopathy, neuropathy and angiopathy. The rate at which these complications develop is unpredictable in individual patients. Indeed, some of these complications may actually precede the development of clinical diabetes.

The mechanism for the development of these complications is poorly understood. There is some suggestive evidence that it may be related to the effects of insulin lack in altering the glycoprotein metabolism in the different target organs.

HYPOGLYCAEMIA

Normal blood glucose concentrations are maintained when the rate of glucose removal from the blood equals the rate at which glucose is added to the circulation. Thus, hypoglycaemia may be caused by excessive removal, by inadequate production of glucose, or by a combination of both mechanisms. The most common causes of hypoglycaemia are insulin overdosage, certain other drugs, neonatal hypoglycaemia and 'functional hypoglycaemia'. Less common causes are deficiencies of pituitary hormones (particularly ACTH and GH), insulinomas, malignancies and genetic defects involving hepatic glycogenolysis and gluconeogenesis.

Excessive insulin secretion

Excessive insulin secretion results in hypoglycaemia because it increases peripheral metabolism of glucose and inhibits the formation of glucose in liver by inhibiting hepatic glycogenolysis and gluconeogenesis. In addition, it decreases the production of glucogenic amino acids, largely by skeletal muscle, thereby depriving the liver of its most important quantitative supply of gluconeogenic precursors. A relative excess of insulin resulting in hypoglycaemia occurs most frequently with excessive insulin therapy. It also occurs commonly by overzealous treatment of maturity onset diabetes with sulphonylureas, particularly chlorpropamide. Because of chlorpropamide's long half life, it continues to stimulate insulin secretion for days. If food intake is stopped, for whatever reason, the chlorpropamide induced insulin secretion results in hypoglycaemia.

The physiologically important stimuli for insulin secretion are glucose, and amino acids, particularly arginine and leucine. A small number of children have an excessive sensitivity to the stimulatory effects of leucine. This results in excessive insulin secretion with normal protein ingestion and development of hypoglycaemia.

An uncommon cause of hypoglycaemia is the excessive secretion of insulin by an insulinoma (β-cell islet tumour). Because the tumour secretes insulin intermittently, the diagnosis may be difficult. Insulinomas frequently are associated with extremely high plasma insulin levels even when overt hypoglycaemia is not present. This occurs because insulinomas frequently secrete the majority of the 'insulin' as proinsulin, the β-cell peptide precursor of insulin, which is 10 per cent as potent as insulin as a hypoglycaemic agent but binds as avidly as insulin with the insulin antibodies used in many insulin assays (p. 542).

Functional hypoglycaemia

Functional hypoglycaemia is frequently used to describe those cases of hypoglycaemia where the causes have (can) not yet be determined. In a significant proportion of cases the hypoglycaemia develops about 3–5 hours after meals and probably is due to a delayed and 'overshoot' response of insulin secretion to glucose. Thus, when the blood glucose is decreasing after a meal the insulin secretion remains relatively high, resulting in hypoglycaemia. This may be an early manifestation of maturity onset diabetes, but generally is unmasked by a diet high in refined carbohydrates.

Defective hepatic glucose production

Defective hepatic glucose production may be secondary to defective glycogenolysis and/or gluconeogenesis. In *glucose-6-phosphatase deficiency* (von Gierke's disease) the enzyme forming glucose-6-phosphate is missing. Thus, the glucose-6-phosphate formed from glycogen breakdown or gluconeogenesis cannot be converted to blood glucose. Since glycogenolysis and gluconeogenesis are important in maintaining a normal blood glucose concentration in the post-absorptive state, hypoglycaemia invariably develops during fasting. In hepatic *phosphorylase deficiency* and hepatic *debrancher enzyme deficiency*, glycogenolysis is impaired resulting in fasting hypoglycaemia. In hepatic *fructose diphosphatase deficiency* (a critical enzyme in the gluconeogenic pathway) gluconeogenesis is impaired, which also results in fasting hypoglycaemia. Deficiencies of all the key gluconeogenic enzymes have been described.

Fasting hypoglycaemia can also occur in hepatitis and in chronic liver disease. Although the mechanisms have not yet been clearly elucidated, it seems probable that the defect will be with decreased hepatic glycogenolysis and/or gluconeogenesis due to effects of liver disease on some of the key enzymes of these pathways.

Hypopituitarism and adrenal cortical insufficiency

In hypopituitarism and in adrenal cortical insufficiency the fasting hypoglycaemia is probably secondary to insufficient rates of hepatic gluconeogenesis due to a deficiency of glucocorticoid secretion. Two contributing factors are probably responsible, although others may also play some role. Firstly, in the absence of glucocorticoids the liver responds poorly, or not at all, to normal gluconeogenic and glycogenolytic stimuli such as an increase in the glucagon : insulin ratio or catecholamines. Secondly, glucocorticoid lack results in decreased rate of amino-acid release from the periphery. Thus, the main substrates for gluconeogenesis are supplied in inadequate amounts. The net result of both effects is decreased hepatic gluconeogenesis and subsequent fasting hypoglycaemia.

Clinical effects

The confusion, abnormal behaviour, coma and convulsions which may all occur with hypoglycaemia are due to decreased blood glucose concentration. Because glucose metabolism normally is the major energy supply to brain and since the rate of glucose utilization by brain is closely related to its blood concentration, any rapid decrease in blood glucose concentration produces a 'metabolic hypoxia'. Initially this results in increased catecholamine secretion by the autonomic nervous system and the adrenal medulla which produces apprehension, tachycardia and sweating. The beneficial metabolic effect of increased catecholamine secretion is the stimulation of hepatic glycogenolysis. This raises the blood glucose level and counteracts the cerebral effects of hypoglycaemia. Although the increased catecholamine levels also stimulate lipolysis and increase the plasma FFA levels this has no direct beneficial effect for brain since it cannot use FFA as a significant source of energy. The increased FFA levels could have an indirect beneficial effect on brain by raising the blood glucose level due to the postulated effects of FFA oxidation on stimulating hepatic gluconeogenesis and inhibiting peripheral glycolysis. However, both of these effects have not been demonstrable in humans, despite the presence of these effects in isolated liver and muscle preparations.

   In certain abnormal states hypoglycaemia may not result in symptoms and signs of cerebral 'metabolic hypoxia'. This is particularly true in the glycogen storage diseases and in starvation. In both cases the blood levels of the ketone bodies are elevated. These can be metabolized by brain as a source of energy. If these concentrations are sufficiently high marked hypoglycaemia may occur without hypoglycaemic symptoms.

OBESITY

Normal weight is maintained when the energy content of ingested food equals the total energy dissipation of the body. Obesity develops when the energy intake exceeds energy dissipation. This can occur with excessive food intake, decreased body energy requirements, or with a combination of both factors.

   The simplistic view of the control of weight is that the appropriate appetite and hunger centres control food ingestion. Any weight gain would then be viewed as reflecting excessive food intake. Since the weight of normal lean subjects generally varies by less than 5 lb (2·3 kg) over a 10 to 20 year period, this would mean exceedingly accurate determination of the energy needs of the body by the appetite and hunger centres. Indeed, a weight gain of 5 lb over a 10 year period would reflect only an average dietary increase of 0·2 per cent above a normal isocaloric diet. While normal weight maintenance by controlling food intake is attractive conceptually, it is difficult to see how a control system mediating the appetite and hunger centres could be capable of responding to such small changes in food intake. In fact, the weight of most people is probably more rigorously controlled than any other biological variable, including the plasma Na^+ concentration. Although these considerations do not rule out regulation of food intake as the primary determinant in controlling body weight, they, nevertheless, suggest that other factors are probably also important. The energy and metabolic intermediates derived from the metabolism of food can be dissipated only in a limited number of ways. They can be used for synthesis of new protein, glycogen, triglycerides, nucleic acids, sterols and some compounds of minimal quantitative significance. The concentration of all these products is fairly constant in the body except for triglycerides which can accumulate to massive amounts in adipose tissue. Thus, the excessive storage of metabolic fuels is essentially limited to triglycerides. Unless the body is capable of altering the efficiency of triglyceride and other end product synthesis, this could not be of any use in regulating body weight.

The energy generated by the metabolism of food is used largely for locomotion, maintenance of ionic gradients across membranes, synthesis of new molecules and thermogenesis. Of these energy requiring processes, thermogenesis and locomotion are quantitatively the most important. When physical activity is increased more energy is dissipated. Thus a significant increase or decrease in physical activity will require appropriate adjustments in food intake to prevent changes in body weight. Thermogenesis varies considerably between individuals with 'normal' thyroid function. Theoretically, it is possible to dissipate the energy generated from the metabolism of surplus food intake by oxidation and dissipation as heat. Such changes in the rate of thermogenesis as a regulatory mechanism in weight control have been suggested and considerable evidence for such a mechanism is now emerging.

A variety of metabolic alterations in carbohydrate and fat metabolism have been demonstrated in obesity. Generally, these abnormalities have been demonstrated in people who are already obese. Therefore, it is not clear whether these alterations are the result of or the contributing factor towards the development of obesity. In any event abnormalities demonstrated in obesity would only be germane to the development of obesity if such abnormalities explained alterations in energy metabolism in a way that would result in a more efficient metabolic system, i.e. more fat for less energy.

For the moment at least, one aspect in the development of obesity is clear. The energy derived from metabolism of dietary intake exceeds the energy needs of the body. Thus, in the treatment of obesity any alteration in this relationship which increases body energy dissipation or decreases intake of food will decrease the body weight. However, these observations, unfortunately, do not reveal very much about other possible control mechanisms involved in weight control.

Obesity for long periods of time is associated with increased frequency of diabetes, hypertension and arteriosclerosis.

STARVATION

The cessation of food intake causes alterations in the body's metabolism which are designed to provide metabolic fuels for energy generation from the energy stores of the body. For practical purposes, energy is stored as glycogen, triglycerides and protein. The hepatic glycogen stores are depleted within 1 or 2 days of starvation (Fig. 14.6). Body lipid provides most of the fuel for energy metabolism for at least a few weeks. Protein degradation provides a relatively small proportion of the body's energy needs as long as adequate lipid stores are present. Terminally, when the body's fat stores are depleted protein degradation probably provides an increasing proportion of the body's energy needs.

The transition from the fed to the fasted state seems to be mediated largely by changes in plasma insulin levels. In the fed state insulin levels are high due to the stimulation of insulin secretion by glucose and amino acids. When no food is ingested the plasma concentration of glucose and appropriate amino acids decreases. Insulin

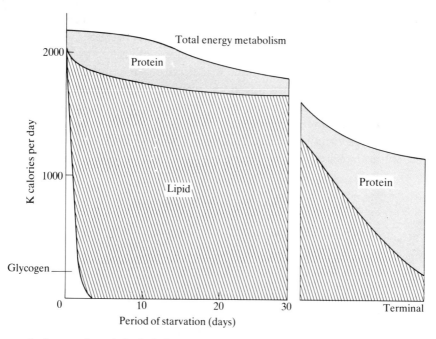

Figure 14.6. *Source of metabolic fuels during starvation.*
The 'protein sparing' observed in human starvation is apparently not present in some smaller animals. The increase rise of protein stores as an energy fuel terminally is observed in animals but the extrapolation indicated here is somewhat speculative in humans.

levels drop producing appropriate changes in different tissues not unlike the changes that occur in the development of diabetic ketoacidosis. Muscle and adipose tissue glucose utilization decreases. Lipolysis, largely in adipose tissue, increases, raising the plasma FFA levels. This results in increased oxidation of FFA in muscle compensating for its decreased rate of glucose utilization. Despite some decrease in the peripheral glucose utilization, considerable glucose is still required to maintain normal oxidative metabolism in brain. To supply this glucose, endogenous glucose production must increase. This is achieved in two ways. Hepatic glycogenolysis is increased, probably mediated by the decreased insulin levels. However, since the hepatic supply of glycogen is limited (muscle glycogen is not mobilized) an additional source of glucose is required. This is achieved by the effects of decreased insulin concentrations in stimulating hepatic gluconeogenesis and muscle proteolysis. The net effect is an increased supply of gluconeogenic precursors to liver and an increased hepatic capacity for converting these amino acids to glucose.

If starvation continues for more than a few days the effect of decreased insulin levels produces additional effects on lipid metabolism. Hepatic uptake of FFA is considerably increased largely due to the increased FFA levels. Because the decreased insulin has inhibited hepatic lipogenesis the increased hepatic FFA uptake is largely diverted to ketone body production, increasing the plasma ketone body concentration.

Brain, then, increases its rate of ketone body oxidation, probably largely because of the increased plasma levels of ketone bodies. It now requires less glucose for oxidative metabolism and decreases its rate of glucose oxidation. The decreased cerebral demand for glucose, coupled with the decreased rates of glucose utilization by muscle decreases the body's demand for glucose. Since the hepatic glycogen has already been depleted this decreased demand is reflected by a decreased requirement for gluconeogenesis. By mechanisms, as yet poorly understood, muscle proteolysis, hepatic gluconeogenesis and urinary nitrogen excretion decrease. This presumably is all related to the body's demand for glucose and is probably the basis for the 'nitrogen sparing' that is observed, at least in obese subjects, undergoing prolonged starvation.

Principles of tests and measurements

MEASUREMENT OF ENERGY EXPENDITURE

Direct calorimetry

Direct calorimetry involves the measurement of total heat loss by the body. This is achieved by placing the person either in an insulated chamber or an insulated suit where heat loss can be accurately measured. Since about one third of the total body heat production is respiratory evaporative heat loss, it is important that the respiratory water loss is known. Direct calorimetry was of importance historically in establishing the biological energy content of food. On the basis of these studies the caloric content of food was found to be 4 kcalories*/g carbohydrate, 9 kcalories/g fat and 4 kcalories/g protein. The values obtained for fat and carbohydrate are very close to the heat content of food determined by direct combustion. However, for proteins the biological caloric content is considerably less than the value obtained by direct combustion. This difference is referred to as the *specific dynamic action* (frequently abbreviated as SDA). This difference, particularly for proteins, is probably due to the additional energy required for the synthesis of urea from the nitrogen moiety of amino acids. The total body heat loss will equal the energy content of the ingested food in persons whose energy metabolism is in a steady state (i.e. there is no change in body weight and alteration in the nature of body fuels stored).

Indirect calorimetry

When the total body heat loss was compared to the oxygen consumption it became apparent that a very close relationship existed between the two. Oxygen consumption, therefore, became an easier and more convenient method of indirectly measuring the total body heat loss. If the inherent assumptions of indirect calorimetry are kept in mind (p. 000), the determination of O_2 consumption, CO_2 production and urinary nitrogen excretion may be used to determine the approximate percentage of total body energy metabolism that is derived from fat, carbohydrate and protein. By assuming that 1·0 g urinary nitrogen represents the oxidation of 6·25 g of protein (about 15 per cent of protein is nitrogen), and that the RQ of protein oxidation equals 0·89, the percentage of non-nitrogenous O_2 consumption and CO_2 production may be calculated. Since the RQ is 0·7 for fat and 1·0 for carbohydrate oxidation (Table 14·1, p. 521), the per cent energy metabolism from fat and carbohydrate can be calculated. In similar fashion, if it is assumed that 1 mole of O_2 consumption generates 108 kcalories of energy (1 litre O_2 = 4·83 kcalories), the actual amounts of carbohydrate and fat oxidized can be estimated.

Basal metabolic rate

The basal metabolic rate (BMR) is the oxygen consumption after an overnight fast in a recumbent subject who is mentally and physically relaxed. The results are usually expressed as per cent of normal values obtained for subjects the same age, sex and body surface area. A major determinant of the BMR is the state of thyroid function. In hyperthyroidism the BMR is increased and in hypothyroidism it is decreased. However, many factors other than thyroid function, including anxiety, catecholamine secretion, anaemia, fever and others affect the BMR. Therefore, its usefulness in detecting abnor-

---

* One calorie is the amount of heat required to raise the temperature of 1 g of water 1° C at 15° C. A kcalorie (kilocalorie) is 1000 calories. However, by common usage, 1 Calorie (capitalized) is frequently used as the equivalent of 1 kcalorie.

malities of thyroid hormone secretion is limited. Many more specific tests of thyroid function are currently available (p. 604).

BLOOD LACTIC AND PYRUVIC ACID

The concentration of these acids in blood is determined by the rate of their addition and removal from the bloodstream. Lactate is produced largely by skeletal muscle and the erythrocytes, although any organ is capable of producing large quantities, particularly during hypoxia. Normally liver and muscle are important in the removal of lactate from blood. In humans, it is not clear which organs are most important in determining the blood pyruvate levels, though muscle and liver are probably the major organs involved.

Lactate and pyruvate are interrelated by the lactate dehydrogenase reaction:

$$\text{Lactate} + \text{NAD}^+ \underset{\substack{\text{lactate} \\ \text{dehydrogenase}}}{\rightleftharpoons} \text{Pyruvate} + \text{NADH} + \text{H}^+$$

This reaction, for practical purposes, is always near equilibrium. Therefore, the ratio of lactate to pyruvate reflects the tissue free NADH/NAD^+ ratio. Since the enzyme's catalytic activity is present only in the cytoplasm of the cell, the lactate : pyruvate ratio reflects the NADH/NAD^+ ratio in this cell compartment (cytosol redox state). Any alterations of the blood lactate or pyruvate concentrations must take account of these interrelationships.

An increase in blood lactate (normal 1–1·5 mM) concentration could represent increased production or decreased removal from blood. However, if the blood pyruvate concentration (normal 0·08 mM) is also determined additional information is obtained about the cytosol redox state of the body. For example, a tripling of blood lactate concentration with a normal lactate : pyruvate ratio (normal 5–8) would indicate either increased addition or decreased removal of lactate and pyruvate from blood. However, it would not suggest an abnormal cytosol redox state. If the blood lactate concentration is increased but the pyruvate concentration is unchanged or decreased it would suggest an increase in the cytosol $\text{NADH} : \text{NAD}^+$ ratio. Under most circumstances this indicates cellular hypoxia. However, an increased L/P ratio may also represent generation of cytosol reducing equivalents that exceed the oxidative capacity of the cell (e.g. excess alcohol intake). In most situations where the blood lactic acid level exceeds 5·0 mM the L/P ratio is increased and cellular hypoxia may be present. Unfortunately the organ affected by hypoxia cannot be determined from the blood L/P ratio.

CARBOHYDRATE METABOLISM

Glucose tolerance test (GTT)

The test involves the administration of a standard oral glucose dose (usually 100 g glucose or 1 g glucose/kg body wt) after an overnight fast. Venous or capillary blood is collected at frequent intervals (usually 0, $\frac{1}{2}$, 1 and 2 h; occasionally also 3, 4 and 5 h) and analysed for blood glucose concentration.

The rise and fall of blood glucose during the GTT is determined by many factors. Among the most important ones are the rate of glucose absorption from the gut, the rate of its utilization by the major organs such as muscle, liver and adipose tissue and the effect of the glucose load on the continued glucose production by glycogenolysis and gluconeogenesis. Probably the most important parameter controlling these processes is the effect of glucose on insulin secretion. Normally the increase in blood glucose results in a rapid stimulation of insulin secretion by the β-cells, usually within 10 minutes of the glucose administration. The increased insulin levels stimulate peripheral utilization of glucose and decrease hepatic glycogenolysis and gluconeogenesis. This contributes to the decrease in blood glucose concentration 30–60 min after administration of glucose. As the blood glucose levels decrease, the stimulus for insulin secretion decreases and the blood glucose levels stabilize again.

Because abnormal glucose homeostasis is one of the hallmarks of diabetes, the GTT is widely used as a tool for the diagnosis of diabetes. The normal fasting blood glucose concentration is less than 95 mg/100 ml (5·2 mmol/l). The peak value is usually less than 160 mg/100 ml (8·8 mmol/l) and occurs between 15 to 30 min after glucose administration. The 2 h value is usually less than 120 mg/100 ml (6·6 mmol/l). However, many factors can affect both the response of insulin secretion to glucose and the response of different tissues to insulin itself. The GTT therefore becomes a very unreliable tool in detecting diabetes unless the factors altering the GTT are taken into account. A low carbohydrate diet or starvation may produce a diabetic-type GTT. Similarly, many serious illnesses (e.g. myocardial infarction), treatment with steroids, thiazides, antiovulatory agents and other drugs may produce a diabetic type GTT. An abnormal GTT in these situations may represent 'prediabetes' since the underlying disease or treatment may have merely unmasked the diabetic trait already present. However, such a conclusion is usually premature.

The most useful test in the diagnosis of diabetes mellitus is the fasting blood glucose concentration which is almost always elevated in diabetes, and the 2 h blood glucose level after a standard carbohydrate load. Particular attention must be attached to 'normal' values for blood glucose since the values vary somewhat from hospital to hospital. While the GTT is widely used for diabetes testing it must be recognized that it rarely adds any information not already provided by the fasting and 2 h post meal blood glucose.

Steroid-augmented GTT

The steroid-augmented GTT is identical to the GTT except that corticosteroids are administered for 1 day prior to the test. It is designed to produce a metabolic 'stress' for detection of the 'prediabetic' state. However, an abnormal steroid-augmented GTT does not necessarily indicate prediabetes and its use is very limited.

Intravenous GTT

The intravenous GTT is similar to the oral GTT, except that it eliminates intestinal absorption of glucose as a variable. Normally 0·3 g glucose/kg body wt is given intravenously over a 3 min period. The exponential decrease in blood glucose levels is plotted semilogarithmically to produce a straight line. The slope of this line is an estimate of the rate of body glucose disposal under standard conditions. An increased slope indicates increased body glucose metabolism which may occur in hyperthyroidism and other hypermetabolic states. Decreased capacity to metabolize glucose is reflected by a decreased slope and is usually obtained in diabetes mellitus.

FASTING

Fasting is a useful physiological perturbation to determine the body's capability of maintaining normal glucose homeostasis. During the first 24 h hepatic glycogenolysis and gluconeogenesis are critical in generating body glucose. After 24 h hepatic gluconeogenesis is the most important. In disorders of hepatic glucose production (p. 534) hypoglycaemia usually develops within 18–24 h. In patients with insulinomas hypoglycaemia invariably develops within 72 h. The hypoglycaemia is due to the excess insulin secretion increasing peripheral glucose utilization and probably decreasing hepatic gluconeogenesis and peripheral release of gluconeogenic amino acids.

GLUCAGON TEST

The glucagon test involves the rapid intravenous administration of glucagon followed by frequent measurements of blood glucose. In normal subjects it produces a maximal rise of blood glucose at 30–60 min, largely due to stimulation of hepatic glycogenolysis. Glucagon also stimulates hepatic gluconeogenesis. However, in humans, it is not clear how much of the rise in blood glucose is due to increased gluconeogenesis. The glucagon test produces no rise in blood glucose in some of the glycogen storage diseases. However, frequently it produces such equivocal results that its usefulness is limited. Since the actual measurements are blood glucose concentration and not rates of hepatic glycogenolysis or gluconeogenesis, it is not surprising that equivocal results are frequently attained. The increased blood glucose levels stimulate insulin secretion and tend to decrease the blood glucose levels. Thus, very transient rises in blood glucose may be observed in a significant proportion of normal subjects.

PLASMA INSULIN CONCENTRATION

Since plasma insulin is rapidly degraded in liver and kidney, changes in the plasma insulin levels largely reflect changes in the rate of pancreatic secretion. Changes in blood glucose concentration change the insulin secretion rate. In normal subjects, despite major changes in glucose loads, there is a fairly constant glucose : insulin ratio in plasma. When insulin is assayed with any of the glucose tolerance tests, a delayed insulin response and an increased glucose : insulin ratio is observed in adult-type diabetes. In juvenile-type diabetes a marked decrease or a total absence of plasma insulin is observed with the GTT. If the GTT is abnormal but the glucose : insulin ratio is low it would suggest peripheral insulin resistance. It is important to remember that the radioimmunoassay for insulin may occasionally not reflect the plasma concentration of the biological activity of insulin. In certain situations, especially insulinomas, this can lead to significant errors in determining the plasma insulin level. The availability of antibodies specific for proinsulin now permits the measurement of proinsulin. This is particularly useful in the diagnosis of insulinomas which frequently secrete abnormally large amounts of proinsulin (p. 534).

FAT METABOLISM

Plasma triglyceride

The plasma triglyceride concentration is a dynamic balance between the rate of very low density lipoprotein production by liver and the rate of its uptake by adipose tissue. Its concentration is not directly related to the rate of energy metabolism of the body.

Plasma FFA

Because the rate of FFA utilization by different organs is largely a function of its plasma level, the plasma FFA concentration can be taken to reflect approximately the rate of adipose tissue lipolysis. It is elevated by catecholamines, starvation, uncontrolled diabetes and frequently in obesity. In all these situations, except possibly excess catecholamine secretion, the elevated FFA levels are probably related to increased lipolysis due to decreased insulin levels.

Plasma ketone bodies

Elevated concentrations of plasma ketone bodies usually reflect excess production by the liver. Altered rates of utilization may also contribute to increased plasma levels in certain situations. Excess production is usually due to a combination of increased plasma FFA levels and decreased rates of hepatic triglyceride synthesis. In diabetic ketoacidosis and starvation a relative lack of insulin is probably responsible.

Since β-hydroxybutyrate and acetoacetate are reactants of a mitochondrial enzyme, β-hydroxybutyrate dehydrogenase, the ratio of β-hydroxybutyrate : acetoacetate can be viewed as a reflection of the mitochondrial redox state in a manner similar to the lactate : pyruvate ratio (p. 540).

PROTEIN METABOLISM

Plasma amino acids

The concentration of each of 20 or so amino acids in plasma is determined by the balance of their addition and removal from plasma. The concentrations of specific amino acids are markedly elevated in plasma in genetic disorders where specific enzymes involving amino acid metabolism are lacking.

Practical assessment

DIABETES MELLITUS

Clinical observations

Symptoms of hyperglycaemia: polyuria, polydipsia and temporary visual disturbance.

Symptoms of water and electrolyte loss: thirst and eventually dehydration with low plasma volume.

Recognition of known complications: retinopathy, nephropathy and neuropathy, and a high incidence of mature cataracts and peripheral vascular disease.

Evidence of causative disease: acromegaly, Cushing's syndrome, phaeochromocytoma, chronic pancreatitis, haemochromatosis.

ROUTINE URINE TESTS

Glucose oxidase tests (Clinistix and Testape), for detection of glycosuria and copper reduction tests (Clinitest and Benedict's test) for measurement of glycosuria.

For ketonuria, see below.

ROUTINE LABORATORY METHODS

Blood samples: venous whole blood.

Fasting blood glucose: Normal less than 100 mg/100 ml (5·5 mmol/l). Diabetic greater than 120 mg/100ml (7·0 mmol/l). Intermediate values indicate impaired glucose tolerance and need for the evaluation.

Two-hours after glucose: (75 g oral load) Normal less than 120 mg/100 ml (7·0 mmol/l). Diabetic above 180 mg/100 ml (10·0 mmol/l). Intermediate values interpreted as above.

Home Blood Glucose: A number of glucose oxidase based strips and reflectance meters permit quite accurate blood glucose determination at home.

Tests for diagnosis of primary causative disease: acromegaly, Cushing's syndrome, phaeochromocytoma, chronic pancreatitis, haemochromatosis.

KETOSIS

Clinical observations

Increased depth and rate of breathing when associated with significant acidosis. The detection of acetone in the breath is greatly overemphasized.

Routine methods

Urine tests: Acetest for acetoacetate and acetone. Ferric chloride test: less sensitive but useful to perform if Acetest strongly positive. Most of the urinary ketone bodies in diabetic ketoacidosis is β-hydroxybutyrate, which is undetected by either test.

Blood: blood pH, P_{CO_2} and $HCO_3{}^-$ measurements. Plasma ketone bodies normally less than 0·2 mM, may be crudely estimated by using Acetest and serially diluting plasma with water until the test is no longer positive. Mild ketosis (4–6 mM) will give a positive Acetest at a 1:2 dilution. The severity of acidosis is most rapidly determined by calculating the plasma $[HCO_3{}^-]$ from the blood pH and P_{CO_2}. The approximate plasma ketone levels can be calculated from the anion gap. Direct quantitative measurements for β-hydroxybutyrate and acetoacetate may be obtained using enzymic fluorimetric techniques. It is important to remember that most of the plasma ketone bodies are usually β-hydroxybutyric acid, which is not detected by the Acetest. Even severe acidosis may be overlooked if measurements other than Acetest are not done.

PREDIABETES

Clinical observations

Obesity: family history of diabetes; hyperglycaemia during stress of pregnancy, severe infection, injury or corticosteroid therapy; in women, birth of large babies.

Routine tests

Urine tests: glycosuria when present may be due to low renal threshold (renal glycosuria). In the last trimester of pregnancy lactose (a reducing sugar) is often present in the urine.

Laboratory tests

Glucose tolerance test: normal.

HYPOGLYCAEMIA

Clinical observations

Episodes of weakness, palpitation, tremor, sweating, abnormal behaviour. Coma and convulsions occur chiefly after fasting with insulinomas. In functional hyperinsulinism symptoms develop 2–5 h after meals. Symptoms relieved by glucose administration.

Signs of causative disease: hypopituitarism, Addison's disease, history of gastric operation, glycogen storage disease, severe liver damage.

Routine methods

Blood glucose: usually less than 40 mg/100 ml (2·2 mmol/l) during episode.

Tests for causative disease

Insulinoma: starvation with glucose and insulin measurements up to 3 days. In normal subjects administration of insulin from a different species (fish insulin is used) will lower blood sugar and suppress insulin secretion. With insulinomas suppression does not occur. Insulin levels are measured by radioimmunoassay using an antibody that does not measure fish insulin. Alternatively, a routine insulin tolerance test can be performed with measurement of c-peptide. Hypopituitarism, adrenal cortical insufficiency, glycogen storage disease.

References

GARROW, J.S. (1981) *Treat Obesity Seriously—a clinical manual.* Edinburgh, London, Melbourne & New York: Churchill Livingstone.

JOHNSTON, D.G. & ALBERTI, K.G.M.M. (1982) New aspects of diabetes. *Clinics in Endocrinology and Metabolism*, 11, No. 2.

NATTRASS, M. & SANTIAGO, J.V. (1984) *Recent Advances in Diabetes*. Edinburgh, London, Melbourne & New York: Churchill Livingstone.

STRYER, L. (1981) *Biochemistry*. 2nd ed. San Francisco, W.H. Freeman & Co.

15

The Pituitary Gland

The human pituitary gland (hypophysis) consists of two parts, the anterior lobe (adenohypophysis) and the posterior lobe (neurohypophysis), which occupy the sella turcica of the sphenoid bone and are connected by the pituitary stalk to the hypothalamus. The optic chiasma lies just above the sella turcica and the cavernous sinuses and the internal carotid arteries lie laterally. The two parts of the pituitary have different origins. The anterior pituitary is derived from an upgrowth from the somatic ectoderm of the posterior nasopharynx (Rathke's pouch). The posterior pituitary develops as a downgrowth from the neural ectoderm of the floor of the diencephalon.

The pituitary is functionally connected to the hypothalamus by the pituitary stalk, but the two lobes are connected in different ways. The connection between the hypothalamus and the anterior pituitary is vascular. It consists of a group of venules, the hypophysial portal vessels, which arise from the capillaries in the median eminence of the hypothalamus, and terminate in a secondary capillary plexus in the anterior pituitary. The hypothalamus is connected to the posterior lobe by nerve fibres whose cell bodies are in the supraoptic, paraventricular and supra-chiasmatic nuclei of the hypothalamus; the nerve fibres pass down the pituitary stalk and terminate in the posterior pituitary lobe.

ANTERIOR PITUITARY GLAND

Normal function

HORMONES

Several types of hormone are secreted by the cells of the anterior pituitary.
1   Growth or somatotrophic hormone (GH).
2   Prolactin or lactogenic hormone.
3   Thyrotrophin or thyroid stimulating hormone (TSH).
4   Follicle stimulating hormone (FSH).
5   Luteinizing hormone (LH).
6   Corticotrophin or adrenocorticotrophic hormone (ACTH).
7   The normal human pituitary does not contain significant amounts of melanocyte stimulating hormone (MSH). The 22 amino acid peptide sequence of human beta

melanocyte stimulating hormone (β-MSH) is now known to be part of a larger molecule (β-lipotrophin, β-LPH) which can be degraded to β-MSH by harsh extraction procedures or post-mortem autolysis. In addition to β-LPH the human pituitary contains a family of other peptides related to ACTH which include γ-LPH, corticotrophin like intermediate lobe peptide (CLIP), β-endorphin, a peptide similar to α-MSH and γ-MSH (see below). The role of many of these peptides is unknown.

CELL TYPES

The question whether each anterior pituitary hormone is synthesized by one cell type is not fully answered. Using conventional histological staining techniques, the cells of the anterior pituitary can be divided into acidophils, basophils and chromophobes, depending on the staining of the cytoplasm. GH and prolactin are secreted by acidophils; TSH, FSH and LH by basophils. ACTH is secreted by basophils, although ACTH-producing tumours of the pituitary sometimes consists of chromophobes. ACTH and the different species of melanocyte stimulating hormone are thought to be secreted by the same cell. The intensity of staining of the cytoplasm depends upon the amount of hormone stored in the form of granules. A chromophobe cell may be staining poorly either because it is synthesizing very little hormone, because it is secreting all the hormone as soon as it is synthesized, or because the stored hormone is not in granule form. Using more refined histological and immunohistochemical techniques it is possible to identify more than the three traditional cell types. Because the cells of the anterior pituitary cannot regenerate, any structural damage is usually permanent.

CHEMICAL PROPERTIES

Pituitary extracts contain hundreds of proteins and peptides; some are breakdown products of hormones, others are biosynthetic precursors. Attempts to identify anterior pituitary hormones in addition to those listed above are made frequently, but none of them are fully accepted. The pituitary content of GH (5–15 mg/gland) is 1000-fold greater than the other hormones.

The chemical properties and structure of the major anterior pituitary hormones are known (Table 15.1), and information on structure-activity relationships is available. There is considerable variation in chemical structure in the same hormone from different animal species. Some of the hormones of the anterior pituitary are chemically related and may have evolved from a single hormone. For example, TSH, FSH and LH share a common alpha chain. ACTH and β-lipotrophin are derived from a common precursor molecule which has three MSH sequences (Fig. 15.1). Alpha-MSH is the same as the first 13 amino acids of ACTH. In the Nakanishi numbering system referring to the whole ACTH/β-LPH precursor β-MSH is 84–101 and is part of the β-LPH molecule. Gamma-MSH is the third melanotrophin core sequence contained within the 77 amino acid NH$_2$-terminal part of the precursor molecule. The detailed structure of human

Table 15.1.

Hormone	Structure	Approximate Molecular weight	Number of Amino acid residues
GH	Single chain polypeptide	21,000	191
Prolactin	Single chain polypeptide	22,000	198
TSH	Double chain glycoprotein	28,000	209
FSH	Double chain glycoprotein	28,000	210
ACTH	Single chain polypeptide	4,500	39
LH	Double chain glycoprotein	28,000	207
β-LPH	Single chain polypeptide	10,000	91

prolactin is now known, a number of amino acid sequences being similar to growth hormone.

It is interesting to compare the structures and activities of the hormones secreted by the anterior pituitary with the peptide hormones secreted by the placenta. Placental lactogen resembles pituitary growth hormone and prolactin; the lactogen and growth hormone, for example, have 85 per cent of the positions in the two hormones occupied by identical amino acids. Placental gonadotrophin (chorionic gonadotrophin) resembles pituitary LH.

PHYSIOLOGICAL EFFECTS

The effects of the trophic hormones TSH (p. 594), LH and FSH (p. 656), and ACTH (p. 573) are discussed in the chapters concerned with their respective target organs. All these

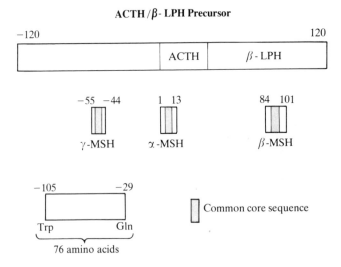

Figure 15.1
The structure of pro-opiocortin, the ACTH/β-LPH precursor molecule, indicating the positions of the three MSH sequences, γ-MSH is contained within the 76 amino acial N-terminal sequence which is released into the circulation at the same time as ACTH and β-LPH.

hormones act through specific sites in the membrane which then activates the adenyl cyclase on the inner surface, to produce cyclic adenosine mono-phosphate (cAMP) within the cytoplasm. cAMP then activates enzyme systems to generate the hormone for which the target cell is biochemically adapted. Recent work has shown that part of the ACTH precursor molecule, pro-gamma MSH, has no direct effect on corticosteroid production by the adrenal but markedly potentiates the effect of ACTH. This action unlike that of ACTH involves DNA transcription. Any effects which these trophic hormones may have on other tissues are probably without physiological significance. An exception to this is the hypogonadism associated with high levels of prolactin. In man this appears to be due at least in part to an antigonadotrophic role of prolactin at the gonadal level which interferes with steroidogenesis by the ovary or testis and thus affects hypothalamic-pituitary feedback control. The role of β-LPH in man is uncertain.

Growth hormone increases the transport of amino acids into cells and their incorporation into protein. It inhibits fat synthesis and facilitates release of fatty acids from adipose tissue. The uptake of glucose into cells is inhibited, and there is a decreased sensitivity to the hypoglycaemic effect of insulin. In man the administration of human GH leads to the retention of nitrogen, phosphorous, sodium, potassium and chloride. Despite the diverse sites of action of growth hormone its precise role in intermediary metabolism is uncertain. The adult hypopituitary patient, given replacement therapy with adrenal corticosteroids, thyroid hormone and the appropriate sex hormones, does not show signs of growth hormone deficiency.

The study of growth responses *in vitro* has shown that some responses are not promoted directly by pituitary GH, but are stimulated by serum from normal subjects. Such responses are greater when serum from acromegalic patients is used and reduced by serum from hypopituitary patients, unless the patient has been given GH treatment. This serum factor is not GH but is stimulated by GH, and has been named somatomedin. The effect most commonly studied has been the *in vitro* uptake of isotopically-labelled sulphate by cartilage (the sulphation factor). Using different isolation procedures two somatomedins active on cartilage have been identified. Somatomedin A is a neutral polypeptide, molecular weight 7000 with 59 amino acid residues. This has been isolated from human plasma and can be measured by bioassay using $^{35}SO_4$ uptake into embryonic chick cartilage. Somatomedin C is a basic polypeptide isolated from human plasma and assayed using either $^{35}SO_4$ or ^3H-thymidine uptake into rat cartilage. Radioimmunoassay for both these somatomedins have now been developed. Somatomedins are produced by the liver.

Prolactin is required for human lactation. Hormonal regulation of lactation has been worked out most fully in the rat, in which it appears that mammary duct growth requires the combined actions of adrenocorticoids, growth hormone and oestrogens, whereas progesterone and prolactin are needed in addition to bring about the growth of the aveolar lobules. Milk secretion is initiated when progesterone and oestrogen are simultaneously withdrawn, but only in the continued presence of prolactin and adrenocorticoids. In vertebrates more than 100 physiological actions have been ascribed to prolactin. These have been classified into effects on reproduction, water and electrolyte balance, somatic growth, ectodermal development and synergistic actions with gonadal and adrenal steroids. In man the function of prolactin appears to be much more limited to lactation and reproduction. There is little evidence for effects on water and electrolyte balance but it may play a role in vitamin D metabolism and foetal lung maturation, and high levels may be associated with abnormal adrenal steroid production.

REGULATION OF ANTERIOR PITUITARY FUNCTION

The hypothalamus is the main regulator of anterior pituitary function. In addition there are a few regulatory mechanisms which act directly on the anterior pituitary.

Hypothalamus

The hypothalamus plays an important role in regulating anterior pituitary function by producing stimulating and inhibitory hormones (Table 15.2). These hormones are thought to be liberated by hypothalamic nerve endings close to the capillaries of the

Table 15.2. Hypothalamic hormones

Releasing hormones	Inhibitory hormones
GH-releasing factor (GHRF)	GH-release inhibiting hormone (GH-RIH somatostatin)
TSH-releasing hormone (TRH), a tripeptide	MSH-inhibitory factor (MIF), a tripeptide
Gonadotrophin-releasing hormone (GnRH) a decapeptide	Prolactin inhibitory factor (PIF) (Dopamine)
ACTH-releasing factor (CRF)	

median eminence, and are then carried in the hypophysial portal vessels into the anterior pituitary. Several of the hypothalamic hormones have been isolated; their structure has been determined and they have been synthesized. These are thyrotrophin releasing hormone (TRH), gonadotrophin releasing hormone (GnRH), somatostatin and the prolactin inhibitory factor, dopamine. Recently a peptide with 41 amino acid residues has been isolated from ovine hypothalami and shown to be a potent corticotrophin releasing factor. A peptide with 40 amino acids has been isolated from a pancreatic tumour in a patient with acromegaly and somatotroph lymperplasia. This peptide is immunologically similar to one present in the hypothalamus and has been shown to release GM in normal subjects and some patients with isolated GH deficiency (see below).

The activity of the hypothalamus may in turn be modified by neural factors, particularly from the reticular activating system and the limbic area, by drugs modifying the neurotransmitters of the hypothalamus and by feedback from target gland secretion or from the anterior pituitary itself. Endogenously produced opiates may also play a role in the control of hypothalamic-pituitary function. Administration of an enkephalin analogue has been shown to raise serum prolactin and growth hormone levels but lower serum levels of luteinizing hormone, follicle stimulatory hormone and ACTH. Prior administration of the opiate antagonist naloxone attenuates these hormonal responses. Naloxone alone has no effect on the basal levels of prolactin or growth hormone but lowers TSH and elevates LH and FSH. The pulsatility of LH release is increased, suggesting that opioid peptides modulate pulsatile LHRH release.

Target organ feedback

For the trophic hormones (ACTH, TSH, FSH, LH) the feedback from the target gland helps to stabilize their plasma level. The main effect is a negative feedback mechanism where an increase in target gland hormone level inhibits secretion of the corresponding trophic hormone, e.g. cortisol inhibits the release of ACTH and thyroid hormones inhibit the release of TSH. The major sites of negative feedback control are believed to be hypothalamic in the case of ACTH and the gonadotrophins, but may be in the pituitary itself in the case of TSH. The feedback control of gonadotrophins is more complex. Despite the fact that there appears to be only one gonadotrophin releasing hormone the levels of LH and FSH are dissociated in certain pathological states. When the germ cells are damaged by irradiation, for example, LH levels are normal but FSH levels rise. In the male there is a reciprocal relationship between spermatogenesis and FSH secretion. Abundant evidence suggests that a substance is produced during spermatogenesis that inhibits FSH secretion. This substance was first suspected in 1932 by McCullagh who called it inhibin. Fractionation studies suggest that it is a protein but it has not been fully isolated or characterized. In the female inhibin appears to be produced by the ovary and has been found in follicular fluid. In women LH secretion may be stimulated by low but inhibited by high oestrogen levels. In men LH secretion is inhibited by testosterone administration. However, this may not be a direct effect but mediated via peripheral conversion of testosterone to oestrogen.

Regulation of growth hormone

Growth hormone secretion is regulated by a hypothalamic centre which controls the activity of the pituitary somatotrophic cells by means of a GH-releasing hormone secreted into the portal system. The hypothalamic centre responds to a variety of both chemical and neural influences (Table 15.3). It is difficult to provide a unifying concept to explain why such a variety of factors influence GH secretion. The most consistent responses are those following changes in blood sugar levels and following the onset of sleep. It must be appreciated that the pattern of GH secretion is usually as intermittent peaks. GH secretion is not normally continued after being stimulated even though the

Chapter 15

Table 15.3. Factors affecting GH secretion

Stimulation	Inhibition
Exercise	Hyperglycaemia
Onset of deep sleep	Emotional deprivation in children
Hypoglycaemia	
Protein meal	
Aminoacids intravenously	
Stress, e.g. arterial puncture	
surgery	
bacterial endotoxin	
vasopressin	
Glucagon	
Dopamine agonists	
GH-releasing factor	
Other factors tending to increase growth hormone concentrations	*Other factors tending to decrease concentrations*
Malnutrition	Obesity
Chronic hepatic and renal disease	Hypothyroidism
Oestrogens	3rd trimester of pregnancy
β-adrenergic blockade	High dose glucocorticoids
	Dopamine antagonists
	Medroxyprogesterone
	α-adrenergic blockade

factor affecting GH secretion is still present, e.g. exercise. Other physiological and pathological conditions have been identified where the change in serum GH level in response to hyperglycaemia is sometimes a paradoxical rise instead of suppression. The response to other stimuli may also be paradoxical. In acromegaly dopamine agonists inhibit rather than stimulate GH secretion. Thus long acting dopamine agonists such as bromocriptine can be used in the treatment of this condition.

Regulation of prolactin

The secretion of prolactin is under the control of the hypothalamus. Unlike the other major anterior pituitary hormones, the predominant control is inhibition via a prolactin inhibitory hormone. Current evidence suggests that dopamine is the most important inhibitor of prolactin secretion. Thus dopamine agonist drugs inhibit prolactin secretion and can be used to lower elevated prolactin levels. Conversely dopamine antagonists stimulate prolactin secretion. There is also evidence for a short feedback loop whereby elevation of prolactin levels stimulates hypothalamic dopamine secretion and thus inhibits prolactin secretion. The factors affecting prolactin secretion are outlined in Table 15.4.

Table 15.4. Factors and conditions affecting prolactin secretion

Stimulation	Inhibition
Breast stimulation	Dopamine agonists
Pregnancy	
Oestrogens	
Renal failure	
Dopamine antagonists	
Thyrotrophin releasing hormone	
Hypoglycaemia	
L-tryptophan	
Primary hypothyroidism	
Stress	
Hypothalamic-pituitary disease	

Disordered function

Any disordered function must be considered in terms of hypothalamic as well as anterior pituitary function. Deficient or excessive secretion of anterior pituitary hormones may be due to a defect at either or both of these levels.

UNDERSECRETION

Isolated deficiencies of gonadotrophins, TSH, GH and ACTH do occur, usually as a congenital deficiency. It is likely that the majority of these isolated deficiencies result from a lack of the relevant releasing hormone rather than an inability of the pituitary to secrete the trophic hormone. Thus in isolated gonadotrophin deficiency administration of the gonadotrophin releasing hormone stimulates anterior pituitary synthesis and release of LH and FSH. Interestingly this defect is commonly associated with anosmia. It will be appreciated from the anatomy that acquired lesions of the hypothalamus or pituitary will usually result in more than one hormone deficiency. Loss of the trophic hormones results in loss of function of their target organs, which are considered in their respective chapters. Loss of GH in children results in short stature.

OVERSECRETION

The effect produced by excessive secretion of GH from an adenoma of the pituitary depends upon the age at which it occurs. Pituitary gigantism results if epiphyses are open. Acromegaly results if epiphyses are closed, with the overgrowth of the skeleton and soft tissues without increase in length. Excessive GH secretion causing gigantism or acromegaly is almost always associated with a pituitary tumour. Whether the tumour is the primary defect or a secondary development from excessive stimulation by the hypothalamic GH releasing factor is not known.

Excessive secretion of ACTH occurs as a normal response to primary adrenal failure. The effect of excessive secretion of ACTH on normally functioning adrenals is an excessive secretion of cortisol, producing the symptoms and signs of Cushing's syndrome. When plasma ACTH concentrations in pituitary dependent Cushing's syndrome are increased, this points to an abnormality in the negative feedback control of pituitary ACTH; in this situation the normal circadian rhythm of ACTH secretion is lost. Sometimes the source of the increased ACTH secretion is a pituitary tumour which may grow after the adrenals have been removed. Such tumours may produce β-LPH and other melanocyte stimulating peptides as well as ACTH, thus leading to intense skin pigmentation Nelson's Syndrome. Sometimes the source of increased ACTH secretion is from hyperplasia of the corticotrophs of the anterior pituitary. The role of the hypothalamus in the pathogenesis of this condition is probably important but the details are not known. ACTH may also be produced by certain neoplasms such as oat cell carcinoma of bronchus, carcinoid tumours, medullary carcinomas of the thyroid, islet cell and thymic tumours. The clinical condition that may result from this is referred to as the ectopic ACTH syndrome. If the tumour is rapidly growing such as an oat cell carcinoma the patient may have none of the usual symptoms or signs of Cushing's syndrome. In such cases skin pigmentation, dependent oedema, hyperglycaemia and hypokalaemic alkalosis are frequently found.

Excessive secretion of prolactin results in breast secretion, galactorrhoea, in both women and men. However, the incidence of galactorrhoea with hyperprolactinaemia is variable (30–80 per cent) depending to a large extent on the care taken to examine the breasts. The major clinical manifestation is usually hypogonadism with menstrual disturbance and infertility in women and impotence in men. The menstrual abnormality may be of any type including amenhorrhoea, oligomenorrhoea, dysfunctional uterine bleeding, regular anovulatory cycles or cycles deficient in luteal phase progesterone production. Hyperprolactinaemia is commonly caused by a prolactin secreting pituitary adenoma (prolactinoma). In some cases these are macro-adenomas that enlarge the pituitary fossa and may produce visual field loss by suprasellar extension and pressure on the optic chiasma. Frequently, however, the pituitary tumours are very small (micro-adenomas) and the pituitary fossa is not obviously enlarged. In addition to prolactinomas any type of hypothalamic or pituitary disease may produce hyperprolactinaemia by preventing dopamine from reaching the normal lactotrophs. The other causes of elevated prolactin levels are listed in Table 15.4.

Excessive secretion of TSH occurs as a normal response to primary thyroid failure. Theoretically some pituitary tumours might be associated with an increased secretion of TSH and hyperthyroidism, but in practice this is very rare. Trophoblastic neoplasms such as choriocarcinoma can be associated with hyperthyroidism. The thyroid stimulating substance differs from TSH or thyroid stimulating immunoglobulins and is now known to be human chorionic gonadotrophin (HCG). Excessive secretion of LH and FSH occurs as a normal response to primary gonadal failure. Very rarely pituitary and other tumours have been documented *in vivo* as secreting either LH or FSH or both. Recently, however, apparently functionless pituitary tumours have been shown *in vitro* to produce LH, FSH and prolactin. Elevated levels of LH with low or normal serum FSH are found in some patients with the polycystic ovary syndrome.

Principles and tests of measurements

MEASUREMENT OF ANTERIOR PITUITARY HORMONES IN BLOOD AND URINE

In the last few years the two main difficulties of sensitivity and specificity in the assay of anterior pituitary hormones have been partly overcome by improved techniques. The concentrations of anterior pituitary hormones in body fluids under most physiological conditions range between 10^{-10} and 10^{-12} M., i.e. picograms or nanograms per ml. These concentrations are usually beyond the sensitivity of bioassay techniques and their application to clinical work is limited. Bioassay of total gonadotrophin in a concentrated extract of urine has been used, but even this is limited both in sensitivity (it cannot distinguish clearly between a low normal result and true deficiency), and in specificity (it measures both FSH and LH activity).

Anterior pituitary hormones are now measured using the principle of saturation analysis. The most commonly employed technique is radioimmunoassay. An antibody is prepared against the particular hormone and, under appropriate conditions of incubation, the binding of the hormone to this antibody is inversely proportional to the amount of unlabelled hormone present, according to the Law of Mass Action. Radioactively-labelled hormone is added to the system to measure the proportion found. One of the advantages of this technique is the greater sensitivity compared with bioassays. Measurements can be made on unextracted plasma for most of the pituitary hormones and such measurements have provided much useful clinical information. Measurement of the rate of excretion of anterior pituitary peptides could provide useful information, but, usually, their rate of renal clearance is low and extracts of urine are required. The immunological pregnancy tests on urine are successful because of the greatly increased rate of secretion of chorionic gonadotrophin.

Problems of specificity have not been eliminated by these techniques. Although the antigenic determinants in the hormone may react specifically with the antibodies, these antigenic determinants may be only a small part of the hormone molecule and may be different from the part of the molecule on which biological activity depends. Hence the assay may measure inactive precursors, inactive degradation fragments or even other hormones with identical antigenic sites, as well as the biologically active molecules of hormone.

TESTS

Pituitary hormone assays are usually made on blood samples collected under well-defined conditions. The tests can be divided into those increasing and those suppress-

Chapter 15

ing the secretion of pituitary hormones. Changes in the plasma concentrations of a hormone may not be only due to a change in secretion rate. In practice, however, there are few examples of a significant discrepancy.

In selecting appropriate stimuli, a rapid reliable response with minimum side-effects is needed. In interpreting the results, it is important to consider whether the stimulus is acting via the hypothalamus or directly on the anterior pituitary. For example, ACTH secretion is increased by stimulation of the hypothalamus by acute hypoglycaemia, and by stimulation of the anterior pituitary by lysine vasopressin.

Growth hormone

Many stimulation tests have been used to assess growth hormone secretory reserve. Of these the most reliable has been the insulin hypoglycaemia test. In this test it is essential that the blood glucose should fall to below 2 mmol/l and the patient should have hypoglycaemic symptoms such as sweating. An impaired growth hormone response to hypoglycaemia is not necessarily associated with impaired growth. Such a response is commonly found in peri-pubertal children. Conversely high levels of growth hormone may be found in children of short stature with the Laron syndrome in which somatomedin production is impaired. In patients with epilepsy or ischaemic heart disease insulin hypoglycaemia is contraindicated. In this situation growth hormone reserve can be assessed following subcutaneous glucagon, intravenous arginine or the oral administration of 'Bovril' or L-dopa.

Prolactin

Prolactin is released in a pulsatile way with marked peaks occurring during sleep. Stress such as venepuncture also stimulates prolactin secretion and to assess basal prolactin levels more than one plasma sample should be taken. In normal subjects prolactin levels are usually below 360 mU/l. Levels consistently in excess of 2000 mU/l are strongly suggestive of a prolactin secreting pituitary tumour. Dynamic function tests can be useful in distinguishing tumours from other causes of hyperprolactinaemia. The prolactin responses to TRH or dopamine antagonists such as metoclopramide are much smaller in patients with prolactinomas. Suppression tests are not helpful. Dopamine or dopamine agonists such as bromocriptine will lower prolactin levels regardless of cause.

ACTH

ACTH secretion is normally assessed indirectly by measuring plasma cortisol levels (p. 581). In non-stressed normal subjects there is a circadian rhythm of ACTH and cortisol with levels being highest on waking in the morning and lowest on going to bed at night. ACTH and hence cortisol secretion is stimulated by hypoglycaemia and indirectly by metyrapone. This drug acts by inhibiting adrenal 11β-hydroxylase activity and thus blocks cortisol secretion. This activates the negative feedback control of ACTH. Increased levels of ACTH stimulate the production of adrenal corticosteroids prior to the enzyme block. The level of 11-deoxycortisol, in particular, rises and can be measured either directly in the blood or by the increase in 17-oxogenic steroids excreted in the urine. These stimulation tests are usually used in patients with suspected hypopituitarism. In patients with suspected cortisol excess glucocorticoid suppression tests are performed. Patients with Cushing's syndrome show no suppression of plasma cortisol with low doses of a glucocorticoid such as dexamethasone. In Cushing's disease (pituitary dependent Cushing's syndrome) the plasma cortisol level usually suppresses when high doses of dexamethasone are given (p. 583).

TSH

The intravenous administration of TRH releases TSH, the levels being highest at about 20 minutes after the injection. Because the releasing hormone controls both synthesis and release of TSH, the TRH test is not a useful way to distinguish between hypothalamic and pituitary disease. The greatest value of the test is in the diagnosis of borderline hyperthyroidism. The high levels of thyroid hormone suppress TSH secretion and when TRH is given there is no TSH response. In primary hypothyroidism basal TSH levels are elevated and there is an exaggerated TSH response to TRH. Suppression tests involving the administration of triiodothyronine or thyroxine used to be used in the diagnosis of hyperthyroidism but have been largely replaced by the TRH test.

Gonadotrophins

The gonadotrophin releasing hormone releases both LH and FSH. Before puberty the LH response is much lower than that of FSH. In the adult the LH rise is exaggerated compared to FSH. The antioestrogen clomiphene stimulates gonadotrophin releasing hormone and hence LH and FSH secretion. Suppression tests are not often used.

Practical assessment

UNDERSECRETION OF GH

Clinical observations

Children: retarded growth; normal size at birth; slowing of growth noted after age 2 to 4 years; no craniofacial anomalies or other congenital defects; delayed bone age and sexual maturation; gross clinical evidence of thyroid and adrenal deficiency usually absent; craniopharyngioma or pituitary tumour may produce space-occupying symptoms and signs.

Adults usually show evidence of multiple endocrine deficiencies; skin is pale, fine and soft; body hair sparse or absent; external genitalia atrophic.

Routine methods

Insulin test: plasma GH and cortisol levels can be measured; test may be dangerous if patient has severe hypopituitarism. Test should not be performed if patient has ischaemic heart disease or epilepsy. *Other tests* include exercise, intravenous arginine, oral meat extract (Bovril), subcutaneous glucagon. Other pituitary hormone tests; radiology of skull and visual field charting.

OVERSECRETION OF GH

Clinical observations

Children and adolescents: gigantism—excessive growth in height often with minor features of acromegaly.

Adults: acromegaly with enlarged hands and feet; coarse facial features; prognathism; large tongue; thick skin with excessive sweating; diabetes mellitus may be found.

Routine methods

Plasma GH levels increased and do not suppress during an oral glucose tolerance test. In some patients GH levels may rise during GTT. Other pituitary function tests should be performed.

    Radiology of skull usually shows evidence of pituitary tumour.

    Visual field charting.

POSTERIOR PITUITARY GLAND

Normal function

POSTERIOR PITUITARY HORMONES

The typical mammalian posterior pituitary contains two active cyclic octapeptides, oxytocin and vasopressin. Oxytocin has the same amino acid sequence in all species; vasopressin has either arginine or lysine in the eighth position. Human vasopressin is in the arginine form. Only two amino acids are different in oxytocin and vasopressin, otherwise they have similar chemical structures. They are synthesized in the hypo-thalamus where they are reversibly bound to a specific protein, called neurophysin, which is released into the circulation together with the active peptides. In some animals, such as the cow, there is clear evidence that specific neurophysins exist for vasopressin and oxytocin. In man the situation is less clear. There seem to be two major neurophysins, one being referred to as nicotine-stimulated neurophysin and the other as oestrogen-stimulated neurophysin. Neurophysins have not been shown to have any specific biological action apart from their possible role in the storage of peptides in the neurosecretory granules of the posterior pituitary.

The main function of vasopressin in man is to permit increased reabsorption of water by the kidney (p. 180). Therefore it is often referred to as the antidiuretic hormone (ADH). Continued administration of the hormone to animals or to man causes a progressive retention of water so that body fluids become hyptonic. After two or three days there is an increase in the rate of renal excretion of sodium which continues as long as water is taken freely; body weight levels off at a new steady state. The rise in the rate of sodium excretion is not due to a direct effect of vasopressin on the rate of renal tubular sodium reabsorption, but is probably related to an increase in the rate of glomerular filtration, a decrease in proximal tubular reabsorption of sodium and to a decrease in the rate of aldosterone secretion. Vasopressin also makes smooth muscle contract but it is unlikely that vasopressin, in physiological amounts, has any role in the maintenance of normal blood pressure. Corticotrophin releasing factor appears to consist of at least two components, one of which is probably arginine vasopressin. A variety of other physiological roles for vasopressin have been suggested. Of these one of the most interesting is its postulated action in the central nervous system. There is convincing evidence that it plays an important role in memory in rats. A useful syn-thetic analogue has been prepared (DDAVP—1-desamino-8D-arginine vasopressin) which has a high ADH activity but no significant pressor effect. It also has a much longer duration of action than natural arginine vasopressin.

Oxytocin causes contraction of the parturient uterus. Its other major function in mammals is to cause ejection of milk from the breasts. The stimulus of suckling excites a reflex discharge of oxytocin which in turn causes contraction of myoepithelial fibres surrounding the mammary ducts. Animals in whom posterior pituitary function has been destroyed can no longer rear their young. In some species, including the rat, oxytocin has a powerful natriuretic action, but in man this effect appears to be insig-nificant. In view of their structural similarities, it is hardly surprising that oxytocin and vasopressin share some biological effects.

Regulation of posterior pituitary function

Posterior pituitary hormones are formed in the hypothalamus by the cells of the supraoptic and paraventricular nuclei. They are transported as neuro-secretory granu-les along the axons of the hypothalamo-hypophysial tract and stored in dilated nerve endings within the posterior lobe. These endings are closely applied to the capillaries into which the hormones can be released rapidly in response to impulses propagated by the supraoptic nerve fibres.

Vasopressin release can be elicited (1) by an increase in plasma osmolality—probably detected directly by the supraoptic cells; (2) by a reduction in circulating blood volume—possibly detected by stretch receptors in the atria and great veins; (3) by sensory or emotional stimuli—pain, fear, rage etc; (4) by muscular excercise; (5) by fainting; (6) by surgical trauma; (7) by suckling; (8) by certain drugs—acetylcholine, nicotine, morphine, barbituates, vincristine, clofibrate and some anaesthetic agents. Release of vasopressin can be inhibited (1) by a decrease in plasma osmolality; (2) by expansion of the circulating blood volume; (3) occasionally by emotional stimuli; (4) by ethanol, atropine and alpha agonists. A variety of drugs may alter the effects of vasopressin on the kidney. Chlorpropamide and carbamazepine potentiate the peripheral action of vasopressin while lithium and demethylchlortetracycline block the anti-diuretic effect.

Oxytocin secretion is increased during parturition, suckling, coitus and certain stimuli which excite vasopressin release—e.g. hypertonic saline infusion. Despite major advances in the assays for oxytocin it is still not clear whether this hormone plays an important role in the initiation or maintenance of human labour. Animal evidence where raised oxytocin levels have been found during labour and the widespread use of oxytocin in the induction of labour have led many people to assume that oxytocin must be an important factor. Specific assays have shown low levels of oxytocin in human maternal blood. Higher levels are found in the umbilical circulation, the hormone being released by the foetus. The importance of this observation is unknown.

Disordered function

UNDERSECRETION (see Table 15.5)

The manifestations of posterior pituitary insufficiency are essentially those of lack of antidiuretic hormone; that is, diabetes insipidus. As a result of diminished reabsorption of water by the renal tubules the urine becomes copious and dilute. Plasma osmolality rises and intense thirst develops.

A genetic defect in the synthesis of vasopressin with normal synthesis of oxytocin has been shown in rats with hereditary diabetes insipidus. This may also obtain in human hereditary hypothalamic diabetes insipidus.

Over 85 per cent of the hypothalamo-hypophysial system must be destroyed to cause permanent diabetes insipidus. For example, the polyuria following surgical division of the pituitary stalk may be transient. When anterior as well as posterior pituitary function is deficient polyuria is slight or absent but frank diabetes insipidus may appear subsequently if cortisone or one of its derivatives is given.

Table 15.5. Causes of diabetes insipidus

Cranial Di	Nephrogenic Di
1. *Primary:*	1. *Hereditary:*
Hereditary (Dominant or Recessive)	Sex-linked recessive form
Idiopathic	Cystinosis
	Sickle cell disease
	Medullary cystic disease
2. *Secondary:*	2. *Acquired:*
Post-traumatic (head injury or neurosurgery)	Hypercalcaemia
Tumours (primary or secondary)	Hypokalaemia
Miscellaneous	Pyelonephritis
(Histiocytosis-X,	Obstructive uropathy
Tuberculosis, Sarcoidosis,	Heavy metal poisoning
Meningitis, Encephalitis)	Lithium therapy

Lack of oxytocin has not been described as an isolated defect. Oxytocin lack does not seem to affect parturition in human diabetes insipidus; the effect on suckling is not clear. Difficulties with both functions have been observed in animals with diabetes insipidus.

OVERSECRETION

A sustained and inappropriate secretion of ADH is held responsible for the hyponatremia and renal 'sodium-wasting' which occur in some patients with malignant disease, especially oat cell cárcinoma of the bronchus, some cases of pneumonia, lung abscess, pulmonary tuberculosis, any type of neoplastic, traumatic or inflammatory cerebral disease or acute porphyria. In certain patients with malignant disease the source of the antidiuretic activity has been shown to be the tumour itself. Secondary hypersecretion of ADH occurs in oedematous states and may occur in adrenocortical deficiency (p. 575) and myxoedema. In oedematous states an increased rate of secretion is an appropriate response to increased sodium retention. In some cases, however, where water is retained in excess of sodium, the posterior pituitary may be responding to a volume stimulus as well as to an osmotic one. In cirrhosis of the liver decreased hepatic inactivation of ADH may occur but probably does not lead to water retention. The cause of the impaired diuretic response to water in adrenocortical insufficiency is considered on page 575. A number of drugs have been shown to enhance ADH activity. The best studied is chlorpropamide which potentiates the effect of ADH on the distal renal tubule. Clofibrate may stimulate endogenous ADH release.

Primary hypersecretion of oxytocin has been found with certain tumours such as oat cell carcinoma of the bronchus and adenocarcinomas of the pancreas. These tumours also secrete vasopressin.

Principles of tests and measurements

MEASUREMENT OF POSTERIOR PITUITARY HORMONES
IN BLOOD AND URINE

Vasopressin in urine or plasma can be bioassayed by preparing extracts which are injected intravenously into ethanol-anaesthetised hydrated rats. The changes in urine volume and electrical conductivity are compared with those produced by arginine vasopressin. Non-specific antidiuretic activity is excluded by establishing that the activity is destroyed by thioglycollate, and that the response is identical with that to arginine vasopressin. During partial hydration of normal human subjects plasma vasopressin levels measured by bioassay range from 1–5 ng/l. During water loading levels fall to about 0·5 ng/l and with water deprivation rise to about 15 ng/l. Oxytocin can be bioassayed in blood using the contraction of the isolated guinea pig uterus or milk ejection in the cannulated rabbit mammary duct. In animals such as the goat bioassay results correlate well with radioimmunoassay, the levels in the second stage of labour being about 200 ng/l. In the human very discrepant results have been obtained with bioassayists suggesting levels from 160–15,000 ng/l in the second stage of labour where radioimmunoassay results have been less than 2 ng/l.

Highly specific and sensitive radioimmunoassays have now been developed for both oxytocin and vasopressin. Antibodies have been produced either by immunisation with the peptide alone, emulsified with Freund's adjuvant or with the peptide conjugated to a carrier such as albumin. However, despite the fact that satisfactory antisera have been developed, the very low circulating levels of vasopressin and oxytocin and the presence in plasma of substances that non-specifically inhibit antigen-antibody binding have necessitated the development of methods of extracting the hormones from plasma. The results of the vasopressin radioimmunoassay correlate well with those obtained by bioassay.

PLASMA OSMOLALITY

In diabetes insipidus underhydration is usual so that the plasma osmolality tends to be above normal (i.e. greater than 284 ± 4 (SD) mosm/kg). In compulsive water drinking (primary polydipsia) overhydration is usual so that plasma osmolality is decreased.

WATER DEPRIVATION

Normally, water deprivation stimulates ADH secretion so that the urine specific gravity is greater than 1020 and the urine osmolality is greater than 800 mosm/kg after 8 hours. The retention of water prevents dehydration and plasma osmolality alters very little. In diabetes insipidus water deprivation fails to release ADH so that the urine specific gravity usually does not exceed 1·014 and the osmolality does not exceed 300 mosm/kg. Water loss continues and severe thirst and dehydration develop with rapid loss of body weight. Water deprivation should be terminated when the body weight has fallen by 3 per cent or more. In some patients with partial diabetes insipidus a more concentrated urine may be produced. This condition can be detected by giving vasopressin after urine osmolality has reached a plateau with water deprivation. In normal subjects vasopressin produces no further increase in urine osmolality whereas there is a further rise in patients with partial diabetes insipidus.

HYPERTONIC SALINE INFUSION

A hypertonic solution of sodium chloride is infused intravenously in the hydrated subject. The normal response to the increased plasma osmolality is a release of ADH, a fall in urine volume and a rise in urine osmolality. Sometimes the antidiuretic effects of the released vasopressin may be masked by a solute diuresis. This potential error in interpretation can be avoided by calculating the osmolar and free water clearances. This test is difficult to perform and is rarely necessary. It is essential that it is not carried out in a dehydrated subject as it may then produce a dangerous rise in plasma osmolality. The procedure is contraindicated in patients with heart disease.

ADMINISTRATION OF NICOTINE

This is an unpleasant test especially for the non-smoker and the results obtained must be interpreted with caution. After hydration the smoke of 1–3 cigarettes is rapidly inhaled or nicotine base given intravenously. Side effects include nausea and hypotension. The consequent reduction in glomerular filtration rate may then produce an antidiuresis which may mistakenly be thought to be due to the release of vasopressin. As with the hypertonic saline test a change in free water clearance rather than urine flow rate should be used as an index of response. The mechanism of action of nicotine is poorly understood. For many years it was thought to release vasopressin via the cholinergic innervation of the posterior pituitary but this is probably not true.

RESPONSE TO VASOPRESSIN

If there is a subnormal rise in urine osmolality with water deprivation the response of the renal tubules to vasopressin should be tested. This is most conveniently done by giving the long acting vasopressin analogue DDAVP (2 μg by intramuscular injection or 20 μg intranasally). In cranial diabetes insipidus this will produce a marked antidiuresis in comparison to the lack of response found in nephrogenic diabetes insipidus. In compulsive water drinking the renal response to exogenous vasopressin may be poor as the long standing overhydration impairs the counter current multiplier system.

Other tests

A variety of non-osmotic tests apart from nicotine stimulate vasopressin release. These include hypotension produced by agents such as trimetaphan, nausea or emesis following apomorphine, exercise, and insulin-induced hypoglycaemia. These tests may not produce the same results as the osmotic ones and thus should not be thought to be equivalent to the standard stimuli. Very occasionally patients may have a selective osmoreceptor defect and fail to respond to osmolar stimuli but respond well to others.

Practical assessment

Clinical observations

Polydipsia and polyuria with dilute urine are cardinal features; in infants fever, dehydration or simply failure to thrive may be the presenting symptoms. In adults the main difficulty lies in distinguishing between true diabetes insipidus and compulsive water drinking (primary polydipsia).

Routine methods

Plasma and urine osmolality measurements before and during a standardized water deprivation test.

Response to vasopressin

Bioassay and immunoassay of vasopressin and oxytocin are specialized procedures. Search for possible cause of diabetes insipidus (Table 15.5).

In patients with suspected cranial diabetes insipidus it is essential to test anterior pituitary function before that of the posterior pituitary. Diabetes insipidus may be masked by ACTH and hence cortisol deficiency. Renal handling of water is also impaired in hypothyroidism. In such patients replacement therapy with hydrocortisone and/or triiodothyronine should be given before water deprivation testing.

References

BEARDWELL, C. & ROBERTSON, G.L. (eds.) (1981) *Clinical Endocrinology*. London, Butterworth.
BESSER, G.M. (ed.) (1977) The hypothalamus and pituitary. *Clinics in Endocrinology and Metabolism*. Philadelphia, W.B. Saunders.
GRAY, C.H. & JAMES, V.H.T. (eds.) (1979) *Hormones in Blood*, 3rd edn. New York, Academic Press.
WILLIAMS, ROBERT (ed.) (1981) *Textbook of Endocrinology*, 6th edn. Philadelphia, W.B. Saunders.

16

The Adrenal Glands

ADRENAL CORTEX

Normal function

The cortex of the adrenal glands is the source of all the known steroid hormones elaborated in the body except for those derived from the gonads and placenta. The steroid hormones of the adrenal cortex play an important role in the regulation of the body's metabolic processes and without them life is hazardous. In particular, they play a major part in the regulation of salt and water metabolism, renal function, carbohydrate and protein metabolism and in the response of the body to such noxious stimuli as severe injury, surgical operations and infections (i.e. 'resistance to stress').

THE ADRENOCORTICAL STEROID HORMONES

The steroid hormones mediating these functions are usually referred to as adrenocorticosteroids. Like all steroid hormones, their chemical structure is based on the reduced cyclopentenophenanthrene ring (carbon atoms 1 to 17 of Fig. 16.1 (i)) but in addition all those which have biological activity also have (a) a double bond between carbon atoms 3 and 4 in the A ring together with a ketone group in the 3 position (the Δ^4, 3-ketone grouping, Fig. 16.1 (ii)) and (b) a 2-carbon side chain at carbon 17 with a ketone group in the 20 position and a hydroxyl group in the 21 position. This α-ketol side chain (Fig. 16.1 (iii)) has strong reducing properties and is the basis of much chemical and immunological specificity. Thus the adrenocorticosteroids have 21 carbon atoms, in contrast to other steroids, such as oestrogens and androgens, which have 18 and 19 carbon atoms respectively.

The important chemical differences between the various adrenocorticosteroids consist mainly of varying substitutions with OH or O in the C-11, C-17 or C-18 positions, and aldosterone is unique in having an aldehyde grouping at C-18. Of the many C-21 steroids with this type of basic chemical structure that have been isolated from adrenal tissue, probably only cortisol (11β-OH, 17α-OH), corticosterone (11β-OH) and aldosterone (11β-OH, 18-aldehyde or the 11,18 hemiacetyl form) are important physiologically. It is only these three which have been identified in the adrenal venous blood in sufficient amounts likely to have a significant biological effect. The proportions of each of these steroids vary considerably from species to species, but in man cortisol predominates. Cortisone (11-ketone, 17α-OH) and desoxycorticosterone

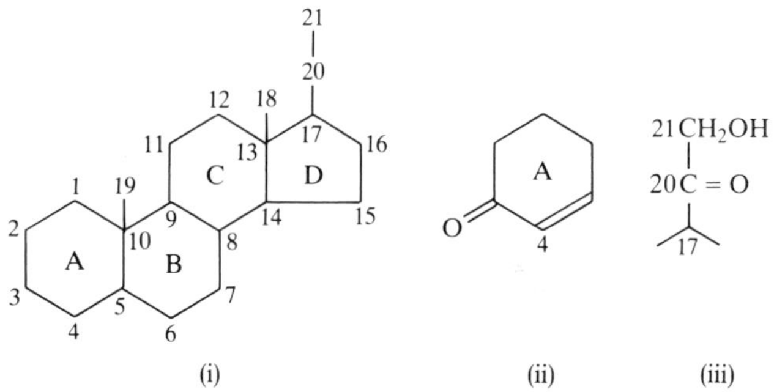

Figure 16.1. *Corticosteroid configurations*

(no substitutions on C-11 or C-17) have also been identified in adrenal venous blood but only in amounts which are unlikely to be important physiologically.

The adrenal cortex also produces weak androgens which are physiologically important as the principal source of androgens in women throughout their life and, by conversion in peripheral tissues, as the principal source of oestrogens after the menopause. The main secretory products are the C-19 steroids dehydroepiandrosterone (and its sulphate ester) and Δ^4-androstenedione (and its 11-oxygenated derivative, 11β-hydroxy-androstenedione).

Functional anatomy

The adrenocorticosteroids are synthesized from cholesterol, which mainly comes from low-density lipoproteins of plasma origin. It is stored in the esterified form in the cytosol as lipid droplets and it is these which give the adrenal cortex its characteristic vivid yellow colour. The cortex is divided into three zones, viz., from outside inwards, the glomerulosa, the fasciculata and the reticularis. Under resting conditions, each has its own characteristic microscopic and electronmicroscopic appearance but, for example, ACTH transforms the 'clear' lipid-rich cells of the fasciculata into the 'compact' lipid-poor cells of the reticularis. Aldosterone is only produced by the glomerulosa, since only glomerulosa cells contain the enzyme, 18-dehydrogenase, necessary for the last stage of its synthesis. 18-hydroxylation occurs, however, in all zones, so that 18-hydroxycorticosterone and 18-hydrodesoxycorticosterone are produced by all three zones. Only the fasciculata/reticularis contains 17-hydroxylase, so that the production of cortisol is limited to these inner zones. There is good evidence that adrenal androgen production is largely a function of the reticularis.

Adrenal vasculature

The adrenal cortex is one of the most richly vascularized tissues in the body. Arteries enter the capsule from the aorta, renal and inferior phrenic arteries to form an extensive subcapsular plexus. Capillary loops surround the outer cells of the glomerulosa, and from them further capillaries pass directly through the deeper layers of the cortex, between the columns of fasciculata cells, to open into a relatively wide plexus of capillary sinusoids surrounding the cells of the reticularis. Thus, there is no direct arterial supply to the inner two zones of the adrenal cortex. Venous blood from both cortex and medulla is collected into a large central vein which has curious, unique, eccentric longitudinal muscle bundles in its wall. The functional significance of these muscle bundles is obscure, although they are clearly so strategically placed that they may modulate blood flow through the adrenal gland.

Pathways of steroid biosynthesis

Free cholesterol, formed by hydrolysis of the cholesterol esters in the cytosol, is converted into pregnenolone, which has reduced ring A with a double bond between C5 and 6, a hydroxyl group attached to C3 and a methyl group at C21. A series of enzymic steps are involved which are rate-limiting in corticosteroid synthesis. This process takes place in the mitochondria, and it is at this point in the biosynthetic sequence that ACTH acts primarily. Pregnenolone then leaves the mitochondria and is transferred to the endoplasmic reticulum, where it is hydroxylated at C-17 to form 17-OH pregnenolone. It is then oxidized with the formation of the characteristic Δ^4, 3-ketone grouping in ring A into 17α-hydroxyprogesterone by 3β-hydroxysteroid dehydrogenase and an isomerase. The action of 21-hydroxylase then leads to the formation of 11-deoxy-cortisol. This then returns to the mitochondria for 11β-hydroxylation to form cortisol. In the glomerulosa, where there is no 17α-hydroxylase, 11-desoxycorticosterone is formed in the endoplasmic reticulum, which then returns to the mitochondria for 11β-hydroxylation into corticosterone and thence, by 18-hydroxylation, followed by 18-dehydrogenation, it is converted into aldosterone. In the endoplasmic reticulum some of the 17α-hydroxypregnenolone is converted, via a C20,22-lyase, into dehydro-epiandrosterone, which is then converted by the 3β-hydroxysteroid dehydrogenase/isomerase system into androstenedione. Some of this returns to the mitochondria for 11β-hydroxylation to yield 11β-hydroxyandrostenedione. Approximately half of the dehydroepiandrosterone formed is secreted in the form of its sulphate ester.

Thus, in the fasciculata/reticularis, 17α-hydroxylation is followed by 21-hydroxylation which, in turn, is followed by 11β-hydroxylation, although 17α-hydroxylation may sometimes take place after—rather than before—the action of the 3β-hydroxysteroid dehydrogenase/isomerase system.

The fetal adrenal cortex

A remarkable feature of the development of the fetal adrenal cortex, which as in adults is under the control of ACTH, is the fact that by mid-pregnancy it is ten to twenty times the size of the adult gland. It rapidly regresses during the first year of life, but the weight of the neonatal adrenal is not regained until puberty. Most of the fetal adrenal is composed of a specific fetal zone which is characterized functionally by an inability to utilize cholesterol as its substrate for steroidogenesis and the absence of the 3β-hydroxysteroid dehydrogenase/isomerase system. The placenta, however, converts cholesterol into progesterone, which is the precursor for the lmited quantities of corticosteroids produced. The placenta also synthesizes pregnenolone, and this is the substrate for the formation of the large quantities of dehydroepiandrosterone and its sulphate ester which are a major product of the fetal adrenal cortex. Together with 16α-hydroxy dehydroepiandrosterone, formed in the fetal liver, these steroids are converted to oestrogens by the placenta (p. 668, Chapter 19). Aromatization of 16α-hydroxydehydroepiandrosterone yields oestriol, the main maternal oestrogen in late pregnancy. Maternal oestriol concentrations are used, clinically, as an index of the integrity of the feto-placental unit.

BIOLOGICAL ACTIVITIES OF THE ADRENAL STEROIDS

The adrenocorticosteroids

The 21-carbon atom adrenal steroids are historically classified as having predominant effects on carbohydrate metabolism (the 'glucocorticoids') or sodium and potassium metabolism (the 'mineralocorticoids'). Although this classification is convenient, it is important to recognize that, depending on the concentration of the steroid hormone actually exposed to its target cell, considerable overlap is likely to exist. For example, in experimental animals in a variety of bioassay systems, aldosterone is about 500 times more potent than cortisol in promoting [Na]/[K] exchanges in secretions emanating from extracellular fluid. Yet, when we consider the relative circulating concentrations of these hormones in people and their likely relative affinities for circulating binding proteins, virtually all of the 'glucocorticoid' activity of the circulating blood is subserved by cortisol, whereas only about 80 per cent of the 'mineralocorticoid' activity is subserved by aldosterone; the rest is probably largely subserved by cortisol, but our knowledge may well be incomplete on this critical point.

'*Glucocorticoids*' influence carbohydrate metabolism by promoting the conversion of protein to glucose (gluconeogenesis), by inhibiting the peripheral utilization of glucose (probably because they antagonize insulin) and by increasing glycogen deposition in the liver. Deficiency of glucocorticoids causes a lowered fasting blood sugar which is largely the result of depressed gluconeogenesis.

Apart from the qualifications mentioned above, the term 'glucocorticoid' is additionally deceptive because glucocorticoids have other very important physiological actions which are unrelated to carbohydrate metabolism. For example, the potency ratios for 'glucocorticoid' and 'anti-inflammatory' activities in a wide range of animal models are approximately equal.

The resistance to the 'stress' caused by non-specific harmful influences such as mechanical or thermal injuries, severe infections, or the harmful effects of antibody-antigen reactions is reduced in the absence of adequate glucocorticoid activity. An increased concentration of glucocorticoids in the blood causes a fall of eosinophils and lymphocytes and an increase of neutrophil granulocytes: erythropoiesis is enhanced. The inflammatory response to tissue irritants is suppressed when pharmacological doses of glucocorticoids are given and wound healing may be delayed. Large amounts of glucocorticoids also reduce the rate of antibody formation; infections spread easily, and the survival of tissue transplants is prolonged. Glucocorticoids probably help to maintain a normal rate of glomerular filtration and they promote free water excretion. Mental changes are frequent. The type of psychiatric manifestation depends largely upon the personality of the individual patient, but euphoria is especially common. More specific actions include stimulation of acid and pepsin secretion from the gastric mucosa and, most importantly, suppression of the secretion of corticotrophin from the pituitary and hence adrenal cortical atrophy.

"*Mineralocorticoids*" primarily influence the rate of sodium and potassium transport across cell membranes, promoting entry of sodium into cells, and extrusion of potassium from them. This is most easily seen in the case of the renal tubule because the characteristic changes are quickly reflected in the urine. With increased mineralocortocoid activity the rate of sodium excretion falls and that of potassium rises. Conversely, lack of mineralocorticoids leads to sodium loss and potassium retention. Similar effects can be demonstrated on the ratio of sodium to potassium concentration in saliva and faeces and on the concentration of sodium in the sweat. Probably many other tissue cells (if not all of them) respond to mineralocorticoids in a similar fashion.

Compelling studies with isolated segments of rabbit nephrons, microperfused with various solutions close to the composition of extracellular fluid indicate that, contrary to classical studies which suggested an exchange mechanism between sodium and potassium/hydrogen, mineralocorticoid action on sodium reabsorption is dissociated anatomically from the effect on tubular potassium secretion. None the less, the net physiological effect is as if there was a simple exchange between sodium and potassium/hydrogen in the distal part of the distal nephron and collecting duct, although it is always greatly modified by the quantitatively more dominant events in the proximal, non-steroid sensitive, parts of the nephron.

Sex hormones of the adrenal cortex

Adrenal androgens are responsible for the early development of pubic hair in the male and for the growth of body hair in women but they cannot maintain secondary sexual characteristics of castrated male animals even if the gland is under the influence of prolonged maximal ACTH stimulation. Conversely, although they are unable to prevent the effects of loss of ovarian function, the adrenals may give rise, indirectly, to enough oestrogen to stimulate the growth of 'oestrogen-dependent' breast carcinoma after the main sources of oestrogen has been removed by oophorectomy. The adrenals probably contribute little to normal sex hormone activity, particularly in males, but over-production of androgens by the adrenal is important in pathological conditions.

Cellular basis of steroid action

Tissue receptor proteins for sex steroid hormones are now well recognized. The uterus, vagina and breasts respond specifically to oestradiol and the seminal vesicles, prostate and hair follicles respond specifically to dihydrotestosterone. Otherwise androgens and oestrogens have relatively few effects. In contrast, the glucocorticoids (and probably also the mineralocorticoids) appear to act on almost every cell of the body. "Recep-tors" are found (i.e. proteins which bind the relevant steroid with considerable affinity yet the steroid is displaced readily within the physiopathological range) in muscle, cartilage, kidney, fat, leucocytes, pituitary, liver and many areas of the brain. This is in keeping with the ubiquitous nature of adrenocortical steroid action.

Steroids, in contrast to polypeptides, penetrate the cell membrane, probably because of their relatively high lipid solubility. Combination with receptors in the cytosol promotes transport to the nucleus, where translation of the code for the synthe-sis of the specific enzymes relevant to the appropriate steroid action takes place. This process takes about 30–90 minutes, so that any pathophysiological event which happens within a few minutes of the application of a stimulatory or inhibitory pertur-bation is unlikely to be steroid dependent. For example, an intravenous injection of hydrocortisone in the treatment of status asthmaticus cannot be expected to have much therapeutic effect in less than about an hour.

THE METABOLISM OF ADRENOCORTICOSTEROIDS

Biologically active adrenocortical steroid hormones are relatively insoluble in water, but protein solutions (e.g. albumin) increase their solubility by nonspecific binding. In plasma, corticosteroids are loosely bound to albumin, but cortisol, corticosterone, desoxycorticosterone and other adrenal steroids are also more strongly bound to an α-globulin (the corticosteroid-binding globulin or transcortin). At the temperature and hydrogen-ion concentration of the extracellular fluid about 90 per cent of cortisol is protein-bound. This is considerably weaker than the binding of thyroxine to its specific binding globulin.

Cortisol has a biological half-life of about $1\frac{1}{2}$ hr but aldosterone has one of only about 20 min, which is probably a reflection of the fact that corticosteroid-binding globulin has little or no affinity for aldosterone.

The liver is largely responsible for the degradation of adrenocorticosteroids into biologically inactive metabolites, mainly their tetrahydro derivatives. These are then conjugated with glucuronic acid and sulphate. This makes them easily soluble in water and the conjugates are excreted into the urine. The urinary steroids consist largely of conjugated steroids of adrenal origin, although in adult males about half of the total metabolites of androgen metabolism come from the testes. The rate of urinary cortico-steroid excretion is considerably influenced by alterations of hepatic metabolism, renal function and plasma protein steroid-binding capacity.

There is now good evidence that, in the metabolism of C-19 steroids, their metabolites not only possess special biological activities of their own, but also act as substrates for hormone production in other tissues. For example, dehydroisoandrosterone sulphate is an important substrate for the production of oestrogens by the placenta.

CONTROL OF ADRENOCORTICAL HORMONE PRODUCTION

The rate of cortisol, corticosterone, and desoxycorticosterone (and their 18-hydroxy derivatives) production virtually ceases when the stimulus of corticotrophin (ACTH) is removed. Without the pituitary hormone adrenal production of these steroids falls to vanishingly small levels. In turn the rate of ACTH production by the pituitary is influenced mainly by the level of cortisol in the extracellular fluids so providing a negative feed-back mechanism via the hypothalamus and pituitary to maintain cortisol levels in the peripheral blood. The hypothalamus stimulates ACTH release by means of a neuro-hormone (the corticotrophin-releasing factor), which is carried to the anterior pituitary gland from the median eminence of the hypothalamus via the hypothalamo-hypophysial portal vessels. This hypothalamic system is stimulated by many factors, both nervous and humoral. It is responsible for the increased ACTH secretion which results from stressful stimuli such as fear, pain, or severe injuries, and it also provides the mechanism by which adrenaline causes ACTH release.

In contrast to cortisol, aldosterone production is only slightly affected by ACTH and it continues largely unchanged following removal of the pituitary. It is an old observation that in experimental animals hypophysectomy is followed by atrophy of the two inner zones of the adrenal cortex (the zona fasciculata and the zona reticularis) but not by the outer zona glomerulosa where aldosterone is produced. Clinically the independence of aldosterone from pituitary control is well illustrated by the contrast between patients with panhypopituitarism and those with Addison's disease. Patients without adrenal glands may die within a few days from sodium depletion and potassium intoxication, whereas those without pituitary function linger on for much longer before finally succumbing not to any electrolyte disturbance but to an ill-defined failure to respond to some stressful influence.

The rate of aldosterone secretion depends on the blood volume and on the potassium concentration in the plasma. Aldosterone secretion is increased by reduction of the circulating blood volume, as for example, in sodium depletion and also by an increase of plasma potassium concentration. It is decreased by expansion of the blood

Chapter 16

volume, as, for example, in sodium loading and also by potassium depletion. These changes occur in the absence of the pituitary except that they may be somewhat sluggish. Another system, therefore, must be responsible and there is now good evidence that some humoral agent is involved. The dominant candidate for this role is a small octapeptide, angiotensin II. The first step in production comes from the action of the renal enzyme renin on circulating renin-substrate, an α-2 globulin of hepatic origin (p. 167). Usually changes of renin-substrate concentration are not rate-limiting, so that, unless otherwise specified, changes of plasma renin concentration or plasma renin activity can be equated, approximately, with changes of plasma angiotensin II concentration. Angiotensin, in small amounts, selectively increases the rate of aldosterone production from the adrenal glomerulosa and, in dogs, haemorrhage does not increase the rate of aldosterone production in the absence of the kidneys. A raised plasma renin concentration is seen in many situations associated with hypovolaemia, such as sodium depletion, and a low one is found in situations associated with hypervolaemia, such as sodium loading. The way that changes of blood volume influence renin output is still conjectural. In some situations, changes of mean pressure in the renal artery are likely to be relevant but alterations in the delicate changes of sodium concentration in the region of the macula densa (p. 166) are also important. In human beings undergoing physiological adjustments to the upright posture or to acute hypovolaemia, alterations in the activity of the sympathetic nerves are dominant (p. 175).

Changes of plasma aldosterone concentration generally follow those of plasma renin activity with a delay of about 15 minutes, but, in both sheep and man, there are many important physiological situations, such as bilateral nephrectomy, sodium depletion and hypoxia where the change of aldosterone concentration (or secretion) is considerably more than that which would be predicted from the changes of angiotensin II concentration, even allowing for concomitant changes of plasma potassium concentration. This means either that aldosterone secretion is considerably more sensitive to changes of plasma potassium concentration than had been anticipated or that there is some, undiscovered, circulating agent capable of stimulating aldosterone secretion. Some evidence is given for the latter view by both *in vitro* and *in vivo* studies in animals which indicate that sodium depletion stimulates aldosterone secretion late in the biosynthetic pathway (between corticosterone and aldosterone); other known factors such as potassium, angiotensin II and adrenocorticotrophic hormone appear to act more proximally.

Disordered function

Functional disorders of the adrenal cortex manifest themselves clinically either as the result of too little or too much secretion of the usual physiological major products, viz., cortisol and aldosterone, or alternatively, because of a biosynthetic defect, steroid hormones, viz., the adrenal androgens, are produced in amounts which have important physiological effects.

HYPOFUNCTION OF THE ADRENAL CORTEX

Since the adrenal cortex is neither functionally nor anatomically homogeneous, it is difficult to describe under a single heading the effects of 'hypofunction' of the gland. A pathological process which diffusely and slowly destroys the adrenal cortex without respect for functional zones (e.g. Addison's disease) will understandably produce a different clinical picture from one which causes differential atrophy of the inner two zones (e.g. anterior pituitary failure). Similarly, a failure of synthesis of physiologically important hormones such as cortisol, due to enzymic defects in the biosynthetic pathway (e.g. congenital adrenal hyperplasia) will impair cortisol secretion (and hence produce the signs of cortisol deficiency) but will cause hyperplasia of the gland and the overproduction of other steroids because the normal inhibition of ACTH secretion is absent.

Clinically two important and interdependent syndromes are recognized. One is due, primarily, to an acute failure of cortisol secretion and the other to a more chronic failure of both cortisol and aldosterone secretion. The first can be completely reversed by adequate glucocorticoid therapy but the second requires the replacement of salt and water as well. Once the initial sodium depletion is made good, an adequate body content of sodium can then be maintained by treatment with mineralocorticoids.

Acute failure of cortisol secretion (acute adrenal insufficiency) usually occurs when patients whose adrenals are incapable of secreting more than basal amounts of gluco-corticoids, are exposed to a relatively sudden stress. The adrenal cortex is incapable of increasing the secretion of glucocorticoid in response to a greater demand. This unresponsiveness may be the result of primary disease in the adrenal cortex itself, as in Addison's disease and congenital adrenal hyperplasia, or it may develop as a result of a secondary atrophy due to chronic failure of corticotrophin secretion. This occurs, of course, in panhypopituitarism; but it most commonly develops as a result of prolonged treatment with suppressive doses of some glucocorticoid type of preparation such as cortisone, prednisone or dexamethasone. Sometimes, especially in children, the disease responsible for the 'stress' may also itself damage the adrenal cortex by causing adrenal haemorrhage. This is seen in severe meningococcal septicaemia, in diphtheria and occasionally after severe abdominal injuries.

The clinical picture is one of profound shock. The blood pressure falls and severe oliguria develops; there is urea retention and the plasma potassium concentration rises. The body temperature is often raised initially but becomes subnormal later. Vomiting is common and the patient may become dehydrated, but the external losses of sodium are modest. Nevertheless, the plasma sodium concentration often falls not because of overall sodium depletion but because the distribution of water and sodium between cells and the extracellular fluid is altered. Superficially, therefore, the clinical picture resembles that of severe water and salt loss, but treatment with saline solutions is not enough. All the changes, including the hyponatraemia, can be reversed by giving cortisol, cortisone or one of their synthetic analogues.

However, when acute adrenal insufficiency occurs in patients already the victims of prolonged mineralocorticoid deficiency, as in Addison's disease or when vomiting has been severe, then correction of the circulatory and electrolyte disturbance requires sodium repletion, often urgently.

Chronic insufficiency of both glucocorticoid and mineralocorticoid secretion occurs in Addison's disease. This condition may be the result of a diffusely destructive pathological process affecting the whole adrenal gland, such as tuberculosis, haemochromatosis or carcinomatosis. Usually, however, it is the result of an autoimmune process which selectively destroys the cortex, leaving the medulla intact. Appropriate antibodies can be detected in the circulation and some patients may also suffer from Hashimoto's thyroiditis. Clinically the condition is characterized by progressive weakness, loss of weight, thirst and polyuria, and pigmentation of the skin and mucosae. The pigmentation tends to accumulate in areas such as flexor surfaces (e.g. palmar creases) and the buccal mucosa. Vitiligo is common and suggests an autoimmune cause.

Lack of aldosterone leads to characteristic changes in the plasma concentrations of sodium, potassium and urea. The kidneys fail to conserve sodium adequately and tubular secretion of potassium is impaired so that the plasma level of sodium tends to fall and that of potassium to rise. In mild cases the urinary sodium loss is at least partially counterbalanced by increased salt consumption, because many patients with Addison's disease have an excessive taste for it. Obvious hyponatraemia is, therefore, a late sign but an elevation of the plasma potassium level occurs relatively early. When sodium depletion does supervene it causes a reduction of the plasma volume with haemoconcentration, a fall in the systemic blood pressure, and urea retention as a result of impaired renal blood flow. However, lack of glucocorticoid activity is primarily responsible for most of the other manifestations of chronic adrenal insufficiency. Although progressive weakness may be partially due to sodium loss, only treatment with glucocortoids will relieve it completely. The impairment of renal function and particularly the impairment of free water excretion as a result of glucocorticoid deficiency leads to delayed water excretion, nocturia and urea retention. Since gluconeogenesis from protein is impaired, glucocorticoid deficiency tends to cause fasting hypoglycaemia and the hypoglycaemic effects of insulin are potentiated. Patients with acute adrenal insufficiency should always be given glucose.

The fall of plasma cortisol concentration abolishes the normal suppressive effect of cortisol on the release of pituitary corticotrophin. Corticotrophin secretion becomes excessive and can be detected easily in peripheral blood. In the absence of the usual target organ one would not expect this to have any particular biological effect, but corticotrophin, which is a polypeptide containing 39 amino acids, shares a sequence of 7 amino acids contained within β-lipotropin which accounts for most of the melanocyte-stimulating activity produced by the anterior pituitary. Even the best corticotrophin preparations have slight melanocyte-stimulating activity, and suppression of the pituitary by treatment with glucocorticoids will slowly clear Addisonian pigmentation. It is, therefore, certain that an excessive secretion of β-lipotropin is a secondary result of cortisol deficiency and that this causes the excessive pigmentation seen in chronic adrenal insufficiency.

ADRENAL INSUFFICIENCY IN PITUITARY FAILURE

As mentioned above, corticotrophin lack causes failure of cortisol secretion from the zona fasciculata and the zona reticularis, both of which eventually atrophy. Mineralocorticoid function, on the other hand, is preserved because aldosterone is produced by the zona glomerulosa which remains relatively unaffected.

Therefore, patients with adrenal insufficiency secondary to a pituitary lesion differ clinically from those with a primary lesion of the adrenal gland itself by having normal electrolyte balance, reduced skin pigmentation and often evidence of hypofunction of the other endocrine glands controlled by pituitary trophic hormones, e.g. the gonads and the thyroid. Otherwise they suffer from the effects of cortisol lack as described above, particularly an inability to respond adequately to infections and other stressful stimuli.

Nowadays the cause of such pituitary failure is likely to be due to a tumour, a granuloma or ineffective follow-up of patients who have had pituitary irradiation or a trans-frontal or a trans-sphenoidal hypophysectomy. Postpartum necrosis (Sheehan's syndrome) is now very rare in the Western world, although 20 years ago it was the commonest cause.

CONGENITAL ADRENAL HYPERPLASIA

Congenital adrenal hyperplasia is caused by an inborn deficiency of one or more of the enzymes responsible for adequate hydroxylation of the steroid nucleus; the production of cortisol is usually defective. In the most common form hydroxylation at the C-21 position is impaired so that 17-hydroxyprogesterone accumulates and its urinary metabolite, pregnanetriol, appears in excessive amount in the urine. The low level of circulating cortisol stimulates ACTH release from the pituitary which then, in turn, causes hyperplasia of the adrenal cortex and over-production of other steroids, particularly androgens. The clinical result of this is virilism, excessive growth and sometimes excessive pigmentation. Boys show precocious sexual development and the girls are masculinized. If the defect is severe, genetic female children are born with male-looking external genitalia (pseudo-hermaphroditism) and they are often mistakenly reared as boys. Clinically these children suffer from a relative lack of cortisol. In boys the endocrine disorder may be first suspected only because of a failure to respond normally to the stress of a minor infection or injury.

About one-third of the cases also show evidence of excessive sodium loss. Vomiting is common and sodium depletion occurs rapidly unless treated. In boys the condition is often unsuspected until the correct diagnosis is made by finding hyperkalaemia.

A number of different mechanisms have been suggested to explain this salt-wasting syndrome. Cortisol deficiency is rather more pronounced than in children without salt-wasting but lack of the weak mineralocorticoid action of cortisol is insufficient to explain it. It is now clear that these children have a low basal rate of aldosterone secretion which does not increase normally in response to sodium deprivation. Since the administration of ACTH often produces a large increase in the rate of renal sodium loss whereas normally it produces a decrease, some, unidentified sodium-losing steroid hormone, was hitherto thought to be responsible. Hypoaldosteronism may well be an adequate explanation for the general tendency to sodium loss but the increased rate of sodium excretion following ACTH requires a further explanation. The biological effect of aldosterone is probably directly antagonized. We know that a number of natural steroids, such as progesterone and 17-OH progesterone, will compete with aldosterone (and also other minerelo-corticoids) for receptor sites on the renal tubule, so causing sodium loss and potassium retention. Synthetic steroids with a spironolactone grouping attached to C-17 have been prepared which also have this effect, and they are useful adjuncts to diuretic therapy because they do not cause potassium loss.

More rarely, children with congenital adrenal hyperplasia develop severe arterial hypertension which may be associated with oversecretion of desoxycorticosterone, a mineralocorticoid which regularly produces hypertension experimentally. In these cases there is a predominant defect of 11β-hydroxylation as when 11β-hydroxylase is inhibited artificially with metyrapone. When compounds with this type of action are administered to normal people the adrenal production of 11-deoxygenated steroids is increased, especially 17α-hydroxy-11-deoxycorticosterone (Reichstein's compound S), and 11-deoxycorticosterone. Rarely, a defect of 17-hydroxylase may lead to hypogonadism and hypertension.

Administration of some form of glucocorticoid will reduce the adrenal androgen output to normal if given in a way which produces a sustained suppression of the pituitary and yet does not give rise to signs of overdosage. All the manifestations of congenital adrenal hyperplasia *can* be controlled if adequate treatment is started early enough; girls can undergo a normal puberty and even become pregnant. But this is difficult and continuous monitoring of the appropriate hormone levels is essential.

HYPERFUNCTION OF THE ADRENAL CORTEX

As with hypofunction, the effects of adrenocortical overactivity depend upon which steroids are primarily affected.

CUSHING'S SYNDROME

Over-production of cortisol leads to Cushing's syndrome if sustained for long enough.

Clinically, most of the effects can be interpreted in terms of the known actions of cortisol. Carbohydrate metabolism is impaired largely because there is excessive glucose formation from protein. Diabetes mellitus is common and even those patients without overt glucosuria usually show a diabetic type of glucose tolerance curve. Whether or not there is a demonstrable abnormality of glucose handling is largely a function of the reserve of insulin within the pancreatic β-cells; but, characteristically, the diabetes when it occurs is of the non-ketotic adult type. The excessive protein breakdown is responsible for the atrophied skin which bruises easily and readily cracks at stress points to form striae. It is also responsible for the wasted, weak muscles and contributes both to the classical picture of central obesity with relatively thin limbs and to the development of osteoporosis. Excess cortisol leads to arterial hypertension by some unknown mechanism and also causes erythrocytosis by what appears to be a direct stimulating effect on the bone marrow. The cause of the striking central obesity of Cushing's syndrome is obscure. Women develop oligomenorrhea or amenorrhea and become hirsute as a result of excess androgen production. An abnormally high output of the androgen, dehydroepiandrosterone, is particularly characteristic of adrenal carcinomata.

Since cortisol does have some mineralocorticoid activity it might be expected that manifestations of this would appear in Cushing's syndrome even though aldosterone production is not elevated. In mild cases, particularly those due to pituitary-dependent adrenal hyperplasia (Cushing's disease) no abnormality of electrolyte balance is usually demonstrable, but with the ectopic ACTH both syndrome hypokalaemia and a raised plasma bicarbonate concentration are common. These are useful differential diagnostic points.

Cushing's syndrome may be due to a primary tumour of the adrenal or to bilateral hyperplasia of the adrenal cortex. Hyperplasia may result from pituitary ACTH hypersecretion or from ACTH secreted ectopically by malignant tumours of other tissues, e.g. oat-cell carcinoma of the bronchus. In children an adrenal carcinoma is usually responsible, whilst in adults hyperplasia is more common. In patients with Cushing's disease, the defect lies in the pituitary. Pituitary adenomata—either small tumours which do not enlarge the pituitary fosse or rarely larger ones which do—are found in about 75 per cent of these patients at trans-sphenoidal surgery. They are particularly likely to show themselves when the restraint of high circulating cortisol levels on pituitary ACTH production has been removed by total adrenalectomy (Nelson's syndrome). In this syndrome circulating plasma ACTH concentrations are high when considered in relation to the prevailing cortisol concentration and relative autonomy of ACTH production is indicated by the fact that the normal circadian rhythm of plasma ACTH concentration is attenuated or abolished. Whether the primary defect in Cushing's disease lies in the pituitary itself (e.g. the development of a hyperactive clone of ACTH-producing cells) or in the hypothalamus (e.g. a primary overproduction of corticotrophin-releasing factor) is not yet clear, but the current evidence is in favour of a primary pituitary over-production of ACTH. Thus many patients can be cured by selective removal of an adenoma usually performed via the trans-sphenoidal route.

HYPERALDOSTERONISM

Overproduction of aldosterone, without alteration in the rate of production of other steroid hormones, occurs either because there is a primary abnormality in the adrenal glomerulosa or because there is an extra-adrenal disease process which causes a sustained increase in the concentration of circulating angiotensin II, the dominant known factor which regulates the rate of aldosterone secretion.

In hyperaldosteronism of adrenal origin the plasma renin and circulating angiotensin II concentration is low, and the adrenal glomerulosa either contains an adenoma or is the site of bilateral hyperplasia. The clinical syndrome so produced is characterized by systemic hypertension (which is only rarely malignant) and potassium depletion with alkalosis. Muscle weakness, cramps and carpopedal spasm are frequent in patients with an adenoma but are unusual in patients with hyperplasia. In the absence of cardiac failure, oedema is not seen. This may seem paradoxical because, being a mineralocorticoid hormone, aldosterone promotes sodium retention. However, if the administration of large amounts of aldosterone (or any mineralocorticoid) is continued for more than approximately 5–21 days (depending on the initial body sodium content) the renal sodium-retaining effect disappears and the rate of renal sodium excretion rises to reach equilibrium with sodium intake. This 'escape' phenomenon applies to the rate of renal sodium excretion only; an elevated rate of distal tubular potassium secretion and the effect of aldosterone on other organs such as the colon and salivary glands, continues unchanged. The mechanism is probably related to altered Starling forces in the peritubular capillaries of the proximal nephron (p. 161) as a consequence of the expansion of plasma volume induced by aldosterone. Thus, oedema is not seen although the total body sodium, and hence the plasma volume, is in fact, increased slightly.

In hyperaldosteronism of extra-adrenal origin associated with an elevated plasma renin concentration and hence angiotensin II concentration, two separate clinical syndromes are seen. If the elevated plasma renin concentration is secondary to a reduced effective plasma volume because the plasma colloid osmotic pressure is reduced, then oedema is prominent, arterial pressure is normal and potassium depletion modest in the absence of diuretic therapy. This occurs in the hyperaldosteronism of hypoproteinaemic states such as cirrhosis of the liver with ascites and the nephrotic syndrome; the hyperaldosteronism is a consequence of the process leading to oedema formation and not its cause. If, on the other hand, the elevated plasma renin concentration is due to a primary vascular disorder which produces renal ischaemia, such as accelerated arterial hypertension or renal artery stenosis, then the clinical syndrome of hypertension and potassium depletion without oedema develops. However, in contrast to hyperaldosteronism due to a primary adrenal lesion, the hypertension is more severe and the plasma sodium concentration tends to be low rather than high.

Thus, there are two main varieties of *secondary* hyperaldosteronism, one is seen when oedema is due to hypoproteinaemia and the other when renal ischaemia mimics the hypertension and potassium depletion of a primary adrenal lesion, i.e. the syndrome of so called *primary* hyperaldosteronism.

Hyperaldosteronism without hypertension can very rarely be due to a primary resistance to the vascular effects of angiotensin (Bartter's syndrome).

The ion-exchange mechanism in the distal tubules of the kidney probably explains why the syndrome of primary aldosteronism is characterized by hypokalaemia whilst that of secondary aldosteronism is usually not. The increased proximal tubular sodium reabsorption seen in oedematous states will reduce the amount of sodium delivered to the distal tubule and so reduce the potassium loss which would otherwise occur in exchange for sodium. The fact that the hypokalaemia of primary aldosteronism can be corrected by feeding a low sodium diet probably reflects this process.

Ten to forty per cent of patients with essential hypertension show a low plasma renin concentration and normal rate of aldosterone secretion. A primary adrenal overproduction of some other mineralocorticoid has been postulated. These patients, however, have an increased sensitivity of the zone glomeratosa to angiotensin II and this is probably the correct explanation for their hyporeninaemia.

IDIOPATHIC HIRSUTISM IN WOMEN

This is a psychologically distressing condition which, since it is often associated with irregular or infrequent menstruation and reduced fertility, suggests that some endocrine abnormality is present rather than an abnormal sensitivity of the hair follicles to normal amounts of circulating androgen. The plasma testosterone concentration or the output of testosterone in the urine is increased in about half the patients. Some patients show an enhanced response to exogenous ACTH and it is probably those who show some response to adrenal suppressive therapy. Others respond to ovarian suppression, e.g. when contraceptive steroids are administered (p. 692). The fundamental defect is therefore, either an enhanced sensitivity of the hair-follicles or overproduction of androgens, particularly testosterone, from the adrenal glands or from the ovaries.

Principles of tests and measurements

In principle procedures for the assessment of adrenal cortical function, and indeed for the function of any endocrine gland, fall into three groups: (i) there are those which are designed to measure the actual rate of hormone secretion, (ii) there are others designed to assess the responsiveness to relevant physiological stimuli and (iii) there are those which depend upon the actual physiological effect of the hormone, whatever the intermediate mechanisms.

In the case of the adrenal cortex, assessment of the 'basal' rate of hormone secretion can now be carried out by the method of double-isotope dilution-derivative analysis, at least for cortisol, aldosterone and testosterone. But the methods are time-consuming and, although specific, depend upon assumptions about the pathways of metabolism which may not always be well validated. The use of methods depending on some index of hormone response, such as the response to a water load, the change of the sodium-dependent transrectal potential difference or the Na/K ration in saliva or urine are in principle excellent, but the variation in response in normal people and their cumbersome nature preclude their use today.

The pattern of plasma electrolyte and urea concentrations can but provide a useful guide. For example, in primary adrenal insufficiency (such as Addison's disease) the plasma concentrations of urea and potassium rise and the plasma sodium concentration tends to fall. The potassium and urea changes do not occur in adrenal insufficiency secondary to anterior hypopituitarism because aldosterone production is barely affected; the plasma sodium concentration may be, however, even lower than in primary adrenal insufficiency because of the impairment of water excretory capacity so characteristic of anterior pituitary failure. By contrast, primary hyperaldosteronism is associated with hypokalaemia, a raised plasma bicarbonate and a high normal plasma sodium concentration. The similar clinical syndrome of arterial hypertension and hypokalaemia, due to diuretic treatment of a hypertension patient or renovascular disease, is associated with a low normal plasma sodium concentration. In patients with Cushing's syndrome the presence of a hypokalaemic alkalosis suggests either an adrenal carcinoma or a carcinoma elsewhere (such as an oat-cell carcinoma of the bronchus) which produces ACTH in large amounts ectopically.

For definitive diagnosis we are therefore left with chemical techniques which, when combined with recognition of the relevant physiological context, provide a powerful diagnostic tool. In the past chemical analyses of urine steroids, such as 17-oxogenic steroids (largely products of cortisol metabolism) and 17-oxosteroids (largely products of androgen metabolism) have been used, but increasingly, immunologically specific measurements of particular adrenal steroids in plasma and urine by radioimmunoassay have found favour because they are so convenient.

The interpretation of plasma levels of adrenocortical steroid hormones is dominated by three main principles: (i) hormone secretion is usually pulsatile and thus any one plasma level may well be misrepresentative; (ii) the plasma level only represents the balance between the rate of hormone secretion and the rate of its degradation, so that alterations of, for example, hepatic and renal function may well affect plasma levels and (iii) the total plasma level itself is usually influenced by the concentration of a relevant "binding-protein". It is assumed, but not fully proven, that it is the concentration of unbound or "free" hormone to which the target cells respond. Free, unbound steroid concentrations can be measured by appropriate techniques, involving dialysis or ultra-filtration. The recent recognition of the value of salivary levels promises to greatly simplify the assessment of plasma free steroid hormone levels.

The interpretation of the rate of renal excretion of free steroids depends mainly on (i) the vagaries of renal function and (ii) the extent to which the plasma binding protein is saturable within the physiological range. With cortisol, for example, as the total plasma level rises, an increasing proportion is free, unbound hormone. This is then filtered at the glomerulus and so, after a modest amount of tubular reabsorption, appears in the urine. Hence urine free cortisol can be regarded, in the presence of normal renal function, as a reasonable index of the circulating free cortisol concentration.

DEFINITIVE DIAGNOSTIC TESTS OF ADRENAL FUNCTION

These now increasingly depend on the recognition of the major physiological factor which controls the adrenocortical hormone concerned, which, when both are measured in the same plasma sample, usually gives a good indication of the correct diagnosis, e.g. simultaneous plasma ACTH and cortisol levels in the diagnosis of Cushing's syndrome or plasma renin activity and plasma aldosterone levels in the diagnosis of Conn's syndrome (primary hyperaldosteronism) or Addison's disease.

Hyperfunction

The study of circadian rhythms and the use of suppression and stimulation tests provide indices of the autonomy· of adrenocortical function, which remains the main means by which a tumour is distinguished from hyperplasia.

In the diagnosis of possible adrenocortical disease, the problem is first whether the clinical suspicion is warranted, secondly whether an adrenal tumour rather than hyper-plasia is present and thirdly, if so, on which side it is.

In people suspected of having an over-production of *cortisol,* the plasma cortisol concentration and, if possible, ACTH concentration at 8 a.m. after an overnight sleep is contrasted with the levels seen at midnight. A fall of about 4-fold in both measure-ments is expected in normal people, that of cortisol being less than 200 nmol/l (7 μg/100 ml) at midnight. A plasma sample taken at about 5 p.m. would be expected to show at least a 2- to 3-fold drop of plasma cortisol concentration in normal people. The use of salivary measurements is likely to increase these differences. The 24 hour urine free cortisol output is a most discriminating test. The administration of 2 mg dexamethazone the previous evening will, in normal people, obliterate the usual rise of ACTH concentration in the early morning and lead to a vanishingly low plasma cortisol concentration at that time. If this does not happen, then more formal testing is required—first a plasma and urine free cortisol response to a moderate suppressive influence (2 mg dexamethazone daily for 3 days) and then the response to a massive suppressive influence (8 mg dexamethazone daily for 3 days). The distinction between a pituitary lesion and an adrenal lesion can usually be made in this way (in pituitary dependent Cushing's disease plasma cortisol levels usually suppress in contrast to the lack of cortisol suppression found with adrenal adenomas or carcinomas), but the possibility of an ectopic source of ACTH production must always be borne in mind.

The distinction between ACTH-dependent and non-ACTH dependent causes of Cushing's syndrome is best made by the measurement of ACTH. In Cushing's disease plasma ACTH levels are often within the normal 8 a.m. range but are clearly elevated at midnight. Of interest is the finding that ACTH levels in normal subjects are already falling at 9 a.m. and that there is little or no overlap with the normal range at this time. The time when the blood sample is taken should thus be accurately recorded.

In patients with primary adrenal disease such as an adrenal adenoma causing Cushing's syndrome ACTH levels are low or undetectable. In the ectopic ACTH syndrome ACTH levels range from those found in Cushing's disease to very much higher levels. Lesions such as bronchial carcinoid tumours often produce low levels and may thus mimic Cushing's disease. Oat cell carcinomas more frequently secrete high levels. Tumours producing ACTH ectopically usually secrete a high molecular weight form of ACTH in addition to $^{1-39}$ACTH.

Another common test involves the use of the 11β-hydroxylase inhibitor, metyrapone, which therefore reduces plasma cortisol levels. ACTH levels increase and normal adrenals respond by producing those steroids, particularly 11-deoxycortisol (Reichstein's compound S) which do not require 11β-hydroxylation. In bilateral hyperplasia the response is excessive, whereas, if there is an adrenal tumour or an ectopic ACTH producing tumour, the response is minimal.

With patients suspected of having primary over-production of *aldosterone* the first step after measuring plasma aldosterone is to ascertain the level of plasma renin activity. If this is high then a primary adrenal abnormality can be ruled out. If the plasma aldosterone level is high and plasma renin activity is low, a diagnosis of primary hyperaldosteronism becomes very likely and the problem becomes one of deciding whether it is due to an adenoma or to bilateral adrenal zona glomerulosa hyperplasia. It has been shown that, on a high sodium intake, if the plasma aldosterone concentration rises when the patient is up and about during the morning, then glomerulosa hyperplasia is almost certain, whereas if the plasma aldosterone concentration falls despite the normal activation of the renin-aldosterone induced by upright ambulation, then an adenoma is very likely to be present. The principle behind this is that a glomerulosa adenoma is relatively insensitive to its usual major physiological stimuli (viz. plasma renin and plasma potassium, both of which are low) so that changes of the plasma concentration of a normally rather minor stimulus, viz., ACTH (whose plasma concentration falls sharply during the morning, irrespective of posture) becomes dominant.

In the future it is likely that anatomical tests will outweigh physiological tests in the localization of an adrenal tumour, particularly a Conn's adenoma. But the adrenals lie in a difficult area deep above the kidneys in close proximity to ribs; artefacts are common with any imaging technique and non-functioning adrenal adenomata are common. Even those which take into account functional criteria such as radionuclide scanning following an injection of radiolabelled cholesterol, the substrate for steroid biosynthesis, are liable to misinterpretation. Adrenal venous catheterization is a relatively inexpensive, if mildly invasive, method of lateralization. With this technique recognition is essential (i) that the anatomy of adrenal venous drainage on the right is often anomalous and (ii) that an index of dilution by non-adrenal effluent, such as the plasma cortisol concentration, is required on both sides. Catheterization data can nonetheless be extremely valuable.

The principles of tests concerned with the diagnosis of adrenal androgen overproduction are extensively considered in Chapter 19.

Hypofunction

Addison's disease is suspected clinically, tentatively confirmed by the pattern of plasma electrolyte concentrations (high potassium and urea concentration and a rather low sodium concentration) and definitively established by showing that plasma cortisol does not change, or changes only minimally, in response to maximal doses of ACTH. The latter should be performed whilst the patient is on physiological steroid replacement therapy.

A high ACTH and plasma renin level in the face of low plasma cortisol and aldosterone levels will clinch the diagnosis. The finding of circulating adrenal antibodies indicates that the cause is autoimmune adrenal disease while adrenal calcification suggests tuberculosis.

Practical assessment

ADDISON'S DISEASE – PRIMARY ADRENAL INSUFFICIENCY

Clinical observations

Lethargy, weight loss, anorexia, vomiting, increased skin pigmentation, appearance of buccal pigmentation in white-skinned people, symptoms of hypotension (particularly on standing) and poor response to stress. Family history of autoimmune disease; presence of vitiligo. History of tuberculosis or contact with it.

Clinical investigations

To confirm the diagnosis: plasma K^+ and blood urea concentrations (both raised) and plasma sodium concentration (low normal). Plasma ACTH (considerably raised) and plasma cortisol levels (low normal). Plasma cortisol after ACTH stimulation (minimal response—less than an increment of 200 nmol/litre or 7 μgm/100 ml): this test can be performed during steroid hormone replacement therapy providing the glucocorticoid being given is not measured by the cortisol assay.

To ascertain cause: straight X-ray of upper abdomen or ultrasound for evidence of tuberculous adrenal calcification. Determination of circulating adrenal antibodies.

HYPOPITUITARISM – SECONDARY ADRENAL INSUFFICIENCY

Clinical observations

Lethargy, skin pallor, poor response to stress, amenorrhoea, impotence, sensitivity to cold, scanty pubic and body hair. Symptoms and signs of an intracranial tumour or a history of poor lactation following a haemorrhagic childbirth. History of corticosteroid therapy.

Clinical investigations

To confirm diagnosis: plasma ACTH (low) and cortisol (low) with evidence of hypothyroidism (low plasms thyroxine with low TSH) and gonadal insufficiency (low plasma testosterone or oestrogen with low LH and FSH). If the basal plasma cortisol is normal then dynamic function tests such as the insulin tolerance test will usually be needed.

To ascertain cause: skull X-rays; if there is a visual field defect then computer-assisted tomographic (CT) scanning or air-encephalography or metrizamide cisternography; plotting of visual fields (see p. 560).

CUSHING'S SYNDROME – GLUCOCORTICOID EXCESS

Clinical observations

Central obesity, weak muscles, round, red face (plethoric, "moon" face), atrophic skin with bruises and purple striae (usually abdomen, upper thighs and arms), alterations of mood, modest hypertension, perhaps glycosuria. In women, excessive body and facial hair, and disordered menstruation.

Clinical investigations

To confirm the diagnosis: circadian variation of plasma cortisol. Effect of 2 mg dexamethazone the evening before plasma sampling at 8 a.m. for cortisol or 2 mg daily for 3 days, and measuring renal excretion of urinary free cortisol (as an index of the plasma free cortisol integrated over 24 h).

To ascertain the cause: measurement of plasma ACTH at 8 a.m. and midnight. Hypokalaemia suggests a tumour, ectopic or adrenal, dexamethazone suppression test (8 mg per day for 3 days), metapyrone test to distinguish between tumour and hyperplasia. Recent imaging techniques (ultrasound, computer-assisted tomography, radio-iodinated 19-nor, 6α-methyl cholesterol scanning) together with selective adrenal venous catheterization obviate the need for the earlier techniques of presacral gas insufflation and angiography. Measurement of plasma dehydroepiandrosterone or of the urinary excretion products of androgen metabolism (the 17-oxosteroids) will help to diagnose an adrenal carcinoma.

To assess the effects of glucocorticoid excess: glucose tolerance test for diabetes, radiology of the spine for osteoporosis.

CONN'S SYNDROME – MINERALCORTICOID EXCESS

Clinical observations

Hypertension with unprovoked hypokalaemia, muscle weakness, tetany, polyuria.

Clinical investigations

To confirm the diagnosis: plasma renin activity in both the recumbent and erect posture (low), plasma aldosterone concentration recumbent at 8 a.m. (high).

To ascertain the cause: plasma aldosterone concentration at 12 noon on a high salt diet after a normal morning's ambulant activity—a fall in plasma aldosterone concentration indicates the presence of bilateral adrenal glomerulosa adenoma (and hence the need for surgery); a rise in plasma aldosterone concentration indicates the presence of bilateral adrenal glomerulosa hyperplasia (and hence the need for medical treatment).

To ascertain the site (and to confirm the diagnosis of adenoma or hyperplasia) collection of adrenal venous blood under radiological control and the estimation of the ratio of aldosterone to cortisol concentration both in adrenal venous effluent and peripheral blood. The estimation of cortisol is necessary because it is inevitable that so-called adrenal venous blood will be to some extent mixed with blood which does not come from the adrenal glands. High resolution computer assisted tomographic scanning will probably obviate the need for adrenal venous sampling in the future.

ADRENAL VIRILISM

This may be congenital due to an enzyme defect or it may be acquired as a result of the development of an adrenal adenoma or carcinoma.

Clinical observations

Precocious secondary sexual development in boys but the testes remain infantile, pseudohermaphroditism in girls or early virilization (masculine body build and hair distribution, thin head hair, greasy skin with acne, deep voice, enlarged clitoris). In infants, associated in some with a tendancy to sodium depletion (usually 21-hydroxylase deficiency) and in others with hypertension (11β-hydroxylase or 17α-hydroxylase deficiency).

Clinical investigations (see Chapter 18)

Determination of chromosomal sex by examination of buccal epithelial cells or circulating leucocytes. Plasma concentrations of ACTH, dehydroepiandrosterone, Δ^4-androstenedione, and 17α-hydroxy progesterone. Imaging techniques for the detection of an adrenal tumour as described above.

Table 16.1. Useful normal values for adrenal steroid measurement

Plasma cortisol	8–9 a.m.	100–550 nmol/l
	Midnight	less than 200 nmol/l
Plasma ACTH	8 a.m.	less than 80 ng/l
Urine free cortisol	—	less than 400 nmol/24 h
Urine 17-oxogenic steroids	—	15–60 μmol/24 h
Plasma aldosterone	8–9 a.m. (recumbent)	100–400 pmol/l
Plasma renin activity	8–9 a.m. (recumbent)	1–3 pmol h^{-1} ml^{-1}
	9–10 a.m. (ambulant)	2–4 pmol h^{-1} ml^{-1}
Urine aldosterone (as the 18-glucuronide)	—	10–50 nmol/24 h
Plasma testosterone	Men	10–40 nmol/l
	Women	1–3 nmol/l
	Prepubertal girls	less than 2 nmol/l
Plasma 17α-OH progesterone	Children (prepuberty)	less than 15 nmol/l
Plasma Δ^4-androstenedione	Children (prepuberty)	less than 3·5 nmol/l
	Adults	3–8 nmol/l
Urine 17-oxosteroids	Men	35–70 μmol/24 h
	Women	15–50 μmol/24 h

References

JAMES, V.H.T. (ed.) (1979) *The Adrenal Gland:* L. Martini (series ed.) *Comprehensive Endocrinology.* New York, Raven Press.

NELSON, D.H. (ed.) (1980) *The Adrenal Cortex: Physiological Function and Disease,* Vol. XVIII Major Problems in Internal Medicine, L.H. Smith (series ed.). Philadelphia, W.B. Saunders.

THE ADRENAL MEDULLA

Normal function

The adrenal medulla is a specialized part of the sympathetic nervous system. It represents the postaganglionic neurones and, like them, releases sympathetic amines in response to stimulation by preganglionic sympathetic neurones. The adrenal medulla differs from postganglionic sympathetic neurones in two ways; firstly adrenaline as well as noradrenaline is liberated in adults but not in children, and secondly, the amines are discharged into the blood stream (thus making the adrenal medulla an endocrine or ductless gland). They act on their effector organs via the circulation rather than by neuronal transmission.

In contrast to other endocrine glands, abolition of adrenal medullary function does not apparently interfere with the body's metabolic processes and medullary secretion only seems to have physiological significance in response to specific stimulation by the hypothalamus. The gland, like other postganglionic neurones of the sympathetic nervous system, provides the basis for many reactions to a sudden stressful stimulus. This underlines the fact that, in contrast to other hormones, those of the adrenal medulla act very swiftly. Traditionally this response has been epitomized as the response to 'flight or fright'. An increased rate of discharge of noradrenaline from sympathetic nerve endings is essential for this and it is doubtful whether the adrenaline released from the adrenal medulla is necessary. However, despite the low concentration of adrenaline in the circulation, adrenaline released in response to stress may be useful because it will increase blood flow through striated muscle. But this can only be a small effect because blood flow through striated muscle is largely regulated by a change of vasoconstrictor tone mediated by noradrenaline.

Biological activities of noradrenaline and adrenaline

Adrenaline given by injection constricts the arterioles in the skin and splanchnic circulation but it dilates the arterioles in striated muscle; the net effect is usually to reduce the overall peripheral resistance. It increases the force of myocardial contraction and accelerates the heart; in larger quantities, it increases myocardial excitability. Because of this cardiac action, cardiac output rises provided venous return is maintained. The systolic blood pressure rises but the effect on the diastolic blood pressure is dose-dependent; small quantities lower it but large quantities elevate it. The pulse pressure invariably rises. It also relaxes the smooth muscle of the uterus, the bronchioles, the intestine and the bladder. Sweating is not seen following a single injection of adrenaline although sweating is a characteristic feature of phaeochromocytoma. However, the sweat glands and hair follicles are stimulated by adrenaline injected directly into the skin.

Adrenaline also has metabolic effects. It produces a rise of the blood sugar concentration by stimulating the breakdown of liver glycogen to glucose and it increases the metabolic rate.

In contrast, infusions of *noradrenaline* cause generalized vasoconstriction so that both systolic and diastolic blood pressures rise. This causes a reflex bradycardia which can be blocked by atropine. However, when noradrenaline is released in the heart as a neuromuscular transmitter, it has an effect similar to that given by adrenaline. It increases the rate and force of myocardial contraction and, in large quantities, it increases mycardial excitability. Cardiac output tends to rise but because of the reflex bradycardia, the effect is much less than that given by adrenaline. In a healthy heart, cardiac output is maintained by an increase of stroke volume despite the bradycardia but, in patients with latent cardiac decompensation, the bradycardia will cause a reduction of cardiac output.

Noradrenaline does not relax bronchiolar muscle and it is approximately eight times weaker than adrenaline in causing hepatic glycogen breakdown. It is, however, more potent than adrenaline in causing the release of free fatty acids from adipose tissue.

Biosynthesis and metabolism of the adrenal medullary hormones

Noradrenaline and adrenaline are synthesized from tyrosine by a series of enzymic steps. The benzene ring is hydroxylated to form 3,4-dihydroxyphenylalanine (dopa), a process which is controlled by a specific enzyme, tyrosine hydroxylase. The activity of this enzyme is the main factor governing the rate of noradrenaline synthesis. It is interesting that the enzyme is inhibited by noradrenaline so that, by end-product inhibition, the rate of noradrenaline synthesis appears to be self-limiting. Dopa is decarboxylated by a dopa decarboxylase to form dopamine; the rate of this reaction can be reduced by alpha-methyl dopa, a hypotensive drug, which is treated as an alternative substrate. A hydroxyl group is then added to the side-chain to form noradrenaline, which is methylated by N-methyl transferase to form adrenaline. This step is governed by phenylethanolamine N-methyltransferase whose activity is critically dependent on the presence of glucocorticoid hormones secreted by the adrenal cortex.

In contrast to other endocrine glands the secretion of noradrenaline and adrenaline into the blood stream seems to occur only in response to neuronal stimulation; there is little secretion under basal conditions. The hypothalamus discharges impulses down the preganglionic neurones to the adrenal medulla in response to severe injury, emotion, hypoglycaemia and severe cold. The immediate stimulus to the medullary cells is, of course, the chemical transmitter substance of the sympathetic ganglia, acetylcholine; and the synapses between the preganglionic nerve endings and the medullary cells, like those in the sympathetic ganglia, respond to ganglion-blocking and ganglion-stimulating agents in the same way. For example, parasympathetic blocking agents prevent catecholamine release whereas parasympatheticomimetic drugs stimulate it.

The evidence suggests that the pattern of catecholamine release differs according to the stimulus. For example, fear appears to favour adrenaline release and anger noradrenaline release. This differential effect is made possible by the fact that there are different cell populations within the adrenal medulla, some contain adrenaline storage granules and others contain noradrenaline storage granules. Recent studies have shown that enkephalin and large molecular weight enkephalin precursors are found in the adrenal medulla and are stored in the same cells as the catecholamines. Large amounts of metenkephalin have been found in phaeochromocytomas.

Within a very few minutes of stopping an infusion of noradrenaline its cardiovascular effects have worn off although the effects of adrenaline persist for longer. Noradrenaline is taken up by platelets and then rapidly enters postganglionic nerve terminals as if it had been released locally. Adrenaline will also enter nerve terminals in higher concentrations. In addition, a second uptake mechanism is available in non-neuronal tissue for both noradrenaline and adrenaline.

Superimposed on these uptake mechanisms, both catecholamines are inactivated metabolically either by *o*-methylation (via the largely extracellular *o*-methyl transferase) to 3-methoxy-4-hydroxy mandelic acid (vanillyl mandelic acid or VMA) or they are inactivated by deamination (via the largely intracellular monoamine oxidase) to normet- and metadrenaline. These enzymic steps generally operate sequentially so that the main urinary metabolite is VMA and its conjugates. Only about 5 per cent of the catechol amines are excreted in the urine as such, i.e. in biologically active form. Whereas there is normally about 4 times as much adrenaline as noradrenaline in the adrenal medulla the reverse is true in the urine. Total adrenalectomy barely affects noradrenaline excretion although adrenaline excretion virtually vanishes. It is clear, therefore, that extra-adrenal tissues are normally responsible for most of the noradrenaline found in the plasma and urine.

Disordered function

The only known abnormality of the adrenal medulla is oversecretion. This is nearly always the result of a phaeochromocytoma, a tumour of the chromaffin cells which is usually benign.

The clinical manifestations depend firstly on the relative amounts of noradrenaline and adrenaline oversecreted, and secondly on whether the oversecretion is continuous or intermittent. Since all phaeochromocytomata produce an absolute excess of noradrenaline the dominating feature is nearly always systemic hypertension. This is sustained in most patients and paroxysmal in a minority (25 per cent). Attacks of severe symptoms are not invariable, but when they occur they usually consist of headache, sweating and apprehension, and they are often precipitated by cold, fasting, exercise or palpating the region of the tumour.

Overt diabetes develops in about 10 per cent and many patients with phaeochromocytoma show an impaired glucose tolerance, but whether or not carbohydrate metabolism is affected will depend on the absolute quantity of adrenaline secreted. For example, disordered carbohydrate metabolism has not been a feature of the cases reported in children presumably because their adrenal glands contain noradrenaline predominantly. Increase of the metabolic rate and apprehension sometimes mimics thyrotoxicosis. In patients who continuously oversecrete noradrenaline alone hypertension is sustained and carbohydrate metabolism is not detectably disordered, so that differentiation from essential hypertension can be particularly difficult.

In about 10 per cent of patients the tumour is malignant although histologically it is not possible to differentiate the benign from the malignant with any useful degree of certainty. Metastases to liver, lungs, local lymph nodes or bone are to be expected. The rate of deterioration is extremely variable.

In another 10 per cent of patients there is a familial tendency, usually associated with hyperparathyroidism and medullary carcinoma of the thyroid. Interestingly the tumour is inherited as an autosomal dominant with a high degree of penetrance; genetic counselling may, therefore, be rewarding.

Principles of tests and measurements

Adrenal medullary function may be assessed by the measurement of the rate of excretion of the main metabolites of the two catecholamines in the urine, by the measurement of the free biologically active hormones in the plasma or by observing the effects of drugs which influence hormone secretion or activity. With the development of rapid and reasonably specific methods of chemical assay, the use of pharmacological tests has become rare.

The renal output of vanillyl mandelic acid provides a convenient chemical method for the detection of patients with phaeochromocytoma. Normally, and in essential hypertension, less than 35 pmoles 24^{-1} (7 mg 24 h^{-1}) is found in the urine. Eating bananas (which contain noradrenaline), and foods containing tyramine (such as strong cheeses) or vanilla may give false positive results. The urine output of normet- and metadrenaline can also be determined. Urine noradrenaline and adrenaline can be measured by their effects on the blood pressure of anaesthetized animals or on smooth muscle preparations. Most of these assays cannot differentiate between noradrenaline and adrenaline.

Nowadays, relatively simple and robust non-bioassay methods have become available for the measurement of plasma and urine adrenaline and noradrenaline concentrations. The plasma catecholamine response to pentolinium, a ganglion-blocking drug, is a useful test, particularly in the context of an elevated blood pressure which may be due to anxiety rather than a phaeochromocytoma. A phaeochromocytoma secretes catechol amines post-synaptically and relatively autonomously so that the plasma levels of adrenaline and noradrenaline change little in response to an intravenous injection of pentolinium. If on the other hand elevated plasma catecholamine levels are due to psychological factors then a sharp fall in plasma catecholamine concentration occurs following ganglion blockade, since psychological factors stimulate the adrenal medulla presynaptically. These new methods for the measurement of plasma levels of adrenaline and noradrenaline are much more specific than hitherto and information can now be obtained by venous sampling as to the precise site of catecholamine oversecretion.

Pharmacological tests for the diagnosis of phaemochromocytoma are largely redundant both because of their potential danger and also because of the high incidence of false positive results.

References

AXELROD, JULIAS (1971) Noradrenaline: fate and control of its biosynthesis. *Science* 173, 598–606.
JAMES, V.H.T. (ed.) (1979) *The Adrenal Gland:* L. Martini (series ed.) *Comprehensive Endocrinology.* New York, Raven Press.

The Thyroid Gland

Normal function

Because the hormones made by the thyroid gland—thyroxine and triiodothyronine—contain a high proportion of iodine and are the only compounds of physiological importance known to contain this element, thyroid function is closely related to iodine metabolism. The effects of these two hormones, although qualitatively very similar, differ dramatically in their time course of action in man. In this chapter only these two hormones will be considered in detail but the relatively recently discovered hormone, calcitonin, must also be mentioned because, in man, it is principally a hormone made by a special group of cells (the 'C cells') interspersed between thyroid follicles. The role of calcitonin in human calcium metabolism remains incompletely defined but is discussed on p. 320. In mammals, thyroid hormones play an important role in the regulation of growth and heat production but in lower orders they also influence metamorphosis. Although the function of many different tissues can be affected by thyroid hormones, no single key site of physiological action has been defined. Recent evidence concerning nuclear receptors for thyroid hormones suggests that this is the level at which their main physiological role is mediated. The assessment of thyroid function in patients was, in the first instance, made by the measurement of basal metabolic rate, which depends on the peripheral utilization of thyroid hormones, then, by studying aspects of iodine metabolism in thyroid itself, which reflected the rate at which thyroxine and triiodothyronine were being synthesized, and, in recent times, assessment now includes simple tests which examine the relationship between peripheral thyroid hormone levels and the release of pituitary thyroid stimulating hormone after the injection of hypothalamic thyrotrophin releasing hormone. These methods, along with sensitive radiochemical techniques for the measurement of thyroxine and triiodothyronine in body fluids, enable a very precise assessment of thyroid function to be made without greatly inconveniencing patients or exposing them to significant risk.

THE THYROID HORMONES

Chemically these hormones are iodinated forms of thyronine, the formula of which is shown below.

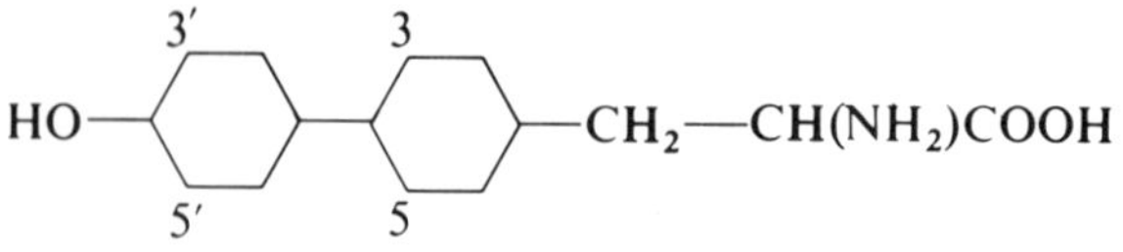

Thyroxine has 4 iodine atoms, attached to the 3 : 5 and 3′ : 5′ positions. Triiodo-thyronine has 3 iodine atoms, attached to the 3 : 5 and 3′ positions.

Synthetic analogues have been prepared by substitution of other halogens for iodine in these 2 molecules; they possess some physiological activity but are much less potent than the naturally occurring compounds. In common with many other physio-logically active compounds, the naturally occurring forms of thyroxine and triiodothy-ronine are both laevorotatory.

Within the thyroid gland the greater part of its store of iodinated compounds is in the colloid as mono- and di-iodotynosine which are precursors of the fully formed hormones made from them by oxidative coupling. Thyroxine, tri-iodothyronine and the iodinated tyrosines are all bound to thyroglobulin by peptide bonds which are broken by the action of thyroidal-protease. In the circulating blood thyroxine is present in very much greater concentration than triiodothyronine and is mostly bound to a specific carrier protein known as thyroxine-binding globulin. This carrier protein migrates in the inter-α-region on electrophoresis and is only partially saturated in health (about 20 per cent); the metabolically active free-thyroxine is minute compared with the total amount of hormone bound to protein. Thyroxine is also bound to a lesser extent to albumin and pre-albumin. Similar protein binding sites exist for tri-iodothyronine and for both hormones a stable equilibrium exists between the free and bound hormone. The concentrations of free-thyroxine and triiodothyronine are minute compared with the total quantities present but these probably determine the amount of hormone available to the tissues.

The relative concentrations of thyroxine and triiodothyronine in blood does not reflect their relative importance in the mediation of the peripheral actions of the thyroid through these hormones as, when due account is taken of the potency of the two hormones and of their half lives, it is highly probable that the more important thyroid hormone is triiodothyronine. Indeed, it has even been postulated that thy-roxine may be regarded as a pro-hormone converted to the more potent triiodothyro-nine by a thyroxine de-iodinase. The total levels of these hormones are both elevated by any drug which increases the concentration of thyroxine binding globulin. The most common agent which does this is an oestrogen-containing oral contraceptive and, under these circumstances, the normal range of thyroxine and triiodothyronine is no longer relevant to assessing thyroid status as the total hormone levels will be elevated, although the free levels remain essentially unchanged.

The de-iodination of thyroxine, which may well be of critical importance in thyroid hormone action at a cellular level, can also occur through an alternative thyroidal de-iodinase which generates 33′ : 5′ triiodothyronine commonly known as reverse tri-iodothyronine (rT_3). This compound is metabolically inactive.

Production of thyroid hormones

The outstanding physiological property of thyroid cells is their ability to concentrate iodide by an active transport system which normally results in the gland concentration being 25 times greater than in serum. In states of iodine deficiency, or when the thyroid is stimulated, this gradient becomes even greater, thus conserving body iodine.

Dietary iodine intake is mainly derived from some species of fish, iodized salt and cow's milk which may contain substantial quantities if the cattle are fed on artificial cake to which supplemental iodine has been added. In urban communities without evidence of iodine deficiency, most measurements show an intake of the order of 100 μg/day.

In health the concentration of plasma free inorganic iodide is quite low (less than 1 μg/100 ml) and its clearance is by the thyroid or kidneys. Other tissues such as salivary or mammary glands also actively transport iodide from the blood but only the thyroid is capable of synthesizing its hormones.

Certain anions—notably perchlorate and thiocyanate—inhibit the transport of iodide by the thyroid and may be used as antithyroid drugs in some circumstances. The artificial radionucleide, technetium ^{99m}Tc, incorporated into the pertechnetate ion, is also transported, as is iodide, into the thyroid cell and this artificial radionucleide is commonly used in modern techniques for thyroid scanning (see page 606).

The iodide in the thyroid is oxidized and becomes incorporated into tyrosine radicals to form mono- and diiodotyrosine. This step is controlled by an enzyme system not yet identified. The tyrosine radicals which are thus iodinated are already part of the protein molecule which is to become thyroglobulin.

The formation of thyroxine is accomplished by the coupling together of the diiodotyrosine radicals:

$$ \text{HO} - \overset{\displaystyle I}{\underset{\displaystyle I}{\bigcirc}} - CH_2 - CH(NH_2)COOH $$

with the extrusion of alanine. This step is also an oxidative process. The formation of triiodothyronine is probably by coupling 1 monoiodotyrosine radical and 1 diiodotyrosine radical.

Both the iodination of tyrosines and the coupling of iodotyrosines can be prevented by small quantities of antithyroid substances of the thiourea series. Thus, it is possible to do a 'pharmacological dissection' of thyroid hormone synthesis; perchlorate prevents iodide concentration whereas thiourea prevents the iodination of protein without affecting the iodide concentrating mechanism.

Thyroglobulin is stored within the colloid of the thyroid acini. When hormone is required, thyroglobulin is broken down by a protease to release thyroxine and triiodothyronine which pass out into the blood. The same proteolytic process also releases mono- and di-iodotyrosine, which are then acted upon by de-iodinases which release the iodide they contain. This iodide is mostly used again within the thyroid. After thyroxine and triiodothyronine are broken down in the peripheral tissues part of their iodine is also reutilized by the thyroid. Thus, both these methods are employed by the body to conserve its store of iodine.

CONTROL OF THYROID FUNCTION

The anterior pituitary gland secretes a thyroid stimulating hormone (TSH or thyrotrophin) which stimulates all known aspects of thyroid function: iodide concentration, iodination of protein, proteolysis of thyroglobulin and release of hormone into the blood. Thyroid function persists after complete hypophysectomy, albeit at a greatly diminished level compared with intact animals. Thus, there is no single metabolic step which is wholly dependent upon TSH and evidence is accumulating that, in common with many other hormones, the effects of TSH on the thyroid are mediated by cyclic adenosine $3' : 5'$ monophosphate (cyclic AMP). Cyclic AMP is formed from adenosine triphosphate (ATP) by the action of the enzyme adenyl cyclase which is activated by TSH.

The release of TSH from the pituitary is controlled by the level of thyroid hormones in the circulating blood through a 'negative feed back' mechanism. The release of TSH is stimulated when circulating thyroid hormone levels fall and, vice versa, so that, in health, the output of hormonal iodine from the thyroid (taking account of both thyroxine and triiodothyronine) is approximately 150 μg/day. Because the main mechanism for the release of TSH from the pituitary is almost certainly mediated through the influence of thyrotrophin-releasing hormone (TRH) it is theoretically possible that the negative 'feedback' could operate at either hypothalamic or at pituitary level but the balance of evidence still favours the latter; both triiodothyronine (T_3) and thyroxine (T_4) are effective at the pituitary level and the slight excess of circulating thyroid hormone will quickly shut down the release of TSH. Because it is now possible to use TRH clinically to test the integrity of the pituitary—thyroid axis, simple dynamic tests of the interrelationship between the pituitary and the thyroid are a practical clinical possibility (see below). The structure of the TRH proved to be a relatively simple tripeptide, pyroglutamyl-histidyl-prolineamide, which can be synthesized. The role of inhibitory dopaminergic mechanisms in the release of TSH is not yet wholly clear but they may constitute a balancing influence in the regulation of TSH release. In experimental animals, thyroid function responds to changes in environmental temperature, increasing when it is cold and declining when it is higher than normal. It is not clear whether these adjustments depend chiefly upon hypothalamic influences or upon changes in the rate of peripheral utilization of thyroid hormones.

Thyroid function may be depressed by stresses such as surgical operations, intercurrent illnesses and the administration of adrenal steroids. Whether or not emotional stresses, action through the hypothalamus, influence thyroid function is still uncertain.

The operation of the pituitary-thyroid feedback mechanism is seen clearly in experiments which interfere with hormone synthesis either by depriving an animal of its dietary iodine, or by inhibiting thyroidal transport of iodide with potassium perchlorate or by preventing the iodination of protein by one of the drugs of the thiouracil group. All of these situations result in a decline in circulating thyroid hormone which stimulates the pituitary to release more TSH, which, in turn, increases thyroid function and will, in time, lead to thyroid enlargement. This thyroid enlargement can be prevented by the simultaneous administration of any substance having thyroid hormone-like activity. This is the basis of the goitre prevention test, used for the assessment of the thyroid hormone-like activity of substances chemically related to thyroxine.

The thyroid is able to compensate for low blood iodide levels by increasing the efficiency of the iodide transporting mechanism; the ability to transport iodide is preserved, although in a greatly enfeebled way, after hypophysectomy and there is some evidence from hypophysectomized animals that there is also an internal regulating mechanism in thyroid follicles which limits the accumulation of iodide when colloid stores are adequate.

There is, however, no doubt that, in health, thyroid function is effectively controlled by TSH which reaches the gland in the circulating blood. In patients with thyrotoxicosis, however, the influence of thyroid stimulating immunoglobulins requires separate consideration (see below).

PHYSIOLOGICAL EFFECTS OF THYROID HORMONES

The precise primary mechanism and sites of action of thyroid hormones remain unknown. There is a latent period of many hours between the administration of thyroxine and any measurable effect, even when very large doses are given. Recent interest has centred on nuclear receptors, both for thyroxine and for triiodothyronine, so that the diverse effects of thyroid hormone on cell function may depend on stimulation of nuclear RNA synthesis which follows thyroid hormone administration. The effects of 'uncoupling' of oxidative phosphorylation which can be demonstrated after large doses of thyroxine, is most likely to be related to the general demand for readily available energy consequent upon generalized stimulation of synthetic functions in the cell. Most of the experiments upon which the older hypotheses depended involved doses of thyroid hormones far above the physiological range.

In physiological concentrations, the thyroid hormones are anabolic and failure of growth is one of the most striking effects of removal or destruction of the thyroid in young animals. In some amphibia, metamorphosis cannot occur in the absence of thyroid hormones.

In adult animals, thyroidectomy is followed by a dramatic fall in heat production. Approximately 40 per cent of mammalian heat production is under thyroid control so that caloric output falls by this amount after thyroidectomy, with an accompanying decline in heat production and oxygen consumption and carbon dioxide output.

The changes induced by thyroidectomy can all be reversed by thyroxine or triiodothyronine. If these hormones are given in greater than physiological doses, the metabolic rate of all tissues is increased and, with this, goes a parallel increase in general activity and respiration and pulse rate. The increased caloric output is partially compensated for by an increased food intake but, if the dose is sufficiently high, this compensation is inadequate and weight loss results.

THYROID HORMONE METABOLISM

The effects of thyroxine and triiodothyronine cannot be distinguished; the latter compound acts more rapidly and, when compared on a weight for weight basis, is three to five times more potent. There is a specific thyroxine binding globulin in man which has a higher affinity for thyroxine than for triiodothyronine. Both hormones also bind to thyroxine binding pre-albumen but, again, the affinity is greater for thyroxine. Apart from these specific thyroid hormone binding proteins, serum albumen has a low affinity for binding thyroxine but the capacity is very great. The half life of thyroxine is much longer (6 to 7 days) than of triiodothyronine ($1-1\frac{1}{2}$ days). When account is taken of the greater potency of triiodothyronine, its more rapid turnover and lower affinity for the circulating plasma proteins to which it is bound, it is likely that the ultimate effects of this hormone are at least as great, if not greater, than those of thyroxine.

Disordered function

HYPOTHYROIDISM

Failure of adequate thyroid hormone production arises most commonly because of some defect in the function of the thyroid gland itself, although, rarely, this may be a consequence of a pituitary lesion. The mechanisms underlying thyroid failure may be of various types: congenital deficiency may arise because of absence of various enzymes concerned with thyroid hormone synthesis (sporadic goitrous cretinism); embryological failure of thyroid development; biochemical (as the result of extreme iodine deficiency or ingestion of antithyroid substances); immunological (associated with the presence in the circulation of autoantibodies against thyroglobulin or other thyroid constituents— see page 604); surgical removal; post-irradiation thyroid failure (after treatment with radioactive iodine). When the pituitary functions normally, it reacts to hypothyroidism by secreting abnormally large amounts of TSH which cause enlargement of the thyroid if the gland is capable of growth. This is seen, for example, in genetically determined goitrous cretinism, goitre due to iodine deficiency or the ingestion of antithyroid substances. Thyroid enlargement is also a feature of lymphadenoid goitre associated with autoimmunization (Hashimoto's disease). In the other groups the lesion in the thyroid tends to deprive the gland of the capacity to respond by enlargement. Hypothyroidism due to pituitary failure is only really rarely seen as an isolated effect. The other trophic hormones are also frequently absent so that the clinical picture is complicated by hypogonadism and hypoadrenalism. Atrophy of the thyroid in the absence of TSH is not complete to that some slight residual function persists, although the amounts of thyroid hormones produced are not sufficient for normal requirements. The use of TRH to differentiate between hypothyroidism due to intrinsic thyroid failure and hypothyroidism resulting from pituitary disease is discussed below —page 605.

Compensated hypothyroidism

Many thyroid lesions which are capable of causing complete thyroid failure occur quite frequently in an incomplete form. Under these circumstances, if there is enough thyroid tissue capable of increasing its function under the influence of TSH, the compensative process operated through the 'feedback' mechanism will sometimes be successful in restoring the deficiency. In that event, there may be no signs of thyroid deficiency but the thyroid will usually be enlarged. This mechanism accounts for goitre in areas of iodine deficiency where most of the inhabitants are euthyroid; their thyroid enlargement is the price they pay for sustaining normal levels of thyroid hormone production.

EFFECTS OF HYPOTHYROIDISM

In the adult

The physiological changes are considered here: the clinical manifestations, some of which cannot be fully explained, are listed on page 609.

General metabolic effects. There is a reduction in the basal metabolic rate which may be as great as 40 per cent. Associated with this there is a general slowing down of many processes and a tendency to gain weight which may be partly due to retention of fluid.

Cardiovascular effects. The heart rate is slowed and the cardiac output is decreased. The heart is often said to be enlarged on X-ray; this may be due to a pericardial effusion and effusions into other serous cavities can also occur. The electrocardiograph characteristically shows, in addition to sinus bradycardia, low voltages and flattened or inverted T-waves.

Skin changes. The skin is thickened, dry and shows a tendency to scale. The cutaneous thickening is due to a mucinous infiltration which has histological staining characteristics due to an increased content of mucopolysaccharides; this change in its fully developed state is called 'myxoedema' and gives its name to the syndrome of severe thyroid deficiency in adults.

Neurological effects. Apart from the general slowing down of mental processes which almost invariably accompanies thyroid deficiency, there may be a psychosis. Surprisingly, the severest mental disturbances in myxoedema are often manic states; depressive or apathetic mental disturbances are less commonly seen.

Paraesthesiae and aches and pains which cannot be accounted for, may occur anywhere and patients with myxoedema frequently suffer from a 'carpal tunnel' syndrome. Fine movements are often clumsy and the gait may be unsteady. Muscle tone is increased (myxoedematous myotonia—page 607) and leads to the associated delayed relaxation of tendon reflexes which can usually be elicited during routine physical examination of hypothyroid patients. Increasing deafness may be another presenting symptom.

Hypothyroid coma. If untreated, severe hypothyroidism may ultimately lead to coma, frequently associated with hypothermia—a very low metabolic rate and CO_2 retention (respiratory acidosis). The possibility of underlying hypothyroidism must always be considered in patients presenting with hypothermic coma.

In the child

'Cretinism' is the term used to describe hypothyroidism present from birth; 'juvenile myxoedema' describes hypothyroidism developing in a previously healthy child. In childhood the chief effects are retardation of growth and mental development; there are certain other characteristics (page 609) whose physiological explanation is uncertain.

HYPERTHYROIDISM

In most, but not all, cases of hyperthyroidism, the thyroid is diffusely hyperplastic and when associated with eye signs (see below) the condition is frequently referred to as Graves' disease. Most clinicians also recognize hperthyroidism associated with a multinodular goitre or a solitary 'toxic adenoma' as being separate categories.

In recent years there has accumulated a convincing body of evidence which suggests that the increased thyroid activity of Graves' disease is associated with an abnormal, immunoglobulin-associated, thyroid stimulating activity in the circulation. Many different techniques have been applied to the measurement of this thyroid stimulating activity; each technique depends on a different parameter of thyroid activation and, because it is uncertain whether different methods are actually measuring precisely the same activity, a complicated nomenclature has grown up around this field of study. The original activity is known as the 'long acting thyroid stimulator', abbreviated as LATS, because of its prolonged time course of action in assay animals. The parameter of thyroid activation in the early assays was radioiodine release. Recent studies have also involved the capacity of immunoglobulins from patients with Graves' disease to block the binding of LATS to a human thyroid binding protein, the capacity of immunoglobulins to block the binding of radioiodinated TSH to human thyroid membranes, or several different ways of measuring adenylate cyclase activation *in vitro*. None of the techniques is simple and they have, so far, only been applied in research studies. Nevertheless, there is now enough evidence to relate the presence of these activities to the natural history of Graves' disease and the occurrence of neonatal thyrotoxicosis (see below) for the hypothesis to be taken increasingly seriously in the past decade.

Because LATS and the other immunoglobulin-associated activities are recovered with gamma globulins, it is postulated that the activities are thyroid stimulating immunoglobulins. Chemical fractionation of immunoglobulin from patients with Graves' disease shows that the thyroid stimulating activity of LATS resides within the same region of the molecule as antigen binding sites in antibodies. However, the site of origin and the mechanism of formation of thyroid stimulating immunoglobulins remain uncertain. It would appear, however, that the activities are directed towards regions on the human thyroid membrane near to the TSH receptor, or to part of it, and, although there is no convincing precedent for stimulation of the tissue by an antibody, the present evidence has many analogies to the acetyl choline receptor antibody of myasthenia gravis (see page 359).

Exophthalmos

Patients with hyperthyroidism often (but not invariably) show a number of eye signs of which the best known is exophthalmos, or protrusion of the eye from the orbit. This may be associated with weakness of the muscles which move the eyeball (external ophthalmoplegia) and spasm of the levator palpabrae superioris muscle (causing lid retraction and lid lag). The periorbital tissues may be greatly swollen and oedematous.

The cause of the eye signs associated with thyrotoxicosis is not known but it is clearly not a direct result of the increased production of thyroid hormones, as it cannot be induced by excessive thyroxine treatment in myxoedema and the eye signs are never seen in patients who are thyrotoxic as a result of a solitary toxic adenoma.

Although most commonly, the eye signs and hyperthyroidism develop concomitantly, this is not invariably so. Thus, the eye changes may present in euthyroid subjects who never subsequently develop hyperthyroidism, although thyroid overactivity will commonly occur eventually. Sometimes a period of thyroid overactivity has responded satisfactorily to treatment and, thereafter, eye signs develop or they may appear during the recovery phase. These latter two sequences have led some physicians to consider that treatment of thyroid overactivity will invariably be followed by deterioration in any associated eye signs. This is not necessarily so; indeed, control of thyroid overactivity probably offers the best chance of amelioration of the associated eye involvement.

There is no clear and consistent relationship between the serum level of thyroid stimulating immunoglobulins and the severity of associated eye signs in thyrotoxicosis. Approximately 50 per cent of patients who present with the eye signs of Graves' disease without associated thyroid overactivity (ophthalmic Graves' disease) show the absence of any response to the injection of intravenous TRH which is seen in Graves' disease associated with thyroid overactivity (see below).

In thyrotoxicosis associated with multinodular goitre eye signs are not a common feature as in Graves' disease and there is no evidence to link the pathogenesis of thyroid overactivity with thyroid-stimulating immunoglobulins.

Solitary toxic adenoma

Hyperthyroidism may sometimes result from overproduction of thyroid hormones in the localized part of the thyroid seen pathologically as a discrete encapsulated nodule in a gland which is otherwise inactive. Exophthalmos does not occur in such cases (see above) and the lesion is usually regarded as a benign new growth. Such adenomas may produce disproportionately large quantities of triiodothyronine as compared with thyroxine (T_3 thyrotoxicosis).

Thyroid addiction

The excessive intake of thyroid hormones induces hyperthyroidism. There are a few individuals who appear to relish the effects of elevation of thyroid hormone levels and become 'thyroid addicts'. In such patients (who are very rare) the thyroid is not enlarged and there are no abnormal changes in the eyes. When the uptake of radioiodine is measured in these patients it is totally suppressed (see below).

EFFECTS OF HYPERTHYROIDISM

Specifically clinical manifestations are summarized on page 609.

General metabolic effects

There is an increase in the metabolic rate which may be as great as 80 per cent. Associated with this there is an increase in rate of many body processes and weight loss is very common. Severe cases, particularly following exposure to the stress of an acute infection or surgical operation, may develop a 'thyrotoxic crisis', with extreme tachycardia, hyperpyrexia and collapse. The condition may be fatal unless rapidly treated.

Cardiovascular effects

There is an increase in heart rate and cardiac output. Thyroid hormones may induce atrial fibrillation, or, occasionally, flutter, especially in older patients. Heart failure may ensue in severe cases. These symptoms are related to increased sensitivity to sympathetic nervous activity and to circulating catecholamines and can be largely reversed by treatment with beta adrenergic blocking agents.

Neurological effects

Normal variations in mood are exaggerated and many patients confess to increased nervousness and irritability. In a severe case there may be mania. Tremor is frequently present and is best demonstrated by observing the outstretched fingers. The mechanism for tremor is not understood.

Thyrotoxic myopathy is discussed on page 363. Lid retraction is at least partly caused by the direct action of thyroid hormones on the levator palpebrae superioris.

SIMPLE GOITRE

In this condition the thyroid is enlarged, but the output of thyroid hormones is normal. There are, therefore, by definition, no other consequences remote from the thyroid, although, in some cases, with substantial enlargement mechanical effects, such as tracheal obstruction or displacement, may cause respiratory distress and there may be dysphagia due to pressure on the oesophagus.

IMMUNOLOGICAL REACTIONS OF THE THYROID

It is now well established that auto-antibodies can be formed against normal constituents of the thyroid both in man and in experimental animals deliberately immunized against extracts of their own thyroids. Such antibodies occur in certain thyroid diseases and high titres of antibody to thyroglobulin, for example, are frequently encountered in Hashimoto's disease, and in Graves' disease there may be high titres of antibodies to thyroid microsomes. Although 'kits' are available commercially for measurement of these antibodies and there are some clinical circumstances in which due weight has to be given to their detection, the measurements do not constitute a diagnostic test as there is considerable variation in the pattern of autoantibody formation in different thyroid autoimmune diseases.

The mechanism whereby these antigens are 'exposed' remains uncertain and it is, in general, more likely that, with the exception of thyroid stimulating antibodies, they are not directly related to the disease process. There is, however, a relationship between the presence of high titres of thyroid autoantibodies and the extent to which the thyroid is infiltrated with plasma cells and lymphocytes.

Principles of tests and measurements

There are many different methods of assessing the functional state of the thyroid gland. The first was measurement of the basal metabolic rate which is now seldom used. After an era in which radioactive isotopes of iodine were widely used for the assessment of thyroid function in man, increasingly precise chemical, and radiochemical, techniques now allow both the estimation of circulating thyroid hormone levels as well as measurement of the response of circulating pituitary TSH to the intravenous injection of TRH (see above, page 597) which allows a dynamic assessment of the interrelationship between the thyroid gland and the controlling influence of the anterior pituitary.

Plasma thyroxine and triiodothyronine levels

The measurements depend on the principles of radioimmuno-assay. The concentration of plasma thyroxine varies slightly according to the details of the technique used but has a wide normal range lying usually between 50 and 150 nmol/l. The level of triiodothyronine is much lower and usually lies between 1 and 3 nmol/l in health. Both measurements are now widely available through the marketing of commercial 'kits' which supply the necessary radioactively labelled hormone as well as the appropriate antibodies. These measurements are of total thyroid hormone and do not measure the all important 'free' moiety which, in the case of thyroxine, is less than 0·1 per cent of the total. Both thyroxine and triiodothyronine circulate, for the greater part, bound to plasma proteins (see page 595). Thyroxine binding globulin may vary in certain disease states and is also influenced by drug treatment, particularly the influence of oestrogens which increase the thyroxine binding globulin level. In healthy subjects, however, this does not influence the free thyroxine fraction and thyroid function is normal because the circulating hormone and the binding proteins reach new levels of equilibrium with each other, leaving the free or unbound fraction essentially unchanged.

In order to take account of variable quantities of binding protein in the circulating plasma several different techniques have been developed known collectively as 'T_3 resin uptake measurements'. In these techniques radioactively labelled T_3 is mixed with the patient's serum prior to exposure to a resin which binds the proportion of the radio labelled T_3 which was not bound to unoccupied binding sites in the patient's serum. Measurement of the radioactivity bound to the resin, which can readily be separated, gives, therefore, an inverse measure of the number of unoccupied binding sites in the patient's plasma. Various calculations to give a 'free thyroxine index' can be made by combining the total plasma thyroxine and the T_3 resin uptake. It is probable that, in the future, more precise measurements of TBG will be combined with total plasma thyroxine in the same way to give a simpler measurement of 'free thyroxine'.

Plasma TSH *and the intravenous* TRH *test*

In general, the available radioimmunoassays for TSH are not sufficiently sensitive to measure normal levels. In myxoedema, however, when the plasma TSH is elevated (see page 598), the level can almost invariably be detected by standard radioimmunoassays for TSH. The extent of elevation of basal TSH levels in myxoedema does, however, vary considerably and in doubtful cases the intravenous injection of TRH (200 µg) will discriminate between those patients with thyroid failure by inducing an exaggerated elevation of plasma TSH which reaches a peak approximately 20 minutes after the injection. Normal ranges of response can be defined so that the test provides a speedy method of discriminating between patients who are euthyroid, borderline hypothyroid or hyperthyroid, in whom the response is suppressed.

The tests so far listed allow a comprehensive assessment of the state of thyroid hormone production. They can all be completed either by the withdrawal of a blood sample or, in the case of the TRH test, a simple procedure lasting only 20 minutes. The older chemical methods of assessing thyroid hormone levels, such as the protein bound iodine, are now almost wholly abandoned as they obviously lacked specificity and were subject to many interfering influences from the use of iodine-containing medications or radiological contrast media.

RADIOISOTOPE TRACER TESTS

Similarly radioisotope tracer tests have declined greatly in their importance to the assessment of thyroid function in patients, although their role in isotope scanning of the thyroid remains of some importance. Uptake of radioiodine by the thyroid gland is still measured quantitatively in some centres and may play a role in predicting the radiation dose to the patient's thyroid following a therapeutic dose of radioiodine. In centres, therefore, which measure radiation dosage in radioiodine therapy, uptake measurements are still employed. As in other similar tracer studies, the principle upon which these tests depend is that radioactive isotopes of iodine are handled by the body in exactly the same manner as the stable, natural occurring isotope ^{127}I. Although several radioactive isotopes of iodine are available, most centres which perform such studies use ^{131}I, as the half life is sufficient to allow the thyroidal content of a tracer dose to be estimated at appropriate intervals up to 48 hours after oral administration to fasting subjects.

Although in thyroid overactivity it is possible to measure, after 48 hours, a significant proportion of a tracer dose of ^{131}I circulating bound to protein (because the tracer is partially incorporated in newly released thyroxine and triiodothyronine), this measurement is no longer widely used as a test of thyroid function.

Thyroidal radioisotope scintiscanning

In addition to measurement of thyroidal uptake of radioactive iodine it is possible either with a collimated scintillation counter or, nowadays more commonly, a 'gamma camera' to study the anatomical localization of radioactive iodine in the thyroid. Lingual or intra-thoracic thyroid tissue may be located and nodules of hyper-functioning tissue detected within a normally placed thyroid. This technique may also be used to establish that areas of the thyroid are not functioning and, occasionally, secondary tumours remote from the thyroid, if sufficiently highly differentiated, may be shown to function after normal thyroid tissue has been ablated. Isotopes of elements other than iodine which are concentrated by the thyroid may also be used for this purpose. The commonest is the artificial radionucleide technetium-^{99M}Tc which is trapped by the thyroid cells in a manner very similar to the trapping of iodide, and emits suitable gamma radiation for scintiscanning. The radiation dose from technetium-^{99M}Tc is very low and the quality of scintiscan not inferior to the use of ^{131}I. Radioactive technetium is not bound to protein in the thyroid.

In interpreting radioiodine tracer studies, care must be taken to ensure that the patient has not been taking drugs which modify the results. The list of drugs which may do this is long; apart from obvious culprits like excess iodides (present in many cough medicines), antithyroid drugs and thyroxine, it also includes agents such as phenylbutazone, iodopyrine, resorcinol and para amino salicyclic acid which all possess weak antithyroid effects. Iodine containing opaque media used for radiology can also seriously interfere with these tests which should not be performed within three months of angiography and up to two years after bronchography. The tests are permanently invalidated by myelography.

Basal metabolic rate

This old established method of assessing thyroid function has now been entirely replaced by more recent techniques except for research purposes. Nevertheless, the BMR remains the only way of measuring the effects of thyroid hormones upon the body as a whole. Essentially, it is a measure of heat output of a fasting subject at complete rest (page 539). The average normal figures were obtained by Robertson and Reid (1952) and in normal subjects the results fall within ± 15 per cent of their figures. Results greater than $+15$ per cent indicate hyperthyroidism, provided that certain other disease states have been excluded, whilst results below -15 per cent indicate hypothyroidism in the absence of gross malnutrition or adrenal corticoid deficiency.

'DELAYED-RELAXATION' OF TENDON REFLEXES

Tendon reflex relaxation time has been used as an aspect of the peripheral action of thyroid hormones on a specific type of tissue which can be measured with ease clinically. Thyroid deficiency has a characteristic effect on the time course of muscular relaxation after the induction of a stretch reflex. This is manifested by a delay in the relaxation phase and is not the consequence of any change in the conduction of nervous impulses. An experienced clinician can detect this phenomenon during his physical examination but there are several relatively simple devices which record the duration of the contraction and relaxation phases of the ankle jerk which may be conveniently followed in the movement of the foot. The most usual measurement for differentiating patients with thyroid disturbances from normal subjects is the 'time to half relaxation' which is prolonged in thyroid deficiency and shortened in hyperthyroidism. The test is simple to perform, gives an immediate result and the patient is not subjected to any inconvenience or discomfort. The relaxation time is, of course, also altered in primary muscular disorders and many other diseases so that it cannot be used as the sole basis for confirming the diagnosis of a thyroid disorder. It may, however, be a most useful screening procedure as the changes in thyroid disorders are remarkably consistent and revert to normal after appropriate treatment.

BLOOD CHOLESTEROL

In hypothyroid animals the production and excretion of cholesterol are both decreased, but since the latter is reduced to a greater extent than the former and net result is an increase in blood cholesterol level. In hyperthyroid animals the reverse occurs and blood cholesterol falls. Similar changes occur in patients with hypo- and hyperthyroidism. The elevation of plasma cholesterol in hypothyroidism is usually above the upper limit of normal (approximately 300 mg/100 ml) and is a fairly constant phenomenon. Exceptions occur when the disease is exceptionally advanced (myxoedema coma) probably because malnutrition is also present, and in young children the increase may not be found. In hypothyroidism secondary to a pituitary lesion the rise in blood cholesterol, though it may occur, is less constant. Hypercholesterol-aemia is also encountered in other diseases, such as obstructive jaundice, diabetes mellitus and nephrotic syndrome; it also occurs as an inherited trait associated with premature excessive atheroma.

In a group of hyperthyroid patients it is easy to demonstrate that the mean blood cholesterol concentration is less than normal. Most of the values, however, lie in the lower part of the normal range so that the estimation of serum cholesterol has little practical value in supporting a diagnosis of hyperthyroidism.

OTHER TESTS

The output of creatine in the urine is decreased in hypothyroidism and increased in hyperthyroidism. Many other conditions (e.g. muscular dystrophies) also cause excessive creatinuria (see page 376) so that this cannot be used as a test for hyperthyroidism.

Radiology is useful in assessing the mechanical effects of goitres. For example, X-ray pictures of the thoracic inlet will show the extent to which the trachea is deviated from its normal course and compressed by thyroid enlargement. Similar information for the oesophagus can be obtained on barium swallow.

It is now possible to examine the thyroid by ultrasound and this technique allows the differentiation of solid and cystic swellings in the thyroid gland, a point which can sometimes be of clinical importance.

Estimation of skeletal age by X-ray pictures of appropriate joints provides a good index of the state of thyroid function in childhood in that 'bone age' lags behind chronological age when hypothyroidism is present during childhood.

Practical assessment

CLINICAL OBSERVATIONS

Hypothyroidism in adults

Main difficulties are insidious onset and lack of specific manifestations. Question relatives and ask for old photographs.

Slowing down mentally and physically; dislike of cold; gain in weight; constipation; vague aches and pains; parasthesiae; puffiness of face; localized swellings around eyes and lips; loss of hair; hoarseness of voice; dry, coarse and cold skin particularly in older subjects; slow heart rate; heart sounds quiet; deafness; tendon jerks relax slowly; psychosis of no characteristic type; coma with hypothermia.

Presence of goitre suggests drugs (phenylbutazone, para-amino salicylic acid, resorcinol) or Hashimoto's thyroiditis (page 604).

Hypogonadism or hypoadrenalism suggest pituitary failure.

Hypothyroidism in children

Failure of mental and physical development; large tongue; distended abdomen with umbilical hernia; retarded bone age. Goitrous cretinism may indicate congenital enzyme deficiency (page 599).

Hyperthyroidism

Goitre usually diffuse, may be nodular; occasionally inconspicuous; very rarely solitary hyperactive nodule.

General metabolic effects; loss of weight; increased appetite especially in younger subjects; feels hot and sweaty; warm moist skin.

Other local manifestations probably due to general metabolic effects, tremor; weakness and wasting; diarrhoea; emotional liability; psychosis without characteristic features.

Circulatory effects, often predominant in older patients; palpitations; tachycardia; elevated sleeping pulse rate; atrial fibrillation; wide pulse pressure; heart failure; rarely oedema without heart failure.

Eye manifestations; lid retraction; exophthalmos; ophthalmoplegia (diplopia or squint); periorbital oedema; oedema of conjunctivae.

ROUTINE METHODS

Laboratory confirmation always advisable lest diagnosis later doubted and treatment interrupted.

Hypothyroidism

Plasma thyroxine and triiodothyronine (page 605). Plasma TSH (page 605) relaxation time of tendon jerks (page 607). BMR (page 607) ECG; immunological evidence; ESR; specific antibodies (page 604); radiological estimation of 'bone age'. Exaggerated increase in plasma TSH in response to intravenous TRH (page 605).

Hyperthyroidism

Plasma thyroxine and triiodothyronine (page 605); absence of any increase in TSH on injection of TRH intravenously (page 610); BMR; scintiscan (page 606); estimation of thyroid stimulating immunoglobulins (page 601).

SPECIAL TECHNIQUES

Needle biopsy for histology in lymphadenoid change or for chromatography after radioiodine.

Perchlorate after radioiodine to test for protein binding (page 596); labelled 'T$_3$' resin uptake tests; thyroid stimulating immunoglobulin assay.

References

DeGroot, L.J. & Stanbury, J.B. (1975) *The Thyroid Gland and its Diseases*. 4th Edition. New York, London, Sydney, Toronto, John Wiley & Sons.

Evered, D.C. (1976) *Diseases of the Thyroid*. London, Pitman Medical.

Evered, D.C. & Hall, R. (eds.) (1979) *Hypothyroidism and Goitre. Clinics in Endocrinology and Metabolism*, Vol. 8, No. 1. Philadelphia, W.B. Saunders.

Kendall-Taylor, P. (1977) Thyroid function and disease, pp. 37–60 In *Recent Advances in Endocrinology and Metabolism*. Ed. O'Riordan, J.L.H. Edinburgh, London and New York, Churchill Livingstone.

Volpe R. (ed.) (1978) *Thyrotoxicosis. Clinics in Endocrinology and Metabolism*, Vol. 7, No. 1. Philadelphia, W.B. Saunders.

Robertson, J.D. & Reid, D.D. (1952) Standards for the basal metabolism of normal people in Britain. *Lancet* i, 940–3.

18

Genetics

Normal function

GENES AND CHROMOSOMES

Genes

The transmission of discrete traits from each parent is termed *segregation*. Each discrete unit of transmission is termed a *gene* and each inherited character is determined by the action of two genes, one from each parent.

Chromosomes

The genes are located on the chromosomes in the nucleus of the cell. In man there are 46 chromosomes, comprising two sex chromosomes (XX in the female and XY in the male) and 22 pairs of like (*homologous*) chromosomes known as autosomes. (Fig. 18.1).

The 22 autosomal pairs can be considered in groups according to the size and position of the *centromere* joining the two arms. The acrocentric chromosome pairs numbered 13, 14, 15 and 21, 22 have the centromere near one end and carry small satellites on fine stalks on their short arms, which are involved in the organization of the nucleolus, and are known to carry the genes for ribosomal ribo-nucleic acid (RNA). Those with the centromere at the middle of the chromosome or off-centred are termed metacentric or submetacentric. Normal variations in arm length, centromere position and size of satellite are commonly seen for several of the chromosomes, especially the acrocentrics, the Y and numbers 1, 9 and 16.

Each chromosome pair can now be identified precisely by the use of staining techniques that produce banding patterns. The first technique to be developed was a fluorescent method using quinacrine dyes (Q banding) and this was soon followed by methods using Giemsa stains (G banding, see Fig. 18.1). In general, the bands produced by Q and G banding are similar, with only a few exceptions. Other staining techniques that are less often used include R banding (reverse banding) which produces a 'negative image' of G banding, and C banding which specifically stains the centromeric region of the chromosome.

In order to describe segments of particular chromosomes, a system has been developed to define individual bands. Bands are recognized as parts of the chromosome that are clearly distinguishable from adjacent segments by being either lighter or darker.

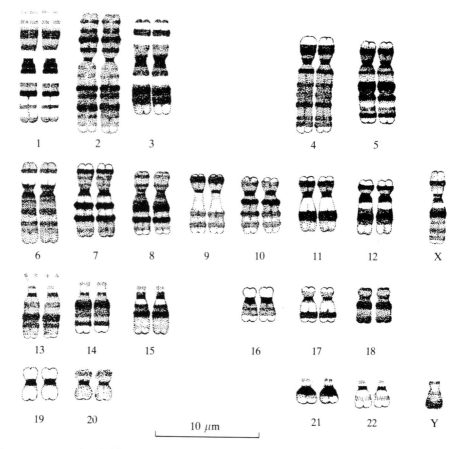

Figure 18.1. *A G-banded karyotype*

The letters p and q are used to designate the short and long arms of a chromosome respectively. Each arm is divided into regions by the presence of landmarks which can either be the centromere, the end of an arm, or particularly prominent and consistent bands. Regions are numbered sequentially from the centromere distally. By convention, a landmark band is taken to be wholly inside the distal region that it demarcates and represents the first band in that region. For example the symbols 6q22 would designate the second band in the second region distal to the centromere on the long arm of chromosome No. 6 (see Fig. 18.2). Each band visible by the light microscope represents a large amount of genetic material containing hundreds of genes. Individual genes are not visible by the light microscope.

Mitosis

Division of somatic cells is known as *mitosis*. The chromosomes that are visualized by normal cytogenetic techniques are arrested at a stage of mitosis known as the *meta-*

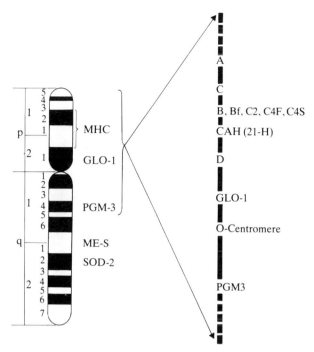

Figure 18.2. *Band classification and gene map of chromosome No. 6.*

Schematic representation of chromosome 6 with its gene assignments (for explanation see text). An expanded map of the HLA region is shown of the right.

Left: MHC = Major Histocompatibility complex; GLO-1 = glyoxylase − 1; PGM-3 = phosphoglucomutase − 3; ME-S = malic enzyme (soluble); SOD-2 = superoxide dismutase − 2.

Right; A, B, C, D = MHC loci; CAH (21 − H) = Congenital adrenal hyperplasia (21-hydroxylase deficiency); B, Bf, C2, C4F, C4S = components of complement.

phase (Fig. 18.3). During the metaphase each chromosome consists of two identical structures, the *chromatids*, joined side by side at the *centromere*. This is because the genetic material at this stage is duplicated, each individual chromatid of a given chromosome being a copy of the other. During the next stage of mitosis, *anaphase*, the chromosomes split longitudinally at the centromere into their constituent chromatids which then begin to move to opposite poles of the cell. The process is completed during *telophase* when two new daughter cells are formed by the creation of a nuclear envelope around each complement of daughter chromatids. Between cell divisions, in *interphase*, the genetic material is very much elongated and individual chromosomes cannot be visualized by the light microscope. The genetic material is duplicated again ready for the next round of cell division at the end of this stage. The term *prophase* is used for the stage when the chromosomes begin to condense and can be seen microscopically using the appropriate stains. After prophase, the renewed onset of metaphase is recognized by the lining up of the chromosomes on the so called metaphase plate.

Meiosis

During the production of gametes the normal diploid complement of 46 chromosomes must be reduced to the haploid number of 23 in the gametes by the process known as *meiosis*. Another important function must also be achieved. This is the independent assortment of maternally and paternally derived genes in the different gametes of an individual. The genes determining a particular trait are situated at identical points or *loci* on each member of an homologous pair of chromosomes. The maternally and

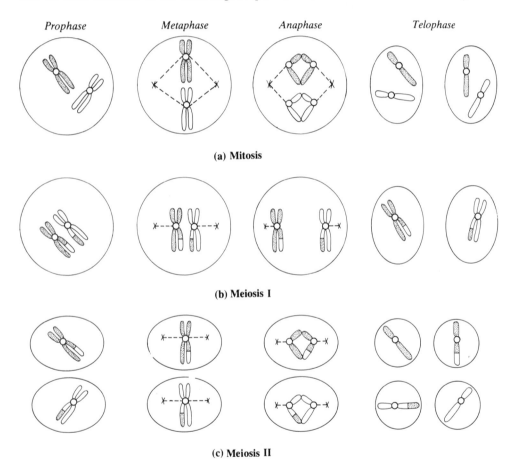

Figure 18.3.
   (a)   Normal Mitosis
   (b)   Normal Meiosis I (illustrating crossing over of genetic material)
   (c)   Normal Meiosis II
For simplicity only one pair of homologous chromosomes is shown.

paternally derived genes at a given locus may not be identical, and are known as *alleles*. During meiosis the independent assortment of alleles is brought about by two means:

(a) Different gametes receive different combinations of maternally and paternally derived chromosomes.

(b) There is an exchange of genetic material between the two members of an homologous pair of chromosomes, known as *crossing over*. This process results in the creation of a chromosome containing part of the maternal and part of the paternal member of an homologous pair, known as a *recombinant* chromosome. Without crossing over, alleles on the same chromosome would always be inherited *en bloc*. Crossing over takes place during the first step of meiosis known as meiosis I which is divided into prophase, metaphase, anaphase and telophase as in mitosis. However, meiosis I differs in that homologous members of a pair of chromosomes line up and exchange genetic material during prophase, and at anaphase each member *chromosome* passes to an opposite pole without division at the centromere. Meiosis I is known as reduction division, because it results in the formation of two daughter cells containing 23 chromosomes, each of which has two chromatids that may not be identical because of crossing over.

The second step, meiosis II, is similar to mitosis except that each 'parent' cell now has only 23 chromosomes and each 'daughter' cell contains 23 chromatids at telophase. Fusion of parental gametes at fertilization restores the diploid number of chromosomes.

Gene maps

The presence of loci close to one another on chromosomes leads to the phenomenon of linkage. The closer loci are together on a chromosome, the lower the probability that a crossover will take place between them during meiosis I and the higher the probability that two alleles inherited from one parent will be passed on *together* to an offspring. In this case the loci are said to be closely *linked*. Linkage between loci can be studied in suitable families by looking at the joint inheritance of specific traits and an estimate can be gained of the distance separating the loci. These family studies are informative for linkage but do not give information about the chromosomes on which the loci are situated. This problem is known as *gene assignment* and can be tackled by many different methods. One method is to study the traits in families in which individuals have recognizable chromosome variants; the first autosomal assignment, that of the Duffy blood group locus to chromosome No. 1 was made in this way. Another technique, known as somatic cell hybridization, makes it possible to study the expression of genes in cultured cells that contain only a few distinguishable human chromosomes, the other chromosomes being from another species such as the mouse. Using these methods, a human gene map is being constructed giving information about the assignment of various loci and their distance from other loci on the same chromosome.

Figure 18.2 is a schematic drawing of a No. 6 metaphase chromosome with the band classification given on the left. On the right the symbols represent some gene loci that have been assigned to chromosome No. 6. On the far right an expanded map is drawn of the HLA region to show the arrangement of the loci. This region is thought to lie somewhere within bands 6p21 and 6p22. Some loci within the HLA region are known to be separated by specific map distances which are proportional to the probability that a cross-over will occur between them. Other loci can be shown to be on chromosome No. 6, although their distance from loci on the same chromosome is not known. Loci that are known to be on the same chromosome are said to be *syntenic* although they may be so far apart that alleles segregate independently, so that they are not strictly linked. At present about 10 per cent of known loci have been assigned to chromosomes and further progress is rapidly being made.

GENES AND DNA

Biochemistry of the gene

The genes carried on individual chromosomes and their linear sequence have been determined in certain plants and animals, such as the fruit fly Drosophila, by observations on the way in which the characters determined by the genes segregate during breeding experiments. These classical observations can now be viewed in the light of more recent knowledge of the biochemical nature of this genetic material. According to the Watson-Crick model the nucleic acid deoxyribonucleic acid (DNA) consists of a double helix each spiral of which consists of a backbone of repeating deoxyribose-phosphate molecules. Attached to each sugar moiety is one of four bases: adenine, guanine, cytosine or thymine. The two helices are held together by hydrogen bonds linking their bases. Since adenine is always paired with thymine, and guanine with cytosine, the two helices are complementary. Because of the linkages of the deoxyribose-phosphate molecules, each DNA strand has a direction one end being designated 5' and the other 3'. The two strands of the helix run antiparallel, that is one strand runs 5' → 3' whilst the complementary strand runs 3' → 5'. Before cell division, the two DNA strands separate and a fresh complementary strand is synthesized on each of the two original strands. This provides a mechanism for the replication of the DNA with an exact reproduction of the original sequence of nucleic acid bases and is the biochemical counterpart of the chromosomal division that follows it.

The replication of DNA can be studied by growing cells in a medium containing tritiated thymidine, or 5-bromo-deoxyuridine (BrdU), which is incorporated into DNA. It is found that replication of the DNA of any one chromosome starts at different points at slightly staggered time intervals and that different chromosomes replicate at different times within the whole synthetic period. In particular, in cells from normal female subjects, one of the two X chromosomes always replicates much later than all the other chromosomes. It is this late replicating X chromosome that subsequently forms the sex chromatin that characterizes the cells of the normal female. Sex chromatin is present as a small densely staining chromatin body, characteristically located at the inner surface of the nuclear membrane in human epithelial cells and fibroblasts, or as the 'drumstick' of polymorphonuclear neutrophil leucocytes.

The presence of the Y chromosome can be demonstrated in interphase or dividing cells, and even in spermatozoa, by staining with a quinacrine dye and examining under ultra-violet light. Y chromatin at the distal end of the long arm of the Y fluoresces brilliantly.

In order to explain the maintenance of correct gene balance between females with two X chromosomes and males with only one it has been postulated that in mammals one of the two X chromosomes of the female is genetically inactivated at an early stage of development of the embryo, and that in each cell the chromosome inactivated is randomly determined. This is the so called Lyon hypothesis. It is also suggested that, once inactivation has occurred, the same X chromosome is subsequently inactivated in all the cells descended from any one particular cell. It has been supposed that it is the late replicating, sex chromatin forming X chromosome that is inactivated. The evidence supporting this hypothesis is now virtually conclusive; however, in man, inactivation apparently does not involve the entire X chromosome because some loci, such as the Xg blood group locus, have been shown not to be inactivated. These loci are probably at the terminal end of the short arms that associate with the short arms of the Y at meiosis.

The genetic code, transcription and translation

Genetic information is carried in the DNA chain by triplets of nucleic acid bases. Each set of three bases constitutes the instruction for a single amino acid and a sequence of base triplets encodes the sequence of amino acids for a complete polypeptide chain. Since there are four bases, giving 64 possible triplets, and only 20 amino acids required for protein synthesis, several amino acids are coded by more than one base triplet or *codon*. In addition to codons for amino acids there are also three codons to indicate the end of a gene, the so called 'stop' or chain termination codons.

The DNA code is *transcribed* by the synthesis of ribo-nucleic acid (RNA) in a 5′ to 3′ direction on a 3′ to 5′ DNA template. This RNA is therefore single stranded and carries a complementary sequence of bases, which are the same as for DNA with the exception of thymine which is replaced by uracil. RNA that codes for a specific sequence of amino acids is known as m-RNA (messenger-RNA); it passes to the cytoplasm where ribosomes together with transfer RNA molecules, one for each of the 20 amino acids, assemble the polypeptide chain in correct amino acid sequence, a process known as *translation*.

A chain of events links the gene and its eventual clinical or biochemical character or phenotype. Nucleic acid base triplets in DNA chain (gene)→ (transcription)→ m-RNA→ (translation)→ amino acid sequence in polypeptide chains→ protein→ biochemical phenotype→ clinical or physiological phenotype.

There is evidence that in eukaryotes the sequence of base triplets coding for a specific protein in the DNA of the genome sometimes may not be continuous, but may be interrupted by one or more sequences of bases that do not code for amino acids. The function of these intervening sequences or 'introns' is not known. The manufacture of the appropriate m-RNA molecule is thought to involve the transcription of the whole DNA sequence into RNA and then the 'splicing out' of the 'introns'.

In addition, m-RNA produced in the nucleus is modified before passage into the cytoplasm. At the 5' end a 'cap' is added consisting of a guanosine triphosphate (GTP) base in a 5'-5' linkage and methylation of the other 5' terminal bases. A the 3' end a sequence of adenylate residues is added, about 50–75 bases in length. Even without these additions the two ends are elongated by transcribed sequences that are eventually untranslated and whose function is unknown.

The traditional dogma of 'one gene, one polypeptide chain', although still valid in principle, now has to be modified with the finding that immunoglobulins, which consist of different variable and constant regions on the same polypeptide chain, are coded for by V and C genes which are separated in the germ line and only come together in a continuous m-RNA molecule in lymphocytes producing a specific antibody.

Gene regulation

Not all genes determine the production of enzymes or other proteins concerned with cell function. One model of gene action in micro-organisms classifies genes as either structural genes (determining the amino acid sequence of particular polypeptide chains) or controlling genes. The action of a group of linked and functionally related structural genes is controlled by an operator gene on the same chromosome, the *operator* and its structural genes constituting an *operon*. The operator in turn is controlled by a regulator gene which may be on another chromosome and is therefore not part of the linkage system of the operon. The regulator gene determines the synthesis of a repressor substance that inhibits the transcription of structural genes in the operon by binding to the operator gene. Transcription of the operon is stimulated by an inducer which either blocks the repressor at its combining site on the operator gene or more probably binds the repressor rendering it inactive. The substrate of an enzyme whose synthesis is determined by the structural genes of the operon acts as the inducer (Fig. 18.4).

The control of protein synthesis in eukaryotes probably differs from that in micro-organisms and may take place during the intranuclear and cytoplasmic 'processing' of m-RNA as well as at the level of DNA transcription.

Chromosome structure

The chromosomes seen by light microscopy consist of chromatin, which contains the continuous DNA strand complexed with a class of basic proteins, the *histones*, as well as with a variety of non-histone proteins and some RNA. Although the naked DNA duplex

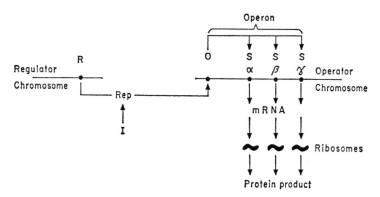

Figure 18.4. *A model for the genetic control of protein synthesis.*
R—Regulator gene; Rep—Repressor substance; I—inducer substance; O—Operator gene;
Sα etc.—Structural gene; mRNA—Messenger RNA.

has a diameter of 22 Å, the basic fibrillar unit of chromatin appears to be a 100 Å
strand consisting of DNA and histones. The structure looks like a string of beads with
repeating 200-base pair units of DNA of which 140 base pairs are wound around a core
of eight histone molecules to form a *nucleosome* whilst the remaining bases link up to
the next unit. More coiling must take place before all the DNA can be packed into a
metaphase chromosome and it is envisaged that the 100 ÅDNA-histone fibril is super-
coiled into a solenoid structure which itself is involved in even higher levels of coiling.

GENES AND FAMILIES

Modes of inheritance

The two genes at the same locus on homologous chromosomes are termed alleles. If
they are identical the individual is homozygous for the gene concerned but if they differ
he is heterozygous. If a gene is present on the X chromosome in the male but not on
the Y it is said to be hemizygous.

The effect of a major gene may be apparent whether it is present in single or double
dose, i.e. in heterozygous or homozygous state, or it may only be apparent when it is
present in the homozygous state. This difference accounts for the two common modes
of inheritance of characters determined by a single gene pair, dominant and recessive
inheritance. Both modes of inheritance may be illustrated by the ABO blood group
system.

In this system there are three genes, A, B and O. The A and B genes determine the
blood group whenever they are present, that is they are codominant. Their expression
is dominant over that of the O gene which is not expressed in their presence. Individuals who have the genotypes AA or AO are group 'A', those who have the genotypes
BB or BO are group 'B' and those who have the genotype AB are group 'AB'. It is

evident from Fig. 18.5 that in dominant inheritance the trait inherited, in this case the A antigen, is passed to half the offspring of a parent carrying the gene in single dose, is present equally in the two sexes and is expressed in each individual carrying the gene in each generation.

On the other hand the inheritance of the O antigen is recessive to the A and B antigens. In recessive inheritance the trait is expressed in only one in four of the children of two heterozygous parents neither of whom shows the trait. The trait is only present in individuals inheriting the gene from both parents. As with dominant inheritance the sexes are equally affected. With a common recessive trait, like blood group O, one or even both parents may also be homozygous for the gene and show the trait, but with rare traits both parents are usually heterozygous.

Dominance and recessivity are qualities not of the gene itself but of its expression. They depend on the method of observation used and are subjective rather than objective. For example, if the O antigen stimulated the production of antibody or if it were possible to recognize the actual protein determined by the ABO genes it would be possible to detect the O antigen in AO or BO individuals. The O gene would then in this sense be expressed in a dominant manner.

Sometimes the expression of a gene is influenced by environmental factors or other genes so that it is manifest in some people carrying the gene but not in others. In this situation it is said to show *incomplete penetrance*. Whether penetrance is complete or incomplete, the degree to which the gene is expressed may vary from maximal expression to a degree that is only just detectable, so called *variable expressivity*.

The pattern of inheritance is also modified when the locus is on a sex chromosome.

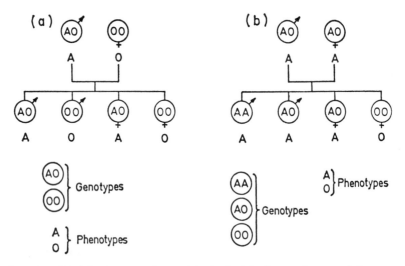

Figure 18.5. *Autosomal dominance and recessivity in the inheritance of a normal character, the ABO blood group.*

(a) Dominant inheritance of group A; (b) Recessive inheritance of group O.

Apart from genes directly involved in sex differentiation, or affecting the morphology of the Y chromosome, there are no certain genes on the Y chromosome of man. Hence for practical purposes sex-linked inheritance in man is X-linked. In recessive X-linked inheritance the gene is expressed only in the hemizygous male, but is transmitted through heterozygous females, whereas in dominant X-linked inheritance it is expressed also in the heterozygous female.

The expression of many characters determined by genes on autosomes may be modified by the sex of the individual, or may be manifested in only one sex. These are termed sex-influenced or sex-limited conditions and are not to be confused with sex-linkage. An example of a sex-limited character is the normal baldness that develops in many men with age. Sex influence on autosomal genes has always to be taken into account in interpreting the sex ratio of any character studied.

Effects of linkage

Although linkage is a term that is applied to loci, the effects of linkage can be studied in families by looking at the joint segregation of particular alleles for genetic abnormalities or for common blood groups and enzyme polymorphisms. For autosomal loci one must find families where at least one of the parents is heterozygous for alleles at both of the loci under investigation, a so called *double heterozygote*. If this individual has an abnormal or marker allele A at the first locus and B at the second locus under study then there are two possible situations.

(i) The A and B alleles are at loci on the *same* chromosome, this is known as linkage in *coupling*;

(ii) The A and B alleles are at loci on opposite members of the homologous pair of chromosomes. This is known as linkage in *repulsion*. The presence of linkage in coupling or repulsion is known as the *linkage phase* and in general can only be determined by studying the parents of the double heterozygote.

If an individual is doubly heterozygous for alleles at loci on separate chromosomes then those alleles will segregate independently in his gametes. This means that if he passes on the A allele to an offspring the probability that that offspring will receive the B allele is 0·5 (see Fig. 18.6). If the loci are close together on the same chromosomes then independent segregation will only occur if an uneven number cross-overs occurs between them at meiosis, the probability of this happening is proportional to the distance between them. If the alleles are in coupling then when the double heterozygote passes the A allele to an offspring the chances will be greater than 0.5 that he will pass on the B allele. For linkage in repulsion the chances would be less than the 0·5 in the same situation. The frequency with which cross-overs occur between two loci is known as the *recombination fraction*, it can be studied by looking at the proportion of children of a double heterozygote in coupling who have received the A allele or the B allele, but not both or neither, or, for double heterozygotes in repulsion, both the A and B alleles or neither, but not one or the other. If loci are so far apart on a chromosome that

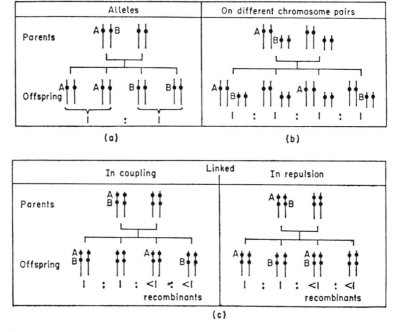

Figure 18.6. *Segregation.*

(a) of alleles at the same locus; (b) of alleles at loci on different chromosomes; (c) of alleles at different loci on the same chromosome.

cross-overs almost always occur between them, or if they are on different chromosomes, then the recombination fraction is 0·5.

Multifactorial traits and polygenic inheritance

Many quantitative traits such as height, blood pressure, intelligence etc. have environmental as well as genetic components and their determination is said to be multifactorial. These traits usually show a 'normal' distribution in the population. Their genetic component is thought to be polygenic, i.e. it is determined by alleles at many different loci. This makes the inheritance of such traits difficult to study, however, it is possible to devise models to predict the distribution of a given trait in a population and in the relatives of a given individual.

One of the simplest models assumes that the trait is determined by the additive effect of genes at various loci, each one having a relatively small effect. A hypothetical example would be a trait determined by 4 loci at each of which there were two possible alleles, one, t, adding 1 unit to the quantitative trait and the other, T, adding 5 units. If the gene frequency of each allele were equal in the population then the distribution of the trait could be calculated by expanding the binomial $(0·5p + 0·5q)^8$ where p = the number of t alleles and q = the number of T, alleles in a given individual (p + q = 8).

The results are shown in Fig. 18.7(a). The distribution of the trait in relatives of an individual with a particular phenotype can also be constructed, using the fact that the first degree relatives share half their genes in common with an individual, and second degree one quarter etc. Thus, the first degree relatives of individuals with 8,T alleles would have on average 6T and 2t alleles, four of the T alleles being present because of common ancestry and 2T alleles coming from the population 'pool'. The distribution of the trait in these first degree relatives is obtained by expanding the binomial $(0.25p + 0.75q)^8$ to give the skewed distribution shown in Fig. 18.7(a). Although the distribution provided by the above model results in nine discrete classes of individual it is not difficult to imagine that the influence of a greater number of loci and varying

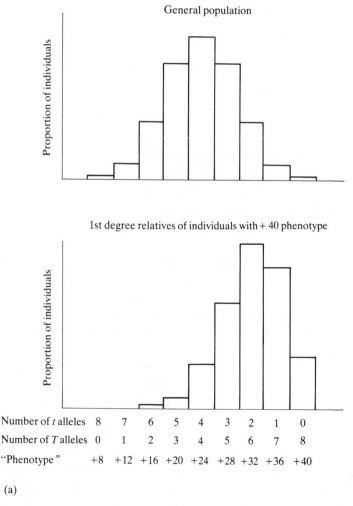

(a)

Figure 18.7. (a) Population distribution and distribution in first degree relatives of a hypothetical quantitative trait;

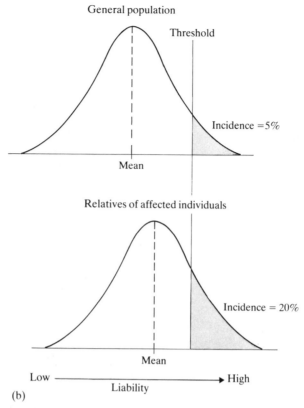

Figure 18.7. (b) Population distribution and distribution in close relatives of affected individuals of "liability" for a qualitative trait. (For explanation see text).

numbers of alleles, together with the effects of the environment and measurement error would produce a continuous normal distribution.

More sophisticated models can also be constructed that take account of the fact that some alleles are dominant to others, the effect of assortative mating (e.g. the tendency of tall men to marry tall women) and the relative contributions of different facets of the environment. These models can be used to predict the similarities between various types of relative within a family. Some qualitative traits such as cleft lip or congenital dislocation of the hip can also be explained by a polygenic model of inheritance. In this case it is envisaged that each individual in the population has a 'liability' to develop the trait, part of which is inherited in a polygenic manner and part of which comes from the environment. It is postulated that the amount of 'liability' is normally distributed in the population and that individuals who exceed a certain threshold level develop the trait. Again, it is possible to derive the liability distribution for first, second degree etc. relatives of an affected individual and to predict how many of these will lie beyond the threshold and therefore be affected. The situation is illustrated in Fig. 18.7(b).

GENES AND POPULATIONS

Hardy–Weinberg equilibrium

Population genetics is the study of the behaviour of genes in populations and is particularly concerned with the factors that tend to increase or decrease the frequency of particular genes and in the conditions under which stable equilibria of gene frequencies are obtained. The findings of population geneticists can be useful in clinical practice; for example, it is sometimes necessary to know the carrier frequency for an autosomal recessive disorder in order to give appropriate genetic advice. If one knows the frequency of the disorder in the population the carrier frequency can be obtained from one of the basic tenets of population genetics, the Hardy–Weinberg law. This law predicts that if there are two possible alleles A and a at a given locus with gene frequencies in the population of p and q respectively (where $p + q = 1$) then the frequencies of AA, Aa and aa individuals are given by p^2, $2pq$ and q^2 respectively ($p^2 + 2pq + q^2 = 1$). For example, in Caucasian populations, the autosomal recessive disorder cystic fibrosis has a frequency of approximately 1 in 1,600; designating the abnormal allele as *a* this means that:

$$q^2 = 1/1,600$$

$$q = 1/40$$

$$p = 39/40$$

$$2pq = 2 \times 1/40 \times 39/40$$

$$\simeq 1/20$$

in other words approximately one in twenty people are heterozygote carriers for the cystic fibrosis gene.

Polymorphism

For many known loci in Man there are up to three or four alleles of relatively frequent occurrence and a larger number of rare alleles. Two genes present at a given locus constitute the genotypé for that locus and the number of possible genotypes depends on the number of alleles. For two alleles there are three genotypes, for three six, and for four ten genotypes.

Alleles and genotypes in man have been most extensively studied for the genes determining blood groups and those for certain serum or red cell enzymes and proteins. In at least a quarter to a third of such systems two or more alleles of high population frequency have been found (i.e. occurring in more than 1 per cent of people); this phenomenon is known as *genetic polymorphism*. Obviously this phenomenon provides the basis for extensive variation of genetic origin. The major alleles within a polymorphic system arise by mutation and their frequencies in any community are determined by either natural selection or by random drift of the frequency of selectively neutral genes by chance.

Haplotype frequencies and
linkage disequilibrium

The Hardy–Weinberg law can be extended to take in situations where there are more than two alleles at a locus or to study the joint distribution of alleles at different loci. When two or more loci are studied together then the proportion of gametes containing different combinations of alleles, *haplotypes*, produced by members of the population can be predicted by constructing a gametic matrix. For example, take two loci with alleles A and a at the first locus with gene frequencies p and q and alleles B and b at the second locus with gene frequencies r and s, then the gametic matrix:

1st locus

	A	a	
B	pr	qr	r
b	ps	qs	s
	p	q	1

2nd locus

predicts that gametes containing allele A at the first locus and B at the second will occur with frequency pr and the other combinations can also be obtained from the matrix. An important point is that with linked loci there is generally no association between alleles in the population because of the effects of crossing over during many generations. For the example above this would mean that double heterozygote individuals in coupling phase for A and B would be present with the same frequency as individuals in repulsion phase. Mathematically this fact is expressed by the value $\Delta = (pr \times qs) - (qr \times ps)$ being equal to zero.

   If the Δ value is non-zero for alleles at linked loci this means that some alleles at the first locus are positively or negatively associated with others at the second locus in gametes produced by members of the population. This phenomenon is known as *linkage disequilibrium* and is sometimes seen with alleles at very closely linked loci. It is an important concept to grasp in order to understand the mechanisms of the association of some genetic diseases with the expression of particular blood group and histocompatibility antigens.

ABNORMAL FUNCTION

CHROMOSOMAL DISORDERS

General types of chromosome anomaly

Chromosome abnormalities may involve loss or increase in the number of chromosomes or alteration in the morphology of individual chromosomes. All the cells of the body may be abnormal or only a proportion of them or only the cells of a particular tissue. Abnormalities may be inherited, they may arise during gametogenesis, they may occur in the first few divisions of the zygote or they may be induced during later fetal or post-natal life.

The last mentioned group includes chromosomal damage induced by radiation, chemicals, viruses or mycoplasma and aberrations observed in neoplastic tissues. They are clearly not genetic. The other classes are either clearly genetic or are possibly due to environmental influence on the very early embryo and give rise to recognizable syndromes of developmental abnormality.

Variation in total chromosome number is due to *non-disjunction* at cell division. If during meiosis one pair of chromosomes fails to separate and both migrate to the same pole the resultant gametes will have either one too many, that is 24, or one too few chromosomes, that is 22. The effect is that in the subsequent zygote one chromosome pair is represented by three homologous chromosomes, (*trisomy*) or by only one (*monosomy*). Apart from monosomy for one of the X chromosomes and very occasionally of chromosome No. 21, it appears that monosomic zygotes are not viable. Trisomy is occasionally due to inheritance of an extra chromosome from a trisomic parent, a phenomenon known as *secondary non-disjunction.*

Abnormal migration can involve one of the chromatids of a chromosome following splitting of the centromere in mitosis. In this case the daughter chromatids may migrate to the same pole with a similar effect of non-disjunction in gametogenesis. Mitotic non-disjunction is associated with *mosaicism*, that is the presence of more than one cell line in the zygote. Non-disjunction at the first mitotic division in the zygote results in two cell lines, one with 47 and one with 45 chromosomes. If it occurs at the second division three cell lines result with 46 chromosomes in half the cells and 45 or 47 chromosomes in a quarter each. Non-disjunction at later divisions results in abnormal cell lines representing progressively smaller proportions of the total cell population so that by about the sixth or seventh division they would no longer be readily detectable.

Abnormalities of chromosome morphology arise in several ways. Breakage occurring simultaneously in two chromosomes, with rejoining of part of each chromosome to a part of the other, *translocation*, results in two abnormal chromosomes (Fig. 18.8). In the commonest stable type of translocation in man, *centric fusion*, two acro-

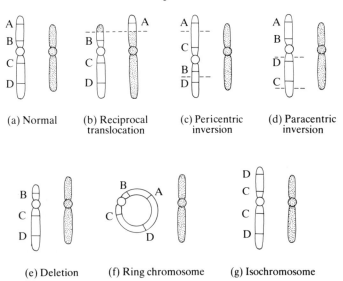

Figure 18.8. *Different types of chromosomal aberration.*
For simplicity only two non-homologous chromosomes are shown. Hypothetical loci A–D are marked on the non-shaded chromosome.

centric chromosomes are involved usually with the loss of one centromere and the short arms of both chromosomes. Such a translocation is balanced in that there is no significant alteration in the total amount of chromosomal material and no clinical abnormality. If however such a translocated chromosome is transmitted to a child it may not be balanced but may result in effective trisomy producing corresponding clinical abnormalities. Reciprocal translocations involving parts of the arms of two chromosomes may also result in an unbalanced chromosome complement in offspring. In this case the offspring may be trisomic for part of one chromosome and monosomic for part of the other.

Horizontal splitting of the centromeres at metaphase instead of the normal vertical splitting is known as *misdivision* and results in two *isochromosomes*, one consisting of two long and the other of two short arms for the chromosomes concerned (Fig. 18.8).

Another fairly common abnormality is a *deletion*, loss of part of an arm of a chromosome following a break with failure to rejoin. Very small deletions may not be detectable by cytological methods and if associated with a gene deficiency can be transmitted as an apparently dominant condition (see p. 638).

Inversions involve two breaks in a chromosome with the inversion of the intervening material and the subsequent reconstitution of the chromosome. They can involve the centromere (*pericentric inversion*), when they sometimes cause a change in the shape of the chromosome or they involve just one arm (*paracentric inversion*) in which case their presence is suspected by an alteration in the banding pattern. *Ring chromosomes* may also be formed by breaks at each end of a chromosome with subsequent end to end rejoining. Inversions which do not involve a significant loss of genetic material rarely cause clinical abnormality although they may result in the production of gametes with an unbalanced karyotype due to crossing-over within the complex pairing configurations of homologous chromosomal regions at meiosis.

If cells of some individuals are grown in specialized media (for example folate-deficient) some chromosomes exhibit a characteristic elongation of chromatin at one point; a so called *fragile site*. The most important of these fragile sites is situated at the bottom end of the long arm of the X (Xq27–28). It has been found that this 'fragile-X' chromosome is present in a proportion of males with moderate to severe mental retardation. Affected males tend to have large testes and prominent ears and jaws. Only a few apparently normal males have been described who express the fragile X. Affected males with 'fragile-X-linked mental retardation', occur in pedigrees with an X-linked pattern of inheritance. Females can carry the abnormal X-chromosomes in a heterozygous form and are occasionally retarded themselves. Techniques are being developed for reliable carrier detection and prenatal diagnosis of the disorder.

In order to describe a specific karyotype a shorthand is used which gives the number of chromosomes, the types of sex chromosomes and the type of any abnormal chromosome. For example 46,XY represents a normal male and 47,XY,+21 represents a male with Down's syndrome due to an extra No. 21. Translocations can also be described, 46,XY,−14,+t(14q21q) represents a male with an extra long arm of a 21 chromosome translocated onto a number 14 by centric fusion. Such an individual would clinically have Down's syndrome (see Fig. 18.9). More complicated rearrangements involving specific chromosome regions may also be described but the conventions used are complex and such cases should be discussed with the cytogeneticist involved.

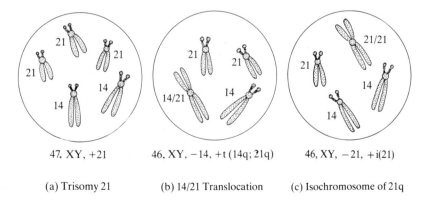

47, XY, +21	46, XY, −14, +t (14q; 21q)	46, XY, −21, +i(21)
(a) Trisomy 21	(b) 14/21 Translocation	(c) Isochromosome of 21q

Figure 18.9. *Different chromosomal types of Down's syndrome.*

For simplicity only chromosomes 14 and 21 are shown. It is assumed that each individual is a male.

Frequency of chromosomal abnormalities

At birth the frequency of sex chromosomes abnormalities is about 3 per 1,000, and of autosomal abnormalities about 4 per 1,000. Of the latter group about 1·5 per 1,000 represent autosomal trisomies (mainly Down's syndrome), the remainder being trans-locations, mainly balanced. The frequency of chromosome abnormalities in mis-carriages is much higher. Studies which include abortuses expelled up to the end of the second trimester show a frequency of 20–30 per cent whilst in those studies including the early stages of pregnancy the figure is closer to 50 per cent.

Down's syndrome

The commonest autosomal anomaly is Down's syndrome, or mongolism, which is present in about 1 in 600–700 live births. In the majority of cases there are 47 chromo-somes, the extra chromosome being a small acrocentric, number 21. The clinical fea-tures are well known.

   This syndrome is usually due to non-disjunction during oogenesis in the mother although a recent study demonstrated that in about 25 per cent of cases the extra chromosome has a paternal origin. The incidence of Down's syndrome due to non-disjunction shows a marked association with maternal age. The overall incidence of Down's syndrome in children born to 18-year old mothers is about 1 in 1,000 while at a maternal age of 46 it is about 1 in 40.

   Occasionally the mother herself is a mosaic of normal and trisomic cells. If the ovaries are trisomic for chromosome 21 then Down's syndrome can arise by secondary non-disjunction.

   In about 3 per cent of cases of Down's syndrome there is the normal number of chromosomes, 46, but an abnormal translocated chromosome is present. The trans-location arises by centric fusion of a number 21 with one of the group 13 to 15, most commonly number 14, with loss of the short arms of each chromosome (see Fig. 18.9). Alternatively there is a centric fusion of a number 21 with a 22 or another 21. Such translocations arise either *de novo* in the affected child or they may be inherited from a clinically normal parent who carries the translocation in a balanced form.

   The presence of the extra chromosome 21, whether as an independent chromosome or as part of a translocated chromosome results in the familiar defects of development of the face and limbs and in mental retardation. There may also be developmental abnormality of the heart or of the alimentary tract. Characteristic alterations of the dermatoglyphic patterns, provide further examples of developmental effects.

Other autosomal trisomic syndromes

Another syndrome, Edwards's syndrome, with an incidence of about 1 in 7,000 live births is characterized by a different spectrum of congenital anomalies, and by frequent arches on the finger tips. It is associated with early death. Chromosomal analysis again shows 47 chromosomes with the extra chromosomes being number 18. Familial translocation cases have also been described.

A third syndrome, Patau's syndrome, with an incidence of less than 1 in 10,000 shows a further set of developmental defects. Most affected infants die within a few months but a few have survived several years. Chromosomal analysis once again shows 47 chromosomes, the trisomic chromosome is a No. 13. Very rarely live-born individuals trisomic for other autosomes have been described particularly involving chromosomes 8, 9 and 22.

Other forms of autosomal imbalance

Partial deletions or duplications of chromosomes also produce recognizable syndromes. A well known example is the Cri du Chat syndrome due to a deletion in the short arm of chromosome number 5. Many other examples have been described involving all the autosomes. Triploidy, in which there is triplication of every chromosome with a total count of 69, is found in aborted fetuses and may be occasionally detected, in mosaic form, in living subjects.

Sex chromosomal anomalies

Genetically determined sexual disorders can be classified into disorders of sex chromosome number and inter-sex states due to a variety of other causes. Most, but not all, of these conditions are associated with sterility or severe infertility and may show some degree of mental retardation and/or psychiatric disorders.

Klinefelter's syndrome. This is a disorder of men characterized by small atrophic tests, azoospermia, low 17-ketosteroid excretion and raised urinary gonadotrophin. Other variable features include eunuchoidal stature, gynaecomastia and (occasionally) mental retardation. Histologically the testes show 'ghost' tubules devoid of elastic tissue. However, the Leydig cells, which secrete androgen, persist so that secondary sexual characters develop. There is no gross distortion of the dermal ridge patterns.

The majority of cases are chromatin positive, i.e. their cells show the presence of sex chromatin. The incidence of chromatin positive males is about one in 800 live births.

Chromatin positive cases of Klinefelter's syndrome have been found to possess 47 chromosomes with an XXY sex chromosome constitution. It is not usually possible to tell whether the non-disjunction leading to the XXY constitution is paternal or maternal but in some instances this can be determined from the Xg blood groups of the patient's family. The results of these studies show that in 60 per cent of Klinefelter's syndrome the extra-X is maternal in origin and in 40 per cent paternal, furthermore the non-disjunctive event most frequently takes place at the first meiotic division.

A minority of chromatin positive cases has been found to have a complex chromosomal constitution with more than two X chromosomes or more than one Y chromosome. They include XXXY and XXXXY, with cells containing two or three sex chromatin bodies respectively and showing more severe mental defects than the XXY cases; and XXYY. Others include mosaics XY/XXY, XX/XXY, XXY/XXXXY and XXY/XXXY.

Turner's syndrome. This is a disorder of women; the features are primary amenorrhoea, sterility and short stature. There is an increased excretion of pituitary gonadotrophin and biopsy shows the gonads to consist of fibrous streaks lacking ovarian follicles. Included in the syndrome as originally described by Turner was neck webbing but this is not an invariable finding. Other features that may be present are an increased carrying angle at the elbow, dyplasia of nails, cardiac defects especially coarctation of the aorta, a low hair line on the neck, renal malformations and lymphoedema at birth. The finger tips show well developed ridge patterns and the palm a distally placed axial triradius. The subjects are of normal intelligence.

The incidence at birth is about 1 in 5,000. This low incidence is at least in part, if not entirely, due to the fact that a high proportion of cases are non-viable and are lost through spontaneous abortion.

Individuals with this syndrome lack sex chromatin in their cells and chromosomal analysis shows 45 chromosomes with only one sex chromosome, an X. Mosaic forms are fairly common and include XO/XX, XO/XY, XO/XYY, XO/XXY and XO/XX/XXX. The XO/XY or XO/XYY cases may show masculinization and in extreme cases present as males with short stature and other stigmata of Turner's syndrome.

A few women have been described with primary amenorrhoea, normal stature but eunuchoidal proportions and an infantile uterus. When laparotomy has been performed small fibrous ovaries with primordial follicles have been found. The cells have either no sex chromatin or small sex chromatin bodies, and chromosomal analysis reveals one normal X chromosome and one X chromosome from which material from the long arms has been deleted. In general, deletion of the short arms of the X-chromosomes leads to individuals who have the somatic features of Turner's syndrome.

Isochromosomes for the long or short arm of the X-chromosome lead to features similar to partial deletions of the X-chromosomes.

Other sex chromosomal anomalies. Another sex chromosomal anomaly is the finding of three X chromosomes in women who may be clinically normal. In some cases there is secondary amenorrhoea or mental retardation. They are chromatin positive with a high total sex chromatin count and have two sex chromatin bodies in some of their cells. Patients with four X chromosomes are infertile and severely mentally retarded. Mosaics, such as XX/XXX, XO/XXX and XO/XX/XXX are also encountered.

The triple-X syndrome has a similar incidence to Klinefelter's syndrome at birth (1 in 800).

In males a double-Y syndrome has been described. Males who are XYY may be fertile and do not necessarily show any abnormalities. However, their average height is greater than that of the general male population. A high proportion of the taller inmates of criminal mental hospitals have been found to have an XYY constitution. However, no increased incidence of XYY males has been found among the inmates of ordinary prisons, Borstals or Approved Schools. Males who are XXYY usually present as tall Klinefelter's but may occasionally be fertile. Surveys of the new-born have shown that as many as 1 in 500 males may have an extra Y chromosome; the vast majority of these display no gross behavioural disturbance.

Chromosomes and cancer

Many cancers have abnormal karyotypes. In most cases the type of chromosomal alteration is not constant nor specific—for example in acute lymphoblastic leukaemia a variety of different chromosomal aberrations occur although chromosomes 8 and 9 tend to be involved more frequently than would occur by chance. In other cases a specific chromosome aberration seems to be characteristic. In chronic myeloid leukaemia, cells of the myeloid series almost invariably exhibit the so called Philadelphia chromosome, a chromosome 22 from which the terminal portion of the long arm has been translocated onto another chromosome (usually the long arm of chromosome 9).

Changes in the chromosome composition of the cells can be used for prognostic purposes, for example, the acquisition of an extra Philadelphia chromosome or an iso-chromosome of the long arm of chromosome 17 usually heralds a blastic crisis.

Certain chromosome constitutions in themselves predispose an individual to specific forms of cancer. A well known example is the prediliction of individuals with Down's syndrome to develop acute lymphoblastic leukaemia. Specific deletions can be associated with characteristic tumours, for example individuals with a deletion of a band on the long arm of chromosome 13 (13q14) have a tendency to develop bilateral retinoblastoma.

Some single gene disorders predispose to an increased frequency of chromosome breakage in the cells of affected individuals. Two disorders, ataxia telangiectasia and Fanconi's anaemia, are probably due to DNA repair defects whilst Bloom's syndrone has a different mechanism and shows an increased frequency of exchange of genetic material between chromatids of the same chromosome, which can be demonstrated by special staining techniques, and is known as sister chromatid exchange. It is of interest that individuals with these disorders also have an increased tendency to develop malignancies.

SINGLE GENE DEFECTS

In the fifth edition of his catalogue McKusick lists 736 autosomal dominant, 521 autosomal recessive and 107 X-linked definite, single gene, traits. Not all of these may be described as "abnormalities", however, these numbers give some idea of the extent of known genetic loci in man.

Molecular basis of single gene disorders

For many disorders, especially those of haemoglobin structure and function, the actual abnormality within the DNA of the genome has been elucidated. Using these disorders to serve as paradigms of the mechanisms of single gene mutation in general, alterations of DNA structure can be classified into different categories:

Point mutations. These consist of the substitution of single nucleotides within a gene resulting in a change in the triplet code for a single amino acid and the formation of a protein with altered structure or charge. Familiar examples are the haemoglobin β-chain mutants resulting in Hbs S, C and E as well as over 200 other point mutations involving α or β haemoglobin chains. A particular type of point mutation is that which results either in the formation of a 'stop' codon in the middle of a gene or in the mutation of the normal 'stop' codon at the end of the gene. An example of the first type is Hb McKees Rocks where the β-chain is terminated early due to a point mutation resulting in a 'stop' codon. The second type of mutation results in an elongated globin chain due to the gene being transcribed past the chain termination ('stop') codon. Some forms of α-thalassaemia, such as Hb Constant Spring, are known to be examples of this type of mutation.

Frame-shift mutation. Insertion or deletion of a number of nucleotides that is not a multiple of three will result in a shift of the reading frame when the DNA is being transcribed. This means that the code for amino acids distal to the mutation will be completely altered. Furthermore, the normal chain termination codon will also be read out of frame resulting in the production of elongated chains. Three such mutations of the α-chain have been described (Hb Cranston, Tak and Wayne).

Gene deletion. Modern methods of molecular biology have been used to demonstrate the fact that in some forms of α-thalasaemia the actual structural genes for the α chains are missing. Although gene deletion has only rarely been shown to be the cause of the β-thalassaemias, a form of gene deletion thought to be due to unequal crossing-over does exist, which results in a "fusion gene" between the β and δ genes and the deletion of normal β and δ genes. Individuals homozygous for the mutation have no β or δ chains, but produce a unique globin chain consisting of the N-terminal residues of the δ chain and the C-terminal residues of the β-chain—the so called Lepore haemoglobin.

Defects of regulation. The β-thalassaemias appear to be very heterogeneous at the molecular level, some individuals producing small amounts of β-chain and some none at all. Some defects in this group of disorders could lie outside the structural gene, involving defects in the stability or processing of the m-RNA.

Functional effects of single gene disorders

The physiological and biochemical abnormalities resulting from the altered structure or quantity or specific proteins reflects their ubiquitous nature as enzymes, carriers, hormones, receptors and basic "building blocks" in animals. The varied types of functional abnormality caused by single gene mutation are summarized below.

Enzyme defects. In born errors of metabolism are due to an abnormal gene causing a deficiency of a specific enzyme or the synthesis of an inactive enzyme with a resultant metabolic block at the step concerned:

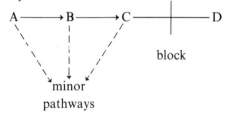

block

minor
pathways

There are several possible consequences of a block. There may be effects arising from the failure to produce the final metabolite D as in the failure to synthesize hydrocortisone in various forms of the adrenogenital syndrome.

Accumulation of the immediate precursor C may produce toxic effects, as does unconjugated bilirubin from lack of glucuronyl transferase in Crigler-Najjar syndrome. The more remote precursors A or B may be harmful factors, as in glucose-6-phosphatase deficiency (von Gierke's disease) in which glycogen accumulates in liver cells. The lysosomal storage disorders are a special group where the precursor accumulates within the lysosomes due to a deficiency of specific lysosomal enzymes. Disorders involving abnormal storage of mucopolysaccharides, oligosaccharides, sphingolipids and glycogen have been described.

Transport defects. Defects in proteins that serve as carriers for specific metabolites are well recognized. Apart from the classical example of haemoglobinopathies, abnormalities of hormone and trace element carrier proteins in plasma have been described. Active transport of metabolites across membranes is also dependent upon carrier proteins. Many examples in the kidney and intestine are known for example, cystinuria, Hartnup Disease, imminoglycinuria and renal glycosuria.

Receptor defects. Many cells possess specialized surface membrane receptors, which interact with circulating hormones or metabolites, including polypeptide and steroid hormones, neurotransmitters and plasma lipoproteins. Defects in the function or turn-over of hormone receptors include diabetes insipidus, pseudohypoparathyroidism and testicular feminization. Type IIa hyperlipoproteinaemia has been shown to be associated with a defect of low density lipoprotein receptors on the cell surface.

Cell surface antigen defects. Cell surface antigens are thought to be of great importance, both in embryological development and in the maintenance of the immune response. It is to be anticipated that mutations of these antigen systems will be shown to be the cause of some developmental defects and immunodeficiency diseases. It is thought that absence of the male specific H-Y antigen is the cause of certain forms of male pseudo-hermaphroditism (i.e., phenotypic females with a 46,XY "male" karyotype). Individuals with combined immunodeficiency who lack HLA-A and B antigens on the surface of their cells have been described.

Structural protein defects. The most plentiful structural protein in animals is collagen. Defects in collagen synthesis cause different forms of the Ehlers-Danlos syndrome and osteogenesis imperfecta, disorders that have been shown to display a marked degree of biochemical heterogeneity. The defect itself may be due to a mutation of one of the genes for a particular collagen chain as in Type IV Ehler's Danlos syndrome where type III collagen is missing; or it may be due to defect in an enzyme that modifies the collagen chain as in Type V Ehler's Danlos syndrome where lysyl oxidase, an enzyme important in cross-linking of collagen, is absent. Defects of the relative production of type I and type III collagen are thought to cause some types of osteogenesis imperfecta.

Contractile protein defects. Although no abnormalities of actin or myosin, the main contractile proteins of muscle, have been described an abnormality of cilia resulting in immotility is known. Individuals with this disorder, known as the immotile cilia syndrome, lack certain proteins that make up the microtubular apparatus which runs the length of the cilium. In one form they exhibit the Kartagener triad (bronchiectasis, sinusitis and situs inversus). Inheritance is probably autosomal recessive with affected males being sterile due to immobile spermatozoa.

Single gene defects in families

Very often the diagnosis of a specific disorder will tell the clinical geneticist which mode of inheritance is operating and hence the genetic prognosis that may be given. At other times the situation may not be so clear, for example, a previously unclassified malformation may be encountered that is manifested by many members of a family and in this case the mode of inheritance must be deduced from the relationship of affected and unaffected members. Alternatively a disorder may exhibit different modes of inheritance, for example retinitis pigmentosa can be inherited in an autosomal recessive, autosomal dominant or X-linked recessive fashion, the family history being at times the only method of distinguishing between the different possibilities. Certain important points regarding each type of inheritance should be made.

Autosomal recessive inheritance. This is usually characterized by the presence of affected sibs with normal parents. There tends to be increased parental consanguinity that becomes more frequent the rarer the condition. This is because if an individual carries a rare detrimental gene, the approximate probability that a first cousin is a carrier is 1 in 8 irrespective of the rarity of the condition, whereas the probability that an unrelated spouse is a carrier is dependent upon the gene frequency in the population. This means that the rarer the gene is in the population, the more likely it will be that couples who are both heterozygotes will be consanguineous. The presence of only one affected sib, even in quite large sibships, does not mean that the disorder cannot be inherited. For example, even for well recognized autosomal recessive conditions, the probability that the clinician will encounter a sibship of five individuals with more than one affected sib is less than fifty per cent. This means that the clinician must be on the alert for a possible genetic cause of obscure conditions, even when there is only one affected member in the sibship, and a detailed family history, with specific reference to possible consanguinity, must be taken.

Autosomal dominant inheritance. Analysis of autosomal dominant pedigrees is made difficult by two factors, mutation and reduced penetrance. It is to be expected that a proportion of cases of an autosomal dominant disorder will have no family history and will be caused by a mutation in one of the gametes passed on by normal parents. In the extreme case, where a particular disorder is a genetic lethal (i.e., affected individuals do not reproduce), all affected individuals will have received new mutations. On the other hand, the phenomenon of reduced penetrance is well recognized in autosomal dominant conditions and can lead to a situation where an affected individual has normal parents, even though one of them carries the gene (which is therefore non-penetrant in that parent). In other cases examination of apparently non-affected family members will sometimes reveal minimal stigmata of the condition. Where there is only one affected person in a family, the assessment of the carrier status of the parents will depend upon the known penetrance of the gene, the percentage of cases that are known to be new mutations and the number and relationship of unaffected individuals in the family.

X-*linked recessive inheritance.* The pattern of inheritance in X-linked recessive pedigrees, with only males being affected and females being phenotypically normal, even when they are carriers, should be familiar. Because males have only one X-chromosome and must pass this on to daughters, *all* daughters of an affected male will be carriers. Because males pass only the Y chromosome on to sons, no son of an affected male will be affected. If a female has one affected son, and there are no other affected males in the family, the probability that she is a carrier must be carefully assessed. J. B. S. Haldane showed that, given certain assumptions, one third of all cases of a lethal X-linked recessive disorder are new mutations (i.e. their mothers are not carriers). Therefore, if a woman has one son who is affected with such a disorder, there being no other affected males in the pedigree, the probability that she is a carrier is two thirds. This probability is reduced if there are normal sons as well, or if there are other unaffected males in the pedigree. (To take a slightly extreme example, it is highly unlikely that a woman with one affected son and ten normal sons is a carrier; it is most likely that the affected son has received a new mutation). Calculation of such risks is complex and a logical theorem known as Bayes' theorem must be used. If reliable carrier detection tests that distinguish between carriers and non-carriers with no false classification are available, then the problem is removed. However, if such tests produce overlapping ranges for carriers and non-carriers, the results must be incorporated into the Bayesian calculations mentioned above.

X-*linked dominant inheritance.* Superficially this can look very much like autosomal dominant inheritance. Distinguishing features include the facts that all the daughters of affected males are affected and that there is no male to male transmission of the disorder.

NON-MENDELIAN FAMILIAL CONDITIONS

It must be appreciated that pedigrees resembling the examples of Mendelian inheritance shown in the last section can be produced by entities other than single gene defects. These entities can be caused by genetic or environmental factors and fall into clearly defined categories:

Chromosomal defects

Small chromosome deletions may be missed on routine cytogenic analysis but may be the cause of multiple affected members of a family. It is to be expected that even with highly refined cytogenetic techniques some chromosomal re-arrangement will be microscopically undetectable. Familial cases of some tumours, for example retinoblastoma and Wilm's tumour have been shown to be associated with very small chromosome deletions.

Multifactorial disorders

The way in which the effects of genes at multiple loci can combine with environmental factors to produce variation in a given quantitative trait has been mentioned and the cause of similarity between close relatives has been outlined. The imposition of an artificial cut off point to define "normality" (e.g. for blood pressure) can lead to the clustering of individuals with "abnormal" values within a family and the appearance that a single gene is operating.

Qualitative traits such as cleft lip and palate or congenital dislocation of the hip, that are thought to be inherited in a polygenic manner, can also show clustering within families. The important point here is that the probability that relatives of an affected individual will be affected themselves drops markedly the more distantly related the relatives are. The risks become almost the same as the population incidence for third degree relatives, whereas for autosomal dominant single gene disorders the risk is only reduced by a half for each degree of relationship. Because more severely affected individuals would be expected to lie further beyond the threshold on the "liability scale" for a polygenic disorder, the probability of their relatives being affected is increased in proportion to the severity of their condition. This is not the case for single gene disorders.

Maternal factors

Disorders that are caused by an abnormal intra-uterine environment can often lead to the presence of many affected members of a sibship. It is well recognized that specific malformation syndromes can be caused by drugs or substances taken by the mother; fetal alcohol, phenytoin, warfarin and aminopterin syndromes have all been described. Uncontrolled phenylketonuria in a mother can lead to retardation in children, even though they are not homozygous for the abnormal gene.

Slow virus infections

It is thought that slow viruses can cause chronic neurological disorders, such as Creutzfeldt–Jakob disease and Kuru, which appear to segregate in a mendelian manner in families. In the case of Kuru the explanation appears to be the custom of eating the brains of relatives at death, but the explanation in the case of Creutzfeldt–Jakob disease is obscure.

DISEASE ASSOCIATION, PLEIOTROPY AND LINKAGE DISEQUILIBRIUM

In recent years, considerable interest has been generated by the discovery that individuals with particular blood groups or HLA types are more susceptible to developing certain diseases than other members of the population. The association of duodenal ulcer with blood group O, ankylosing spondylitis with HLA-B27, and juvenile onset diabetes mellitus with HLA-B8, −BW15 and −DW3 antigens are familiar examples. It must be stressed that these are population associations, they do not imply that all persons with a specific blood group will develop a given disease nor that individuals with the disease must have the particular blood group or HLA antigen.

There are two types of explanation for these phenomena:

(a)   That the susceptibility to the disease is caused in some way by the presence of the actual antigen produced by the blood group or HLA locus (for example, the presence of a specific antigen on a cell could make it more susceptible to particular viral infections). In this case the disease susceptibility is said to be a *pleiotropic effect* of the gene for the specific antigen.

(b)   That the disease is caused by an allele at a locus closely linked to the blood group or HLA locus. This explanation requires the additional postulate that there is linkage disequilibrium between the allele for the disease susceptibility and the allele for the antigen. Support for this hypothesis comes from the discovery that significant linkage disequilibrium between certain alleles of the HLA system does exist. For example in caucasians HLA-A1 has a gene frequency of 0·14 and HLA-B8 of 0·1 so that the haplotype frequency would be expected to be 0·014, in fact the HLA-A1, B8 haplotype occurs with a frequency of 0·064 (a Δ value of 0·05). In addition, certain diseases that are known to be caused by single gene defects, for example, complement C-2 deficiency, 21-hydroxylase deficiency (a form of adrenogenital syndrome) and haemochromatosis have been found to be linked to the HLA region. Furthermore, some of these abnormal genes are in linkage disequilibrium with alleles at the HLA loci. Although specific genes for disease susceptibility have not been discovered in man, comparison with the major histocompatibility region of the mouse suggests that the HLA region might contain loci that control the level of immune response to specific antigens.

The population studies must be followed up by family and immunological studies in order to fully elucidate the mechanism of specific disease associations.

CONGENITAL MALFORMATIONS

Approximately 2 per cent of children are born with significant congenital malformations. A few of these malformations are due to environmental factors such as maternal Rubella infection or to drugs taken by the mother (thalidomide is the prime example). Some malformations, a minority, are caused by single gene defects. Most malformations such as neural tube defects, dislocation of the hip, cardiac defects and pyloric stenosis are thought to have a multifactorial aetiology. The combination of several unrelated abnormalities in a child should raise the possibility of chromosomal imbalance. Because of the possibility of a translocation carrier state in the parents such children should always have their chromosomes analyzed.

The diagnosis of specific patterns of congenital malformation has been made complex with the recognition of many hundreds of syndromes. Every effort should be made to establish a diagnosis because of the importance for prognosis and genetic counselling. Computer data banks and centralized syndrome identification services can be used where available.

PRINCIPLES OF TESTS AND MEASUREMENTS

CHROMOSOME ANALYSIS

Chromosomes are usually analyzed in cells that are in metaphase. This requires the availability of viable dividing cells. In human these are generally derived from four sources:

Peripheral blood lymphocytes

These are cultured for 60–72 hours in the presence of phytohaemagglutinin, a plant extract that has been found to increase cell division. One to two hours before harvest, colchicine is added which arrests the cells in metaphase. A hypotonic solution is added to expand the nuclei and the cells are fixed and dry mounted on a slide before staining. Micro-methods requiring only a few drops of blood, have been developed. Lymphocytes are used for routine karyotyping.

Bone marrow

The cells of the marrow are rapidly dividing so that phytohaemagglutinin is not used. The principles of slide preparation are otherwise similar to those for peripheral blood. Bone marrow specimens are used in the study of leukaemias, and occasionally to obtain a rapid karyotype result.

Fibroblasts

Fibroblasts can be grown in culture from small skin biopsies under sterile conditions. Again colchicine is used to arrest the cells in metaphase. Fibroblasts are used when mosaicism is suspected or when peripheral blood may not be available, for example from early post mortem material.

Amniotic fluid cells

These can be obtained by amniocentesis at 14–16 weeks gestation and grown in culture in the same way as fibroblasts, although amniotic fluid contains different types of cells. There are usually enough cells for chromosome analysis after 2–3 weeks culture. Amniotic fluid cells are used for the prenatal diagnosis of chromosome and biochemical disorders. In the latter case culture times may be longer, in order to obtain enough cells for biochemical assay.

Chromosome staining

At least one banding method is used (usually G-Banding). Each complement of chromosomes is counted and analyzed for structural abnormalities. Specialized banding techniques such as C and T banding may be necessary in difficult cases. Techniques are available for the study of late prophase or early metaphase chromosomes which are longer and show more bands than metaphase chromosomes. This makes it more probable that a small deletion will be detected. However, the greater resolution obtained by using these techniques must be balanced against the greater length of time needed for analysis.

Sex chromatin

The presence of Barr bodies may be demonstrated in cells obtained from the buccal mucosa and directly mounted and stained on a slide. In approximately 30–60 per cent of cells from a normal female Barr bodies will be present. Buccal smears are usually carried out as an initial test to identify sex chromosome anomalies (in this case full chromosome analysis should also be carried out) or in an attempt to demonstrate mosaicism for a sex chromosome anomaly.

BIOCHEMICAL TESTS

Biochemical tests are used to identify specific enzyme deficiencies in affected individuals. The choice of tissue depends upon the particular enzyme deficiency. In suitable disorders biochemical tests on cultured amniotic fluid cells can be used for ante-natal diagnosis.

A variety of tests can be used to detect individuals who carry an abnormal gene but who are clinically normal. These include biochemical tests for heterozygote carriers of autosomal recessive enzyme deficiencies (for example Tay Sachs Disease) and measurement of circulating metabolites (for example phenylalanine loading tests to detect carriers for phenylketonuria).

Measurement of creatine phosphokinase (CPK) levels in female relatives of Duchenne muscular dystrophy patients provides a means of estimating the probability of carrier status. However, in this case, because of the overlap of activities between normal individuals and known carriers, the results of the CPK estimation must be combined with the pedigree data in order to come up with an overall probability that a female is a carrier. Carrier detection in haemophilia A families presents a similar problem, the carrier detection test in this case being combination of the Factor VIII levels and the levels of circulating immunologically cross reacting material.

In X-linked recessive biochemical disorders such as the Lesch-Nyhan syndrome, Hunter syndrome (mucopolysaccharidosis type II), and Fabry's disease carrier detection may be facilitated by estimating enzyme levels in cloned fibroblasts. In carrier females one would expect to find two types of clone, one with normal activity and one with enzyme activities in the male hemizygote range. This is a manifestation of the Lyon hypothesis. Hair roots can be used as an extension of this approach because they are thought to be derived from very few cells.

DERMATOGLYPHICS

The dermal ridges of the palms and finger tips carry the openings of the sweat ducts which are just visible to the naked eye. Their function is still unknown: it may help to give better friction in grip or it may be to prevent obstruction of the sweat ducts on flection of the skin. They form parallel lines covering the surface. Where three systems of parallel lines meet, provided each of the angles between them is at least 90°, they form a central point termed a triradius. There are three main types of pattern that may interrupt the parallel system on the finger tips. These are the simple *arch*, the *loop* in which a ridge turns back on itself through 180° and which is always associated with one triradius and the *whorl* associated with two triradii (Fig. 18.10).

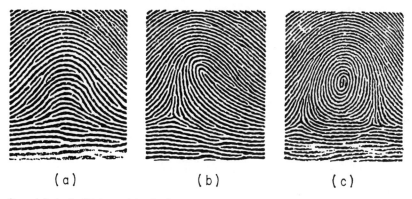

(a)　　　　　(b)　　　　　(c)

Figure 18.10. (a) *Arch*; (b) *loop*; (c) *whorl*.

The ridges are formed during the third and fourth months of fetal life and follow patterns that are retained for the rest of life. The departures from simple parallel ridges to more complex patterns reflect the shape at the time the ridges are laid down, and the patterns formed cover the surface in the most economical way possible following topological principles. Quantitative studies have shown that the ridge patterns are genetically determined in a multifactorial manner.

The major practical application of these studies is obviously forensic, but they do have clinical uses. The ridges can be observed by direct inspection in good light, with, if necessary, a magnifying lens. They are easily recorded by a clean method using a starch impregnated paper containing an iodide which is released by oxidizing agent wiped on the skin.

Characteristic patterns are seen in most of the established chromosomal disorders. Unusual patterns in children with retarded mental development, especially if unlike those of the parents, are suggestive of a developmental insult dating from at least the time the ridges are laid down. Dermatoglyphics are also of value in determining the zygosity of twins.

ASSESSMENT OF GENETIC RISK

There are three essential steps in assessing the risk of a genetic disorder recurring in a family. These are an accurate initial diagnosis; family history, including if necessary family investigation, details of age, sex and medical history of relatives, miscarriages, consanguinity and the ante-natal and labour record of the index patient; and a knowledge either of the mode of inheritance or of the observed frequency of recurrence of the disorder in question. Frequently a specific phenotype may be caused by genes exhibiting different modes of inheritance or solely by environmental factors. The counsellor must be aware of this heterogeneity of aetiology. Such information is available in standard text books and original reports. Only a minority of diseases or malformations are determined by a single major gene with Mendelian inheritance. Most disorders for which genetic advice is sought are either chromosomal or due to a complex mixture of environmental and genetic factors.

Recurrence of chromosomal disorders

Most forms of chromosomal disorder carry a low risk of recurrence. If a normal couple has one child with Down's syndrome due to standard trisomy-21, and if parental mosaicism is excluded, the risk of recurrence will depend on the mother's age. Some families are thought to have a predisposition to non-disjunction and consequently the risk is higher than the usual population risk for the mother's age, especially if any close relatives have had a child with Down's or any other chromosomal anomaly. The average recurrence risk is about 1 in 100. If either parent has Down's syndrome or is mosaic for it then the risk of recurrence are very high, up to 1 in 2. Higher risks are also found in translocation cases, where one parent carries the translocation (about 10 per cent for a female carrier 2 per cent for a male carrier of a 14/21 translocation). A fertile male of XYY sex constitution is liable to have a proportion of XYY and of XXY sons and similarly an XXX woman may have an XXY son or an XXX daughter. These possibilities should, therefore, be excluded if advice is sought as to the recurrence of the triple-X, XYY or Klinefelter anomalies.

A succession of children with multiple congenital anomalies, even if different, could be due to a structural chromosomal rearrangement in one of the parents and can be looked for in such cases. Although recurrent abortion is seldom due to genetic factors it is known that in a third to a half of all spontaneous abortions the foetus has a demonstrable chromosomal abnormality. It is therefore, to be expected that in occasional instances recurrent abortion will be due to a transmissible chromosomal anomaly in a parent.

Recurrence of single gene defects

In disorders due to a single major gene, advice is easy to give and will usually be just a matter of quoting the Mendelian ratios of 1 in 2 in a dominant disorder or 1 in 4 for a recessive. However, even here a number of complicating factors may be present. In a dominant disorder the expression of the gene may not always be complete and then the risk of recurrence will be less than 1 in 2, although the carrier rate will be unchanged. If the parent is not himself affected but some other member of his family is, an estimate may have to be made of the risk of the parent being a carrier who may manifest the disorder late in life. This may be a problem in Huntington's chorea, for example. A parent can also carry a harmful gene without manifesting any disease, as in dystrophia myotonica. If a dominant disorder has appeared in a child for the first time the relative probability of fresh mutation or of prior failure to manifest have to be estimated.

In recessive disorders advice may be sought on the risk of a potential carrier having affected children. Even in the absence of any method of directly detecting carriers it is always possible to calculate a probability based on the known gene frequency. In conditions with intermediate inheritance like the thalassaemias, carriers can usually be directly detected.

X-linked recessive disorders call for the skill of the counsellor in calculating the probability that a given female is a carrier. The results of carrier detection tests may have to be combined with the pedigree data in order to come up with a final probability. Such calculations are by no means straightforward. Programmes for computers and programmable calculators are available as are nomograms and tables for specific situations.

Recurrence of genetically complex disorders

The problem of heterogeneity of aetiology also arises with conditions that are not determined by a single gene. Whatever the overall risk of recurrence, as determined from observations of many families, the individual family history may lead to modification of the empirical risk. Such empirical risks vary from one community to another and that for the community of the family concerned should be used. Reliable risks have been determined for common disorders like the major malformations of the central nervous system, the common cardiac defects, talipes, congenital dislocation of the hip and pyloric stenosis. For rarer congenital disorders the number of recorded instances of familial recurrence is too small for any reliable figure to be given. For all these conditions the likelihood of recurrence is greatly increased if there is already more than one affected child. For example the risk of recurrence of a neural tube defect rises from about 1 in 20 after a single affected child to about 1 in 10 after two such children.

Cousin marriage

Occasionally advice is sought on the risk of a cousin marriage. First cousins have 1 in 8 of their genes in common and hence have a greater risk of having children homozygous for a recessive condition. If there is any evidence of recessive disorder on either side of the family the risk for the particular condition is obviously increased. Because every person in the population is thought to carry 1 or 2 detrimental recessive alleles, first cousins are usually quoted a 2–3 per cent risk of having a child with a recessive disorder even if there is nothing suspicious in the family history.

ANTENATAL DIAGNOSIS

The ability to detect serious developmental defects early in pregnancy, in time for a termination if necessary, has revolutionized the management of families at risk for these disorders. Much ingenuity has been applied to the problem of antenatal diagnosis, resulting in the development of different methods for different problems. These methods are summarized below.

Amniocentesis. The indications for amniocentesis in prenatal diagnosis have to be weighed against the risk of the procedure. The risk of amniocentesis at 14–16 weeks is still being evaluated, however, it appears that there may be up to 1 per cent chance of precipitating a miscarriage and an additional risk, of the same order of magnitude, of causing minor malformations such as talipes or dislocation of the hip. The present indications for amniocentesis are as follows:
(i)   Risk of chromosomal anomalies:
    Maternal age: at present the recommended age is roughly 38 years or over, (this depends upon the assessment of the risk of the procedure and the availability of laboratory and obstetric facilities. In the U.S.A. a cut-off of 35 years is used).
    Previous child with a chromosome anomaly due to non-disjunction.
    Parent a carrier for chromosome re-arrangement.
(ii)   X-linked recessive disorders:
    Women who have a high risk of producing a male affected with an X-linked recessive disorder that cannot be detected in utero may elect to have fetal sexing carried out by chromosome analysis and to have all males aborted. However, in this case, at least 50 per cent of aborted males will be unaffected.
(iii)   Risk of inborn errors of metabolism:
    Biochemical diagnosis of many disorders can be carried out on cells and amniotic fluid. An accurate biochemical diagnosis of previously affected sibs or of carrier status in the parents is necessary.
(iv)   Risk of diseases linked to marker loci:
    Certain conditions may not be expressed in amniotic fluid or cells but the disease

loci may be linked to antigen or enzyme loci that are. In a minority of families genotyping amniotic cells or fluid may enable the geneticist to calculate the probability that a fetus carries the abnormal gene. This approach has been used for dystrophia myotonica utilizing linkage to the secretor locus for the ABO blood group substances, and for 21-hydroxylase deficiency which is linked to the HLA region. Restriction enzyme recognition site polymorphisms (see below) will play an increasingly important rôle in linkage analysis.

(v)   Risk of neural tube defects:

In couples who have had one previous child with a neural tube defect, there is about a 5 per cent risk of recurrence. If one parent is affected or has extensive spina bifida occulta, the risks are about the same. Neural tube defects can be detected by finding an increased level of α-fetoprotein in the amniotic fluid.

Estimation of α-fetoprotein levels in maternal serum appears to be a good indicator of an affected fetus. Population screening of all pregnant mothers between 16 to 18 weeks of pregnancy helps to define an at risk group upon whom amniocentesis can be carried out. Serum screening at present detects about 90 per cent of anencephalic and 80 per cent of open spina bifida cases. Women with a high serum α-fetoprotein level have about a 15 per cent risk of carrying an affected fetus—although this figure varies with the cut off level used.

(vi)   Risk of diseases detectable by analysis of DNA:

Advances in molecular biology have made it possible to detect the presence or absence of specific genes and to recognize certain changes in the base sequence of the genome by directly analyzing the DNA. These techniques are dependent upon the availability of a single stranded, radioactively labelled *complementary DNA (cDNA)* probe for the gene under investigation. At present this cDNA is prepared from purified m-RNA for the appropriate gene by the use of an enzyme known as reverse transcriptase. The m-RNA is obtained from cells actively synthesizing the protein coded for by the specific gene. The cDNA is mixed with denatured single stranded DNA from the cells under investigation. If the specific gene is present the cDNA will hybridize with it to reform a double helix. The kinetics of such hybridization can be studied and the number of genes present can be inferred.

Greater specificity can be obtained by the use of enzymes, known as *restriction enzymes*, that cleave double stranded DNA at specific base sequences. Using different combinations of these enzymes the genome of a cell can be cut up into fragments and analyzed by electrophoresis in an agarose gel. Sections of DNA containing a specific gene can be identified by overlaying the gel with cDNA for that gene. Restriction enzyme analysis not only enables gene deletions to be identified but also provides a means of detecting base changes that result in an alteration of specific restriction enzyme recognition sites. In the latter case an alteration in the size of DNA fragments containing a specific gene would be obtained.

At present these techniques have been used to identify gene deletions, as in certain forms of homozygous α-thalassaemia and to detect alterations in a restriction enzyme recognition site such as the base change that gives rise to the sickle cell β-globin gene. It is envisaged that cDNA for many different types of gene will be available in the future, as more sophisticated techniques for isolating specific m-RNA molecules are developed, and through the development of techniques for the artificial manufacture of specific DNA sequences. This will greatly increase the range of disorders that can be diagnosed by these methods.

Fetoscopy. This is the direct visualization of the fetus using a fibreoptic fetoscope into the amniotic cavity via a transabdominal 2–3 mm cannula. A wide range of disorders can be diagnosed by inspection of specific parts of the fetus. In addition, fetal blood sampling can be carried out either from a placental vein or from an umbilical vein near to the insertion into the placenta. This enables a number of haematological disorders to be diagnosed including thalassaemia, sickle cell anaemia and haemophilia. Skin biopsies may also be obtained from the scalp of the fetus. The range of disorders that can be diagnosed by these methods is limited only by the ingenuity of the investigator in devizing a suitable micro-assay or a means of histological diagnosis for the disorder in question and by the problem of whether the disorder is expressed at the time of fetoscopy.

At present fetoscopy carries a higher risk to the fetus than does amniocentesis, so that the risk of a given disorder must be carefully assessed and weighed against the risk of the procedure.

Ultrasound. Ultrasound is already used in the diagnosis of anencephaly and certain cases of spina bifida as well as in the diagnosis of renal malformations. Technical advances in the future, allowing for greater resolution, should make it possible to diagnose certain bone dysplasias, limb defects, and malformations of the internal organs including the heart and brain.

Radiography. The fact that the long bones, ribs and spine are all largely ossified at 16 weeks gestation makes the diagnosis of certain bone dysplasias and limb defects by radiography feasible at around this time. Special skill is required, both in the technique of positioning the mother and fetus for radiography, and in the interpretation of the subsequent X-rays. Contrast medium injected into the amniotic fluid in order to outline the fetus is occasionally used.

Practical assessment

The affected individual who brings a family to the attention of the medical profession is known as the *proband* (or *propositus*). Individuals seeking genetic advice are known as *consultands*. Assessment of genetic disease within families starts with the investigation of the proband, moves on to enquiries about the pedigree and can include investigation of other family members.

ASSESSMENT OF PROBAND

1 History, examination, routine tests according to nature of the condition.
2 Special tests where appropriate (see text); buccal smear, chromosome analysis of lymphocyte or fibroblast cultures, enzyme studies, dermatoglyphics.
3 Accurate diagnosis; supra-specialist advice may be needed for rare neurological, eye, ear or bone abnormalities etc. For rare congenital malformation syndromes advice can be sought from syndrome identification services and computer data bases.

FAMILY HISTORY

Racial and geographical origin of the family, at least a three generation pedigree of affected and unaffected family members, evidence of consanguinity.

ASSESSMENT OF AFFECTED FAMILY MEMBERS

As for the proband.

ASSESSMENT OF NON-AFFECTED FAMILY MEMBERS

1 Biochemical and haematological carrier detection tests, where applicable.
2 Chromosome analysis—if a translocation carrier or a mosaic individual is suspected.
3 Other tests to identify minimally affected gene carriers; e.g. Wood's Lamp examination of the skin for tuberose sclerosis, slit lamp examination for dystrophia myotonia.

ASSESSMENT OF GENETIC RISK

1 Formulation of an accurate diagnosis including assessment of the mode of inheritance.
2 Estimation of penetrance, expressivity etc, from the actual family and from other reported cases.
3 Pedigree analysis including Bayesian calculations, estimation of degree of inbreeding etc. (computer programmes are available).

ANCILLARY METHODS

1 Linkage analysis.
2 Antenatal diagnosis.

References

HARRIS, HARRY. (1982). *The Principles of Human Biochemical Genetics,* 3rd ed. North-Holland Publishing Company, Amsterdam and London; American Elsevier Publishing Co., New York.

MCKUSICK, V.A. (1983). *Mendelian Inheritance in Man,* 6th ed. The Johns Hopkins Press, Baltimore.

SMITH, D.W. (1982). *Recognizable Patterns of Human Malformation,* 3rd ed. W.B. Saunders Company, Philadelphia.

STANBURY, J.B., WYNGAARDEN, J.B. and FREDRICKSON, D.S., eds. (1983). *The Metabolic Basis of Inherited Disease.* 5th ed. McGraw-Hill Book Company, New York.

STEVENSON, A.C. and DAVISON, B.C.C. (1976). *Genetic Counselling,* 2nd ed. William Heinemann Medical Books, London.

THOMPSON, J.S. and THOMPSON M.W. (1980). *Genetics in Medicine,* 3rd ed. W.B. Saunders Company, Philadelphia.

YUNIS, J.J., ed. (1977). *New Chromosomal Syndromes.* Academic Press, Inc., New York, San Francisco and London.

HARPER, P.S. (1981). *Practical Genetic Counselling.* John Wright, Bristol.

19

Sex and Reproduction

Normal function

The sex of an individual may be described in terms of the genetic constitution (genotype), in terms of the external appearance (phenotype) and in terms of the sexual role assumed in society and personal relationships (gender). Studies of patients reared in a sex that is discordant with their chromosomal, gonadal, hormonal, or even phenotypic sex have shown that there is no single feature that determines a person's gender. There is, instead, a hierarchy of sexual determinants which, when in harmony with each other, produce the differences between men and women that we all recognize so easily. When relationships between the different components break down, various forms of ambiguous sexuality occur. Although many are explained by genetic, embryological and hormonal disturbances, what finally determines how individuals perceive themselves sexually, and are perceived by others, remains poorly understood. It is probable that gender and psychosexual orientation are mainly determined by complex social and psychological influences.

PRE-NATAL SEXUAL DEVELOPMENT

Normal development of the sexual organs involves four distinct processes which follow one another in an orderly sequence.

The first, which is the determination of genetic sex, depends on the complement of chromosomes established at fertilization. Division of cells before the formation of gametes results in a halving of the number of chromosomes (meiosis or reduction division) so that the ovum contains only a single X chromosome and the spermatozoon either an X or a Y chromosome. When fertilization occurs and the maternal and paternal genetic material are united to form the zygote, the original number of chromosomes is restored. If the spermatozoon that fertilizes the ovum carries a Y chromosome the offspring will be genetically male (46XY); if it carries an X chromosome the genetic sex is female (46XX).

When embryonic female cells divide, replication of the DNA of the pair of X chromosomes is asynchronous and the chromosome whose replication is delayed condenses and moves towards the nuclear membrane, where it can be identified as a darkly staining mass, the Barr body. Although cells occasionally acquire several X chromosomes, all but one condense, thus resulting in the formation of several Barr bodies. The number of Barr bodies is therefore always one less than the number of X chromosomes. The condensed chromatin that forms the Barr body is genetically largely inert.

The second process of differentiation is development of the gonads. At about the fourth week of gestation primordial germ cells and celomic epithelial cells coalesce to form a primitive gonad, the sex of which cannot at this stage be determined morphologically. Further development depends upon the individual's chromosomal complement. Formation of a testis is induced by a cell surface protein whose presence can be detected immunologically—the H-Y antigen. The rate of production of this protein is determined by genes located near the centromere of the short arm of the Y chromosome. In the presence of the H-Y antigen, germ cells and epithelial cells form cord-like structures which gradually form a lamina propria and become seminiferous tubules. The tunica albuginea of the testis appears as the lamina propria is completed.

In the absence of the H-Y antigen, the majority of germ cells enlarge to become oogonia. At about 12 weeks the oogonia enter an early stage of meiosis and are then termed oocytes. This event marks the morphological transition of the indifferent primitive gonad into an ovary. Unless oocytes undergo atresia, they remain in this early (diplotene) stage until just before ovulation when meiosis is completed. By the fifth month of the intrauterine life, the ovaries contain about 7 million germ cells but thereafter the number rapidly falls so that at birth there are only 2 million and at the age of 10, about 300,000 oocytes. No new oocytes are formed postnatally. They disappear almost entirely from the ovaries by the age of 50.

The rate of atresia of oocytes is genetically determined. In the absence of the second X chromosome (as occurs, for instance in 45XO Turner syndrome—page 687) the rate is so rapid that, compared with a normal fetus, the reduction in the number of oocytes can be detected as early as 16 weeks of gestation and depletion of the pool of oocytes is usually complete before the age of puberty. The failure of oocytes and of follicles to persist in the ovaries of subjects with this genotype indicates that, normally, inactivation of the late replicating X chromosome is incomplete and that both X chromosomes are required for complete ovarian function. The few oocytes lost during normal ovulation cycles (page 663) contribute so little to the overall rate of atresia that prolonged anovulation (as occurs, for instance, during treatment with oral contraceptives) does not delay the menopause.

The third process of sexual development involves differentiation of one of the pairs of the internal ducts, with regression of the other (Table 19.1). The Müllerian and Wolffian ducts, in contrast to the primitive gonad, are unipotential and the internal genitalia, therefore, develop from specific primordia. By the seventh week of intrauterine life, both duct systems are present. The inherent direction of development is along female lines and, in the absence of a fetal gonad, structures derived from the Müllerian duct (uterus, fallopian tubes and the upper third of the vagina) persist and the Wolffian ducts involute. Masculine differentiation requires the presence of a testis, which causes

Table 19.1. Development of internal genitalia

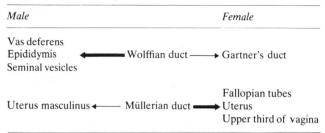

Male		Female
Vas deferens		
Epididymis	◄━━━━ Wolffian duct ━━━►	Gartner's duct
Seminal vesicles		
		Fallopian tubes
Uterus masculinus ◄───	Müllerian duct ━━►	Uterus
		Upper third of vagina

both regression of the Müllerian duct and stimulation of the Wolffian duct. Regression of the Müllerian ducts occurs at about the time the tunica albuginea of the testes develops and is induced by a protein (molecular weight of about 3,000 daltons) secreted by the Sertoli cells of the seminiferous tubules. This *Müllerian Regression Factor* acts locally—thus, in experimental animals, after unilateral orchidectomy of early embryos, Müllerian regression only fails to occur (that is, an oviduct develops) on the side lacking the testis.

Development of structures derived from the Wolffian duct (vas deferens, epididymus and seminal vesicles) occurs after Müllerian regression has occurred. Wolffian development is directly stimulated by high local concentrations of testosterone, secreted by the Leydig cells of the fetal testis under the influence of chorionic gonadotrophin.

The fourth process of sexual differentiation is the development of the external genitalia from structures derived from the urogenital sinus, the genital tubercle and the genital swelling (Table 19.2). Like the gonads, the primordia of the external genitalia are

Table 19.2. Development of external genitalia

Male		Female
(Determinant is presence of dihydrotestosterone)		(Determinant is absence of dihydrotestosterone)
Penis ◄───────	Genital tubercle ─────►	*Clitoris*
(Corpora cavernosa and glans)		(Corpora cavernosa and glans)
Corpora spongiosum ◄──	Urethral folds ───►	*Labia minora*
(Encloses penile urethra)		
Scrotum ◄───────	Labio-scrotal swellings ───►	*Labia majora*
Prostate ◄───────	Urogenital sinus ───►	*Vagina (lower two thirds)*
(Bulbo-urethral glands)		Para-urethral glands; (Bartholin's glands)

bipotential and, in the absence of androgens, they develop along female lines. In contrast to the structures derived from the Wolffian duct, virilization of the external genitalia is induced by dihydrotestosterone (page 681) rather than by testosterone. Dihydrotestosterone is not secreted by the testis but is formed locally in the primordial cells of the external genitalia by 5-α reduction of testosterone. Conversion of testosterone to dihydrotestosterone occurs in these cells before the Wolffian ducts differentiate but, when virilization of the external genitalia is completed, Wolffian duct derivatives themselves acquire 5-α reductase activity and once differentiated they too will respond to dihydrotestosterone.

Virilization of the external genitalia is thus the first embryological manifestation of androgen action. Initially the anogenital distance lengthens and then the urethral folds fuse in the mid-line to form the corpus spongeosium which encloses the phallic urethra. The labio-scrotal swellings, which are lateral to the urethral folds, fuse to form the scrotum and the ventral epidermis of the penis. The corpus cavernosum and glans are formed by enlargement of the genital tubercle (Table 19.2).

Development of the vagina

At about the tenth week of fetal life, the distal portions of the Müllerian ducts fuse into a single cord, the caudal end of which forms the Müllerian tubercle. As the Müllerian tubercle contacts the urogenital sinus it triggers cell division in its epithelium and the two structures fuse to form a solid vaginal plate. Later the central cells of this plate break down and epithelial cells from the urogenital sinus invade the Müllerian duct and extend upwards to form a junction of squamous and columnar cells at the level of the external cervical os.

Exposure of the fetus to stilboestrol (and also to other non-steroidal oestrogens such as dienoestrol and hexoestrol) disrupts the orderly and complete replacement of Müllerian by urogenital sinus epithelium. Thus in a large proportion of girls born to mothers who received stilboestrol during pregnancy, columnar epithelium, normally present only above the cervix, is also found in the vagina. This abnormally sited columnar epithelium gives rise to a variety of anatomical defects, the commonest of which is vaginal adenosis. At about the time of puberty these columnar cells may undergo malignant transformation to produce a multicentric vaginal adenocarcinoma. More commonly squamous metaplasia occurs, thus causing the alternative risk of the development of a squamous carcinoma of the vagina. This hormonal teratogenesis is of particular importance because the malignant change does not become evident until the second or third decade of life. Experiments in animals suggest that it is the increased oestrogen secretion at puberty that provokes malignant change.

In addition to these changes in the vagina, abnormalities of the fallopian tubes have also been reported in women exposed prenatally to stilboestrol. These changes, together with those in the sons of women given synthetic oestrogens during pregnancy (epidydimal cysts, oligospermia), emphasize the wide ranging vulnerability of the fetus

Table 19.3. Embryological Inducers male sexual differentiation

Process	Inducer			
	H-Y antigen	Müllerian regression factor	Testosterone	Dihydrotestosterone
Testicular differentiation	+			
Müllerian duct regression		+		
Wolffian duct virilisation			+	
Virilisation of urogenital sinus and external genitalia				+

at critical stages of development and suggest that current concepts of embryological inducers need to be expanded.

By the twelfth week of fetal life proliferation of the septum between the bladder and the vagina displaces the vagina posteriorly and gives it a separate opening. After this separation has occurred, fusion of the labio-scrotal folds is no longer possible, even under intense androgenic stimulation. Consequently the presence of labial fusion in a subject with female pseudo hermaphroditism (page 688), indicates that exposure to androgens must have occurred before the twelfth week of intrauterine life.

Summary

The inherent direction of pre-natal sexual differentiation is along female lines but in the presence of a male genotype and a functioning testis, masculine differentiation is imposed upon the fetus (Table 19.3). The testis secretes two hormones which act as embryological inducers—Mullerian Regression Factor which causes local regression of the Müllerian ducts, and testosterone, which in high concentrations directly stimulates the development of Wolffian derivatives and which serves as a prehormone for dihydrotestosterone, which masculinizes the external genitalia.

POSTNATAL SEXUAL DEVELOPMENT

PUBERTY

Sexual development is a continuous process which remains incomplete until after puberty when procreation becomes possible.

In the male, during early embryonic life, the Leydig cells are under the influence of chorionic gonadotrophin and towards the end of the first trimester there is a marked increase of testosterone in fetal blood and amniotic fluid (Fig 19.1). As described earlier, testosterone is required by the external genitalia as a prehormone for dihydro-testosterone formation but it also directly stimulates Wolffian duct differentiation. The high levels of testosterone suppress fetal pituitary gonadotrophin secretion, as shown by the marked differences between male and female fetuses in blood and amniotic fluid gonadotrophin concentrations (Fig 19.1). Later, oestrogen produced by the feto-placental unit (page 668) suppresses fetal pituitary gonadotrophin secretion in both sexes. These observations indicate that the long-loop negative feedback relationship of testosterone and LH is present at the very earliest stages of sexual development. More-over, when the fetal pituitary is released from the negative feedback restraint imposed by the oestrogens of pregnancy, there is a striking surge of LH and testosterone secre-tion. The plasma concentrations of both hormones remain elevated for the first few weeks of life. This perinatal increase of LH and testosterone contributes to descent of

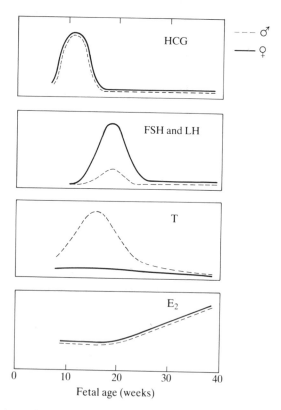

Figure 19.1. *Schematic representation of hormonal patterns (mean serum levels) in both sexes during fetal life in man. T = testosterone; E_2 = oestradiol-17β. Data derived mainly from Reyes et al (1974) and Clements et al (1976).*

the testicles into the scrotum, but in up to 10 per cent of boys descent is delayed until the first few weeks of extrauterine life and, indeed, may not occur at all in subjects with congenital LHRH, and therefore LH, deficiency.

After a few months the Leydig cells involute and in the normal boy there is little overt evidence of gonadal secretion. The Leydig cells do however remain responsive to chorionic gonadotrophin and they do secrete some testosterone, since prepubertal castration causes an increase in gonadotrophin secretion. The spermatic tubules are solid at birth; canalization and spermatogenesis begin at the onset of puberty and until that time the germ cells remain diploid. As the volume of the spermatic tubules increases, the testes enlarge in size—the first physical sign of puberty in boys. Testicular enlargement is also temporally associated with an increase in the secretion of testosterone, whose actions are manifest by the orderly appearance of the secondary sexual characteristics of men.

The earliest detectable endocrine change of puberty in both sexes is an increase in amplitude of pulsatile gonadotrophin secretion (page 676). Initially the increase only occurs at night but during mid-puberty daytime hormone secretion increases too. There is a gradual increase of the mean concentrations of gonadotrophins to reach the levels characteristic of adults.

In the female less is known of the prenatal and early postnatal secretion of sex steroids because of the unimportant role of oestrogens in early sexual development. During late childhood oestradiol secretion increases in a waxing and waning fashion (Fig 19.2), mirroring development and atresia of crops of follicles. There is a gradual increase of mean oestrogen levels, which is manifest phenotypically initially by breast development and subsequently by uterine enlargement and endometrial proliferation. Eventually enough oestrogen is secreted so that withdrawal bleeding occurs when the follicles undergo atresia and the levels of oestrogen wane. The menarche thus represents the uterine response to a rising but fluctuating level of oestrogen and does not indicate the onset of ovulatory cycles. Ovulation, which usually occurs only after 1–2 years of more or less irregular anovulatory menstrual cycles, develops when the neuro-endocrine process underlying oestrogen-mediated positive feedback matures (page 678).

The appearance of secondary sexual hair in girls is caused by androgens rather than oestrogens. The source is predominately the adrenal (hence the term adrenarche) but what causes the change in adrenal secretion at puberty is poorly understood. The change is not regulated by corticotrophin.

Maturation of the skeleton—the bone age—is in part determined by gonadal hormones, which together with thyroxine and growth hormone cause orderly development and later fusion of the epiphyses. Fusion of the epiphyses of long bones ultimately limits growth in stature but, in the absence of gonadal steroids, growth hormone and thyroxine continue to promote growth. In patients with hypogonadism developing in childhood this is reflected by the length of the span exceeding the height. The change

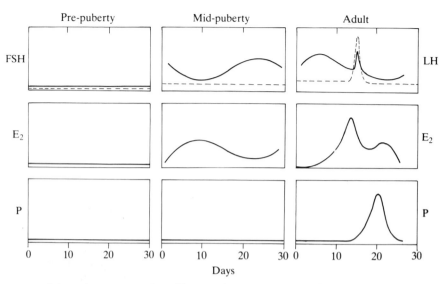

Figure 19.2. *Schematic representation of hormonal patterns in serum during puberty in girls. $E_2 =$ oestradiol-17β; P = progesterone. Note the increasing basal titres of gonadotrophins as pubertal development progresses, the onset of rhythmic FSH fluctuations by mid-puberty, and the establishment of the adult menstrual cycle pattern by one year postmenarche. E_2 levels also begin to show rhythmic changes by mid-puberty.*

in bodily proportions occurs because each arm and leg contains two epiphyses: when the height is measured the growth points of the limbs are measured in parallel but when the span is measured they are measured in series.

The initiation of puberty

The factors reponsible for the initiation of puberty are poorly understood. As mentioned, gonadotrophins can be detected in the pituitary and blood of fetuses from the third month of intra-uterine life. The concentrations are low in the serum of children and rise *pari passu* with the degree of sexual maturity. The concentraion of follicle stimulating hormone in serum rises earlier than that of luteinizing hormone, resulting in a fall of the ratio of FSH to LH with the attainment of sexual maturity. In young children with primary gonadal failure (for instance, 45XO Turner's syndrome) serum concentrations of FSH and LH are higher than those in normal children of the same developmental and chronological age, confirming the existence before puberty of a feedback mechanism for the control of gonadotrophin secretion. Possibly puberty is initiated by an upward 'set' of the feedback mechanism controlling gonadotrophin secretion. Certainly in agonadal children there is a secondary rise of LH and FSH concentrations at the time at which puberty would normally have occurred. Studies in animals have also indicated, however, that a major component of the pubertal process involves a change in gonadal responsiveness to gonadotrophins. Thus, in sexually mature rats, the testis responds to doses of LH that are several thousand times smaller than those required to stimulate testosterone production before puberty. This pubertal change in the response to LH is contingent upon prior exposure of the testis to FSH, indicating the importance of both gonadotrophins in the process of sexual maturation.

The clinical observation of the association of disorders of the timing of puberty with lesions in the posterior hypothalamic region indicates the important role of the hypothalamus in the initiation of puberty. The pineal secretes a hormone—melatonin—which in animals delays the onset of puberty through an action on the brain. In boys pinealomas may be associated with precocious puberty. Many factors, including nutrition, climate, light-dark relationships and the secretion of other endocrine glands are known to influence the pubertal process and it must be accepted that at present no single hypothesis can account for all the mechanisms involved.

FERTILITY

The overall success of reproduction depends upon integration of a variety of physiological mechanisms. Although social and psychological influences may overide them, only the primary functions of the genital organs essential for fusion of the male and female gametes and their genetic material will be discussed here.

Female fertility

The physiology of fertility in women is most conveniently considered at several discrete levels—the uterus and genital tract, the ovary and the hypothalamic-pituitary unit. Though cyclical activity at these levels is described separately they are of course closely linked and interdependent.

The uterine cycle. Menstruation refers to the cyclical shedding of an endometrium previously exposed to stimulation by endogenous oestrogen and progesterone. It is the overt manifestation of a series of morphological changes that occur in response to the fluctuating secretion of oestrogen and progesterone.

The endometrium consists of a surface layer of columnar epithelial cells and an underlying stroma. The surface of the endometrium is invaginated to form crypts ('glands') which are lined by cuboidal epithelial cells. The underlying stroma consists of spindle shaped cells permeated by blood vessels.

At the start of the uterine cycle (that is, after menstruation has occurred) mitotic activity increases in the basal unshed portion of the endometrium and in the stroma. The mucosa becomes thickened, the 'glands' elongated, and the cells columnar. This 'proliferative' endometrium is formed in response to stimulation by oestradiol, secreted in increasing quantities by maturing ovarian follicles. After ovulation, in response to progesterone, mitotic activity in the endometrium diminishes and the glands become coiled and secrete a mucus rich in glycogen. The cells lining the glands take on a characteristic appearance as the mucus displaces the nucleus towards the base of the cell and vacuolation develops beneath it. The stroma becomes increasingly vascular and its cells loosen and enlarge; those most superficial come to resemble the decidual cells of early pregnancy. These changes, which produce the 'secretory' endometrium, represent the response to stimulation by progesterone, secreted by the corpus luteum.

Both oestradiol and progesterone are needed to maintain the secretory endometrium. However, despite the large amounts of oestrogen secreted by the corpus luteum (Fig 19.3), proliferation of the endometrium does not persist during the luteal phase. It is thought that the altered response to oestradiol in the luteal phase is mediated by the action of progesterone which, in addition to its own specific morphological effects, limits oestrogen-induced proliferation of the endometrium. There are two mechanisms by which this occurs—firstly progesterone induces an enzyme in the cytoplasm of endometrial cells (oestradiol 17β-dehydrogenase) which converts oestradiol 17β to oestrone. Oestrone has a lower affinity than oestradiol for the cytoplasmic oestrogen receptor protein (page 683) and is therefore a less potent oestrogen than oestradiol. Secondly, progesterone inhibits synthesis of the cytoplasmic receptor proteins for oestrogen and so reduces the total amount of oestrogen that can enter the cell nucleus. Prolonged exposure of the endometrium to oestrogen in the absence of progesterone leads to continued proliferative activity, with the eventual development of endometrial

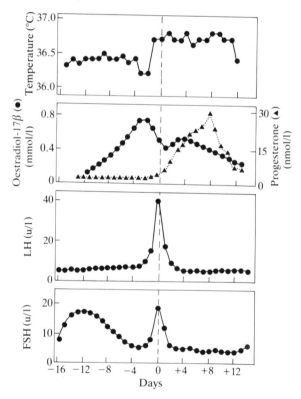

Figure 19.3. *Mean treads in serum hormone levels during the menstrual cycle.*

hyperplasia. This sequence of events, which can be reversed by treatment with progestogens, occurs in women with anovular menstrual cycles and in postmenopausal women who are treated with unopposed oestrogens. Since treatment with progestogens also reverses the hyperplasia induced by synthetic oestrogens, which cannot be converted to oestrone, it is likely that the direct inhibition by progesterone of synthesis of oestrogen receptors is more important than the induction of oestradiol 17β dehydrogenase.

The cyclical changes in the histology and cytology of the endometrium are very stereotyped and their appearance in endometrial biopsies can be used to identify the stage of the cycle (i.e, the relationship to ovulation) with considerable accuracy.

As the function of the corpus luteum wanes (Fig 19.3) and oestradiol and progesterone concentrations fall to critically low levels, the spiral arteries of the endometrium contract and cause an avascular necrosis of the superficial layers of the endometrium. The basal portion of the endometrium is supplied by straight arteries that are not sensitive to hormonal fluctuations and so this portion is left unshed. It appears that hormonally mediated alterations in the rate of prostaglandin production ultimately determine the vascular responses of the endometrium and therefore the pattern of uterine bleeding. Alterations in platelets and clotting factors are also important because heavy menstrual bleeding may complicate bleeding disorders.

Menstrual fluid is composed of blood, cellular debris and mucus. The blood normally clots and undergoes lysis within the uterine cavity, so menstrual fluid has a low content of platelets and fibrinogen. Heavy bleeding may result in rapid evacuation of menstrual fluid before intrauterine clotting and lysis have been completed and so causes clots to be passed.

Menstruation usually lasts between 3–7 days and the average blood loss is about 50 ml (range 10–80 ml). The quantity lost and the duration of the cycle vary with age, the variations being greatest at the extremes of reproductive life. The duration of the menstrual cycle is surprisingly variable, the mean and standard deviation for women aged 20–45 being 29 ± 7 days.

The cervix. Cyclical changes similar to those in the endometrium occur in the epithelium of the endocervix (the endocervical 'glands') and result in cyclical changes in the quantity and physical properties of the cervical mucus. In the first week of the menstrual cycle only small amounts of a viscous mucus are produced. As oestrogen levels rise the quantity of mucus increases 30-fold and it becomes less viscous and more elastic. Its penetrability by sperm is maximal in the periovulatory phase. Progesterone reverses these features so during the luteal phase cervical mucus is viscous and of poor penetrability.

The vagina. The vagina is lined by a statified squamous epithelium in which five cell layers can be recognized cytologically. Oestrogen stimulates proliferation and keratinization of the superficial layers, so that increasing numbers of acidophilic cells (which stain pink with the Papanicolou stain and have pyknotic nuclei) are formed in the pre-ovulatory phase. The ratio of the numbers of superficial to intermediate cells in a smear taken from the vaginal wall can be expressed as a karyopyknotic index (KPI). This index gives an estimate of the degree of oestrogen exposure of the vagina. Progesterone reduces the proportion of acidophilic cells, to the KPI falls during the luteal phase.

The ovary

The ovary consists of follicles in varying stages of maturation and atresia, distributed in a supporting tissue (the stroma).

The *primordial follicle* consists of an oocyte surrounded by a single layer of spindle shaped cells which are united to the plasma membrane of the oocyte by protoplasmic processes. The primordial follicle is separated from the surrounding stroma by a definite but inconspicuous basement membrane. These follicles, located almost exclusively in the cortex of the ovary, form a pool of non-growing follicles from which a number are selected for further development. The factors underlying the selection process are unknown but they do not involve peptide or steroid hormones. The term *primary follicle* is used when the cells surrounding the oocyte become cuboidal. These cuboidal (granulosa) cells divided to form a multilayered stratum granulosum. The granulosa cells secrete a mucoid substance that separates them from the oocyte (the zona pellucida) but they retain physical contact with the oocyte through processes which traverse the zona pellucida. As the granulosa layer is formed and the follicle migrates towards the medulla, adjacent stromal cells become arranged concentrically around the follicle to form the theca. The highly vascular inner layer (theca interna) undergoes epithelioid transformation and at the same time fluid-filled spaces appear between the granulosa cells. When these spaces become confluent and form an antrum, the term *Graaffian* or *antral follicle* is used. Antral fluid contains mucopolysaccharides which are secreted by granulosa cells in response to FSH, together with plasma proteins and steroid and peptide hormones. As the Graaffian follicle enlarges the concentration of oestradiol increases to reach the astonishingly high concentration of 1–2 g/l in the immediate preovulatory phase. Just before ovulation the diameter of the Graaffian follicle, some 200–400 times larger than the primordial follicle from which it was derived, is 1·8–2·2 cms—that is, large enough to be readily detected using modern ultrasound equipment. The follicle is also large enough to be punctured under direct vision through the laparoscope, so permitting the harvesting of human eggs for *in vitro* fertilization.

Ovulation occurs about 30 hours after the beginning of the mid-cycle surge of LH. The preovulatory follicle enlarges and a stigma appears on the surface of the ovary. Cells in this region enlarge and become filled with lysosomal vesicles which contain proteolytic enzymes. As these cells degenerate proteases are released into the surrounding tissue and cause sequential disintegration of the tunica albugenia, the theca externa and then the theca interna. At the same time the follicle expands, capillaries in the theca dilate and fenestrations appear in the vessel walls. Oedema develops in the follicular wall which then bulges through the stigma. The oocyte looses contact with the granulosa cells, meiotic division is reinitiated and, floating freely in the follicular fluid, the oocyte is finally discharged, with its cumulus surrounding it, into the peritoneal cavity.

After the mid-cycle surge of LH, the granulosa cells enlarge and acquire the typical ultrastructural appearances of steroid secreting cells to form a *corpus luteum*. Their high lipid content gives them a characteristic yellow colour, although of course when a fresh corpus luteum is seen at laparoscopy the associated bleeding makes it appear red (corpus haemorrhagicum). The luteinized granulosa cells secrete progesterone in response to LH. Oestradiol and androgens are also secreted during the luteal phase but their cells of origin are uncertain. Steroid secretion by the corpus luteum rapidly increases during the week following ovulation and peak plasma progesterone concentrations occur on day 19–23 of the cycle (Fig. 19.3). Not all the endocrine factors that maintain the corpus luteum are known with certainty but continued pulsatile secretion of LH during the luteal phase is definitely required.

If conception occurs and a syncitiotrophoblast is formed, luteal function is prolonged by stimulation by chorionic gonadotrophin. In the absence of HCG, the population of LH/HCG receptors on the luteinized granulosa cells falls, the blood supply diminishes and the corpus luteum involutes. Progesterone and oestradiol levels fall and menstruation ensues. The luteal phase normally lasts 11–14 days, during which time progesterone concentrations should equal or exceed 25 nmol/l, 7–10 days after ovulation (Fig 19.3).

The ovarian stroma consists of ordinary connective tissue and interstitial cells that are thought to arise from undifferentiated theca cells of atretic follicles. These interstitial cells undergo morphological changes on stimulation with LH and HCG and they are thought to secrete androgens. Hilar cells, which are morphologically indistinguishable from the Leydig cells of the testis, are found in the hilus of the ovary. Their origin and function are unknown, but hyperplastic and neoplastic changes may be associated with virilism.

Ovulation cycle (Fig. 19.3). During the luteal phase, serum gonadotrophin concentrations are low. They rise at the end of the luteal phase as oestrogen and progesterone concentrations fall (activation of negative feedback). The increase of FSH secretion recruits development of some 15–20 primary follicles. As these follicles mature and enlarge they secrete oestradiol which, *inter alia*, suppresses further secretion of FSH. The degree to which the declining amounts of FSH can stimulate further follicular development depends upon the number of receptors for FSH on the granulosa cells. Production of these receptors is stimulated by oestrogen, so FSH is concentrated in the follicles which make the most oestradiol. There is thus an intraovarian autoregulatory mechanism by which the follicle with the maximum steroidogenic potential becomes dominant and is selected for ovulation while the remainder undergo atresia. This autoregulatory mechanism accounts for the characteristic uniovulation of the human. During treatment of amenorrhoeic women with injections of gonadotrophins it may be overidden and multiple ovulations and conceptions then result.

Peripheral blood levels of oestradiol continue to rise in the late preovulatory phase and when they reach a critical level, or perhaps a critical rate of increase, there is an abrupt and massive discharge of LH (oestrogen-mediated positive feedback). This pre-ovulatory discharge of LH causes rupture of the follicle and transformation of the ruptured follicle into a corpus luteum. The corpus luteum secretes progesterone and oestradiol which in turn suppresses gonadotrophin secretion and so LH and FSH concentrations remain low during the luteal phase. If a pregnancy does not occur the corpus luteum involutes and the restraint on gonadotrophin secretion is released. The fall of oestradiol and progesterone causes menstruation as well as activating, by negative feedback, the increase of gonadotrophin secretion that stimulates follicular development for the next cycle.

Studies in women with amenorrhoea, treated with LHRH to induce ovulation, have shown that a normal ovulation cycle can be created if fixed amounts of the releasing hormone are given intermittently every 90–120 minutes throughout the cycle. These, and similar studies in animals, indicate the fundamental importance to normal gonadotrophin secretion of episodic stimulation of the pituitary with LHRH and they also show that the modulating effect of gonadal steroids is predominantly exerted at the pituitary level. In addition these studies make it clear that a major endocrine contribution to the organization of the ovulation cycle is provided by the ovary. Thus the mid-cycle surge of LH secretion which causes ovulation only occurs when the ovarian endocrine signal (rising oestradiol concentrations) indicates that follicular maturation is complete. If this signal does not occur, or is not interpreted correctly (that is, when there is a defect of oestrogen-mediated positive feedback), anovulatory menstrual cycles result.

The above account of hypothalamic-pituitary-ovarian interaction explains many features of the normal ovulation cycle. It does not, however, account for the spontaneous recovery of ovulation cycles which frequently occurs after an episode of amenorrhoea. It is clear that in addition to positive and negative feedback control of gonadotrophin secretion, there must be some other mechanism (perhaps in the ovary, perhaps in the central nervous system) responsible for the initiation and characteristic continuity of ovulation cycles.

The oral contraceptive

Steroid contraceptives act by preventing ovulation, fertilization and nidation. All the preparations cause alterations in the cervical mucus (making it thick and viscous and so impairing sperm penetration), affect the endometrium (by reducing the glycogen production necessary for nutrition of a blastocyst) and interfere with contractions of the uterus and fallopian tubes (so impairing transport of gametes and/or a blastocyst). The most effective contraceptives, however, are those that prevent ovulation.

When given alone in high doses both oestrogens and progestogens inhibit ovulation. Both steroids, however, disrupt the normal pattern of cyclical bleeding and in addition oestrogens in high dose have many systemic adverse effects. Progestogens in high doses disrupt gonadotrophic secretion so severely that amenorrhoea is common in patients given doses sufficient to block ovulation. When the dose is reduced, contraceptive efficiency falls but it can be improved by combining it with an oestrogen.

The combined oral contraceptive. Preparations currently available in the UK contain 50 μg or less of ethinyloestradiol or mestranol and various doses of one of five different progestogens. Ethinyloestradiol has the same chemical structure as oestradiol except for the addition of an ethinyl group at the 17α position. This structural modification protects adjacent portions of the molecule from hydroxylation and subsequent conjugation and so prolongs its biological action. Mestranol is inactive as an oestrogen until it is metabolized in the liver to ethinlyoestradiol. In terms of inhibition of ovulation, the two are equipotent.

The introduction of a 17α ethinyl group into the testosterone molecule produces ethisterone, which is an orally active androgen. Removal of the C19 carbon atom from ethisterone, to produce 19 norethisterone, converts the steroid into an orally active progestogen. The five progestogens currently in use are all derivatives of 19 norethisterone and *in vivo* all of them, with the exception of norgestrol, are metabolized to norethisterone before becoming biologically active. The l-isomer of norgesterol, however, known as D norgesterol, binds directly to progesterone receptors (the d-isomer is inactive). In current formulations androgenic and anabolic activity of these ethisterone derivatives are clinically unimportant. Their progestational potency varies widely but there are too few clinical studies to allow useful comparisons of this activity in the combined formulations presently used.

The combination of an oestrogen and a progestagen, which imitates the steroid milieu of the luteal phase, suppresses gonadotrophin secretion. Follicular development does not occur and endogenous levels of oestradiol 17β are therefore in the postmenopausal range. Since there is no preovulatory rise in oestradiol, the mid-cycle surge of LH secretion does not occur. Suppression of gonadotrophin secretion persists for several days after the preparation is stopped so there is normally little follicular development during the treatment-free week.

There are, currently, two types of progestogen-only contraception. *Low dose progestogens* (the "mini pill") impair fertility by a variety of actions but ovulation is only inhibited in 30–50 per cent of cycles. Contraceptive efficiency is correspondingly impaired but when ovulation is prevented the mechanism is thought to be through interference with oestrogen-mediated positive feedback. Follicular maturation, and therefore endogenous oestrogen secretion, is not inhibited, however, because negative feedback remains intact and there is therefore a high incidence of breakthrough bleeding.

Injectable progestogens such as norethisterone enanthate, or medroxy progesterone acetate in depot form, inhibit ovulation when given at about 3 monthly intervals and so provide effective contraception. Menstrual irregularities are, however, common and up to 80 per cent of subjects develop amenorrhoea during treatment. Return of fertile ovulatory cycles when treatment has been stopped is not as reliable when compared with discontinuing oral contraceptives.

MALE FERTILITY

The testis is composed of a myriad of tubules which drain through the rete testis into the epididymis. Spermatozoa develop' from primordial germ cells (spermatogonia) within the seminiferous tubules. Each tubule is lined by spermatogonia resting on a basal lamina, while spermatocytes, in later stages of development, are found at successively higher layers of the epithelium. Spermatogenesis consists of three major stages: replication of stem cells by mitotic division of spermatogonia, meiotic (reduction) division of groups of spermatogonia to form the haploid spermatocytes (that is, cells with half the chromosomal number) and metamorphosis of the structurally fairly simple spermatids (themselves formed from spermatocytes) into the complex and highly differentiated spermatozoa (spermiogenesis).

The supporting (Sertoli) cells also lie on the basal lamina. They have an elaborate system of thin processes that extend upward and surround spermatids. Large portions of spermatids are actually embedded in the cytoplasm of these cells, which provide mechanical support and protection and also participate in the nutrition of developing germ cells.

Spermatogenesis occurs cyclically, different areas of the testis and of any one seminiferous tubule being at different stages of development at any one time. Each cycle lasts about 16 days. The whole process, culminating the production of spermatozoa, consists of four consecutive cycles and takes approximately $2\frac{1}{2}$ months.

Both gonadotrophins and testosterone are required for the formation of mature spermatozoa. Testosterone stimulates the earliest phases. The influence of LH is indirect and mediated by the Leydig cells. It reflects the need for a high concentration of testosterone *within* the testis. FSH is also required, both for maturation of the later stages of spermatogenesis and for stimulation of the Sertoli cells to produce androgen-binding protein. This protein, which binds testosterone and dihydrotestosterone with high affinity, concentrates androgens within the tubule. The steroid-protein complex is transported to the epididymis where it dissociates to produce very high local concentrations of androgens.

Spermatogenic tissue is very sensitive to changes in temperature, particularly during the long meiotic pro-phase of the primary spermatocytes. Normally the temperature of the scrotal contents is $1°C$ lower than that of the central body core. The difference is maintained by variations in blood flow and in the position of the testis: contraction of the dartos muscle occurs in response to cold. The blood supply helps to maintain the gradient since the testicular artery courses through the pampiniform plexus, which allows pre-cooling of the arterial blood. Obliteration of the temperature difference, as occurs in cryptorchism, and in men with varicoeles, impairs spermatogenesis and fertility.

Spermatozoa are transported to the epididymis by peristaltic movements of the seminiferous tubules and by ciliary movements of the cells lining the ductus efferens. It is only after passage through the head of the epididymus (which takes about 2 weeks) that sperm acquire fertilizing ability. After being stored in the tail of the epididymis further transport takes place during ejaculation. The spermatozoa are then diluted by secretions of the prostate and seminal vesicles. Fructose, provided by the seminal vesicles, provides the energy source required for motility. Considerable numbers of spermatozoa are stored in the seminal vesicles so that, depending upon the frequency of ejaculation, fertility may be preserved for weeks or months after vasectomy.

Fertility in men depends, therefore, upon adequate production, storage, and transport of spermatozoa. In addition to the specific effects of FSH and testosterone on spermatogenesis, all of the other processes involved are sensitive to androgens. Thus, in the absence of testosterone, spermatogenesis is reduced, the accessory glands involute and impotence and lack of libido develop.

TRANSPORT OF SPERMATOZA AND OVA IN THE FEMALE
AND FERTILIZATION OF THE OVUM

At coitus 100–150 million spermatozoa are ejaculated. Spermatozoa have been recovered from the oviducts as early as 30 minutes after coitus, but less than a fraction of 1 per cent of those deposited in the vagina reach the ovum. Transport in the oviduct is achieved by flagellar movements of the spermatozoa and muscular activity of the female reproductive tract. The physical characteristics of cervical mucus in the pre-ovulatory phase of the cycle (page 661) encourage the survival and forward movement of spermatozoa. At other times of the cycle, however, the mucus acts as a barrier.

Within minutes of ovulation, the ovum passes through the ostium of the Fallopian tubes. Contractions of the oviduct and ciliary movements of its epithilium propel the ovum into the ampulla, where, as a result of highly coordinated muscular contractions, it remains for several days. Fertilization takes place there. If fertilization does not occur the ovum then moves into the uterus. If it does occur, there is an interval of approximately one week before implantation, although the blastocyst enters the uterus about 2 days after fertilization.

PREGNANCY

Between fertilization and implantation, the ovum divides repeatedly, to form a solid sphere (the morula) which consists of about 200 cells. Fluid accumulates between the cells, leading to the formation of an outer layer (the trophoblast) and an inner layer (the developing embryo). The embryo with its trophoblast is called the blastocyst.

The trophoblast erodes the surface epithelium of the uterus and allows the blasto-cyst to sink into the underlying stroma. The embryo and its membranes thus become completely encapsulated by the endometrium and contact with the opposite wall of the uterus is prevented: the placenta thus develops on one side of the uterus only. The extent to which the trophoblast invades the uterus is limited by the depth of the decidua of the endometrium. In the fallopian tubes the decidual reaction is very slight, which explains why small tubal pregnancies perforate the tube so easily.

The trophoblast differentiates into two further types of cell: an outer syn-cytiotrophoblast, which burrows into the uterine stroma and comes into direct contact with the maternal blood, and an inner cytotrophoblast which grows into the parts of the outer layer projecting into the endometrium. The placental architecture is complete by about three months, which corresponds, in time, to the maximal concentrations of HCG (secreted by the syncytiotrophoblast) detected in the blood of the mother.

The feto-placental unit

The placenta represents the interface between mother and fetus, and, in addition to its nutritional function, it is rich in enzymes and capable of the synthesis of a variety of proteins and steroids. However, as described below, the placenta is incapable of per-forming the complete sequence of biosynthetic steps necessary to produce the high concentrations of steroid hormones that characterize pregnancy. In order to do this it utilizes steroid precursors which are made in the fetus. This interdependence has led to the concept of a 'feto-placental unit', which produces the high concentrations of oestro-gen and progesterone that circulate during pregnancy (Fig. 19.4).

In contrast to the fetal adrenal cortex, the placenta is rich in the enzyme 3β-hydroxysteroid dehydrogenase (3β-OH) and, as pregnancy advances, it synthesizes increasing quantities of progesterone. The major substrate of this essential precursor of steroids is free (non-esterified) cholesterol from the maternal circulation. In the placen-ta the enzymes required for further hydroxylation of progesterone are lacking so that progesterone is either passed into the maternal circulation or to the fetus. In the mother about 20 per cent of the progesterone is inactivated by conversion to pregnane-diol, whose excretion increases throughout pregnancy.

The fetal adrenal cortex is deficient in 3β-OH steroid dehydrogenase and it uses progesterone (derived from the placenta) as the precursor for the biosynthesis of gluco-corticoids and mineralocorticoids. The fetal adrenal cortex can synthesize pregneno-lone from acetate but, because of the lack of 3β-OH, it cannot oxidize it to progesterone. Pregnenolone is either conjugated in the fetus to form its sulphate or hydroxylated to form 17-hydroxypregnenolone; these are then converted to dehydro-epiandrosterone sulphate (DHAS) or dehydroepiandrosterone (DHA) respectively. DHAS is desulphated in the placenta and converted to androstenedione, which is in equilibrium

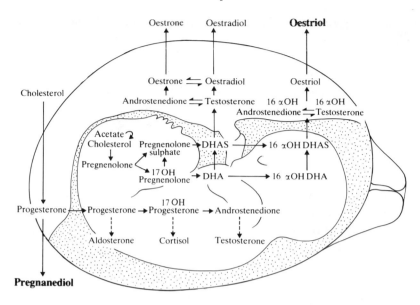

Figure 19.4. *Pathways of steroid biosynthesis in the feto-placental unit.*

The steroids measured in maternal urine are enclosed in boxes. Hydroxylation of dehydroepiandrosterone (DHA) and its sulphate (DHAS) at the C_{15} position occurs in the *fetal* liver. Following desulphation in the placenta these compounds are converted to C_{19} steroids which can be aromatized to form oestrogens, quantitatively the most important of which is oestriol. In contrast, pregnanediol is produced in the *maternal* liver from a precursor (progesterone) which is synthesized by the placenta alone.

with testosterone (Fig. 19.4). The placenta is rich in aromatizing enzymes and converts these compounds into oestrone and oestradiol-17β respectively. Oestriol is quantitatively the main oestrogen in maternal blood and urine: it is formed by the placenta, via 16α-OH DHAS, derived from the fetal adrenal cortex. The amounts of oestrogen in maternal blood or urine therefore reflect both the fetal production of precursors and the metabolic activity of the placenta. The major pathways of steroid biosynthesis by the fetus and the placenta are summarized above.

The lack of 16α-, , 21 and 11β-hydroxylating enzymes in the placenta, and the lack of 3β-hydroxy steroid dehydrogenase in the fetal adrenal cortex, emphasizes that, as far as steroid biosynthesis is concerned, both the fetal adrenal cortex and the placenta are endocrinologically incomplete. However, together, the 'feto-placental unit' needs no steroid precursors (with the exception of cholesterol) from the mother, and, in man, the outcome of pregnancy does not depend on the maternal adrenal or pituitary or, after the placenta is established, on the maternal ovary. However, the fetal adrenal does require the secretion of fetal corticotrophins. When these are absent, as in anencephaly, the concentration of oestrogens in maternal blood and urine is very low.

The role of progesterone and oestrogens during pregnancy. Progesterone (secreted by the placenta) is a precursor for the fetal adrenal cortex; with oestrogen it suppresses the secretion of pituitary gonadotrophins so that ovulation is inhibited during pregnancy. A high concentration of progesterone in maternal blood is also required for development of the breast and initiation of lactation (p. 672). Very high concentrations in the uterus itself are thought to inhibit uterine contractions so that parturition is delayed.

The role of the large amounts of oestrogen produced during pregnancy is less clear. However it is known that in women pregnant with a fetus with aryl sulphatase deficiency (a sex linked enzyme deficiency which prevents desulphation of steroid conjugates, and so results in subnormal oestrogen production (and excretion)) progress through labour is impaired because of failure of the cervix to dilate. This occurs despite normal myometrial growth and contraction, both spontaneously and in response to infusion of oxytocin. Women pregnant with a fetus with aryl sulphatase deficiency usually require delivery by caesarian section. Because of lack of eostrogen priming of the breasts and of the pituitary (page 673) fails to occur and they rarely lactate successfully. In the fetus exposure to oestrogen causes hypertrophy of the breasts and vulva; occasionally in females there may be uterine bleeding due to the sudden withdrawal of oestrogen after birth. In males there may be transitory metaplastic changes in the prostate as well as changes in the breast.

Shortly after birth, through an unknown mechanism, the fetal adrenal cortex involutes and the 3β-steroid dehydroxygenase enzyme appears in the adrenal of the newborn.

Protein hormones of pregnancy

The placenta makes a number of protein hormones of which the most important are human chorionic gonadotrophin (HCG), human placental lactogen (HPL) and relaxin. In addition, the placenta secretes several recently described glycopeptides (SPI, pregnancy plasma proteins A–D) into the maternal circulation. At the present time the biological actions and functions of these compounds are unknown. Human chorionic gonadotrophin is described on page 679.

Human placental lactogen (HPL) is synthesized by the syncytiotrophoblast where it may be identified as early as three weeks after fertilization. It is first measurable in the serum at about the fourth week of pregnancy and rises to a peak at term when about 1 gm is synthesized each day. There is a rapid turnover in the maternal circulation (the half-time of disappearance is only 30 minutes); its rate of metabolic clearance remains constant throughout pregnancy.

In the rabbit and the rat HPL is luteotropic and, in combination with HCG, it prolongs the life of the corpus luteum for longer than either of the hormones given alone. Whether this action has any significance in man is uncertain. In the tibia-width bioassay of growth hormone, HPL has weak growth-promoting activity when given alone, but it potentiates the action of growth hormone. HPL has considerable chemical similarity to growth hormone (163 of its 190 amino acids are shared with growth hormone) and indeed immunologically they cross-react; thus most radioimmunoassays for growth hormone are invalid during pregnancy. HPL shares many biological properties with growth hormone: it reduces the peripheral utilization of glucose, in the fasting state it stimulates the release of free fatty acids, and it increases the secretion of insulin in response to glucose loading. HPL is thought to account for the increasing requirements for insulin of diabetic patients who are pregnant. It also causes retention of nitrogen and calcium. These actions suggest that HPL acts in the mother to ensure the nutritional needs of the fetus. In addition, in animals, HPL is lactogenic: although secretion of milk does not occur during pregnancy it is likely that HPL contributes to the breast development which occurs at this time.

Relaxin is a low molecular weight peptide (about 6,500 daltons) which, first isolated from the ovaries of the pregnant sow and now identified in the corpus luteum of pregnant women, causes relaxation of the interpubic ligament. Its chemical sequence strongly resembles part of the insulin molecule. Concentrations of relaxin rise during pregnancy and fall after removal of the corpus luteum at any stage of pregnancy. With current methods of measurement relaxin cannot be detected in non-pregnant women or in men.

PARTURITION

The normal period of human gestation is terminated some forty weeks after conception by the onset of labour. Labour takes place by a combination of contraction of the uterine muscles and relaxation of the pelvic floor.

The mechanism of the initiation of labour remains uncertain. Throughout pregnancy the uterus undergoes numerous contractions which increase in frequency as the pregnancy matures. There are both hormonal and mechanical factors which favour their development. Oestrogens induce the synthesis of actomyosin and high energy phosphate in myometrial cells; they stimulate the uptake of catecholamines by the myometrium, and bring the membrane potential of the cells into the range in which spontaneous depolarization occurs. In this way they enhance the sensitivity of the uterus to several other stimuli. The most important mechanical factor is the increasing volume of its contents: the uterine muscles contract and hypertrophy in response to being stretched. At the same time, the excitability of uterine muscle is inhibited by progesterone, which, by causing hyperpolarization of the myometrial cell, reduces its membrane potential and blocks electrical propagation and, therefore, coordination of muscular contractions.

Labour is initiated by a change in the balance between stimulation and inhibition of uterine contractions. The balance is altered in favour of increasing contractions by several mechanisms, though which is most critical is uncertain. Mechanical factors, such as the increasing volume and activity of the uterine contents, and stretching of the cervix and vagina, may result, *inter alia*, in activation of an autonomic spinal reflex and also in release of oxytocin and prostaglandins. None of these mechanisms on its own is adequate, since labour is unimpeded by epidural anaesthesia and may be normal in the presence of hypothalamic lesions (such as those causing diabetes insipidus) in which lack of oxytocin may be presumed. Oxytocin is released in a pulsatile fashion, but since in maternal blood it is more frequently measurable after rather than before the onset of labour, it is unlikely to initiate the process. Oxytocin has a higher concentration in fetal than in maternal blood, but an overridingly important fetal contribution to the initiation of labour seems unlikely in view of the spontaneous delivery of stillborn fetuses. The major role of oxytocin appears to be the acceleration of established labour and the reduction of post-partum blood loss.

Prostaglandins are potent stimulators of uterine muscle and their concentration is high in amniotic fluid and maternal blood during—but not before—labour: like oxytocin, it appears that they contribute to but do not initiate labour.

In the later weeks of pregnancy, although the uterus continues to enlarge, the rate of production of progesterone increases less rapidly. More important than the blood concentration however is the uterine concentration, which is very high because of the proximity of the site of progesterone production. Progesterone is therefore well suited, both by its action and source, to exert a local blocking effect on uterine contractions and thereby to exert a potent regulatory role on the process of labour. The administration of exogenous progesterone is however less successful in delaying parturition than is the use of oxytocin in inducing it. This is probably explained by the route by which the hormones normally reach the uterus: oxytocin is secreted into the systemic circulation, whereas progesterone reaches the uterus directly and so achieves local concentrations which greatly exceed those in blood.

Relaxation of the cervix during labour is an active process that involves specific biochemical changes in connective tissue. It is stimulated by prostaglandins and by oestrogens. In pregnancies complicated by aryl sulphatase deficiency (page 670) in which oestrogen production is very low, the cervix dilates very poorly despite normal contractions of the muscles of the upper segment of the uterus. Ripening of the cervix can however be achieved by applying oestrogens and prostaglandins directly to the cervix.

In conclusion, labour occurs in response to a change in the balance between those factors stimulating and those factors inhibiting uterine contractions. The particular event which precipitates labour is uncertain—most likely it is a combination of events, whose complete analysis will be an important milestone in the attempt to control labour and so improve fetal survival.

LACTATION

Milk is a protein-containing fluid of highly specialized composition which is secreted by epithelial (alveolar) cells. The alveolar cells are arranged in acini which drain into specialised collecting ducts. The acini are encircled by myo-epithelial cells which, by contracting, eject the milk into the nipple. The glandular tissue is arranged in lobules and is supported by a stroma of connective tissue and fat.

From *in vitro* studies it appears that initiation of the division of cells destined to form the alveolae depends upon stimulation by insulin, growth hormone and other as yet unidentified factors in serum. The rate of cell division is increased by oestrogen. Differentiation of these cells to form the alveolar lobules requires the presence of hydrocortisone and prolactin. For the secretion of milk, prolactin is essential. *In vitro* studies have shown that it binds to specific high affinity receptors on the plasma membrane of alveolar cells and induces synthesis of several forms of nuclear RNA. New polysomes then enter the cytoplasm and there is an increase of protein (casein and lactalbumin) synthesis and milk is produced. Although prolactin interacts with a membrane-associated receptor it does not stimulate adenylate cyclase, so the signal must be processed by a mechanism which is different from that usually associated with polypeptide hormones. It is probable that prostaglandins contribute to the "second message" transmitted to the nucleus.

There is a specific oestrogen binding protein in the cytoplasm of alveolar cells which is also present in a proportion of cancers of the breast. In animals, however, the presence of the pituitary is required for oestrogen to have its full biological action on the breast. Finally, studies *in vivo* indicate that, in addition to the hormones mentioned, thyroxine and progesterone are required for full mammary development. Progesterone has an important role in stimulating development during pregnancy, but *in vitro* studies have shown that it also inhibits synthesis of messenger RNA in response to stimulation by prolactin. It thus inhibits nuclear transcription and therefore the production of milk proteins. Progesterone also reduces the number of prolactin receptors on the surface of alveolar cells. These various actions inhibit milk secretion during pregnancy in spite of the high levels at that time of all of the other hormones that stimulate the breast.

The breasts enlarge throughout pregnancy under the influence of the rising concentrations of oestrogen and progesterone, prolactin and HPL. Serum prolactin concentrations increase 1–20 fold in response to the increasing production of oestrogens by the feto-placental unit (page 668). They decline after delivery but increase in response to each episode of suckling. The sensory input for this neuro-endocrine reflex originates in nerve terminals in the nipple, stimulation of which even in the nulliparous subject results in release of prolactin. There is a striking impairment of breast sensitivity during pregnancy but within hours of delivery tactile sensitivity increases dramatically. A similar neuroendocrine reflex controls the release of oxytocin, which hormone aids in the ejection of milk by causing contraction of the myo-epithelial cells that surround the alveolar lobules and collecting ducts.

Postpartum initiation of lactation is related to a combination of events of which the most important is the direct effect on the breast of the declining concentrations of progesterone and oestrogen. Adequate production of prolactin is essential for the initiation and maintenance of lactation, and if prolactin secretion is inhibited at any stage milk production promptly ceases.

During postpartum (and inappropriate—see page 698) lactation, normal gonadotrophin secretion is disrupted. The most striking change is a reduction in pulsatile LH release. There is consequently impairment of oestrogen-mediated positive feedback and a fall of oestrogen levels without an increase of gonadotrophin secretion (i.e. a disturbance of negative feedback). This widespread abnormality of hypothalamic-pituitary control of gonadotrophin secretion is directly related to the hyperprolactinaemia and it causes amenorrhoea and therefore infertility. During weaning, as the frequency and duration of suckling falls, the number of episodes of hyperprolactinaemia falls too, gonadotrophin secretion returns to normal, ovulatory menstrual cycles resume and normal fertility returns.

In western societies supplementary feeding at an early stage is so common that lactational infertility is very unreliable but in societies where intensive lactation is practised (that is, when the breast is the sole source of the infant's nutrition) normal breast feeding provides the major method of family planning and population control.

THE CLIMACTERIC

In women there is a gradual decline in ovarian activity towards the end of the fifth decade of life. Initially there is resistance to further development of primordial and primary follicles. Consequently there is a subnormal increase of oestrogen secretion in the late follicular phase and therefore a failure of oestrogen-mediated positive feedback. In the absence of ovulation corpora lutea do not form so there is no increase of progesterone secretion. The uterus is subject to the influence of oestrogen alone, the levels of which may fluctuate considerably but without their normal periodicity. Endometrial hyperplasia may therefore develop (page 661) and cause prolonged and irregular bleeding.

As ovulation ceases plasma oestradiol concentrations fall, eventually reaching levels which cannot sustain the endometrium and menstrual bleeding then ceases. Secretion of FSH and LH increases to give the hormonal profile characteristic of primary gonadal failure. When follicular secretion of oestradiol has ceased, the major unconjugated circulating oestrogen in postmenopausal women becomes oestrone, produced by extraglandular conversion of androstendione, which is secreted by the adrenal cortex. Androstendione secretion is controlled by ACTH, and ACTH is subject to mechanisms of control quite independent of androgen and oestrogen production (page 553). The major determinant, therefore, of oestrone production after the menopause becomes the rate of extraglandular aromatization of androstendione. Aromatization occurs in fat, liver, muscle and brain tissue. Quantitatively the most important site is adipose tissue, so obese postmenopausal women have high rates of oestrone production and a higher than normal risk of endometrial carcinoma. Thin women have low rates of oestrone production and a higher than normal risk of osteoporosis.

The low oestrogen levels characteristic of postmenopausal women result in a variety of changes, such as atrophy of the genitalia and a general decrease in the rate of protein anabolism; the skin loses its youthful elasticity and the bones become osteoporotic. Vaso-motor instability, typified by hot flushes, also develops. The precise cause is uncertain—for instance, although they typically occur in young women after ovariectomy and can be prevented or cured by treatment with oestrogens, flushes are not related to particular concentrations of unconjugated oestrogen (or androgens) in blood. Each flush is associated with an abrupt discharge of LH, a phenomenon which represents a change in hypothalamic activity rather than a direct cause of the flush. Ultimately the patient's oestrogen status must be a fundamental component of the mechanism underlying climacteric flushes but the precise cause of this unpleasant symptom remains elusive.

In men as judged by hormone measurements, gonadal function remains remarkably stable in the healthy elderly population. The suggestion that there is a male counterpart of the gonadal failure of women was based on studies of hospitalized rather than healthy men. Studies of spermatogenesis in the elderly are rare and often based on histological examination of testes removed in the course of treatment of prostatic cancer. These studies are subject to the criticism that such patients represent an unhealthy sample of the population of elderly men. Nonetheless behavioural studies of the ageing male population certainly describe a marked reduction in sexual activity and potency with age but these changes do not seem to be related to concentrations of testosterone (or dihydrotesterone) in plasma.

HORMONES OF REPRODUCTION

Gonadotrophins

The gonadotrophins are carbohydrate-containing polypeptides which stimulate the gonads to secrete sex steroids and to release gametes. In man there are three gonadotrophins: luteinizing hormone (LH) or in the male, interstitial cell stimulating hormone (ICSH), follicle stimulating hormone (FSH) and human chorionic gonadotrophin (HCG). In other species prolactin is a gonadotrophin and is responsible for the maintenance of the corpus luteum—hence the earlier name, luteotrophin.

LH and FSH are secreted by the pituitary glands of both sexes. ICSH and LH are identical and FSH exists in the same form in men and women. Although much is known of the chemical nature of the gonadotrophins, preparations of 100 per cent purity have not been made, nor using classical methods has it been possible to synthesize them. Synthesis by gene transcription can, however, be confidently anticipated.

The gonadotrophins consist of 2 non-covalently linked carbohydrate-containing peptides of comparable size, designated α and β subunits. The α subunits in LH, HCG, FSH and thyrotrophin (TSH) are .very similar. They have no biological activity. The β

subunits of each hormone are quite disimilar and they confer the biological and immunological specificity of the hormones but like the α subunits they are inactive in the dissociated form. For biological potency, association of both subunits in the whole molecule is required. The common α subunit explains the immunological cross-reactions of the parent hormones since most antisera raised against one cross-react with the others.

Luteinizing hormone (LH). LH is synthesized in specific cells of the pituitary which can be identified immunologically and by electron microscopy. Its molecular weight is 32,000 daltons; that of its subunits half of this. Its half-time of disappearance after hypophysectomy (20 minutes) is shorter than that of FSH and HCG. The plasma concentration of LH oscillates throughout the day within a fairly narrow range. Some LH is taken up by the gonads but the majority is removed and metabolized, mainly by the kidneys.

In men, LH stimulates the interstitial (Leydig) cells of the testis to synthesize and to release testosterone, androstenedione and oestratheol. LH acts on interstitial cells via membrane associated adenyl cyclase whose activation causes an accumulation of cyclic AMP. Subsequently there is an increase in protein synthesis and conversion of a number of precursers of testosterone occurs. If protein synthesis is blocked experimentally, synthesis of testosterone does not increase; these data suggest that, *inter alia*, LH induces formation of enzymes required for the synthesis of testosterone, and indeed an increase of the enzyme responsible for conversion of 17-hydroxyprogesterone to androstenedione has been demonstrated in response to LH. In addition to its effects on steroidogenesis LH may also raise serum testosterone levels by increasing testicular blood flow.

LH also plays an important role in spermatogenesis, although this probably represents the requirement for a high concentration of testosterone within the testis, rather than a direct effect of LH on the germinal epithelium itself.

In women, LH stimulates oestrogen production by the ovary during both the follicular and the luteal phases. LH is required for the ovulatory process (p. 662) and it stimulates progesterone secretion by the corpus luteum. In the corpus luteum, as in Leydig cells, LH causes an increase of cyclic AMP. There is again an increase in protein synthesis, and steroidogenesis does not occur if this is blocked. LH increases the hydroxylation of cholesterol to 20-α hydroxycholesterol, with the subsequent production of progesterone.

The synthesis and release of LH is controlled by luteinizing hormone releasing hormone (LHRH). LHRH is a decapeptide which also causes release of FSH, though in acute experiments in adults the dose required is greater than that needed for the release of LH. In men the secretion and/or action of LHRH is controlled by prevailing levels of testosterone. Interruption of this feedback, for instance by castration, results in high levels of LH, which can be suppressed by treatment with testosterone. When plasma LH concentrations are measured at frequent intervals (i.e. every 15–20 minutes) it becomes apparent that the hormone is released episodically, in bursts which normally occur every 90–120 minutes. Following each increase of LH there is a measurable rise of plasma testosterone concentration. Modulation of the pituitary—Leydig cell axis can occur both by alterations in the mean plasma concentration of LH ("amplitude modulation") and by alteration in pulse frequency ("frequency modulation"). The amount of information contained in the hormonal signal represented by LH secretion is thus considerably enhanced by having two modes of communication. Evidence is also accumulating that steroid feedback control of LH involves differential effects on the amplitude and frequency of the episodes of LH release. The most familiar example of this occurs in women: the increase of oestrogen alone in the preovulatory phase of the menstrual cycle causes an increase in the amplitude of the pulses of LH; the addition of progesterone to oestrogen, as occurs during the luteal phase, reduces both of these features of LH secretion.

Episodic release of LH is a fundamental feature of normal LH secretion. It is characteristically lost in certain conditions (page 695). It depends on episodic stimulation of the pituitary by LHRH, an observation with immense therapeutic importance, since episodic treatment with LHRH can repair certain reproductive defects and be used to treat, say, delayed puberty and anovulatory infertility.

It is thought that pulsatile LH secretion is required by the target organ so that LH receptors, once activated by the trophic hormone, can be replenished and so respond to the next stimulatory pulse. If the concentration of the trophic hormone is raised but held at a constant level, the Leydig cell response becomes attenuated (desensitization) in part because the number of membrane-associated LH receptors cannot be maintained (down regulation of the numbers of receptors). Down regulation of receptors is an important mechanism of control in several endocrine systems (see, for instance, insulin secretion—page 525). The most familiar examples in reproductive endocrinology are, firstly, the reduction during pregnancy of progesterone secretion by the corpus luteum, despite the continuously high levels of HCG; and, secondly, the low-normal levels of testosterone found in men with choriocarcinoma, despite the persistently very high concentrations of HCG. The pituitary response to LHRH can be desensitized through down regulation of LHRH receptors by treatment with long-acting analogues of the releasing hormone. LH release in response to both endogenous and exogenous native LHRH is then inhibited and anovulation ensues. This down regulation of LHRH receptors provides the basis for a peptide-based method of contraception. Since the peptide in question is a simple analogue of LHRH and, like the parent molecule, is a decapeptide, it can be absorbed through the nasal mucosa. The possibility of a new form of contraception by a nasally administered superactive LHRH analogue is mentioned here because it demonstrates the fundamental importance of basic physiological knowledge to human progress and health.

The control of LH secretion in women is complex because in addition to episodic release there is evidence of both positive and negative feedback control. In the rat there are two hypothalamic centres which control the release of LH by the pituitary. These centres, termed the tonic and the cyclical centres, are anatomically separated, the tonic centre being found in the arcuate ventomedial area of the hypothalamus, and the cyclical centre more anteriorly in the supra chiasmatic portion. In man, although evidence of anatomical separation is lacking, it is convenient conceptually to consider the control of LH and FSH secretion in these terms. Thus, the first level of hypothalamic control involves the tonic discharge of gonadotrophin releasing hormone in sufficient amounts to maintain LH and FSH release at basal levels. It is this centre which provides negative feedback control of gonadotrophin secretion and which is sensitive to a fall in circulating levels of oestrogen (or, in males, of testosterone). Superimposed on this control mechanism is the cyclical centre, which is responsible for the pre-ovulatory discharge of large amounts of gonadotrophin. This centre does not exist in males. Discharge of the cyclical centre is triggered by an *increase* of oestrogen (oestrogen-mediated positive feedback). Progestogens may modify this response. For instance, when ovariectomized rats are treated with graded doses of oestrogen, there is a fall in serum LH levels in those animals given small doses (negative feedback on tonic centre) but an increase is subsequently found in those animals treated with a larger dose (positive feedback on cyclical centre). When progestins are given there is also release of LH in the animals given the smaller dose of oestrogen, suggesting that progestins lower the threshold for the stimulatory effect of oestrogen on the cyclical centre.

The theory of a dual hypothalamic control of gonadotrophin secretion helps to explain the hormonal events of the normal ovulatory cycle, although it is salutory to note that oestrogen mediated positive feedback can occur in women with organic hypothalamic disease during treatment with LHRH given in pulses of fixed amplitude and frequency. Here it is, however, important to recall that the organization of the central nervous system into cyclical and tonic centres is critically influenced by the gonadal hormones. Thus brief exposure of a female rat to testosterone in the first few days of life prevents the subsequent development of cyclical discharge of gonado-trophins. Moreover, if male rate are castrated at birth and treated with oestrogen, subsequent ovarian transplants will undergo cyclical changes, further emphasizing the hormonal dependence of the organization of the central nervous system.

Follicle stimulating hormone (FSH). FSH is secreted by basophilic cells of the pituitary; although the precise cell of origin has not yet been identified, it is probably distinct from that which produces LH. The synthesis and release of FSH, like that of LH, is under the control of the hypothalamus. LHRH, in doses greater than needed for LH release, causes release of FSH and a separate releasing hormone for FSH has not been described. Following hypophysectomy the half-time of disappearance of FSH is longer than that of LH, being four hours.

In men, FSH contributes to growth of the seminiferous tubules and the maintenance of spermatogenesis. Since most of the volume of the testis is made up of tubular elements, it is therefore FSH which causes the increase in size of the testis at puberty. FSH is required in the preparations of gonadotrophins used to induce spermatogenesis in patients with infertility due to lack of gonadotrophins.

In the male the secretion of FSH in response to LHRH is controlled by two factors. Since castration results in elevated levels of FSH which return to eugonadal levels during treatment with testosterone, it is clear that the latter hormone has a role in the maintenance of normal FSH levels. This effect is, however, not specific for FSH, since LH levels are always suppressed by the amounts of testosterone required to suppress FSH. If spermatogenesis is selectively impaired, leaving Leydig cell function (i.e. testosterone production) intact, FSH levels are elevated but LH remains at normal levels, suggesting the existence of a factor derived from the tubular epithelium responsible for the feedback control of FSH. This hypothetical specific inhibitor of FSH secretion, termed 'inhibin', has not yet been isolated, nor has its cell of origin been identified. Most workers consider that it is a peptide released from Sertoli cells (*cf* Mullerian regression factor, page 653), and that its secretion depends upon normal spermatogenesis. Its characterization is a matter of great interest since it might impair spermatogenesis reversibly (by inhibition of FSH release) while leaving masculinization (i.e. Leydig cell function) unaffected. Theoretically it could then provide the basis for a reversible contraceptive for men.

In women FSH stimulates the ovary, causing enlargement and development of the follicle. It stimulates aromalase activity in the granulosa cells so estrogens are secreted in increasing amounts. The degree of stimulation of the follicles is closely related to the total amount of FSH to which the ovary is exposed.

In women, the control of the secretion of FSH is similar to that of LH. There is tonic release, which is controlled by negative feedback, and there is also a mid-cycle pre-ovulatory discharge of FSH, presumably dependent upon positive feedback by gonadal hormones. However, in contrast to LH, the role of the mid-cycle peak of FSH is unknown: it is not required for ovulation of the ripened follicle. Episodic secretion of FSH does not seem to be as important as it is for LH action. The precise fluctuations of LH and FSH levels during the ovulatory cycle are described on p. 663.

Human chorionic gonadotrophin

HCG is a glycopeptide secreted by the syncytiotrophoblast. Its molecular weight is in the region of 60,000, that of its subunits being half of this. The biological actions of HCG are similar to those of LH and indeed the two hormones bind to and activate the same receptors on the membranes of responsive cells. Moreover in the majority of radioimmunoassays LH and HCG cannot be distinguished. However, important chemical differences, mainly related to carbohydrate composition, do exist: the greater sialic acid content of HCG gives it a longer half-time in the body and therefore makes it a preferable hormone to LH in certain therapeutic regimens.

In subjects in whom the date of conception is accurately known, HCG can be detected in blood and urine as early as 5 days before the expected but missed menstrual period. This corresponds to the time of implantation of the conceptus. When the levels of HCG in the blood of women in the first trimester are related to time, the concentration initially rises logarithmically with a doubling time of just under 3 days. As the metabolic clearance rate of HCG remains constant throughout pregnancy, this doubling time represents the rate of increase of synthesis of HCG. The levels reach a maximum at 50–70 days after the last menstrual period and then decline after about 12 weeks of gestation.

The actions of HCG mimic those of LH, although some FSH activity is observed with high doses. HCG causes ovulation when given to anovulatory patients pretreated with FSH rich preparations of gonadotrophin. HCG causes luteinization of granulosa cells and is luteotrophic—that is, it can prolong the life of the corpus luteum. In the corpus luteum, HCG, like LH, enhances the conversion of most of the precursors of progesterone into progesterone. However, the physiological significance of these actions is uncertain because the endogenous levels of HCG are highest at a stage of pregnancy when corpus luteum function is waning. Moreover the secretion of HCG persists throughout pregnancy. HCG is thought to bridge the transition between the steroidogenic role of the corpus luteum and that of the placenta. It may, in addition, contribute to the mother's local immunological tolerance of her fetus, since experimentally HCG can inhibit blast cell transformation in response to stimulation by phytohaemoglutanin.

When injected into the male, HCG causes hypertrophy of the interstitial cells of the testis and stimulates the adult and fetal gonad to secrete androgen. *In vitro* studies indicate that it is the earlier stages of testosterone biosynthesis which are accelerated by HCG. Testicular production of oestrogens is also increased. In large doses HCG may desensitize the testis, in part through overstimulation of oestrogen production and in part through down-regulation of LH/HCG receptors. Overstimulation of the testis by HCG results in the development of gynaecomastia and impaired secretion and action of testosterone. These effects are reversible by reducing the dose of HCG. There is almost certainly no direct effect of HCG on the normal adrenal. HCG may be used in the treatment of non-testicular causes of hypogonadism.

GONADAL HORMONES

Hormones of the testis

The principal steroids secreted by the testis are testosterone and androstenedione, oestradiol, and the postulated hormone 'inhibin'. 17-hydroxyprogesterone is also secreted by the testis.

Testosterone. In men the major source of testosterone is the Leydig (interstitial) cell, in which the complete synthesis of androstenedione and testosterone from acetate has been demonstrated. There are two main biosynthetic pathways to androstenedione, which is the immediate precursor of testosterone. One pathway, via C21 steroid precursors, is through pregnenolone, the other, via C19 precursors, is through dehydroepiandrosterone (DHA). It is uncertain which is the major route. The final biosynthetic step involves the reduction of the 17-oxo group of androstenedione to form the 17-hydroxyl group of testosterone.

Testosterone secretion by the testis is increased by LH and HCG, which also increases the output of androstenedione and oestrogens. In men about 7 mg of testosterone are produced each day by the testis, there being only a small contribution from the adrenal. The hormone circulates reversibly bound to a specific sex hormone binding (β-2) globulin (SHBG) and to albumin. In men there is a diurnal variation of testosterone levels which is thought to reflect a diurnal variation in its release. Testosterone is inactivated in the liver by reduction (to androsterone and aetiocholanolone) and subsequent conjugation of the products of reduction; these conjugates form about one-third of the 17-oxo steriods which are excreted in the urine of men. A small fraction of testosterone (which as described above is not itself a 17-oxo steroid) is conjugated directly and excreted as the glucuronide. Less than 1 per cent is excreted as the free hormone.

In sensitive target organs, testosterone undergoes further metabolic changes which are related to its biological effects. Firstly, there is a redox type of reaction, whereby testosterone and adrenostenedione are reversibly interconverted. This process also takes place in the liver. It may represent a local control mechanism, since androstenedione is only a weak androgen. In androgen sensitive tissue, testosterone is converted to dihydrotestosterone by reduction of the C4-5 double bond. The dihydrotestosterone is bound to a cytoplasmic protein and then transported to the nucleus where it is bound to a basic protein of the chromatin. The complex causes an increase in template RNA in the nucleus, which results in increased protein synthesis and aminoacid uptake in those cells capable of binding testosterone and converting it to the active form. The rate of conversion of testosterone to dihydrotestosterone varies in tissues obtained from different parts of the body, being highest in the most androgen sensitive areas.

Testosterone has two major biological functions. Firstly, it is an androgen, which means, by definition, that it produces and maintains the secondary sexual characteristics of adult men. Thus it causes growth of the phallus and scrotum and development of the prostate and seminal vesicles. It is also required for full maturation of the seminiferous tubules. Testosterone is responsible for the masculine pattern of skin texture, growth of body hair and baldness, and for sebum secretion. Testosterone also has important actions on the central nervous system. It enhances libido in both sexes, affects the organization of the brain (page 678) and probably has a direct action on the nervous mechanism which controls erection of the penis.

The second major action of testosterone is an anabolic one: it increases muscle mass, causes nitrogen retention and alters the distribution of fat. Interestingly, in muscle the 5α-reductase enzyme is absent and therefore this anabolic activity of testosterone appears not to require formation of dihydrotestosterone.

There are a number of other actions of testosterone which are less well-defined at a biochemical level. In the presence of growth hormone, testosterone stimulates the growth of bones, and, indeed, it may also stimulate the secretion of growth hormone. These actions are largely responsible for the pubertal growth spurt. Subsequently, through an unknown mechanism, testosterone affects closure of the epiphyses. Testosterone also stimulates the production of red blood cells: the details are uncertain but it appears to have a direct action on stem cells and to enhance both the production and response to erythropoietin.

Androstenedione. This steroid, the immediate precursor of testosterone, is synthesized and released by the testis in response to stimulation with LH and HCG. Androstenedione is also produced by the adrenal, a source of particular importance in women (p. 686). Androstenedione is a weak androgen, probably acting in this way only through its conversion to testosterone, which takes place in the gonads, the liver and in androgen sensitive tissues.

Oestrogens. Oestradiol in men is mainly derived by extra gonadal conversion, from testosterone (50 per cent) and from oestrone (20 per cent) which is itself derived from androstenedione. About a third of the circulating oestradiol is directly secreted by the testis, the Leydig cells being the site of origin. The physiological significance of circulating oestradiol in men is not clear. The levels increase at puberty, and this increase may contribute to the transient gynaecomastia which occurs in the majority of boys at that time. Oestrogen regulates the synthesis of sex hormone binding globulin which determines the proportion of the total testosterone in serum that is free and therefore biologically active. Oestrogens also inhibit the conversion of testosterone to DHT. Both of these actions are important effects of treatment with oestrogen of patients with carcinoma of the prostate.

Hormones of the ovary

The principal hormones secreted by the ovaries are oestradiol-17β, progesterone and 17-hydroxyprogesterone. Small amounts of testosterone are also secreted by the normal ovary.

Oestrogens. Oestrogens cause the post-natal development and maintenance of the female sexual organs and secondary sexual characteristics. They have specific actions on the uterus and are required for the maintenance of pregnancy.

Over 30 naturally-occurring oestrogens have been isolated: the major oestrogens are oestradiol-17β, oestriol and oestrone. All of the naturally-occurring oestrogens are steroids but the most potent synthetic oestrogen, stilboestrol, is not. These steriods usually contain 18 carbon atoms and the A ring is phenolic, with a hydroxyl group at the C3 position. The three major oestrogens are named according to the number of hydroxyl groups they contain.

There are several routes in the ovarian biosynthesis of oestradiol, the most potent naturally-occurring oestrogen. The initial pathways are similar to those in the adrenal cortex, but hydroxylation at the 21 and 11 positions does not occur in the ovary. Oestrogens are thus formed by a series of attacks on the cholesterol molecule, forming C19 compounds (androstenedione and testosterone) which are then aromatized to form the C18 oestrogens, by a group of enzymes ('aromatase') which are present in the ovary, the testes, the adrenals, the placenta and the hypothalamus. The source of oestradiol in the follicular phase is the follicle and its surrounding cells. Under the influence of LH the theca cells convert progesterone, via intermediates, to testosterone. Aromatization takes place in the granulosa cells under the influence of FSH. Oestrogen secretion by the ovary in the follicular phase is thus stimulated by both FSH and LH; in the luteal phase, FSH is not required, oestradiol being secreted by the corpus luteum in response to LH.

Oestradiol may be converted to oestrone by many tissues (page 660). Since oestrone is a less potent oestrogen than oestradiol, this relationship is reminiscent of that between testosterone and androstenedione, except that in the case of the oestrogens the equilibrium favours the less potent compound. Oestradiol is also converted via 'oestrone', to oestriol (although it is not the only source of this compound). The metabolic products of the oestrogens are conjugated, predominantly in the liver and gut, to form glucuronisides and sulphates. There is an enterohepatic circulation of these conjugates, which are also excreted in the urine. However, up to 50 per cent of the oestradiol is excreted in the urine in the unconjugated form. The conjugates do not appear to have any direct biological role: the glucuronisides are quite rapidly excreted. However, the sulphates have much longer biological half-lives in the body and may represent an important source of inactive stored hormone, particularly during pregnancy when they can be reconverted to the active form by the placenta.

Oestradiol circulates in the blood partially bound to the globulin which is the same as that which binds testosterone (sex hormone binding globulin). The levels of oestradiol vary with the stages of ovulatory cycle, rise during pregnancy and fall precipitously at the menopause.

Oestrogens have widespread actions throughout the body, some obvious and familiar, others more subtle but probably of great physiological significance. The appearance of female secondary sexual characteristics at puberty is caused by oestrogen: these actions include the development of the internal and external genitalia and of the breasts, and the characteristic female distribution of fat. The vagina enlarges and its epithelium becomes cornified. As oestrogen concentrations rise, vaginal and cervical secretions increase and their pH falls. Oestrogen is also required for the normal increase of vaginal secretions during coitus.

The actions of oestrogen on the uterus have been extensively studied. The most obvious effects are on the endometrium during the pre-ovulatory phase and include proliferation of the glandular and superficial epithelium, the stroma and the blood vessels. The endometrium may grow in width by as much as 5 mm in the two weeks between menstruation and ovulation. Oestrogens also affect the myometrium, which hypertrophies under its influence.

Studies of the mechanism of action of oestrogen have shown that injected labelled oestradiol is taken up very rapidly by the uterus. Binding of oestradiol to a protein in the cytoplasm has been demonstrated within 15 seconds of exposure to the steroid. This cytoplasmic protein binds both steroidal and non-steroidal oestrogens, although with varying affinities, affinities which correspond closely to the known biological potencies of the compounds. The receptor bound oestrogen is transported to the nucleus where there is an early increase in ribosomal and messenger RNA and protein synthesis, and a later increase in DNA dependent RNA synthesis. That increased RNA synthesis is fundamental to the response of the uterus to oestrogen is shown by the effect of intra-luminal application of RNA obtained from oestrogen stimulated uteri. If such material is introduced into an oestrogen deprived uterus, all of the morphological changes associated with the actions of oestrogen can be reproduced.

In the rat, oestrogens are also actively taken up at the anterior portion of the hypothalamus, which in this species is the area of the cyclical centre that controls the preovulatory discharge of LH.

Oestrogens have a variety of other actions. They increase the sensitivity of ovarian follicles to FSH by causing synthesis of FSH receptors on granulosa cells. The motility of the female reproductive tract is partly controlled by oestrogens. The role of these hormones in the control of gonadotrophin secretion is described in detail on page 663 but here it is important to note that oestrogens also stimulate secretion of prolactin by a direct action on the lactotrophs.

Oestrogens increase the hepatic synthesis of a number of specific binding proteins in serum, and during pregnancy or treatment with oral contraceptives serum thyroxine and cortisol concentrations may be raised. Since the rise of the concentration of these hormones is due to a rise of their protein-bound fractions it is not associated with an increase of the biological action of these hormones. Oestrogen also raises the concentration of the sex hormone-binding globulin and thereby in women can elevate the level of serum testosterone. Since the metabolic clearance rate of testosterone depends upon the amount of free testosterone and not upon the total concentration, the assessment of testosterone production may become complex in patients treated with oestrogens. This is of importance in the investigation and management of patients with hirsutism (p. 691).

Other clinically important effects of oestrogen include an increase in the rate of hydroxylation of steroids in the adrenal, so that in patients taking oestrogen (e.g. the oral contraceptive) larger doses of metapyrone are needed to inhibit 11-hydroxylation, in this test of the hypothalamic-pituitary-adrenal axis. Oestrogens promote the release of growth hormone and may impair carbohydrate metabolism. In large doses they may lead to salt and water retention, an important effect when they are used in the treatment of elderly patients with carcinoma of the prostate. Certain of the clotting factors are increased by oestrogens and this is thought to account for the increased incidence of thromboembolism in patients taking oral contraceptives and in those given oestrogens for suppression of lactation and in the chemotherapy of malignant disease.

Progesterone. The term 'progesterone' was originally coined to describe the active principle of the corpus luteum which maintains pregnancy in ovariectomized rabbits. The term is now used to describe a specific C21 steroid, released mainly by the corpus luteum and placenta, which is also an important intermediate in the biosynthetic pathways of all steroid secreting cells. Progesterone has a variety of actions (*vide infra*). The terms progestin or progestogen describe a group of steroids, many of them synthetic, which have the characteristic effects of progesterone on the uterine endometrium (p. 661).

In the ovary progesterone is secreted and released mainly from the corpus luteum, though some progesterone is secreted shortly before ovulation. The cells of the corpus luteum appear to have a relative deficiency of 17α-hydroxylase, thus limiting further metabolism of progesterone by the ovary. The secretion of progesterone increases in response to LH, and if HCG is given during the luteal phase there is a further rise in the concentration of plasma progesterone. Recent evidence indicates that continuing secretion of LH after ovulation is required to maintain progesterone secretion by the corpus luteum. HCG, with its much longer biological half-life, fulfills this function in subfertile patients treated with gonadotrophins.

The largest amounts of progesterone are secreted by the placenta: about 250 mg are produced each day in the last trimester of pregnancy. Some progesterone is also released from the adrenal, although in women this contribution to the total production is small.

Progesterone circulates in blood partially bound to a globulin which is thought to be the same one that binds cortisol. It is also loosely bound to albumen. There are numerous metabolites of progesterone, but the main one is pregnanediol which is still occasionally measured in urine as an indication of its secretion.

The principal effects of progesterone are on the female reproductive tract. Progesterone decreases motility of the fallopian tube and contractility of the uterus. The former action may be important in the process of fertilization and implantation, and the latter in the control of parturition (p. 671). The best known action of progesterone is on the oestrogen-stimulated endometrium, which undergoes further differentiation under the influence of progesterone (p. 661).

Progesterone also has important effects on the central nervous system. In some species it is thought to influence maternal behaviour. It contributes to the regulation of gonadotrophin secretion and, in association with oestrogen, causes oestrous behaviour in certain animals. Through an action on the brain progesterone causes the rise in basal body temperature that occurs after ovulation. Progesterone antagonizes the action of aldosterone on the kidney, by competing for aldosterone-binding proteins in the epithelial cells of the distal renal tubule. This action may explain the relative ineffectiveness and high rate of secretion of aldosterone during pregnancy and may also explain the high aldosterone secretion rate in patients with the non-salt-losing form of congenital adrenal hyperplasia (due to 21-hydroxylase deficiency), in whom circulating levels of progesterone are raised.

Testosterone production in women

In women in the reproductive phase of life, the normal rate of production of testosterone is about one-twentieth that of men. The source of the testosterone is complex, since it is both secreted by the ovaries and the adrenals, and is also derived from extra glandular conversion of androstenedione (Fig. 19.5). This conversion takes place in the liver and in androgen sensitive tissues such as the sebaceous glands and hair follicles of skin. It accounts for about half of the daily production of testosterone. Of the testosterone directly secreted, less than one-fifth comes from the ovary, which also produces only a quarter of the daily production of androstenedione—thus the majority of testosterone is produced, either directly or indirectly, from the adrenal.

In both sexes testosterone is further metabolized to dihydrotestosterone (DHT). In women the amount produced each day is about 15 per cent of that in men. In both sexes DHT is almost entirely derived by extra glandular conversion of precursors: though in men the main precursor is testosterone, in women it is androstenedione, which emphasizes again the important contribution of the adrenal to the formation of

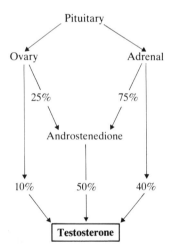

Figure 19.5. *Source of testosterone in women.*
    In normal women during the reproductive phase of life, half of the testosterone produced each day is derived from glandular secretion and half from conversion (mainly, but not exclusively, in the liver) of androstenedione. The adrenal provides 4/5 of directly secreted testosterone; it also provides 3/4 of the amounts of androstenedione produced each day. In women androstenedione is also the major precurser of dihydrotestosterone. These data explain why in women it is adrenal rather than gonadal failure which results in loss of secondary sexual hair.

androgens in women. Clinically these data are reflected in the observation that in women loss of secondary sexual hair indicates adrenal failure, whether primary (as in Addison's disease) or secondary to ACTH deficiency. In contrast, in men such a change indicates gonadal failure.

Abnormal function

DISORDERS OF GONADAL DIFFERENTIATION

Seminiferous tubule dysgenesis—47XXY Klinefelter's syndrome

This syndrome is one of the commonest causes of primary hypogonadism in males and occurs in one in 500 live male births. It is associated with an additional X chromosome which gives a chromatin positive buccal smear and a 47XXY karyotype. Mosaic forms also occur.

The extra X chromosome sterilizes the testis: initially there is a reduction in spermatogonia and after the time at which puberty would normally have occurred, hyalinization and fibrosis of the seminiferous tubules develop. The testes are small (depletion of tubules) and firm (fibrosis of tubules). Testosterone levels are usually less than 50 per cent of normal so symptoms and signs of androgen deficiency are common. LH concentrations are high and the oestrogen production rate is increased (partly through secretion of oestradiol by the testis in response to the high LH concentrations and partly because of extra-testicular aromatization). The raised oestrogen production rate causes gynaecomastia.

Primary gonadal dysgenesis—45XO Turner's syndrome

This syndrome occurs in one in 2,700 live female births but since a 45XO karyotype occurs in 5 per cent of abortuses, it is probably the commonest chromosomal lesion in man. It is not associated with increased maternal age.

Complete sex chromosomal monosomy is associated with:

Sexual infantilism: as indicated on page 652, in the absence of the second chromosome the rate of atresia is so rapid that the ovaries are depleted of germ cells and follicles by the age of puberty. The loss of oocytes causes sterility and the loss of follicles causes oestrogen deficiency, which is responsible for failure of pubertal development. Plasma gonadotrophin concentrations are higher than in children of the same chronological age.

Short stature: 45XO individuals commonly suffer intra-uterine growth retardation, their birth weight is usually subnormal and most are short by school age. They do not experience a pubertal growth spurt; their final height is usually below 5 ft (152 cm).

Growth hormone secretion is normal, as is the production of somatomedin(s), so the defect is thought to be at the level of cartilage.

Associated somatic abnormalities: there are a variety of associated somatic abnormalities (webbed neck, wide carrying angle, etc) the prevalence of which depends on the extent of the chromosomal lesion (i.e. the proportion of somatic cells that are 45XO).

About 20 per cent of cases of otherwise typical Turner's syndrome have chromatin positive buccal smears. These cases usually have a structurally abnormal chromosome (isochromosome) or mosaicism. The ratio of XX : XO (or XY cells) in the gonad determines whether it develops into an ovary, a testis or a hypoplastic streak gonad. In patients with less extensive loss of the second X chromosome, oocytes and follicles may persist for some years, permitting normal pubertal development but causing secondary amenorrhoea and a premature menopause. The ratio of XO : XX cells in extra-gonadal tissues determines whether there is short stature or the other physical abnormalities referred to above.

TRUE HERMAPHRODITISM

True hermaphroditism exists when both ovarian and testicular tissue is present in the same subject. The essential feature is that the ovarian tissue contains follicles and the testicular tissue contains spermatic tubules. The amount of interstitial tissue is variable and although its steroidogenic potential determines the phenotype, its presence is of no diagnostic significance. The testicular and ovarian tissue may be on the same side (forming an ovotestis) or on opposite sides. About one-quarter of cases have chromosomal mosaicism. The condition is rare.

DISORDERS OF SEXUAL DIFFERENTIATION

Sexual ambiguity caused by a defect in differentiation of the external and/or internal genitalia is called pseudohermaphroditism. In these patients the gonadal and chromosomal sex are in accord.

Female pseudohermaphroditism

This refers to masculinization of subjects possessing ovaries and a 46XX karyotype. Since these subjects do not possess testes, masculinization is restricted to the external genitalia. The source of androgens may be fetal (congenital adrenal hyperplasia due to 21 or 11 hydroxylase deficiency), maternal (for instance, pregnancy luteoma), or transfer of maternal medication (for instance, certain androgenic progestogens) to the fetal circulation. The extent of genital ambiguity depends not only on the amount of androgen but also on the timing of fetal exposure (page 655).

Male pseudohermaphroditism

Male pseudohermaphroditism occurs when genetic males differentiate partly or completely as females. There are theoretically 4 basic defects—failure of secretion or action of testosterone and failure of secretion or action of Müllerian Regression Factor. Since Müllerian Regression Factor cannot be measured at present, the latter two conditions cannot yet be distinguished. The phenotype depends upon the severity of the defect and the stage at which it begins to operate.

Testicular dysgenesis: 46XY agonadal subjects

If testes do not develop at all, or if they regress (because of some teratogenic insult) before Müllerian Regression Factor is produced, an agonadal 46XY subject may acquire complete Müllerian derivatives with female external genitalia (Swyer's syndrome). Regression later in embryogenesis causes varying degrees of male pseudohermaphroditism.

Defects of testosterone secretion in 46XY subjects with testes

Reduced biosynthesis of testosterone occurs in subjects with Leydig cell agenesis, fetal pituitary LH deficiency (page 657) and in subjects with specific inherited defects of the enzymes involved in the biosynthesis of testosterone. Reduced production of testosterone results in ambiguous or female genitalia, no Wolffian derivatives and severe gynaecomastia at puberty. In patients with defects of testosterone biosynthesis, the critical finding is a low plasma testosterone concentration with a subnormal response to stimulation by HCG. The pattern of the increased testosterone precursors in blood and urine allows precise diagnosis of the enzymatic defect.

Defects of testosterone metabolism

Failure to convert testosterone to dihydrotestosterone because of an inherited autosomal recessive deficiency of 5α reductase produces a characteristic syndrome. 46XY individuals with cryptorchid testes, fully developed vasa deferentia, epididymes and seminal vesicles and female or ambiguous external genitalia are reared as girls, but at puberty the testes descend and usually develop normal spermatogensis. The subjects then develop virilization, growth of the phallus and scrotum and an increase in muscle mass and deepening of the voice. Oestrogen production is normal so gynaecomastia does not occur. At puberty the subjects change their gender and develop male heterosexual orientation. Plasma testosterone levels are raised but dihydrotestosterone concentrations are low. Impaired conversion of testosterone to dihydrotestosterone can be demonstrated *in vitro* in skin and fibroblasts and *in vivo* by measuring the ratio in urine of 5α (androsterone) to 5β (aetiocholanolone) reduced metabolites of testosterone.

Despite its rarity, the syndrome of 5α reductase deficiency is of immense biological importance because it illustrates which aspects of androgen action depend on testosterone and which require dihydrotestosterone. Thus growth of the prostate and facial hair, recession of scalp hair and development of acne do not occur in these subjects though they can be induced by treatment with dihydrotestosterone.

Defects in the response of androgen sensitive tissue

Resistance to androgens is the commonest cause of male pseudohermaphroditism and occurs in complete and incomplete forms. The cardinal endocrine feature is the association of defective virilization with raised plasma testosterone concentrations.

Complete androgen resistance. This syndrome, also known (misleadingly) as testicular feminization, is inherited as an X linked recessive. It is caused by complete failure of peripheral tissues to respond to testosterone and/or dihydrotestosterone. The cytoplasmic receptor for dihydrotestosterone is usually absent but in the few cases where the cytoplasmic uptake of labelled dihydrotestosterone can be demonstrated the defect is presumably 'distal' to the receptor and involves either the DNA acceptor protein or a primary disorder of nuclear chromatin.

The patients have female external genitalia, are reared as girls, develop breasts at puberty but present with primary amenorrhoea and failure to develop secondary sexual hair. The male genotype results in the formation of testes and suppression of Müllerian development but the lack of response to androgens causes failure of development of the Wolffian derivatives and permits development of female external genitalia. The source of both testosterone (whose levels are normal or slightly raised for a normal *male*) and oestrogens are the intraabdominal testes. These testes are prone to malignancy and are therefore removed, following which the patients develop flushing which can be controlled with oestrogen but not by testosterone.

Incomplete androgen resistance. Reifenstein's syndrome is inherited as an autosomal recessive defect. In the most severe form there is perineoscrotal hypospadias and sterility, with the development of gynaecomastia at puberty. Milder variants exist. Presumably the cause of the phenotypic variation is a variation in the amount of intracellular androgen binding protein—cases have been described in which the only abnormalities have been gynaecomastia and oligospermia and infertility.

Defects of Müllerian regression

Persistence of Müllerian duct derivatives occurs in 46XY individuals with agenesis or very early embryonic regression of the testes. It can also occur as an isolated familial abnormality (uterus and tubes present in an otherwise normal man) due to reduced synthesis or response to Müllerian Regression Factor.

DISORDERS OF PUBERTY

Puberty is a gradual process, sexual and somatic maturation taking place over several years. Sexual maturation begins at about the same age in both sexes but the pubertal growth spurt starts around 2 years later in boys. Consequently boys continue to grow at the pre growth spurt rate for about 2 years longer than girls, which largely accounts for the difference of about 10 cm in the average height of adult men and women.

Precocious puberty

Isosexual precocious puberty occurs when signs of sexual maturation appear before the age of 8 years in girls or 10 years in boys. The term implies premature development of the adult pattern of sexual function: gonadal maturation occurs and spermatogenesis and ovulation may be sufficient for procreation. *Pseudoprecocity* indicates premature development of secondary sexual characteristics due to excessive gonadal steroids, arising for instance from a gonadal or adrenal tumour. In this case normal gonadal function is inhibited. *Heterosexual precocity* is the term used to describe pre-pubertal effects of androgens in girls (congenital adrenal hyperplasia) or the effects of oestrogens in boys (testicular tumour).

In the normal child, the hypothalamic-pituitary-gonadal axis is functional before puberty and is thought to be under inhibitory control from other parts of the brain. Sexual precocity occurs when the inhibition is released prematurely. The condition occurs twice as often in girls than in boys, and in girls only rarely is an associated anatomical lesion discovered. In contrast, in boys, lesions in the brain or adrenals or testes are found in over half the cases; in many of the remainder the electroencephalogram is abnormal. The intracranial lesions are usually in the posterior hypothalamic region and are thought to cause precocious sexual development by damaging the parts of the brain contributing the inhibitory signals which restrain sexual function before puberty.

Delayed puberty

The condition is usually not diagnosed until the patient is aged 17 years. It is commoner in boys and is not usually associated with destructive lesions in the hypothalamic-pituitary region. The patients are often of short stature and of course have not experienced the pubertal growth spurt. Naturally, all of the causes of gonadal failure (see below) may be responsible, and must be considered in the individual case.

MASCULINIZATION OF WOMEN

An excess of androgens in women results in defeminization and virilism. The virilism may be generalized, or restricted to the skin, where it is manifest clinically as hirsutism.

Table 19.4. *Results of an excess of androgen in women*

Cutaneous virilism	Generalized virilism	Defeminization
Coarsening and lengethening of terminal hair	Temporal and occipital baldness	Atrophy of breasts, uterus and vagina
Increased dermal collagen	Masculine body contour (increased muscle bulk and decreased fat)	Amenorrhoea
Increased secretion of sebum and sweat	Deepened voice	
Acne	Clitoral hypertrophy	

As indicated in Table 19.4, hirsutism is usually but one aspect of a generalized cutaneous virilism:

Masculinization of women may occur because of ingestion of androgens, because of an increase in the endogenous production of androgens or their precursors, or because of increased sensitivity of hair follicles to normal amounts of androgens. The latter is rare. In patients with hormone secreting ovarian tumours, direct secretion of large quantities of testosterone is usually responsible. In patients with the common forms of congenital adrenal hyperplasia, increased production of androstendione results in high levels of testosterone and dihydrotestosterone. In some patients with idiopathic hirsutism and with the polycystic ovary syndrome, there may be an increase of both ovarian and adrenal secretion of androgens and precursors, suggesting a general disorder of steroid-producing tissues in these conditions. In most of these patients, however, the increased androgen is produced by direct secretion by the ovaries.

Polycystic ovary syndrome

The cardinal feature of this disorder is the presence of bilaterally enlarged ovaries containing multiple subcapsular cysts. The tunica albugenia is thickened, there is variable luteinization within the cysts together with theca cell hyperplasia. The usual clinical history is of a normal menarche succeeded by several years of progressive menstrual irregularity (based usually on anovular menstrual cycles) culminating in amenorrhoea. Most patients have cutaneous virilization and many complain of infertility. Obesity is a variable but common feature.

This heterogeneous group of patients have been studied from several points of view. The high ratio of testosterone to oestrogen in fluid obtained from the cysts and the results of *in vitro* incubation studies have suggested a deficiency of aromatase activity in the granulosa cells of the follicles. This defect can, however, be readily overcome by treatment with FSH. In a proportion of cases a major contributor to the often six fold elevation of the rate of testosterone production is direct glandular secretion by the adrenal. However, adrenal secretion of testosterone is only partly suppressible by doses of glucocorticoids sufficient to reduce the cortisol secretion rate to immeasurable levels (p. 583). This increase in non-corticotrophin dependent adrenal production of C19 steroids may represent an exaggeration of the increase in adrenal androgens that normally occurs during puberty (the adrenarche). Measurement of gonadotrophins generally reveal low levels of FSH (perhaps responsible for the deficiency of aromatase activity in the granulosa cells—p. 662), but markedly increased levels of LH. These high LH concentrations are variable, they lack the fluctuations characteristic of the normal menstrual cycle and they stimulate testosterone secretion by the theca cells. Aromatization in the granulosa cells of the follicles of the large amounts of androgens produced by the theca is reduced, and the high intra-ovarian levels of testosterone may contribute to the high rate of atresia of follicles that characterizes polycystic ovaries. The androgens are, however, aromatized in extra ovarian tissues, particularly in fat, and so patients with polycystic ovaries may have clinical features of hyperoestrogenization (for instance, endometrial hyperplasia) as well as hyperandrogenization.

Treatment of these patients depends upon their primary complaint. Infertility is usually reversed by treatment with clomiphene and only rarely is treatment with gonadotrophins necessary. Wedge resection of the ovaries causes a transient improvement of fertility through an unknown mechanism, but this operation does not help the virilism and since it may be followed by periovarian adhesions is nowadays rarely advised if hirsutism is the major complaint, the logical treatment is with antiandrogens. Cyproterone acetate is a progestogen with moderate antiandrogen and mild glucocorticoid activity. When used in combination with oestrogen it has the following effects: the combination of oestrogen with the progestational activity of cyproterone acetate suppresses LH secretion and so reduces ovarian secretion of testosterone; the glucocorticoid-like activity of cyproterone acetate suppresses ACTH secretion and so reduces adrenal production of androgens. Oestrogen increases the synthesis of sex hormone binding globulin and so impairs transport of testosterone by reducing free testosterone levels. Oestrogen also reduces 5α reductase activity and so impairs conversion of testosterone to dihydrotestosterone. Finally, cyproterone acetate, being a peripheral antiandrogen, reduces the nuclear uptake of dihydrotestosterone. Thus, this combined treatment impairs production, transport, metabolism and action of androgens. It is, at present, the most effective treatment available for hyperandrogenization of women.

DISORDERS OF FERTILITY

INFERTILITY IN MEN

About 15 per cent of marriages are childless. Reproductive disorders in the male account for approximately half of these. Impairment of fertility may arise because of germinal cell failure, Leydig cell failure, or a failure of transport of normally formed spermatozoa.

Germinal cell failure

Defects of spermatogenesis may be caused by a primary disorder of the germinal epithelium, or because of impaired secretion of gonadotrophins (see below). Primary disorders of spermatogenesis may be isolated or associated with Leydig cell failure; they may be congenital (e.g. Klinefelter's syndrome) or acquired (e.g. orchitis, radiation injury). One cause of increasing importance is the use of antimitotic agents, such as cyclophosphamide, which produce a selective impairment of spermatogenesis. Isolated disorders of the germinal epithelium, of unknown cause, account for 90 per cent of cases of infertility in men. The condition is associated with a low sperm count, and a large proportion of the spermatozoa show defective morphology and impaired motility. LH levels are normal, but FSH levels are raised in those cases with severe arrest of spermatogenesis. In patients with primary lesions of the seminiferous tubules the prognosis for fertility is poor.

Leydig cell failure (hypogonadism)

Failure of Leydig cells results in impaired secretion of androgens. The condition occurs either because of a deficiency of LH or because of a primary testicular disorder ('primary hypogonadism'). In the majority of patients with impaired or absent secretion of gonadotrophins, the levels of both LH and FSH are low and the testis is infantile. The failure of secretion of gonadotrophins may be caused by an acquired destructive lesion in the hypothalamic-pituitary region: in addition to anatomical evidence of the lesion it is often possible to detect failure of other pituitary hormones. Hyperprolactinaemia may cause hypogonadism and impotence by impairing the normal secretion of gonadotrophins.

Isolated gonadotrophin deficiency is usually unaccompanied by overt hypothalamic-pituitary disease ('primary hypogonadotrophic hypogonadism'). Since LH and FSH levels rise in these patients after acute administration of LHRH, this condition is thought to result primarily from a defect in the formation of the releasing hormone. In young patients this disorder may be difficult to distinguish from delayed puberty, but in some cases of hypogonadotrophic hypogonadism there is an associated (genetic) absence of the olfactory bulbs, resulting in hyposmia (Kallman's syndrome). Treatment with HCG (to stimulate Leydig cells) and FSH (to induce spermatogenesis) induces testicular function and the prognosis for fertility is good. These patients may also be treated with HRH.

Very rarely isolated LH deficiency occurs: the patient is eunuchoid and there is some impairment of spermatogenesis. Both the androgen deficiency and fertility are restored by treatment with HCG alone ('the fertile eunuch').

Primary disorders of the testis affecting the Leydig cell may result from lesions which are genetic (p. 689), chromosomal (Klinefelter's syndrome), congenital (anorchia) or acquired (orchitis). In these patients eunuchism is associated with raised levels of both gonadotrophins. The prognosis for fertility is poor and treatment is directed at androgen replacement.

Transport failure

Failure of transport of normally formed spermatozoa may occur because of blockage of the vas or epididymus: azoospermia is associated with normal sized testes and a normal testicular biopsy. Sometimes the problem is due to coital failure, either through impotence, lack of libido or ignorance. Although impotence may be due to androgen deficiency, hyperprolactinaemia, or neurological disease it usually results from psychological problems.

INFERTILITY IN WOMEN

Infertility in women may arise because of a failure to ovulate or because of a mechanical hindrance to fertilization. Recurrent abortion or stillbirth also cause childlessness.

Failure of ovulation

Anovular infertility is usually associated with amenorrhoea and may result from disintegration of any of the stages in the mechanism which underlies ovulation. The widespread interconnections with other parts of the brain of the hypothalamic centres controlling LHRH release provide the neurophysiological background to the many effects that psychological factors have on reproductive function. Indeed 'stress' is the commonest single cause of secondary amenorrhoea, the 'stress' varying from serious psychiatric disorder to temporary environmental change (traveller's amenorrhoea). Anorexia nervosa is characteristically associated with amenorrhoea which is often the presenting feature. There is a reversible hypothalamic lesion in these patients, which is closely related to the degree of weight loss: thus, although in severely affected cases the response of gonadotrophins to clomiphene and to LHRH is lost, it returns (as do ovulation cycles) when the patient gains weight.

Structural lesions in the hypothalamus and pituitary may cause infertility by impairing the secretion of LHRH or of the gonadotrophins. Hypothalamic-pituitary activity may be adversely affected by disorders of other endocrine glands—as seen for instance in the oligomenorrhoea of hyperthyroidism. Hyperprolactinaemia is the cause of amenorrhoea in 20 per cent of cases. Almost half have small (micro) prolactin-secreting pituitary tumours, but drugs (p. 699) and hypothyroidism must also be excluded.

Disorders primarily affecting the ovaries which cause infertility include ovarian dysgenesis and the premature menopause. In the former, streak ovaries may be associated with chromosomal abnormalities and a variety of congenital defects (45XO Turner's syndrome, p. 687) or it may occur as an isolated finding in a subject with an XX karyotype (pure gonadal dysgenesis). In patients with primary ovarian failure the lack of oocytes causes sterility and the lack of follicles causes oestrogen deficiency.

In patients with organically caused gonadotrophin (or LHRH) deficiency, ovulation may be induced by treatment with exogenous gonadotrophins. Treatment with a mixed preparation (usually extracted from the urine of postmenopausal women) is given to induce follicular maturation and ovulation of the ripened follicle is achieved with HCG. Follicular maturation is monitored endocrinologically by measurements of oestrogen production, usually in association with ultrasonic monitoring of the ovaries. If oestrogen levels rise excessively, HCG is withheld in order to avoid multiple ovulations and also to avoid the 'hyperstimulation' syndrome. In this condition, which only occurs when HCG has been given to a subject given too much FSH, there is rapid ovarian enlargement causing pain and in severe cases ascites, pleural effusions and shock.

Clomiphene can be used to induce ovulation in anovulatory patients who have normal ovaries and who do not have a destructive lesion in the hypothalamic pituitary region. This compound causes release of LH and FSH which results in follicular maturation and subsequent ovulation. However, some subjects who ovulate after clomiphene still fail to achieve a pregnancy. A proportion of these have a 'short luteal phase' in which the corpus luteum functions poorly and involutes early, menstruation occurring about a week after ovulation. The short luteal phase may also occur spontaneously. One cause is an inadequate initial increase of FSH and LH secretion, resulting in inadequate follicular maturation and therefore a defective corpus luteum. Hyperprolactinaemic amenorrhea is treated by suppression of excessive prolactin secretion. This can be accomplished by withdrawal of provoking drugs, correction of hypothyroidism or pharmocologically, using dopaminergic agonists such as bromocripture. Only occasionally is destructive treatment (surgicel extirpation or irradiation) necessary.

Infertility with normal ovulation

This usually results from structural disorders which prevent fertilization. Fibrous adhesions within the pelvis may prevent the ovum from gaining entrance to the oviduct by impairing the motility of the fimbria. Spasm of the utero-tubal junction can be demonstrated radiologically in some patients and has been postulated as a cause of infertility. Salpingitis may cause partial or complete blockage of the tubes and so prevent union of the gametes. The treatment of these conditions is surgical. If conservative treatment is unsuccessful, *in vitro* fertilization with embryo transfer is appropriate for these patients.

DISORDERS OF THE BREAST

GYNAECOMASTIA

The term refers to a concentric increase in glandular and stromal tissue in the male breast. It may vary from the size of a button to that of the female breast. While usually bilateral, the condition is often asymmetrical and may be unilateral.

During fetal life, development of the breasts, like that of the other accessory organs of reproduction, is under the influence of testosterone. In the case of the breast, development is suppressed by testosterone, which inhibits embryogenesis of the ducts and the nipple. Consequently when breast development occurs in a previously normal man it is never as complete as in women, which explains why, although gynaecomastia is common, galactorrhoea in men is exceedingly rare.

The breast is very sensitive to oestrogen and indeed in the normal pubertal girl thelarche precedes even the uterine response to oestrogen (p. 657). It is therefore not surprising that development of the male breast occurs when there are quantitatively relatively minor changes in the amount of oestrogen produced. The critical factor appears to be a change in the normal ratio of oestradiol to testosterone. The source of oestradiol in men has been described on p. 682. Here it is necessary to emphasize that about two-thirds of its daily production is derived from peripheral (extragonadal) conversion of predominantly andiogenic precursors. The factors controlling the rate of conversion are unknown, but it increases with obesity, with age, and is raised in patients with cirrhosis of the liver.

Physiological gynaecomastia occurs in the neonate, in the adolescent and in the elderly. In the neonate the cause is exposure to the large amounts of oestrogen produced by the feto-placental unit. In the elderly gynaecomastia may be related to a fall of testosterone production, the effects of which are emphasized by the increased peripheral formation of oestradiol that occurs in the elderly. Adolescent (pubertal) gynaecomastia is common, occurring transiently in 60–80 per cent of boys. The cause is unknown, but may reflect a relatively early maturation of the peripheral aromatizing mechanism before the pubertal increase of testosterone secretion. Alternatively it may result from increasing stimulation of the Leydig cells by the rising concentration of LH (see below).

Pathological gynaecomastia occurs either because of reduction in the secretion of testosterone or an increase of oestradiol production. Oestrogen levels may increase because of oestrogen administration, either in the form of intentional treatment (as in cancer of the prostate) or because of contamination of other compounds. Digoxin behaves like an oestrogen and in addition to causing gynaecomastia occasionally, it can have an oestrogenic effect on the vaginal smear.

The condition in which increased conversion of precursors occurs most commonly as a cause of gynaecomastia is cirrhosis of the liver. Direct testicular secretion of oestradiol is probably the cause of gynaecomastia in patients treated with HCG, because in normal subjects treatment with this gonadotrophin causes a twofold increase in testosterone production but a tenfold increase in oestrogen production. This effect of HCG is probably also responsible for the gynaecomastia seen in patients with carcinomas producing 'ectopic' HCG. It is possible that the rising LH of puberty and the high LH levels that are found in partial Leydig cell failure (for instance Klinefelter's syndrome) similarly result in a relatively greater increase of oestradiol than testosterone production by the testis.

Deficient production of testosterone results in a striking increase of the ratio of oestradiol to testosterone. This occurs in the castrated male because a proportion of oestrogen is produced by precursors arising from the adrenal. In the disorders caused by congenital lesions operating during fetal life (male pseudohermaphroditism) the gynaecomastia is usually very florid because the usual testosterone induced suppression of fetal breast development has not occurred. The most striking example of this occurs in testicular feminization in which, though the increase in the oestrogen testosterone ratio is minor in chemical terms, it is biologically most profound because of the insensitivity to testosterone that characterizes this disorder. Cryptorchidism may be associated with androgen deficiency and gynaecomastia—indeed it appears that the failure of migration may in some cases be the result rather than the cause of the testicular deficiency. About 80 per cent of patients with Klinefelter's syndrome have gynaecomastia. The cause is usually the low testosterone production rate (p. 687) but in some cases an increase in oestrogen production has been implicated. All of the causes of post-pubertal testicular failure (for instance post-infective or traumatic or irradiation testicular atrophy) may be associated with gynaecomastia. A cause seen frequently in general medical practice is therapy with spironolactone: this compound is an anti androgen (like cyproterone acetate, p. 693) but it also produces a pharmacological castration by reducing testosterone secretion through inhibition of the testicular enzyme which converts 17α-hydroxy-progesterone to androstenedione.

Although the final endocrine pathway to gynaecomastia involves an increase of oestrogen relative to testosterone, the lack of symmetry and the occasional patient with unilateral gynaecomastia indicate that in any hormonal milieu, local differences in sensitivity are important in determining the biological response.

Galactorrhoea

Persistent lactation normally only follows childbirth. Inappropriate non-puerperal lactation—galactorrhoea—may be caused by a number of conditions which are thought to act at various levels in the sequence of neuro-humoral events controlling the release of prolactin (page 551). Serum prolactin levels are usually raised in patients with spontaneous galactorrhoea. However, the levels do not correlate with the amount of milk produced and clearly the degree of development of the breast itself is important in determining its response to a particular level of prolactin. The secretion of prolactin is largely controlled by an inhibitory factor produced by the hypothalamus. This prolactin inhibitory factor is now known to be dopamine. Dopamine is secreted into the pituitary portal circulation and inhibits the synthesis and release of prolactin by the lactotrophs. Inflammatory, traumatic and neoplastic lesions in the region of the hypothalamus or pituitary stalk may cause hyperprolactinaemia and therefore galactorrhoea and amenorrhoea, either by disrupting the site of dopamine synthesis or preventing dopamine from reaching the pituitary. Such lesions include granulomas, tuberuculosis and sarcoidosis and tumours such as craniopharyngiomas. Many chromophobe adenomas of the pituitary are now known to secrete prolactin and they typically present with amenorrhoea and galactorrhoea. Drugs which block the secretion or action of dopamine, which include reserpine, α methyldopa and the phenothiazine tranquillizers cause hyperprolactinaemia. Conversely prolactin levels can be lowered by stimulation of pituitary dopamine receptors. This can be achieved by the administration of l-dopa which is converted in the brain to dopamine. A variety of longer acting and better tolerated dopamine receptor agonists (bromocriptine, mertergoline) have, however, been developed for clinical use.

As discussed on page 672, physiological lactation is associated with impaired fertility, the mechanism operating by a prolactin-mediated disruption of normal gonadotrophin secretion. The same mechanism is thought to operate in patients with inappropriate hyperprolactinaemia and when prolactin concentrations are lowered to normal, whether by elimination of the provoking cause (e.g. drugs or hypothyroidism), by activation of dopamine receptors (e.g. treatment with bromocriptine), or by surgical extirpation of a prolactin-secreting pituitary adenoma, normal gonadotrophin secretion resumes, ovulation cycles return and fertility is restored to normal.

Cancer of the breast

This disease is the commonest malignancy of women, and has been the subject of intense endocrinological investigation. In about half of the patients the urinary excretion of androgen metabolites is reduced, and patients in this group experience more rapid recurrences and do not benefit from major endocrine ablation. Prospective studies have shown that this abnormality in steroid excretion precedes the development of cancer. Some breast cancers can convert steroids *in vitro* from one form into another, and others contain a high affinity receptor which binds oestradiol. The presence of this receptor can be used to predict the response to oestrogen deprival (for instance, by ovariectomy) with an encouraging degree of precision.

Cancer of the prostate

The prostate is an organ whose androgen dependance is so striking that its rate of growth in rats is utilized as an endpoint for bioassays of testosterone and LH. The majority of malignant tumours arising in the prostate of man retain a degree of androgen dependence and consequently may be treated by androgen deprival. This can be achieved by castration (or by hypophysectomy) but often patients with prostatic cancer are treated with oestrogens. Recently desensitisation of the pituitary with superactive analogues of LHRH have been shown to be very effective in inhibiting LH release.

The sites of action of oestrogen are shown in Fig. 19.6 which indicates that oestrogen reduces testosterone secretion indirectly, by inhibiting LH release, and directly, by impairing the activity of several enzymes within the Leydig cell. By increasing sex hormone binding globulin, oestrogen lowers the concentration of free testosterone, and therefore reduces the amount of testosterone available to bind to androgen receptors. *In vitro* studies have shown that oestrogen inhibits 5α reductase activity and

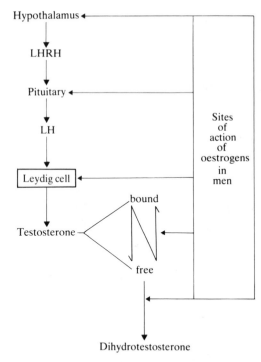

Figure 19.6. *Actions of oestrogen on secretion and metabolism of testosterone in men.*
    Oestrogens reduce testosterone secretion (a) indirectly, by reducing the secretion of LH (by inhibiting the production and action of LHRH), and (b) directly, by inhibiting enzymes within the Leydig cell. In addition, oestrogens reduce free testosterone concentration by increasing the amounts of sex hormone binding globulin. Finally, oestrogens inhibit the 5α-reductase system which converts testosterone to dihydrotestosterone.

so reduces conversion of testosterone to dihydrotestosterone. Finally, oestrogens have a direct action on the prostate, inhibiting protein synthesis and thereby causing a fall in the seminal concentration of acid phosphatase.

Principles of tests and measurements

GONADOTROPHINS

Radioimmunoassay

There are a number of problems to be considered in the interpretation of the results of radioimmunoassay of gonadotrophins. Firstly, as mentioned earlier, preparations of gonadotrophins of complete purity have not been made and so the assays have to be standardized against reference preparations which are of agreed potency, are widely available and are stable enough to last for many years. The potencies of these preparations have been defined in terms of their biological activity but radioimmunoassay does not measure biological activity—it depends upon an immunological reaction between a hormone (antigen) and an antiserum. Moreover with many hormones, biological and immunological activities involve different parts of the molecule—for instance, the carbohydrate moiety of HCG is essential for biological activity *in vivo* but it is not involved in immunological potency. There is thus an inherent difficulty in expressing radioimmunoassay results in biological units, and not surprisingly there is considerable confusion over the terminology of the units used in this field.

Secondly, because of the shared α subunits of the glycopeptides (p. 675) radioimmunoassays for one hormone often cross-react with those for the others. Indeed most radioimmunoassays for LH utilize antisera raised against HCG. The degree of specificity of radioimmunoassays for gonadotrophins must therefore be determined with considerable care. When particular problems arise, as for instance in the measurement of HCG in the presence of LH (needed for the very early diagnosis of pregnancy), antisera raised against the β subunits are used (p. 675). Thus measurement of HCG in early pregnancy utilizes antisera raised against HCG β. In addition to these fundamental problems, it is now clear that reference preparations derived from extracts of urine are inappropriate standards for assays performed on serum. In the past the only widely available internationally agreed reference preparations were of urinary origin but most workers now use one of various pituitary gonadotrophins to standardize their radioimmunoassays. There is still however no universal agreement about the optimum preparation.

Luteinizing hormone

Since in men the release of LH is controlled by feedback inhibition by testosterone, an elevated level of LH indicates primary gonadal failure. In certain conditions there is not a simple inverse relationship between LH and testosterone and it is thought that in partial Leydig cell failure normal levels of testosterone may be maintained by increased secretion of LH—"compensated partial hypogonadism". Most radioimmunoassays of LH are insufficiently sensitive to distinguish normal from low levels. When the LH concentration is not raised in a patient with a low level of testosterone, stimulation tests can be performed to help define the defect.

In women LH (and FSH) concentrations rise after ovariectomy and therefore elevated gonadotrophin concentrations characterize the postmenopausal woman. An isolated increase of LH (usually associated with an increase of testosterone) is seen in women with the polycystic ovary syndrome.

Follicle stimulating hormone

In primary ovarian dysgenesis, FSH levels rise earlier than LH levels and so if the disorder is suspected in a child below the age of 5, though LH levels may be normal, FSH will be raised. In women with the polycystic ovary syndrome and amenorrhoea, occasionally the differential diagnosis includes the premature menopause. LH levels are raised in both. FSH levels are greatly elevated in the former, but low in the polycystic ovary syndrome.

In men LH levels are raised in most conditions in which FSH levels are elevated, with the exception of selective defects of the germinal epithelium. In patients with this condition (for instance, after anti-cancer chemotherapy) FSH levels may be raised in the presence of a normal Leydig cell—LH axis.

The clomiphene test

In the adult male LH and FSH concentrations rise after several days treatment with clomiphene. The mechanism of this response is not understood but one factor is an increase of sex hormone binding globulin which causes a fall in the free testosterone concentration. The gonadotrophin response can be blocked by concurrent adminis-tration of testosterone (or oestrogen). Failure to respond to clomiphene is seen in patients with disorders of the hypothalamus or pituitary and before puberty.

In well-oestrogenized women with amenorrhoea, clomiphene causes an increase of LH and FSH secretion. The mechanism is thought to operate by blockade of hypo-thalamic oestrogen receptors. The simulated oestrogen deficiency activates negative feedback and so causes an increase of gonadotrophin secretion which in turn causes follicular development with the evolution of an ovulation cycle. The action of clomi-phene can be assessed by measuring the increase of gonadotrophin secretion during treatment, or by determining whether the patient ovulates (p. 703) after the treatment is discontinued.

The LHRH test

When given as a single intravenous injection, LHRH causes release of LH and FSH within a few minutes in both sexes. The quantities of these hormones released depends upon the pituitary content of LH and FSH, which in turn depends upon the endogenous production of LHRH. Exaggerated responses are therefore seen in patients with primary gonadal failure, since the gonadal steroids modulate gonadotrophin secretion in part by their action on the hypothalamic production of endogenous LHRH. A failure to respond (or an impaired response) to LHRH occurs in patients with pituitary failure and, because of low endogenous LHRH levels, in patients with hypothalamic disorders. Sub-normal responses occur in hypogonadotrophic hypogonadism (for instance, Kallman's syndrome, p. 694) and in certain patients with anovulation caused by hypothalamic disorders. Subnormal responses, which return to normal as the patient gains weight, are seen in anorexia nervosa.

GONADAL STEROIDS

Testosterone

In the male testosterone measurements are useful to confirm hypogonadism, particu-larly when only minor degrees are present. Low testosterone levels are caused by Leydig cell failure, LH deficiency or hyperprolactinaemia. Testosterone measurements are therefore of more value when taken in conjunction with estimations of LH and prolactin.

In women testosterone measurements are required for the diagnosis of virilism. The total serum testosterone concentration may, however, be normal in many patients with idiopathic hirsutism despite the almost universally observed increase in the testosterone production rate. This occurs because of an associated increase in the metabolic clearance rate of testosterone, which is thought to result from a (testosterone induced) fall of sex hormone binding globulin (SHBG). It is important to remember that treatment with oestrogen raises SHBG and so oestrogen therapy may alter the plasma concentrations (and metabolic clearance rate) of testosterone by an effect on its transport rather than on its secretion.

Very high levels of testosterone suggest production by a tumour; if the urinary 17-oxosteroids are normal the source is usually ovarian.

HCG *test*

In men testosterone (and oestrogen) levels rise after injection of HCG. Measurements of testosterone after injection of standard amounts of HCG may be used to diagnose the presence of intra-abdominal testes, and the quantitative response may be used both as an index of Leydig cell reserve and of the adequacy of therapy with exogenous gonadotrophins.

Oestrogens

In men measurements of oestrogen are usually made in the investigation of gynaecomastia. High levels may be seen with testicular and adrenal tumours. In women oestrogen measurements are useful in the diagnosis and management of amenorhoea. The oestrogen may either be measured in a blood or urine sample or assessed by its biological effect on the patient (vaginal smear, endometrial biopsy, progestogen challenge). The main value of oestrogen measurements is in the selection of patients for therapy: patients with low oestrogen levels rarely respond to clomiphene. Oestrogen measurements are also required to monitor induction of ovulation with gonadotrophin injections.

DIAGNOSIS OF OVULATION

The only certain proof of ovulation is pregnancy. In its absence ovulation is usually diagnosed by one (or a combination) of the following methods:

1 Basal body temperature. Progesterone causes the basal body temperature to rise in the luteal phase by about 1°C. A biphasic temperature chart is presumptive evidence of ovulation. It is also used to diagnose the short luteal phase, a cause of ovulatory infertility in which the time between the rise of temperature and menstruation is reduced to less than 10 days.

2   Endometrial biopsy. The presence of a secretory endometrium indicates progesterone secretion and therefore corpus luteum formation.

3   Serum progesterone (or urine pregnanediol). An increase in the second half of the cycle, or about 15 days after starting a course of clomiphene, are the most convenient indices of corpus luteum function.

4   Laparoscopic visualization. This can give evidence of ovulation by direct observation of a recently formed corpus luteum.

Recently advances in the resolution of diagnostic ultrasound equipment has led to the introduction of serial ultrasonic monitoring of follicular development. This method now provides the most reliable assessment of the time of ovulation.

ASSESSMENT OF FETUS AND PLACENTA

A large number of tests are used to evaluate feto-placental function and this suggests that none is completely satisfactory. Some assess mainly the function of the placenta (measurement of progesterone and its metabolites, measurement of HPL and placental enzymes), some are directed mainly at determining fetal dimensions and alterations in their size (measurement of the biparietal diameter of the fetal skull by ultrasound) and others attempt to assess the well being of the feto-placental unit (oestriol measurements). Since these tests evaluate different aspects of the same process, it is to be expected that the information they give is complementary and that none is exclusively the best.

STEROID ASSAYS

Progesterone and pregnanediol

After the first trimester, progesterone is synthesized by the placenta, using cholesterol from the maternal circulation as substrate (Fig. 19.4). Progesterone cannot be utilized for further steroid biosynthesis by the placenta (p. 668) and a proportion passes into the fetal circulation to be used as a substrate by the fetal adrenal cortex; the remainder passes into the maternal circulation and is converted by the liver into many metabolites, which are then excreted in the urine. Pregnanediol is one of these metabolites. Its measurement in maternal urine therefore gives an index of the steroid biosynthetic capacity of the placenta. These measurements have now been largely replaced in clinical practice by measurements of oestriol.

Oestriol

The production of oestriol increases a thousand-fold by the last weeks of pregnancy, an increase which exceeds that of the other two major oestrogens by a factor of ten. Oestriol is produced by the placental conversion of a precursor, which is synthesized by the fetal adrenal cortex under the influence of the fetal pituitary. Reference to Fig. 19.4 indicates that hydroxylation of DHAS at the 16 position in the fetal liver is followed by placental aromatization to form oestriol. Oestriol production is therefore influenced by the availability of a substrate derived from the fetus and the efficiency of placental aromatization. Since the latter is oxygen-dependent it can be appreciated how uterine blood flow and placental well-being contribute to the levels of oestriol observed in the mother. However, the major determinant is the production of the substrate and its delivery to the placenta by the fetal circulation.

The excretion of oestriol is subject to a day to day variation of 20 per cent, so serial measurements are obligatory. In patients with recurrent abortions oestriol levels have prognostic significance, the levels being higher in those subjects who carry their pregnancies to term. Oestriol excretion is low in severe toxaemia and when growth of the fetus is retarded. A series of low levels, or a sudden sharp reduction, suggests the fetus is in danger. The levels are very low in anencephaly, fetal adrenal hypoplasia, aryl sulphatase deficiency (p. 670) and after intra-uterine death.

Blood oestriol levels can now be measured and, since in pregnancy oestriol forms 80–90 per cent of the total oestrogen produced, assay specificity in this situation is of minor importance.

PEPTIDE AND PROTEIN ASSAYS

Human chorionic gonadotrophin (p. 679)

Measurements of HCG are used to confirm the diagnosis of pregnancy, but they are of no value in predicting its outcome. The immunological methods which are used are designed to be insensitive and therefore they detect only high levels. When HCG measurements are used to monitor the results of treatment of trophoblastic disease, more sensitive radioimmunoassays must be used, as large quantities of HCG may be produced in the presence of a negative 'immuno' pregnancy test.

Human placental lactogen (HPL) (p. 670)

HPL levels reflect the capacity of the placenta for biosynthesis of proteins. The information gained from measurements of HPL during pregnancy is of more value than that gained from a knowledge of HCG concentrations. The reason is partly due to the difference in the normal pattern of production of the two hormones and partly due to the difference in their rates of metabolism. The concentration of HCG rises to a peak in early pregnancy and then declines, whereas that of HPL continues to rise and does not reach a plateau until the last weeks of pregnancy. Thus a fall of HPL concentration is compared with a level that is normally rising throughout pregnancy and changes are therefore emphasized. However, in spite of this, there is considerable overlap of HPL levels in normal and abnormal pregnancies and serial estimations greatly increase the information that can be derived from the measurements.

In patients with vaginal bleeding in early pregnancy, levels of HPL that are lower than normal are associated with inevitable abortion and they therefore have prognostic value. Later in pregnancy, a series of low levels of HPL particularly when occurring in a patient with hypertension, is of considerable value in the prediction of fetal distress and therefore the deployment of facilities for fetal monitoring and resuscitation.

Measurements of enzymes released by the placenta (heat stable alkaline phosphatase, cystine amino peptidase) have been used as a guide to placental function, but in general the results have been disappointing.

ULTRASOUND

Ultrasound can be used to obtain very accurate measurements of various dimensions of the fetus, including the size of the fetal head (the biparietal diameter is measured). Using serial measurements, the rate of growth can be determined and a distinction made between retardation of fetal growth and mistaken maturity. Ultrasound can also be used to localize the placenta, is helpful in the diagnosis of trophoblastic disease and is used increasingly for antenatal diagnosis of defects in fetal development.

Practical Assessment

IN WOMEN

CLINICAL OBSERVATIONS

Menstrual history; appearance of external genitalia and vagina; presence and size of internal genitalia (by palpation); body hair and other secondary sexual characteristics. Isolated failure of breast development suggests local resistance to circulating oestrogens. Similarly, failure to menstruate (primary amenorrhoea) with normal secondary sexual characteristics suggests lack of uterine development (or imperforate hymen). Cessation of menstruation indicates ovarian failure, whether primary or secondary. Complete loss of secondary sexual hair indicates adrenal failure, which may be primary or secondary. Galactonhoea suggests hyperprolactinaemia.

ROUTINE TESTS

Occurrence of ovulation is established by the methods listed on p. 703. Measurement of urinary HCG is used to confirm pregnancy. Measurement of urinary 17-oxosteroids establishes whether an increase of testosterone secretion is of adrenal origin.

SPECIAL TESTS

Measurements of LH, and LH response to LHRH and clomiphene. Diagnosis of ovulation following clomiphene. In children with ambiguous genitalia, establishment of chromosomal sex and ratio of 11-oxo to 11-deoxo 17-oxosteroids. Hysterosalpingography is used to establish the presence of tubal obstruction. Laparoscopy allows ovarian visualization and biopsy without formal laparotomy. Tubal patency can be established at laperoscopy by observing the passage of dye through the tubes.

IN MEN

CLINICAL OBSERVATIONS

Age at puberty; sexual potency; appearance of external genitalia, particularly the consistency and size of the testes; body hair; breast development. Undescended testes may be normal ('retractile testes'). Less commonly they may indicate a true failure of pituitary gonadotrophin. Treatment with HCG often results in descent so long as the position is not ectopic.

ROUTINE METHODS

Spermatogenesis can be assessed by examination of seminal fluid, collected by masturbation after 5 days' abstinence. Criteria employed are the volume of the specimen, the concentration of spermatozoa per unit volume of fluid, the degree of motility of the spermatozoa, and the morphology of the spermatozoa. The most important of these criteria are the motility of the spermatozoa. A count of 10–20 million/ml is adequate, provided that motility is unimpaired. The specimen must be warm and fresh when examined and it must be collected in a dry container and not in a sheath, since cooling, moisture and rubber tend to kill the sperms. The variability of the sperm count in any individual may be considerable; at least three specimens need to be examined if the first count is low.

SPECIAL TECHNIQUES

Testicular biopsy: biopsy of the testes is sometimes performed in cases of hypogonadism. The procedure adds little to the routine assessment of gonadal function; in cases of azoospermia a high FSH indicates tubular destruction, a low level suggests testicular obstruction. Spermatogenesis is most easily assessed by examination of the seminal fluid. Interstitial cell function is readily assessed by clinical observation of the secondary sex characteristics and by measurement of testosterone and LH.

Response to HCG

Injections of chorionic gonadotrophin increase the output of testosterone in patients with testicular deficiency due to pituitary gonadotrophin failure and may result in clinical improvement. Those with primary testicular failure show an impaired response. Patients with pituitary infantilism develop secondary sex characters and show a spurt of growth, but the response continues only as long as the treatment is maintained. On the other hand, those with constitutional stunting of growth and delayed puberty continue to improve spontaneously after an initial course of treatment lasting up to six months. An increase of testosterone (or oestrogen) after stimulation with HCG indicates the presence of testes in patients in whom they cannot be detected in the scrotum (e.g. ectopic testes).

TESTS IN EITHER SEX

BONE AGE

This can be estimated from radiographs of various epiphyses whose time of appearance and fusion is compared with that of the general population. Bone age is a more reliable index of gonodal function during adolescence than is height age. The wrist and hand is usually assessed.

MEASUREMENT OF BODY PROPORTIONS

This is a useful test in the assessment of primary failure of the gonads, or selective gonadotrophic failure, especially in males. Owing to the failure of epiphyses to mature in the absence of androgenic hormones, growth of the long bones persists unduly because the secretion of pituitary growth hormone continues. The span of the out-stretched arms exceeds the total height, and the distance from floor to pubis exceeds that of pubis to vertex.

PITUITARY FUNCTION

Pituitary function is assessed by measurement of gonadotrophins in the basal state and after stimulation tests. Association with other endocrine disorders suggests hypo-gonadism is of pituitary origin.

References

GRUMBACK, M. & CONTE, F. (1981) Disorders of sex differentiation. In WILLIAMS R.H. (ed.) *Textbook of Endocrinology*, 6th edn, Chapter 9. Philadelphia, W.B. Saunders.

ROSS, GRIFF T. & VAN DE WIELE, RAYMOND L. (1981) The ovaries and the breasts. In WILLIAMS R.H. (ed.) *Textbook of Endocrinology*, 6th edn, Chapter 7. Philadelphia, W.B. Saunders.

TULCHINSKY, D. & RYAN, K.J. (1980) *Maternal–Fetal Endocrinology*. Philadelphia, W.B. Saunders.

YEN, S.S.C. & JAFFE, R.B. (eds.) (1978) *Reproductive Endocrinology: physiology, pathophysiology and clinical management*. Philadelphia, W.B. Saunders.

Index